VOLUME

II COMPUTER AND ROBOT VISION

Robert M. Haralick
University of Washington

Linda G. Shapiro
University of Washington

 ADDISON-WESLEY PUBLISHING COMPANY

Reading, Massachusetts • Menlo Park, California • New York
Don Mills, Ontario • Wokingham, England • Amsterdam • Bonn
Sydney • Singapore • Tokyo • Madrid • San Juan • Milan • Paris

Library of Congress Cataloging-in-Publication Data
(Revised for vol. 2)

Haralick, Robert M.
    Computer and robot vision.

    Includes bibliographical references and index.
    1. Computer vision.   2. Robot vision.   2. Image
Processing.   I. Shapiro, Linda G.   II. Title.
TA1632.H37   1993   621.39'9   90-25550
ISBN 0-201-56943-4 (v. 2)

1 2 3 4 5 6 7 8 9 10-HA-95949392

CONTENTS

14   Analytic Photogrammetry   125

15   Motion and Surface Structure from Time Varying Image Sequences   187

18   Object Models and Matching   427

21   Glossary of Computer Vision Terms   571

12 ILLUMINATION

12.1 Introduction

The two key questions in understanding three-dimensional image formation are: (1) *What determines where some point on the object will appear on the image?* (2) *What determines how bright the image of some surface on the object will be?* The first may be answered through a geometric perspective projection model (Chapter 13). The second involves radiometry, general illumination models, and surfaces having both diffuse and specular reflection components. In this chapter we briefly describe how this model can be used in photometric stereo and shape from shading. We also show how to extend the illumination model to include polarization.

Different points on the objects in front of the imaging system will have different intensity values on the image, depending on the amount of incident radiance, how they are illuminated, how they reflect light, how the reflected light is collected by a lens system, and how the sensor responds to the incoming light. Figure 12.1 shows the basic reflection phenomenon. The image intensity I is proportional to the scene radiance. The scene radiance depends on (1) the amount of light that falls on a surface, (2) the fraction of the incident light that is reflected, and (3) the geometry of light reflection, that is, the direction from which it is viewed as well as the direction from which it is illuminated. Mathematically the image intensity can be written as

$$I = g\, J^i\, f_r\, C\, S + b \qquad (12.1)$$

where J^i is the incident radiance, f_r is the bidirectional reflectance function, C is the lens collection, S is the sensor responsivity, g is the sensor gain, and b is the sensor offset.

In this chapter we discuss each component of Eq. (12.1) in detail and present an illumination model that can be used for incident light that is arbitrarily polarized. Section 2 provides an elementary overview of radiometric calculations, the bidirectional reflection distribution function, the lens collection, and the image intensity.

1

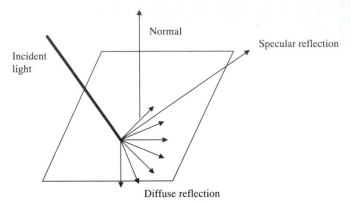

Figure 12.1 Refraction of light bouncing off a surface patch.

In section 3 we discuss photometric stereo and in section 4 we discuss the determination of shape from shading. Section 5 discusses the polarization of light. Section 6 describes the Fresnel equation. Section 7 derives a bi-directional reflectance function which includes polarization. Lens collection for polarized light is discussed in section 8, and section 9 rewrites Eq. (12.1) in terms of the polarized bi-directional reflectance function.

12.2   Radiometry

Radiometry is the measurement of the flow and transfer of radiant energy in terms of both the power emitted from or incident upon an area and the power radiated within a small solid angle about a given direction. Here we define the basic terms of radiometry and illustrate their use in calculations that determine how much light falls on a surface, and subsequently in calculations that determine the brightness of an illuminated surface.

The amount of light falling on a surface is called *irradiance*, and the amount of light emitted from a surface is called *radiance*. More formally, irradiance is the power per unit area of radiant energy falling on a surface. It is measured in units of watts per square meter. Radiance is the power per unit foreshortened area emitted into a unit solid angle. It is measured in units of watts per square meter per steradian. The *radiant intensity* of a point illumination source is the power per steradian the source radiates. It is measured in units of watts per steradian. The radiant intensity may be a function of polar and azimuth angles.

Radiant energy may leave an infinitesimal area on a surface in any direction. To specify the direction, we may place a hemisphere around the area. A direction is specified by the angle from the north and an angle from the east. The angle from North is called the polar angle, and the angle from East is called the azimuth angle (see Fig. 12.2).

The solid angle subtended by a surface patch is defined by the cone whose vertex is at the point of radiation and whose axis is the line segment going from

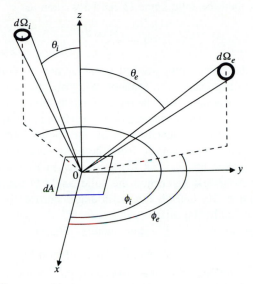

Figure 12.2 Geometry of incident and reflected elementary beams. The z-axis is chosen along the normal to the surface element dA at 0. The polar angle is measured from the z-axis, which we can think of as pointing north. The azimuth angle is measured from the x-axis, which we can think of as pointing east.

the point of radiation to the center of the surface patch. The size of the solid angle is equal to the area intercepted by the cone on a unit radius sphere centered at the point of radiation. The solid angle is measured in steradians. The total solid angle about a point in space is 4π steradians.

As Fig. 12.3 shows, if a surface of area A is at a distance d from a point of radiation and the surface normal makes an angle of θ with respect to the cone axis,

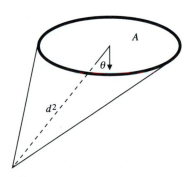

Figure 12.3 Determination of the solid angle Ω subtended by a small surface patch; $\Omega = \frac{\cos\theta A}{d^2}$.

and, if $d^2 >> A$, then the solid angle Ω can be written as

$$\Omega = \frac{A \cos \theta}{d^2}$$

To illustrate these concepts, we first consider how to compute the irradiance at a surface due to a point source radiator. Suppose that a surface patch of area A is illuminated by a point source having a constant radiant intensity of I_o (w/sr). Relative to a coordinate system centered on the surface patch and whose polar angle is oriented with respect to the surface normal, the (polar, azimuth) direction of the point source is (θ_o, ϕ_o). The distance from the point source to the surface patch is d meters. To determine the incident irradiance (w/m²) on the surface due to the point source, we must first determine the solid angle Ω the surface patch makes with the point source. As before, this solid angle satisfies $\Omega = A \cos \theta_o / d^2$. The total power intercepted by the surface patch is then $I_o A \cos \theta_o / d^2$ watts. This power is distributed over an area of A square meters. Therefore the surface irradiance is $\dfrac{I_o A \cos \theta_o / d^2}{A} = I_o \cos \theta_o / d^2$ w/m². This relationship shows that irradiance varies inversely as the square of the distance from the illuminated surface to the source (see Fig. 12.4) directly with the intensity of the illumination and directly with the cosine of the angle between the point source and the surface normal of the illuminated surface patch. This relationship is known as the Lambertian cosine law of incidence.

Now suppose that this point source radiator is illuminating a surface patch on the horizontal plane a distance h below the point source and a horizontal distance a from the point source. Then $\cos \theta_o = h / \sqrt{h^2 + a^2}$ and $d^2 = h^2 + a^2$. Hence the surface irradiance is

$$\frac{I_o \cos \theta_o}{d^2} = I_o \frac{h}{\sqrt{h^2 + a^2}} \frac{1}{h^2 + a^2}$$

$$= \frac{I_o}{h^2} \cos^3 \theta_o$$

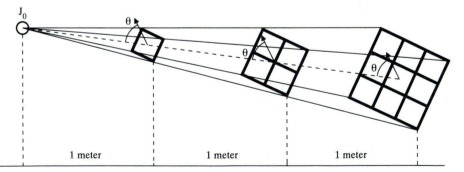

Figure 12.4 Law of inverse squares. At 1 meter the light in the solid angle illuminates 1 unit of area. At 2 meters the light in the solid angle illuminates 4 units of area. At 3 meters the light in the solid angle illuminates 9 units of area. The incident irradiance at distance d is given by $J_o A \cos \theta_o / d^2$, where the source has isotropic radiance of J_o w/m²-sr.

Thus the irradiance at a horizontal surface produced by a point source a fixed height h above the surface varies as the cube of the cosine of the angle between the point source and the surface normal of the horizontal surface.

Suppose that a small surface patch is illuminated by a surface radiator having an isotropic (independent of direction) radiance J_o w/m²-sr. This means that each square meter of the surface of the illuminating source radiates J_o w/sr. Relative to the small illuminated surface patch, the (polar, azimuth) direction of the differential illuminating surface is (θ_o, ϕ_o) (see Fig. 12.2). The distance from the differential illuminating surface patch of dS to the small illuminated surface patch is d meters. To determine the incident irradiance (w/m²) at the small illuminated surface patch, we must first determine the differential solid angle $d\Omega$ that the surface patch makes with the illumination differential surface patch. This solid angle satisfies $d\Omega = dS \cos\theta_o / d^2$ (see Fig. 12.3). Next suppose that the axis of the differential solid angle makes an angle of θ_1 with respect to the normal of the illuminated surface. Then the differential incident irradiance can be computed as the product of the radiance of the illuminating surface patch, the solid angle, and $\cos\theta_1$; that is, $dJ = J_o \cos\theta_1 dS \cos\theta_o / d^2$ (w/m²), and it is all coming from direction (θ_o, ϕ_o).

To compute the total power radiated, we determine the differential power $d\Phi$ in the direction (θ, ϕ) emitted by a point source and having radiant intensity $I(\theta, \phi)$ (w/sr). To do this, we first put a sphere of radius r centered around the differential area. On the unit sphere we put an annulus concentric around the north pole at a latitude of θ from the north. Consider a slice of the annulus that is $d\phi$ wide. The radius of the annulus is $r \sin\theta$. The length of the differential slice of the annulus is $r \sin\theta d\phi$. The width of the annulus is $rd\theta$. The differential area of the slice is then $dS = r^2 \sin\theta d\theta d\phi$. This differential area is everywhere normal to the radius of the sphere. The differential solid angle $d\Omega$ subtended by this area is

$$d\Omega = r^2 \sin\theta d\theta d\phi / r^2$$
$$= \sin\theta d\theta d\phi \tag{12.2}$$

The radiant power passing through the solid angle $d\Omega$ is always flowing in a direction orthogonal to the area dS. Hence the differential radiant power $d\Phi$ passing through the differential solid angle $d\Omega$ is

$$d\Phi = I(\theta, \phi)d\Omega$$
$$= I(\theta, \phi) \sin\theta d\theta d\phi \tag{12.3}$$

The total radiant power radiated by all the differential surface patches dS can then be determined by integrating $d\Phi$ over the hemisphere of 2π steradians.

$$\Phi = \int_{\theta=0}^{\pi/2} \int_{\phi=0}^{2\pi} I(\theta, \phi) \sin\theta d\theta d\phi \quad \text{watts}$$

The differential radiance dJ in the direction (θ, ϕ) is defined by

$$dJ = \frac{d\Phi}{d\Omega dS \cos\theta} \tag{12.4}$$

Substituting Eq. (12.3) into (12.4) results in

$$dJ = \frac{I(\theta, \phi)}{dS \cos \theta}$$

Radiating surfaces for which the radiance is independent of direction are called diffuse radiators. They are often associated with Lambertian surfaces. For them $I(\theta, \phi)/\cos \theta$ must be a constant. Therefore we must be able to write the radiant intensity $I(\theta, \phi)$ of a diffuse radiator as $I(\theta, \phi) = I_o \cos \theta$.

The total radiant power radiated by a point diffuse radiator or a diffusely radiating differential surface patch of radiant intensity $I_o \cos \theta$ (w/sr) is then

$$\Phi = \int_{\theta=0}^{\pi/2} \int_{\phi=0}^{2\pi} I_o \cos \theta \sin \theta d\theta d\phi$$

$$= 2\pi I_o \int_{\theta=0}^{\pi/2} \frac{1}{2} \sin 2\theta d\theta$$

$$= -\pi I_o \frac{\cos 2\theta}{2} \Big|_{\theta=0}^{\pi/2} = \frac{-\pi I_o}{2}(-1)$$

$$= \pi I_o \text{ watts}$$

For our next problem we suppose that a large circular disk of radius a meters is a diffuse radiator having a radiance of J_o(w/m²-sr). This disk illuminates a small surface patch that is parallel to the disk and lying at a distance r_o meters down the axis of the disk. To determine the irradiance of the illuminated surface patch, we can reason as follows: A differential area on the disk is $rdrd\phi$. Consider the line segment from this differential area to the illuminated surface patch. It makes an angle of θ with the normal to the disk and has length $r_o/\cos \theta$. The foreshortened differential area is $\cos \theta rdrd\phi$. The solid angle made by the foreshortened differential area is $\cos \theta rdrd\phi/(r_o/\cos \theta)^2$ steradians. The differential irradiance dJ at the illuminated surface patch produced by a radiance of J_o (w/m²-sr) flowing into a solid angle of $\cos \theta rdrd\theta/(r_o/\cos \theta)^2$ (sr), where the angle at which this radiant energy hits the surface with respect to the surface normal is θ, is $J_o \cos^4 \theta rdrd\phi/r_o^2$ (w/m²). Integrating this differential irradiance over all the differential areas of the disk that produce it results in an irradiance J, where

$$J = \int_{r=0}^{a} \int_{\phi=0}^{2\pi} \frac{J_o \cos^4 \theta \, r \, dr \, d\phi}{r_o^2}$$

But $\cos^4 \theta = r_o^4/(r^2 + r_o^2)^2$. Hence

$$J = \frac{2\pi J_o}{r_o^2} \int_{r=0}^{a} \frac{r_o^4}{(r^2 + r_o^2)^2} r \, dr$$

$$= \pi J_o r_o^2 \int_{r=0}^{a} \frac{2r \, dr}{(r^2 + r_o^2)^2} = \frac{-\pi J_o r_o^2}{r^2 + r_o^2} \Big|_{r=0}^{a}$$

$$= \pi J_o \left(1 - \frac{r_0^2}{a^2 + r_0^2}\right) = \frac{\pi a^2 J_o}{r_o^2 + a^2} \text{ (w/m}^2)$$

This result shows that the irradiance depends directly on the area πa^2 of the illuminating disk and the radiance $J_\circ$ of the illuminating disk, and inversely depends on the squared distance of the illuminated surface patch to the periphery of the disk.

A second way to work this problem is first to center a unit hemisphere around the illuminated surface patch, and then integrate over all the energy flowing into the differential surface area of that portion of the hemisphere cut by a cone whose apex is the center of the hemisphere and whose base is the illuminating disk. In this situation we then integrate ϕ from 0 to 2π and θ from 0 to $\theta_\circ$, where $\cos\theta_\circ = r_\circ/\sqrt{r_\circ^2 + a^2}$. The differential surface area is $\sin\theta d\theta d\phi$. The radiance flowing into this surface area is $J_\circ(\text{w/m}^2\text{-sr})$. The energy reaches the illuminated surface at an angle θ with respect to the surface normal. Hence the differential irradiance dJ at the illuminated surface patch produced by the radiance flowing through the differential surface area is

$$dJ = J_\circ \cos\theta \sin\theta d\theta d\phi$$

Integrating this produces

$$J = \int_{\phi=0}^{2\pi} \int_{\theta=0}^{\theta_\circ} J_\circ \cos\theta \sin\theta d\theta d\phi$$
$$= 2\pi J_\circ \left(\frac{-\cos^2\theta}{2} \right) \Big|_{\theta=0}^{\theta=\theta_\circ}$$
$$= -\pi J_\circ (\cos^2\theta_\circ - 1) = \pi J_\circ \sin^2\theta_\circ$$
$$= \pi J_\circ \frac{a^2}{a^2 + r_\circ^2}$$

This solution illustrates an important fact about diffuse radiators. Essentially the constant radiance $J_\circ$ is moving into the solid angle defined by the disk and the center of the small illuminated surface patch. A diffuse illuminator of any shape having radiance $J_\circ(\text{w/m}^2\text{-sr})$ that illuminates exactly this solid angle must produce exactly the same irradiance $\pi J_\circ a^2/(a^2+r_\circ^2)$ (w/m^2) at the illuminated surface patch. Hence illuminating shapes such as tilted ellipses and spheres whose total solid angle oriented around the normal to the illuminated surface patch is the same as the one the disk produced must yield the same irradiance.

12.2.1 Bidirectional Reflectance Function

The bidirectional reflectance distribution function f_r is the fraction of incident light emitted in one direction when the surface is illuminated from another direction. The direction of incident and reflected light rays can be specified in a local coordinate system by using polar and azimuth angles. Figure 12.2 illustrates the specification of light source and sensor positions in spherical coordinates. Let θ be the polar angle between the surface normal and the lens center. Let ϕ be the azimuth angle of the sensor. We use a subscript or superscript e to denote emitting from, and a subscript or superscript i to denote incident to. Let J^i be the irradiance of the incident light at the illuminated surface, and J^r the radiance of the reflected light. Since f_r is the ratio of the scene radiance to the scene irradiance, the differential reflectance model is

$$dJ^r(\theta_e, \phi_e, \theta_i, \phi_i) = dJ^i(\theta_i, \phi_i) f_r(\theta_i, \phi_i, \theta_e, \phi_e)$$

This equation states that the differential emitted radiance (w/m²-sr) in the direction (θ_e, ϕ_e) due to the incident differential irradiance in the direction (θ_i, ϕ_i) is equal to the incident differential irradiance $dJ^i(\theta_i, \phi_i)$ (w/m²) times the bidirectional reflectance distribution function $f_r(\theta_i, \phi_i, \theta_e, \phi_e)$ (1/sr).

To determine the radiance from this differential form, we must integrate.

$$J^r(\theta_e, \phi_e) = \int_{\theta_i=0}^{\pi/2} \int_{\phi_i=0}^{2\pi} dJ^r(\theta_e, \phi_e, \theta_i, \phi_i) \sin\theta_i d\theta_i d\phi_i$$

The integral form of the differential reflectance relationship is

$$J^r(\theta_e, \phi_e) = \int_{\theta_i=0}^{\pi/2} \int_{\phi_i=0}^{2\pi} dJ^i(\theta_i, \phi_i) f_r(\theta_i, \phi_i, \theta_e, \phi_e) \sin\theta_i d\theta_i d\phi_i$$

The bidirectional reflectance distribution function must satisfy the symmetry condition

$$f_r(\theta_i, \phi_i, \theta_e, \phi_e) = f_r(\theta_e, \phi_e, \theta_i, \phi_i)$$

For many surfaces the dependence of f_r on the azimuth angles ϕ_i and ϕ_e is only a dependence on their difference. Thus we may write

$$f_r(\theta_i, \phi_i, \theta_e, \phi_e) = f_r(\theta_i, \theta_e; \phi_e - \phi_i)$$

If the incident source irradiance is isotropic, as it might be if it came from an external source, such as the sky, that radiates light of the same power in all directions, then $dJ^i(\theta_i, \phi_i) = J^i_o$, and we then have

$$J^r(\theta_e, \phi_e) = J^i_o \int_{\theta_i=0}^{\pi/2} \int_{\phi_i=0}^{2\pi} f_r(\theta_i, \phi_i, \theta_e, \phi_e) \sin\theta_i d\theta_i d\phi_i$$

The integral $\int_{\theta_i=0}^{\pi/2} \int_{\phi_i=0}^{2\pi} f_r(\theta_i, \phi_i, \theta_e, \phi_e) \sin\theta_i d\theta_i d\phi_i$ is called the hemispherical reflectance.

If the radiance comes from only one direction, (θ_o, ϕ_o), then we can write the radiance as $J^i_o \delta(\theta_i - \theta_o)\delta(\phi_i - \phi_o)/\sin\theta_o$ (w/m²-sr), where δ is the Dirac delta distribution function. It is defined by $\int_{x=-\infty}^{\infty} \delta(x-x_o)f(x)dx = f(x_o)$, from which it follows that $\int_{x=-\infty}^{\infty} \delta(x-x_o)dx = 1$.

In the expression $J^i_o \delta(\theta_i - \theta_o)\delta(\phi_i - \phi_o)/\sin\theta_o$, J^i_o has units of w/m², and $\delta(\theta_i - \theta_o)\delta(\phi_i - \phi_o)/\sin\theta_o$ has units of sr⁻¹.

Writing the radiance as $J_o^i \delta(\theta_i - \theta_o)\delta(\phi_i - \phi_o)/\sin\theta_o$ may seem a bit counterintuitive. Integrating this over the solid angle of 4π steradian, however, produces

$$\int_{\phi_i=0}^{\pi/2} \int_{\phi_i=0}^{2\pi} \frac{J_o^i \delta(\theta_i - \theta_o)\delta(\phi_i - \phi_o)}{\sin\theta_o} \sin\theta_i d\theta_i d\phi_i$$

$$= \frac{J_o^i}{\sin\theta_o} \int_{\theta_i=0}^{\pi} \delta(\theta_i - \theta_o)\sin\theta_i d\phi_i$$

$$= \frac{J_o^i}{\sin\theta_o} \sin\theta_o = J_o^i$$

indicating that J_o^i is the irradiance and the term $\delta(\theta_i - \theta_o)\delta(\phi_i - \phi_o)/\sin\theta_o$ simply gives the proper angular dependence. If the radiance is $J_o^i \delta(\theta_i - \theta_o)\delta(\phi_i - \phi_o)/\sin\theta_o$, then the incident irradiance is $J_o^i \delta(\theta_i - \theta_o)\delta(\phi_i - \phi_o)\cos\theta_o/\sin\theta_o$. If the incident irradiance comes from only one direction, the reflected radiance can be written as

$$J^r(\theta_e, \phi_e) = \int_{\theta_i=0}^{\pi/2} \int_{\phi_i=0}^{2\pi} J_o^i \frac{\delta(\theta_i - \theta_o)\delta(\phi_i - \phi_o)\cos\theta_o}{\sin\theta_o} f_r(\theta_i, \phi_i, \theta_e, \phi_e) \sin\theta_i d\theta_i d\phi_i$$

$$= \frac{J_o^i \sin\theta_o \cos\theta_o f_r(\theta_o, \phi_o, \theta_e, \phi_e)}{\sin\theta_o} = J_o^i f_r(\theta_o, \phi_o, \theta_e, \phi_e)\cos\theta_o$$

A Lambertian surface is a perfectly diffusing surface, with a matte appearance. The bidirectional reflectance distribution function for it is $f_r(\theta_i, \phi_i, \theta_e, \phi_e) = r/\pi$. Therefore the differential relationship for the emitted radiance for a Lambertian surface is

$$dJ^r(\theta_e, \phi_e) = \frac{rdJ^i}{\pi} \quad \text{w/m}^2 - \text{sr}$$

A Lambertian surface has the property of consistent brightness no matter from what direction it is viewed. Hence in any direction the power radiated into a fixed solid angle is the same. This does not mean, however, that each differential area of the Lambertian surface radiates the same amount of power in all directions. To understand this difference, think of a sensor that, through a viewing cone of fixed radial angle Ω, measures the power emitted into the viewing cone. The output of the sensor is a number that is the power emitted per steradian. Suppose the distance between the sensor and the surface is d meters and the direction of the sensor relative to the illuminated surface patch is (θ_e, ϕ_e). The area of the footprint of the sensor's viewing cone on the surface patch is $\Omega d^2/\cos\theta_e$ square meters.

The viewing cone then integrates all the radiant power that is radiated from the area $\Omega d^2/\cos\theta_e$ in the direction (θ_e, ϕ_e). The radiated power (w/m²-sr) that moves through or from an area in a particular direction is called the specific intensity (Chandrasekhar, 1960). The difference between it and radiance is that radiance is the radiated power (w/m²-sr) that moves through an area that must be perpendicular to the direction of flow. Specific intensity does not require the area to be the perpendicular area associated with the solid angle.

So from the point of view of the Lambertian radiator, the power radiated (w/m²-sr) from the area $\Omega d^2/\cos\theta_e$—that is, the specific intensity—must be $J^r\cos\theta_e$. The product of $\Omega d^2/\cos\theta_e$ and $J'\cos\theta_e$ is the power per steradian flowing into the viewing cone, and it does not depend on the viewing angle.

Consider now the ratio of the total power radiated onto a Lambertian surface to the total power reflected from the surface. Let dJ^i be the incident irradiance w/m^2 on the Lambertian surface having area dA. Then the total power incident on the Lambertian surface is $dJ^i dA$ watts. Let dJ^r be the reflected radiance (w/m^2-sr). For a Lambertian surface, $dJ^r = rdJ^i/\pi$. The flux density of the radiated power is $\int_{\theta_r=0}^{\pi}\int_{\phi=0}^{2\pi} dJ^r\cos\theta_r\sin\theta_r d\theta_r d\phi_r = \pi dJ^r w/m^2$. The total power radiated is $\pi dJ^r dA$ watts. The ratio of the power radiated to the power incident is then $\pi dJ^r dA/dJ^i dA = r$. The unitless fraction r is called the reflectance factor of the Lambertian surface.

Values of the reflectance factors for some common materials whose dominant reflection characteristic is Lambertian include the following: For white blotting paper, $r = .8$; for white writing paper, $r = .68$; for white ceilings or yellow paper, $r = .6$; for dark brown paper, $r = .13$; and for dark velvet, $r = .004$.

■ EXAMPLE 12.1

Consider a uniform point light source having a radiant power of 300 watts per steradian. A surface patch of 1 cm^2 is illuminated by this light source. The surface patch is 2 meters from the light source, and the cosine of the angle between its surface normal and the incident vector of the light source is .866. The reflectance of the surface patch is Lambertian with reflectivity of .7.

A 1-cm^2 patch at 2 meters, with the cosine of the angle between its surface normal and the incident light being .866, subtends a solid angle of $.866\frac{10^{-4}}{4}$ steradians. Therefore

$$\text{Irradiance} = 300\frac{\text{watts}}{\text{sr}} \times \frac{10^{-4}}{4}\text{sr} \times .866 \times \frac{1}{1\text{ cm}^2} = 6.495 \text{ milliwatts}/\text{cm}^2$$

To determine the radiance produced by this illuminated Lambertian surface patch, we note that because the surface is Lambertian, to convert from irradiance to radiance we must divide by π:

$$\text{Radiance} = 6.495\frac{\text{milliwatts}}{\text{cm}^2} \times .7\frac{1}{\pi} = 1.447\frac{\text{milliwatts}}{\text{cm}^2 - \text{sr}}$$

A surface that is a perfect reflector has a bidirectional reflection distribution function $f_r(\theta_i,\phi_i,\theta_e,\phi_e) = \delta(\theta_i-\theta_e)\delta(\phi_e-\phi_i-\pi)/(\sin\theta_i\cos\theta_i)$. Suppose that the normal incident irradiance is $J(\theta_i,\phi_i)$. Then the incident irradiance at the surface is $J(\theta,\phi_i)\cos\theta_i$. If the surface is a perfect reflector, then the reflected radiance

$J^r(\theta_e, \phi_e)$ can be computed from

$$J^r(\theta_e, \phi_e) = \int_{\phi_i=0}^{2\pi} \int_{\theta_i=0}^{\pi/2} \frac{\delta(\theta_i - \theta_e)\delta(\phi_e - \phi_i - \pi)}{\sin\theta_i \cos\theta_i} J(\theta_i, \phi_i) \cos\theta_i \sin\theta_i d\theta_i d\phi_i$$

$$= J(\theta_e, \phi_e - \pi)$$

This illustrates what we expect: The reflected radiance from a perfect radiator is related to the mathematical reflection of the normal incident irradiance. If the source radiance is $J^i = J^i_o[\delta(\theta_i - \theta_o)\delta(\phi_e - \phi_o)]/\sin\theta_o$, then the reflected radiance is

$$J^r(\theta_e, \phi_e) = \int_{\theta_i=0}^{\pi/2} \int_{\phi_i=0}^{2\pi} J^i_o \frac{\delta(\theta_i - \theta_o)\delta(\phi_e - \phi_o)}{\sin\theta_o}$$

$$\frac{\delta(\theta_i - \theta_e)\delta(\phi_e - \phi_i - \pi)}{\sin\theta_i \cos\theta_i} \sin\theta_i \cos\theta_i d\theta_i d\phi_i$$

$$= J^i_o \frac{\delta(\theta_e - \theta_o)\delta(\phi_e - \phi_o - \pi)}{\sin\theta_e}$$

12.2.2 Photometry

Photometry is the study of radiant light energy specifically as it results in physical sensation in the human eye. From this point of view, brightness can be defined as that attribute of sensation by which an observer is aware of differences of observed radiant energy. Photometry has a set of concepts parallel to those of radiometry. Where radiometry refers to radiant energy, photometry refers to luminous energy. Where radiometry refers to power, photometry refers to luminous flux. The unit of luminous flux is the lumen. The luminous flux leaving a point source per unit solid angle is called the luminous intensity. It has units of lumens per steradian. By definition, one lumen per steradian is equal to one candela. The luminous flux per unit area incident upon a surface is called illuminance, and its units are lumens per square meter. One lumen per square meter is, by definition, one lux. One lumen per square foot is called a foot-candle. Since one foot equals .3048 meters,

$$1\text{lux} = \frac{1}{(.3048)^2} \text{ foot-candles}$$

$$= 10.76 \text{ foot-candles}$$

Finally, the luminous flux per unit solid angle per unit of projected area is called luminance, and its units are lumens per square meter per steradian.

As a frame of reference, the sun observed at zenith from the earth's surface has a luminance of 1.6×10^9 candelas per square meter. A frosted 60-watt bulb has a luminance of 1.2×10^5 candelas per square meter. The average brightness of a clear sky is 8×10^3 candelas per square meter. The ground on an overcast day has a brightness of between 34 and 100 candelas per square meter. Luminances greater than 5 candelas per square inch are unsafe for long-term direct human viewing.

The smallest noticeable luminance is about .001 candelas per square meter. The luminous intensity of a candle is generally between 1 and 2 candelas.

12.2.3 Torrance-Sparrow Model

The Torrance-Sparrow model is the basis for a reflectance model that can be generalized for all states of polarization of light: completely unpolarized, partially polarized, or completely polarized.

The Torrance-Sparrow model (1965, 1966, 1967) assumes that the intensity of the reflected light J^r from a roughened surface consists of two parts: the specular reflection J^r_s from the mirrorlike surface facets and the diffuse reflection J^r_d. The specular part is a function of the angle of reflection, and the diffuse part is independent of the angle of reflection. Therefore J^r_s is dependent on the view point, whereas J^r_d is not.

Considering surfaces for which $f_r(\theta_i, \phi_i, \theta_e, \phi_e) = f_r(\theta_i, \theta_e; \phi_e - \phi_i)$, we write $\phi = \phi_e - \phi_i$. The angles are illustrated in Fig. 12.5.

The Torrance-Sparrow model states that

$$J^r(\theta_i; \theta_e, \phi; \lambda) = sJ^r_s(\theta_i; \theta_e, \phi; \lambda) + (1 - s)J^r_d(\theta_i; \lambda) \qquad (12.5)$$

where $s(0 \leq s \leq 1)$ denotes the proportion of specular reflection that depends on the photometric properties of the object surface and λ denotes the wavelength of light. If a source of constant radiance J^i is oriented at various angles of incidence θ_i, the diffuse component of reflected flux will vary directly with $\cos\theta_i$. In other words,

$$J^r_d(\theta_i; \lambda) = r_d J^i \cos\theta_i d\omega \qquad (12.6)$$

for some constant r_d, which is the diffuse reflectance factor of the surface.

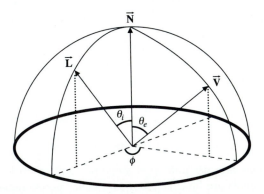

Figure 12.5 Light source and sensor positions as specified by a pair of angles: The polar angle is measured from the north pole and the azimuth angle is measured from a reference line along the equator. For surfaces whose reflection function depends only on the differences between the incident and emitted azimuth angles, we use the following convention: We choose the reference line as the line passing through the origin and the projected point of the light source position on the xy-plane.

The reflected radiance J_s^r is the specularly reflected radiant flux per unit projected surface area per unit solid angle. The detail derivation of J_s^r is given in Torrance and Sparrow (1967). Briefly,

$$J_s^r(\theta_i; \theta_e, \phi; \lambda) = J^i d\Omega \frac{FGD}{\pi \cos\theta_e} \qquad (12.7)$$

F is called the Fresnel term and is a function of the complex index of refraction of the material and the incidence angle to the microfacet. Since the Fresnel term comes from the reflection of specially polarized light, we will focus on it in Section 12.4.

G is called the geometric attenuation factor and takes shadowing and masking of one microfacet by another into consideration. Let $\vec{N}$ be a unit surface normal and $\vec{L}$ and $\vec{V}$ be unit positional vectors of the light source and the sensor, respectively (see Fig. 12.5). Let $\vec{H}$ be the unit vector that bisects the angle between $\vec{L}$ and $\vec{V}$; $\vec{H} = \vec{L} + \vec{V}/\|\vec{L} + \vec{V}\|$. Then G can be written as

$$G = \min\left\{1, \frac{2(\vec{N}\cdot\vec{H})(\vec{N}\cdot\vec{V})}{\vec{V}\cdot\vec{H}}, \frac{2(\vec{N}\cdot\vec{H})(\vec{N}\cdot\vec{L})}{\vec{V}\cdot\vec{H}}\right\} \qquad (12.8)$$

D is considered the surface roughness index that is modeled by a facet slope distribution. For isotropic surfaces a normal distribution for the number of facets whose surface normals are within a small solid angle in the direction of $\vec{H}$ per unit surface area is assumed.

If we substitute relations (12.6) and (12.7), Eq. (12.5) results in

$$J^r(\theta_i; \theta_e, \phi; \lambda) = J^i d\Omega \cos\theta_i [sR_s + (1-s)R_d] \qquad (12.9)$$

where

$$R_s = \frac{FDG}{\pi \cos\theta_i \cos\theta_e} \qquad (12.10)$$

It is easy to see that angles θ_i, θ, and ϕ can be represented by the three vectors $\vec{N}$, $\vec{L}$, and $\vec{V}$, where

$$\cos\theta_i = \vec{N}\cdot\vec{L}$$
$$\cos\theta_e = \vec{N}\cdot\vec{V}$$
$$\cos\phi = \frac{[\vec{L} - (\vec{L}\cdot\vec{N})\vec{N}]\cdot[\vec{V} - (\vec{V}\cdot\vec{N})\vec{N}]}{\|[\vec{L} - (\vec{L}\cdot\vec{N})\vec{N}]\|\,\|[\vec{V} - (\vec{V}\cdot\vec{N})\vec{N}]\|}$$

With these identities we can rewrite Eq. (12.10) as

$$R_s = \frac{FDG}{\pi(\vec{N}\cdot\vec{L})(\vec{N}\cdot\vec{V})} \qquad (12.11)$$

12.2.4 Lens Collection

Not all the light reflected from the object surface is captured by the lens system. Only a portion of the reflected light comes through the lens and affects the camera film. We call that portion of the reflected light the *lens collection*, and it can be derived by considering how the image irradiance is related to the scene radiance. Let r_1 be the distance between the object and the lens, and r_2 the distance between the lens and the image of the object. Let f be the distance between the image plane and the lens, a the diameter of the lens, and α the angle between the ray from the object patch to the center of the lens (see Fig. 12.6).

To find the irradiance incident on differential area da_2 coming from differential area da_1, having radiance dJ^i, and passing through a lens having aperture area $A = \pi a^2/4$, we can reason as follows: The foreshortened area of the aperture stop as seen by the differential area da_1 is $A \cos \alpha$. The distance from the area da_1 to the aperture is $s_1/\cos \alpha$. The solid angle Ω subtended by the aperture stop as seen from da_1 is

$$\Omega = \frac{A \cos \alpha}{r_1^2} = \frac{A \cos^3 \alpha}{s_1^2}$$

The differential radiant power $d\Phi$ passing through the aperture due to the energy radiating from da_1 is

$$d\Phi = dJ^i \Omega da_1 \cos \alpha$$

Therefore the radiant power passing through the aperture from da_1 is

$$d\Phi = \frac{dJ^i A \cos^4 \alpha da_1}{s_1^2}$$

If we neglect transmission losses through the optical system, the radiant power reaching da_2 is $d\Phi$. The irradiance incident to da_2 is

$$dJ^r = \frac{d\Phi}{da_2} = \frac{dJ^i A \cos^4 \alpha da_1}{s_1^2 da_2}$$

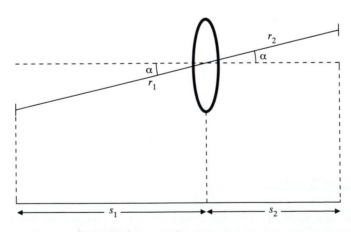

Figure 12.6 Geometry for lens collection.

If we assume $s_1 \gg s_2$, then $s_2 = f$ and the magnification provided by the lens is s_1/s_2. Hence $da_2/da_1 = (s_1/s_2)^2 = s_1^2/f^2$. Therefore

$$dJ^r = \frac{dJ^i A \cos^4 \alpha}{s_1^2} \frac{s_1^2}{f^2} = \frac{dJ^i A \cos^4 \alpha}{f^2}$$

Since $A = \pi a^2/4$,

$$dJ^r = \frac{\pi dJ^i \cos^4 \alpha}{4} \left(\frac{a}{f}\right)^2$$

Then the lens collection C is given by:

$$C = \frac{\pi}{4} \left(\frac{a}{f}\right)^2 \cos^4 \alpha \qquad (12.12)$$

Lens collection is discussed in detail in Horn (1986).

12.2.5 Image Intensity

The image intensity gray level I associated with some small area of the image plane can then be represented as the integral of all light collected at the given pixel position coming from the observed surface patch, modified by sensor gain g and bias b. If $J^r(\theta, \phi; \lambda)$ is the radiance of the observed surface patch (watts/m²-sr-m), if the viewing cone of the camera for the pixel position subtends a solid angle of Ω, and if the distance to the observed patch is r, then the power received for the pixel position is $J^r(\theta, \phi; \lambda)\Omega(\Omega r^2)$, and we can write

$$I(\theta, \phi) = g \int_\lambda C\, S(\lambda) J^r(\theta, \phi; \lambda)\Omega^2 r^2 d\lambda + b$$

12.3   Photometric Stereo

In photometric stereo there is one camera but K light sources having known intensitites $i_1, \ldots, i_K$ and incident vectors $v_1, \ldots, v_K$ to a given surface patch. Let n be the surface normal vector of the surface patch having Lambertian reflectance with reflectivity r.

In photometric stereo the camera sees the surface patch K times, one time when each light source is activated and the remaining ones are deactivated. This produces the observed gray levels $f_1, \ldots, f_K$. By the model of Lambertian reflectance,

$$f_k = g i_k v_k \cdot n + b, \quad k = 1, \ldots, K \qquad (12.13)$$

where g is the gain of the sensor and b is its offset.

If the camera has been photometrically calibrated, g and b are known. So letting $f_k^* = \dfrac{f_k - b}{g i_k}$ and

$$f^* = \begin{pmatrix} f_1^* \\ \vdots \\ f_K^* \end{pmatrix} \qquad V = \begin{pmatrix} v_1' \\ \vdots \\ v_K' \end{pmatrix}$$

we can write Eq. (12.13) in matrix form as

$$f^* = rVn \tag{12.14}$$

If the surface normal n is known, the least-squares solution for the reflectivity r is given by

$$r = \frac{f^{*\prime}V_n}{(Vn)'(Vn)}$$

If $K = 3$, a solution for the surface normal n is given by

$$n = \frac{V^{-1}f^*}{\|V^{-1}f^*\|}$$

This is essentially the form Woodham (1978) gives.

If $K > 3$, a least-squares solution can be obtained from

$$n = \frac{(V'V)^{-1}V'f^*}{\|(V'V)^{-1}V'f^*\|}$$

If the g and b are not known, the camera must first be calibrated. This can be done by using a geometric setup where the incident angle of the light source relative to the surface normal is known. The surfaces with known reflectivities are illuminated with a light source of known intensities.

Let i_k be the known intensity of the light source for the kth trial; V_k, the known incident direction of the light source for the kth trial; n, the known unit length surface normal vector; r_k, the known reflectivity of the surface illuminated for the kth trial; and y_k, the observed value from the camera. Let $X_k = i_k r_k v_k n$. Then the unknown gain g and offset b satisfy

$$\begin{pmatrix} x_1 & 1 \\ x_2 & 1 \\ & \vdots \\ x_k & 1 \end{pmatrix} \begin{pmatrix} g \\ b \end{pmatrix} = \begin{pmatrix} y_1 \\ y_2 \\ \vdots \\ y_k \end{pmatrix}$$

This leads to the least-squares solution for (g, b):

$$\begin{pmatrix} g \\ b \end{pmatrix} = \begin{pmatrix} \sum_{k=1}^{K} x_k^2 & \sum_{k=1}^{K} x_k \\ \sum_{k=1}^{K} x_k & K \end{pmatrix}^{-1} \begin{pmatrix} \sum_{k=1}^{K} x_k y_k \\ \sum_{k=1}^{K} y_k \end{pmatrix}$$

Ikeuchi (1981) used a line light source and derived a photometric stereo method for specular reflection. Ray, Birk, and Kelley (1983) did an error analysis and found that the surface normal could be determined to within better than $5°$ in zenith angle and $10°$ in azimuth angle. Ikeuchi (1987) used a dual camera photometric stereo that permits depth determination as well as surface normal calculation.

12.4   Shape from Shading

Nonplanar Lambertian surfaces of constant reflectance factor appear shaded on an image. The shading gives a secondary clue to the shape of the observed surface. In this section we briefly discuss how the shape of the Lambertian surface may be recovered from the image shading.

Suppose that a Lambertian surface is illuminated by a distant point source of light located in a direction having unit vector (a,b,c) from the surface. Suppose also that this surface is viewed by a distant camera so that the perspective projection can be approximated by an orthographic projection. This means that a surface point having position (x,y,z) projects to the image at position (x,y). If the surface can be expressed as $z = g(x,y)$, then the unit vector normal to the surface at (x,y) is

$$\frac{1}{\sqrt{\left(\frac{\partial g}{\partial x}\right)^2 + \left(\frac{\partial g}{\partial y}\right)^2 + 1}} \begin{pmatrix} \frac{\partial g}{\partial x} \\ \frac{\partial g}{\partial y} \\ -1 \end{pmatrix}$$

The gray level at the image position (x,y) can, within a multiplicative constant, be written as

$$I(x,y) = \frac{ap(x+y) + bq(x+y) - c}{\sqrt{p^2(x,y) + q^2(x,y) + 1}} \tag{12.15}$$

and $p = (\partial g/\partial x)$ and $q = (\partial g/\partial y)$.

This immediately shows that planar surfaces have uniform brightness because $p(x,y)$ and $q(x,y)$ are constant for such surfaces. Surfaces that have some curvature are surfaces for which the partial derivatives $p(x,y)$ and $q(x,y)$ are not constant. Hence, because of Eq. (12.15), the variation in surface brightness has some connection with the variation expressed by the partial derivatives $p(x,y)$ and $q(x,y)$. These partial derivatives, in turn, provide information about surface height $g(x,y)$.

Since $p(x,y)$ and $q(x,y)$ are partial derivatives of g, we can write a first-order Taylor expression for g as follows:

$$g(x,y) = g(x+1,y) - p(x,y)$$
$$g(x,y) = g(x,y+1) - q(x,y) \tag{12.16}$$

Equations (12.15) and (12.16) constitute a system of nonlinear equations that, with boundary conditions on $g(x,y)$, can be solved for the unknown surface height $g(x,y)$ and partial derivatives $p(x,y)$ and $q(x,y)$. However, the system is not an overconstrained system, and without any further assumptions the solution, by whatever technique, is likely to be very sensitive to noise.

Horn (1975) was the first to solve Eq. (12.15) and (12.16) by a characteristic strip method; however, the solution is quite sensitive to noise. To help reduce this sensitivity, Ikeuchi and Horn (1981) solve the system of equations (12.15) by constraining $p(x,y)$ and $q(x,y)$ to be smooth functions. This solution does not make use of the system of equations (12.16). Ikeuchi and Horn proceed as follows: Define the reflectance map $R(p,q)$ by

$$R(p,q) = \frac{ap+bq+c}{\sqrt{p^2+q^2+1}}$$

Let a penalty constant λ be chosen. Define the criterion function to be minimized by the choice of p and q as

$$\epsilon^2 = \sum_r \sum_c \left\{ I(r,c) - R\left[p(r,c),q(r,c)\right] \right\}^2$$
$$+ \lambda \left\{ [p(r+1,c) - p(r,c)]^2 + [p(r,c+1) - p(r,c)]^2 \right.$$
$$\left. + [q(r+1,c) - q(r,c)]^2 + [q(r,c+1) - q(r,c)]^2 \right\}$$

The minimizing p and q must make the partial derivatives of ϵ^2 with respect to p and q at each pixel (r,c) equal to zero.

Now

$$\frac{\partial \epsilon^2}{\partial p(i,j)} = -2\left\{ I(i,j) - R\left[p(i,j),q(i,j)\right] \right\} \frac{\partial R}{\partial p}\left[p(i,j),q(i,j)\right]$$
$$+ \lambda \left\{ 2\left[p(i+1,j) - p(i,j)\right](-1) + 2\left[p(i,j+1) - p(i,j)\right](-1) \right.$$
$$\left. + 2\left[p(i,j) - p(i-1,j)\right] + 2\left[p(i,j) - p(i,j-1)\right] \right\}$$

Setting the partial derivative to zero and rearranging produces

$$p(i,j) = \bar{p}(i,j) + \eta \left\{ I(i,j) - R\left[p(i,j),q(i,j)\right] \right\} \frac{\partial R}{\partial p}\left[p(i,j),q(i,j)\right]$$

where

$$\bar{p}(i,j) = \frac{1}{4}\left[p(i+1,j) + p(i-1,j) + p(i,j+1) + p(i,j-1)\right] \qquad (12.17)$$

and $\eta = 4/\lambda$.

Similarly, taking the partial derivative of ϵ^2 with respect to $q(i,j)$ and setting it to zero results in

$$q(i,j) = \bar{q}(i,j) + \eta \left\{ I(i,j) - R\left[p(i,j),q(i,j)\right] \right\} \frac{\partial R}{\partial q}\left[p(i,j),q(i,j)\right] \qquad (12.18)$$

The pair of equations (12.17) and (12.18) then provide the basis of an iterative solution technique. For each (i,j), let $p_o(i,j)$ and $q_o(i,j)$ be given. Then define the iteration

$$p_{t+1}(i,j) = \bar{p}_t(i,j) + \eta \left\{ I(i,j) - R\left[\bar{p}_t(i,j), \bar{q}_t(i,j)\right] \right\} \frac{\partial R}{\partial p} \left[p_t(i,j), q_t(i,j)\right]$$

$$q_{t+1}(i,j) = \bar{q}_t(i,j) + \eta \left\{ I(i,j) - R\left[\bar{p}_t(i,j), \bar{q}_t(i,j)\right] \right\} \frac{\partial R}{\partial q} \left[p_t(i,j), q_t(i,j)\right]$$

A major difficulty with the Ikeuchi and Horn procedure is that the system of equations (12.16) is not used. This system has two related kinds of information. The first is that, given boundary values such as a first row and column for g, it is possible to determine g everywhere else. The second is a relationship between p and q:

$$p(r,c+1) - p(r,c) = q(r+1,c) - q(r,c) \qquad (12.19)$$

In problems that are essentially unconstrained, it is vital to use all constraints. In this case, not using Eq. (12.16) produces p and q that do not satisfy (12.16). Hence, there is no guarantee that there is any function with p and q as its first partial derivatives.

Smith (1982) uses a similar idea to Ikeuchi and Horn's. However, the penalty function Smith uses is

$$\sum_{(r,c)} [\tilde{p}(r,c)]^2 + [\tilde{q}(r,c)]^2$$

where $\tilde{p}(r,c) = p(r,c+1) + p(r,c-1) + p(r+1,c) + p(r-1,c) - 4p(r,c)$ and $\tilde{q}(r,c) = q(r,c+1) + q(r,c-1) + q(r+1,c) + q(r-1,c) - 4q(r,c)$.

Stratt (1979) makes use of the relationship (12.19). We decide here to use a slight modification of the Strat approach. The criterion function is to be minimized by a choice of $p(r,c)$ and $q(r,c)$ in

$$\epsilon^2 = \sum_{(r,c)} \left\{ I(r,c) - R\left[p(r,c), q(r,c)\right] \right\}^2$$

$$+ \lambda \sum_{(r,c)} \left[p(r,c+1) - p(r,c) - q(r+1,c) + q(r,c)\right]^2$$

where λ is the given constant that weights the second error term relative to the first. Taking the partial derivative of ϵ^2 with respect to $p(i,j)$ produces

$$\frac{\partial \epsilon^2}{\partial p(i,j)} = -2\left\{ I(i,j) - R\left[p(i,j), q(i,j)\right] \right\} \frac{\partial R}{\partial p}\left[p(i,j), q(i,j)\right]$$

$$+ 2\lambda \left\{ (p(i,j) - p(i,j-1) - q(i+1,j-1) + q(i,j)(-1) \right.$$

$$\left. + [p(i,j+1) - p(i,j) - q(i+1,j) + q(i,j)](-1) \right\}$$

Setting the partial derivative to zero and rearranging results in

$$p(i,j) = \bar{p}(i,j) - \tilde{q}(i,j) + \eta \{I(i,j) - R[p(i,j),q(i,j)]\} \frac{\partial R}{\partial p} [p(i,j),q(i,j)]$$

where

$$\bar{p}(i,j) = \frac{1}{4} [p(i,j-1) + p(i,j+1) + 2p(i,j)]$$

$$\tilde{q}(i,j) = \frac{1}{4} [q(i,j-1) - q(i+1,j-1) + q(i+1,j) - q(i,j)]$$

$$\eta = \frac{1}{4\lambda}$$

Similarly, taking the partial derivatives of ϵ^2 with respect to $q(i,j)$ produces

$$q(i,j) = \bar{q}(i,j) - \tilde{p}(i,j) + \eta \{I(i,j) - R[p(i,j),q(i,j)]\} \frac{\partial R}{\partial q} [p(i,j),q(i,j)]$$

These relationships then suggest the following iterative solution. Let $p_o(i,j)$ and $q_o(i,j)$ be the initial values. Then define the next values in each iteration by

$$p_{t+1}(i,j) = \bar{p}_t(i,j) - \tilde{q}_t(i,j) + \eta \{I(i,j) - R[p_t(i,j),q_t(i,j)]\} \frac{\partial R}{\partial p}$$

$$\times [p_t(i,j),q_t(i,j)]$$

$$q_{t+1}(i,j) = \bar{q}_t(i,j) - \tilde{p}_t(i,j) + \eta \{I(i,j) - R[p_t(i,j),q_t(i,j)]\} \frac{\partial R}{\partial q}$$

$$\times [p_t(i,j),q_t(i,j)]$$

The methods we have discussed up to now for the inference of shape from shading are global methods. Pentland (1984) addresses the problem of what can be inferred on the basis of local intensity information, such as first and second derivatives, when the local intensity value comes from a spherically shaped object surface patch. Pentland shows that under this spherical-surface-patch constraint, there are two possible interpretations for any local intensity information pattern: one a spherical convex surface and the other a spherical concave surface.

One way to make the shape-from-shading problem easier to solve is to provide more constraints. These constraints must come from additional assumptions. Pentland made the assumption of spherical object surface. This assumption can be relaxed some. The shape-from-shading problem can be formulated by assuming that the observed surface is quadratic. Under this constraint,

$$g(r,c) = k_1 + k_2 r + k_3 c + k_4 r^2 + k_5 rc + k_6 c^2$$

$$p(r,c) = k_2 + 2k_4 r = k_5 c$$

$$q(r,c) = k_3 + k_5 r + 2k_6 c$$

Now we can write for the criterion function

$$\epsilon^2(k_2,k_3,k_4,k_5,k_6) = \sum_{(r,c)} \left[I(r,c) - \frac{ap(r,c) + bq(r,c) + c}{\sqrt{1 + p^2(r,c) + q^2(r,c)}} \right]^2$$

To minimize the criterion function, an optimization technique such as the Marquardt (Draper and Smith, 1987) method can be used.

12.4.1 Shape from Focus

So far we have discussed shape recovery based on the shading produced by Lambertian surfaces. It is also possible to recover shape from the shading profile of object edges. The basic idea is that cameras do not have infinite depth of field. The degree to which edges may be defocused is related to how far the three-dimensional edge is away from the depths at which the edges are sharply in focus. Pentland (1984, 1987), Grossman (1987), Subbarao (1987), and Subbarao and Gurumoorthy (1988) all have made contributions to this technique.

Subbarao and Gurumoorthy compute the line spread function $\theta(r)$ by

$$\theta(r) = \frac{g'_\alpha(r)}{\int_{-\infty}^{\infty} g'_\alpha(r)dr}$$

where α is the gradient direction and g'_α is the first directional derivative taken in the direction of the gradient. The location of the edge is then placed at $\bar{r}$, where

$$\bar{r} = \int_{-\infty}^{\infty} r\theta(r)dr$$

The spread or defocusing of the edge can then be obtained from

$$\sigma^2 = \int_{-\infty}^{\infty} (r - \bar{r})^2 \theta(r)dr$$

Subbarao argues that there is a relationship between σ and μ, the distance between the three-dimensional edge and the camera. He argues that

$$\sigma = m\mu^{-1} + c$$

where m and c are camera constants.

12.5 Polarization†

An illumination source can be characterized by four factors: directionality, intensity, spectral distribution, and polarization. The *directionality* of the illumination source is a term relative to the surface normal and is included in the definition of the bidirectional reflectance function which was discussed in Section 12.2.1. The *intensity* is the energy coming out from the source per unit foreshortened area per unit solid angle. The *spectral distribution* is a function of wavelength λ. Any

---

† The material in section 12.5 to 12.10 is contributed by S. K. Yi and is based on a chapter of his dissertation (Yi, 1989).

light source has its own spectral distribution. *Polarization* is not a measure of the direction, intensity, or divergence of the light and in fact is not a measure of the propagation of the light at all. It describes the time-varying vibration of the light energy in a certain direction. Linearly polarized light changes its direction by 180° every other instant of time. Circularly polarized light changes its direction so that the endpoints of the vibrating vector form a circle. Elliptically polarized light changes its direction so that the endpoints of the vibrating vector form an ellipse. Elliptically or circularly polarized light propagates, making a spine curve. Since every light is at least partially polarized, we begin with a detailed discussion of polarization.

Polarization of light was first discovered by Huygens in 1690. He found that a ray of light is divided into two separate rays of equal intensity when it traverses a crystal of Iceland spar in a direction not parallel to the crystallographic axis. More than a century later Malus detected this phenomenon when light undergoes reflection. The former is called polarization by refraction; the latter, polarization by reflection.

Mathematical Meaning of Polarization

Here we discuss the polarization of light mathematically by using wave theory. Let $\vec{r}(x, y, z)$ be a position vector of a point in space. According to Maxwell's equations, in a homogeneous medium with regions free of currents and charges, each rectangular component $V(\vec{r}, t)$ of the field vectors satisfies the homogeneous wave equation

$$\nabla^2 V - \frac{1}{v^2}\frac{\partial^2 V}{\partial t^2} = 0 \qquad (12.20)$$

where v is the velocity. Let $\vec{s} = (s_x, s_y, s_z)$ be a unit vector in a fixed direction. Then $\vec{r} \cdot \vec{s}$ is the projection of the position onto a line passing through origin and in a direction $\vec{s}$. The points at which $\vec{r} \cdot \vec{s}$ is a given constant are planes where the normal is $\vec{s}$. Therefore any solution of Eq. (12.20) of the form

$$V = V(\vec{r} \cdot \vec{s}, t)$$

is said to represent a *plane wave*. Among the plane waves, a *harmonic plane wave* is suitable for describing the polarization. A harmonic plane wave propagating in the direction $\vec{s}$ with velocity v has the form

$$A \cos\left[\omega\left(t - \frac{\vec{r} \cdot \vec{s}}{v}\right) + \delta\right]$$

where $A(>0)$ is called the *amplitude*, ω is called the *angular frequency*, and δ is a constant. The argument in the cosine term is called the *phase factor*.

When the plane wave is time harmonic, each Cartesian component of the electric vector $\vec{E}$ of the wave has the form

$$A \cos(\tau + \delta)$$

where τ is the variable part of the phase factor. Now we consider how $\vec{E}$ varies with time. Figure 12.7 illustrates propagation of the x- and y-components of light $\vec{E}$ in the z-direction. Let the z-axis be in the $\vec{s}$ direction, and let $\vec{i}$, $\vec{j}$, $\vec{k}$ be the unit vectors in the x-, y-, and z-directions, respectively. Then $\vec{E}$ can be written as

$$\vec{E} = E_x\vec{i} + E_y\vec{j} + E_z\vec{k}$$

where the x-, y-, and z-components of the light $\vec{E}$ are

$$E_x = A\cos(\tau + \delta_x)$$
$$E_y = B\cos(\tau + \delta_y) \qquad (12.21)$$
$$E_z = 0$$

The curve that the endpoint of the electric vector describes at a typical point in space is the locus of the points whose coordinates (E_x, E_y) are given by Eq. (12.21). Eliminating τ from Eq. (12.21), we have

$$\left(\frac{E_x}{A}\right)^2 + \left(\frac{E_y}{B}\right)^2 - \frac{2\cos\delta}{ABE_xE_y} - \sin^2\delta = 0 \qquad (12.22)$$

where $\delta = \delta_y - \delta_x$. This is an ellipse because the corresponding determinant of Eq. (12.22)

$$\begin{vmatrix} \dfrac{1}{A^2} & -\dfrac{\cos\delta}{AB} \\ -\dfrac{\cos\delta}{AB} & \dfrac{1}{B^2} \end{vmatrix} = \frac{\sin^2\delta}{A^2B^2}$$

is nonnegative. This ellipse is inscribed in a rectangle whose sides are parallel to the coordinate axes and whose lengths are $2A$ and $2B$ (See Fig. 12.8). Hence the light $\vec{E}$ is said to be *elliptically polarized*. This means that $\vec{E}$ changes its vibration direction along the perimeter of the ellipse in one cycle.

Let A' and B' be the lengths of the major and minor axes of the ellipse rotated

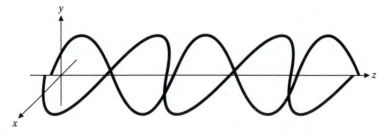

Figure 12.7 Propagation of the x- and y-components of the light $\vec{E}$ in the z-direction. The endpoint of the light $\vec{E}$ changes as it propagates.

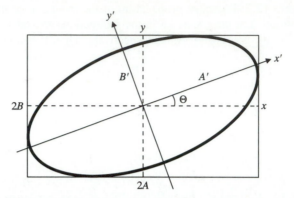

Figure 12.8 Locus of the endpoint of the light $\vec{E}$ forming an ellipse bounded by a rectangle whose sides are parallel to the coordinate axes and whose lengths are $2A$ and $2B$.

by an angle Θ, respectively. Then we have

$$A'^2 = A^2 \cos^2 \Theta + B^2 \sin^2 \Theta + 2AB \cos \Theta \sin \Theta \cos \delta$$
$$B'^2 = A^2 \sin^2 \Theta + B^2 \cos^2 \Theta - 2AB \cos \Theta \sin \Theta \cos \delta$$
$$\tan 2\Theta = \frac{2AB}{A^2 - B^2} \cos \delta$$

It is easy to see that this ellipse degenerates into a straight line or a circle when the parameters A, B, and δ have special values. Figure 12.9 illustrates how polarization direction changes according to the phase difference. Let n be any integer. When the phase difference of the two components is a multiple of π, the ellipse becomes a straight line. More precisely, if

$$\delta = \delta_y - \delta_x = n\pi$$

then Eq. (12.22) reads

$$E_y = (-1)^n \frac{B}{A} E_x$$

This means that $\vec{E}$ changes its vibration direction 180° every other instant of time. Therefore $\vec{E}$ is said to be *linearly polarized*. When the amplitudes of the x- and y-components are equal and the phase difference of the two components is a multiple of $\pi/2$, the ellipse becomes a circle. Precisely, if

$$\delta = n\pi \pm \frac{\pi}{2} \quad \text{and} \quad A = B$$

then Eq. (12.22) becomes

$$E_x^2 + E_y^2 - A^2 = 0$$

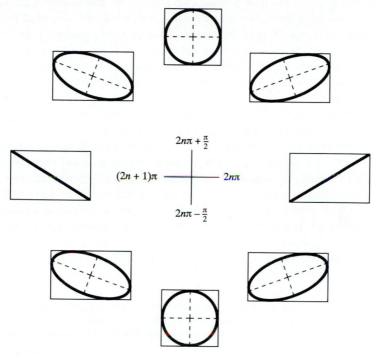

Figure 12.9 Polarizations of light with various phase differences δ. In general, the endpoint of the light $\vec{E}$ forms an ellipse, and the ellipse degenerates into a straight line or a circle when δ and the lengths of two sides of the bounding rectangle have special values. When $δ = nπ$, $\vec{E}$ is said to be linearly polarized. When $δ = 2nπ \pm π/2$ and the two sides of the bounding rectangle are of equal length, $\vec{E}$ is said to be circularly polarized.

Hence $\vec{E}$ is said to be *circularly polarized*. This means that $\vec{E}$ changes its vibration direction along the perimeter of the circle in one cycle. Moreover, if $δ = 2nπ + \frac{π}{2}$, $\vec{E}$ is said to be *right-handed* circular polarization, and if $δ = 2nπ - \frac{π}{2}$, $\vec{E}$ is said to be *left-handed* circular polarization. The former makes a circle in the clockwise direction; the latter, in the counterclockwise direction.

Usefulness of Polarization in Machine Vision

Many sources in the literature confirm the importance and usefulness of polarization of light in machine vision (Batchelor, 1985; Beckmann, 1968; Beckmann and Spizzichino, 1963; Browne and Norton-Wayne, 1986; Egan, 1985; Harding, 1987, 1988; Kingslake, 1983; Koshikawa, 1979; Nicodemus et al., 1977; Schroeder, 1984; Torrance, Sparrow, and Birkebak, 1966; and Wolfe and Wang, 1982). Now we briefly discuss how polarization of light can be used in machine vision.

Any polarization of light can be regarded as a sum of two linearly polarized waves of light with some constant phase difference between them. If a linear polarizer is placed at 45° to the polarization of a linearly polarized beam of light, only some fraction of light will be transmitted. The amount of transmitted light is 1 over the square root of 2 times the total intensity, which can be computed from a 45° right triangle decomposition of the vector of the linearly polarized light. If another polarizer is placed after the first at an additional 45° from the first, again the same fraction of the beam transmitted by the first polarizer is transmitted. Now if the first polarizer is removed from the beam path, the light will be totally blocked by the second polarizer because it is perpendicular to the linearly polarized light, and therefore it only passes light that is orthogonally polarized from the input beam. This effect is due to the vector property of the polarization.

The light wave can be considered a vector sum of two separate and orthogonal waves of light—for example, components polarized parallel and perpendicularly to the incidence plane. By changing the phase difference between the two separate waves, the vectors of polarization can be further affected. One wave of light may be delayed with respect to the other if the light passes through a medium whose index of refraction is different for a parallel polarized wave of light from that for a perpendicularly polarized wave. Such materials are called *birefringent*. The delay means that one wave may see a slightly longer path than the other. When these materials are arranged such that one wave sees an optical path half a wavelength longer than that seen by the other wave, the resulting optic system is called a half-wave plate. A half-wave plate maintains the linear polarization but rotates the orientation of a linear polarization. This effect is important because many sensors have different responses to different polarizations.

A quarter-wave plate changes linear polarization to circular polarization, or circular polarization back to linear polarization. When illuminating an object that has a different reflection coefficient depending on the state of polarization, a circularly polarized light can be used to suppress glints while minimizing the effect of illumination incidence angle.

At Brewster's angle, the parallel polarized light is totally transmitted and the perpendicularly polarized light is partially transmitted and partially reflected. This effect can be used to remove the specular reflections from the window or metal surfaces by looking through them at Brewster's angle.

12.5.1   Representation of Light Using the Coherency Matrix

Monochromatic light is always polarized. In other words, the endpoint of the electric vector at each point in space moves periodically with increasing time. In the case of unpolarized light, the endpoint may be assumed to move quite irregularly, and the light shows no preferential directional properties when resolved in different directions at right angles to the direction of propagation. In general, the variation of the field vectors is neither completely regular nor completely irregular, and we may say that the light is *partially polarized*. Such light arises usually from unpolarized light by reflection or scattering. The effects of a partially polarized light wave depend on the intensities of any two mutually orthogonal components of the electric vector at right angles to the direction of propagation (Born and Wolf, 1980). The characterization of partially polarized light requires four parameters.

Here we discuss the representation method of polarization called the *coherency matrix*. Other representation methods, such as the Stokes parameter, can be found in any standard optics texts, including Born and Wolf (1980) and Möller (1988).

Suppose a monochromatic light wave $\vec{E}$ of mean frequency ν is propagating in the z-direction.

$$\vec{E} = A(t)e^{j[\phi_x(t) - 2\pi\nu t]}\vec{i} + B(t)e^{j[\phi_y(t) - 2\pi\nu t]}\vec{j} \tag{12.23}$$

The x- and y-components of this equation represent the components of the electric vector at a point O in two mutually orthogonal directions at right angles to the direction of propagation. For notational convenience let E_x and E_y denote the x- and y-components, respectively. For a light wave $\vec{E}$, a two-by-two matrix Ψ can be defined as follows:

$$\Psi = \begin{bmatrix} \langle E_x E_x^\star \rangle & \langle E_x E_y^\star \rangle \\ \langle E_y E_x^\star \rangle & \langle E_y E_y^\star \rangle \end{bmatrix} = \begin{bmatrix} \langle A^2 \rangle & \langle AB e^{j(\phi_x - \phi_y)} \rangle \\ \langle AB e^{-j(\phi_x - \phi_y)} \rangle & \langle B^2 \rangle \end{bmatrix} \tag{12.24}$$

where $\langle X \rangle$ denotes the expected value of X and E^* denotes the conjugate of E. This matrix Ψ is called the *coherency matrix* of the light wave $\vec{E}$. Let $\Psi_{i,j}$ denote the row i, column j, element of Ψ. Note that Ψ is Hermitian and the total intensity of the light is equal to the trace of the matrix Ψ,

$$\mathbf{Tr}(\Psi) = \Psi_{1,1} + \Psi_{2,2}$$

It is well known that the trace and the determinant of the matrix are invariant under the rotation of the axes. This means we can choose the coordinate system arbitrarily.

The coherency matrix can be used to represent the polarization of the light wave uniquely. Suppose E_y is subjected to a delay ε with respect to E_x and consider the intensity $J(\theta, \varepsilon)$ of the light vibrations in the direction that makes an angle θ with the positive x-direction (see Fig. 12.10). The component of the electric vector in the θ-direction, after the retardation ε has been introduced, is

$$E(t; \theta, \varepsilon) = E_x \cos\theta + E_y e^{j\varepsilon} \sin\theta \tag{12.25}$$

so that the intensity of the light in the θ-direction becomes

$$
\begin{aligned}
J(\theta, \varepsilon) &= \langle E(t; \theta, \varepsilon) E^\star(t; \theta, \varepsilon) \rangle \\
&= \langle E_x E_x^\star \rangle \cos^2\theta + \langle E_y E_y^\star \rangle \sin^2\theta + (\langle E_x E_y^\star \rangle e^{-j\varepsilon} + \langle E_y E_x^\star \rangle e^{j\varepsilon}) \cos\theta \sin\theta \\
&= \Psi_{1,1} \cos^2\theta + \Psi_{2,2} \sin^2\theta + (\Psi_{1,2} e^{-j\varepsilon} + \Psi_{2,1} e^{j\varepsilon}) \cos\theta \sin\theta
\end{aligned}
$$

Light that is most frequently encountered in nature has the property that the intensity of its components in any direction perpendicular to the direction of propagation is the same. Moreover, the intensity is not affected by any previous retardation of one of the rectangular components relative to the other, into which the light may have been resolved. In other words, $J(\theta, \varepsilon)$ is constant for all values of θ and ε. Such light may be said to be *completely unpolarized* and is often also called *natural*

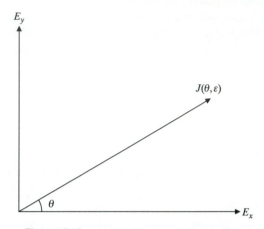

Figure 12.10 Intensity of light in the θ-direction.

light. Since $J(\theta, \varepsilon)$ is independent of θ and ε if and only if

$$\Psi_{1,2} = \Psi_{2,1} = 0 \quad \text{and} \quad \Psi_{1,1} = \Psi_{2,2}$$

the coherency matrix of natural light of intensity J can be written as

$$\Psi = \begin{bmatrix} \frac{J}{2} & 0 \\ 0 & \frac{J}{2} \end{bmatrix} = \frac{1}{2}J \begin{bmatrix} 1 & 0 \\ 0 & 1 \end{bmatrix}$$

To represent polarized light, first suppose the light is strictly monochromatic (monochromatic light is always polarized). Since the amplitude A and B and the phase factors ϕ_x and ϕ_y do not depend on time, the coherency matrix can be written as

$$\Psi = \begin{bmatrix} A^2 & ABe^{j\delta} \\ ABe^{-j\delta} & B^2 \end{bmatrix} \tag{12.26}$$

where $\delta = \phi_x - \phi_y$. Let **Det**(Ψ) denote the determinant of the matrix Ψ. One can easily observe that **Det**(Ψ) $= 0$. In other words, if the determinant of the coherency matrix is zero, then the light is said to be *completely polarized*. This is also true when the light is not monochromatic.

   In the case of partially polarized light, we use the following well-known fact: If several independent light waves propagating in the same direction are combined, the resulting wave has a coherency matrix that is equal to the sum of the coherency matrices of the individual waves. Conversely, any light wave can be represented as the sum of several independent light waves. Any quasi-monochromatic light wave is considered to be the sum of two independent waves: a completely unpolarized wave and a completely polarized wave. Hence the coherency matrix Ψ can be uniquely decomposed into two matrices, Ψ^u for the completely unpolarized part and Ψ^p for the completely polarized part:

$$\Psi = \Psi^u + \Psi^p = \begin{bmatrix} \alpha & 0 \\ 0 & \alpha \end{bmatrix} + \begin{bmatrix} \beta & \eta \\ \eta^\star & \gamma \end{bmatrix} \tag{12.27}$$

where $\alpha, \beta,$ and γ are all nonnegative, and $\beta\gamma = \eta\eta^\star$. One can easily observe the following identities:

$$2\alpha = \Psi_{1,1} + \Psi_{2,2} - D$$
$$2\beta = \Psi_{1,1} - \Psi_{2,2} + D$$
$$2\gamma = \Psi_{2,2} - \Psi_{1,1} + D$$
$$\eta = \Psi_{1,2}$$

where

$$D = \sqrt{[\text{Tr}(\Psi)]^2 - 4\text{Det}(\Psi)}$$

The total intensity of the light is

$$J_{\text{tot}} = \text{Tr}(\Psi) = \Psi_{1,1} + \Psi_{2,2}$$

and the total intensity of the polarized part is

$$J_{\text{pol}} = \text{Tr}(\Psi^p) = \beta + \gamma = D$$

The ratio of the intensity of the polarized part to the total intensity is called the *degree of polarization* of the light and is defined by

$$P = \frac{J_{\text{pol}}}{J_{\text{tot}}} = \sqrt{1 - \frac{4(\Psi_{1,1}\Psi_{2,2} - \Psi_{1,2}\Psi_{2,1})}{(\Psi_{1,1} + \Psi_{2,2})^2}} \tag{12.28}$$

It can be shown that $0 \leq P \leq 1$. When $P = 1$, the light is completely polarized and the determinant of its coherency matrix is zero. When $P = 0$, the light is completely unpolarized, and we have

$$(\Psi_{1,1} - \Psi_{2,2})^2 + 4\Psi_{1,2}\Psi_{2,1} = 0$$

Since $\Psi_{2,1} = \Psi_{1,2}^\star$, we must have

$$\Psi_{1,1} = \Psi_{2,2} \quad \text{and} \quad \Psi_{1,2} = \Psi_{2,1} = 0$$

This is in accordance with the previous argument on natural light. In all other cases where $0 < P < 1$, the light is said to be *partially polarized*.

12.5.2 Representation of Light Intensity

As we have seen so far, the intensity of any light can be represented as a sum of the intensities of two orthogonal polarization components. There are two plane polarizations that are frequently used in optics: S-pol and P-pol. S-pol is the component polarized perpendicularly to the incidence plane, and P-pol is the component po-larized parallel to the incidence plane. We will use the notations $\perp$ for S-pol and $\parallel$ for P-pol to emphasize the direction of polarization.

Since any polarized light can be divided into these two plane-polarized compo-nents, the illumination intensity can be written as the sum of the intensities of the S-pol and P-pol components:

$$J = J_{\parallel} + J_{\perp} \tag{12.29}$$

Note that the intensity J is a function of wavelength λ, and let $Q(\lambda)$ be the spectral distribution function of the illumination source. Then the overall intensity of an illumination source can be computed by

$$
\begin{aligned}
J &= \int J(\lambda)Q(\lambda)d\lambda \\
&= \int J_{\parallel}(\lambda)Q(\lambda)d\lambda + \int J_{\perp}(\lambda)Q(\lambda)d\lambda
\end{aligned}
\tag{12.30}
$$

12.6   Fresnel Equation

When a plane harmonic wave is incident upon a planar boundary separating two optical media, there will be a reflected wave and a transmitted wave. The amplitudes of the reflected and transmitted waves are dependent on the refractive indices of the two optical media and the incident angle. Fresnel and Arago investigated the interference of polarized rays of light and found that two rays polarized at right angles to each other never interfere. Using this fact, Fresnel derived the amplitude ratios of two orthogonally polarized light waves. These two special polarizations are P-pol and S-pol, as we have just discussed. The Fresnel equation can be derived by using Maxwell's equation and the energy conservation law. Maxwell's equation states that the tangential component of the field vector must be the same on both sides of the interface. The energy conservation law states that the incident energy is equal to the sum of the reflected energy and the transmitted energy. The Fresnel equation can be found in most standard optics texts, including Fowles (1975), Born and Wolf (1980), and Möller (1988). Here we show only the result.

As in Fig. 12.11, let ψ_i and ψ_t be the incident angle and the refracted angle, respectively. Let A^i, A^r, and A^t be the amplitude of the incident light, the reflected light, and the transmitted light, respectively. Since in our case the incidence medium is always air and the refractive index of air is 1, the Fresnel equation for reflection is

$$r_{\perp} = \frac{A^i_{\perp}}{A^r_{\perp}} = \frac{\cos \psi_i - n \cos \psi_t}{\cos \psi_i + n \cos \psi_t} \tag{12.31}$$

$$r_{\parallel} = \frac{A^i_{\parallel}}{A^r_{\parallel}} = \frac{n \cos \psi_t - \cos \psi_i}{\cos \psi_i + n \cos \psi_t} \tag{12.32}$$

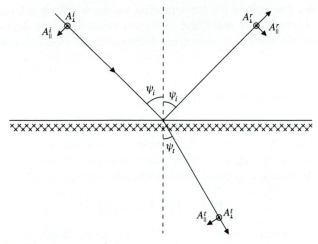

Figure 12.11 Fresnel law of reflectance. A P-pol element is denoted by an arrow; an S-pol element, by a circle. In general, some of the incident light is reflected, and the other part is transmitted. ψ_i is the incidence angle, and ψ_t is the refraction angle. The P-pol component and the S-pol component have different reflectance and transmittance coefficients.

The reflection coefficient ρ is the ratio of the incident energy to the reflected energy. Since the energy of a harmonic wave is equal to the square of the amplitude,

$$\rho_\perp = |r_\perp|^2, \qquad \rho_{||} = |r_{||}|^2$$

From Eqs. (12.31) and (12.32), eliminating ψ_t by using the law of refraction and setting $\psi = \psi_i$, we have (Sparrow, 1978)

$$\rho_\perp(\psi) = \frac{a^2 + b^2 - 2a\cos\psi + \cos^2\psi}{a^2 + b^2 + 2a\cos\psi + \cos^2\psi} \tag{12.33}$$

$$\rho_{||}(\psi) = \rho_\perp(\psi)\frac{a^2 + b^2 - 2a\sin\psi\tan\psi + \sin^2\psi\cos^2\psi}{a^2 + b^2 + 2a\sin\psi\tan\psi + \sin^2\psi\tan^2\psi} \tag{12.34}$$

where ψ is the incidence angle and a and b are quantities related to the incidence angle ψ, the refractive index n, and the extinction coefficient k of the material:

$$a = \sqrt{\frac{\sqrt{c^2 + 4n^2k^2} + c}{2}}$$

$$b = \sqrt{\frac{\sqrt{c^2 + 4n^2k^2} - c}{2}}$$

where

$$c = n^2 - k^2 - \sin^2\psi$$

One can see that $\rho_\parallel$ and $\rho_\perp$ are dependent on the wavelength λ of the incident light because the refractive index and extinction coefficients are dependent on the wavelength, and the Fresnel equations are derived for the monochromatic wave.

12.7   Reflection of Polarized Light

To understand how arbitrarily polarized light is reflected, we consider a small patch on an object that is perfectly flat and smooth and is modeled in microfacets. We assume that the light is *ergodic*, which means that the time average of the light is equivalent to its ensemble average. Mathematically, suppose a light $\vec{E}$ is given by

$$\vec{E}(t) = E_x(t)\vec{i} + E_y(t)\vec{j}$$

where $\vec{i}$ and $\vec{j}$ are unit vectors in the x- and y-directions, respectively. Let

$$\vec{E}(\omega) = \int_{-\infty}^{\infty} \vec{E}(t)e^{-j\omega t} dt \qquad (12.35)$$

$$= E_x(\omega)\vec{i} + E_y(\omega)\vec{j} \qquad (12.36)$$

The ergodic assumption tells us that

$$\langle E_x E_y^{\star} \rangle_t = \int_{-\infty}^{\infty} E_x(t)E_y^{\star}(t)dt \qquad (12.37)$$

$$= \frac{1}{2\pi} \int_{-\infty}^{\infty} E_x(\omega)E_y^{\star}(\omega)d\omega \qquad (12.38)$$

$$= \langle E_x E_y^{\star} \rangle_\omega \qquad (12.39)$$

Therefore the coherency matrix can be written in terms of the ensemble average instead of the time average.

$$\Psi = \begin{bmatrix} \langle E_x E_x^{\star} \rangle_t & \langle E_x E_y^{\star} \rangle_t \\ \langle E_x^{\star} E_y \rangle_t & \langle E_y E_y^{\star} \rangle_t \end{bmatrix} = \begin{bmatrix} \langle E_x E_x^{\star} \rangle_\omega & \langle E_x E_y^{\star} \rangle_\omega \\ \langle E_x^{\star} E_y \rangle_\omega & \langle E_y E_y^{\star} \rangle_\omega \end{bmatrix}$$

Suppose the incident light $\vec{E}^i$ is given by

$$\vec{E}^i(\omega) = E_x^i(\omega)\vec{i} + E_y^i(\omega)\vec{j}$$

where $\vec{i}$ and $\vec{j}$ are unit vectors in the x- and y-directions, respectively. Let the ratio of the amplitude of the reflected light to that of the incident light in each direction be $r_x(\omega)$ and $r_y(\omega)$. Then the reflected light can be written as

$$\vec{E}^r(\omega) = r_x(\omega)E_x^i(\omega)\vec{i} + r_y(\omega)E_y^i(\omega)\vec{j} \qquad (12.40)$$

$$= E_x^r(\omega)\vec{i} + E_y^r(\omega)\vec{j} \qquad (12.41)$$

Since

$$\langle E_x^r(\omega) E_x^r{}^\star(\omega) \rangle = \frac{1}{2\pi} \int_{-\infty}^{\infty} E_x^r(\omega) E_x^r{}^\star(\omega) d\omega \qquad (12.42)$$

$$= \frac{1}{2\pi} \int_{-\infty}^{\infty} |r_x(\omega)|^2 E_x^i(\omega) E_x^i{}^\star(\omega) d\omega \qquad (12.43)$$

$$= \langle |r_x(\omega)|^2 E_x^i E_x^i{}^\star \rangle \qquad (12.44)$$

the coherency matrix of the reflected light is

$$\Psi^r = \begin{bmatrix} \langle |r_x(\omega)|^2 E_x^i E_x^i{}^\star \rangle & \langle r_x(\omega) r_y^\star(\omega) E_x^i E_y^i{}^\star \rangle \\ \langle r_x^\star(\omega) r_y(\omega) E_x^i{}^\star E_y^i \rangle & \langle |r_y(\omega)|^2 E_y^i E_y^i{}^\star \rangle \end{bmatrix}$$

If the incident light is monochromatic, $r_x(\omega)$ and $r_y(\omega)$ can be written without the ω term. Let ρ_x and ρ_y denote $|r_x|^2$ and $|r_y|^2$, respectively. Then the intensity of the reflected light is

$$J^r = \mathbf{Tr}(\Psi^r) = \rho_x \langle E_x^i E_x^i{}^\star \rangle + \rho_y \langle E_y^i E_y^i{}^\star \rangle \qquad (12.45)$$

If E_x^i and E_y^i are given by

$$E_x^i = A e^{j\phi_x}, \qquad E_y^i = B e^{j\phi_y}$$

then the intensity of the reflected light is

$$J^r = \rho_x \frac{A^2}{2} + \rho_y \frac{B^2}{2}$$

Therefore the reflectance coefficient is

$$\rho = \frac{J^r}{J^i} = \frac{\rho_x A^2 + \rho_y B^2}{A^2 + B^2} \qquad (12.46)$$

Here there has been no assumption on the polarization of the incident light. Hence this equation is valid for completely unpolarized light as well as for completely polarized light. Note that Eq. (12.46) conforms to Jones's calculus (Fowles, 1975) and the reflectance coefficient matrix (Tsang, Kong, and Shin, 1985). It is also valid for the partially polarized light. To see this, consider the coherency matrix of partially polarized light given by Eq. (12.27). From Eq. (12.46) the reflectance coefficient of a partially polarized light can be written

$$\rho = \frac{\rho_x(\alpha + \beta) + \rho_y(\alpha + \gamma)}{(\alpha + \beta) + (\alpha + \gamma)} = \frac{\alpha(\rho_x + \rho_y) + (\beta\rho_x + \gamma\rho_y)}{2\alpha + (\beta + \gamma)} \qquad (12.47)$$

Since the degree of polarization is

$$P = \frac{J_{pol}}{J_{tot}} = \frac{\beta + \gamma}{2\alpha + \beta + \gamma}$$

Eq. (12.47) becomes

$$\rho = (1 - P)\frac{\rho_x + \rho_y}{2} + P\frac{\rho_x\beta + \rho_y\gamma}{\beta + \gamma}$$

This means that the reflectance coefficient of the partially polarized light is the weighted sum of the reflectance coefficients of the unpolarized part and the polarized part.

Equation (12.46) holds for any pair of mutually orthogonal components of the electric vector at right angles to the direction of propagation. Let the x-direction be parallel to the plane of incidence (P-pol) and y-direction be perpendicular to the plane of incidence (S-pol). Let $J^i_{||}$ and $J^i_{\perp}$ denote the intensity of the P-pol and S-pol components of the incident light, respectively. Then Eq. (12.46) can be rewritten as

$$\rho = \rho_{||} \cos^2 \varphi + \rho_{\perp} \sin^2 \varphi \qquad (12.48)$$

where

$$\tan^2 \varphi = \frac{J^i_{\perp}}{J^i_{||}}$$

Note that this equation is derived for monochromatic light and therefore depends on the wavelength λ.

12.8   A New Bidirectional Reflectance Function

Using Eq. (12.48), we can now rewrite the specular term R_s in the Torrance-Sparrow model as

$$R_s = \frac{\rho_{||}(\lambda)\cos^2 \varphi + \rho_{\perp}(\lambda)\sin^2 \varphi}{\pi} \frac{DG}{(\vec{N} \cdot \vec{L})(\vec{N} \cdot \vec{V})} \qquad (12.49)$$

For notational convenience we let

$$K = \frac{DG}{\pi(\vec{N} \cdot \vec{L})(\vec{N} \cdot \vec{V})}$$

and remove λ from Eq. (12.49). From Eqs. (12.5), (12.6), and (12.49) we have a bidirectional reflectance function of the form

$$\begin{aligned}
R &= sR_s + (1-s)R_d \\
&= sK(\rho_{||}\cos^2 \varphi + \rho_{\perp}\sin^2 \varphi) + (1-s)R_d \qquad (12.50)
\end{aligned}$$

Using Eqs. (12.9) and (12.50), we can write the intensity of the reflected light J^r as

$$\begin{aligned}
J^r &= J^i d\omega \cos \psi R \\
&= sK(\rho_{||}\cos^2 \varphi + \rho_{\perp}\sin^2 \varphi)J^i d\omega \cos \psi + (1-s)R_d J^i d\omega \cos \psi \\
&= sK(\rho_{||}J^i_{||} + \rho_{\perp}J^i_{\perp})d\omega \cos \psi + (1-s)R_d(J^i_{||} + J^i_{\perp})d\omega \cos \psi \\
&= [sK\rho_{||} + (1-s)R_d]J^i_{||}d\omega \cos \psi + [sK\rho_{\perp} + (1-s)R_d]J^i_{\perp}d\omega \cos \psi \\
&= [sK\rho_{||} + (1-s)R_d]J^i_{||}d\omega \vec{N} \cdot \vec{L} + [sK\rho_{\perp} + (1-s)R_d]J^i_{\perp}d\omega \vec{N} \cdot \vec{L} \qquad (12.51)
\end{aligned}$$

If we let

$$R_{||} = sK\rho_{||} + (1 - s)R_d \tag{12.52}$$

$$R_{\perp} = sK\rho_{\perp} + (1 - s)R_d \tag{12.53}$$

then the reflected light can be written in a matrix form:

$$\begin{bmatrix} J^r_{||} \\ J^r_{\perp} \end{bmatrix} = d\omega \vec{\mathbf{N}} \cdot \vec{\mathbf{L}} \begin{bmatrix} R_{||} & 0 \\ 0 & R_{\perp} \end{bmatrix} \begin{bmatrix} J^i_{||} \\ J^i_{\perp} \end{bmatrix} \tag{12.54}$$

12.9   Image Intensity

With the knowledge we have obtained so far, the image intensity in Eq. (12.1) can be written in terms of the illumination parameters, the sensor parameters, and the bidirectional reflectance function. If the light is represented with a column vector whose elements denote the intensities of P-pol and S-pol components at a certain wavelength λ, from Eq. (12.54) the total intensity of the reflected light at that wavelength is

$$J^r(\lambda) = d\omega \vec{\mathbf{N}} \cdot \vec{\mathbf{L}}[R_{||}(\lambda)J^i_{||}(\lambda) + R_{\perp}(\lambda)J^i_{\perp}(\lambda)] \tag{12.55}$$

In fact, one can measure the intensity of the P-pol and S-pol components of light by a polarizer (Born and Wolf, 1980; Wolff, 1987) using Eq. (12.26). The image intensity I now can be written as

$$I = \int CSJ^r(\lambda)Q(\lambda)d\lambda$$

$$= \int CSQ(\lambda)d\omega \vec{\mathbf{N}} \cdot \vec{\mathbf{L}}[R_{||}(\lambda)J^i_{||}(\lambda) + R_{\perp}(\lambda)J^i_{\perp}(\lambda)]d\lambda \tag{12.56}$$

The relation given in Eq. (12.56) can also be extended to the case of multiple light sources and an extended light source, like the sun.

12.10   Related Work

Reflectance models have been used in computer graphics (Phong, 1975; Blinn and Newell, 1976; Blinn, 1977; Whitted, 1980; Cook and Torrance, 1982) and image analysis (Brelstaff, and Blake, 1988; Healey, 1988; Healey and Binford, 1988a, 1988b; Healey and Blanz, 1988; and Klinker, Shafer, and Kanade, 1988). The Torrance-Sparrow model (1967) is the first theoretical reflectance model in the literature that takes specular reflection as well as diffusion into consideration. This model was built from the viewpoint of geometric optics and validated by conducting a number of experiments. Blinn (1977) rewrote the Torrance-Sparrow model using vector notation, and Cook extended the model by including ambient light, multiple light sources, and spectral dependencies (Cook and Torrance, 1982). These models have been successfully applied to computer-synthesized picture generation.) There also have been efforts to evaluate reflectance models (Buchanan 1986). Cook's model is regarded as the best. The Torrance-Sparrow model (1965) is good only if the incident light is completely unpolarized. In other words, it will not work when the incident light is polarized. In general, light is polarized, at least partially.

In computer vision research the properties of the polarization of light have been used in shape understanding and material classification. Koshikawa (1979) proposed a polarimetric method to find surface normals. In his research the stokes parameters are used to represent the state of polarization. Wolff (1988a, 1988b, 1989a, 1989b; Wolff and Boult, 1989) used polarized incident light for material classification. Drawing on the theory of the polarization of quasi-monochromatic light radiation developed by Emil Wolf (1970; Born and Wolf, 1980) and the Torrance-Sparrow model, Wolff derived a set of equations from which the Fresnel reflectance coefficients can be computed by measuring the intensities of the reflected light. With the Fresnel reflectance coefficients, one can determine the orientation of a given surface (Wolff, 1987). Wolff's result and derivation are similar to the ones given here; however, no reflectance model is presented in his work.

■ Exercises

12.1.   The half-angle ψ of a right circular cone is defined by the plane angle formed between the cone axis and any line on the cone passing through its vertex. Show that the relationship between the solid angle Ω of the cone and the half-angle ψ of the cone is given by

$$\Omega = 2\pi(1 - \cos\psi)$$

12.2.   Using the relationship of Exercise (12.1), show that for small angles, $\Omega = \pi\psi^2$.

12.3.   A point source of I_o (w/sr) illuminates a vertical surface located a distance h below the point source and a horizontal distance a from the point source. Show that the irradiance at the vertical surface due to the point source is given by $I_o \sin\theta_o \cos^2\theta_o/h^2$, where $\cos\theta_o = h/\sqrt{h^2 + a^2}$.

12.4.   Suppose that two point sources, each having a radiant intensity of I_o (w/sr), are each a distance h meters above the horizontal plane and a distance kh apart. Determine the irradiance produced by these point sources on a small horizontal patch located on the horizontal plane (which is a distance h meters below the sources) at a position exactly between the two sources. Then determine the irradiance produced by these point sources on a small horizontal patch located on the horizontal plane a distance of h meters immediately below one of the sources. Show that when $k = 1$, the ratio of the irradiance at the point between the source to a point immediately below one of the sources is about .65. Show that when $k = 2$, this ratio is about .18.

12.5.   Suppose a uniform point source radiator of I_o w/sr is illuminating a surface patch on the horizontal plane a distance h below the point sources and a horizontal distance a from the point source. Show that as h increases from small values to higher values, the irradiance at the illuminated surface patch increases and then decreases, reaching a maximum value when $h = a/\sqrt{2}$, and at the height $h = a/\sqrt{2}$, the irradiance is $2I_o/2\sqrt{3}a^2$w/m^2.

12.6. Assume that the sky can be represented as an extended horizontal plane having radiance of E w/m^2-sr independent of direction. Show that the irradiance on a horizontal surface is $E\pi$ w/m^2.

12.7. A uniformly diffusing circular-disk fluorescent-light panel of 40 m^2 area is placed 200 m directly above the center of a surface patch that is parallel to the panel. If the light panel has radiance E w/m^2-sr independent of direction, show that the irradiance at the center of the surface patch is just short of 10^{-3} E w/m^2.

12.8. A rectangle in the $z = 0$ plane, $0 \le x \le a$, $0 \le y \le b$, is a Lambertian illuminator having a constant radiance of J (w/m^2-sr). Show that the irradiance on a small surface patch centered at (x_o, y_o, z_o) in the $z = z_o$ plane is given by

$$\int_{x=0}^{a} \int_{y=0}^{b} \frac{Jz_o^2}{[(x - x_o)^2 + (y - y_o)^2 + z_o^2]^2} dx dy$$

12.9. Suppose that a cylinder given by

$$g(x,y) = d - \sqrt{r^2 - y^2}, \qquad -r \le y \le r$$

has a Lambertian surface and is illuminated by a distant point light source in a direction having unit vector (a,b,c). Show that the image intensity on an orthographic projection is proportional to

$$\frac{by + c\sqrt{r^2 - y^2}}{r}$$

12.10. Suppose that a sphere given by

$$g(x,y) = d - \sqrt{r^2 - x^2 - y^2}, \qquad -r \le x, y \le r$$

has a Lambertian surface and is illuminated by a distant point light source in a direction having unit vector (a,b,c). Show that the image intensity on an orthographic projection is proportional to

$$ax + by + c\sqrt{\frac{r^2 - x^2 - y^2}{r}}$$

Bibliography

Arnsprang, J., "Depth from Photometric Motion," *Proceedings of the Fifth Scandinavian Conference on Image Analysis,* Vol. 1, Stockholm, 1987, pp. 505–512.

Babu, M. D. R., C.-H. Lee, and A. Rosenfeld, "Determining Plane Orientation from Specular Reflectance," *Pattern Recognition,* Vol. 18, 1985, pp. 53–62.

Batchelor, B. G., D. A. Hill, and D. C. Hodgson (eds.), *Automated Visual Inspection,* IFS, Bedford, United Kingdom, 1985.

Beckmann, P., *The Depolarization of Electromagnetic Waves,* Golem Press, Boulder, CO, 1968.

Beckmann, P., and A. Spizzichino, *The Scattering of Electromagnetic Waves from Rough Surfaces,* Pergamon Press, Oxford, 1963.

Blake, A., and G. Brelstaff, "Geometry from Specularities," *IEEE Proceedings of the Second International Conference on Computer Vision,* Tarpon Springs, MD, 1988, pp. 394–403.

Blinn, J. F., "Models of Light Reflection for Computer Synthesized Pictures," *Proceedings of Computer Graphics,* Vol. 11, 1977, pp. 192–198.

Blinn, J. F., and M. E. Newell, "Texture and Reflection in Computer Generated Images," *Communications of the ACM,* Vol. 19, 1976, pp. 542–547.

Bolle, R. M., "Information Extraction about Complex Three-Dimensional Objects from Visual Data," Brown University Technical Report No. LEMS-6, 1984.

Bolle, R. M., and D. B. Cooper, "Bayesian Recognition of Local 3-D Shape by Approximating Image Intensity Functions with Quadric Polynomials," *IEEE Transactions on Pattern Analysis and Machine Intelligence,* Vol. PAMI-6, 1984, pp. 418–429.

Born, M., and E. Wolf, *Principles of Optics,* 6th ed., Pergamon Press, Oxford, 1980.

Brelstaff, G., and A. Blake, "Detecting Specular Reflections Using Lambertian Constraints," *Proceedings of the Second International Conference on Computer Vision,* Tampa, FL, 1988, pp. 297–302.

Brown, C. M., D. H. Ballard, and O. A. Kimball, "Constraint Interaction in Shape-from-Shading Algorithms," *Computer Science and Computer Engineering Research Review,* C.S. Dept. Annual Report, University of Rochester, Rochester, NY, 1982, pp. 3–21.

Browne, A., and L. Norton-Wayne, *Vision and Information Processing for Automation,* Plenum Press, New York, 1986.

Bruckstein, A. M., "On Shape from Shading," *Computer Vision, Graphics, and Image Processing,* Vol. 44, 1988, pp. 139–154.

Buchanan, C. G., "Determining Surface Orientation from Specular Highlights," Master's thesis, Computer Science Department, University of Toronto, 1986.

Cernuschi-Frias, B., and D. B. Cooper, "3-D Space Location and Orientation Parameter Estimation of Lambertian Spheres and Cylinders from a Single 2-D Image by Fitting Lines and Ellipses to Thresholded Data," *IEEE Transactions on Pattern Analysis and Machine Intelligence,* Vol. PAMI-6, 1984, pp. 430–441.

Chandrasekhar, S., *Radiative Transfer,* Dover, New York, 1960.

Chiaradia, M. T., et al., "A Photometric Approach to Determine Surface Objects Orientations for Vision Systems," in *Advances in Image Processing and Pattern Recognition,* V. Cappellini and R. Marconi (eds.), Elsevier, North-Holland, 1986.

Cohen, F. S., and J.-F. P. Cayula, "3-D Object Recognition from a Single Image," *SPIE,* Vol. 521, *Intelligent Robots and Computer Vision* 1984, pp. 7–13.

Cook, R. L., and K. E. Torrance, "A Reflectance Model for Computer Graphics," *ACM Transactions on Graphics,* Vol. 1, pp. 7–24.

Draper, N. R., and H. Smith, *Applied Regression Analysis,* 2d ed., Wiley, New York, 1987.

Egan, W. G., *Photometry and Polarization in Remote Sensing,* Elsevier, New York, 1985.

Ferriem, F. P., and M. D. Levine, "Where and Why Local Shading Analysis Works," *IEEE Transactions on Pattern Analysis and Machine Intelligence,* Vol. 11, 1989, pp. 198–206.

Forsyth, D., and A. Zisserman, "Shape from Shading in the Light of Mutual Illumination," *Image and Vision Computing,* Vol. 8, 1990, pp. 43-49.

Fowles, G. R., *Introduction to Modern Optics,* Holt, Rinehart and Winston, New York, 1975.

Frankot, R. T., and R. Chellappa, "A Method for Enforcing Integrability in Shape from Shading Algorithms," *IEEE Transactions on Pattern Analysis and Machine Intelligence,* Vol. 10, 1988, pp. 439–451.

Grossman, P., "Depth from Focus," *Pattern Recognition Letters,* Vol. 5, 1987, pp. 63–69.

Hambrick, L. N., M. H. Loew, and R. L. Carroll, "The Entry-Exit Method of Shadow Boundary Segmentation," *IEEE Transactions on Pattern Analysis and Machine Intelligence,* Vol. PAMI-9, 1987, pp. 597–607.

Harding, K. G., "Optical Considerations for Machine Vision," Industrial Technology Institute, Ann Arbor, MI, 1987.

——, "Advanced Optical Considerations for Machine Vision Applications," Industrial Technology Institute, Ann Arbor, MI, 1988.

Healey, G., "A Color Reflectance Model and Its Use for Segmentation," *Proceedings of the Second International Conference on Computer Vision,* Tampa, FL, 1988, pp. 460–466.

Healey, G., and T. O. Binford, "Local Shape from Specularity," *Computer Vision, Graphics, and Image Processing,* Vol. 42, 1988a, pp. 62–86.

——, "Predicting Material Classes," *Proceedings of ARPA Image Understanding Workshop,* Boston, 1988b.

Healey, G., and W. E. Blanz, "Identifying Metal Surfaces in Color Images," *Proceedings of the SPIE,* Vol. 939, Hybrid Image and Signal Processing, pp. 71–74, Orlando, FL, 1988.

Horn, B. K. P., "Obtaining Shape from Shading Information," in *The Psychology of Machine Vision,* P. H. Winston (ed.), McGraw-Hill, New York, 1975, pp. 115–155.

——, "Understanding Image Intensities," *Artificial Intelligence,* Vol. 8, 1977, pp. 201–231.

——, *Robot Vision,* MIT Press, Cambridge, MA, 1986.

Horn, B. K. P., and M. J. Brooks (eds.), *Shape from Shading,* MIT Press, Cambridge, MA, 1989.

Ikeuchi, K., "Determining Surface Orientations of Specular Surfaces by Using the Photometric Stereo Method," *IEEE Transactions on Pattern Analysis and Machine Intelligence,* Vol. PAMI-3, 1981, pp. 661–669.

——, "Reconstructing a Depth Map from Intensity Maps," *Proceedings of the Seventh International Conference on Pattern Recognition,* Montreal, 1984, pp. 736–738.

——, "Determining a Depth Map Using a Dual Photometric Stereo," *International Journal of Robotics Research,* Vol. 6, 1987, pp. 15–31.

Ikeuchi, K., and B. K. P. Horn, "Numerical Shape from Shading and Occluding Boundaries," *Artificial Intelligence,* Vol. 17, 1981, pp. 141–184.

Kanade, T., "Recovery of the Three-Dimensional Shape of an Object from a Single View," *Artificial Intelligence,* Vol. 17, 1981, pp. 409–460.

Kingslake, R., *Optical Systit Design,* Academic Press, New York, 1983.

Klinker, G. J., S. A. Shafer, and T. Kanada, "Color Image Analysis with an Intrinsic Reflection Model," *Proceedings of the Second International Conference on Computer Vision,* Tampa, FL, 1988, pp. 292–296.

Koshikawa, K., "A Polarimetric Approach to Shape Understanding of Glossy Objects," *Proceedings of the Sixth International Joint Conference on Artificial Intelligence,* Tokyo, 1979, pp. 493–495.

Lee, H. C., E. J. Breneman, and C. P. Schulte, "Modeling Light Reflection for Computer Color Vision," *IEEE Transactions on Pattern Analysis and Machine Intelligence,* Vol. 12, 1990, pp. 402–416.

Malik, J., and D. Maydan, "Recovering Three-Dimensional Shape from a Single Image of Curved Objects," *IEEE Transactions on Pattern Analysis and Machine Intelligence,* Vol. 11, 1989, pp. 555–566.

McPherson, C. A., "Three-Dimensional Robot Vision," *Proceedings of the SPIE, Intelligent Robots: Third International Conference on Robot Vision and Sensory Controls,* Cambridge, MA, 1983, pp. 116–126.

Möller, K. D., *Optics,* University Science Books, Mill Valley, CA, 1988.

Mundy, J. L., and G. B. Porter III, "Visual Inspection of Metal Surfaces," *Fifth International Conference on Pattern Recognition,* Vol. 1, Miami Beach, FL, 1980 pp. 232–237.

Nagel, H.-H., "Representation of Moving Rigid Objects Based on Visual Observations," *Computer,* Vol. 14, 1981, pp. 29–39.

Nicodemus, F., et al., *Geometric Considerations and Nomenclature for Reflectance,* Monograph No. 160, U.S. Department of Commerce, National Bureau of Standards, Washington, DC, 1977.

Penna, M. A., "Local and Semi-Local Shape from Shading for a Single-Perspective Image of a Smooth Object," *Computer Vision, Graphics, and Image Processing,* Vol. 46, 1989a, pp. 346–366.

———, "A Shape from Shading Analysis for a Single Perspective Image of a Polyhedron," *IEEE Transactions on Pattern Analysis and Machine Intelligence,* Vol. 11, 1989b, pp. 545–554.

Pentland, A. P., "Local Shading Analysis," *IEEE Transactions on Pattern Analysis and Machine Intelligence,* Vol. PAMI-6, 1984, pp. 170–187.

———, "A New Sense for Depth of Field," *IEEE Transactions on Pattern Analysis and Machine Intelligence,* Vol. PAMI-9, 1987, pp. 523–531.

Phong, B. T., "Illumination for Computer-Generated Pictures," *Communications of the ACM,* Vol. 18, 1975, pp. 311–317.

Pong, T. C., R. M. Haralick, and L. G. Shapiro, "Shape from Shading Using the Facet Model," *Pattern Recognition,* Vol. 22, 1989, pp. 683–696.

Ray, R., J. Birk, and R. B. Kelley, "Error Analysis of Surface Normals Determined by Radiometry," *IEEE Transactions on Pattern Analysis and Machine Intelligence,* Vol. PAMI-5, 1983, pp. 631–645.

Schroeder, H. E., "Practical Illumination Concept and Technique for Machine Vision Applications," *Proceedings of Robots 8 Conference,* Detroit, 1984.

Shatt, T. M., "A Numerical Method for Shape from Shading from a Single Image," Master Thesis, Artificial Intelligence Laboratory, MIT, Cambridge, MA, 1979.

Simchiny, T. R. Chellappa, and M. Shao, "Direct Analytical Methods for Solving Poisson Equations in Computer Vision Problems," *IEEE Transactions on Pattern Analysis & Machine Intelligence*, Vol. 12, 1990, pp. 44–50.

Smith, G. B., "Shape From Shading: An Assessment" SRI Tech. Note. 287, Menlo Park, CA, 1983.

Sjoberg, R. W., and B. K. P. Horn, "Atmospheric Modeling for the Generation of Albedo Images," *Proceedings of the Image Understanding Workshop,* College Park, MD, 1980, pp. 58–70.

Sparrow, E. M., and R. D. Cess, *Radiation Heat Transfer,* McGraw-Hill, New York, 1978.

Subbarao, M., "Direct Proving of Depth Maps: Differential Methods," *IEEE Computer Society Conference on Computer Vision,* Miami Beach, FL, 1987, pp. 58–65.

Subbarao, M., and N. Gurumoorthy, "Depth Recovery from Blurred Edges," *Proceedings, of the Conference on Computer Vision and Pattern Recognition,* Ann Arbor, MI, 1988, pp. 498–503.

Torrance, K. E., and E. M. Sparrow, "Biangular Reflectance of an Electric Nonconductor as a Function of Wavelength and Surface Roughness," *Journal of Heat Transfer,* Vol. 87, Series C, 1965, pp. 283–292.

——, "Theory for Off-Specular Reflection from Roughened Surfaces," *Journal of the Optical Society of America,* Vol. 57, 1967, pp. 1105–14.

Torrance, K. E., E. M. Sparrow, and R. C. Birkebak, "Polarization, Directional Distribution, and Off-Specular Peak Phenomena in Light Reflected from Roughened Surfaces," *Journal of the Optical Society of America,* Vol. 56, 1966, pp. 916–925.

Tsang, L., J. A. Kong, and R. T. Shin, *Theory of Microwave Remote Sensing,* Wiley, New York, 1985.

Westphal, H., "Photometric Stereo Considering Diffuse Illumination," in *Proceedings of the Sixth International Conference on Pattern Recognition,* Munich, 1982, pp. 310–312.

Whitted, T., "An Improved Illumination Model for Shaded Display," *Communications of the ACM,* Vol. 23, 1980, pp. 343–349.

Wolf, E., "Coherence Properties of Polarized Electromagnetic Radiation," in *Coherence and Fluctuations of Light,* L. Mandel and E. Wolf (eds.), Dover, New York, 1970, pp. 364–380.

Wolfe, W. L., and Y.-J. Wang, "Comparison of Theory and Experiments for Bidirectional Reflectance Distribution Function (BRDF) of Microrough Surfaces," *Proceedings of the SPIE,* Vol. 362, *Scattering in Optical Materials,* San Diego, August 1982, pp. 40–45.

Wolff, L. B., "Surface Orientation from Polarization Images," *Proceedings of the SPIE,* Vol. 850, *Optics, Illumination, and Image Sensing for Machine Vision II,* Cambridge, MA, 1987, pp. 110–121.

——, "Classification of Material Surfaces from the Polarization of Specular Highlights," *Proceedings of the SPIE,* Vol. 850, *Optics, Illumination, and Image Sensing for Machine Vision III,* Cambridge, MA, 1988a, pp. 206–213.

——, "Segmentation of Specular Highlights from Object Surfaces," *Proceedings of the SPIE,* Vol. 850, *Optics, Illumination, and Image Sensing for Machine Vision III,* Cambridge, MA, 1988b, pp. 198–205.

——, "Material Classification and Separation of Reflection Components Using Polarization/Radiometric Information," *Proceedings of the DARPA Image Understanding Workshop,* Palo Alto, CA, 1989a, pp. 232–244.

——, "Using Polarization to Separate Reflection Components," *Proceedings of the IEEE Computer Society Conference on Computer Vision and Pattern Recognition,* San Diego, 1989b, pp. 363–369.

Wolff, L. B., and T. E. Boult, "Polarization/Radiometric-Based Material Classification," *Proceedings of the IEEE Computer Society Conference on Computer Vision and Pattern Recognition,* San Diego, 1989, pp. 383–395.

Woodham, R. J., "A Cooperative Algorithm for Determining Surface Orientation from a Single View," *Proceedings of the International Joint Conference on Artificial Intelligence,* Cambridge, MA, 1977, pp. 635–641.

——, "Photometric Stereo: A Reflectance Map Technique for Determining Object Relief from Image Intensity," *Proceedings of the Second National Conference of the Canadian Society for Computational Studies of Intelligence,* Toronto, 1978, pp. 117–124.

Woodham, R. J., "Relating Properties of Surface Curvature to Image Intensity," International Joint Conference on Artificial Intelligence, Tokyo, 1979, pp. 971–977.

——, "Analyzing Images of Curved Surfaces," *Artificial Intelligence,* Vol. 17, 1981, pp. 117–140.

Wu, Z., and L. Li, "A Line-Integration Method for Depth Recovery from Surface Normals," *Computer Vision, Graphics, and Image Processing,* Vol. 43, 1988, pp. 53-66.

Yi, S. K., "Illumination Control Expert for Machine Vision, A Goal Driver's Approach," Ph.D. Dissertation. Computer Science Dept., University of Washington, Seattle, WA, 1990.

13 PERSPECTIVE PROJECTION GEOMETRY

13.1 Introduction

Computer vision problems often involve interpreting the information on a two-dimensional (2D) image of a three-dimensional (3D) world in order to determine the placement of the 3D objects portrayed in the image. To do this requires understanding the perspective transformation governing the geometric way 3D information is projected onto the 2D image. This chapter reviews the basic and well-known concepts and properties of the perspective transformation. Then we show how partial 3D-world-model knowledge about distances between points, parallel lines, perpendicular lines, intersecting lines, direction cosines of lines, or lines in the same plane can be used with their corresponding perspective projections on the image in order to determine the location of these points, lines, or planes in the 3D world.

The idea of using the geometry in the 2D-perspective projection to infer 3D information about the objects being viewed is certainly not new. Aspects of this problem have been discussed in the vision literature beginning with the classic paper of Roberts (1965). Indeed, how to relate points in 2D-perspective projections to their corresponding 3D points has long been one of the principal concerns in photogrammetry resection procedures. The three-point resection problem was first solved by the German mathematician Grunert (1841). Finsterwalder and Scheufele (1903) and Müller (1925) are among the early papers in the photogrammetric literature. More recently published procedures for camera calibration include Abidi and Eason (1985); El-Hakim (1985); Haralick and Chu (1984); Hung, Lee, and Fang (1986); Isaguirre and Summers (1985); Magee and Aggarwal (1985); Mansbach (1986); Martins, Birk, and Kelley (1981); Sobel (1974), Tsai (1987); and Yakimovsky and Cunningham (1976). Chapter 14 treats the calibration problem in detail. First, however, we review the basic principles of perspective projection geometry.

Geometry contains more than just 3D-to-2D-point corrrespondences. The object being viewed, especially if man-made, may have collinear distinguished points, lines, line segments, planes, coplanar points, and coplanar lines. These types of geometric entities and their relationships are so prevalent in the world which must be interpreted and understood through vision that the prior probabilities for these kinds of entities and their relationships are high enough that in most situations people have a natural tendency to prefer interpretations using the largest number of such geometric relationships which are consistent with the observed perspective imagery.

Sections 2 and 3 are largely tutorial. They define the one- and two-dimensional perspective projection and for completeness give some of its well-known properties. Section 4 derives a variety of relationships, some of which are known and others of which are new, by which the observed perspective projection information and a small amount of model information about a 3D object's points and lines can be combined to infer other 3D information. Some of the inverse projection ideas presented in this section were first introduced into the computer vision literature by Haralick (1970) in the world reference frame rather than the camera reference frame. Horaud (1987) develops a couple of the relationships found in this section. Section 5 discusses how to derive the 3D position and orientation of a known circle from its perspective projection. Section 6 discusses how range information can be determined from structured light. Section 7 discusses the well-known invariance of cross-ratio to perspective projection, first in one-dimensional perspective projection and then in two-dimensional perspective projection. We show how any two cross-ratios in a five-point coplanar configuration, together with the topology of the configuration determine the remaining three cross-ratios. By the invariance of the cross-ratio to perspective projection, these relationships hold both on the plane in 3D in which the observed five-point configuration lies, and in any 2D perspective projection. Section 7 closes with some remarks on the use of the cross-ratio in matching.

13.2   One-Dimensional Perspective Projection

Suppose we have a camera taking one-dimensional pictures in a two-dimensional world. As shown in Fig. 13.1, the camera lens is at the origin and points directly down the y-axis. In order to keep the image in a positive orientation, we assume that the image line is at a distance f in front of the camera lens and that the lens projects forward to it. This eliminates the problems of left–right reversal in an image behind the lens. The image line for this first example is parallel to the x-axis.

According to the geometric ray optics model for the lens, the lens will focus a point (r, s) onto the image line, which is a line parallel to the x-axis and at a distance f directly in front of the lens. The distance f is called the *camera constant*. The position on the line is determined by where the line from (r, s) to the origin intersects the image line. Hence the perspective projection has coordinates $(rf/s, f)$ in the original two-dimensional coordinate system. Relative to the one-dimensional coordinate system of the image line, the coordinate is rf/s.

Note that both the numerator and the denominator of rf/s are linear combinations of r and s. This suggests that if the numerator and the denominator were computed by an appropriate linear transformation, the one-dimensional perspective coordinates could be computed by taking ratios of components of the transformed

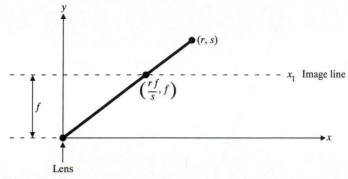

Figure 13.1 For a lens oriented along the y-axis and an image plane parallel to the x-axis, the perspective projection gives the point (r,s) the coordinates $(rf/s, f)$ on the one-dimensional image.

vector. We illustrate this by using homogeneous coordinates. The point (r,s) can be represented as $(r,s,1)$ in the homogeneous coordinate system. The first linear transformation translates the point $(r,s,1)$ down the y-axis by a distance of f. The second transformation takes the perspective transformation to the image line. Hence,

$$\begin{pmatrix} u \\ v \end{pmatrix} = \begin{pmatrix} 1 & 0 & 0 \\ 0 & 1/f & 1 \end{pmatrix} \begin{pmatrix} 1 & 0 & 0 \\ 0 & 1 & -f \\ 0 & 0 & 1 \end{pmatrix} \begin{pmatrix} r \\ s \\ 1 \end{pmatrix}$$

The one-dimensional image line coordinates for the point are then given by $x_I = u/v = rf/s$.

Figure 13.2 illustrates a more complex example. The lens is still at the origin but is pointing down the y'-axis. The x'- and y'-axes are the x- and y-axes rotated by an angle of θ. The projection (r,s) of the point (p,q) can be determined as the intersection of the image line with the line from the origin to (p,q). The image line is given by the equation $(-\sin\theta \ \cos\theta)\begin{pmatrix} x \\ y \end{pmatrix} = f$.

The line between the origin and the point (p,q) is given by $(-q \ p)\begin{pmatrix} x \\ y \end{pmatrix} = 0$.

Hence the perspective projection (r,s) is determined by $\begin{pmatrix} -\sin\theta & \cos\theta \\ -q & p \end{pmatrix}\begin{pmatrix} r \\ s \end{pmatrix} = \begin{pmatrix} f \\ 0 \end{pmatrix}$. This system of equations has the solution

$$\begin{pmatrix} r \\ s \end{pmatrix} = \left(\frac{f}{-p\sin\theta + q\cos\theta} \right) \begin{pmatrix} p \\ q \end{pmatrix}$$

To represent this in a coordinate system relative to the image line, we must first rotate the x-y-axes to the x'-y'-axes. Let us call the coordinates of (r,s) in the x'-y' reference frame (r',s'). Then $\begin{pmatrix} \cos\theta & \sin\theta \\ -\sin\theta & \cos\theta \end{pmatrix}\begin{pmatrix} r \\ s \end{pmatrix} = \begin{pmatrix} r' \\ s' \end{pmatrix}$.

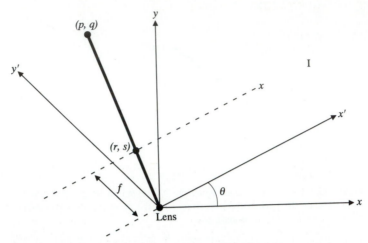

Figure 13.2 Geometry of a general one-dimensional perspective projection in a two-dimensional world. The lens is at the origin and looks down the y'-axis. The image line is a distance f in front of the lens and it is parallel to the x'-axis. The x'-y' axes are the x-y axes rotated anticlockwise by an angle θ.

Hence

$$\begin{pmatrix} r' \\ s' \end{pmatrix} = \frac{f}{-p\sin\theta + q\cos\theta} \begin{pmatrix} p\cos\theta + q\sin\theta \\ -p\sin\theta + q\cos\theta \end{pmatrix}$$

$$= f \begin{pmatrix} [p\cos\theta + q\sin\theta]/[-p\sin\theta + q\cos\theta] \\ 1 \end{pmatrix}$$

Note that the y'-coordinate is f, exactly as we expect it to be, since the image line is at a distance f down the y'-axis from the origin. The x'-coordinate is the coordinate with respect to the image line. Hence $x_I = f\frac{p\cos\theta + q\sin\theta}{-p\sin\theta + q\cos\theta}$. This relationship also has a numerator and a denominator that are linear combinations of the original coordinates (p, q). Hence the point x_I can be written as the ratio of two linear combinations.

Rewriting the relationships in terms of a homogeneous coordinate system, we obtain

$$\begin{pmatrix} u \\ v \end{pmatrix} = \begin{pmatrix} 1 & 0 & 0 \\ 0 & 1/f & 1 \end{pmatrix} \begin{pmatrix} 1 & 0 & 0 \\ 0 & 1 & -f \\ 0 & 0 & 1 \end{pmatrix} \begin{pmatrix} \cos\theta & \sin\theta & 0 \\ -\sin\theta & \cos\theta & 0 \\ 0 & 0 & 1 \end{pmatrix} \begin{pmatrix} p \\ q \\ 1 \end{pmatrix}$$

| Perspective projection | Translation to image line | Rotation of axes |

Dividing u by v results in x_I :

$$x_I = \frac{u}{v} = f\frac{p\cos\theta + q\sin\theta}{-p\sin\theta + q\cos\theta}$$

13.3 The Perspective Projection in 3D

Now suppose the optic axis of our camera lens is along a line parallel to the z-axis. To obtain the image frame coordinates for a given point in 3D space, we first translate this point to a 3D coordinate system centered at the lens of the camera. Then we translate along the z-axis by a distance f to the desired location of the projection image plane, and finally we take the perspective transformation.

We do this by using a homogeneous coordinate system that assumes an arbitrary position of the lens. Let (x, y, z) be the original coordinates of a point in 3D space. Let (x_0, y_0, z_0) be the position of the lens (called the *center of perspectivity*), and let (u, v) be the coordinates of the perspective projection of (x, y, z) on the image projection plane. Then $u = x*/t*$ and $v = y*/t*$, where

$$
\begin{pmatrix} x* \\ y* \\ t* \end{pmatrix} = \underbrace{\begin{pmatrix} 1 & 0 & 0 & 0 \\ 0 & 1 & 0 & 0 \\ 0 & 0 & 1/f & 1 \end{pmatrix}}_{\substack{\text{Perspective} \\ \text{projection}}} \underbrace{\begin{pmatrix} 1 & 0 & 0 & 0 \\ 0 & 1 & 0 & 0 \\ 0 & 0 & 1 & -f \\ 0 & 0 & 0 & 1 \end{pmatrix}}_{\substack{\text{Translation to} \\ \text{projection}}} \underbrace{\begin{pmatrix} 1 & 0 & 0 & -x_0 \\ 0 & 1 & 0 & -y_0 \\ 0 & 0 & 1 & -z_0 \\ 0 & 0 & 0 & 1 \end{pmatrix}}_{\substack{\text{Translation} \\ \text{to lens}}} \begin{pmatrix} x \\ y \\ z \\ 1 \end{pmatrix}
$$

$$
= \begin{pmatrix} x - x_0 \\ y - y_0 \\ (z - z_0)/f \end{pmatrix}
$$

Thus

$$
u = f\frac{x - x_0}{z - z_0} \quad \text{and} \quad v = f\frac{y - y_0}{z - z_0}
$$

13.3.1 Smaller Appearance of Farther Objects

We can, without loss of generality take the center of perspectivity to be the origin. The first property we notice about perspective projections is that objects appear smaller the farther they are from the center of perspectivity (Fig. 13.3). To understand this algebraically, assume that the center of perspectivity is at the origin and that we are observing the perspective projection of a line segment one of whose endpoints is at

$$
\begin{pmatrix} x_0 \\ y_0 \\ z_0 \end{pmatrix} - \frac{\Lambda}{2} \begin{pmatrix} a \\ b \\ c \end{pmatrix}
$$

and the other at

$$
\begin{pmatrix} x_0 \\ y_0 \\ z_0 \end{pmatrix} + \frac{\Lambda}{2} \begin{pmatrix} a \\ b \\ c \end{pmatrix}, \quad \text{where} \quad \begin{pmatrix} a \\ b \\ c \end{pmatrix}
$$

Figure 13.3 In perspective projection the farther objects appear smaller. Courtesy of Joseph D'Amelio, 1984 © with permission of the publisher Van Nostrand Reinhold, N.Y.

is the vector of direction cosines of the line segment, the constant Λ is the length

of the line segment, and $\begin{pmatrix} x_0 \\ y_0 \\ z_0 \end{pmatrix}$ is the center of the line segment.

Letting w denote the horizontal width of the projected line segment and h denote its vertical height, we can obtain

$$w = \left| f \frac{x_0 + \frac{\Lambda}{2}a}{z_0 + \frac{\Lambda}{2}c} - f \frac{x_0 - \frac{\Lambda}{2}a}{z_0 - \frac{\Lambda}{2}c} \right|$$

$$= f\Lambda \left| \frac{az_0 - cx_0}{z_0^2 - \frac{\Lambda^2}{4}c^2} \right|$$

$$h = \left| f\frac{y_0 + \frac{\Lambda}{2}b}{z_0 + \frac{\Lambda}{2}c} - f\frac{y_0 - \frac{\Lambda}{2}b}{z_0 - \frac{\Lambda}{2}c} \right|$$

$$= f\Lambda \left| \frac{bz_0 - cy_0}{z_0^2 - \frac{\Lambda^2}{4}c^2} \right|$$

As z_0 gets larger and larger, both w and h must get smaller and smaller. In the special case when the line segment lies in a plane parallel to the image plane ($c = 0$), the horizontal width and the vertical height of the projected line segment are proportional to the horizontal width and vertical height of the 3D line segment and inversely proportional to the distance between the image plane and the $z = z_0$ plane in which the line segment lies.

$$w = \frac{f\Lambda|a|}{z_0}$$

$$h = \frac{f\Lambda|b|}{z_0}$$

The second noticeable property about perspective projection images is that centrally located line segments lying in a plane parallel to the image plane show their maximum size. The more the centrally located line segment rotates away from a plane parallel to the image plane, the shorter the line segment will appear in its perspective image. This phenomenon, illustrated in Fig. 13.4, is called *foreshortening*.

The foreshortening phenomenon can be understood algebraically by writing out an expression for the projected length L of the line segment:

$$L = \sqrt{h^2 + w^2}$$

$$= \frac{f\Lambda}{z_0^2 - \frac{\Lambda}{4}c^2}\sqrt{(az_0 - cx_0)^2 + (bz_0 - cy_0)^2}$$

$$= \frac{f\Lambda}{z_0^2 - \frac{\Lambda^2}{4}c^2}\sqrt{(1 - c^2)z_0^2 + c^2(x_0^2 + y_0^2) - 2cz_0(ax_0 + by_0)}$$

For centrally located line segments, $x_0 \approx 0$ and $y_0 \approx 0$. Hence

$$L \approx \frac{f\Lambda}{z_0^2 - \frac{\Lambda^2}{4}c^2}z_0(1 - c^2)^{\frac{1}{2}}$$

Considering L as a function of c, we find the slope of L to be

$$\frac{\partial L}{\partial c} = \frac{-f\Lambda z_0 c}{(1 - c^2)^{\frac{1}{2}}(z_0^2 - \frac{\Lambda^2}{4}c^2)^2}\left[z_0^2 + \frac{c^2\Lambda^2}{4} - \frac{\Lambda^2}{2}\right]$$

We assume z_0 is large enough so that $z_0^2 + (c^2\Lambda^2)/4 - \Lambda^2/2 > 0$. For $c < 0$, the slope is positive, and for $c > 0$, it is negative. A relative maximum occurs when $c = 0$. As the slope of the line segment in the direction of the z-axis increases, $|c|$ increases to 1, which makes L decrease to zero. As the slope of the line segment decreases in the direction of the z-axis, $|c|$ decreases to zero, which makes L increase to $f\Lambda/z_0$.

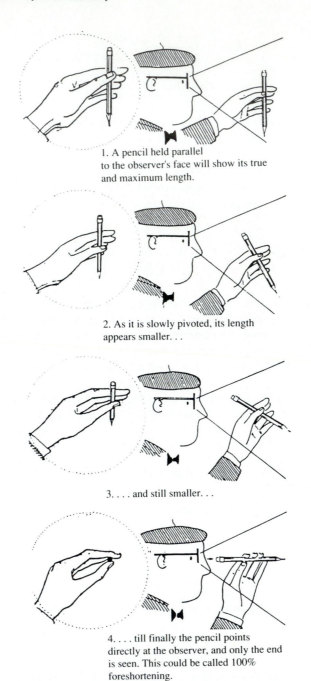

1. A pencil held parallel to the observer's face will show its true and maximum length.

2. As it is slowly pivoted, its length appears smaller. . .

3. . . . and still smaller. . .

4. . . . till finally the pencil points directly at the observer, and only the end is seen. This could be called 100% foreshortening.

Figure 13.4 In a perspective projection image, lengths appear foreshortened the more the 3D line segment is oriented away from a plane parallel to the image plane. Courtesy of Joseph D'Amelio, 1984 © with permission of the publisher Van Nostrand Reinhold, N.Y.

13.3.2 Lines to Lines

The two most important properties of the 2D-perspective transformation is the way lines in the 3D world relate to lines in the image plane: Lines in the 3D world transform to lines in the image plane; parallel lines in the 3D world, having nonzero slope along the z-axis, meet in a vanishing point in the image plane.

To see that lines in the original 3D world correspond to lines in the image plane, we let p_1 and p_2 be two points in the 3D world represented in this homogeneous coordinate system. The line passing through p_1 and p_2 consists of all points having the form $\lambda p_1 + (1 - \lambda)p_2$ for some constant λ. We let the perspective projections of p_1 and p_2 be $\binom{u_1/w_1}{v_1/w_1}$ and $\binom{u_2/w_2}{v_2/w_2}$, respectively, where for the perspective transformation matrix T,

$$\begin{pmatrix} u_1 \\ v_1 \\ w_1 \end{pmatrix} = T p_1 \quad \text{and} \quad \begin{pmatrix} u_2 \\ v_2 \\ w_2 \end{pmatrix} = T p_2.$$

The exact nature of the matrix T is not important in this discussion. We need only use the fact that T is a linear operator. Hence

$$T[\lambda p_1 + (1 - \lambda)p_2] = \lambda T p_1 + (1 - \lambda)T p_2$$

$$= \lambda \begin{pmatrix} u_1 \\ v_1 \\ w_1 \end{pmatrix} + (1 - \lambda) \begin{pmatrix} u_2 \\ v_2 \\ w_2 \end{pmatrix}$$

The line segment on the image plane passing through $\binom{u_1/w_1}{v_1/w_1}$ and $\binom{u_2/w_2}{v_2/w_2}$ consists of all points of the form $\eta \binom{u_1/w_1}{v_1/w_1} + (1 - \eta)\binom{u_2/w_2}{v_2/w_2}$, for some constant η. To show that the perspective transformation of every line in the 3D world is a line in the image plane and every line in the image plane corresponds to a line in the 3D model, we need to show that for every λ there exists an η and for every η there exists a λ such that the following relationship is satisfied:

$$\begin{pmatrix} [\lambda u_1 + (1 - \lambda)u_2]/[\lambda w_1 + (1 - \lambda)w_2] \\ [\lambda v_1 + (1 - \lambda)v_2]/[\lambda w_1 + (1 - \lambda)w_2] \end{pmatrix} = \eta \begin{pmatrix} u_1/w_1 \\ v_1/w_1 \end{pmatrix} + (1 - \eta) \begin{pmatrix} u_2/w_2 \\ v_2/w_2 \end{pmatrix}$$

It is easily verified that if λ is given, then

$$\eta = \frac{\lambda w_1}{\lambda w_1 + (1 - \lambda)w_2}$$

And if η is given, then

$$\lambda = \frac{\eta w_2}{\eta w_2 + (1 - \eta)w_1}$$

satisfies the required relationship. Therefore lines correspond to lines.

13.3.3   Perspective Projections of Convex Polyhedra are Convex

Let

$$\left\{ \begin{pmatrix} x_n \\ y_n \\ z_n \end{pmatrix} \middle| \; n = 1, \ldots, N \right\}$$

constitute the vertices of a convex polyhedron. Let

$$\left\{ \begin{pmatrix} u_n \\ v_n \end{pmatrix} \middle| \; n = 1, \ldots, N \right\}$$

be the corresponding perspective projections of the polyhedron vertices. Then by the perspective projection equations

$$u_n = f \, \frac{x_n}{z_n}$$

$$v_n = f \, \frac{y_n}{z_n}$$

Let $\begin{pmatrix} x \\ y \\ z \end{pmatrix}$ be any point in the 3D convex polyhedron. Then there exist constants

$$\lambda_1, \ldots, \lambda_N, \; \lambda_n > 0, \; \sum_{n=1}^{N} \lambda_n = 1$$

such that

$$\begin{pmatrix} x \\ y \\ z \end{pmatrix} = \sum_{n=1}^{N} \lambda_n \begin{pmatrix} x_n \\ y_n \\ z_n \end{pmatrix}$$

Let $\begin{pmatrix} u \\ v \end{pmatrix}$ be the perspective projection of $\begin{pmatrix} x \\ y \\ z \end{pmatrix}$. Then

$$u = f \, \frac{x}{z}$$

$$v = f \, \frac{y}{z}$$

Hence

$$u = f \, \frac{\sum_{n=1}^{N} \lambda_n x_n}{\sum_{n=1}^{N} \lambda_n z_n}$$

$$v = f \, \frac{\sum_{n=1}^{N} \lambda_n y_n}{\sum_{n=1}^{N} \lambda_n z_n}$$

But

$$x_n = \frac{u_n z_n}{f} \qquad \text{and} \qquad y_n = \frac{v_n z_n}{f}$$

So

$$u = f \frac{\sum_{n=1}^{N} \lambda_n u_n z_n / f}{\sum_{m=1}^{N} \lambda_m z_m}$$

$$= \sum_{n=1}^{N} \left(\frac{\lambda_n z_n}{\sum_{m=1}^{N} \lambda_m z_m} \right) u_n.$$

Similarly,

$$v = f \frac{\sum_{n=1}^{N} \lambda_n v_n z_n / f}{\sum_{m=1}^{N} \lambda_m z_m}$$

$$= \sum_{n=1}^{N} \left(\frac{\lambda_n z_n}{\sum_{m=1}^{N} \lambda_m z_m} \right) v_n$$

Since $z_n > 0$ and $\lambda_n > 0$,

$$\frac{\lambda_n z_n}{\sum_{m=1}^{N} \lambda_m z_m} > 0.$$

Also

$$\sum_{n=1}^{N} \frac{\lambda_n z_n}{\sum_{m=1}^{N} \lambda_m z_m} = 1.$$

This means that the perspective projection $\begin{pmatrix} u \\ v \end{pmatrix}$ of $\begin{pmatrix} x \\ y \\ z \end{pmatrix}$ is a convex combination

of the perspective projections of the vertices of the polyhedron.

13.3.4 Vanishing Point

Perspective projections of parallel 3D lines having nonzero slope along the optic z-axis meet, in general, in a vanishing point on the image projection plane. To see this, consider a 3D line L represented by its direction cosines. The line L consists of the set of points

$$L = \left\{ \begin{pmatrix} x \\ y \\ z \end{pmatrix} \middle| \text{ for some } \lambda, \begin{pmatrix} x \\ y \\ z \end{pmatrix} = \begin{pmatrix} a_1 \\ a_2 \\ a_3 \end{pmatrix} + \lambda \begin{pmatrix} b_1 \\ b_2 \\ b_3 \end{pmatrix} \right\}$$

The line L passes through the point $\begin{pmatrix} a_1 \\ a_2 \\ a_3 \end{pmatrix}$ and has direction cosines $\begin{pmatrix} b_1 \\ b_2 \\ b_3 \end{pmatrix}$. In this

representation the direction vector $\begin{pmatrix} b_1 \\ b_2 \\ b_3 \end{pmatrix}$ has unit length; that is, $b_1^2 + b_2^2 + b_3^2 = 1$.

Parallel lines here are lines that have the same direction vectors.

The perspective projection $\binom{u}{v}$ of any point on the line L is given by

$$u = f\frac{a_1 + \lambda b_1 - x_0}{a_3 + \lambda b_3 - z_0}, \qquad v = f\frac{a_2 + \lambda b_2 - y_0}{a_3 + \lambda b_3 - z_0}$$

where (x_0, y_0, z_0) is the center of perspectivity. For lines that have some slope along the optic axis, $b_3 \neq 0$. Points that are on the line and infinitely far from the center of the lens will have perspective projection

$$u_\infty = \lim_{\lambda \to \infty} f\frac{a_1 + \lambda b_1 - x_0}{a_3 + \lambda b_3 - z_0} = f\frac{b_1}{b_3}$$

(13.1)

$$v_\infty = \lim_{\lambda \to \infty} f\frac{a_2 + \lambda b_2 - y_0}{a_3 + \lambda b_3 - z_0} = f\frac{b_2}{b_3}$$

Since the point $\binom{u_\infty}{v_\infty}$ is independent of a_1, a_2, a_3, all parallel 3D lines that have nonzero slope along the optic axis have perspective projections that meet at the same point, called the *vanishing point,* on the perspective projection image plane.

The limit on $\lambda \to -\infty$ gives the same result as the limit on $\lambda \to \infty$. Why is this, and what does it mean? One limit corresponds to the vanishing point arising from the ends of parallel lines in front of the camera lens, and the other corresponds to the ends of the parallel lines behind the camera lens. Both vanishing points are the same.

For lines that have no slope along the optic axis, $b_3 = 0$. Such a line is on a plane parallel to the image projection plane. These lines have perspective projections that are parallel to the 3D line. We can easily understand this by observing that when $b_3 = 0$,

$$u = f\frac{a_1 + \lambda b_1 - x_0}{a_3 - z_0}$$

$$v = f\frac{a_2 + \lambda b_2 - y_0}{a_3 - z_0}$$

Multiplying the first equation by b_2, the second equation by b_1, and subtracting yields

$$b_2 u - b_1 v = \frac{f}{a_3 - z_0}[(a_1 - x_0)b_2 - (a_2 - y_0)b_1]$$

Notice that the slope of the line depends solely on b_1 and b_2, and that the right-hand side of the equation is a constant with respect to u and v. This means that all lines that are parallel to the image projection plane will have perspective projection lines that have the same slope and differ only in their intercept.

What makes vanishing points important is that from the position of a vanishing point on the perspective projection image plane, it is possible to infer the direction cosine vector for the 3D parallel lines whose perspective projections meet at the vanishing point. To see this, just collect together all the constraints: two for the position of the vanishing point and one because the direction cosine vector has length 1. Solving for $b_1, b_2,$ and b_3 in terms of $f, u_\infty,$ and v_∞ yields

$$\begin{pmatrix} b_1 \\ b_2 \\ b_3 \end{pmatrix} = \frac{1}{\sqrt{u_\infty^2 + v_\infty^2 + f_\infty^2}} \begin{pmatrix} u_\infty \\ v_\infty \\ f \end{pmatrix} \tag{13.2}$$

Finally, we develop one more vanishing-point relation. The vanishing points for all 3D lines that have nonzero slope along the optic axis direction and lie in parallel planes (planes with identical normals) are constrained to lie along a line in the perspective projection image plane. To see this, consider any line having direction cosine vector $\begin{pmatrix} b_1 \\ b_2 \\ b_3 \end{pmatrix}$, where $b_3 \neq 0$. Its vanishing point will be

$$u_\infty = f \frac{b_1}{b_3}$$
$$v_\infty = f \frac{b_2}{b_3}$$

Points on the 3D line have the form

$$\begin{pmatrix} x \\ y \\ z \end{pmatrix} = \begin{pmatrix} a_1 \\ a_2 \\ a_3 \end{pmatrix} + \lambda \begin{pmatrix} b_1 \\ b_2 \\ b_3 \end{pmatrix}$$

If the line is perpendicular to the vector $\begin{pmatrix} A \\ B \\ C \end{pmatrix}$, then its direction cosine vector will be perpendicular to $\begin{pmatrix} A \\ B \\ C \end{pmatrix}$. Hence $Ab_1 + Bb_2 + Cb_3 = 0$. Substituting $u_\infty b_3/f$ for b_1 and $v_\infty b_3/f$ for b_2 results in

$$\frac{Au_\infty b_3}{f} + \frac{Bv_\infty b_3}{f} + Cb_3 = 0$$

Since $b_3 \neq 0$, we finally obtain

$$Au_\infty + Bv_\infty + Cf = 0$$

This shows that the vanishing point for any line perpendicular to the vector $\begin{pmatrix} A \\ B \\ C \end{pmatrix}$ must satisfy $Au_\infty + Bv_\infty + Cf = 0$ on the perspective projection image plane. Thus vanishing points arising from lines all of which are perpendicular to the same vector must lie along a line in the image plane.

These relations imply that it is quite easy to infer the normal to a plane by observing the vanishing points resulting from two pairs of parallel lines each lying in some plane with the same surface normal. Let $\binom{u_1}{v_1}$ be the first vanishing point and $\binom{u_2}{v_2}$ the second vanishing point. Then

$$\begin{pmatrix} u_1 & v_1 \\ u_2 & v_2 \end{pmatrix} \begin{pmatrix} A \\ B \end{pmatrix} = -Cf \begin{pmatrix} 1 \\ 1 \end{pmatrix}$$

so that

$$A = Cf(v_1 - v_2)/(u_1v_2 - v_1u_2)$$
$$B = Cf(u_2 - u_1)/(u_1v_2 - v_1u_2)$$

Since the plane normal $\begin{pmatrix} A \\ B \\ C \end{pmatrix}$ has length 1, we can use the constraint $A^2 + B^2 +$

$C^2 = 1$ to obtain

$$A = (v_1 - v_2)f \Big/ \sqrt{(v_1 - v_2)^2 f^2 + (u_2 - u_1)^2 f^2 + (u_1v_2 - u_2v_1)^2}$$

$$B = (u_2 - u_1)f \Big/ \sqrt{(v_1 - v_2)^2 f^2 + (u_2 - u_1)^2 f^2 + (u_1v_2 - u_2v_1)^2}$$

$$C = (u_1v_2 - u_2v_1) \Big/ \sqrt{(v_1 - v_2)^2 f^2 + (u_2 - u_1)^2 f^2 + (u_1v_2 - u_2v_1)^2} \quad (13.3)$$

Of course, $(-A, -B, -C)$ is also a plane normal, and the choice between (A, B, C) and $(-A, -B, -C)$ is arbitrary.

EXAMPLE 13.1

Take the focal length of the lens to be $f = 3$ and the plane normal to be $\frac{1}{5\sqrt{2}} \begin{pmatrix} 3 \\ 4 \\ 5 \end{pmatrix}$. Take the direction cosines for one pair of parallel lines to be $\frac{1}{\sqrt{6}} \begin{pmatrix} 1 \\ -2 \\ 1 \end{pmatrix}$ and for the other pair of parallel lines to be $\frac{1}{\sqrt{19}} \begin{pmatrix} 1 \\ 3 \\ -3 \end{pmatrix}$. The observed vanishing point for the first pair of parallel lines will be $\binom{u_1}{v_1} = \binom{3}{-6}$. The observed vanishing point for the second pair of parallel lines will be $\binom{u_2}{v_2} = \binom{-1}{-3}$. In this case from the observed vanishing point and the known focal length of the lens, we obtain

$$(v_1 - v_2)f = -9$$
$$(u_2 - u_1)f = -12$$
$$(u_1v_2 - u_2v_1) = -15$$

Hence

$$A = (v_1 - v_2)f \Big/ \sqrt{(v_1 - v_2)^2 f^2 + (u_2 - u_1)^2 f^2 + (u_1 v_2 - u_2 v_1)^2}$$

$$= (-9/\sqrt{450}) = (-3/5\sqrt{2})$$

$$B = (u_2 - u_1)f \Big/ \sqrt{(v_1 - v_2)^2 f^2 + (u_2 - u_1)^2 f^2 + (u_1 v_2 - u_2 v_1)^2}$$

$$= (-12/\sqrt{450}) = (-4/5\sqrt{2})$$

$$C = (u_1 v_2 - u_2 v_1) \Big/ \sqrt{(v_1 - v_2)^2 f^2 + (u_2 - u_1)^2 f^2 + (u_1 v_2 - u_2 v_1)^2}$$

$$= (-15/\sqrt{450}) = (-5/5\sqrt{2})$$

■

13.3.5 Vanishing Line

Consider any plane with surface normal (A, B, C). Points in such a plane satisfy $Ax + By + Cz + D = 0$ for some D. The set of all vanishing points for lines in any such plane is the line L_∞, called the vanishing line, $L_\infty = \left\{ \begin{pmatrix} u \\ v \end{pmatrix} \middle| Au + Bv + Cf = 0 \right\}$; and every point in L_∞ is a vanishing point for some line in such a plane (Fig. 13.5).

To see this vanishing line property algebraically, let D be given and let (u_∞, v_∞) be a vanishing point arising from a line in the plane $Ax + By + Cz + D = 0$. Then for some line with direction cosine (b_1, b_2, b_3),

$$\begin{pmatrix} u_\infty \\ v_\infty \end{pmatrix} = \frac{f}{b_3} \begin{pmatrix} b_1 \\ b_2 \end{pmatrix}$$

and $Ab_1 + Bb_2 + Cb_3 = 0$.

Consider $Au_\infty + Bv_\infty + Cf$.

$$Au_\infty + Bv_\infty + Cf = A\frac{fb_1}{b_3} + B\frac{fb_2}{b_3} + Cf$$

$$\frac{f}{b_3}(Ab_1 + Bb_2 + Cb_3) = 0$$

Hence $(u_\infty, v_\infty) \in L_\infty$.

Conversely, suppose $(u_o, v_o) \in L_\infty$. Since $(u_o, v_o) \in L_\infty$, $Au_o + Bv_o + Cf = 0$. Take any line with direction cosines

$$\frac{1}{\sqrt{u_o^2 + v_o^2 + f^2}} \begin{pmatrix} u_o \\ v_o \\ f \end{pmatrix}$$

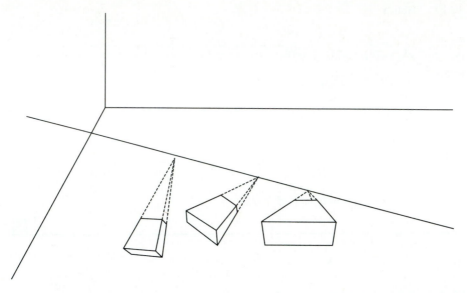

Figure 13.5 All lines lying in planes parallel to the slanted floor have vanishing points that lie along a vanishing line.

Such a line has vanishing point

$$
\begin{pmatrix} u_\infty \\ v_\infty \end{pmatrix} = \frac{f}{f/\sqrt{u_o^2 = v_o^2 + f^2}} \begin{pmatrix} u_o/\sqrt{u_o^2 + v_o^2 + f^2} \\ v_o/\sqrt{u_o^2 + v_o^2 + f^2} \end{pmatrix}
$$
$$
= \begin{pmatrix} u_o \\ v_o \end{pmatrix}
$$

Therefore any line

$$
L = \left\{ \begin{pmatrix} x \\ y \\ z \end{pmatrix} \middle| \begin{pmatrix} x \\ y \\ z \end{pmatrix} = \begin{pmatrix} a_1 \\ a_2 \\ a_3 \end{pmatrix} + \frac{\lambda}{\sqrt{u_o^2 + v_o^2 + f^2}} \begin{pmatrix} u_o \\ v_o \\ f \end{pmatrix} \right\}
$$

will lie in the plane $Ax + By + Cz + D = 0$, where $D = (Aa_1 + Ba_2 + Ca_3)$, and have vanishing point (u_o, v_o).

The vanishing line for the set of planes with surface normal (A, B, C) has an easy-to-visualize geometric interpretation. Select from this set of planes the one going through the origin. This plane satisfies $Ax + By + Cz = 0$. Consider the intersection of this plane with image plane $z = f$. The intersection is a line L on the image plane given by

$$
L = \left\{ \begin{pmatrix} x \\ y \end{pmatrix} \middle| Ax + By + Cf = 0 \right\}
$$

The line L is the vanishing line L_∞ for the planes having surface normal (A, B, C).

13.3.6 3D Lines–2D Perspective Projection Lines

There is a relationship between the parameters of a 3D line and the parameters of the perspective projection of the line that we will use repeatedly. Suppose that a 3D line L is given as the set

$$L = \left\{ \begin{pmatrix} x \\ y \\ z \end{pmatrix} \;\middle|\; \text{for some } \lambda, \; \begin{pmatrix} x \\ y \\ z \end{pmatrix} = \begin{pmatrix} a_1 \\ a_2 \\ a_3 \end{pmatrix} + \lambda \begin{pmatrix} b_1 \\ b_2 \\ b_3 \end{pmatrix} \right\}$$

and the perspective projection of the line L is given as the set

$$M = \left\{ \begin{pmatrix} u \\ v \end{pmatrix} \;\middle|\; \text{for some } \eta, \; \begin{pmatrix} u \\ v \end{pmatrix} = \begin{pmatrix} c_1 \\ c_2 \end{pmatrix} + \eta \begin{pmatrix} d_1 \\ d_2 \end{pmatrix} \right\}$$

We take the camera lens as the origin of the coordinate system, so that

$$u = f \, \frac{x}{z}$$
$$v = f \, \frac{y}{z}$$

Hence for any given λ, η must satisfy

$$c_1 + \eta d_1 = f(a_1 + \lambda b_1)/(a_3 + \lambda b_3)$$
$$c_2 + \eta d_2 = f(a_2 + \lambda b_2)/(a_3 + \lambda b_3)$$

Multiplying the first equation by d_2, the second equation by d_1, and subtracting one from the other results in the elimination of η:

$$c_1 d_2 - c_2 d_1 = \frac{d_2 f(a_1 + \lambda b_1) - d_1 f(a_2 + \lambda b_2)}{a_3 + \lambda b_3}$$

This equation must hold for any given λ. Rewriting this as a part that depends on λ and a part that is independent of λ results in

$$(c_1 d_2 - c_2 d_1)a_3 - (d_2 a_1 - d_1 a_2)f + \lambda \, [b_3(c_1 d_2 - c_2 d_1) - (d_2 b_1 - d_1 b_2)f] = 0$$

Since this relation must hold for every λ, we obtain two equations that relate parameters d_1, d_2, c_1, and c_2 of the observed perspective projection line with the unknown parameters a_1, a_2, a_3, b_1, b_2, and b_3 of the 3D line:

$$d_2 f a_1 - d_1 f a_2 + (c_2 d_1 - c_1 d_2)a_3 = 0$$

$$\tag{13.4}$$

$$d_2 f b_1 - d_1 f b_2 + (c_2 d_1 - c_1 d_2)b_3 = 0$$

So there is a total of six unknown parameters of a 3D line, and from the perspective projection of a 3D line we obtain two constraints. However, one constraint is inherent in the parametrization of the line: $b_1^2 + b_2^2 + b_3^2 = 1$. Finally, the point $\begin{pmatrix} a_1 \\ a_2 \\ a_3 \end{pmatrix}$ through which the line passes is not unique. We can make it unique by requiring the vector from the origin to it to be perpendicular to $\begin{pmatrix} b_1 \\ b_2 \\ b_3 \end{pmatrix}$. In this case $a_1b_1 + a_2b_2 + a_3b_3 = 0$. Now knowledge of the perspective projection of an unknown line provides only four of the six constraints needed to determine the line.

EXAMPLE 13.2

Suppose the perspective projection of the 3D line given by

$$L = \left\{ \begin{pmatrix} x \\ y \\ z \end{pmatrix} \; \middle| \; \begin{pmatrix} x \\ y \\ z \end{pmatrix} = \begin{pmatrix} 3 \\ 4 \\ 5 \end{pmatrix} + \lambda \begin{pmatrix} \frac{1}{\sqrt{2}} \\ 0 \\ \frac{1}{\sqrt{2}} \end{pmatrix} \right\}$$

is observed. If $f = 10$, then the line on the perspective projection image will be given by

$$M = \left\{ \begin{pmatrix} u \\ v \end{pmatrix} \; \middle| \; \begin{pmatrix} u \\ v \end{pmatrix} = \begin{pmatrix} 6 \\ 8 \end{pmatrix} + \eta \frac{1}{\sqrt{5}} \begin{pmatrix} -1 \\ 2 \end{pmatrix} \right\}.$$

Here

$$\begin{pmatrix} a_1 \\ a_2 \\ a_3 \end{pmatrix} = \begin{pmatrix} 3 \\ 4 \\ 5 \end{pmatrix}, \quad \begin{pmatrix} b_1 \\ b_2 \\ b_3 \end{pmatrix} = \frac{1}{\sqrt{2}} \begin{pmatrix} 1 \\ 0 \\ 1 \end{pmatrix}$$

$$\begin{pmatrix} c_1 \\ c_2 \end{pmatrix} = \begin{pmatrix} 6 \\ 8 \end{pmatrix}, \quad \begin{pmatrix} d_1 \\ d_2 \end{pmatrix} = \frac{1}{\sqrt{5}} \begin{pmatrix} -1 \\ 2 \end{pmatrix}$$

Notice that

$$d_2 f a_1 - d_1 f a_2 + (c_2 d_1 - c_1 d_2) a_3 = \left(\frac{2}{\sqrt{5}} \right)(10)(3) - \left(\frac{-1}{\sqrt{5}} \right)(10)(4) +$$

$$\left[(8) \left(\frac{-1}{\sqrt{5}} \right) - (6) \left(\frac{2}{\sqrt{5}} \right) \right] (5)$$

$$= \frac{60 + 40 - 100}{\sqrt{5}} = 0$$

and that

$$d_2 f b_1 - d_1 f b_2 + (c_2 d_1 - c_1 d_2) b_3 = \left(\frac{2}{\sqrt{5}} \right)(10) \left(\frac{1}{\sqrt{2}} \right) - \left(\frac{-1}{\sqrt{5}} \right)(10)(0)$$

$$+ \left[(8) \left(\frac{-1}{\sqrt{5}} \right) - (6) \left(\frac{2}{\sqrt{5}} \right) \right] \frac{1}{\sqrt{2}}$$

$$= \frac{20 + 0 - 20}{\sqrt{10}} = 0$$

∎

13.4   2D to 3D Inference Using Perspective Projection

In the previous section we indicated that knowledge of the perspective projection of an unknown 3D line provides four of the six constraints required to specify the line. To find additional constraints, we can use 3D-world-model information about the points and lines whose perspective projection we observe. Then such relations with the perspective geometry constraints can often provide enough information to uniquely determine everything in 3D. Such knowledge can come about when we have a model of the object being viewed in the perspective projection. In this section we show how a variety of kinds of relations between lines, their direction cosines, distances along lines, and the planes the lines may lie in all lead to the formation of additional constraint equations.

13.4.1   Inverse Perspective Projection

A point whose perspective projection is (u,v) in the coordinate system of the image projection plane that is a distance f in front of the camera lens has 3D coordinates (u,v,f). Since the camera lens is the origin, a ray passing through the point (u,v,f) and the origin consists of all multiples of (u,v,f). Furthermore, since the origin is the center of perspectivity, any 3D point on the ray consisting of a multiple of (u,v,f) has perspective projection (u,v). We call the line L,

$$L = \left\{ \begin{pmatrix} x \\ y \\ z \end{pmatrix} \middle| \begin{pmatrix} x \\ y \\ z \end{pmatrix} = \lambda \begin{pmatrix} u \\ v \\ f \end{pmatrix} \right\}$$

the inverse perspective projection of the point (u,v).

13.4.2   Line Segment with Known Direction Cosines and Known Length

Suppose we know that we are observing the perspective projection of a 3D line hav-

ing known direction cosines $\begin{pmatrix} b_1 \\ b_2 \\ b_3 \end{pmatrix}$ and known line length of δ. We know everything

but the position of the line segment in three dimensions. The 3D line segment L can be represented in the form

$$L = \left\{ \begin{pmatrix} x \\ y \\ z \end{pmatrix} \middle| \begin{pmatrix} x \\ y \\ z \end{pmatrix} = \begin{pmatrix} a_1 \\ a_2 \\ a_3 \end{pmatrix} + \lambda \begin{pmatrix} b_1 \\ b_2 \\ b_3 \end{pmatrix}, \frac{-\delta}{2} \leq \lambda \leq \frac{\delta}{2} \right\}$$

In this case one end of the line segment is at

$$p_1 = \begin{pmatrix} a_1 \\ a_2 \\ a_3 \end{pmatrix} + \frac{\delta}{2} \begin{pmatrix} b_1 \\ b_2 \\ b_3 \end{pmatrix}$$

and the other end is at

$$p_2 = \begin{pmatrix} a_1 \\ a_2 \\ a_3 \end{pmatrix} - \frac{\delta}{2} \begin{pmatrix} b_1 \\ b_2 \\ b_3 \end{pmatrix}.$$

Assuming a coordinate system where the camera lens is the origin, suppose the perspective projections of p_1 and p_2 are observed to be $\begin{pmatrix} u_1 \\ v_1 \end{pmatrix}$ and $\begin{pmatrix} u_2 \\ v_2 \end{pmatrix}$, respectively.

By the geometry of the perspective projection, p_1 must lie someplace along the ray defined by the line passing through the center of the lens and the point $\begin{pmatrix} u_1 \\ v_1 \\ f \end{pmatrix}$ on the perspective projection image plane. Hence for some η_1,

$$p_1 = \eta_1 \begin{pmatrix} u_1 \\ v_1 \\ f \end{pmatrix}.$$

Similarly, for some η_2,

$$p_2 = \eta_2 \begin{pmatrix} u_2 \\ v_2 \\ f \end{pmatrix}.$$

The vector from p_2 to p_1 is $p_1 - p_2$, which can be written as $p_1 - p_2 = \delta \begin{pmatrix} b_1 \\ b_2 \\ b_3 \end{pmatrix}$.

This implies that η_1 and η_2 are constrained to satisfy

$$\eta_1 \begin{pmatrix} u_1 \\ v_1 \\ f \end{pmatrix} - \eta_2 \begin{pmatrix} u_2 \\ v_2 \\ f \end{pmatrix} = \delta \begin{pmatrix} b_1 \\ b_2 \\ b_3 \end{pmatrix}.$$

This is an overconstrained equation for η_1 and η_2:

$$\begin{pmatrix} u_1 & -u_2 \\ v_1 & -v_2 \\ f & -f \end{pmatrix} \begin{pmatrix} \eta_1 \\ \eta_2 \end{pmatrix} = \delta \begin{pmatrix} b_1 \\ b_2 \\ b_3 \end{pmatrix}.$$

Solving the overconstrained equation, we obtain

$$\begin{pmatrix} \eta_1 \\ \eta_2 \end{pmatrix} = \delta \begin{pmatrix} \frac{(u_1 v_2 - u_2 v_1)(b_1 v_2 - b_2 u_2) + (u_2 - u_1) f (u_2 b_3 - f b_1) + (v_2 - v_1) f (v_2 b_3 - f b_2)}{(u_1 v_2 - u_2 v_1)^2 + [(u_1 - u_2)^2 + (v_1 - v_2)^2] f^2} \\ \frac{(u_1 v_2 - u_2 v_1)(b_1 v_1 - b_2 u_1) + (u_2 - u_1) f (u_1 b_3 - f b_1) + (v_2 - v_1) f (v_1 b_3 - f b_2)}{(u_1 v_2 - u_2 v_1)^2 + [(u_1 - u_2)^2 + (v_1 - v_2)^2] f^2} \end{pmatrix}.$$

Having the values for η_1 and η_2, we can substitute and immediately write down the

3D position of the line segment endpoints:

$$p_1 = \frac{\delta[(u_1v_2-u_2v_1)(b_1v_2-b_2u_2)+(u_2-u_1)f(u_2b_3-fb_1)+(v_2-v_1)f(v_2b_3-fb_2)]}{(u_1v_2-u_2v_1)^2+[(u_1-u_2)^2+(v_1-v_2)^2]f^2} \begin{pmatrix} u_1 \\ v_1 \\ f \end{pmatrix}$$

$$(13.5)$$

$$p_2 = \frac{\delta[(u_1v_2-u_2v_1)(b_1v_1-b_2u_1)+(u_2-u_1)f(u_1b_3-fb_1)+(v_2-v_1)f(v_1b_3-fb_2)]}{(u_1v_2-u_2v_1)^2+[(u_1-u_2)^2+(v_1-v_2)^2]f^2} \begin{pmatrix} u_2 \\ v_2 \\ f \end{pmatrix}$$

In this development, δ is really the directed distance from p_2 to p_1. The knowledge we actually have is about $|\delta|$. This means that we must try $+|\delta|$ and $-|\delta|$ in place of δ in the equations determining p_1 and p_2. Only one of the alternatives will provide values for p_1 and p_2 that are consistent with the points being in front of the lens. An easier way to proceed in this kind of problem is as follows:

Let $\begin{pmatrix} a \\ b \\ c \end{pmatrix}$ be the position of the first point and $\begin{pmatrix} a \\ b \\ c \end{pmatrix} + \rho \begin{pmatrix} m_1 \\ m_2 \\ m_3 \end{pmatrix}$ the position

of the second point. The term ρ is the directed distance between two points and $\begin{pmatrix} m_1 \\ m_2 \\ m_3 \end{pmatrix}$ is the set of direction cosines for the line segment defined by the two

points. Let $\begin{pmatrix} u_1 \\ v_1 \end{pmatrix}$ and $\begin{pmatrix} u_2 \\ v_2 \end{pmatrix}$ be their perspective projections. From the perspective projection equations

$$u_1 = f\frac{a}{c}, \qquad u_2 = f\frac{a+\rho m_1}{c+\rho m_3}$$

$$v_1 = f\frac{b}{c}, \qquad v_2 = f\frac{b+\rho m_2}{c+\rho m_3}$$

Solving for a and b in the first pair of equations and substituting in the second yields

$$\begin{pmatrix} u_2 - u_1 \\ v_2 - v_1 \end{pmatrix} c = \rho \begin{pmatrix} fm_1 - u_2m_3 \\ fm_2 - v_2m_3 \end{pmatrix}$$

Since ρ is known, the least-squares solution for a, b, and c can be given by

$$c = \frac{\rho[(u_2 - u_1)(fm_1 - u_2m_3) + (v_2 - v_1)(fm_2 - v_2m_3)]}{(u_2 - u_1)^2 + (v_1 - v_2)^2}$$

$$a = \frac{u_1}{f}c \qquad\qquad (13.6)$$

$$b = \frac{v_1}{f}c$$

■ **EXAMPLE 13.3**

Suppose we know that we are observing the perspective projection of two parallel 3D line segments of known length. The extended lines of which the observed perspective projection line segments are a part must meet at a vanishing point. Once the coordinates of the vanishing point are determined, the direction cosines for the 3D line segment can be determined. Since the length of the line segment and its direction cosine are known, the 3D position of the line segment can be determined.

In the example, the first line segment has endpoints

$$\begin{pmatrix} 10 \\ 12 \\ 14 \end{pmatrix} \text{ and } \begin{pmatrix} 13 \\ 18 \\ 23 \end{pmatrix}$$

The second line segment has endpoints

$$\begin{pmatrix} 21 \\ 11 \\ 14 \end{pmatrix} \text{ and } \begin{pmatrix} 28 \\ 25 \\ 35 \end{pmatrix}$$

The distance between the projective image plane and the center of perspectivity is 5. The observed perspective projections of the endpoints for the first line segment are

$$\begin{pmatrix} u_1 \\ v_1 \end{pmatrix} = \begin{pmatrix} 3.571428 \\ 4.285714 \end{pmatrix}, \qquad \begin{pmatrix} u_2 \\ v_2 \end{pmatrix} = \begin{pmatrix} 2.826087 \\ 3.913043 \end{pmatrix}$$

and for the second line segment

$$\begin{pmatrix} u_3 \\ v_3 \end{pmatrix} = \begin{pmatrix} 7.5 \\ 3.92857 \end{pmatrix}, \qquad \begin{pmatrix} u_4 \\ v_4 \end{pmatrix} = \begin{pmatrix} 4.0 \\ 3.571428 \end{pmatrix}$$

The point $\begin{pmatrix} u_\infty \\ v_\infty \end{pmatrix}$ of intersection of the lines is the vanishing point, and it must satisfy

$$\begin{pmatrix} v_2 - v_1 & u_1 - u_2 \\ v_4 - v_3 & u_3 - u_4 \end{pmatrix} \begin{pmatrix} u_\infty \\ v_\infty \end{pmatrix} = \begin{pmatrix} u_1 v_2 - u_2 v_1 \\ u_3 v_4 - u_4 v_3 \end{pmatrix}$$

Hence

$$u_\infty = \frac{\begin{vmatrix} u_1 v_2 - u_2 v_1 & u_1 - u_2 \\ u_3 v_4 - u_4 v_3 & u_3 - u_4 \end{vmatrix}}{\begin{vmatrix} v_2 - v_1 & u_1 - u_2 \\ v_4 - v_3 & u_3 - u_4 \end{vmatrix}}$$

$$v_\infty = \frac{\begin{vmatrix} v_2 - v_1 & u_1 v_2 - u_2 v_1 \\ v_4 - v_3 & u_3 v_4 - u_4 v_3 \end{vmatrix}}{\begin{vmatrix} v_2 - v_1 & u_1 - u_2 \\ v_4 - v_3 & u_3 - u_4 \end{vmatrix}}$$

Then

$$v_2 - v_1 = -.372671$$
$$u_1 - u_2 = .745348$$
$$u_1 v_2 - u_2 v_1 = 1.8633807$$
$$v_4 - v_3 = -.357142$$
$$u_3 - u_4 = 3.5$$
$$u_3 v_4 - u_4 v_3 = 11.07143$$
$$u_\infty = 1.6666$$
$$v_\infty = 3.3333$$

Having the vanishing point, we can determine the direction cosines of the 3D lines:

$$\begin{pmatrix} b_1 \\ b_2 \\ b_3 \end{pmatrix} = \frac{1}{\sqrt{u_\infty^2 + v_\infty^2 + f^2}} \begin{pmatrix} u_\infty \\ v_\infty \\ f \end{pmatrix}$$

Substituting the values for u, v, and f results in

$$\begin{pmatrix} b_1 \\ b_2 \\ b_3 \end{pmatrix} = \begin{pmatrix} .267261 \\ .534522 \\ .801784 \end{pmatrix}$$

Finally, substituting the values for $u_1, u_2, v_1, v_2, b_1, b_2, b_3, f$, and δ into the equation determining p_1 and p_2, we obtain

$$p_1 = \begin{pmatrix} 10 \\ 12 \\ 14 \end{pmatrix}, \begin{pmatrix} -10 \\ -12 \\ -14 \end{pmatrix}$$

$$p_2 = \begin{pmatrix} 13 \\ 18 \\ 23 \end{pmatrix}, \begin{pmatrix} -13 \\ -18 \\ -23 \end{pmatrix}$$

The alternatives having negative z-values can be eliminated because they represent points behind, not in front of, the lens.

■ **EXAMPLE 13.4**

Let the directed line segment have direction cosines

$$\begin{pmatrix} m_1 \\ m_2 \\ m_3 \end{pmatrix} = \begin{pmatrix} 1 \\ 2 \\ 3 \end{pmatrix} \frac{1}{\sqrt{14}}$$

and length $= 3\sqrt{14}$. Let the distance between the image plane and the front of the camera lens be $f = 5$ and the observed perspective projection be

$$\begin{pmatrix} u_1 \\ v_1 \end{pmatrix} = \frac{5}{14}\begin{pmatrix} 10 \\ 12 \end{pmatrix} \text{ and } \begin{pmatrix} u_2 \\ v_2 \end{pmatrix} = \frac{5}{23}\begin{pmatrix} 13 \\ 18 \end{pmatrix}$$

Then

$$u_2 - u_1 = -5\left(\frac{48}{14.23}\right)$$

$$v_2 - v_1 = -5\left(\frac{24}{14.23}\right)$$

$$fm_1 - u_2 m_3 = -\frac{16.5\sqrt{14}}{23.14}$$

$$fm_2 - v_2 m_3 = -\frac{8.5\sqrt{14}}{14.23}$$

Hence

$$c = \frac{3\sqrt{14}\left[-5\frac{48}{14.23}\left(\frac{-16}{14.23}5\sqrt{14}\right) + \left(5\frac{-24}{14.23}\right)\left(\frac{-8}{14.23}5\sqrt{14}\right)\right]}{\left(-5\frac{24}{14.23}\right)^2 + \left(-5\frac{48}{14.23}\right)^2} = 14$$

$$a = \frac{5\frac{10}{14}14}{5} = 10$$

$$c = \frac{5\frac{12}{14}14}{5} = 12$$

■

13.4.3 Collinear Points with Known Interpoint Distances

Suppose we know that we are observing the perspective projection of a real or virtual 3D line whose position and slope are not known. On this line there are N distinguished points with known interpoint distances. When $N > 2$, this constitutes sufficient information to determine all the parameters of the line as well as the 3D position of the distinguished points.

Let the unknown direction cosines of the line be $b = \begin{pmatrix} b_1 \\ b_2 \\ b_3 \end{pmatrix}$. Let the N distinguished points be $p, p + \delta_1 b, \ldots, p + \delta_{N-1}b$, where δ_n represents the distance between the $(n+1)$th and nth distinguished points. Let the first distinguished point

$p = \begin{pmatrix} p_1 \\ p_2 \\ p_3 \end{pmatrix}$. Then the perspective projection $\begin{pmatrix} u_n \\ v_n \end{pmatrix}$ of the nth distinguished point

is given by

$$u_n = f\,\frac{p_1 + \delta_n b_1}{p_3 + \delta_n b_3}$$

$$v_n = f\,\frac{p_2 + \delta_n b_2}{p_3 + \delta_n b_3}, \qquad n = 0,\ldots,N-1$$

where we use the convention that $\delta_0 = 0$. Multiplying both sides by $p_3 + \delta_n b_3$ and rewriting produces

$$fp_1 - u_n p_3 + \delta_n f b_1 - \delta_n u_n b_3 = 0$$

$$fp_2 - v_n p_3 + \delta_n f b_2 - \delta_n v_n b_3 = 0, \qquad n = 0,\ldots,N-1$$

This may be written in matrix form:

$$\underbrace{\begin{pmatrix} \delta_0 f & 0 & -\delta_0 u_0 \\ & \vdots & \\ \delta_{N-1}f & 0 & -\delta_{N-1}u_{N-1} \\ 0 & \delta_0 f & -\delta_0 v_0 \\ & \vdots & \\ 0 & \delta_{N-1}f & -\delta_{N-1}v_{N-1} \end{pmatrix}}_{A} \begin{pmatrix} b_1 \\ b_2 \\ b_3 \end{pmatrix} + \underbrace{\begin{pmatrix} f & 0 & -u_0 \\ & \vdots & \\ f & 0 & -u_{N-1} \\ 0 & f & -v_0 \\ & \vdots & \\ 0 & f & -v_{N-1} \end{pmatrix}}_{B} \begin{pmatrix} p_1 \\ p_2 \\ p_3 \end{pmatrix} = \begin{pmatrix} 0 \\ \vdots \\ 0 \end{pmatrix}$$

$$Ab + Bp = 0$$

where both A and B are $2N \times 3$ matrices. Since there are $2N$ equations plus the constant equation $b'b = 1$, and since there are six unknowns, the smallest N permitting a solution is $N = 3$.

The overconstrained system can be solved by finding b and p that minimize $\epsilon^2 = (Ab + Bp)'(Ab + Bp)$, subject to the constraint $b'b = 1$. Taking a partial derivative of ϵ^2 with respect to each of the components of p produces

$$\frac{\partial \epsilon^2}{\partial p} = 2B'(Ab + Bp)$$

Setting this to zero and solving for p, we obtain

$$B'(Ab + Bp) = 0$$

$$p = -(B'B)^{-1}B'Ab$$

Now we may substitute the above expression for p back in the expression for ϵ^2 :

$$
\begin{aligned}
\epsilon^2 &= (Ab + Bp)'(Ab + Bp) \\
&= [Ab - B(B'B)^{-1}B'Ab]'[Ab - B(B'B)^{-1}B'Ab] \\
&= \{[I - B(B'B)^{-1}B']Ab\}'\{[I - B(B'B)^{-1}B']Ab\}
\end{aligned}
$$

Noticing that $I - B(B'B)^{-1}B'$ is idempotent, we can immediately write that b is the eigenvector of $A'[I - B(B'B)^{-1}B']A$ having smallest eigenvalue. Equivalently, b is the right singular vector of $[I - B(B'B)^{-1}B']A$ having smallest singular value.

13.4.4 N Parallel Lines

Suppose we know that we are observing the perspective projection of N parallel 3D lines each of which has unknown direction cosines $\begin{pmatrix} b_1 \\ b_2 \\ b_3 \end{pmatrix}$. Then we can determine the direction cosines. We represent the perspective projection of the nth line by

$$
L_n = \left\{ \begin{pmatrix} u \\ v \end{pmatrix} \middle| \begin{pmatrix} u \\ v \end{pmatrix} = \begin{pmatrix} c_n \\ d_n \end{pmatrix} + \eta \begin{pmatrix} g_n \\ h_n \end{pmatrix} \right\}, \qquad n = 1, \ldots, N
$$

The line parameters on the perspective projection image plane and the direction cosines $\begin{pmatrix} b_1 \\ b_2 \\ b_3 \end{pmatrix}$ of the parallel 3D lines are related by

$$
h_n f b_1 - g_n f b_2 + (d_n g_n - c_n h_n)b_3 = 0, \qquad n = 1, \ldots, N
$$

This overconstrained system may be written in matrix form

$$
\underbrace{\begin{pmatrix} h_1 f & -g_1 f & d_1 g_1 - c_1 h_1 \\ \vdots & \vdots & \vdots \\ h_N f & -g_N f & d_N g_N - c_N h_N \end{pmatrix}}_{A} \begin{pmatrix} b_1 \\ b_2 \\ b_3 \end{pmatrix} = \begin{pmatrix} 0 \\ \vdots \\ 0 \end{pmatrix}
$$

$$
Ab = 0
$$

The least-squares solution for b that minimizes $\|Ab\|^2$ is given by the right singular vector of A having smallest singular value.

In the special case when $N = 2$, an explicit expression for the direction cosines may be given. From

$$
\begin{aligned}
h_1 f b_1 - g_1 f b_2 + (d_1 g_1 - c_1 h_1)b_3 &= 0 \\
h_2 f b_1 - g_2 f b_2 + (d_2 g_2 - c_2 h_2)b_3 &= 0
\end{aligned}
$$

we see that $\begin{pmatrix} b_1 \\ b_2 \\ b_3 \end{pmatrix}$ must be perpendicular to $\begin{pmatrix} h_1 f \\ -g_1 f \\ d_1 g_1 - c_1 h_1 \end{pmatrix}$ and

$\begin{pmatrix} h_2 f \\ -g_2 f \\ d_2 g_2 - c_2 h_2 \end{pmatrix}$. Hence we may express $\begin{pmatrix} b_1 \\ b_2 \\ b_3 \end{pmatrix}$ in terms of the vector cross-

product operator:

$$
\begin{pmatrix} b_1 \\ b_2 \\ b_3 \end{pmatrix} = \pm \frac{\begin{pmatrix} h_1 f \\ -g_1 f \\ d_1 g_1 - c_1 h_1 \end{pmatrix} \times \begin{pmatrix} h_2 f \\ -g_2 f \\ d_2 g_2 - c_2 h_2 \end{pmatrix}}{\left\| \begin{pmatrix} h_1 f \\ -g_1 f \\ d_1 g_1 - c_1 h_1 \end{pmatrix} \times \begin{pmatrix} h_2 f \\ -g_2 f \\ d_2 g_2 - c_2 h_2 \end{pmatrix} \right\|} \tag{13.7}
$$

13.4.5 N Lines Intersecting at a Point with Known Angles

Suppose we know that we are observing the perspective projection of N lines, $N \geq 3$, that intersect at the same point, and we know the angles between every pair of the 3D lines. From this it is possible to determine the orientation of the 3D lines.

Let the three 3D lines be

$$
L_n = \left\{ \begin{pmatrix} x \\ y \\ z \end{pmatrix} \middle| \begin{pmatrix} x \\ y \\ z \end{pmatrix} = \begin{pmatrix} a \\ b \\ c \end{pmatrix} + \lambda \begin{pmatrix} i_n \\ j_n \\ k_n \end{pmatrix} \right\}, \qquad n = 1, \ldots, N
$$

and it is understood that $i_n^2 + j_n^2 + k_n^2 = 1$. Suppose the corresponding observed perspective projection lines are

$$
M_n = \left\{ \begin{pmatrix} u \\ v \end{pmatrix} \middle| \begin{pmatrix} u \\ v \end{pmatrix} = \begin{pmatrix} g \\ h \end{pmatrix} + \eta \begin{pmatrix} \alpha_n \\ \beta_n \end{pmatrix} \right\}, \qquad n = 1, \ldots, N
$$

Then by the relation between the 3D lines and their perspective projection we have

$$
\beta_n f i_n - \alpha_n f j_n + (h \alpha_n - g \beta_n) k_n = 0, \qquad n = 1, \ldots, N
$$

From our knowledge about the angle between every pair of 3D lines, we can write

$$
i_p i_q + j_p j_q + k_p k_q = \cos \theta_{pq}, \qquad p, q = 1, \ldots, N \quad \text{and} \quad p < q
$$

where θ_{pq} is the known angle between L_p and L_q.

Each linear constraint $\beta_n f i_n - \alpha_n f j_n + (h \alpha_n - g \beta_n) k_n = 0$ forces the vector $\begin{pmatrix} i_n \\ j_n \\ k_n \end{pmatrix}$ to lie in a particular two-dimensional subspace. We let an orthonormal basis for this subspace be $\{u_n, v_n\}$. Then for some coefficients a_n and b_n we can represent

$$\begin{pmatrix} i_n \\ j_n \\ k_n \end{pmatrix} \quad \text{by} \quad \begin{pmatrix} i_n \\ j_n \\ k_n \end{pmatrix} = a_n u_n + b_n v_n$$

Substituting this into the dot product for θ_{pq}, we obtain

$$\cos \theta_{pq} = i_p i_q + j_p j_q + k_p k_q = (a_p u_p + b_p v_p)' (a_q u_q + b_q v_q)$$

$$= (a_p b_p) \begin{pmatrix} u_p' u_q & u_p' v_q \\ v_p' u_q & v_p' v_q \end{pmatrix} \begin{pmatrix} a_q \\ b_q \end{pmatrix}$$

Since

$$\begin{pmatrix} i_n \\ j_n \\ k_n \end{pmatrix}' \begin{pmatrix} i_n \\ j_n \\ k_n \end{pmatrix} = 1,$$

$$1 = (a_n u_n + b_n v_n)'(a_n u_n + b_n v_n)$$
$$= a_n^2 u_n' u_n + a_n b_n u_n' v_n + a_n b_n v_n' u_n + b_n^2 v_n' v_n$$
$$= a_n^2 + b_n^2$$

Let

$$W_{pq} = \begin{pmatrix} u_p' u_q & u_p' v_q \\ v_p' u_q & v_p' v_q \end{pmatrix}$$

Then the system of equations to be solved for (a_p, b_p), $p = 1, \ldots, N$ is

$$(a_p \ b_p) W_{pq} \begin{pmatrix} a_q \\ b_q \end{pmatrix} = \cos \theta_{pq}, \qquad p < q; \qquad p, q = 1, \ldots, N$$

$$a_p^2 + b_p^2 = 1, \qquad p = 1, \ldots, N \qquad\qquad (13.8)$$

This is a nonlinear system of equations, and for $N = 3$ it has multiple solutions. Since the function

$$\sum_p \sum_q \left[(a_p \ b_p) W_{pq} \begin{pmatrix} a_q \\ b_q \end{pmatrix} - \cos \theta_{pq} \right]^2$$

is a sum of squares, for N large enough it has a minimum that can be found by an iterative technique if a good enough initial approximation to the desired answer is available. Choose a_p^0, b_p^0 to be an initial approximation to the correct solution satisfying $a_p^2 + b_p^2 = 1$, $p = 1, \ldots, N$. Suppose a_p^n, b_p^n have been determined. For each iteration we choose a p different from the ones we chose in the last $N - 1$ iterations. Let

$$\begin{pmatrix} c_{pq}^n \\ d_{pq}^n \end{pmatrix} = W_{pq} \begin{pmatrix} a_q^n \\ b_q^n \end{pmatrix}$$

At iteration $n+1$ we choose (a_p^{n+1}, b_p^{n+1}) to minimize

$$\epsilon^2 = \sum_{\substack{q \\ q \neq p}} \left[(a_p^{n+1} b_p^{n+1}) \, W_{pq} \begin{pmatrix} a_q^n \\ b_q^n \end{pmatrix} - \cos\theta_{pq} \right]^2$$

$$= \sum_{\substack{q \\ q \neq p}} \left(a_p^{n+1} c_{pq}^n + b_p^{n+1} d_{pq}^n - \cos\theta_{pq} \right)^2$$

subject to

$$\left(a_p^{n+1} \right)^2 + \left(b_p^{n+1} \right)^2 = 1$$

Using the method of Lagrange multipliers leads to the requirement that

$$\sum_{\substack{q \\ q \neq p}} \left[a_p^{n+1} c_{pq}^n + b_p^{n+1} d_{pq}^n - \cos\theta_{pq} \right] c_{pq}^n + \lambda a_p^{n+1} = 0$$

$$\sum_{\substack{q \\ q \neq p}} \left[a_p^{n+1} c_{pq}^n + b_p^{n+1} d_{pq}^n - \cos\theta_{pq} \right] d_{pq}^n + \lambda a_p^{n+1} = 0$$

Let

$$A = \sum_{\substack{q \\ q \neq p}} \left(c_{pq}^n \right)^2$$

$$B = \sum_{\substack{q \\ q \neq p}} c_{pq}^n d_{pq}^n$$

$$C = \sum_{\substack{q \\ q \neq p}} \left(d_{pq}^n \right)^2$$

$$E = \sum_{\substack{q \\ q \neq p}} c_{pq}^n \cos\theta_{pq}$$

$$F = \sum_{\substack{q \\ q \neq p}} d_{pq}^n \cos\theta_{pq}$$

$$\begin{pmatrix} \lambda + A & B \\ B & \lambda + C \end{pmatrix} \begin{pmatrix} a_p^{n+1} \\ b_p^{n+1} \end{pmatrix} = \begin{pmatrix} E \\ F \end{pmatrix}$$

This implies that

$$a_p^{n+1} = \frac{\begin{vmatrix} E & B \\ F & \lambda + C \end{vmatrix}}{\begin{vmatrix} \lambda + A & B \\ B & \lambda + C \end{vmatrix}}$$

$$b_p^{n+1} = \frac{\begin{vmatrix} \lambda + A & E \\ B & F \end{vmatrix}}{\begin{vmatrix} \lambda + A & B \\ B & \lambda + C \end{vmatrix}}$$

The value for λ can be determined by using the constraint $(a_p^{n+1})^2 + (b_p^{n+1})^2 = 1$. This leads to the quartic equation

$$\lambda^4 + 2(A + C)\lambda^3 + (A^2 + 4AC + C^2 - 2B^2 - E^2 - F^2)\lambda^2$$
$$+ 2[(AC - B^2)(A + C) - (E^2C - 2EFB + F^2A)]\lambda$$
$$+ (AC)^2 - 2AB^2C + B^4 - (EC - FB)^2 - (FA - EB)^2 = 0$$

which is solved for λ by using any one of the standard root finders. Each of the possible four different roots will determine a value for a_p^{n+1} and b_p^{n+1}. At each iteration we take the value of λ to be selected as the value producing values for a_p^{n+1} and b_p^{n+1} that minimize ϵ^2.

13.4.6 N Intersecting Lines in a Known Plane

Suppose we know that we are observing the perspective projection of N lines all of which lie in the same known plane and which intersect at the same point. Then the 3D intersection point can be determined.

Let the 3D lines be given by

$$L_n = \left\{ \begin{pmatrix} x \\ y \\ z \end{pmatrix} \Big| \begin{pmatrix} x \\ y \\ z \end{pmatrix} = \begin{pmatrix} a \\ b \\ c \end{pmatrix} + \lambda \begin{pmatrix} i_n \\ j_n \\ k_n \end{pmatrix} \right\}, \qquad n = 1, \ldots, N$$

and the corresponding perspective projections be given by

$$M_n = \left\{ \begin{pmatrix} u \\ v \end{pmatrix} \Big| \begin{pmatrix} u \\ v \end{pmatrix} = \begin{pmatrix} g \\ h \end{pmatrix} + \eta \begin{pmatrix} \alpha_n \\ \beta_n \end{pmatrix} \right\}, \qquad n = 1, \ldots, N$$

By the 2D-to-3D-line relationships we have

$$\beta_n f a - \alpha_n f b + (h_n \alpha_n - g_n \beta_n)c = 0, \qquad n = 1, \ldots, N$$

Since the point of intersection $\begin{pmatrix} a \\ b \\ c \end{pmatrix}$ must lie in the known plane, we can write

$Aa + Bb + Cc + D = 0$. This overconstrained system can be written in matrix form as

$$\underbrace{\begin{pmatrix} \beta_1 f & -\alpha_1 f & h_1\alpha_1 - g_1\beta_1 \\ \vdots & \vdots & \vdots \\ \beta_N f & -\alpha_N f & h_N\alpha_N - g_N\beta_N \\ A & B & C \end{pmatrix}}_{E} \begin{pmatrix} a \\ b \\ c \end{pmatrix} = \begin{pmatrix} 0 \\ \vdots \\ 0 \\ -D \end{pmatrix}$$

The solution can then be given in terms of the singular value decomposition of E.

Let $E = USV'$,   where U and V are orthonormal and S is diagonal. Then

$$\begin{pmatrix} a \\ b \\ c \end{pmatrix} = VS^{-1}U' \begin{pmatrix} 0 \\ \vdots \\ 0 \\ -D \end{pmatrix} = -DVS^{-1}u_N \tag{13.9}$$

where u'_N is the bottom row of U.

13.4.7 Three Lines in a Plane with One Perpendicular to the Other Two

Suppose we know that we are observing the perspective projection of three lines that lie in the same plane, one of which is perpendicular to the other two. In this case two of the lines must be parallel and therefore have the same direction cosines.

Let the 3D lines be given by

$$L_1 = \left\{ \begin{pmatrix} x \\ y \\ z \end{pmatrix} \middle| \begin{pmatrix} x \\ y \\ z \end{pmatrix} = \begin{pmatrix} a_1 \\ a_2 \\ a_3 \end{pmatrix} + \lambda \begin{pmatrix} k_1 \\ k_2 \\ k_3 \end{pmatrix} \right\}$$

$$L_2 = \left\{ \begin{pmatrix} x \\ y \\ z \end{pmatrix} \middle| \begin{pmatrix} x \\ y \\ z \end{pmatrix} = \begin{pmatrix} b_1 \\ b_2 \\ b_3 \end{pmatrix} + \lambda \begin{pmatrix} k_1 \\ k_2 \\ k_3 \end{pmatrix} \right\}$$

$$L_3 = \left\{ \begin{pmatrix} x \\ y \\ z \end{pmatrix} \middle| \begin{pmatrix} x \\ y \\ z \end{pmatrix} = \begin{pmatrix} c_1 \\ c_2 \\ c_3 \end{pmatrix} + \lambda \begin{pmatrix} m_1 \\ m_2 \\ m_3 \end{pmatrix} \right\}$$

Let the corresponding perspective projections be given by

$$M_1 = \left\{ \begin{pmatrix} u \\ v \end{pmatrix} \middle| \begin{pmatrix} u \\ v \end{pmatrix} = \begin{pmatrix} g_1 \\ h_1 \end{pmatrix} + \eta \begin{pmatrix} \alpha_1 \\ \beta_1 \end{pmatrix} \right\}$$

$$M_2 = \left\{ \begin{pmatrix} u \\ v \end{pmatrix} \middle| \begin{pmatrix} u \\ v \end{pmatrix} = \begin{pmatrix} g_2 \\ h_2 \end{pmatrix} + \eta \begin{pmatrix} \alpha_2 \\ \beta_2 \end{pmatrix} \right\}$$

$$M_3 = \left\{ \begin{pmatrix} u \\ v \end{pmatrix} \middle| \begin{pmatrix} u \\ v \end{pmatrix} = \begin{pmatrix} g_3 \\ h_3 \end{pmatrix} + \eta \begin{pmatrix} \alpha_3 \\ \beta_3 \end{pmatrix} \right\}$$

Then by the relation between the 3D lines and their perspective projections we have

$$\beta_1 f k_1 - \alpha_1 f k_2 + (h_1 \alpha_1 - g_1 \beta_1) k_3 = 0$$
$$\beta_2 f k_1 - \alpha_2 f k_2 + (h_2 \alpha_2 - g_2 \beta_2) k_3 = 0$$
$$\beta_3 f m_1 - \alpha_3 f m_2 + (h_3 \alpha_3 - g_3 \beta_3) m_3 = 0$$

Then

$$\begin{pmatrix} k_1 \\ k_2 \\ k_3 \end{pmatrix}$$

must be perpendicular to

$$
\begin{pmatrix} \beta_1 f \\ -\alpha_1 f \\ h_1\alpha_1 - g_1\beta_1 \end{pmatrix} \quad \text{and} \quad \begin{pmatrix} \beta_2 f \\ -\alpha_2 f \\ h_2\alpha_2 - g_2\beta_2 \end{pmatrix}
$$

Hence, using vector cross-products, we may express $\begin{pmatrix} k_1 \\ k_2 \\ k_3 \end{pmatrix}$ by

$$
\begin{pmatrix} k_1 \\ k_2 \\ k_3 \end{pmatrix} = \pm \frac{\begin{pmatrix} \beta_1 f \\ -\alpha_1 f \\ h_1\alpha_1 - g_1\beta_1 \end{pmatrix} \times \begin{pmatrix} \beta_2 f \\ -\alpha_2 f \\ h_2\alpha_2 - g_2\beta_2 \end{pmatrix}}{\left\| \begin{pmatrix} \beta_1 f \\ -\alpha_1 f \\ h_1\alpha_1 - g_1\beta_1 \end{pmatrix} \times \begin{pmatrix} \beta_2 f \\ -\alpha_2 f \\ h_2\alpha_2 - g_2\beta_2 \end{pmatrix} \right\|}
\tag{13.10}
$$

Since L_3 is perpendicular to L_1 and L_2,

$$
k_1 m_1 + k_2 m_2 + k_3 m_3 = 0
$$

Now we have $\begin{pmatrix} m_1 \\ m_2 \\ m_3 \end{pmatrix}$ perpendicular to

$$
\begin{pmatrix} k_1 \\ k_2 \\ k_3 \end{pmatrix} \quad \text{and} \quad \begin{pmatrix} \beta_3 f \\ -\alpha_3 f \\ h_3\alpha_3 - g_3\beta_3 \end{pmatrix}
$$

Using vector cross-products again, we may write

$$
\begin{pmatrix} m_1 \\ m_2 \\ m_3 \end{pmatrix} = \pm \frac{\begin{pmatrix} k_1 \\ k_2 \\ k_3 \end{pmatrix} \times \begin{pmatrix} \beta_3 f \\ -\alpha_3 f \\ h_3\alpha_3 - g_3\beta_3 \end{pmatrix}}{\left\| \begin{pmatrix} k_1 \\ k_2 \\ k_3 \end{pmatrix} \times \begin{pmatrix} \beta_3 f \\ -\alpha_3 f \\ h_3\alpha_3 - g_3\beta_3 \end{pmatrix} \right\|}
\tag{13.11}
$$

13.4.8 Point with Given Distance to a Known Point

Suppose we observe the perspective projection of a 3D point whose distance to a known 3D point is given. Then the unknown 3D point can be in only two possible positions, and they can be simply determined.

Let the position of the known 3D point be $\begin{pmatrix} a \\ b \\ c \end{pmatrix}$, and let the position of the unknown 3D point be represented as $\begin{pmatrix} a \\ b \\ c \end{pmatrix} + \rho \begin{pmatrix} i \\ j \\ k \end{pmatrix}$, where $\begin{pmatrix} i \\ j \\ k \end{pmatrix}$ is the vector of unknown direction cosines of the line segment defined by the two points and ρ is the known distance between the two points. Let $\binom{u}{v}$ be the perspective projection of the unknown 3D point. Then

$$u = f \frac{a + \rho i}{c + \rho k}$$

$$v = f \frac{a + \rho j}{c + \rho k}$$

Hence

$$i = \frac{u\rho k + uc - fa}{f\rho}$$

$$j = \frac{v\rho k + vc - fb}{f\rho}$$

Since $i^2 + j^2 + k^2 = 1$, we obtain after substitution for i and j the relation

$$\rho^2(u^2 + v^2 + f^2)k^2 + 2\rho[u(uc - fa) + v(vc - fb)]k$$

$$\text{(13.12)}$$

$$+ (uc - fa)^2 + (vc - fb)^2 - (\rho f)^2 = 0$$

which is easy to solve for the two possible values for k. Once having k, the values for i and j are determined, and once i, j, and k have been determined, the 3D point $\begin{pmatrix} a \\ b \\ c \end{pmatrix} + \rho \begin{pmatrix} i \\ j \\ k \end{pmatrix}$ is determined.

■ EXAMPLE 13.5

If $\begin{pmatrix} a \\ b \\ c \end{pmatrix} = \begin{pmatrix} 3 \\ 4 \\ 5 \end{pmatrix}$, $\rho = 5\sqrt{3}$, $\begin{pmatrix} i \\ j \\ k \end{pmatrix} = \frac{1}{\sqrt{3}} \begin{pmatrix} 1 \\ 1 \\ 1 \end{pmatrix}$, and $f = 20$. Then the perspective projection of $\begin{pmatrix} 3 \\ 4 \\ 5 \end{pmatrix} + 5 \begin{pmatrix} 1 \\ 1 \\ 1 \end{pmatrix}$ is

$$u = 20 \frac{3 + 5}{5 + 5} = 16$$

$$v = 20 \frac{4 + 5}{5 + 5} = 18$$

The quadratic equation for k in terms of a, b, c, u, v, ρ, and f is

$$73500k^2 + 8660.25k - 24500 = 0$$

from which $k = .5773, -.6952$. When $k = .5773$, then $i = j = .5773$, and the 3D-point's position is

$$\begin{pmatrix} 3 \\ 4 \\ 5 \end{pmatrix} + 5\sqrt{3} \begin{pmatrix} .5773 \\ .5773 \\ .5773 \end{pmatrix} = \begin{pmatrix} 8 \\ 9 \\ 10 \end{pmatrix}$$

When $k = -.6952$, then $i = -.4407$ and $j = .5679$. The position of the 3D point is then

$$\begin{pmatrix} 3 \\ 4 \\ 5 \end{pmatrix} + 5\sqrt{3} \begin{pmatrix} -.4407 \\ -.5679 \\ -.6952 \end{pmatrix} = \begin{pmatrix} -.81657 \\ -.918158 \\ -1.0206 \end{pmatrix}$$

■

13.4.9 Point in a Known Plane

Suppose we observe the perspective projection of a 3D point and we know that the point is constrained to be in a given plane. Then the 3D position of the point can be uniquely determined.

Let $\begin{pmatrix} a \\ b \\ c \end{pmatrix}$ be the 3D point and $\begin{pmatrix} u \\ v \end{pmatrix}$ be its perspective projection. Then

$$u = f \frac{a}{c}$$
$$v = f \frac{b}{c}$$

Suppose the plane is given by

$$\left\{ \begin{pmatrix} x \\ y \\ z \end{pmatrix} \middle| Ax + By + Cz + D = 0 \right\}$$

Since the point lies in the plane, $Aa + Bb + Cc + D = 0$. But from the perspective equation

$$a = \frac{uc}{f}$$
$$b = \frac{vc}{f}$$

Hence

$$A\left(\frac{uc}{f}\right) + B\left(\frac{vc}{f}\right) + Cc + D = 0$$

from which

$$c = \frac{-Df}{Au + Bv + Cf}$$

$$b = \frac{-Dv}{Au + Bv + Cf}$$ (13.13)

$$a = \frac{-Du}{Au + Bv + Cf}$$

13.4.10 Line in a Known Plane

Suppose we know that we are observing the perspective projection of a line that lies in a known plane. Then the direction cosines for the line can be determined.
Let the 3D line be given by

$$L = \left\{ \begin{pmatrix} x \\ y \\ z \end{pmatrix} \Bigg| \begin{pmatrix} x \\ y \\ z \end{pmatrix} = \begin{pmatrix} a \\ b \\ c \end{pmatrix} + \lambda \begin{pmatrix} i \\ j \\ k \end{pmatrix} \right\}$$

and the corresponding perspective projection line be given by

$$M = \left\{ \begin{pmatrix} u \\ v \end{pmatrix} \Bigg| \begin{pmatrix} u \\ v \end{pmatrix} = \begin{pmatrix} g \\ h \end{pmatrix} + \eta \begin{pmatrix} \alpha \\ \beta \end{pmatrix} \right\}$$

Suppose the known plane in which the 3D line lies is given by

$$\left\{ \begin{pmatrix} x \\ y \\ z \end{pmatrix} \Bigg| Ax + By + Cz + D = 0 \right\}$$

By the 2D-to-3D perspective projection line relationship we have

$$\beta f i - \alpha f j + (h\alpha - g\beta)k = 0$$

Since the line lies in the plane, we have

$$Ai + Bj + Ck = 0$$

Hence

$$\begin{pmatrix} i \\ j \\ k \end{pmatrix} = \frac{\begin{pmatrix} \beta f \\ -\alpha f \\ h\alpha - g\beta \end{pmatrix} \times \begin{pmatrix} A \\ B \\ C \end{pmatrix}}{\left\| \begin{pmatrix} \beta f \\ -\alpha f \\ h\alpha - g\beta \end{pmatrix} \times \begin{pmatrix} A \\ B \\ C \end{pmatrix} \right\|}$$ (13.14)

To determine a point $\begin{pmatrix} a \\ b \\ c \end{pmatrix}$ on the line, we have from the 2D-to-3D perspective projection line relationship

$$\beta f a - \alpha f b + (h\alpha - g\beta)c = 0$$

and we are free to take $\begin{pmatrix} a \\ b \\ c \end{pmatrix}$ to be perpendicular to $\begin{pmatrix} i \\ j \\ k \end{pmatrix}$. Hence

$$\begin{pmatrix} a \\ b \\ c \end{pmatrix} = \lambda \begin{pmatrix} i \\ j \\ k \end{pmatrix} \times \begin{pmatrix} \beta f \\ -\alpha f \\ h\alpha - g\beta \end{pmatrix}$$

Since the line lies in the plane, we have

$$Aa + Bb + Cc + D = 0$$

Hence

$$\lambda = \frac{-D}{(A\ B\ C)\left[\begin{pmatrix} i \\ j \\ k \end{pmatrix} \times \begin{pmatrix} \beta f \\ -\alpha f \\ h\alpha - g\beta \end{pmatrix} \right]}$$

so that

$$\begin{pmatrix} a \\ b \\ c \end{pmatrix} = \frac{-D \begin{pmatrix} i \\ j \\ k \end{pmatrix} \times \begin{pmatrix} \beta f \\ -\alpha f \\ h\alpha - g\beta \end{pmatrix}}{(A\ B\ C)\left[\begin{pmatrix} i \\ j \\ k \end{pmatrix} \times \begin{pmatrix} \beta f \\ -\alpha f \\ h\alpha - g\beta \end{pmatrix} \right]} \tag{13.15}$$

13.4.11 Angle

Suppose we know that we are observing the perspective projection of two lines and that the angle between the lines is known. If we know in addition the direction cosines for one of the lines, then the other line can have only two possible direction cosine values, and they are easily determined.

Let the direction cosines for the known line be $\begin{pmatrix} i_0 \\ j_0 \\ k_0 \end{pmatrix}$ and the direction cosines

for the unknown line be $\begin{pmatrix} i_1 \\ j_1 \\ k_1 \end{pmatrix}$. Suppose the perspective projection of the unknown line is observed to be

$$M = \left\{ \begin{pmatrix} u \\ v \end{pmatrix} \middle| \begin{pmatrix} u \\ v \end{pmatrix} = \begin{pmatrix} g \\ h \end{pmatrix} + \eta \begin{pmatrix} \alpha \\ \beta \end{pmatrix} \right\}$$

By the 2D-to-3D perspective projection line relationship, we have

$$\beta f i_1 - \alpha f j_1 + (h\alpha - g\beta)k_1 = 0$$

Let θ be the angle between the 3D lines. Then

$$i_0 i_1 + j_0 j_1 + k_0 k_1 = \cos \theta$$

Let USV' be the singular value decomposition of the matrix A, where

$$A = \begin{pmatrix} \beta f & -\alpha f & h\alpha - g\beta \\ i_0 & j_0 & k_0 \end{pmatrix}$$

If

$$S = \begin{pmatrix} s_1 & 0 & 0 \\ 0 & s_2 & 0 \end{pmatrix}$$

$$V = (f v_1 \quad v_2 \quad v_3)$$

$$v_3 = \begin{pmatrix} v_{31} \\ v_{32} \\ v_{33} \end{pmatrix}$$

$$V_1 = (f v_1 \quad v_2)$$

$$\begin{pmatrix} b_1 \\ b_2 \\ b_3 \end{pmatrix} = V_1 \begin{pmatrix} s_1 & 0 \\ 0 & s_2 \end{pmatrix}^{-1} U' \begin{pmatrix} 0 \\ \cos \theta \end{pmatrix}$$

then the general solution to the system

$$A \begin{pmatrix} i_1 \\ j_1 \\ k_1 \end{pmatrix} = \begin{pmatrix} 0 \\ \cos \theta \end{pmatrix}$$

is given by

$$\begin{pmatrix} i_1 \\ j_1 \\ k_1 \end{pmatrix} = \begin{pmatrix} b_1 \\ b_2 \\ b_3 \end{pmatrix} + \lambda v_3$$

Since $i_1^2 + j_1^2 + k_1^2 = 1$, we can solve the quadratic equation

$$(b_1 + \lambda v_{31})^2 + (b_2 + \lambda v_{32})^2 + (b_3 + \lambda v_{33})^2 = 1$$

for the two possible values of λ.

13.4.12   Parallelogram

Suppose we know that we are observing the perspective projection of the four corner points of a parallelogram. Then we can determine the normal to the plane on which the parallelogram lies. If in addition we know the length or width of the parallelogram, we can determine the 3D position of each corner point.

Let the four corner points of the parallelogram be

$$
\begin{pmatrix} p_1 \\ p_2 \\ p_3 \end{pmatrix}, \quad
\begin{pmatrix} p_1 \\ p_2 \\ p_3 \end{pmatrix} + W \begin{pmatrix} m_1 \\ m_2 \\ m_3 \end{pmatrix} =
\begin{pmatrix} q_1 \\ q_2 \\ q_3 \end{pmatrix} + H \begin{pmatrix} n_1 \\ n_2 \\ n_3 \end{pmatrix},
$$

$$
\begin{pmatrix} q_1 \\ q_2 \\ q_3 \end{pmatrix}, \quad
\begin{pmatrix} q_1 \\ q_2 \\ q_3 \end{pmatrix} - W \begin{pmatrix} m_1 \\ m_2 \\ m_3 \end{pmatrix} =
\begin{pmatrix} p_1 \\ p_2 \\ p_3 \end{pmatrix} - H \begin{pmatrix} n_1 \\ n_2 \\ n_3 \end{pmatrix},
$$

They are illustrated in Fig. 13.6.

Since opposite sides of a parallelogram are parallel, using the techniques given in Section 13.4.4 we can infer $\begin{pmatrix} m_1 \\ m_2 \\ m_3 \end{pmatrix}$ and $\begin{pmatrix} n_1 \\ n_2 \\ n_3 \end{pmatrix}$ on the basis of the

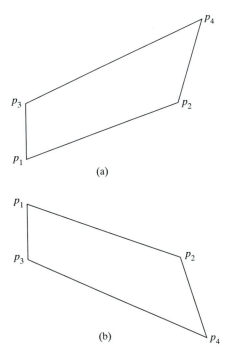

(a)

(b)

Figure 13.6 (a) Perspective projection of the rectangle when viewed from a position above the rectangle. Pan angle $= 30°$, tilt angle $= 40°$, swing angle $= 0°$. (b) Perspective projection of the rectangle when viewed from a position below the rectangle. Pan angle $= 30°$, tilt angle $= 40°$, swing angle $= 0°$.

observed perspective projections of the sides. Let the unknown plane in which the parallelogram lies be given as

$$\left\{ \begin{pmatrix} x \\ y \\ z \end{pmatrix} \middle| Ax + By + Cz + D = 0 \right\}$$

A line with direction cosines $\begin{pmatrix} m_1 \\ m_2 \\ m_3 \end{pmatrix}$ and that lies in the plane must satisfy $Am_1 + Bm_2 + Cm_3 = 0$. Similarly, $An_1 + Bn_2 + Cn_3 = 0$. This immediately implies

$$\begin{pmatrix} A \\ B \\ C \end{pmatrix} = \frac{\begin{pmatrix} m_1 \\ m_2 \\ m_3 \end{pmatrix} \times \begin{pmatrix} n_1 \\ n_2 \\ n_3 \end{pmatrix}}{\left\| \begin{pmatrix} m_1 \\ m_2 \\ m_3 \end{pmatrix} \times \begin{pmatrix} n_1 \\ n_2 \\ n_3 \end{pmatrix} \right\|}$$

If one of W or H is known, then from the perspective projection of the four corner points and the inferred direction cosines, we can use the techniques of Section 13.4.2 to determine the 3D position of each of the corner points.

13.4.13 Triangle with One Vertex Known

Suppose we are observing the perspective projection of a triangle with known side lengths. If the 3D position of one vertex of the triangle is known, then the position of the other vertices can be uniquely determined.

Let

$$\begin{pmatrix} x_1 \\ y_1 \\ z_1 \end{pmatrix}, \begin{pmatrix} x_2 \\ y_2 \\ z_2 \end{pmatrix}, \text{ and } \begin{pmatrix} x_3 \\ y_3 \\ z_3 \end{pmatrix}$$

be the 3D positions of vertices of the triangle with known lengths S_{12}, S_{13}, and S_{23}. Let the corresponding observed perspective projection be

$$u_n = f \frac{x_n}{z_n}$$

$$n = 1, 2, 3$$

$$v_n = f \frac{y_n}{z_n}$$

Suppose that $\begin{pmatrix} x_1 \\ y_1 \\ z_1 \end{pmatrix}$ is known. Then since the side lengths are known,

$$(x_2 - x_1)^2 + (y_2 - y_1)^2 + (z_2 - z_1)^2 = S_{12}^2$$
$$(x_3 - x_1)^2 + (y_3 - y_1)^2 + (z_3 - z_1)^2 = S_{13}^2$$
$$(x_2 - x_3)^2 + (y_2 - y_3)^2 + (z_2 - z_3)^2 = S_{23}^2$$

Since

$$x_2 = \frac{u_2 z_2}{f} \quad \text{and} \quad y_2 = \frac{v_2 z_2}{f}$$
$$x_3 = \frac{u_3 z_3}{f} \quad \text{and} \quad y_3 = \frac{v_3 z_3}{f}$$

the quadratic equation

$$z_2^2(u_2^2 + v_2^2 + f^2) - 2(u_2 x_1 + v_2 y_1 + f z_1) f z_2 + f^2(x_1^2 + y_1^2 + z_1^2) - S_{12}^2 f^2 = 0$$
$$z_3^2(u_3^2 + v_3^2 + f^2) - 2(u_3 x_1 + v_3 y_1 + f z_1) f z_3 + f^2(x_1^2 + y_1^2 + z_1^2) - S_{13}^2 f^2 = 0$$

results. In general there are two possible values for z_2. A similar situation permits two possible values for the solution for z_3 to be determined. Once z_2 and z_3 are known, x_2, y_2 and x_3, y_3 are easily determined from the perspective equations. Only one of the resulting four possible total solutions will satisfy the third equation

$$(x_2 - x_3)^2 + (y_2 - y_3)^2 + (z_2 - z_3)^2 = S_{23}^2$$

EXAMPLE 13.6

Suppose the vertices of the triangle are $\begin{pmatrix} 10 \\ 11 \\ 12 \end{pmatrix}$, $\begin{pmatrix} 11 \\ 12 \\ 14 \end{pmatrix}$, and $\begin{pmatrix} 11 \\ 13 \\ 13 \end{pmatrix}$. Then $S_{12} = \sqrt{6}, S_{13} = \sqrt{6}$, and $S_{23} = \sqrt{5}$. If the focal length $f = 2$, then the perspective projection will be

$$u_1 = 2\frac{10}{12} = 1.6667$$

$$v_1 = 2\frac{11}{12} = 1.8333$$

$$u_2 = 2\frac{11}{14} = 1.5714$$

$$v_2 = 2\frac{12}{14} = 1.7143$$

$$u_3 = 2\frac{11}{13} = 1.1923$$

$$v_3 = 2\frac{13}{13} = 2.0000$$

Taking $\begin{pmatrix} x_1 \\ y_1 \\ z_1 \end{pmatrix} = \begin{pmatrix} 10 \\ 11 \\ 12 \end{pmatrix}$ as known, we can compute

$$u_2^2 + v_2^2 + f^2 = 9.4082$$
$$u_3^2 + v_3^2 + f^2 = 10.8639$$
$$f^2(x_1^2 + y_1^2 + z_1^2) = 1460$$
$$S_{12}^2 f^2 = 24$$
$$S_{13}^2 f^2 = 24$$
$$-2f(u_2 x_1 + v_2 y_1 + f z_1) = -234.2852$$
$$-2f(u_3 x_1 + v_3 y_1 + f z_1) = -251.692$$

Hence

$$9.4082 z_2^2 - 234.2852 z_2 + 1436 = 0$$
$$10.8639 z_3^2 - 251.692 z_3 + 1436 = 0$$

from which

$$z_2 = \frac{234.2852 \pm 29.135}{18.8164} = 14,\ 10.903$$
$$z_3 = \frac{251.692 \pm 30.767}{21.7278} = 13,\ 10.168$$

13.4.14   Triangle with Orientation of One Leg Known

Suppose we are observing the perspective projection of a triangle with known side lengths. If the direction cosines for one leg are known, then the position of all vertices can be uniquely determined.

Let $\begin{pmatrix} x_1 \\ x_2 \\ x_3 \end{pmatrix}$, $\begin{pmatrix} x_2 \\ y_2 \\ z_2 \end{pmatrix}$, and $\begin{pmatrix} x_3 \\ y_3 \\ z_3 \end{pmatrix}$ be the 3D positions of the vertices of the triangle with known side lengths S_{12}, S_{13}, and S_{23}. Then

$$(x_1 - x_2)^2 + (y_1 - y_2)^2 + (z_1 - z_2)^2 = S_{12}^2$$
$$(x_1 - x_3)^2 + (y_1 - y_3)^2 + (z_1 - z_3)^2 = S_{13}^2$$
$$(x_2 - x_3)^2 + (y_2 - y_3)^2 + (z_2 - z_3)^2 = S_{23}^2$$

Suppose that $\begin{pmatrix} i_1 \\ j_1 \\ k_1 \end{pmatrix}$ are the known direction cosines of the line segment defined between the first two vertices. Then

$$\begin{pmatrix} x_1 \\ y_1 \\ z_1 \end{pmatrix} = \begin{pmatrix} x_2 \\ y_2 \\ z_2 \end{pmatrix} + S_{12} \begin{pmatrix} i_1 \\ j_1 \\ k_1 \end{pmatrix}$$

Let the corresponding observed perspective projections of the vertices be $u_n = fx_n/z_n$, $v_n = fy_n/z_n$, $n = 1,2,3$. Then by the technique discussed in Section 13.4.2, the points $\begin{pmatrix} x_1 \\ y_1 \\ z_1 \end{pmatrix}$ and $\begin{pmatrix} x_2 \\ y_2 \\ z_2 \end{pmatrix}$ can be uniquely determined by

$$\begin{pmatrix} x_1 \\ y_1 \\ z_1 \end{pmatrix} = S_{12}[(u_1v_2 - u_2v_1)(i_1v_2 - j_1u_2) +$$

$$(u_2 - u_1)f(u_2k_1 - fi_1) + (v_2 - v_1)f(v_2k_1 - fj_1)] \begin{pmatrix} u_1 \\ v_1 \\ f \end{pmatrix}$$

and

$$\begin{pmatrix} x_2 \\ y_2 \\ z_2 \end{pmatrix} = S_{12}[(u_1v_2 - u_2v_1)(i_1v_1 - j_1u_1) +$$

$$(u_2 - u_1)f(u_1k_1 - fi_1) + (v_2 - v_1)f(v_1k_1 - fj_1)] \begin{pmatrix} u_2 \\ v_2 \\ f \end{pmatrix}$$

Suppose that the unknown direction cosines of the line segment from

$$\begin{pmatrix} x_1 \\ y_1 \\ z_1 \end{pmatrix} \text{ to } \begin{pmatrix} x_3 \\ y_3 \\ z_3 \end{pmatrix} \text{ is } \begin{pmatrix} i_3 \\ j_3 \\ k_3 \end{pmatrix} \text{ and from}$$

$$\begin{pmatrix} x_2 \\ y_2 \\ z_2 \end{pmatrix} \text{ to } \begin{pmatrix} x_3 \\ y_3 \\ z_3 \end{pmatrix} \text{ is } \begin{pmatrix} i_2 \\ j_2 \\ k_2 \end{pmatrix} \text{ so that}$$

$$\begin{pmatrix} x_3 \\ y_3 \\ z_3 \end{pmatrix} = \begin{pmatrix} x_1 \\ y_1 \\ z_1 \end{pmatrix} + S_{13} \begin{pmatrix} i_3 \\ j_3 \\ k_3 \end{pmatrix} = \begin{pmatrix} x_2 \\ y_2 \\ z_2 \end{pmatrix} + S_{23} \begin{pmatrix} i_2 \\ j_2 \\ k_2 \end{pmatrix}$$

Now by the techniques discussed in Section 13.4.8, we must have

$$S_{13}^2(u_3^2 + v_3^2 + f^2)k_3^2 + 2S_{13}[u_3(u_3z_1 - fx_1) + v_3(v_3z_1 - fy_1)]k_3 +$$
$$(u_3z_1 - fx_1)^2 + (v_3z_1 - fy_1)^2 - (S_{13}f)^2 = 0$$

$$(13.16)$$

$$S_{23}^2(u_3^2 + v_3^2 + f^2)k_2^2 + 2S_{23}[u_3(u_3z_2 - fx_2) + v_3(v_3z_2 - fy_2)]k_2 +$$
$$(u_3z_2 - fx_2)^2 + (v_3z_2 - fy_2)^2 - (S_{23}f)^2 = 0$$

These equations can be solved for the two possible values for k_2 and k_3. Then the corresponding values for i_2, i_3, j_2, and j_3 can be determined from

$$i_2 = \frac{u_3 S_{23} k_2 + u_3 z_2 - f x_2}{f S_{23}}$$

$$j_2 = \frac{v_3 S_{23} k_2 + v_3 z_2 - f y_2}{f S_{23}}$$

$$(13.17)$$

$$i_3 = \frac{u_3 S_{13} k_3 + u_3 z_1 - f x_1}{f S_{13}}$$

$$j_3 = \frac{v_3 S_{13} k_3 + v_3 z_1 - f y_1}{f S_{13}}$$

The extraneous solutions can be eliminated by keeping only the solution that satisfies

$$S_{12} \begin{pmatrix} i_1 \\ j_1 \\ k_1 \end{pmatrix} = S_{23} \begin{pmatrix} i_2 \\ j_2 \\ k_2 \end{pmatrix} - S_{13} \begin{pmatrix} i_3 \\ j_3 \\ k_3 \end{pmatrix}$$

13.4.15 Triangle

Suppose we are observing the perspective projection of a triangle with known side lengths. The 3D position of the vertices of the triangle can be determined. However, there are multiple solutions. In photogrammetry this is called the three-point spatial resection problem. Grunert (1841) appears to have been the first to solve it. Müller (1925) describes the Grunert solution. Finsterwalder (1897); Merritt (1949); Fischler and Bolles (1981); Grafarend, Lohse, and Schaffrin (1989); Linainmaa, Harwood, and Davis (1988) all give direct solutions. We follow the Finsterwalder solution.

Let the unknown positions of the triangle vertices be

$$\begin{pmatrix} x_1 \\ y_1 \\ z_1 \end{pmatrix}, \begin{pmatrix} x_2 \\ y_2 \\ z_2 \end{pmatrix}, \text{ and } \begin{pmatrix} x_3 \\ y_3 \\ z_3 \end{pmatrix}$$

The known side lengths S_{12}, S_{13}, and S_{23} satisfy

$$(x_1 - x_2)^2 + (y_1 - y_2)^2 + (z_1 - z_2)^2 = S_{12}^2$$

$$(x_1 - x_3)^2 + (y_1 - y_3)^2 + (z_1 - z_3)^2 = S_{13}^2$$

$$(x_2 - x_3)^2 + (y_2 - y_3)^2 + (z_2 - z_3)^2 = S_{23}^2$$

Let the observed perspective projection of $\begin{pmatrix} x_n \\ y_n \\ z_n \end{pmatrix}$ be $\begin{pmatrix} u_n \\ v_n \end{pmatrix}$. Then

$$u_n = f\frac{x_n}{z_n}$$

$$n = 1, 2, 3$$

$$v_n = f\frac{y_n}{z_n}$$

By substituting for x_n and y_n, we obtain

$$\left(\frac{u_1 z_1}{f} - \frac{u_2 z_2}{f}\right)^2 + \left(\frac{v_1 z_1}{f} - \frac{v_2 z_2}{f}\right)^2 + (z_1 - z_2)^2 = S_{12}^2$$

$$\left(\frac{u_1 z_1}{f} - \frac{u_3 z_3}{f}\right)^2 + \left(\frac{v_1 z_1}{f} - \frac{v_3 z_3}{f}\right)^2 + (z_1 - z_3)^2 = S_{13}^2$$

$$\left(\frac{u_2 z_2}{f} - \frac{u_3 z_3}{f}\right)^2 + \left(\frac{v_2 z_2}{f} - \frac{v_3 z_3}{f}\right)^2 + (z_2 - z_3)^2 = S_{23}^2$$

After some rearranging,

$$(u_1^2 + v_1^2 + f^2)z_1^2 - 2(u_1 u_2 + v_1 v_2 + f^2)z_1 z_2 + (u_2^2 + v_2^2 + f^2)z_2^2 = (fS_{12})^2$$

$$(u_1^2 + v_1^2 + f^2)z_1^2 - 2(u_1 u_3 + v_1 v_3 + f^2)z_1 z_3 + (u_3^2 + v_3^2 + f^2)z_3^2 = (fS_{13})^2$$

$$(u_2^2 + v_2^2 + f^2)z_2^2 - 2(u_2 u_3 + v_2 v_3 + f^2)z_2 z_3 + (u_3^2 + v_3^2 + f^2)z_3^2 = (fS_{23})^2$$

If the z-coordinate of one vertex were known, say z_1, then the two possible solutions for z_2 could be obtained from the first of the three equations, and the two possible solutions for z_3 could be obtained from the second of the three equations. The extraneous solution for z_2, z_3 could be eliminated by keeping only those solutions pairs that satisfy the third equation. Once having $z_1, z_2,$ and z_3, we could readily determine the corresponding values for x_n and y_n from the perspective projection equations. However, if none of the z-coordinates are known, the problem is more complex.

Each quadratic equation in two variables is the equation of an ellipse. The first equation does not involve the variable z_3. Hence it is unconstrained by the first equation. This means that in the space of the three variables $z_1, z_2,$ and z_3, the first equation represents an elliptical cylinder, the axis of the cylinder being parallel to the z_3-axis. Similarly, the second equation represents an elliptical cylinder whose axis is parallel to the z_2-axis, and the third equation represents an elliptical cylinder whose axis is parallel to the z_1-axis. Geometrically, then, the solution to the system corresponds to the intersection of three elliptical cylinders each having mutually orthogonal axes. The intersection of three such cylinders can result in at most eight points, which means there may be as many as eight possible solutions. Let

$$A = u_1^2 + v_1^2 + f^2$$
$$B = u_2^2 + v_2^2 + f^2$$
$$C = u_3^2 + v_3^2 + f^2$$
$$D = -(u_1 u_2 + v_1 v_2 + f^2)$$
$$E = -(u_1 u_3 + v_1 v_3 + f^2)$$
$$F = -(u_2 u_3 + v_2 v_3 + f^2)$$
$$K_{12} = (fS_{12})^2$$
$$K_{13} = (fS_{13})^2$$
$$K_{23} = (fS_{23})^2$$

Then we can rewrite the three equations as

$$Az_1^2 + 2Dz_1 z_2 + Bz_2^2 = K_{12}$$
$$Az_1^2 + 2Ez_1 z_3 + Cz_3^2 = K_{13}$$
$$Bz_2^2 + 2Fz_2 z_3 + Cz_3^2 = K_{23}$$

Finsterwalder (1897), summarized in Finsterwalder and Scheufele (1903, 1937), proceeded as follows to solve this system. Multiply the first equation by K_{13} and the second equation by K_{12} and subtract. This yields

$$A(K_{13} - K_{12})z_1^2 + 2DK_{13}z_1 z_2 + BK_{13}z_2^2 - 2EK_{12}z_1 z_3 - CK_{12}z_3^2 = 0$$

Then multiply the first equation by K_{23} and the third equation by K_{12} and subtract. This yields

$$AK_{23}z_1^2 + 2DK_{23}z_1 z_2 + B(K_{23} - K_{12})z_2^2 - 2FK_{12}z_2 z_3 - CK_{12}z_3^2 = 0$$

Now divide both equations by z_3^2. Letting $s = z_1/z_3$ and $t = z_2/z_3$, we obtain

$$A(K_{13} - K_{12})s^2 + 2DK_{13}st + BK_{13}t^2 - 2EK_{12}s - CK_{12} = 0$$
$$AK_{23}s^2 + 2DK_{23}st + B(K_{23} - K_{12})t^2 - 2FK_{12}t - CK_{12} = 0$$

Multiply the second equation by the free variable λ and add the result to the first equation.

$$A(K_{13} - K_{12} + \lambda K_{23})s^2 + 2D(K_{13} + \lambda K_{23})st + B(K_{13} + \lambda(K_{23} - K_{12}))t^2$$
$$- 2EK_{12}s - 2FK_{12}\lambda t - CK_{12}(1 + \lambda) = 0$$

or

$$A^* s^2 + 2B^* st + C^* t^2 + 2D^* s + 2E^* t + F^* = 0$$

where

$$A^* = A(K_{13} - K_{12} + \lambda K_{23})$$
$$B^* = D(K_{13} + \lambda K_{23})$$
$$C^* = B(K_{13} + \lambda(K_{23} - K_{12}))$$
$$D^* = EK_{12}$$
$$E^* = FK_{12}\lambda$$
$$F^* = CK_{12}(1 + \lambda)$$

Consider this a quadratic equation in

$$A^*s^2 + 2(B^*t + D^*)s + C^*t^2 + 2E^*t + F^* = 0$$

If we were to solve for s in terms of t, we would write

$$s = \frac{-(B^*t + D^*) \pm \sqrt{(B^*t + D^*)^2 - A^*(C^*t^2 + 2E^*t + F^*)}}{A^*}$$

What we desire to do is choose a value for λ that makes the discriminant $(B^*t + D^*)^2 - A^*(C^*t^2 + 2E^*t + F^*)$ a perfect square. That is, we seek to find a value for λ such that for some constant p and q we have

$$(B^*t + D^*)^2 - A^*(C^*t^2 + 2E^*t + F^*) = (pt + q)^2$$

Equating coefficients of t^2, t, and 1 results in

$$B^{*2} - A^*C^* = p^2$$
$$2B^*D^* - 2A^*E^* = 2pq$$
$$D^{*2} - A^*F^* = q^2$$

Since $p^2q^2 = (pq)^2$ is an identity, we must have

$$(B^{*2} - A^*C^*)(D^{*2} - A^*F^*) = (B^*D^* - A^*E^*)^2$$

Multiplying this out, we obtain

$$B^{*2}D^{*2} - B^{*2}A^*F^* - A^*C^*D^{*2} + A^{*2}C^*F^* =$$

$$B^{*2}D^{*2} - 2A^*B^*D^*E^* + A^{*2}E^{*2}$$

Canceling the $B^{*2}D^{*2}$ terms and dividing by A^* yields

$$A^*(C^*F^* - E^{*2}) + B^*(2D^*E^* - B^*F^*) - C^*D^{*2} = 0$$

which is equivalent to the determinant

$$\begin{vmatrix} A^* & B^* & D^* \\ B^* & C^* & E^* \\ D^* & E^* & F^* \end{vmatrix} = 0$$

Hence λ must satisfy the cubic equation

$$A(K_{13} - K_{12} + \lambda K_{23})\{B[K_{13} + \lambda(K_{23} - K_{12})]CK_{12}(1 + \lambda) - (FK_{12}\lambda)^2\}$$
$$+ [D(K_{13} + \lambda K_{23})]\{2EK_{12}FK_{12}\lambda - D(K_{13} + \lambda K_{23})C[K_{12}(1 + \lambda)]\}$$
$$- B[K_{13} + \lambda(K_{23} - K_{12})]E^2K_{12}^2 = 0$$

After dividing out K_{12} and rearranging, we obtain

$$A'\lambda^3 + B'\lambda^2 + C'\lambda + D' = 0$$

where
$$A' = K_{23}[ABC(K_{23} - K_{12}) + AF^2K_{12} - CD^2K_{23}]$$
$$B' = ABC[(K_{23} - K_{12})(K_{13} - K_{12}) - K_{23}K_{12} + K_{23}K_{13} + K_{23}K_{23}]$$
$$+ K_{23}D[2EFK_{12} - CD(2K_{13} + K_{23})] - AF^2K_{12}(K_{13} - K_{12})$$
$$C' = ABC[(K_{13} - K_{12})(K_{23} - K_{12}) + K_{13}K_{23} + (K_{13} - K_{12})K_{13}]$$
$$- D^2C[K_{13}^2 + 2K_{13}K_{23}] - BE^2(K_{12}^2 - K_{23}K_{12}) + 2DEFK_{12}K_{13}$$
$$D' = ABCK_{13}(K_{13} - K_{12}) - D^2CK_{13}^2 - E^2BK_{12}K_{13}$$

The cubic equation in λ is solved for any one of its roots, say λ^*. This value of λ^* is then substituted for λ in the equations determining $A^*, B^*, C^*, D^*, E^*,$ and F^*. Then p and q are easily determined from

$$p = \sqrt{B^{*2} - A^*C^*}$$
$$q = \text{sgn}(B^*D^* - A^*E^*)\sqrt{D^{*2} - A^*F^*}$$

The two solutions for s in terms of t are

$$s = \frac{-(B^*t + D^*) \pm (pt + q)}{A^*}$$

Each solution is a first-order polynomial in t. Thus we can write for the two solution s_1 and s_2

$$s_i = m_it + n_i, \qquad i = 1, 2$$

where

$$m_1 = \frac{p - B^*}{A^*}, \qquad n_1 = \frac{q - D^*}{A^*}$$
$$m_2 = \frac{-p - B^*}{A^*}, \qquad n_2 = \frac{-q - D^*}{A^*}$$

Substituting, in turn, each s_i back into

$$A^*s^2 + 2B^*st + C^*t^2 + 2D^*s + 2E^*t + F^* = 0$$

yields two quadratic equations in t:

$$(A^*m_i^2 + 2B^*m_i + C^*)t^2 + 2(A^*m_in_i + Bn_i + Dm_i + E)t + 2Dn_i + F + A^*n_i^2 = 0$$

where $i = 1, 2$. Each is solved, thereby providing two values of t associated with s_1 and two values of t associated with s_2.

Since $sz_3 = z_1$ and $tz_3 = z_2$, each one of the four solutions (s,t) can be substituted back into the original three equations

$$Az_1^2 + 2Dz_1z_2 + Bz_2^2 = K_{12}$$
$$Az_1^2 + 2Ez_1z_3 + Cz_3^2 = K_{13}$$
$$Bz_2^2 + 2Fz_2z_3 + Cz_3^2 = K_{23}$$

to produce

$$(As^2 + 2Dst + Bt^2)z_3^2 = K_{12}$$
$$(As^2 + 2Es + C)z_3^2 = K_{13}$$
$$(Bt^2 + 2Ft + C)z_3^2 = K_{23}$$

which can be solved in a least-squares sense for z_3^2:

$$z_3^2 = \frac{K_{12}(As^2 + 2Dst + Bt^2) + K_{13}(As^2 + 2Es + C) + K_{23}(Bt^2 + 2Ft + C)}{(As^2 + 2Dst + Bt^2)^2 + (As^2 + 2Es + C)^2 + (Bt^2 + 2Ft + C)^2}$$

Solving this for z_3 then produces two solutions for z_3 for each of the four solutions for (s,t). Then using

$$z_1 = sz_3$$
$$z_2 = tz_3$$

we can completely determine the eight solutions for z_1, z_2, and z_3.

■ **EXAMPLE 13.7**

The perspective projection of a triangle is observed to be

$$\begin{pmatrix} u_1 \\ v_1 \end{pmatrix} = \begin{pmatrix} 0.00645 \\ -0.0105 \end{pmatrix}$$

$$\begin{pmatrix} u_2 \\ v_2 \end{pmatrix} = \begin{pmatrix} 0.0243 \\ -0.0126 \end{pmatrix}$$

$$\begin{pmatrix} u_3 \\ v_3 \end{pmatrix} = \begin{pmatrix} 0.00961 \\ 0.0156 \end{pmatrix}$$

The distance the image plane is in front of the center of perspectivity is 0.075. The side lengths of the triangle are

$$S_{12} = 3, \qquad S_{13} = 4, \qquad S_{23} = 5$$

The four solutions are given by

Solution	Point 1	Point 2	Point 3
1	$\begin{pmatrix} 0.932698 \\ -1.518346 \\ 10.845331 \end{pmatrix}$	$\begin{pmatrix} 3.424111 \\ -1.775465 \\ 10.568245 \end{pmatrix}$	$\begin{pmatrix} 1.102255 \\ 1.789300 \\ 8.602403 \end{pmatrix}$
2	$\begin{pmatrix} 0.924198 \\ -1.504508 \\ 10.746488 \end{pmatrix}$	$\begin{pmatrix} 3.424218 \\ -1.780706 \\ 10.599438 \end{pmatrix}$	$\begin{pmatrix} 1.471471 \\ 2.388653 \\ 11.4830908 \end{pmatrix}$
3	$\begin{pmatrix} 0.960641 \\ -1.563834 \\ 11.170240 \end{pmatrix}$	$\begin{pmatrix} 3.845932 \\ -1.994187 \\ 11.870161 \end{pmatrix}$	$\begin{pmatrix} 1.472456 \\ 2.390250 \\ 11.491589 \end{pmatrix}$
4	$\begin{pmatrix} 0.940637 \\ -1.531269 \\ 10.937637 \end{pmatrix}$	$\begin{pmatrix} 2.775335 \\ -1.439063 \\ 8.565849 \end{pmatrix}$	$\begin{pmatrix} 1.473875 \\ 2.392555 \\ 11.502668 \end{pmatrix}$

■

13.4.16 Determining the Principal Point by Using Parallel Lines

In all our work so far, we have assumed that the origin of the image reference frame is the principal point, the point through which the optic axis passes. This would be true for a calibrated camera. In this section we assume an uncalibrated camera and show how to determine the location of the principal point by using an observation of parallel 3D lines of known orientation.

Suppose we observe the perspective projection of a collection of N 3D lines having the same known direction cosines (b_1, b_2, b_3). In this case, however, the origin of the perspective projection image is not assumed to be the principal point. Let (u, v) represent the coordinates in the perspective projection image whose origin is the principal point, and let $(\hat{u}, \hat{v})$ represent the coordinates in the measurement reference frame. We take the relationship between (u, v) and $(\hat{u}, \hat{v})$ to be a simple translation

$$\begin{pmatrix} \hat{u} \\ \hat{v} \end{pmatrix} = \begin{pmatrix} u \\ v \end{pmatrix} + \begin{pmatrix} \Delta u \\ \Delta v \end{pmatrix}$$

This means that in the measurement reference frame of perspective projection image, the coordinates of the principal point are $(\Delta u, \Delta v)$.

The perspective projection of the nth parallel 3D line is observed to consist of the points in the set

$$\hat{M}_n = \left\{ \begin{pmatrix} \hat{u} \\ \hat{v} \end{pmatrix} \middle| \begin{pmatrix} \hat{u} \\ \hat{v} \end{pmatrix} = \begin{pmatrix} g_n \\ h_n \end{pmatrix} + \eta \begin{pmatrix} c_n \\ d_n \end{pmatrix} \right\}$$

expressed relative to the observed reference frame. Hence in the reference frame where the principal point is the origin, the nth line will consist of the points in the set

$$M_n = \left\{ \begin{pmatrix} u \\ v \end{pmatrix} \middle| \begin{pmatrix} u \\ v \end{pmatrix} = \begin{pmatrix} g_n - \Delta u \\ h_n - \Delta v \end{pmatrix} + \eta \begin{pmatrix} c_n \\ d_n \end{pmatrix} \right\}$$

The relationship between the direction cosines of the nth 3D line and the parameters of its perspective projection line expressed relative to the perspective projection plane coordinate system whose origin is the principal point is given by

$$d_n f b_1 - c_n f b_2 + [(h_n - \Delta v)c_n - (g_n - \Delta u)d_n]b_3 = 0$$

To find Δu and Δv, find those values for Δu, and Δv that minimize the squared error

$$\epsilon^2 = \sum_{n=1}^{N} \{d_n f b_1 - c_n f b_2 + [(h_n - \Delta v)c_n - (g_n - \Delta u)d_n]b_3\}^2$$

Given (b_1, b_2, b_3), we can obtain the minimizing solution for $(\Delta u, \Delta v)$ by taking the partial derivative of ϵ^2 with respect to Δu and Δv, setting the partial derivative to zero, and solving for Δu and Δv. Let

$$\alpha_n = d_n f b_1 - c_n f b_2 + (h_n c_n - g_n d_n)b_3$$

$$\epsilon^2 = \sum_{n=1}^{N} [\alpha_n + b_3(d_n \Delta u - c_n \Delta v)]^2$$

Then

$$\frac{\partial \epsilon^2}{\partial \Delta u} = \sum_{n=1}^{N} 2[\alpha_n + b_3(d_n \Delta u - c_n \Delta v)]d_n b_3$$

$$\frac{\partial \epsilon^2}{\partial \Delta v} = \sum_{n=1}^{N} 2[\alpha_n + b_3(d_n \Delta u - c_n \Delta v)](-c_n b_3)$$

Setting the partial derivative to zero yields the matrix equation

$$\begin{pmatrix} \sum_{n=1}^{N} d_n^2 & -\sum_{n=1}^{N} c_n d_n \\ -\sum_{n=1}^{N} c_n d_n & \sum_{n=1}^{N} c_n^2 \end{pmatrix} \begin{pmatrix} \Delta u \\ \Delta v \end{pmatrix} = \begin{pmatrix} -\sum_{n=1}^{N} \alpha_n d_n / b_3 \\ \sum_{n=1}^{N} \alpha_n c_n / b_3 \end{pmatrix}$$

And its solution is

$$
\begin{pmatrix} \Delta u \\ \Delta v \end{pmatrix} = \frac{\begin{pmatrix} \displaystyle\sum_{n=1}^{N} c_n^2 & \displaystyle\sum_{n=1}^{N} c_n d_n \\[2mm] \displaystyle\sum_{n=1}^{N} c_n d_n & \displaystyle\sum_{n=1}^{N} d_n^2 \end{pmatrix} \begin{pmatrix} -\displaystyle\sum_{n=1}^{N} \alpha_n d_n \\[2mm] \displaystyle\sum_{n=1}^{N} \alpha_n c_n \end{pmatrix}}{b_3 \left[\displaystyle\sum_{n=1}^{N} d_n^2 \displaystyle\sum_{n=1}^{N} c_n^2 - \left(\displaystyle\sum_{n=1}^{N} c_n d_n \right)^2 \right]}
\tag{13.18}
$$

13.5 Circles

From the observed perspective projection of a circle having known radius d_0, it is possible to infer the plane on which the circle lies as well as where the center of the circle lies. Relative to the unknown plane on which the circle lies, the circle center is (r_0, s_0). We can therefore represent the circle by the set

$$
\left\{ \begin{pmatrix} r \\ s \end{pmatrix} \middle| (r - r_0)^2 + (s - s_0)^2 = d_0^2 \right\}
\tag{13.19}
$$

The plane on which the circle lies has an unknown orientation with respect to the camera reference frame and an unknown perpendicular distance δ to the origin of the camera reference frame. The 3D coordinates of any point (x, y, z) on the unknown plane can then be represented in terms of the plane coordinate frame by

$$
\begin{pmatrix} x \\ y \\ z \end{pmatrix} = \begin{pmatrix} a_1 \\ a_2 \\ a_3 \end{pmatrix} r + \begin{pmatrix} b_1 \\ b_2 \\ b_3 \end{pmatrix} s + \begin{pmatrix} c_1 \\ c_2 \\ c_3 \end{pmatrix} \delta
\tag{13.20}
$$

where $(a_1, a_2, a_3)'$ and $(b_1, b_2, b_3)'$ are unit vectors orthogonal to each other whose span is a subspace parallel to the plane, and $(c_1, c_2, c_3)'$ is a unit vector that is normal to the plane. The matrix R,

$$
R = \begin{pmatrix} a_1 & b_1 & c_1 \\ a_2 & b_2 & c_2 \\ a_3 & b_3 & c_3 \end{pmatrix}
$$

is then an orthonormal matrix that, with the correct sign on $(c_1, c_2, c_3)'$, rotates the coordinate system of the unknown plane to the camera coordinate system. We can therefore write the coordinates of any point on the plane in terms of its camera reference frame coordinates as

$$
\begin{pmatrix} r \\ s \\ \delta \end{pmatrix} = \begin{pmatrix} a_1 & a_2 & a_3 \\ b_1 & b_2 & b_3 \\ c_1 & c_2 & c_3 \end{pmatrix} \begin{pmatrix} x \\ y \\ z \end{pmatrix}
\tag{13.21}
$$

The relation between the perspective projection coordinates of a point and its 3D coordinates is given by

$$\begin{pmatrix} u \\ v \end{pmatrix} = \frac{f}{z} \begin{pmatrix} x \\ y \end{pmatrix} \tag{13.22}$$

Equations (13.19), (13.21), and (13.22) algebraically represent all the relations involved in the problem. Solve Eq. (13.22) for x and y in terms of u, v, and z and substitute this back into Eq. (13.21), which then becomes

$$\begin{pmatrix} r \\ s \\ \delta \end{pmatrix} = \frac{1}{f} \begin{pmatrix} a_1 & a_2 & a_3 \\ b_1 & b_2 & b_3 \\ c_1 & c_2 & c_3 \end{pmatrix} \begin{pmatrix} u \\ v \\ f \end{pmatrix} z \tag{13.23}$$

From Eq. (13.23)

$$z = \frac{\delta f}{c_1 u + c_2 v + c_3 f}$$

so that

$$\begin{pmatrix} r \\ s \end{pmatrix} = \begin{pmatrix} a_1 & a_2 & a_3 \\ b_1 & b_2 & b_3 \end{pmatrix} \begin{pmatrix} u \\ v \\ f \end{pmatrix} \frac{\delta}{c_1 u + c_2 v + c_3 f} \tag{13.24}$$

Now substitute the expressions for r and s of Eq. (13.24) into the circle equation (13.19) and simplify. This results in

$$\begin{aligned} &\left[(a_1\delta - r_0 c_1)u + (a_2\delta - r_0 c_2)v + (a_3\delta - r_0 c_3)f\right]^2 + \\ &\left[(b_1\delta - s_0 c_1)u + (b_2\delta - s_0 c_2)v + (b_3\delta - s_0 c_3)f\right]^2 \\ &= d_0^2(c_1 u + c_2 v + c_3 f)^2 \end{aligned} \tag{13.25}$$

Collecting together the terms involving like powers of u and v results in

$$A'u^2 + 2B'uv + C'v^2 + 2fD'u + 2fE'v + f^2 F' = 0$$

where

$$A' = (a_1^2 + b_1^2)\delta^2 - 2\delta c_1(a_1 r_0 + b_1 s_0) + c_1^2(r_0^2 + s_0^2 - d_0^2)$$
$$B' = (a_1 a_2 + b_1 b_2)\delta^2 - \delta c_2(a_1 r_0 + b_1 s_0) - \delta c_1(a_2 r_0 + b_2 s_0) + c_1 c_2(r_0^2 + s_0^2 - d_0^2)$$
$$C' = (a_2^2 + b_2^2)\delta^2 - 2\delta c_2(a_2 r_0 + b_2 s_0) + c_2^2(r_0^2 + s_0^2 - d_0^2)$$
$$D' = (a_1 a_3 + b_1 b_3)\delta^2 - \delta c_3(a_1 r_0 + b_1 s_0) - c_1\delta(a_3 r_0 + b_3 s_0) + c_1 c_3(r_0^2 + s_0^2 - d_0^2)$$
$$E' = (a_2 a_3 + b_2 b_3)\delta^2 - \delta c_3(a_2 r_0 + b_2 s_0) - c_2\delta(a_3 r_0 + b_3 s_0) + c_2 c_3(r_0^2 + s_0^2 - d_0^2)$$
$$F' = (a_3^2 + b_3^2)\delta^2 - 2\delta c_3(a_3 r_0 + b_3 s_0) + c_3^2(r_0^2 + s_0^2 - d_0^2) \tag{13.26}$$

What is observed on the perspective projection image is the projection of the circle, and this can be represented by the set

$$\left\{ \begin{pmatrix} u \\ v \end{pmatrix} \middle| Au^2 + 2Buv + Cv^2 + 2fDu + 2fEv + f^2F = 0 \right\} \qquad (13.27)$$

The relation between the coefficients of the equation of the observed projected circle and the derived projected circle is simply a multiplicative constant k:

$$A = kA'$$
$$B = kB'$$
$$C = kC'$$
$$D = kD'$$
$$E = kE'$$
$$F = kF'$$

From the observed coefficients A, B, C, D, E, F and the known orthonormality of R, the problem now becomes: Solve for the desired unknowns, which are $r_0, s_0, \delta, k, c_1, c_2$, and c_3. To do so, we perform one more step of simplification. We eliminate $a_1, a_2, a_3, b_1, b_2, b_3, r_0$, and s_0. Notice that

$$a_1^2 + b_1^2 = 1 - c_1^2$$
$$a_2^2 + b_2^2 = 1 - c_2^2$$
$$a_3^2 + b_3^2 = 1 - c_3^2$$

$$(13.28)$$

$$a_1 a_2 + b_1 b_2 = -c_1 c_2$$
$$a_1 a_3 + b_1 b_3 = -c_1 c_3$$
$$a_2 a_3 + b_2 b_3 = -c_2 c_3$$

Letting

$$\begin{pmatrix} \lambda_1 \\ \lambda_2 \\ \lambda_3 \end{pmatrix} = \begin{pmatrix} a_1 \\ a_2 \\ a_3 \end{pmatrix} r_0 + \begin{pmatrix} b_1 \\ b_2 \\ b_3 \end{pmatrix} s_0$$

and using the orthonormality of R again results in

$$\lambda_1^2 + \lambda_2^2 + \lambda_3^2 = r_0^2 + s_0^2$$

and

$$\lambda_1 c_1 + \lambda_2 c_2 + \lambda_3 c_3 = 0$$

After using these relations to eliminate $a_1, a_2, a_3, b_1, b_2, b_3, r_0$, and s_0 from Eq. (13.26) and putting together all we have, we obtain a set of relations between the observed coefficients and the unknowns $c_1, c_2, c_3, \lambda_1, \lambda_2, \lambda_3, \delta$, and k:

$$A = k\left[(1 - c_1^2)\delta^2 - 2c_1\delta\lambda_1 + c_1^2(\lambda_1^2 + \lambda_2^2 + \lambda_3^2 - d_0^2)\right]$$

$$B = k\left[-\delta c_1\lambda_2 - \delta c_2\lambda_1 - c_1 c_2\delta^2 + c_1 c_2(\lambda_1^2 + \lambda_2^2 + \lambda_3^2 - d_0^2)\right]$$

$$C = k\left[(1 - c_2^2)\delta^2 - 2\delta c_2\lambda 2 + c_2^2(\lambda_1^2 + \lambda_2^2 + \lambda_3^2 - d_0^2)\right]$$

$$D = k\left[-\delta c_3\lambda_1 - \delta c_1\lambda_3 - c_1 c_3\delta^2 + c_1 c_3(\lambda_1^2 + \lambda_2^2 + \lambda_3^2 - d_0^2)\right]$$

$$E = k\left[-\delta c_3\lambda_2 - \delta c_2\lambda_3 - c_2 c_3\delta^2 + c_2 c_3(\lambda_1^2 + \lambda_2^2 + \lambda_3^2 - d_0^2)\right]$$

$$F = k\left[(1 - c_3^2)\delta^2 - 2\delta c_3\lambda_3 + c_3^2(\lambda_1^2 + \lambda_2^2 + \lambda_3^2 - d_0^2)\right]$$

The solution to this set of nonlinear equations cannot be obtained directly. In the special case when $r_0 = s_0 = 0$ so that $\lambda_1 = \lambda_2 = \lambda_3 = 0$, a direct solution can be obtained. This is the case for which the center of the circle is located on the point of the plane closest to the camera lens. Also in the special case when $c_1 = c_2 = 0$ so that $c_3 = 1$, a direct solution can be obtained. This is the case for which the circle is located on a plane parallel to the image projection plane. Depending on whether the actual situation to be solved is the closest to the first case or to the second case, a direct approximate solution can be obtained by using the closest case. This approximate solution can then be used as the initial approximate solution for the technique described in the next chapter, Section 14.3, which illustrates how a non-linear problem can be linearized and solved in an iterative manner. Here we set up the case for which

$$A = k[\delta^2 - c_1^2(\delta^2 + d_0^2)]$$

$$B = -kc_1 c_2(\delta^2 + d_0^2)$$

$$C = k[\delta^2 - c_2^2(\delta^2 + d_0^2)]$$

$$D = -kc_1 c_3(\delta^2 + d_0^2)$$

$$E = -kc_2 c_3(\delta^2 + d_0^2)$$

$$F = k[\delta^2 - c_3^2(\delta^2 + d_0^2)]$$

$$c_1^2 + c_2^2 + c_3^2 = 1$$

To solve this system of equations, we assume D and $E \neq 0$. The cases when $D = 0$ or $E = 0$ must be worked out separately. Notice that

$$c_2 = \frac{B}{D}c_3$$

$$c_1 = \frac{B}{E}c_3$$

Hence

$$c_3 = \pm \frac{1}{\sqrt{\left(\frac{B}{D}\right)^2 + \left(\frac{B}{E}\right)^2 + 1}}$$

There are two solutions to the plane normal, one of which is the negation of the other. One will correspond to a $\delta > 0$ and the other to a $\delta < 0$.

Next notice that

$$\frac{DE}{B} = -kc_3^2(\delta^2 + d_0^2)$$

Substituting this into the expression for F, results in

$$k = \frac{BF - DE}{B\delta^2}$$

Now

$$A + C + F = k(2\delta^2 - d_0^2)$$

Substituting the expression for k and solving for δ^2, we have

$$\delta^2 = \frac{(DE - BF)d_0^2}{B(A + C + F) + 2(DE - BF)} \tag{13.29}$$

so that

$$\delta = \pm d_0 \sqrt{\frac{DE - BF}{B(A + C + F) + 2(DE - BF)}}$$

To locate the center $(x_0, y_0, z_0)'$ of the circle in 3D space, we use Eq. (13.19):

$$\begin{pmatrix} x_0 \\ y_0 \\ z_0 \end{pmatrix} = \begin{pmatrix} a_1 \\ a_2 \\ a_3 \end{pmatrix} r_0 + \begin{pmatrix} b_1 \\ b_2 \\ b_3 \end{pmatrix} s_0 + \begin{pmatrix} c_1 \\ c_2 \\ c_3 \end{pmatrix} \delta = \begin{pmatrix} \lambda_1 \\ \lambda_2 \\ \lambda_3 \end{pmatrix} + \begin{pmatrix} c_1 \\ c_2 \\ c_3 \end{pmatrix} \delta$$

Since in the case we have worked $\lambda_1 = \lambda_2 = \lambda_3 = 0$, the center of the circle is given by

$$\begin{pmatrix} x_0 \\ y_0 \\ z_0 \end{pmatrix} = \begin{pmatrix} c_1 \\ c_2 \\ c_3 \end{pmatrix} \delta$$

EXAMPLE 13.8

Consider the situation in which the camera constant $f = 7$, the circle has radius 2, and the observed ellipse satisfies

$$Au^2 + 2Buv + Cv^2 + 2fDu + 2fEv + f^2F = 0$$

where

$$A = 321$$
$$B = -58$$
$$C = 234$$
$$D = -87$$
$$E = -174$$
$$F = 89$$

Then $B/D = 2/3$ and $B/E = 1/3$, so that

$$c_3 = \pm \frac{1}{\sqrt{\frac{4}{9} + \frac{1}{9} + 1}} = \pm \frac{3}{\sqrt{14}}$$

Using the positive value of $3/\sqrt{14}$ for c_3, we have

$$c_2 = \frac{B}{D}c_3 = \frac{2}{\sqrt{14}}$$

$$c_1 = \frac{B}{E}c_3 = \frac{1}{\sqrt{14}}$$

From Eq. (13.29) we can determine δ^2:

$$\delta^2 = \frac{(-87)(-174) - (-58)(89)4}{(-58)(321 + 234 + 89) + 2[(-87)(-174) - (-58)(89)]}$$

$$= 25$$

So $\delta = \pm 5$. The center of the circle is then given by

$$\begin{pmatrix} x_0 \\ y_0 \\ z_0 \end{pmatrix} = 5 \begin{pmatrix} 1 \\ 2 \\ 3 \end{pmatrix} \frac{1}{\sqrt{14}}$$

If the circle's center is not the point in the plane closest to the camera lens or if the circle does not lie in a plane parallel to the image plane, we can determine an approximate solution by using the direct, but incorrect, solution for the case believed to be the closest of these two cases. Then we can solve the nonlinear problem by linearizing the problem around the current approximate solution and iterating. For the case of the circle, we have the following eight functions, which depend on the unknown variables $k, \delta, \lambda_1, \lambda_2, \lambda_3, c_1, c_2,$ and c_3:

$$A^* = k\left[\delta^2 - 2c_1\lambda_1\delta + c_1^2(\lambda_1^2 + \lambda_2^2 + \lambda_3^2 - \delta^2 - d_0^2)\right]$$

$$B^* = k\left[-\delta c_1\lambda_2 - \delta c_2\lambda_1 + c_2^2(\lambda_1^2 + \lambda_2^2 + \lambda_3^2 - \delta^2 - d_0^2)\right]$$

$$C^* = k\left[\delta^2 - 2c_2\lambda_2\delta + c_2^2(\lambda_1^2 + \lambda_2^2 + \lambda_3^2 - \delta^2 - d_0^2)\right]$$

$$D^* = k\left[-\delta c_3\lambda_1 - \delta c_1\lambda_3 + c_1c_3(\lambda_1^2 + \lambda_2^2 + \lambda_3^2 - \delta^2 - d_0^2)\right]$$

$$E^* = k\left[-\delta c_3\lambda_2 - \delta c_2\lambda_3 + c_2c_3(\lambda_1^2 + \lambda_2^2 + \lambda_3^2 - \delta^2 - d_0^2)\right]$$

$$F^* = k\left[\delta^2 - 2c_3\lambda_3\delta + c_3^2(\lambda_1^2 + \lambda_2^2 + \lambda_3^2 - \delta^2 - d_0^2)\right]$$

$$G^* = c_1^2 + c_2^2 + c_3^2$$

$$H^* = \lambda_1 c_1 + \lambda_2 c_2 + \lambda_3 c_3$$

The observed or known fixed values of these functions are A for A^*, B for B^*, C for C^*, D for D^*, E for E^*, F for F^*, 1 for G^*, and 0 for H^*. To linearize the nonlinear problem, we expand each function in a first-order Taylor series around the current approximate solution $(k, c_1, c_2, c_3, \delta, \lambda_1, \lambda_2, \lambda_3)$. Then

$$
\begin{pmatrix}
A - A^*(k, c_1, c_2, c_3, \delta, \lambda_1, \lambda_2, \lambda_3) \\
B - B^*(k, c_1, c_2, c_3, \delta, \lambda_1, \lambda_2, \lambda_3) \\
C - C^*(k, c_1, c_2, c_3, \delta, \lambda_1, \lambda_2, \lambda_3) \\
D - D^*(k, c_1, c_2, c_3, \delta, \lambda_1, \lambda_2, \lambda_3) \\
E - E^*(k, c_1, c_2, c_3, \delta, \lambda_1, \lambda_2, \lambda_3) \\
F - F^*(k, c_1, c_2, c_3, \delta, \lambda_1, \lambda_2, \lambda_3) \\
G - G^*(k, c_1, c_2, c_3, \delta, \lambda_1, \lambda_2, \lambda_3) \\
H - H^*(k, c_1, c_2, c_3, \delta, \lambda_1, \lambda_2, \lambda_3)
\end{pmatrix}
= Q
\begin{pmatrix}
\Delta k \\ \Delta c_1 \\ \Delta c_2 \\ \Delta c_3 \\ \Delta \delta \\ \Delta \lambda_1 \\ \Delta \lambda_2 \\ \Delta \lambda_3
\end{pmatrix}
$$

where

$$
Q = \begin{pmatrix}
\dfrac{\partial A^*}{\partial k} & \dfrac{\partial A^*}{\partial c_1} & \dfrac{\partial A^*}{\partial c_2} & \dfrac{\partial A^*}{\partial c_3} & \dfrac{\partial A^*}{\partial \delta} & \dfrac{\partial A^*}{\partial \lambda_1} & \dfrac{\partial A^*}{\partial \lambda_2} & \dfrac{\partial A^*}{\partial \lambda_3} \\
\vdots & & & & & & & \vdots \\
\dfrac{\partial H^*}{\partial k} & \dfrac{\partial H^*}{\partial c_1} & \dfrac{\partial H^*}{\partial c_2} & \dfrac{\partial H^*}{\partial c_3} & \dfrac{\partial H^*}{\partial \delta} & \dfrac{\partial H^*}{\partial \lambda_1} & \dfrac{\partial H^*}{\partial \lambda_2} & \dfrac{\partial H^*}{\partial \lambda_3}
\end{pmatrix}
$$

and all the partial derivatives are evaluated at the current approximate solution.

Solve the system of equations for the adjustments $(\Delta k, \Delta c_1, \Delta c_2, \Delta c_3, \Delta \delta, \Delta l_1, \Delta l_2, \Delta l_3)$. Then update the current approximate solution by using

$$
\begin{pmatrix}
k_1 \\ c_1 \\ c_2 \\ c_3 \\ \delta \\ \lambda_1 \\ \lambda_2 \\ \lambda_3
\end{pmatrix}
\leftarrow
\begin{pmatrix}
k_1 \\ c_1 \\ c_2 \\ c_3 \\ \delta \\ \lambda_1 \\ \lambda_2 \\ \lambda_3
\end{pmatrix}
+
\begin{pmatrix}
\Delta k \\ \Delta c_1 \\ \Delta c_2 \\ \Delta c_3 \\ \Delta \delta \\ \Delta \lambda_1 \\ \Delta \lambda_2 \\ \Delta \lambda_3
\end{pmatrix}
$$

Iterate until convergence.

13.6   Range from Structured Light

Structured light is an active visual sensing technique introduced by Pennington and Will (1970; Pennington, Will, and Shelton, 1970) that relies totally on perspective geometry. It can be used to determine range maps from which reliable object recognition, location, and orientation may be derived. It works by employing a controlled light source to illuminate the scene with a regular pattern of light, such as stripes or a grid. Then from observing the perspective projection image of the projected light pattern reflecting from the object surface, it is possible to recover the 3D position of the illuminated object surface patches. One of the first successful systems of this type to be employed in manufacturing is CONSIGHT (Ward, Rheaume, and Holland, 1982).

To see how this works, we consider a simple situation in which the projected light pattern is a single beam of light illuminating a single point in the scene. Suppose we observe the perspective projection (u,v) of an unknown 3D point (x_0,y_0,z_0) illuminated by a known beam of light

$$\left\{ \begin{pmatrix} x \\ y \\ z \end{pmatrix} \middle| \begin{pmatrix} x \\ y \\ z \end{pmatrix} = \begin{pmatrix} x_1 \\ y_1 \\ z_1 \end{pmatrix} + \lambda \begin{pmatrix} a_1 \\ a_2 \\ a_3 \end{pmatrix} \right\}$$

where (a_1, a_2, a_3) is the direction cosine of the beam.

Since the beam illuminates (x_0, y_0, z_0), we must have

$$\begin{pmatrix} x_0 \\ y_0 \\ z_0 \end{pmatrix} = \begin{pmatrix} x_1 \\ y_1 \\ z_1 \end{pmatrix} + \lambda \begin{pmatrix} a_1 \\ a_2 \\ a_3 \end{pmatrix}$$

Since the point (x_0, y_0, z_0) has perspective projection (u, v), we must also have

$$\begin{pmatrix} u \\ v \end{pmatrix} = \frac{f}{z_0} \begin{pmatrix} x_0 \\ y_0 \end{pmatrix}$$

The unknown point (x_0, y_0, z_0) can then be determined by solving in the least-squares sense the following overconstrained equation for z_0:

$$\begin{pmatrix} \frac{u}{f} & -a_1 \\ \frac{v}{f} & -a_2 \\ 1 & -a_3 \end{pmatrix} \begin{pmatrix} z_0 \\ \lambda \end{pmatrix} = \begin{pmatrix} x_1 \\ y_1 \\ z_1 \end{pmatrix}$$

Hence

$$z_0 = \frac{\frac{ux_1}{f} + \frac{vy_1}{f} + z_1 - \left(\frac{a_1 u}{f} + \frac{a_2 v}{f} + a_3 \right)(a_1 x_1 + a_2 y_1 + a_3 z_1)}{\left(\frac{u}{f} \right)^2 + \left(\frac{v}{f} \right)^2 + 1 - \left(\frac{a_1 u}{f} + \frac{a_2 v}{f} + a_3 \right)^2} \qquad (13.30)$$

Once z_0 is determined, the x_0, y_0 are easily recovered from

$$\begin{pmatrix} x_0 \\ y_0 \end{pmatrix} = \frac{z_0}{f} \begin{pmatrix} u \\ v \end{pmatrix} \qquad (13.31)$$

EXAMPLE 13.9

Let the camera constant $f = 5$, the observed perspective projection $(u, v) = (1.9444, 1.1111)$, and the light beam passing through point $(x_1, y_1, z_1) = (10, 20, 30)$ have direction cosines $(a_1, a_2, a_3) = (5/13, 0, 12/13)$. Then by Eqs. (13.30) and (13.31)

$$z_0 = \frac{38.3333 - 33.8297}{1.20061 - 1.15057} = 90$$

$$x_0 = 35$$

$$y_0 = 20$$

More often than not, the light pattern projected by the laser beam of a structured light ranging system is a stripe or a grid. In the case of a light stripe pattern (Will and Pennington, 1971; Agin and Binford, 1973; Potmesil, 1979; Oshima and Shinai, 1983; Posdamer and Altschuler, 1982; Sato, Kitagawa, and Fujita, 1982; and Asada, Ichikawa, and Tsuji, 1988), the projected light can be thought of as a plane of light. This light plane intersects the illuminated scene object, causing a 3D curve of bright light to be observed on the perspective projection image. Since the light plane is in a known position and orientation with respect to the camera, it is possible to determine the 3D coordinates of each point on the 3D curve: Just intersect the light plane with the inverse perspective projection of the observed stripe on the perspective projection image. This intersection can be done point by point, and it is accomplished exactly as was the point-in-known-plane problem of Section 13.4.9.

A review of range-finding techniques can be found in Nitzan, Brain, and Duda (1977); Jarvis (1983); and Yang and Kak (1989). Hu and Stockman (1989) discuss how to solve for range by using a structured grid pattern. Shrikhande and Stockman (1989); Wang, Mitiche, and Aggarwal (1987); and Will and Pennington (1972) show how to construct a surface orientation map by projecting a grid pattern. Morita, Yajima, and Sakata (1988); Yeung and Lawrence (1986); and Boyer and Kak (1987) discuss possibilities for projecting multiple planes of light, each uniquely coded to permit its unambiguous recognition.

13.7   Cross-Ratio

Under certain conditions, two different one-dimensional perspective projections of the same object in a 2D world are similar in almost a metric sense. A quantity, called the cross-ratio of four perspective projection points, takes the same value anytime the four 2D points are collinear (Fig. 13.7).

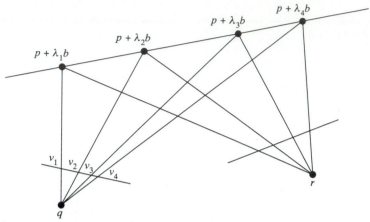

Figure 13.7 Two arbitrary perspective views of the same set of four collinear points lying along some arbitrary line.

13.7.1 Cross-Ratio Definitions and Invariance

Let the four collinear points be $p + \lambda_1 b$, $p + \lambda_2 b$, $p + \lambda_3 b$, and $p + \lambda_4 b$, where λ_1, λ_2, λ_3, and λ_4 are scalars and b is the vector of direction cosines of the line along which they lie. Let q and r be the centers of perspectivity for the two projection images. We will illustrate the invariance of the cross-ratio for the image whose center of perspectivity is q. Thus we may without loss of generality assume that the parameters p and b of the line are oriented with respect to the natural reference frame of the one-dimensional camera whose lens is at q, which we take to be the origin. Let $p = \begin{pmatrix} p_1 \\ p_2 \end{pmatrix}$.

By the perspective projection equations

$$u_n = f \frac{p_1 + \lambda_n b_1}{p_2 + \lambda_n b_2} \qquad n = 1, 2, 3, 4$$

Consider

$$u_i - u_j = f \frac{p_1 + \lambda_i b_1}{p_2 + \lambda_i b_2} - f \frac{p_1 + \lambda_j b_1}{p_2 + \lambda_j b_2}$$

Put this in a common denominator and simplify:

$$u_i - u_j = \frac{f(p_2 b_1 - p_1 b_2)(\lambda_i - \lambda_j)}{(p_2 + \lambda_i b_2)(p_2 + \lambda_j b_2)}$$

The cross-ratio of the perspective projection points u_1, u_2, u_3, and u_4 is defined by

$$\gamma(u_1, u_2, u_3, u_4) = \frac{(u_1 - u_2)(u_3 - u_4)}{(u_1 - u_3)(u_2 - u_4)}$$

Substitute for each difference $(u_i - u_j)$ the expression above. Notice that each of the differences involves the term $f(p_2 b_1 - p_1 b_2)$. Hence the term $f(p_2 b_1 - p_1 b_2)$ will appear twice in the numerator and twice in the denominator and thereby cancel itself out. Also notice that $(u_1 - u_2)(u_3 - u_4)$ contains the product $(p_2 + \lambda_1 b_2)(p_2 + \lambda_2 b)(p_2 + \lambda_3 b_2)(p_2 + \lambda_4 b_2)$, as does $(u_1 - u_3)(u_2 - u_4)$. Hence this product will cancel itself out too. What remains is

$$\gamma(u_1, u_2, u_3, u_4) = \frac{(u_1 - u_2)(u_3 - u_4)}{(u_1 - u_3)(u_2 - u_4)} = \frac{(\lambda_1 - \lambda_2)(\lambda_3 - \lambda_4)}{(\lambda_1 - \lambda_3)(\lambda_2 - \lambda_4)}$$

This illustrates that the cross-ratio is independent of the reference frame, the point p through which the line goes, and the direction cosines of the line. It depends only on the directed distances characterizing the configuration of the four collinear points.

13.7.2 Only One Cross-Ratio

There are 4! ways of permuting the arguments u_1, u_2, u_3, and u_4, and one might quickly think that there are 4! different cross-ratios. However, they are all functionally dependent. To see this, recall that any permutation of u_1, u_2, u_3, and u_4 can be achieved by composing an appropriate sequence of pairwise interchanges. There are six possible pairwise interchanges for four elements: $3 \leftrightarrow 4$, $2 \leftrightarrow 4$, $1 \leftrightarrow 4$, $2 \leftrightarrow 3$, $1 \leftrightarrow 3$, $1 \leftrightarrow 2$. By direct calculation we can verify that the cross-ratios for the six permutations of the arguments u_1, u_2, u_3, and u_4 induced by these six interchanges satisfy

$$\gamma(u_1, u_2, u_4, u_3) = \gamma(u_2, u_1, u_3, u_4,) = \frac{-\gamma(u_1, u_2, u_3, u_4)}{1 - \gamma(u_1, u_2, u_3, u_4)}$$

$$\gamma(u_1, u_4, u_3, u_2) = \gamma(u_3, u_2, u_1, u_4) = 1 - \gamma(u_1, u_2, u_3, u_4)$$

$$\gamma(u_4, u_3, u_2, u_1) = \gamma(u_1, u_3, u_2, u_4) = \frac{1}{\gamma(u_1, u_2, u_3, u_4)}$$

Hence we can deduce that each of the 4! cross-ratios is a function of the cross-ratio $\gamma(u_1, u_2, u_3, u_4)$

13.7.3 Cross-Ratio in Three Dimensions

The cross-ratio derived from one-dimensional perspective projections in a two-dimensional world can be generalized to two-dimensional perspective projections in a three-dimensional world. The generalization relies on the fact that lines in the 3D world project to lines on the 2-D-image perspective projection plane. Since dimensions have increased by one, the observed 3D points must be co-planar, and the invariance will be achieved by two independent cross-ratios.

Consider, for example, the star configuration of five points shown in Fig. 13.8. Let γ_{23} denote the cross-ratio for the line segment $p_2 p_3$, and let γ_{25} denote the cross-ratio for the line segment $p_2 p_5$. Take γ_{23} and γ_{25} to be fixed. Let λ_1, λ_2, and $L_1, 0 < \lambda_1 < \lambda_2 < L_1$, be any numbers for which

$$\gamma_{23} = \frac{\lambda_1(L_1 - \lambda_2)}{\lambda_2(L_1 - \lambda_1)}$$

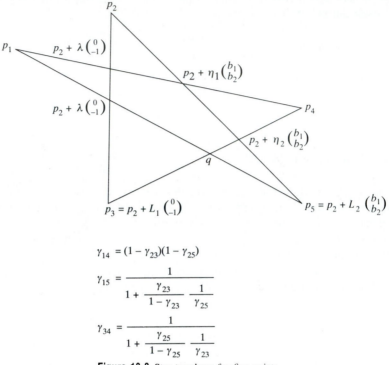

$$\gamma_{14} = (1 - \gamma_{23})(1 - \gamma_{25})$$

$$\gamma_{15} = \cfrac{1}{1 + \cfrac{\gamma_{23}}{1 - \gamma_{23}} \cfrac{1}{\gamma_{25}}}$$

$$\gamma_{34} = \cfrac{1}{1 + \cfrac{\gamma_{25}}{1 - \gamma_{25}} \cfrac{1}{\gamma_{23}}}$$

Figure 13.8 Star topology for five points.

and let η_1, η_2, and $L_2, 0 < \eta_1 < \eta_2 < L_2$, be any number for which

$$\gamma_{25} = \frac{\eta_1(L_2 - \eta_2)}{\eta_2(L_2 - \eta_1)}$$

Once the values for $\lambda_1, \lambda_2, L_1, \eta_1, \eta_2$, and L_2 are given and the angle $p_3 p_2 p_5$ is given, the line segments $p_1 p_4$, $p_1 p_5$, and $p_3 p_4$ are determined, so the configuration becomes fixed. The question is whether the configuration explicitly depends on the particular values $\lambda_1, \lambda_2, L_1, \eta_1, \eta_2$, and L_2 and the angle $p_3 p_2 p_5$ or whether it depends only on the given values of the cross-ratio γ_{23} and γ_{25}. We now show that the latter is indeed the case.

First we demonstrate that γ_{14} is a function only of γ_{23} and γ_{25}. We use a coordinate system where p_2 is the origin and the direction of p_3 from p_2 is $\begin{pmatrix} 0 \\ -1 \end{pmatrix}$.

Then the two points $\begin{pmatrix} 0 \\ -\lambda_1 \end{pmatrix}$ and $\eta_1 \begin{pmatrix} b_1 \\ b_2 \end{pmatrix}$ determine the line $p_1 p_4$. The two points $\begin{pmatrix} 0 \\ -\lambda_2 \end{pmatrix}$ and $L_2 \begin{pmatrix} b_1 \\ b_2 \end{pmatrix}$ determine the line $p_1 p_5$. The intersection of these two lines determines p_1. Hence

$$p_1 = \begin{pmatrix} p_{11} \\ p_{12} \end{pmatrix} = \frac{1}{\begin{vmatrix} \eta_1 b_2 + \lambda_1 & -\eta_1 b_1 \\ L_2 b_2 + \lambda_2 & -L_2 b_1 \end{vmatrix}} \left(\begin{matrix} \begin{vmatrix} \lambda_1 \eta_1 b_1 & -\eta_1 b_1 \\ \lambda_2 L_2 b_1 & -L_2 b_1 \end{vmatrix} \\ \begin{vmatrix} \eta_1 b_2 + \lambda_1 & \lambda_1 \eta_1 b_1 \\ L_2 b_2 + \lambda_2 & \lambda_2 L_2 b_1 \end{vmatrix} \end{matrix} \right)$$

$$p_1 = \frac{1}{\eta_1 \lambda_2 - \lambda_1 L_2} \left(\begin{matrix} \eta_1 b_1 L_2 (\lambda_2 - \lambda_1) \\ \eta_1 L_2 b_2 (\lambda_2 - \lambda_1) + \lambda_1 \lambda_2 (L_2 - \eta_1) \end{matrix} \right) \tag{13.32}$$

The two points $\begin{pmatrix} 0 \\ -\lambda_1 \end{pmatrix}$ and $\eta_1 \begin{pmatrix} b_1 \\ b_2 \end{pmatrix}$ determine the line $p_1 p_4$, and the two points

$\begin{pmatrix} 0 \\ -L_1 \end{pmatrix}$ and $\eta_2 \begin{pmatrix} b_1 \\ b_2 \end{pmatrix}$ determine the line $p_3 p_4$. The intersection of these two lines
determines p_4. Hence

$$p_4 = \frac{1}{\eta_1 L_1 - \lambda_1 \eta_2} \left(\begin{matrix} \eta_1 \eta_2 b_1 (L_1 - \lambda_1) \\ \eta_1 \eta_2 b_2 (L_1 - \lambda_1) + \lambda_1 L_1 (\eta_2 - \eta_1) \end{matrix} \right) \tag{13.33}$$

The four collinear points on line $p_1 p_4$

$$p_1, \qquad \begin{pmatrix} 0 \\ -\lambda_1 \end{pmatrix}, \qquad \eta_1 \begin{pmatrix} b_1 \\ b_2 \end{pmatrix}, \qquad p_4$$

determine the cross-ratio γ_{14}. Calculating the difference between each pair of successive points results in

$$p_1 - \begin{pmatrix} 0 \\ -\lambda_1 \end{pmatrix} = \frac{(\lambda_2 - \lambda_1) L_2}{\eta_1 \lambda_2 - \lambda_1 L_2} \left(\begin{matrix} \eta_1 b_1 \\ \eta_1 b_2 + \lambda_1 \end{matrix} \right)$$

$$\begin{pmatrix} 0 \\ -\lambda_1 \end{pmatrix} - \eta_1 \begin{pmatrix} b_1 \\ b_2 \end{pmatrix} = - \left(\begin{matrix} \eta_1 b_1 \\ \eta_1 b_2 + \lambda_1 \end{matrix} \right)$$

$$\eta_1 \begin{pmatrix} b_1 \\ b_2 \end{pmatrix} - p_4 = \frac{L_1 (\eta_1 - \eta_2)}{\eta_1 L_1 - \lambda_1 \eta_2} \left(\begin{matrix} \eta_1 b_1 \\ \eta_1 b_2 + \lambda_1 \end{matrix} \right)$$

Recall that

$$\gamma(u_1, u_2, u_3, u_4) = \frac{(u_1 - u_2)(u_3 - u_4)}{(u_1 - u_3)(u_2 - u_4)}$$

Letting $\delta_1 = u_1 - u_2$, $\delta_2 = u_2 - u_3$, $\delta_3 = u_3 - u_4$,

$$\gamma = \frac{\delta_1 \delta_3}{(\delta_1 + \delta_2)(\delta_2 + \delta_3)}$$

Then by letting

$$\delta_1 = \frac{(\lambda_2 - \lambda_1)L_2}{\eta_1\lambda_2 - \lambda_1 L_2}$$

$$\delta_2 = -1$$

$$\delta_3 = \frac{L_1(\eta_1 - \eta_2)}{\eta_1 L_1 - \lambda_1\eta_2}$$

we have

$$\gamma_{14} = \frac{\delta_1\delta_3}{(\delta_1 + \delta_2)(\delta_2 + \delta_3)}$$

Since

$$\delta_1 + \delta_2 = \frac{\lambda_2(L_2 - \eta_1)}{\eta_1\lambda_2 - \lambda_1 L_2} \quad \text{and} \quad \delta_2 + \delta_3 = \frac{\eta_2(L_1 - \lambda_1)}{\lambda_1\eta_2 - \eta_1 L_1}$$

their results

$$\gamma_{14} = \frac{[(\lambda_2 - \lambda_1)L_2/(\eta_1\lambda_2 - \lambda_1 L_2)]}{[\lambda_2(L_2 - \eta_1)/(\eta_1\lambda_2 - \lambda_1 L_2)]} \quad \frac{[L_1(\eta_1 - \eta_2)/(\eta_1 L_1 - \lambda_1\eta_2)]}{[\eta_2(L_1 - \lambda_1)/(\lambda_1\eta_2 - \eta_1 L_1)]}$$

Upon simplifying, we obtain

$$\gamma_{14} = \frac{(\lambda_2 - \lambda_1)L_1}{\lambda_2(L_1 - \lambda_1)} \quad \frac{(\eta_2 - \eta_1)L_2}{\eta_2(L_2 - \eta_1)}$$

Identifying 0 with u, λ_1 with u_2, λ_2 with u_3, and L_1 with u_4, we have

$$\frac{(\lambda_2 - \lambda_1)L_1}{\lambda_2(L_1 - \lambda_1)} = \frac{(u_3 - u_2)(u_4 - u_1)}{(u_3 - u_1)(u_4 - u_2)}$$

$$= \gamma(u_2, u_3, u_4, u_1)$$

Using the permutation relationships yields

$$\gamma(u_2, u_3, u_4, u_1) = \frac{1}{\gamma(u_1, u_3, u_4, u_2)}$$

$$= \frac{1}{1 - \gamma(u_1, u_2, u_4, u_3)}$$

$$= \frac{1}{[1 + \gamma(u_1, u_2, u_3, u_4)]/[1 - \gamma(u_1, u_2, u_3, u_4)]}$$

$$= 1 - \gamma(u_1, u_2, u_3, u_4)$$

Therefore

$$\gamma_{14} = (1 - \gamma_{23})(1 - \gamma_{25})$$

To determine γ_{15} we need first to determine the point q of intersection between the line segments p_1p_5 and p_3p_4. We have already determined that

$$p_1 = \frac{1}{\eta_1\lambda_2 - \lambda_1 L_2} \left(\begin{array}{c} \eta_1 b_1 L_2(\lambda_2 - \lambda_1) \\ \eta_1 b_2 L_2(\lambda_2 - \lambda_1) + \lambda_1\lambda_2(L_2 - \eta_1) \end{array} \right)$$

$$p_5 = L_2 \left(\begin{array}{c} b_1 \\ b_2 \end{array} \right)$$

$$p_3 = \left(\begin{array}{c} 0 \\ -L_1 \end{array} \right)$$

$$p_4 = \frac{1}{\eta_1 L_1 - \lambda_1\eta_2} \left(\begin{array}{c} \eta_1\eta_2 b_1(L_1 - \lambda_1) \\ \eta_1\eta_2 b_2(L_1 - \lambda_1) + \lambda_1 L_1(\eta_2 - \eta_1) \end{array} \right)$$

The intersection point q is then given by

$$q = \frac{1}{\lambda_2\eta_2 - L_1 L_2} \left(\begin{array}{c} -b_1\eta_2 L_2(L_1 - \lambda_2) \\ -b_2\eta_2 L_2(L_1 - \lambda_2) + L_1\lambda_2(L_2 - \eta_2) \end{array} \right)$$

The four collinear points on line p_1p_5

$$p_1, \qquad \left(\begin{array}{c} 0 \\ -\lambda_2 \end{array} \right), \qquad q, \qquad p_5$$

determine the cross-ratio γ_{15}. Calculating the difference between each pair of successive points, we find

$$p_1 - \left(\begin{array}{c} 0 \\ -\lambda_2 \end{array} \right) = \frac{\eta_1(\lambda_2 - \lambda_1)}{\eta_1\lambda_2 - \lambda_1 L_2} \left(\begin{array}{c} b_1 L_2 \\ b_2 L_2 + \lambda_2 \end{array} \right)$$

$$\left(\begin{array}{c} 0 \\ -\lambda_2 \end{array} \right) - q = \frac{\eta_2(L_1 - \lambda_2)}{\lambda_2\eta_2 - L_1 L_2} \left(\begin{array}{c} b_1 L_2 \\ b_2 L_2 + \lambda_2 \end{array} \right)$$

$$q - p_5 = \frac{L_1(L_2 - \eta_2)}{\lambda_2\eta_2 - L_1 L_2} \left(\begin{array}{c} b_1 L_2 \\ b_2 L_2 + \lambda_2 \end{array} \right)$$

Letting

$$\delta_1 = \frac{\eta_1(\lambda_2 - \lambda_1)}{\eta_1\lambda_2 - \lambda_1 L_2}$$

$$\delta_2 = \frac{\eta_2(L_1 - \lambda_2)}{\lambda_2\eta_2 - L_1 L_2}$$

$$\delta_3 = \frac{L_1(L_2 - \eta_2)}{\lambda_2\eta_2 - L_1 L_2}$$

we have

$$\gamma_{15} = \frac{\delta_1\delta_3}{(\delta_1 + \delta_2)(\delta_2 + \delta_3)}$$

Since

$$\delta_1 + \delta_2 = \frac{\lambda_1\eta_2(L_1 - \lambda_2)(L_2 - \eta_1) + \eta_1 L_1(L_2 - \eta_2)(\lambda_2 - \lambda_1)}{(\eta_1\lambda_2 - \lambda_1 L_2)(\lambda_2\eta_2 - L_1 L_2)}$$

and $\delta_2 + \delta_3 = 1$, upon substituting and simplifying their results

$$\gamma_{15} = \frac{1}{1 + [\lambda_1(L_1 - \lambda_2)/L_1(\lambda_2 - \lambda_1)][\eta_2(L_2 - \eta_1)/\eta_1(L_2 - \eta_2)]}.$$

Identifying 0 with u_1, λ_1 with u_2, λ_2 with u_3, and L_1 with u_4, we obtain

$$\frac{\lambda_1(L_1 - \lambda_2)}{L_1(\lambda_2 - \lambda_1)} = \frac{(u_2 - u_1)(u_4 - u_3)}{(u_4 - u_1)(u_3 - u_2)}$$
$$= -\gamma(u_2, u_1, u_3, u_4)$$

Using the permutation relationships produces

$$\gamma(u_2, u_1, u_3, u_4) = \frac{-\gamma(u_1, u_2, u_3, u_4)}{1 - \gamma(u_1, u_2, u_3, u_4)}$$

Hence

$$\frac{\lambda_1(L_1 - \lambda_2)}{L_1(\lambda_2 - \lambda_1)} = \frac{\lambda_1(L_1 - \lambda_2)/\lambda_2(L_1 - \lambda_1)}{1 - [\lambda_1(L_1 - \lambda_2)/\lambda_2(L_1 - \lambda_1)]} = \frac{\gamma_{23}}{1 - \gamma_{23}}$$

so that

$$\gamma_{15} = \frac{1}{1 + [\gamma_{23}/(1 - \gamma_{23})][1/\gamma_{25}]}$$

By the symmetry of the situation, it is no surprise that an exactly similar calculation produces

$$\gamma_{34} = \frac{1}{1 + [\gamma_{25}/(1 - \gamma_{25})][1/\gamma_{23}]}$$

Other relationships are easily derived from the ones we have determined. For example, from the relationship $\gamma_{14} = (1 - \gamma_{23})(1 - \gamma_{25})$, it follows that

$$\frac{\gamma_{23}}{1 - \gamma_{23}} = \frac{1 - \gamma_{25}}{\gamma_{14}} - 1$$

Substituting this into the relation

$$\gamma_{15} = \frac{1}{1 + [\gamma_{23}/(1 - \gamma_{23})][1/(\gamma_{25})]}$$

and simplifying produces

$$\gamma_{15} = \left(\frac{\gamma_{14}}{1 - \gamma_{14}}\right)\left(\frac{\gamma_{25}}{1 - \gamma_{25}}\right)$$

From the relation $\gamma_{14} = (1 - \gamma_{23})(1 - \gamma_{25})$ it follows that

$$\frac{1}{\gamma_{25}} = \frac{1}{1 - [\gamma_{14}/(1 - \gamma_{23})]}$$

Substituting this into the relation

$$\gamma_{15} = \frac{1}{1 + [(\gamma_{23})/(1 - \gamma_{23})](1/\gamma_{25})}$$

and simplifying produces

$$\gamma_{23} = (1 - \gamma_{14})(1 - \gamma_{15})$$

Finally, from the relation

$$\gamma_{34} = \frac{1}{1 + [\gamma_{25}/(1 - \gamma_{25})](1/\gamma_{23})}$$

it follows that

$$\frac{1}{\gamma_{25}} = 1 + \frac{\gamma_{34}}{\gamma_{23}(1 - \gamma_{34})}$$

Substituting this into the relation

$$\gamma_{15} = \frac{1}{1 + [\gamma_{23}/(1 - \gamma_{23})](1/\gamma_{25})}$$

and simplifying produces

$$\gamma_{15} = (1 - \gamma_{23})(1 - \gamma_{34})$$

The topology of the point configuration does make a difference. For example, for the tower configuration shown in Figure 13.9, $\gamma_{23} = \gamma_{14}\gamma_{15}$. This relationship can easily be obtained from the known relation $\gamma_{23} = (1 - \gamma_{14})(1 - \gamma_{15})$ of the star configuration. Consider the line $p_1 p_4$. In the star configuration the order of the four collinear points is

$$p_1, \qquad \begin{pmatrix} 0 \\ -\lambda_1 \end{pmatrix}, \qquad \eta_1 \begin{pmatrix} b_1 \\ b_2 \end{pmatrix}, \qquad p_4$$

whereas on the tower configuration it is

$$p_1, \qquad p_4, \qquad \eta_1 \begin{pmatrix} b_1 \\ b_2 \end{pmatrix}, \qquad \begin{pmatrix} 0 \\ -\lambda_1 \end{pmatrix}$$

Notice that the tower configuration order can be obtained from the star configuration order by interchanging the second and fourth points. But from the ratio interchange relationship we have $\gamma(u_1, u_4, u_3, u_2) = 1 - \gamma(u_1, u_2, u_3, u_4)$. Hence in going from the star to the tower configuration, γ_{14} must change to $1 - \gamma_{14}$. Similarly, γ_{15} must change to $1 - \gamma_{15}$. Therefore the relation $\gamma_{23} = (1 - \gamma_{14})(1 - \gamma_{15})$ for the star configuration must transform to $\gamma_{23} = [1 - (1 - \gamma_{14})][1 - (1 - \gamma_{15})] = \gamma_{14}\gamma_{15}$ in

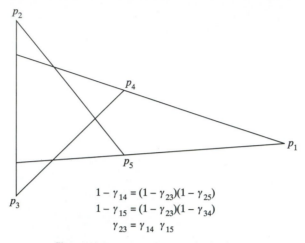

$$1 - \gamma_{14} = (1 - \gamma_{23})(1 - \gamma_{25})$$
$$1 - \gamma_{15} = (1 - \gamma_{23})(1 - \gamma_{34})$$
$$\gamma_{23} = \gamma_{14} \, \gamma_{15}$$

Figure 13.9 Tower topology for five points.

the tower configuration, and the relation $\gamma_{15} = (1 - \gamma_{23})(1 - \gamma_{34})$ for the star configuration must transform to $(1 - \gamma_{15}) = (1 - \gamma_{23})(1 - \gamma_{34})$ in the tower configuration.

■ EXAMPLE 13.10

Let

$$p_2 = \begin{pmatrix} 0 \\ 0 \end{pmatrix} \qquad p_3 = \begin{pmatrix} 0 \\ -6 \end{pmatrix} \qquad b = \frac{1}{5}\begin{pmatrix} 3 \\ -4 \end{pmatrix}$$

$$\lambda_1 = 1 \qquad\qquad \lambda_2 = 3 \qquad\qquad L_1 = 6$$

$$\eta_1 = 2 \qquad\qquad \eta_2 = 4 \qquad\qquad L_2 = 5$$

Then $p_5 = \begin{pmatrix} 3 \\ -4 \end{pmatrix}$. Point p_1 is determined by the intersection between the line
defined by $\begin{pmatrix} 0 \\ -1 \end{pmatrix}$ and $\frac{2}{5}\begin{pmatrix} 3 \\ -4 \end{pmatrix}$ and the line defined by $\begin{pmatrix} 0 \\ -3 \end{pmatrix}$ and $\begin{pmatrix} 3 \\ -4 \end{pmatrix}$.
Using the equation for the intersection of lines each defined by two points, we
obtain

$$p_1 = \begin{pmatrix} p_{11} \\ p_{12} \end{pmatrix} = \frac{1}{\begin{vmatrix} -6/5 & -3/5 \\ -3 & -1 \end{vmatrix}} \left(\begin{pmatrix} \begin{vmatrix} -6/5 & 6/5 \\ -3 & 9 \end{vmatrix} \\ \begin{vmatrix} 6/5 & -3/5 \\ 9 & -1 \end{vmatrix} \end{pmatrix} \right)$$

$$= \begin{pmatrix} 12 \\ -7 \end{pmatrix}$$

Equivalently, we can use the relations we have developed. From Eq. (13.32)

$$p_1 = \begin{pmatrix} p_{11} \\ p_{11} \end{pmatrix} = \frac{1}{\eta_1\lambda_2 - \lambda_1 L_2} \begin{pmatrix} \eta_1 b_1 L_2(\lambda_2 - \lambda_1) \\ \eta_1 b_2 L_2(\lambda_2 - \lambda_1) + \lambda_1\lambda_2(L_2 - \eta_1) \end{pmatrix}$$

The same answer must result. The point p_4 can be determined from Eq. (13.33):

$$p_4 = \begin{pmatrix} p_{41} \\ p_{42} \end{pmatrix} = \frac{1}{\eta_1 L_1 - \lambda_1\eta_2} \begin{pmatrix} \eta_1\eta_2 b_1(L_1 - \lambda_1) \\ \eta_1\eta_2 b_2(L_1 - \lambda_1) + \lambda_1 L_1(\eta_2 - \eta_1) \end{pmatrix}$$

$$= \begin{pmatrix} 3 \\ -2.5 \end{pmatrix}$$

The resulting geometry is shown in Fig. 13.9. The points $p_1, p_4, \eta_1 b$, and $\begin{pmatrix} 0 \\ -\lambda_1 \end{pmatrix}$ lie on the line segment whose endpoints are p_1 and $\begin{pmatrix} 0 \\ -\lambda_1 \end{pmatrix}$, and they divide the line into three segments. The segment lengths are

$$\|p_1 - p_4\| = \delta_1 = 10.062305$$
$$\|p_4 - \eta_1 b\| = \delta_2 = 2.012461$$
$$\|\eta_1 b - \begin{pmatrix} 0 \\ -\lambda_1 \end{pmatrix}\| = \delta_3 = 1.341641$$

from which

$$\gamma_{14} = \frac{\delta_1\delta_3}{(\delta_1 + \delta_2)(\delta_2 + \delta_3)}$$
$$= \frac{1}{3}$$

For the tower topology of this configuration, the relation for γ_{14} in terms of γ_{23} and γ_{25} is given by

$$\gamma_{14} = 1 - (1 - \gamma_{23})(1 - \gamma_{25})$$
$$= 1 - (1 - \frac{1}{5})(1 - \frac{1}{6})$$
$$= 1 - \frac{4}{5} \times \frac{5}{6} = \frac{1}{3}$$

13.7.4 Using Cross-Ratios

There are several ways to use cross-ratios to aid in establishing correspondences. Suppose that the 3D object of interest has five distinguished coplanar points, no three of which are collinear. Then regardless of the topological configuration of the points and the line segments defined by them, there will be five line segments that are cut by other line segments in two distinct places. Each such line segment has an associated given cross-ratio that will be invariant under perspective projection. Since two such cross-ratios determine the other three, we need only two cross-ratios. Suppose it is known that a specific two of these five line segments will show up in the perspective projection image. Then a matching algorithm could locate all line segments in the image and eliminate those whose associated cross-ratio is not the same as one of the two given cross-ratio values. Finally, among the remaining line segments, two that are part of the same five-point configuration and have the required cross-ratio values are selected for the match.

Another use of cross-ratio occurs in the following situation. Suppose that on a 3D object of interest there are N distinguishable coplanar points, at least four of which are noncollinear among themselves and no two of which are collinear with the other $N - 4$ points. Select any four such noncollinear points to constitute a perspective basis. These four points form a quadrilateral, with the points being its vertices. Taking the vertices around the quadrilateral in clockwise order, we name them p_1, p_2, p_3, and p_4. The diagonals of the quadrilateral are the line segments $p_1 p_3$ and $p_2 p_4$, and they intersect at some point z within the quadrilateral.

As shown in Fig. 13.10, the configuration of five points p_1, p_2, p_3, p_4, and z divide into four quadrants the plane on which all the points lie. One quadrant is determined by p_1, z, and p_4; the second by p_2, z, and p_1; the third by p_3, z, and p_2; and the fourth by p_4, z, and p_3. Let q be any of the other distinguished points. If q lies in the quadrant determined by p_2, z, and p_1, the line segment determined by $q p_4$ will intersect the diagonal $p_1 p_3$, and the line segment determined by $q p_3$ will intersect the diagonal $p_2 p_4$. Since q is noncollinear with every pair of $p_1 p_2 p_3 p_4$, q cannot lie along the line defined by either diagonal, and the intersection that the line defined by q makes with the diagonal $p_1 p_3$ will be distinct from p_1, z, and p_3. Similarly, the intersection that the line defined by $q p_3$ makes with the diagonal $p_2 p_4$ will be distinct from p_2, z, and p_4. In this way each diagonal will be partitioned into three segments from which a cross-ratio γ_{13} and γ_{24} can be calculated. We associate with the point q the perspective invariant coordinates $(\gamma_{13}, \gamma_{24})$.

In an analogous manner, if q lies in the quadrant determined by p_3, z, and p_2, the lines determined by $q p_1$ and $q p_4$ will intersect each diagonal in a distinct place. If q lies in the quadrant determined by p_4, z, and p_3, the lines determined by $q p_1$ and $q p_2$ will intersect each diagonal in a distinct plane. Finally, if q lies in the quadrant determined by p_1, z, and p_4, the lines determined by $q p_2$ and $q p_3$ will intersect each diagonal in a distinct plane. Thus no matter where q lies, by using the appropriate lines each diagonal will be partitioned into three segments from which the cross-ratio $\gamma_{13}(q)$ and $\gamma_{24}(q)$ can be calculated. So for each distinguished point q we are able to associate its perspective invariant coordinates $[\gamma_{13}(q), \gamma_{24}(q)]$.

Now suppose we observe a perspective projection of these N distinguished points and are able to match the 3D basic points p_1, p_2, p_3, and p_4 to the corresponding 2D-perspective projection points u_1, u_2, u_3, and u_4. Then the four points u_1, u_2, u_3, and u_4, in the same manner as p_1, p_2, p_3, and p_4, determine a quadrilateral with two diagonals. Any distinguished 3D point q has a corresponding 2D-perspective projection v. By constructing the appropriate lines between v and two of the vertices of the quadrilateral, we can, in the same manner we did for the 3D point q, associate a cross-ratio $[\gamma_{13}(v), \gamma_{24}(v)]$. By the invariance of the

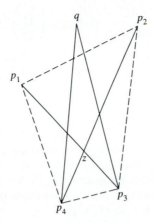

Figure 13.10 Two diagonals of a quadrilateral cut into three segments by two points of intersection on each diagonal. One point is the point of intersection of the two diagonals; the other is the point of intersection determined by the diagonal and the line determined by q and an opposite vertex.

cross-ratio to perspective projection, $[\gamma_{13}(q), \gamma_{24}(q)] = [\gamma_{13}(v), \gamma_{24}(v)]$. This means that after having matched the 3D basis points p_1, p_2, p_3, and p_4 to the corresponding 2D perspective points u_1, u_2, u_3, and u_4, we can use the known perspective invariant coordinates $[\gamma_{13}(q), \gamma_{24}(q)]$ for any 3D point q, and by working backward locate where the corresponding 2D perspective projection v for q is. First we determine the appropriate point on each diagonal on the perspective projection image that divides the diagonals into three segments with the given cross-ratios. Then we construct the two lines determined by the point on each diagonal and the opposite quadrilateral vertex. These two lines intersect, and the point of intersection is exactly the location of the perspective projection of the point q.

This same scheme not only can help in the establishment of correspondences between a 2D-perspective projection image and a set of observed 2D coplanar points, but in exactly the same manner can help in the establishment of correspondences between two 2D-perspective projection images of the same set of 3D coplanar points.

■ Exercises

13.1. Suppose that the distance f the image plane is in front of the camera lens is given by $f = 5$. Determine the perspective projection of the points

$$a. \quad \begin{pmatrix} 10 \\ 12 \\ 14 \end{pmatrix}$$

$$b. \quad \begin{pmatrix} 13 \\ 18 \\ 23 \end{pmatrix}$$

$$c. \quad \begin{pmatrix} 21 \\ 11 \\ 14 \end{pmatrix}$$

$$d. \quad \begin{pmatrix} 28 \\ 25 \\ 35 \end{pmatrix}$$

13.2.   Show that the projected heights of vertical line segments are proportional to their 3D heights when the optic axis of the camera is parallel to the ground plane. Suppose that the camera lens is at a height h_0 above the ground, the vertical height of an observed 3D vertical line segment is h above the ground, and the bottom of the 3D line segment touches the ground. Let V_E designate the y-coordinate of the eye level, V_H the y-coordinate of the top of the projected line segment, and V_G the y-coordinate of the bottom of the projected line segment. If the optic axis of the camera is parallel to the ground plane, show that

$$\frac{h_0}{h} = \frac{V_G - V_H}{V_G}$$

13.3.   Suppose that the distance f the image plane is in front of the camera lens is given by $f = 5$. Suppose that the vanishing point $\begin{pmatrix} u_\infty \\ v_\infty \end{pmatrix}$ for a set of parallel lines is given by

$$\begin{pmatrix} u_\infty \\ v_\infty \end{pmatrix} = \frac{5}{3} \begin{pmatrix} 1 \\ 2 \end{pmatrix}$$

Determine the direction cosines for the set of parallel lines.

13.4.   Show that the vanishing points for all 3D horizontal lines must lie on the line $v = 0$ of the perspective projection image.

13.5.   Show that the vanishing points for all 3D vertical lines must lie on the line $u = 0$ of the perspective projection image.

13.6.   Show that vanishing points arising from 3D lines, all of whose direction cosines are at a given angle θ with respect to the same vector, must lie along a line in the image plane.

13.7.   Suppose $\begin{pmatrix} u_1 \\ v_1 \end{pmatrix}$ and $\begin{pmatrix} u_2 \\ v_2 \end{pmatrix}$ are the perspective projections of two points on a line and that $\begin{pmatrix} u_3 \\ v_3 \end{pmatrix}$ and $\begin{pmatrix} u_4 \\ v_4 \end{pmatrix}$ are the perspective projections of two points on a line different from and parallel to the first line. Show that the vanishing point $\begin{pmatrix} u_\infty \\ v_\infty \end{pmatrix}$ of the parallel lines is determined by

$$u_\infty = \frac{\begin{vmatrix} u_1 v_2 - u_2 v_1 & u_1 - u_2 \\ u_3 v_4 - u_4 v_3 & u_3 - u_4 \end{vmatrix}}{\begin{vmatrix} v_2 - v_1 & u_1 - u_2 \\ v_4 - v_3 & u_3 - u_4 \end{vmatrix}}$$

$$v_\infty = \frac{\begin{vmatrix} v_2 - v_1 & u_1 v_2 - u_2 v_1 \\ v_4 - v_3 & u_3 v_4 - u_4 v_3 \end{vmatrix}}{\begin{vmatrix} v_2 - v_1 & u_1 - u_2 \\ v_4 - v_3 & u_3 - u_4 \end{vmatrix}}$$

13.8. Suppose that $\frac{1}{7}\begin{pmatrix} 25 \\ 30 \end{pmatrix}$ and $\frac{1}{23}\begin{pmatrix} 65 \\ 90 \end{pmatrix}$ are the perspective projections of two points

on a line and that $\frac{1}{14}\begin{pmatrix} 105 \\ 55 \end{pmatrix}$ and $\frac{1}{7}\begin{pmatrix} 28 \\ 25 \end{pmatrix}$ are the perspective projections of two

points on a different but parallel line. Show that the vanishing point $\begin{pmatrix} u_\infty \\ v_\infty \end{pmatrix}$ for

these two parallel lines is given by

$$\begin{pmatrix} u_\infty \\ v_\infty \end{pmatrix} = \frac{1}{3}\begin{pmatrix} 5 \\ 10 \end{pmatrix}$$

13.9. Suppose that a 3D line L is represented by

$$L = \left\{ \begin{pmatrix} x \\ y \\ z \end{pmatrix} \middle| \begin{pmatrix} x \\ y \\ z \end{pmatrix} = \begin{pmatrix} a_1 \\ a_2 \\ a_3 \end{pmatrix} + \lambda \begin{pmatrix} b_1 \\ b_2 \\ b_3 \end{pmatrix} \right\}$$

where $b_1^2 + b_2^2 + b_3^2 = 1$. Let the perspective projection of this line be represented by

$$M = \left\{ \begin{pmatrix} u \\ v \end{pmatrix} \middle| \begin{pmatrix} u \\ v \end{pmatrix} = \begin{pmatrix} c_1 \\ c_2 \end{pmatrix} + \eta \begin{pmatrix} d_1 \\ d_2 \end{pmatrix} \right\}$$

where $d_1^2 + d_2^2 + d_3^2 = 1$. Show that one way to determine $\begin{pmatrix} c_1 \\ c_2 \end{pmatrix}$ and $\begin{pmatrix} d_1 \\ d_2 \end{pmatrix}$ from

$\begin{pmatrix} a_1 \\ a_2 \\ a_3 \end{pmatrix}$ and $\begin{pmatrix} b_1 \\ b_2 \\ b_3 \end{pmatrix}$ is given by

$$\begin{pmatrix} c_1 \\ c_2 \end{pmatrix} = f \begin{pmatrix} a_1/a_3 \\ a_2/a_3 \end{pmatrix}$$

$$\begin{pmatrix} d_1 \\ d_2 \end{pmatrix} = \begin{pmatrix} a_3 b_1 - a_1 b_3 \\ a_3 b_2 - a_2 b_3 \end{pmatrix} \frac{1}{\sqrt{(a_3 b_1 - a_1 b_3)^2 + (a_3 b_2 - a_2 b_3)^2}}$$

13.10. Suppose there are three 3D lines whose unknown direction cosines are $(a_x, a_y, a_z), (b_x, b_y, b_z)$, and (c_x, c_y, c_z). The observed vanishing point for the first line is (u_1, v_1) on a perspective image where f is the camera constant. Suppose that an observed line $\alpha u + \beta v + \gamma$ on the perspective projection image is known to pass through the vanishing point (u_2, v_2) corresponding to 3D lines having unknown direction cosines (b_x, b_y, b_z). Show that if the cosines of the angle between each pair of 3D lines is known—that is, if

$$\lambda_{12} = a_x b_x + a_y b_y + a_z b_z$$
$$\lambda_{13} = a_x c_x + a_y c_y + a_z c_z$$
$$\lambda_{23} = b_x c_x + b_y c_y + b_z c_z$$

are known—then it is possible to recover the unknown direction cosines (b_x, b_y, b_z), and (c_x, c_y, c_z), as well as the unknown vanishing points (u_2, v_2) and (u_3, v_3).

13.11. Consider the star configuration of Fig. 13.8. Show that

a. $$\gamma_{23} = 1 - \frac{\gamma_{14}}{1 - \gamma_{25}}$$

b. $$\gamma_{15} = \frac{\gamma_{14}}{1 - \gamma_{14}} \frac{\gamma_{25}}{1 - \gamma_{25}}$$

c. $$\gamma_{34} = 1 - \frac{\gamma_{25}}{1 - \gamma_{14}}$$

d. $$\gamma_{14} = \frac{\gamma_{15}(1 - \gamma_{25})}{\gamma_{25} + \gamma_{15}(1 - \gamma_{25})}$$

e. $$\gamma_{34} = (1 - \gamma_{25})(1 - \gamma_{15})$$

f. $$\gamma_{23} = \frac{\gamma_{25}(1 - \gamma_{15})}{\gamma_{15} + \gamma_{25}(1 - \gamma_{15})}$$

g. $$\gamma_{23} = \frac{\gamma_{34}\gamma_{25}}{(1 - \gamma_{34})(1 - \gamma_{25})}$$

h. $$\gamma_{15} = 1 - \frac{\gamma_{34}}{1 - \gamma_{25}}$$

i. $$\gamma_{14} = 1 - \frac{\gamma_{25}}{1 - \gamma_{34}}$$

13.12. Use the cross-ratio permutation relationships to show that

a. $$\gamma(u_2, u_3, u_1, u_4) = \frac{\gamma(u_1, u_2, u_3, u_4) - 1}{\gamma(u_1, u_2, u_3, u_4)}$$

b. $$\gamma(u_2, u_1, u_4, u_3) = \gamma(u_1, u_2, u_3, u_4)$$

c. $$\gamma(u_2, u_4, u_3, u_1) = \frac{1}{(u_1, u_2, u_3, u_4)}$$

13.13. Determine all the permutations (i_1, i_2, i_3, i_4) of 1, 2, 3, 4 for which

a. $$\gamma(u_1, u_2, u_3, u_4) = \gamma(u_{i1}, u_{i2}, u_{i3}, u_{i4})$$

b. $$\gamma(u_2, u_3, u_1, u_4) = \gamma(u_{i1}, u_{i2}, u_{i3}, u_{i4})$$

c. $$\gamma(u_2, u_4, u_3, u_1) = \gamma(u_{i1}, u_{i2}, u_{i3}, u_{i4})$$

d. $$\gamma(u_1, u_2, u_4, u_3) = \gamma(u_{i1}, u_{i2}, u_{i3}, u_{i4})$$

e. $$\gamma(u_1, u_4, u_3, u_2) = \gamma(u_{i1}, u_{i2}, u_{i3}, u_{i4})$$

f. $$\gamma(u_4, u_2, u_3, u_1) = \gamma(u_{i1}, u_{i2}, u_{i3}, u_{i4})$$

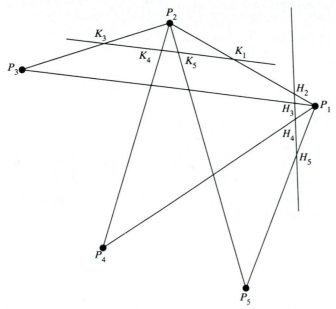

Figure 13.11 Geometric relation of five points in a plane for Exercise 13.14.

13.14. Consider the geometry of the points in a plane in Fig. 13.11. Suppose that the (x, y) position of points P_1, P_2, P_3, and P_4 are known and the (x, y) position of point P_5 is not known. Determine the position of P_5, given the values of the cross-ratios $\gamma(H_2, H_3, H_4, H_5)$ and $\gamma(K_3, K_4, K_5, K_1)$.

13.15. Let the perspective projection (u, v) of a point (x, y, z) be given by

$$u = f\frac{x - x_0}{z - z_0} \qquad v = f\frac{y - y_0}{z - z_0}$$

Suppose that the point (x, y, z) is known to lie in the plane given by $ax + by + cz = 0$
Show that

$$x = \frac{-(by_0 + cz_0 + d)u + bx_0 v + cfx_0}{au + bvf + cf}$$

$$y = \frac{ay_0 u - (ax_0 + cz_0 + d)v + cfy_0}{au + bv + cf}$$

13.16. Suppose that a planar conic is represented as the intersection of a plane $ax + by + cz + d = 0$ with a conic cylinder $Ax^2 + 2Bxy + Cy^2 + 2Dx + 2Ey + F = 0$. Let (u, v) be the perspective projection of a point (x, y, z) on the planar conic. Substitute the relation of the previous problem, expressing x and y in terms of u and v into the conic cylinder equation to show that the perspective projection of a conic curve in 3D space results in a conic in the 2D-perspective projection.

13.17. Suppose a conic $Au^2 + 2Buv + C^2 + 2Du + 2Ev + F = 0$ is observed on the perspective projection image plane. The 3D curve of which this is the projection must lie in the 3D cone

$$\left\{ \begin{pmatrix} x \\ y \\ z \end{pmatrix} \middle| \text{ for some } \lambda, u, v \begin{pmatrix} x \\ y \\ z \end{pmatrix} = \lambda \begin{pmatrix} u \\ v \\ f \end{pmatrix} \right.$$

$$\left. \text{where } Au^2 + 2Buv + C^2 + 2Du + 2Ev + F = 0 \right\}$$

Suppose the 3D curve is known to lie in a plane

$$\left\{ \begin{pmatrix} x \\ y \\ z \end{pmatrix} \middle| ax + by + cz + d = 0 \right\}$$

Intersect the cone and plane and show that the 3D curve must be the intersection of the plane with the conic cylinder

$$\left\{ \begin{pmatrix} x \\ y \\ z \end{pmatrix} \middle| \left(A - \frac{2aD}{cf} + \frac{Fa^2}{c^2f^2} \right) x^2 + \left(2B - \frac{2Db}{cf} - \frac{2Ea}{cf} + \frac{2Fab}{c^2f^2} \right) xy + \right.$$

$$\left. \left(C - \frac{2Eb}{cf} + \frac{Fb^2}{c^2f^2} \right) y^2 + \left(\frac{-2Dd}{cf} + \frac{2ad}{c^2f^2} \right) x + \left(-\frac{2Ed}{cf} + \frac{2bdF}{c^2f^2} \right) + \frac{Fd^2}{c^2f^2} = 0 \right\}$$

13.18. Suppose that a circle is known to lie in a plane parallel to the image projection. Show that the perspective projection of such a circle is a circle.

13.19. Suppose that a circle of known radius d_0 lies in an unknown plane $z = \delta$ parallel to the image projection plane. Suppose that the observed circle on the image projection plane satisfies

$$Cu^2 + Cv^2 + 2fDfu + 2fEv + f^2F = 0$$

where f is the focal length of the camera. Show that

$$\delta = d_0 \sqrt{\frac{C^2}{D^2 + E^2 - FC}}$$

and that the center of the circle in 3D space is given by

$$\begin{pmatrix} x_0 \\ y_0 \\ z_0 \end{pmatrix} = \frac{\delta}{C} \begin{pmatrix} D \\ E \\ C \end{pmatrix}$$

▪ Bibliography

Abidi, M. A., and R. O. Eason, "Camera Calibration in Robot Vision," *Proceedings of the Fourth Scandinavian Conference on Image Analysis,*, Trondheim, Norway, 1985, pp. 471–478.

Agin, G., and T. Binford, "Computer Description of Curved Objects," *Proceedings of the Third International Joint Conference on Artificial Intelligence,* Stanford, CA, 1973, pp. 629–640.

Altschuler, B. R., J. Tabaoda, and M. D. Altschuler, "Laser Electro-Optic System for Three-Dimensional Topographic Mensuration," *Imaging Applications for Automated Industrial Inspection and Assembly,* Vol. 182, 1979, pp. 192–196.

Altschuler, M. D., B. R. Altschuler, and J. Tabaoda, "Measuring Surfaces Space-Coded by a Laser Projected Dot Matrix," *Imaging Applications for Automated Industrial Inspection and Assembly,* Vol. 182, 1979, pp. 187–191.

Alvertos, N., E. L., Hall, and R. L. Anderson, "Omnidirectional Viewing for Robot Vision," *Proceedings of the SPIE Third International Conference on Robot Vision and Sensory Controls,* Cambridge, MA, 1983, pp. 230–239.

Asada, M., H. Ichikawa, and S. Tsuji, "Determining Surface Orientation by Projecting a Stripe Pattern," *IEEE Transactions on Pattern Analysis and Machine Intelligence,* Vol. PAMI-10, 1988, pp. 749–754.

Barnard, S. T., and M. A. Fischler, "Computational Stereo," *Computing Surveys*, Vol. 14, 1982, pp. 553–572.

Barry, M., et al., "A Multi-Level Geometric Reasoning System for Vision," *Artificial Intelligence,* Vol. 37, 1988, pp. 291–332.

Blais, J. A., "Three-Dimensional Similarity," *Canadian Surveyor*, Vol. 1, 1972, pp. 71–76.

Bopp, H., and H. Krauss, "Ein Orientierungs- und Kalibrierungsverfahren für nichttopographische Anwendungen der Photogrammetrie," *ANV*, Vol. 5, 1978, pp. 182–188.

Boyer, K. L., and A. C. Kak, "Color-Encoded Structured Light for Rapid Active Ranging," *IEEE Transactions on Pattern Analysis and Machine Intelligence,* Vol. PAMI-9, 1987, pp. 14–28.

Brill, M., and E. B. Barrett, "Closed-Form Extension of the Anharmonic Ratio to N-Space," *Computer Vision, Graphics, and Image Processing*, Vol. 23, 1983, pp. 92–98.

Case, S. K., J. A. Jalkio, and R. Kim, "Specification and Design of Custom 3D Inspection Systems," *Proceedings of Vision '86,* Detroit, 1986, pp. 31–60.

Cheng, S. K., Y. Y. Hung, and N. K. Loh, "A Computer Vision Technique for Surface Curvature Gaging with Projected Grating," *Intelligent Robots and Computer Vision,* Vol. 521, 1984, pp. 331–336.

Chien, C. H., and J. K. Aggarwal, "Model Construction and Shape Recognition from Occluding Contours," *IEEE Transactions on Pattern Analysis and Machine Intelligence,* Vol. 11, 1989, pp. 372–389.

Chou, H. L., and W. H. Tsai, "A New Approach to Robot Location by House Corners," *Pattern Recognition*, Vol. 19, 1986, pp. 439–451.

Courtney, J. W., M. J., Magee, and J. K. Aggarwal, "Robot Guidance Using Computer Vision," *Pattern Recognition*, Vol. 17, 1984, pp. 585–592.

Crombie, M. A., and W. A. Baracat, "Applying Photogrammetry to Real Time Collection of Digital Image Data," *U.S. Army Corps of Engineers ETL-0275*, Fort Belvoir, VA, 1981, pp. 4–38.

Draper, S. W., "The Use of Gradient and Dual Space in Line Drawing Interpretation," *Artificial Intelligence*, Vol. 17, 1981, pp. 461–508.

Duda, R. O., D. Nitzan, and P. Barret, "Use of Range and Reflectance Data to Find Planar Surface Regions," *IEEE Transactions on Pattern Analysis and Machine Intelligence,* Vol. PAMI-1, 1979, pp. 259–271.

El-Hakim, S. F., "A Photogrammetric Vision System for Robots," *Photogrammetric Engineering and Remote Sensing*, Vol. 51, 1985, pp. 545–552.

Fang, T. J., et al., "Three-Dimensional Object Recognition Using a Transformation Clustering Technique," *Proceedings of the Sixth International Conference on Pattern Recognition,* Munich, 1982, pp. 678–681.

Faugeras, O. D., and F. Lustman, "Inferring Planes by Hypothesis Prediction and Testing for a Mobile Robot," in *Advances in Image Processing and Pattern Recognition,* V. Cappellini and R. Marconi (eds.), Elsevier, North-Holland, 1986, pp. 211–223.

Finsterwalder, S., "Die geometrischen Grundlagen der Photogrammetrie," *Annual Report of the Mathematics Association,* Vol. 6, 1897, pp. 26 ff.

Finsterwalder, S., and W. Scheufele, "Das Rückwärtseinschneiden im Raum," *Class Report from the Mathematical Physics Class,* Bavarian Academy of Science, Munich, Vol. 33, 1903, pp. 591–614; and also in *Sebastian Finsterwalder zum 75 Geburtstage,* Die Deutsche Gesellschaft für Photogrammetrie, Verlag Herbert Wichmann, Berlin, 1937, pp. 86–100.

Fischler, M. A., and R. C. Bolles, "Random Sample Consensus: A Paradigm for Model Fitting with Applications to Image Analysis and Automated Cartography," *Communications of the ACM,* Vol. 24, 1981, pp. 381–395.

Fisher, R. B., "Geometric Constraints from Planar Surface Patch Matching," *Image and Vision Computing,* Vol. 8, 1990, pp. 148–154.

Förstner, W., "The Reliability of Block Triangulation," *Photogrammetric Engineering and Remote Sensing,* Vol. 51, 1985, pp. 1137–49.

Ganapathy, S., "Decomposition of Transformation Matrices for Robot Vision," *IEEE Conference on Robotics,* Atlanta, 1984, pp. 130–139.

Gil, B., A. Mitiche, and J. K. Aggarwal, "Experiments in Combining Intensity and Range Edge Maps," *Computer Vision, Graphics, and Image Processing,* Vol. 21, 1983, pp. 395–411.

Gordon, S. J., and W. P. Seering, "Real-Time Part Position Sensing," *IEEE Transactions on Pattern Analysis and Machine Intelligence,* Vol. 10, 1988, pp. 374–386.

Grafarend, E. W., P. Lohse, and B. Schaffrin, "Dreidimensionaler Rückwärtsschnitt," *Zeitschrift für Vermessungswesen,* Vol. 114, 1989, pp. 61–67, 127–137, 172–175, 225–234, and 278–287.

Grunert, J. A., "Das Pothenotische Problem in erweiterter Gestalt nebst über seine Anwendungen in der Geodäsie, *Grunerts Archiv für Mathematik und Physik,* Band 1, 1841, pp. 238–248.

Halioua, M., and H. C. Liu, "Optical Sensing Techniques for 3D Machine Vision," *Optical Techniques for Industrial Inspection,* Vol. 665, 1986, pp. 150–161.

Haralick, R. M., "Using Perspective Transformations in Scene Analysis," *Computer Graphics and Image Processing*, Vol. 13, 1980, pp. 191–221.

——, "Determining Camera Parameters from the Perspective Projection of a Rectangle," *Pattern Recognition,* Vol. 22, 1988, pp. 225–230.

Haralick, R. M., and Y. H. Chu, "Solving Camera Parameters from the Perspective Projection of a Parameterized Curve," *Pattern Recognition*, Vol. 17, 1984, pp. 637–645.

Hogg, R. V., *An Introduction to Robust Estimation, Robustness in Statistics*, R. L. Launer and G. N. Wilkinson (eds.), Academic Press, San Diego, 1979.

Horaud, R., "New Methods for Matching 3-D Objects with Single Perspective Views," *IEEE Transactions on Pattern Analysis and Machine Intelligence*, Vol. PAMI-9, 1987, pp. 401–412.

Hu, G., and G. Stockman, "3D Surface Solution Using Structured Light and Constraint Propagation," *IEEE Transactions on Pattern Analysis and Machine Intelligence,* Vol. 11, 1989, pp. 390–402.

Huber, P. J., *Robust Statistics*, Wiley, New York, 1981.

Hung, K. C., J. Y., Lee, and Y. K. Fang, "Solving Camera Parameters from Straight Line Coefficients with the Application of Homotopy Method: A New Approach,"

National Cheng Kung University, Department of Electrical Engineering, Taiwan, 1986, pp. 1–28.

Hung, Y., P. S. Yeh, and D. Harwood, "Passive Ranging to Known Planar Point Sets," *Proceedings of the International Conference on Robotics and Automation,* St. Louis, MO, 1985, pp. 80–85.

Isaguirre, A., P. Pu, and J. Summers, "A New Development in Camera Calibration: Calibrating a Pair of Mobile Cameras," *Proceedings of the International Conference on Robotics and Automation,* St. Louis, MO, 1985, pp. 74–79.

Jarvis, R. A., "A Perspective on Range Finding Techniques for Computer Vision," *IEEE Transactions on Pattern Analysis and Machine Intelligence,* Vol. PAMI-5, 1983, pp. 122–139.

Kabuka, M., and E. S. McVey, "A Position Sensing Method Using Images," *Proceedings of the 14th Southeastern Symposium on System Theory,* Blacksburg, VA, 1982, pp. 191–194.

Kamata, S., S. Ishikawa, and K. Kato, "Reconstructing an Edge on a Polyhedron Using an Optimization Method," *Computer Vision, Graphics, and Image Processing,* Vol. 47, 1989, pp. 92–104.

Kanatani, K., "3D Euclidean or 2D Non-Euclidean? Methodology of Image Understanding," Gunma University, Department of Computer Science, Gunma, Japan, 1986, pp. 1–16.

———, "Constraints on Length and Angle," *Computer Vision, Graphics, and Image Processing,* Vol. 41, 1988, pp. 28–42.

Kapur, D., and J. Mundy, "Wu's Method and Its Application to Perspective Viewing," *Artificial Intelligence,* Vol. 37, 1988, pp. 15–36.

Kapur, D., et al., "Reasoning about Three Dimensional Space," Proceedings of the International Conference on Robotics and Automation, St. Louis, MO, 1985, pp. 405–410.

Kitahashi, T., and E. Hiroyuki, "A New Method of 3D Motion Analysis Using a Fundamental Concept of Projective Geometry," Department of Information and Computer Sciences, Toyohashi University of Technology, Toyohashi, Japan, pp. 1–8.

Lee, H. C., and K. S. Fu, "Generating Object Descriptions for Model Retrieval," *IEEE Transactions on Pattern Analysis and Machine Intelligence,* Vol. PAMI-5, 1983, pp. 462–471.

Lee, H. J., and Z. Chen, "Determination of 3D Human Body Postures from a Single View," *Computer Vision, Graphics, and Image Processing,* Vol. 30, 1985, pp. 148–168.

Lin, W. C., K. S. Fu, and T. Sederberg, "Estimation of Three-Dimensional Object Orientation for Computer Vision Systems with Feedback," *Journal of Robotic Systems,* Vol. 1, 1984, pp. 59–82.

Linnainmaa, S., D. Harwood, and L. S. Davis, "Pose Estimation of a Three-Dimensional Object Using Triangle Pairs," *IEEE Transactions on Pattern Analysis and Machine Intelligence,* Vol. 10, 1988, pp. 634–647.

Luh, J. Y. S., and J. A. Klaasen, "A Real-Time 3D Multi-Camera Vision System," *Proceedings of the SPIE International Conference on Robotic Vision and Sensory Controls,* Cambridge, MA, 1983, pp. 400–408.

Magee, M. J., and J. K. Aggarwal, "Determining Vanishing Points from Perspective Images," *Computer Vision, Graphics, and Image Processing,* Vol. 26, 1984, pp. 256–267.

———, "Determining the Position of a Robot Using a Single Calibration Object," *Proceedings of the International Conference on Robotics and Automation,* St. Louis, MO, 1985, pp. 140–149.

Mansbach, P., "Calibration of a Camera and Light Source by Fitting to a Physical Model," *Computer Vision, Graphics, and Image Processing*, Vol. 35, 1986, pp. 200–219.

Markowsky, G., and M. A. Wesley, "Fleshing out Wire Frames," *IBM Journal of Research and Development*, Vol. 24, 1980, pp. 582–597.

Martins, H. A., J. R. Birk, and R. B. Kelley, "Camera Models Based on Data from Two Calibration Planes," *Computer Vision, Graphics, and Image Processing*, Vol. 17, 1981, pp. 173–180.

Mayhew, J. E. W., and H. C. Longuet-Higgins, "A Computational Model of Binocular Depth Perception," *Nature*, Vol. 297, 1982, pp. 376–379.

McVey, E., and J. W. Lee, "Some Accuracy and Resolution Aspects of Computer Vision Distance Measurements," *IEEE Transactions on Pattern Analysis and Machine Intelligence*, Vol. PAMI-4, 1982, pp. 646–649.

Merritt, E. L., "Explicitly Three-Point Resection in Space," *Photogrammetric Engineering*, Vol. 15, 1949, pp. 649–655.

——, "General Explicit Equations for a Single Photograph," *Analytical Photogrammetry*, Pitman, New York, 1958, pp. 43–79.

Mitiche, A., and G. Habelrih, "Interpretation of Straight Line Correspondences Using Angular Relations," *Pattern Recognition*, Vol. 22, 1989, pp. 299–308.

Mitiche, A., S. Seida, and J. K. Aggarwal, "Determining Position and Displacement in Space from Images," *Proceedings of the IEEE Computer Society Conference on Computer Vision and Pattern Recognition*, San Francisco, 1985, pp. 1–6.

Moon, C. W., and E. S. McVey, "Computer Vision Distance Measurement Error Analysis for Imperfect Camera Alignment," *Proceedings of the 14th Southeastern Symposium on System Theory*, 1982, pp. 187–190.

Morita, H., K. Yajima, and S. Sakata, "Reconstruction of Surfaces of 3D Objects by M-array Pattern Projection Method," *Proceedings of the Second International Conference on Pattern Recognition*, Tampa, FL, 1988, pp. 468–473.

Müller, F. J., "Direkte (exakte) Lösung des einfachen Rückwärtseinschneidens im Raume," *Algemeine Vermessungs-Nachrichten*, 1925.

Naeve, A., and J. O. Eklundh, "Projective Geometry and the Recovery of 3D Structure," Laboratory of Computer Vision and Associative Pattern Processing, Stockholm, pp. 1–31.

Nakatani, H., et al., "Extraction of Vanishing Point and Its Application to Scene Analysis on Image Sequence," *Proceedings of the Fifth International Conference on Pattern Recognition*, Miami Beach, FL, 1980, pp. 370–372.

Nitzan, D., A. E. Brain, and R. O. Duda, "The Measurement and Use of Registered Reflectance and Range Data in Scene Analysis," *Proceedings of the IEEE*, Vol. 65, 1977, pp. 206–220.

Parrish, E. A., and A. K. Goksel, "A Camera Model for Natural Scene Processing," *Pattern Recognition*, Vol. 9, 1987, pp. 131–136.

Pennington, K. S., and P. M. Will, "A Grid-Coded Technique for Recording 3-Dimensional Scenes Illuminated with Ambient Light," *Optics Communications*, Vol. 2, 1970, pp. 167–169.

Pennington, K. S., P. M. Will, and G. L. Shelton, "Grid-Coding: A Technique for Extraction of Differences from Scenes," *Optics Communications*, Vol. 2, 1970, pp. 113–119.

Pope, J. A., "An Advantageous, Alternative Parameterization of Rotations for Analytical Photogrammetry," ESSA Technical Report, C and GS 39.

Posdamer, J. L., and M. D. Altschuler, "Surface Measurement by Space-Encoded Projected Beam Systems," *Computer Graphics and Image Processing*, Vol. 18, 1982, pp. 1–17.

Potmesil, M., "Generation of 3D Surface Descriptions from Images of Pattern-Illuminated Objects," *Proceedings of the IEEE Conference on Pattern Recognition and Image Processing,* Chicago, 1979, pp. 553–559.

Preiss, K., "Algorithms for Automatic Conversion of a 3-View Drawing of a Plane-Faced Part to the 3-D Representation," *Computers in Industry,* Vol. 2, 1981, pp. 133–139.

Roberts, L. G., "Machine Perception of Three-Dimensional Solids," in *Optical and Electro-Optical Information Processing,* J.T. Tippett et al. (eds.), M.I.T. Press, Cambridge, MA, 1965, pp. 159–197.

Sansò, F., "An Exact Solution of the Roto-Translation Problem," *Photogrammetria,* Vol. 29, 1973, pp. 203–216.

Sato, Y., H. Kitagawa, and H. Fujita, "Shape Measurement of Curved Objects Using Multiple Slit-Ray Projections," *IEEE Transactions on Pattern Analysis and Machine Intelligence,* Vol. PAMI-4, 1982, pp. 641–646.

Schmidt, R. Q., "Inspection and Adaptive Robot Applications Based on Three-Dimensional Vision Measurements," *SPIE Intelligent Robots and Computer Vision,* Vol. 521, 1984, pp. 346–351.

Schut, G. H. "On Exact Linear Equations for the Computation of the Rotational Elements of Absolute Orientation," *Photogrammetria,* Vol. 15, 1960, pp. 34–37.

Shafer, S. A. and T. Kanade, "Using Shadows in Finding Surface Orientations," Department of Computer Science, Carnegie-Mellon University, Pittsburgh, 1982, pp. 1–61.

——, "Gradient Space under Orthography and Perspective," *Computer Vision, Graphics, and Image Processing,* Vol. 24, 1983, pp. 182–199.

Shakunaga, T., and H. Kaneko, "Shape from Angles under Perspective Projection," *Proceedings of the Second IEEE International Conference on Computer Vision,* Tampa, FL, 1988, pp. 671–678.

——, "Perspective Angle Transform: Principle of Shape from Angles," *International Journal of Computer Vision,* Vol. 3, 1989, pp. 239–254.

Shirai, Y., "Recognition of Polyhedrons with a Range Finder," *Pattern Recognition,* Vol. 4, 1972, pp. 243–250.

Shrikhande, N., and G. Stockman, "Surface Orientation from a Projected Grid," *IEEE Transactions on Pattern Analysis and Machine Intelligence,* Vol. 11, 1989, pp. 650–654.

Silberberg, T., D. Harwood, and L. Davis, "Object Recognition Using Oriented Model Points, Center for Automation Research, University of Maryland, College Park, pp. 1–34.

Sobel, I., "On Calibrating Computer Controlled Cameras for Perceiving 3D Scenes," *Artificial Intelligence,* Vol. 5, 1974, pp. 185–198.

Stockman, G. C., "Use of Geometrical Constraints and Clustering to Determine 3D Object Pose," *Proceedings of the Seventh International Conference on Pattern Recognition,* Montreal, 1984, pp. 742–744.

——, "Three-Dimensional Pose Computations from Multiple Views," *Pattern Recognition in Practice II,* Elsevier, North-Holland, 1985.

Stockman, G. C., and J. C. Esteva, "3D Object Pose from Clustering with Multiple Views," *Pattern Recognition Letters 3,* Elsevier, North-Holland, 1985, pp. 279–286.

Stockman, G. C., et al., "Sensing and Recognition of Rigid Objects Using Structured Light," *IEEE Control Systems,* Vol. 8, 1988, pp. 14–22.

Szczepanski, W., "Die Lösungsvorschläge für den räumlichen Rückwärts-einschnitt," Deutsche Geodätische Kommission, Reihe C: Dissertationen–Heft Nr. 29, 1958, pp. 1–144.

Taboada, J., and B. R. Altschuler, "Rectangular Grid Fringe Pattern for Topographic Applications," *Applied Optics,* Vol. 15, 1976, pp. 597–599.

Thompson, E. H., "On Exact Linear Solution of the Problem of Absolute Orientation," *Photogrammetria,* Vol. 13, 1958, pp. 163–178.

Thompson, M. (ed.), *Manual of Photogrammetry,* 3rd Edition, Falls Church, VA, American Society of Photogrammetry, 1966, pp. 77–101.

Thorpe, C., and S. Shafer, "Topological Correspondence in Line Drawings of Multiple Views of Objects," Department of Computer Science, Carnegie-Mellon University, Pittsburgh, 1983, pp. 1–24.

Tienstra, J. M., "Calculation of Orthogonal Matrices," I.T.C. Delft Series, A 48.

Tsai, R. Y., "A Versatile Camera Calibration Technique for High-Accuracy 3D Machine Vision Metrology Using Off-the-Shelf TV Cameras and Lenses," *IEEE Journal of Robotics and Automation,* Vol. RA-3, 1987, pp. 323–344.

Tsui, H. T., "3D Orientation Estimation of Machine Parts for Automatic Inspection," *Proceedings of the Fourth Scandinavian Conference on Image Analysis,* Trondheim, Norway, 1985, pp. 455–461.

Walker, E. L., and M. Herman, "Geometric Reasoning for Constructing 3D Scene Descriptions from Images," *Artificial Intelligence,* Vol. 37, 1988, pp. 275–290.

Wang, Y. F., A. Mitiche, and J. K. Aggarwal, "Computation of Surface Orientation and Structure of Objects Using Grid Coding," *IEEE Transactions on Pattern Analysis and Machine Intelligence,* Vol. PAMI-9, 1987, pp. 129–137.

Ward, M. R., D. P. Rheaume, and S. W. Holland, "Production Plant CONSIGHT Installations, *Proceedings of the SPIE 26th Annual Technical Symposium on Robotics and Industrial Inspection,* San Diego, Vol. 360, 1982, pp. 297–305.

Watson, L. T., and L. G. Shapiro, "Identification of Space Curves from Two-Dimensional Perspective Views," *IEEE Transactions on Pattern Analysis and Machine Intelligence,* Vol. PAMI-4, 1982, pp. 469–475.

Will, P. M., and K. S. Pennington, "Grid Coding: A Preprocessing Technique for Robot and Machine Vision," *Proceedings of the Second International Joint Conference on Artificial Intelligence,* Imperial College, London, 1971, pp. 66–70.

——— , "Grid Coding: A Novel Technique for Image Processing," *Proceedings of the IEEE,* Vol. 60, 1972, pp. 669–680.

Wrobel, B., and D. Klemm, "Über der Berechnung allgemeiner räumlicher Drehungen," *International Archives of Photogrammetry and Remote Sensing,* Vol. 25, pp. 1153–1163.

Wu, C. K., D. Q. Wang, and R. K. Bajcsy, "Acquiring 3D Spatial Data of a Real Object," *Computer Vision, Graphics, and Image Processing,* Vol. 28, 1984, pp. 126–133.

Yakimovsky, Y., and R. Cunningham, "A System for Extracting 3-Dimensional Measurements from a Stereo Pair of TV Cameras," National Aeronautics and Space Administration, Technical Memorandum 33-769, California Institute of Technology, Pasadena, 1976, pp. 1–15.

Yang, H. S., and A. C. Kak, "Edge Extraction and Labelling from Structured Light 3D Vision Data," in *Selected Topics in Signal Processing,* S. Haykin (Ed.), Prentice-Hall, Englewood Cliffs, NJ, 1989.

Yang, H. S., K. L. Boyer, and A. C. Kak, "Range Data Extraction and Interpretation by Structured Light," *Proceedings of the First International Conference on Artificial Intelligence Applications,* Denver, 1984, pp. 199–205.

Yeung, K. K., and P. D. Lawrence, "A Low Cost Three-Dimensional Vision System Using Space-Encoded Spot Projections," *SPIE Optics, Illumination, and Image Sensing for Machine Vision,* Vol. 728, 1986, pp. 160–172.

14 ANALYTIC PHOTOGRAMMETRY

14.1 Introduction

Analytic photogrammetry, as we deal with it, includes the body of techniques by which, from measurements of one or more 2D-perspective projections of a 3D object, one can make inferences about the 3D position, orientation, and lengths of the observed 3D object parts in a world reference frame. These inference problems can be construed as nonlinear least-squares problems. To solve such problems, photogrammetrists begin with an initially given approximate solution, then iteratively linearize the nonlinear functions around the current approximate solution, and solve the linearized problem for the adjustments to the current solution. The iterative solution for the nonlinear least-squares problem, the absolute orientation problem, the relative orientation problem, and the camera calibration problem, as well as stereo triangulation, are all discussed here from the algebraic point of view. This chapter also explains how to estimate rotation and translation to put a set of 2D-to-2D data points in correspondence, as well as a set of 3D-to-3D data points. The chapter concludes with a short discussion of error propagation, using both the explicit and the implicit forms.

Techniques of analytic photogrammetry were initially used in close-range applications; since the beginning of the 20th century to make topographic maps in scales of 1:5,000 to 1:250,000, until recently the main way by which elevation of different areas of the earth have been represented. Now the elevation information for most new maps in technologically developed countries is stored in digital format. The topographic maps are then produced from these digital terrain models.

In Chapter 13 we approached the inference of object position, orientation, and part lengths from the point of view of the camera reference frame. In this chapter we discuss how to determine one camera reference frame with respect to a world reference frame, given the camera coordinates of the observed perspective projection points and the world coordinates of the corresponding distinguished-object points. We also explain how to determine one camera reference frame with respect to another, how to do stereo triangulation, and how to calibrate a camera by the determination of its interior orientation.

In photogrammetric terminology, the *exterior orientation* of a camera is specified by all the parameters that determine the pose of the camera in the world reference frame. The parameters consist of the position of the center of perspectivity and the direction of the optical axis. Specification of the exterior orientation therefore requires three rotation angles and three translation parameters and is accomplished by obtaining the 3D coordinates of some control points whose corresponding position on the image is known. The *interior orientation* of a camera is specified by all the parameters that determine the geometry of a bundle of 3D rays from the measured image coordinates. The parameters of interior orientation relate the geometry of ideal perspective projection to the physics of a camera. The parameters include the camera constant, the principal point, and the specification of the lens distortion. Complete specification of the orientation of a camera is given by the interior and exterior orientations.

The *relative orientation* of one camera relative to another constitutes a stereo model and is specified by five parameters: three rotation angles and two translations. When two cameras are in relative orientation, each pair of corresponding rays from the two cameras intersect in 3D space. The scale cannot be determined by relative orientation. The process of determining relative orientation assumes that the interior orientation of each camera is known. *Absolute orientation* involves the orientation of a stereo model in a world reference frame. Absolute orientation requires the determination of seven parameters: the scale, the three translation parameters, and the three rotation parameters. It is accomplished by obtaining the 3D coordinates of some central points whose position on the stereo image can be determined. Complete specification of the orientation of a pair of cameras is given by specifying the twelve parameters determined from the relative and absolute orientations. This chapter explores techniques for the determination of the exterior, interior, and relative orientations.

To set the stage for these discussions, we need to develop the perspective projection equations by including all the transformations between the various reference frames. The relation between the camera and world reference frames is given by a translation and rotation. A point $(x, y, z)'$ in the world reference frame must first be expressed relative to the position $(x_0, y_0, z_0)'$ of the camera lens. Thus, relative to an origin that is the camera lens, the point $(x, y, z)'$ has coordinates

$$\begin{pmatrix} x \\ y \\ z \end{pmatrix} - \begin{pmatrix} x_0 \\ y_0 \\ z_0 \end{pmatrix}$$

The perspective projection of the camera is obtained with respect to the z-axis, which we take to be pointing outward. We take the x-axis to be horizontally oriented, pointing to the right, and the y-axis to be vertically oriented, pointing above. Thus we have a left-hand coordinate system. But the directions of the x-, y-, and z-axes of the camera reference frame differ from those of the world reference frame. We represent the rotation by which the world reference frame is brought into correspondence with the camera reference frame as a sequence of three rotations, the first one being around the x axis of the world reference frame; the second, around the y-axis of the once-rotated system; and the third, around the z-axis of the twice-rotated system.

In the photogrammetric literature, especially in relative orientation problems, one often finds ω to be the angle of rotation around the x-axis of the camera reference frame; ϕ, the angle of rotation around the y-axis; and κ, the angle of rotation around the z-axis. We use these conventions here. We call the clockwise angle of rotation ω around the x-axis the tilt angle; ϕ around the y-axis, the pan angle; and κ around the z-axis, the swing angle. Hence, if $R = R(\omega, \phi, \kappa)$ is the 3×3 rotation matrix, we have $R(\omega, \phi, \kappa) = R(\kappa)R(\phi)R(\omega)$, where

$$R(\omega) = \begin{pmatrix} 1 & 0 & 0 \\ 0 & \cos\omega & \sin\omega \\ 0 & -\sin\omega & \cos\omega \end{pmatrix}$$

$$R(\phi) = \begin{pmatrix} \cos\phi & 0 & -\sin\phi \\ 0 & 1 & 0 \\ \sin\phi & 0 & \cos\phi \end{pmatrix}$$

$$R(\kappa) = \begin{pmatrix} \cos\kappa & \sin\kappa & 0 \\ -\sin\kappa & \cos\kappa & 0 \\ 0 & 0 & 1 \end{pmatrix}$$

so that

$R(\omega, \phi, \kappa) =$

$$\begin{pmatrix} \cos\phi\cos\kappa & \sin\omega\sin\phi\cos\kappa + \cos\omega\sin\kappa & -\cos\omega\sin\phi\cos\kappa + \sin\omega\sin\kappa \\ -\cos\phi\sin\kappa & -\sin\omega\sin\phi\sin\kappa + \cos\omega\cos\kappa & \cos\omega\sin\phi\sin\kappa + \sin\omega\cos\kappa \\ \sin\phi & -\sin\omega\cos\phi & \cos\omega\cos\phi \end{pmatrix}$$

$$(14.1)$$

In order to represent relations involving the entries of the rotation matrix R and to avoid long trigonometric expressions, we establish the following convention:

$$R(\omega, \phi, \kappa) = \begin{pmatrix} r_{11} & r_{12} & r_{13} \\ r_{21} & r_{22} & r_{23} \\ r_{31} & r_{32} & r_{33} \end{pmatrix}$$

The angles ω, ϕ, and κ can be obtained directly from the values r_{ij} :

$$\sin\phi = r_{31}$$
$$\tan\omega = (-r_{32})/r_{33}$$
$$\tan\kappa = (-r_{21})/r_{11} \qquad (14.2)$$

Care must be taken to determine the correct quadrants. The choice of quadrants must be consistent with the observed signs of $r_{11}, r_{12}, r_{13}, r_{23}$, and r_{33}.

The point $(x, y, z)'$ in the world reference frame is then represented by the point $(p, q, s)'$ in the camera reference frame, where

$$\begin{pmatrix} p \\ q \\ s \end{pmatrix} = R(\omega, \phi, \kappa) \begin{pmatrix} x - x_0 \\ y - y_0 \\ z - z_0 \end{pmatrix} \tag{14.3}$$

Having a representation for the 3D point in the camera reference frame, we can then take its perspective projection. Assuming a pinhole camera and an image projection plane that is a distance f in front of the camera lens, we have the projection coordinates given by

$$\begin{pmatrix} u \\ v \end{pmatrix} = \frac{f}{s} \begin{pmatrix} p \\ q \end{pmatrix} \tag{14.4}$$

This form assumes that the optic axis of the camera is the z-axis.

In the photogrammetric literature, f is referred to as the *camera constant*. It is of course related to the focal length of the lens. But as the lens must be moved closer to or farther from the image projection plane depending on the distance of the objects focused on, the focal length of the lens must be considered as only the nominal value of f. The actual value of f must be determined by a calibration procedure applied in the exact circumstances in which the camera is used. We discuss this calibration procedure in Section 14.5.

Measurements are made in the image projection plane. The origin of the measurement image plane coordinate system is not necessarily the point where the z-axis of the camera intersects the image plane. This point is called the *principal point*. If its location is (u_0, v_0) in the image measurement plane coordinate system, then the coordinates of the projected point must be written as

$$\begin{pmatrix} u \\ v \end{pmatrix} = \begin{pmatrix} u_0 \\ v_0 \end{pmatrix} + \frac{f}{s} \begin{pmatrix} p \\ q \end{pmatrix} \tag{14.5}$$

Rewriting this in terms of the rotational dependencies directly, we have

$$\frac{u - u_0}{f} = \frac{r_{11}(x - x_0) + r_{12}(y - y_0) + r_{13}(z - z_0)}{r_{31}(x - x_0) + r_{32}(y - y_0) + r_{33}(z - z_0)}$$

$$\tag{14.6}$$

$$\frac{v - v_0}{f} = \frac{r_{21}(x - x_0) + r_{22}(y - y_0) + r_{23}(z - z_0)}{r_{31}(x - x_0) + r_{32}(y - y_0) + r_{33}(z - z_0)}$$

This pair of equations expresses a ratio of camera coordinates in terms of a ratio involving the rotated and translated world coordinates. It is known as the general or *fundamental perspective projection equation* or as the *collinearity* equation.

This pair of equations can be inverted so that a ratio of world coordinates can be expressed in terms of a ratio of rotated camera coordinates. To see this, we simply need to add the identity

$$1 = \frac{r_{31}(x - x_0) + r_{32}(y - y_0) + r_{33}(z - z_0)}{r_{31}(x - x_0) + r_{32}(y - y_0) + r_{33}(z - z_0)}$$

to the two equations we already have and multiply both sides by R^{-1}. Since R is a rotation matrix, $R^{-1} = R'$. So

$$\frac{1}{f} \begin{pmatrix} u - u_0 \\ v - v_0 \\ f \end{pmatrix} = \frac{R \begin{pmatrix} x - x_0 \\ y - y_0 \\ z - z_0 \end{pmatrix}}{r_{31}(x - x_0) + r_{32}(y - y_0) + r_{33}(z - z_0)}$$

$$\frac{1}{f} R' \begin{pmatrix} u - u_0 \\ v - v_0 \\ f \end{pmatrix} = \frac{R'R \begin{pmatrix} x - x_0 \\ y - y_0 \\ z - z_0 \end{pmatrix}}{r_{31}(x - x_0) + r_{32}(y - y_0) + r_{33}(z - z_0)}$$

$$= \frac{\begin{pmatrix} x - x_0 \\ y - y_0 \\ z - z_0 \end{pmatrix}}{r_{31}(x - x_0) + r_{32}(y - y_0) + r_{33}(z - z_0)}$$

Now by taking the ratios of component 1 to component 3 and component 2 to component 3, the denominator $r_{31}(x - x_0) + r_{32}(y - y_0) + r_{33}(z - z_0)$ cancels, and we obtain

$$\frac{x - x_0}{z - z_0} = \frac{r_{11}(u - u_0) + r_{21}(v - v_0) + r_{31}f}{r_{13}(u - u_0) + r_{23}(v - v_0) + r_{33}f}$$

(14.7)

$$\frac{y - y_0}{z - z_0} = \frac{r_{12}(u - u_0) + r_{22}(v - v_0) + r_{32}f}{r_{13}(u - u_0) + r_{23}(v - v_0) + r_{33}f}$$

Equations (14.5) and (14.7) are also called the general fundamental perspective projection equations. They show that the relationship between the measured 2D-perspective projection coordinates and the 3D coordinates is a nonlinear function of $u_o, v_o, x_0, y_0, z_0, \omega, \phi$, and κ. Given enough pairs of corresponding 2D and 3D points and an initial approximate solution, we can solve for the unknown parameters $u_o, v_o, x_0, y_0, z_0, \omega, \phi$, and κ by a nonlinear least-squares technique, which we next discuss.

14.2   Nonlinear Least-Squares Solutions

In those cases when a close-enough approximate solution to the nonlinear least-squares problem is known or given, the exact solution to the nonlinear least-squares problem can be obtained by iteratively solving the linearized problem in which the linearization is taken around the current approximate solution.

Suppose $\beta_1, \ldots, \beta_M$ are the unknown parameters governing each of the nonlinear transformations $g_1, \ldots, g_K$ and that $\alpha_1, \ldots, \alpha_K$ are the observed values of $g_1, \ldots, g_K$. The underlying noise model is given by

$$\alpha_k = g_k(\beta_1, \ldots, \beta_M) + \xi_k, \quad k = 1, \ldots, K \tag{14.8}$$

where $\xi_1, \ldots, \xi_K$ are additive mean zero Gaussian random variables having covariance matrix Σ. Under these assumptions the $\beta_1, \ldots, \beta_M$ that maximize $Prob(\alpha_1, \ldots, \alpha_K | \beta_1, \ldots, \beta_M)$, and that therefore constitute the maximum likelihood solution, are the $\beta_1, \ldots, \beta_M$ that minimize the least-squares criterion

$$\epsilon^2 = (\alpha - g)' \Sigma^{-1} (\alpha - g) \tag{14.9}$$

where

$$\alpha = \begin{pmatrix} \alpha_1 \\ \alpha_2 \\ \vdots \\ \alpha_K \end{pmatrix} \quad \text{and} \quad g = \begin{pmatrix} g_1(\beta_1, \cdots, \beta_M) \\ g_2(\beta_1, \cdots, \beta_M) \\ \vdots \\ g_K(\beta_1, \cdots, \beta_M) \end{pmatrix}$$

We assume that we are not in the underconstrained case. Hence the number K of observations exceeds the number M of unknown parameters by a sufficient margin so that the problem is overconstrained and has a unique solution.

If the transformations $g_1, \ldots, g_K$ were linear functions of $\beta_1, \ldots, \beta_M$, the problem would be a linear least-squares problem whose solution could be written as the normal equation found in standard regression analysis. To solve the case when $g_1, \ldots, g_K$ are nonlinear functions and there is a given approximate solution $\beta^0 = (\beta_1^0, \ldots, \beta_M^0)'$, we begin by linearizing the nonlinear transformations around β^0 and solve for the adjustments $\Delta\beta = (\Delta\beta_1, \ldots, \Delta\beta_M)'$, which when added to β constitute a better approximate solution. We perform this linearization and adjustment iteratively. In most cases five to ten iterations are required to produce the solution to the desired accuracy.

At the tth iteration let $\beta^t = (\beta_1^t, \ldots, \beta_M^t)'$ be the current approximate solution. The linearization proceeds by representing each $g_k(\beta^t + \Delta\beta)$ by a first-order Taylor series expansion of g_k taken around β^t :

$$g_k(\beta^t + \Delta\beta) = g_k(\beta^t) + \Delta g_k(\Delta\beta; \beta^t) \tag{14.10}$$

where Δg_k, the total derivative of g_k, is a linear function of the vector of adjustments $\Delta\beta$ given by

$$\Delta g_k(\Delta\beta; \beta^t) = \left(\frac{\partial g_k}{\partial \beta_1}(\beta^t) \cdots \frac{\partial g_k}{\partial \beta_M}(\beta^t) \right) \Delta\beta$$

The total derivative Δg_k has a direct interpretation apparent from Eq. (14.9). If the parameter vector β is perturbed by an amount $\Delta\beta$, then the resulting value of g_k will be perturbed by a value of Δg_k.

Substituting the linearized expressions into the least-squares criterion, we obtain

$$\epsilon^2 = (\alpha - g^t - G^t \Delta\beta)' \Sigma^{-1} (\alpha - g^t - G^t \Delta\beta)$$

where

$$g^t = \begin{pmatrix} g_1(\beta_1^t, \ldots, \beta_M^t) \\ g_2(\beta_1^t, \ldots, \beta_M^t) \\ \vdots \\ g_K(\beta_1^t, \ldots, \beta_M^t) \end{pmatrix}$$

and G^t is the Jacobian given by

$$G^t = \begin{pmatrix} \frac{\partial g_1}{\partial \beta_1} & \frac{\partial g_1}{\partial \beta_2} & \cdots & \frac{\partial g_1}{\partial \beta_M} \\ \vdots & & & \vdots \\ \frac{\partial g_K}{\partial \beta_1} & \frac{\partial g_K}{\partial \beta_2} & \cdots & \frac{\partial g_K}{\partial \beta_M} \end{pmatrix}$$

and each partial derivative of G^t is evaluated at $(\beta_1^t, \ldots, \beta_M^t)$. Now we see that the nonlinear least-squares problem has become a linear least-squares problem in the adjustments $\Delta\beta$. To solve it, we take partial derivatives of ϵ^2 with respect to each $\Delta\beta_m$ and set each partial derivative to zero. Denoting the vector of such partial derivatives by $\frac{\partial \epsilon^2}{\partial \Delta\beta}$, we have

$$\frac{\partial \epsilon^2}{\partial \Delta\beta} = 0 = -2G^{t'} \Sigma^{-1} (\alpha - g^t - G^t \Delta\beta)$$

From this relation it immediately follows that

$$\Delta\beta = (G^{t'} \Sigma^{-1} G^t)^{-1} G^{t'} \Sigma^{-1} (\alpha - g^t) \tag{14.11}$$

The adjustments are then added to the current β to form the next β:

$$\beta^{t+1} = \beta^t + \Delta\beta$$

When the initial β^0 is close to the true β, five or ten iterations of adjustments are sufficient to estimate the unknown β.

In the case when $K = M$ and the Jacobian matrix G^t is of full rank, the solution for $\Delta\beta$ simplifies:

$$\Delta\beta = G^{t-1} \Sigma (G^{t'})^{-1} G^{t'} \Sigma^{-1} (\alpha - g^t)$$
$$= G^{t-1} (\alpha - g^t) \tag{14.12}$$

14.3   The Exterior Orientation Problem

In the one-camera exterior orientation problem, a set of N 3D points having known positions $(x_n, y_n, z_n)'$, $n = 1, \ldots, N$, in an object reference frame and the corresponding set of 2D-perspective projections $(u_n, v_n)'$, $n = 1, \ldots, N$, are given. The exterior orientation problem is to determine the unknown rotation and translation that put the camera reference frame in the world reference frame.

This problem has sometimes been referred to in the computer vision literature as the *camera calibration problem* when the unknown is where the camera is in the object reference frame, and as the *object pose estimation problem* when the unknown is the object in the camera reference frame. Photogrammetrists call the estimation of the position of 3D points of a known model from their 2D-perspective orientations the *spatial resection problem*.

Let the unknown translation be given by $(x_0, y_0, z_0)'$. The point $(x_n, y_n, z_n)'$ of the object reference frame becomes the point (p_n, q_n, s_n) of the camera reference frame, where

$$\begin{pmatrix} p_n \\ q_n \\ s_n \end{pmatrix} = R(\omega, \phi, \kappa) \begin{pmatrix} x_n - x_0 \\ y_n - y_0 \\ z_n - z_0 \end{pmatrix}, \quad n = 1, \ldots, N \tag{14.13}$$

The observed perspective projection is then given by

$$\begin{pmatrix} u_n \\ v_n \end{pmatrix} = \frac{f}{s_n} \begin{pmatrix} p_n \\ q_n \end{pmatrix}, \quad n = 1, \ldots, N \tag{14.14}$$

This problem can be set up as a nonlinear least-squares problem. We will solve it assuming we have an initial approximate solution so that we can linearize the problem. Our initial approximate solution can come from prior knowledge or from one of the techniques discussed in the previous chapter, such as a solution from the three-point-triangle-section problem. Initial approximate solutions need to be within 10% of scale for the translation parameters and within 15° for the rotational parameters for the linearized solution technique to succeed. In the notation of the previous section, $\beta = (x_0, y_0, z_0, \omega, \phi, \kappa)'$ and

$$g_{2n-1}(\beta) = u_n = f p_n / s_n$$

$$\tag{14.15}$$

$$g_{2n}(\beta) = v_n = f q_n / s_n$$

To develop the standard technique for linearization, we employ, for convenience, a slight change in notation. In the perspective projection equation, we think of $\begin{pmatrix} u_n \\ v_n \end{pmatrix}$ as a function of $(x_0, y_0, z_0, \omega, \phi, \kappa)$ in order to write out the required differentiation:

$$\begin{pmatrix} u_n \\ v_n \end{pmatrix} = \begin{pmatrix} u_n(x_0, y_0, z_0, \omega, \phi, \kappa) \\ v_n(x_0, y_0, z_0, \omega, \phi, \kappa) \end{pmatrix} = \frac{f}{s_n(x_0, y_0, z_0, \omega, \phi, \kappa)} \begin{pmatrix} p_n(x_0, y_0, z_0, \omega, \phi, \kappa) \\ q_n(x_0, y_0, z_0, \omega, \phi, \kappa) \end{pmatrix}$$

where

$$\begin{pmatrix} p_n(x_0, y_0, z_0, \omega, \phi, \kappa) \\ q_n(x_0, y_0, z_0, \omega, \phi, \kappa) \\ s_n(x_0, y_0, z_0, \omega, \phi, \kappa) \end{pmatrix} = R(\omega, \phi, \kappa) \begin{pmatrix} x_n - x_0 \\ y_n - y_0 \\ z_n - z_0 \end{pmatrix}$$

Then Δu_n and Δv_n denote the change in the value of the function u_n and v_n due to a change $(\Delta x_0, \Delta y_0, \Delta z_0, \Delta \omega, \Delta \phi, \Delta \kappa)$ in the value of the parameters $(x_0, y_0, z_0, \omega, \phi, \kappa)$.

At each iteration the principal difficulty is to determine the vector $[\Delta g_1(\Delta \beta; \beta'), \ldots, \Delta g_{2N}(\Delta \beta; \beta')]' = G' \Delta \beta$.

Because all the odd components are really Δu_n for a corresponding n, and all the even components are really Δv_n for a corresponding n,

$$G' \Delta \beta = \begin{pmatrix} \Delta u_1 \\ \Delta v_1 \\ \Delta u_2 \\ \Delta v_2 \\ \vdots \\ \Delta u_N \\ \Delta v_N \end{pmatrix}$$

Therefore to determine the explicit form of G' we need only show how to do it for a pair $(\Delta u_n, \Delta v_n)'$.

Next we show how to do this by using the standard solution. Then we give two alternative forms using adjustments associated with a skew symmetric matrix and then using a rotation represented in terms of its quaternion parameters. Classic papers in this area include Rosenfield (1959) and Thompson (1968). Our development follows Hinsken (1988).

14.3.1 Standard Solution

We proceed by using the chain rule for partial differentiation:

$$\Delta u_n = f \frac{s_n \Delta p_n - p_n \Delta s_n}{s_n^2}$$

$$\Delta v_n = f \frac{s_n \Delta q_n - q_n \Delta s_n}{s_n^2}$$

or in matrix form,

$$\begin{pmatrix} \Delta u_n \\ \Delta v_n \end{pmatrix} = \frac{f}{s_n} \begin{pmatrix} 1 & 0 & -p_n/s_n \\ 0 & 1 & -q_n/s_n \end{pmatrix} \begin{pmatrix} \Delta p_n \\ \Delta q_n \\ \Delta s_n \end{pmatrix} \tag{14.16}$$

Then

$$\begin{pmatrix} \Delta p_n \\ \Delta q_n \\ \Delta s_n \end{pmatrix} = R(\omega, \phi, \kappa) \begin{pmatrix} -\Delta x_0 \\ -\Delta y_0 \\ -\Delta z_0 \end{pmatrix} + \Delta R(\omega, \phi, \kappa) \begin{pmatrix} x_n - x_0 \\ y_n - y_0 \\ z_n - z_0 \end{pmatrix} \tag{14.17}$$

where

$$\Delta R(\omega, \phi, \kappa) = \frac{\partial R(\omega, \phi, \kappa)}{\partial \omega} \Delta \omega + \frac{\partial R(\omega, \phi, \kappa)}{\partial \phi} \Delta \phi + \frac{\partial R(\omega, \phi, \kappa)}{\partial \kappa} \Delta \kappa$$

Since $R(\omega, \phi, \kappa) = R(\kappa)R(\phi)R(\omega)$,

$$\frac{\partial R(\omega, \phi, \kappa)}{\partial \omega} = R(\kappa)R(\phi)\frac{\partial R(\omega)}{\partial \omega}$$

$$\frac{\partial R(\omega, \phi, \kappa)}{\partial \phi} = R(\kappa)\frac{\partial R(\phi)}{\partial \phi}R(\omega)$$

and

$$\frac{\partial R(\omega, \phi, \kappa)}{\partial \kappa} = \frac{\partial R(\kappa)}{\partial \kappa}R(\phi)R(\omega)$$

Hence

$$\frac{\partial R}{\partial \omega}(\omega, \phi, \kappa)$$

$$= \begin{pmatrix} 0 & \cos\omega\sin\phi\cos\kappa - \sin\omega\sin\kappa & \sin\omega\sin\phi\cos\kappa + \cos\omega\sin\kappa \\ 0 & -\cos\omega\sin\phi\sin\kappa - \sin\omega\cos\kappa & \cos\omega\cos\kappa - \sin\omega\sin\phi\sin\kappa \\ 0 & -\cos\omega\cos\phi & -\sin\omega\cos\phi \end{pmatrix}$$

$$\frac{\partial R}{\partial \phi}(\omega, \phi, \kappa)$$

$$= \begin{pmatrix} -\sin\phi\cos\kappa & \sin\omega\cos\phi\cos\kappa & -\cos\omega\cos\phi\cos\kappa \\ \sin\phi\sin\kappa & -\sin\omega\cos\phi\sin\kappa & \cos\omega\cos\phi\sin\kappa \\ \cos\phi & \sin\omega\sin\phi & -\cos\omega\sin\phi \end{pmatrix}$$

$$\frac{\partial R}{\partial \kappa}(\omega, \phi, \kappa)$$

$$= \begin{pmatrix} -\cos\phi\sin\kappa & -\sin\omega\sin\phi\sin\kappa + \cos\omega\cos\kappa & \cos\omega\sin\phi\sin\kappa + \sin\omega\cos\kappa \\ -\cos\phi\cos\kappa & -\sin\omega\sin\phi\cos\kappa - \cos\omega\sin\kappa & \cos\omega\sin\phi\cos\kappa - \sin\omega\sin\kappa \\ 0 & 0 & 0 \end{pmatrix}$$

For the tth iteration of the procedure, we then have

$$\begin{pmatrix} \Delta u_n \\ \Delta v_n \end{pmatrix} = \frac{f}{s_n^t}\begin{pmatrix} 1 & 0 & -p_n^t/s_n^t \\ 0 & 1 & -q_n^t/s_n^t \end{pmatrix}\left[-R(\omega^t, \phi^t, \kappa^t)\begin{pmatrix} \Delta x_0 \\ \Delta y_0 \\ \Delta z_0 \end{pmatrix}\right.$$

$$+ \left.\left(\frac{\partial R}{\partial \omega}(\omega^t, \phi^t, \kappa^t)\Delta\omega + \frac{\partial R}{\partial \phi}(\omega^t, \phi^t, \kappa^t)\Delta\phi + \frac{\partial R}{\partial \kappa}(\omega^t, \phi^t, \kappa^t)\Delta\kappa \right)\begin{pmatrix} x_n - x_0^t \\ y_n - y_0^t \\ z_n - z_0^t \end{pmatrix}\right]$$

where

$$\begin{pmatrix} p_n^t \\ q_n^t \\ s_n^t \end{pmatrix} = R(\omega^t, \phi^t, \kappa^t)\begin{pmatrix} x_n \\ y_n \\ z_n \end{pmatrix}$$

Or putting this in a compact linear form, we have

$$\begin{pmatrix} \Delta u_n \\ \Delta v_n \end{pmatrix} = A_n^t B_n^t \begin{pmatrix} \Delta x_0 \\ \Delta y_0 \\ \Delta z_0 \\ \Delta \omega \\ \Delta \phi \\ \Delta \kappa \end{pmatrix}$$

where

$$A_n^t = \frac{f}{s_n^t} \begin{pmatrix} 1 & 0 & -p_n^t/s_n^t \\ 0 & 1 & -q_n^t/s_n^t \end{pmatrix}$$

and

$$B_n^t = \left(-R(\omega', \phi', \kappa') \ \ Q(\omega', \phi', \kappa', x_0', y_0', z_0') \right)$$

(14.18)

and where Q is a 3×3 matrix given by

$$Q(\omega', \phi', \kappa', x_0', y_0', z_0') =$$
$$\left[\frac{\partial R}{\partial \omega}(\omega', \phi', \kappa') \begin{pmatrix} x_n - x_0' \\ y_n - y_0' \\ z_n - z_0' \end{pmatrix} \ \ \frac{\partial R}{\partial \phi}(\omega', \phi', \kappa') \begin{pmatrix} x_n - x_0' \\ y_n - y_0' \\ z_n - z_0' \end{pmatrix} \ \ \frac{\partial R}{\partial \kappa}(\omega', \phi', \kappa') \begin{pmatrix} x_n - x_0' \\ y_n - y_0' \\ z_n - z_0' \end{pmatrix} \right]$$

This leads to the following algorithm for the solution of the absolute orientation problem. At iteration t, the available solution is $\beta^t = (x_0', y_0', z_0', \omega', \phi', \kappa')$. The nth pair of observed coordinates on the image is (u_n, v_n), and they arise from the perspective projection of (x_n, y_n, z_n). On the basis of the approximate solution, we could infer that the perspective projection of (x_n, y_n, z_n) would have image coordinates (u_n', v_n'), where

$$\begin{pmatrix} u_n' \\ v_n' \end{pmatrix} = \frac{f}{s_n^t} \begin{pmatrix} p_n^t \\ q_n^t \end{pmatrix}$$

$$\begin{pmatrix} p_n^t \\ q_n^t \\ s_n^t \end{pmatrix} = R(\omega', \phi', \kappa') \begin{pmatrix} x_n - x_0' \\ y_n - y_0' \\ z_n - z_0' \end{pmatrix}$$

The vector α and g' of least-squares equation (14.11) is then

$$\alpha = \begin{pmatrix} u_1 \\ v_1 \\ \vdots \\ u_N \\ v_N \end{pmatrix} \quad \text{and} \quad g' = \begin{pmatrix} u_1' \\ v_1' \\ \vdots \\ u_N' \\ v_N' \end{pmatrix}$$

where $u_1, v_1, \ldots, u_N, v_N$ are the observed perspective projection coordinates. The

$2N \times 6$ matrix G' of Eq. (14.11) is given by

$$
G' = \begin{pmatrix} A'_1 B'_1 \\ \vdots \\ A'_N B'_N \end{pmatrix}
$$

Equation (14.11) is used to solve for $\Delta\beta = (\Delta x_0, \Delta y_0, \Delta z_0, \Delta\omega, \Delta\phi, \Delta\kappa)'$. The new approximate solution is then given by

$$
\begin{pmatrix} x_0^{t+1} \\ y_0^{t+1} \\ z_0^{t+1} \\ \omega^{t+1} \\ \phi^{t+1} \\ \kappa^{t+1} \end{pmatrix} = \begin{pmatrix} x_0^t \\ y_0^t \\ z_0^t \\ \omega^t \\ \phi^t \\ \kappa^t \end{pmatrix} + \begin{pmatrix} \Delta x_0 \\ \Delta y_0 \\ \Delta z_0 \\ \Delta\omega \\ \Delta\phi \\ \Delta\kappa \end{pmatrix}
\tag{14.19}
$$

EXAMPLE 14.1

The vertices of three points in the world coordinate system are known to be

$$
\begin{pmatrix} x_1 \\ y_1 \\ z_1 \end{pmatrix} = \begin{pmatrix} 0 \\ 0 \\ 6 \end{pmatrix}, \quad \begin{pmatrix} x_2 \\ y_2 \\ z_2 \end{pmatrix} = \begin{pmatrix} 3 \\ 0 \\ 6 \end{pmatrix}, \quad \begin{pmatrix} x_3 \\ y_3 \\ z_3 \end{pmatrix} = \begin{pmatrix} 0 \\ 4 \\ 6 \end{pmatrix}
$$

These points produce their respective perspective projections in the camera coordinate system of

$$
\begin{pmatrix} u_1 \\ v_1 \end{pmatrix} = \begin{pmatrix} 0.00645 \\ -0.0105 \end{pmatrix}, \quad \begin{pmatrix} u_2 \\ v_2 \end{pmatrix} = \begin{pmatrix} 0.0243 \\ -0.0126 \end{pmatrix}, \quad \begin{pmatrix} u_3 \\ v_3 \end{pmatrix} = \begin{pmatrix} 0.00961 \\ 0.0156 \end{pmatrix}
$$

Starting with an initial approximation solution of 0 for the unknowns x_0, y_0, z_0, ω, ϕ, κ and using Eq. (14.19), we calculate after nine iterations

$$
\begin{pmatrix} x_0 \\ y_0 \\ z_0 \end{pmatrix} = \begin{pmatrix} -3.754 \\ .5256 \\ -4.666 \end{pmatrix}, \quad \begin{pmatrix} \omega \\ \phi \\ \kappa \end{pmatrix} = \begin{pmatrix} -0.08271 \\ 2.355 \\ 0.1481 \end{pmatrix} \text{ radians}
$$

14.3.2 Auxiliary Solution

Instead of iteratively adjusting the angles directly, we can reorganize the calculation such that we iteratively adjust the three auxiliary parameters of a skew symmetric matrix associated with the rotation matrix. Then we determine the adjustments of the angles in terms of the adjustments of the parameters of the skew symmetric matrix associated with the rotation matrix. The first step of our derivation will illustrate how the skew symmetric matrix arises and how it relates to the adjustments $(\Delta\omega, \Delta\phi, \Delta\kappa)$. Then we will reorganize the calculations using these relationships and finally summarize the reorganized algorithm.

Since $R'R = I$, we can use Eq. (14.13) to rewrite Eq. (14.17) in the form

$$\begin{pmatrix} \Delta p_n \\ \Delta q_n \\ \Delta s_n \end{pmatrix} = R(\omega, \phi, \kappa) \begin{pmatrix} \Delta x_0 \\ \Delta y_0 \\ \Delta z_0 \end{pmatrix} + \Delta R(\omega, \phi, \kappa) R'(\omega, \phi, \kappa) \begin{pmatrix} p_n \\ q_n \\ s_n \end{pmatrix} \qquad (14.20)$$

The matrix $\Delta R(\omega, \phi, \kappa) R'(\omega, \phi, \kappa)$ has an especially nice form. It must be skew symmetric, from the relation $RR' = I$. Taking total derivatives on both sides, we obtain

$$R\Delta R' + \Delta R R' = 0$$

Now notice that $R\Delta R' = (\Delta RR')'$. Hence $\Delta RR' = -(\Delta RR')'$. Therefore the matrix $\Delta S_w = \Delta RR'$ must have the form

$$\Delta S_w = \Delta RR' = \begin{pmatrix} 0 & -\Delta w_3 & \Delta w_2 \\ \Delta w_3 & 0 & -\Delta w_1 \\ -\Delta w_2 & \Delta w_1 & 0 \end{pmatrix} \qquad (14.21)$$

and Eq. (14.20) can be rewritten as

$$\begin{pmatrix} \Delta p_n \\ \Delta q_n \\ \Delta s_n \end{pmatrix} = -R(\omega, \phi, \kappa) \begin{pmatrix} \Delta x_0 \\ \Delta y_0 \\ \Delta z_0 \end{pmatrix} + \Delta S_w \begin{pmatrix} p_n \\ q_n \\ s_n \end{pmatrix} \qquad (14.22)$$

The relation between $\Delta w_1, \Delta w_2, \Delta w_3$ and $\Delta\omega, \Delta\phi, \Delta\kappa$ can be determined directly:

$$\begin{aligned} \Delta RR' &= \left(\frac{\partial R}{\partial\omega}\Delta\omega + \frac{\partial R}{\partial\phi}\Delta\phi + \frac{\partial R}{\partial\kappa}\Delta\kappa \right) [R(\kappa)R(\phi)R(\omega)]' \\ &= R(\kappa)R(\phi)\frac{\partial R(\omega)}{\partial\omega}R(\omega)'R(\phi)'R(\kappa)'\Delta\omega \\ &\quad + R(\kappa)\frac{\partial R(\phi)}{\partial\phi}R(\omega)R(\omega)'R(\phi)'R(\kappa)'\Delta\phi \\ &\quad + \frac{\partial R(\kappa)}{\partial\kappa}R(\phi)R(\omega)R(\omega)'R(\phi)'R(\kappa)'\Delta\kappa \\ &= R(\kappa)R(\phi)\frac{\partial R(\omega)}{\partial\omega}R(\omega)'R(\phi)'R(\kappa)'\Delta\omega \\ &\quad + R(\kappa)\frac{\partial R(\phi)}{\partial\phi}R(\phi)'R(\kappa)'\Delta\phi \\ &\quad + \frac{\partial R(\kappa)}{\partial\kappa}R(\kappa)'\Delta\kappa \end{aligned}$$

Since

$$\frac{\partial R(\omega)}{\partial \omega} R(\omega)' = \begin{pmatrix} 0 & 0 & 0 \\ 0 & 0 & 1 \\ 0 & -1 & 0 \end{pmatrix}$$

$$\frac{\partial R(\phi)}{\partial \phi} R(\phi)' = \begin{pmatrix} 0 & 0 & -1 \\ 0 & 0 & 0 \\ 1 & 0 & 0 \end{pmatrix}$$

$$\frac{\partial R(\kappa)}{\partial \kappa} R(\kappa)' = \begin{pmatrix} 0 & 1 & 0 \\ -1 & 0 & 0 \\ 0 & 0 & 0 \end{pmatrix}$$

upon multiplying out there results

$$\Delta R R' = \begin{pmatrix} 0 & \sin\phi & \cos\phi\sin\kappa \\ -\sin\phi & 0 & \cos\phi\cos\kappa \\ -\cos\phi\sin\kappa & -\cos\phi\cos\kappa & 0 \end{pmatrix} \Delta\omega$$

$$+ \begin{pmatrix} 0 & 0 & -\cos\kappa \\ 0 & 0 & \sin\kappa \\ \cos\kappa & -\sin\kappa & 0 \end{pmatrix} \Delta\phi + \begin{pmatrix} 0 & 1 & 0 \\ -1 & 0 & 0 \\ 0 & 0 & 0 \end{pmatrix} \Delta\kappa$$

$$= \begin{pmatrix} 0 & -\Delta w_3 & \Delta w_2 \\ \Delta w_3 & 0 & -\Delta w_1 \\ -\Delta w_2 & \Delta w_1 & 0 \end{pmatrix} \tag{14.23}$$

Upon identifying terms of Eqs. (14.21) and (14.23),

$$\begin{pmatrix} \Delta w_1 \\ \Delta w_2 \\ \Delta w_3 \end{pmatrix} = \begin{pmatrix} -\cos\phi\cos\kappa & -\sin\kappa & 0 \\ \cos\phi\sin\kappa & -\cos\kappa & 0 \\ -\sin\phi & 0 & -1 \end{pmatrix} \begin{pmatrix} \Delta\omega \\ \Delta\phi \\ \Delta\kappa \end{pmatrix} \tag{14.24}$$

from which

$$\begin{pmatrix} \Delta\omega \\ \Delta\phi \\ \Delta\kappa \end{pmatrix} = \begin{pmatrix} -\cos\kappa/\cos\phi & \sin\kappa/\cos\phi & 0 \\ -\sin\kappa & -\cos\kappa & 0 \\ \cos\kappa\tan\phi & -\sin\kappa\tan\phi & -1 \end{pmatrix} \begin{pmatrix} \Delta w_1 \\ \Delta w_2 \\ \Delta w_3 \end{pmatrix} \tag{14.25}$$

Now the product $\Delta S_w \begin{pmatrix} p_n \\ q_n \\ s_n \end{pmatrix}$ is nothing more than the cross-product

$$\begin{pmatrix} \Delta w_1 \\ \Delta w_2 \\ \Delta w_3 \end{pmatrix} \times \begin{pmatrix} p_n \\ q_n \\ s_n \end{pmatrix} = -\begin{pmatrix} p_n \\ q_n \\ s_n \end{pmatrix} \times \begin{pmatrix} \Delta w_1 \\ \Delta w_2 \\ \Delta w_3 \end{pmatrix}$$

Letting

$$S_n = \begin{pmatrix} 0 & s_n & -q_n \\ -s_n & 0 & p_n \\ q_n & -p_n & 0 \end{pmatrix}$$

we can rewrite Eq. (14.22) as

$$\begin{pmatrix} \Delta p_n \\ \Delta q_n \\ \Delta s_n \end{pmatrix} = -R(\omega, \phi, \kappa) \begin{pmatrix} \Delta x_0 \\ \Delta y_0 \\ \Delta z_0 \end{pmatrix} + S_n \begin{pmatrix} \Delta w_1 \\ \Delta w_2 \\ \Delta w_3 \end{pmatrix} \tag{14.26}$$

Putting Eq. (14.26) into Eq. (14.16), we obtain

$$\begin{pmatrix} \Delta u_n \\ \Delta v_n \end{pmatrix} = \frac{-f}{s_n} \begin{pmatrix} 1 & 0 & -p_n/s_n \\ 0 & 1 & -q_n/s_n \end{pmatrix} \left[-R(\omega, \phi, \kappa) \begin{pmatrix} \Delta x_0 \\ \Delta y_0 \\ \Delta z_0 \end{pmatrix} + S_n \begin{pmatrix} \Delta w_1 \\ \Delta w_2 \\ \Delta w_3 \end{pmatrix} \right] \tag{14.27}$$

So for the tth iteration we then have

$$\begin{pmatrix} \Delta u_n \\ \Delta v_n \end{pmatrix} = A_n^t B_n^t \begin{pmatrix} \Delta x_0 \\ \Delta y_0 \\ \Delta z_0 \\ \Delta w_1 \\ \Delta w_2 \\ \Delta w_3 \end{pmatrix} \tag{14.28}$$

where

$$A_n^t = \frac{-f}{s_n^t} \begin{pmatrix} 1 & 0 & -p_n^t/s_n^t \\ 0 & 1 & -q_n^t/s_n^t \end{pmatrix}$$

$$B_n^t = [-R(\omega^t, \phi^t, \kappa^t) S_n^t]$$

$$\begin{pmatrix} p_n^t \\ q_n^t \\ s_n^t \end{pmatrix} = R(\omega^t, \phi^t, \kappa^t) \begin{pmatrix} x_n - x_0 \\ y_n - y_0 \\ z_n - z_0 \end{pmatrix}$$

$$S_n^t = \begin{pmatrix} 0 & s_n^t & -q_n^t \\ -s_n^t & 0 & p_n^t \\ q_n^t & -p_n^t & 0 \end{pmatrix}$$

This leads to the following algorithm for the solution of the absolute orientation problem. At iteration t, the available solution is $\beta^t = (x_0^t, y_0^t, z_0^t, \omega^t, \phi^t, \kappa^t)'$. The nth pair of observed coordinates on the image is (u_n, v_n), and they arise from the perspective projection of (x_n, y_n, z_n). On the basis of the approximate solution, we could infer that the perspective projection of (x_n, y_n, z_n) would have image coordinates (u_n^t, v_n^t), where

$$\begin{pmatrix} u_n^t \\ v_n^t \end{pmatrix} = \frac{f}{s_n^t} \begin{pmatrix} p_n^t \\ q_n^t \end{pmatrix}$$

$$\begin{pmatrix} p_n^t \\ q_n^t \\ s_n^t \end{pmatrix} = R(\omega^t, \phi^t, \kappa^t) \begin{pmatrix} x_n - x_0^t \\ y_n - y_0^t \\ z_n - z_0^t \end{pmatrix}$$

The vector α and g' of the least-squares equation (14.11) is then

$$
\alpha = \begin{pmatrix} u_1 \\ v_1 \\ \vdots \\ u_N \\ v_N \end{pmatrix} \quad \text{and } g' = \begin{pmatrix} u'_1 \\ v'_1 \\ \vdots \\ u'_N \\ v'_N \end{pmatrix}
$$

The $2N \times 6$ matrix G' of Eq. (14.11) is given by

$$
G' = \begin{pmatrix} A'_1 B'_1 \\ \vdots \\ A'_N B'_N \end{pmatrix}
$$

Equation (14.11) is then used to solve for $\Delta\beta = (\Delta x_0, \Delta y_0, \Delta z_0, \Delta w_1, \Delta w_2, \Delta w_3)'$. Then the adjustments $(\Delta w_1, \Delta w_2, \Delta w_3)$ determine the adjustments $(\Delta\omega, \Delta\phi, \Delta\kappa)$ by

$$
\begin{pmatrix} \Delta\omega \\ \Delta\phi \\ \Delta\kappa \end{pmatrix} = \begin{pmatrix} -\cos\kappa'/\cos\phi' & \sin\kappa'/\cos\phi' & 0 \\ -\sin\kappa' & -\cos\kappa' & 0 \\ \cos\kappa'\tan\phi' & -\sin\kappa'\tan\phi' & -1 \end{pmatrix} \begin{pmatrix} \Delta w_1 \\ \Delta w_2 \\ \Delta w_3 \end{pmatrix}
$$

Then the new angles ω^{t+1}, ϕ^{t+1}, and κ^{t+1} are defined by

$$
\begin{pmatrix} \omega^{t+1} \\ \phi^{t+1} \\ \kappa^{t+1} \end{pmatrix} = \begin{pmatrix} \omega^t \\ \phi^t \\ \kappa^t \end{pmatrix} + \begin{pmatrix} \Delta\omega \\ \Delta\phi \\ \Delta\kappa \end{pmatrix}
$$

And the new translation $(x_0^{t+1}, y_0^{t+1}, z_0^{t+1})$ is defined by

$$
\begin{pmatrix} x_0^{t+1} \\ y_0^{t+1} \\ z_0^{t+1} \end{pmatrix} = \begin{pmatrix} x_0^t \\ y_0^t \\ z_0^t \end{pmatrix} + \begin{pmatrix} \Delta x_0 \\ \Delta y_0 \\ \Delta z_0 \end{pmatrix}
$$

14.3.3 Quaternion Representation

There are representations for rotation matrices other than the one we have used. Schut (1958–59) devotes a whole paper to various representations, among which is the quaternion representation.

Schut (1962–63) has shown how, from any skew symmetric matrix

$$
S = \begin{pmatrix} 0 & -c & b \\ c & 0 & -a \\ -b & a & 0 \end{pmatrix}
$$

a rotation matrix R can be constructed. For any scalar d, define the rotation matrix R by

$$
R = (dI + S)(dI - S)^{-1} \tag{14.29}
$$

Note that this definition guarantees that $R'R = I$:

$$\begin{aligned}
R'R &= [(dI + S)(dI - S)^{-1}]'[(dI + S)(dI - S)^{-1}] \\
&= [(dI - S)^{-1}]'(dI + S)'(dI + S)(dI - S)^{-1} \\
&= (dI + S)^{-1}(dI - S)(dI + S)(dI - S)^{-1} \\
&= [(dI + S)^{-1}(dI + S)][(dI - S)(dI - S)^{-1}] \\
&= I
\end{aligned}$$

If

$$S = \begin{pmatrix} 0 & -c & b \\ c & 0 & -a \\ -b & a & 0 \end{pmatrix}$$

then, using Eq. (14.29), we have

$$R = \begin{pmatrix} d^2 + a^2 - b^2 - c^2 & 2(ab - cd) & 2(ac + bd) \\ 2(ab + cd) & d^2 - a^2 + b^2 - c^2 & 2(bc - ad) \\ 2(ac - bd) & 2(bc + ad) & d^2 - a^2 - b^2 + c^2 \end{pmatrix} \frac{1}{a^2 + b^2 + c^2 + d^2}$$

Schut (1962–63), seeing how to make things even simpler than Thompson (1958–59) did, used the quaternion representation to show how, by a solution of a linear system of equations, a 3D rotation matrix can be determined from corresponding 3D point sets. Pope (1970) then used the quaternion representation in the determination of the rotation matrix in the absolute orientation problem. Hinsken (1988), using the quaternion representation, developed a new computationally simpler solution for the determination of the unknown rotation and translation parameters in the absolute orientation problem. Our derivation is based on his. Hinsken uses the quaternion form of the rotation matrix given by Schut (1958–59). Horn (1987) and Horn, Hildren, and Negahdaripour (1988) give related derivations. Hinsken's rotations matrix is specified by

$$R = \begin{pmatrix} d^2 + a^2 - b^2 - c^2 & 2(ab - cd) & 2(ac + bd) \\ 2(ab + cd) & d^2 - a^2 + b^2 - c^2 & 2(bc - ad) \\ 2(ac - bd) & 2(bc + ad) & d^2 - a^2 - b^2 + c^2 \end{pmatrix} \tag{14.30}$$

where the parameters $a, b, c,$ and d are constrained to satisfy

$$a^2 + b^2 + c^2 + d^2 = 1 \tag{14.31}$$

In terms of the quaternion parameters,

$$\Delta R = \frac{\partial R}{\partial a} \Delta a + \frac{\partial R}{\partial b} \Delta b + \frac{\partial R}{\partial c} \Delta c + \frac{\partial R}{\partial d} \Delta d$$

so that

$$\Delta R R' = \frac{\partial R}{\partial a} R' \Delta a + \frac{\partial R}{\partial b} R' \Delta b + \frac{\partial R}{\partial c} R' \Delta c + \frac{\partial R}{\partial d} R' \Delta d \tag{14.32}$$

Calculating the matrices directly, we obtain

$$\frac{\partial R}{\partial a}R' = 2\begin{pmatrix} a & b & c \\ -b & a & -d \\ -c & d & a \end{pmatrix}$$

$$\frac{\partial R}{\partial b}R' = 2\begin{pmatrix} b & -a & d \\ a & b & c \\ -d & -c & b \end{pmatrix}$$

$$\frac{\partial R}{\partial c}R' = 2\begin{pmatrix} c & -d & -a \\ d & c & -b \\ a & b & c \end{pmatrix}$$

$$\frac{\partial R}{\partial d}R' = 2\begin{pmatrix} d & c & -b \\ -c & d & a \\ b & -a & d \end{pmatrix}$$

Identifying the terms of Eq. (14.32) with Eq. (14.21), we obtain

$$\begin{pmatrix} \Delta w_1 \\ \Delta w_2 \\ \Delta w_3 \end{pmatrix} = 2\begin{pmatrix} d & -c & b & -a \\ c & d & -a & -b \\ -b & a & d & -c \end{pmatrix}\begin{pmatrix} \Delta a \\ \Delta b \\ \Delta c \\ \Delta d \end{pmatrix} \tag{14.33}$$

Taking the total derivative of Eq. (14.31), we have

$$2a\,\Delta a + 2b\,\Delta b + 2c\,\Delta c + 2d\,\Delta d = 0$$

so that

$$\Delta d = -\frac{a\,\Delta a + b\,\Delta b + c\,\Delta c}{d} \tag{14.34}$$

Substituting the expression for Δd of Eq. (14.34) into Eq. (14.33) results in

$$\begin{pmatrix} \Delta w_1 \\ \Delta w_2 \\ \Delta w_3 \end{pmatrix} = \frac{2}{d}\begin{bmatrix} d^2 + a^2 & ab - cd & ac + bd \\ ab + cd & d^2 + b^2 & bc - ad \\ ac - bd & bc + ad & d^2 + c^2 \end{bmatrix}\begin{pmatrix} \Delta a \\ \Delta b \\ \Delta c \end{pmatrix} \tag{14.35}$$

Solving Eq. (14.35) for $\begin{pmatrix} \Delta a \\ \Delta b \\ \Delta c \end{pmatrix}$, we obtain

$$\begin{pmatrix} \Delta a \\ \Delta b \\ \Delta c \end{pmatrix} = \frac{1}{2}\begin{pmatrix} d & c & -b \\ -c & d & a \\ b & -a & d \end{pmatrix}\begin{pmatrix} \Delta w_1 \\ \Delta w_2 \\ \Delta w_3 \end{pmatrix} \tag{14.36}$$

This leads to the following algorithm for the solution of the absolute orientation problem. At iteration t, the available solution is $\beta^t = (x_0^t, y_0^t, z_0^t, a^t, b^t, c^t, d^t)$. As before, the nth pair of observed coordinates on the image is (u_n, v_n), and they arise from the perspective projection of (x_n, y_n, z_n). On the basis of the approximate solution, we could infer that the 3D point (x_n, y_n, z_n) would have image coordinates (u'_n, v'_n), where

$$\begin{pmatrix} u'_n \\ v'_n \end{pmatrix} = \frac{f}{s'_n} \begin{pmatrix} p'_n \\ q'_n \end{pmatrix}$$

$$\begin{pmatrix} p'_n \\ q'_n \\ s'_n \end{pmatrix} = R(a', b', c', d') \begin{pmatrix} x_n - x'_0 \\ y_n - y'_0 \\ z_n - z'_0 \end{pmatrix}$$

The vector α and g' of least-squares equation (14.11) is then

$$\alpha = \begin{pmatrix} u_1 \\ v_1 \\ \vdots \\ u_N \\ v_N \end{pmatrix} \quad \text{and} \quad g' = \begin{pmatrix} u'_1 \\ v'_1 \\ \vdots \\ u'_N \\ v'_N \end{pmatrix}$$

The $2N \times 6$ matrix G' is given by

$$G' = \begin{pmatrix} A'_1 B'_1 \\ \vdots \\ A'_N B'_N \end{pmatrix}$$

where A_n and B_n are defined in eq. (14.27).

The normal equation (14.11) provides the least-squares solution to determine $\Delta\beta = (\Delta x_0, \Delta y_0, \Delta z_0, \Delta w_1, \Delta w_2, \Delta w_3)$. The adjustments $(\Delta w_1, \Delta w_2, \Delta w_3)$ determine the adjustments $(\Delta a, \Delta b, \Delta c, \Delta d)$ by

$$\begin{pmatrix} \Delta a \\ \Delta b \\ \Delta c \end{pmatrix} = \frac{1}{2} \begin{pmatrix} d' & c' & -b' \\ -c' & d' & a' \\ b' & -a' & d' \end{pmatrix} \begin{pmatrix} \Delta w_1 \\ \Delta w_2 \\ \Delta w_3 \end{pmatrix}$$

$$\Delta d = -\frac{a' \Delta a + b' \Delta b + c' \Delta c}{d'}$$

The new quaternion parameters $a^{t+1}, b^{t+1}, c^{t+1}$, and d^{t+1} are defined by

$$\begin{pmatrix} a^{t+1} \\ b^{t+1} \\ c^{t+1} \\ d^{t+1} \end{pmatrix} = k \left[\begin{pmatrix} a^t \\ b^t \\ c^t \\ d^t \end{pmatrix} + \begin{pmatrix} \Delta a \\ \Delta b \\ \Delta c \\ \Delta d \end{pmatrix} \right]$$

where $k = \sqrt{(a^t + \Delta a)^2 + (b^t + \Delta b)^2 + (c^t + \Delta c)^2 + (d^t + \Delta d)^2}$. And as before, the new translation parameters $(x_0^{t+1}, y_0^{t+1}, z_0^{t+1})$ are defined by

$$\begin{pmatrix} x_0^{t+1} \\ y_0^{t+1} \\ z_0^{t+1} \end{pmatrix} = \begin{pmatrix} x_0^t \\ y_0^t \\ z_0^t \end{pmatrix} + \begin{pmatrix} \Delta x_0 \\ \Delta y_0 \\ \Delta z_0 \end{pmatrix}$$

14.4   Relative Orientation

Relative orientation is typically concerned with the determination of the position and orientation of one photograph with respect to another, given a set of corresponding image points. After one camera is in relative orientation with another, corresponding points from the two photographs intersect in 3D space. Stereo and multiview imageries require solving the relative orientation problem. We illustrate the solution of the relative orientation problem for the case of stereo imagery. Classic papers on this topic include Thompson (1959, 1968), Shmid (1956), and Schut (1957).

Let (x_L, y_L, z_L) and (x_R, y_R, z_R) be the 3D positions of the left and right camera lenses, respectively. Let $(\omega_L, \phi_L, \kappa_L)$ and $(\omega_R, \phi_R, \kappa_R)$ be the rotation angles specifying the exterior orientation of the left and right camera reference frames, respectively. Let a set of N image points $\{(u_{Ln}, v_{Ln})\}_{n=1}^{N}$ from the left image and the corresponding set of N points $\{(u_{Rn}, v_{Rn})\}_{n=1}^{N}$ from the right image be given. The point (u_{Ln}, v_{Ln}) on the left image corresponds to the point (u_{Rn}, v_{Rn}) on the right image if there is some 3D point (x_n, y_n, z_n) such that (u_{Ln}, v_{Ln}) is the perspective projection of (x_n, y_n, z_n) on the left image and (u_{Rn}, v_{Rn}) is the perspective projection of (x_n, y_n, z_n) on the right image.

The photogrammetrists take the separation of the two camera lenses along the x-axis as a constant that controls the scale. It is not a parameter of the relative orientation. Therefore the relative orientation is specified by the five parameters $(y_R - y_L), (z_R - z_L), (w_R - w_L), (\phi_R - \phi_L)$, and $(\kappa_R - \kappa_L)$. Our solution assumes that the interior orientation of the cameras is known and that all image positions are expressed to identical scale and with respect to their principal points.

14.4.1   Standard Solution

To set up the problem, let

$$Q_L' = R(\omega_L, \phi_L, \kappa_L)$$

be the rotation matrix associated with the exterior orientation of the left image and

$$Q_R' = R(\omega_R, \phi_R, \kappa_R)$$

be the rotation matrix associated with the exterior orientation of the right image. Let f_R be the distance between the right image plane and the front of the right lens, and f_L the distance between the left image plane and the front of the left lens. Then from the perspective projection collinearity equation,

$$\begin{pmatrix} u_{Ln} \\ v_{Ln} \\ f_L \end{pmatrix} = \frac{1}{\lambda_{Ln}} Q_L' \begin{pmatrix} x_n - x_L \\ y_n - y_L \\ z_n - z_L \end{pmatrix} \quad \text{and} \quad \begin{pmatrix} u_{Rn} \\ v_{Rn} \\ f_R \end{pmatrix} = \frac{1}{\lambda_{Rn}} Q_R' \begin{pmatrix} x_n - x_R \\ y_n - y_R \\ z_n - z_R \end{pmatrix}$$

Hence

$$\begin{pmatrix} x_n \\ y_n \\ z_n \end{pmatrix} = \begin{pmatrix} x_L \\ y_L \\ z_L \end{pmatrix} + \lambda_{Ln} Q_L \begin{pmatrix} u_{Ln} \\ v_{Ln} \\ f_L \end{pmatrix} = \begin{pmatrix} x_R \\ y_R \\ z_R \end{pmatrix} + \lambda_{Rn} Q_L \begin{pmatrix} u_{Rn} \\ v_{Rn} \\ f_R \end{pmatrix} \qquad (14.37)$$

where $\lambda_{Ln} = \dfrac{s_{Ln}}{f_L}$ and $\lambda_{Rn} = \dfrac{s_{Rn}}{f_R}$.

Upon subtracting one equation from the other, we have

$$0 = \begin{pmatrix} x_R \\ y_R \\ z_R \end{pmatrix} - \begin{pmatrix} x_L \\ y_L \\ z_L \end{pmatrix} + \lambda_{Rn} Q_R \begin{pmatrix} u_{Rn} \\ v_{Rn} \\ f_R \end{pmatrix} - \lambda_{Ln} Q_L \begin{pmatrix} u_{Ln} \\ v_{Ln} \\ f_L \end{pmatrix}$$

Since the vector

$$Q_R \begin{pmatrix} u_{Rn} \\ v_{Rn} \\ f_R \end{pmatrix} \times Q_L \begin{pmatrix} u_{Ln} \\ v_{Ln} \\ f_L \end{pmatrix}$$

is orthogonal to both

$$Q_R \begin{pmatrix} u_{Rn} \\ v_{Rn} \\ f_R \end{pmatrix} \text{ and } Q_L \begin{pmatrix} u_{Ln} \\ v_{Ln} \\ f_L \end{pmatrix}$$

we obtain the coplanarity equation (14.38), which each pair of corresponding image points must satisfy:

$$\begin{pmatrix} x_R - x_L \\ y_R - y_L \\ z_R - z_L \end{pmatrix}' \left[Q_R \begin{pmatrix} u_{Rn} \\ v_{Rn} \\ f_R \end{pmatrix} \times Q_L \begin{pmatrix} u_{Ln} \\ v_{Ln} \\ f_L \end{pmatrix} \right] = 0, \quad n = 1, \dots, N \qquad (14.38)$$

Schut (1957) discusses the relation between the coplanarity constraint and other ways of setting up the problem. He shows that they can all be reduced to the coplanarity constraint.

Since the relative orientation is the reference frame of the right image specified with respect to the reference frame of the left image, we take $\begin{pmatrix} x_L \\ y_L \\ z_L \end{pmatrix} = 0$ and $Q_L = I$, the identity matrix. Also we take x_R to be known. From this it is clear that there must be at least five point correspondences to establish five coplanarity equations in order to solve for the five unknown parameters $y_R, z_R, \omega_R, \phi_R$, and κ_R. In this formulation the nonlinear system of equations that has to be solved in the least-squares sense for $y_R, z_R, \omega_R, \phi_R$, and κ_R is

$$\begin{pmatrix} x_R \\ y_R \\ z_R \end{pmatrix}' \left[Q_R \begin{pmatrix} u_{Rn} \\ v_{Rn} \\ f_R \end{pmatrix} \times \begin{pmatrix} u_{Ln} \\ v_{Ln} \\ f_L \end{pmatrix} \right] = 0, \quad n = 1, \dots, N \qquad (14.39)$$

To solve this system, we assume an initial approximate solution $y_R^0, z_R^0, \omega_R^0, \phi_R^0, \kappa_R^0$. At each iteration we linearize the nonlinear equations around the current solution and solve the linearized system in the least-squares sense. Letting $R(\omega_R, \phi_R, \kappa_R) = Q_R$, we have at iteration t

$$0 = \begin{pmatrix} x_R \\ y_R^t + \Delta y \\ z_R^t + \Delta z \end{pmatrix}' \left[R(\omega_R^t + \Delta\omega, \phi_R^t + \Delta\phi, \kappa_R^t + \Delta\kappa) \begin{pmatrix} u_{Rn} \\ v_{Rn} \\ f_R \end{pmatrix} \times \begin{pmatrix} u_{Ln} \\ v_{Ln} \\ f_L \end{pmatrix} \right]$$

$$= \begin{pmatrix} x_R \\ y_R^t \\ z_R^t \end{pmatrix}' \left[R(\omega_R^t, \phi_R^t, \kappa_R^t) \begin{pmatrix} u_{Rn} \\ v_{Rn} \\ f_R \end{pmatrix} \times \begin{pmatrix} u_{Ln} \\ v_{Ln} \\ f_L \end{pmatrix} \right] +$$

$$\begin{pmatrix} x_R \\ y_R^t \\ z_R^t \end{pmatrix}' \left[\Delta R(\omega_R^t, \phi_R^t, \kappa_R^t) \begin{pmatrix} u_{Rn} \\ v_{Rn} \\ f_R \end{pmatrix} \times \begin{pmatrix} u_{Ln} \\ v_{Ln} \\ f_L \end{pmatrix} \right] +$$

$$\begin{pmatrix} 0 \\ \Delta y \\ \Delta z \end{pmatrix}' \left[R(\omega_R^t, \phi_R^t, \kappa_R^t) \begin{pmatrix} u_{Rn} \\ v_{Rn} \\ f_R \end{pmatrix} \times \begin{pmatrix} u_{Ln} \\ v_{Ln} \\ f_L \end{pmatrix} \right]$$

Since

$$\Delta R(\omega_R^t, \phi_R^t, \kappa_R^t) = \frac{\partial R}{\partial \omega}(\omega_R^t, \phi_R^t, \kappa_R^t)\Delta\omega + \frac{\partial R}{\partial \phi}(\omega_R^t, \phi_R^t, \kappa_R^t)\Delta\phi + \frac{\partial R}{\partial \kappa}(\omega_R^t, \phi_R^t, \kappa_R^t)\Delta\kappa$$

we can rewrite the system as the linear overconstrained system

$$\begin{pmatrix} a_{11}^t & a_{21}^t & a_{31}^t & a_{51}^t & a_{61}^t \\ & & \vdots & & \\ a_{1N}^t & a_{2N}^t & a_{3N}^t & a_{5N}^t & a_{6N}^t \end{pmatrix} \begin{pmatrix} \Delta\omega \\ \Delta\phi \\ \Delta\kappa \\ \Delta y \\ \Delta z \end{pmatrix} = -x_R \begin{pmatrix} a_{41}^t \\ \vdots \\ a_{4N}^t \end{pmatrix} - y_R \begin{pmatrix} a_{51}^t \\ \vdots \\ a_{5N}^t \end{pmatrix} - z_R \begin{pmatrix} a_{61}^t \\ \vdots \\ a_{6N}^t \end{pmatrix}$$

$$(14.40)$$

where

$$a_{1n}^t = \begin{pmatrix} x_R \\ y_R^t \\ z_R^t \end{pmatrix}' \left[\frac{\partial R}{\partial \omega}(\omega_R^t, \phi_R^t, \kappa_R^t) \begin{pmatrix} u_{Rn} \\ v_{Rn} \\ f_R \end{pmatrix} \times \begin{pmatrix} u_{Ln} \\ v_{Ln} \\ f_L \end{pmatrix} \right]$$

$$a_{2n}^t = \begin{pmatrix} x_R \\ y_R^t \\ z_R^t \end{pmatrix}' \left[\frac{\partial R}{\partial \phi}(\omega_R^t, \phi_R^t, \kappa_R^t) \begin{pmatrix} u_{Rn} \\ v_{Rn} \\ f_R \end{pmatrix} \times \begin{pmatrix} u_{Ln} \\ v_{Ln} \\ f_L \end{pmatrix} \right]$$

$$a_{3n}^t = \begin{pmatrix} x_R \\ y_R^t \\ z_R^t \end{pmatrix}' \left[\frac{\partial R}{\partial \kappa}(\omega_R^t, \phi_R^t, \kappa_R^t) \begin{pmatrix} u_{Rn} \\ v_{Rn} \\ f_R \end{pmatrix} \times \begin{pmatrix} u_{Ln} \\ v_{Ln} \\ f_L \end{pmatrix} \right]$$

$$\begin{pmatrix} a_{4n}^t \\ a_{5n}^t \\ a_{6n}^t \end{pmatrix} = R(\omega_R^t, \phi_R^t, \kappa_R^t) \begin{pmatrix} u_{Rn} \\ v_{Rn} \\ f_R \end{pmatrix} \times \begin{pmatrix} u_{Ln} \\ v_{Ln} \\ f_L \end{pmatrix}, \quad n = 1, \ldots, N$$

This overconstrained linear system is solved in the least-squares sense, and the adjusted values for the next iteration are given by

$$
\begin{pmatrix} \omega_R^{t+1} \\ \phi_R^{t+1} \\ \kappa_R^{t+1} \\ y_R^{t+1} \\ z_R^{t+1} \end{pmatrix} = \begin{pmatrix} \omega_R^t \\ \phi_R^t \\ \kappa_R^t \\ y_R^t \\ z_R^t \end{pmatrix} + \begin{pmatrix} \Delta\omega \\ \Delta\phi \\ \Delta\kappa \\ \Delta y \\ \Delta z \end{pmatrix}
\tag{14.41}
$$

■ EXAMPLE 14.2

Suppose the center of perspective for the left camera is the origin, and for the right camera it is at (x_R, y_R, z_R), where $x_R = 5$. Suppose that the six corresponding pairs of perspective projection points are as given below.

	1	2	3	4	5	6
u_{L_n}	0.00157	0.0228	0.0067	0.0779	-0.00801	-0.054
v_{L_n}	0.0119	0.00833	0.0406	0.0405	-0.0327	-0.0327
u_{R_n}	-0.0347	-0.00655	-0.0273	0.101	-0.0297	-0.0749
v_{R_n}	0.0610	0.0581	0.109	0.106	-0.0129	-0.00889

(table column header: n)

Then after 10 iterations of Eqs. (14.41) and (14.40), we compute

$$\omega_R = 0.0506 \text{ radians}$$
$$\phi_R = -0.127 \text{ radians}$$
$$\kappa_R = 0.0811 \text{ radians}$$
$$y_R = -3.638$$
$$z_R = 3.605$$

■

14.4.2 Quaternian Solution

Hinsken (1988) sets up the problem in a slightly different way. Instead of determining the relative orientation of the right image with respect to the left image, he aligns a reference frame having its x-axis along the line from the left image lens to the right image lens. The relative orientation is then determined by the angles $(\omega_R, \phi_R, \kappa_R)$, which rotate the right image into this reference frame, and the angles $(\omega_L, \phi_L, \kappa_L)$, which rotate the left image into this reference frame.

At first glance it appears that this setup has six parameters. But we will shortly see that there are really only five linearly independent parameters, just as there were in the first way we set up the relative orientation problem. If we let $b_x = x_R - x_L$, eq. (14.38) becomes

$$
\begin{pmatrix} b_x \\ 0 \\ 0 \end{pmatrix}' \left[R(\omega_R, \phi_R, \kappa_R) \begin{pmatrix} u_{Rn} \\ v_{Rn} \\ f_R \end{pmatrix} \times R(\omega_L, \phi_L, \kappa_L) \begin{pmatrix} u_{Ln} \\ v_{Ln} \\ f_L \end{pmatrix} \right] = 0, \quad n = 1, \dots, N
$$

To solve this system of overconstrained equations, we proceed as before and linearize:

$$0 = \begin{pmatrix} b_x \\ 0 \\ 0 \end{pmatrix}' \left[R(\omega_R + \Delta\omega_R, \phi_r + \Delta\phi_R, \kappa_R + \Delta\kappa_R) \begin{pmatrix} u_{Rn} \\ v_{Rn} \\ f_R \end{pmatrix} \right.$$

$$\left. \times R(\omega_L + \Delta\omega_L, \phi_r + \Delta\phi_L, \kappa_L + \Delta\kappa_L) \begin{pmatrix} u_{Ln} \\ v_{Ln} \\ f_L \end{pmatrix} \right]$$

$$= \begin{pmatrix} b_x \\ 0 \\ 0 \end{pmatrix}' \left[R(\omega_R, \phi_R, \kappa_R) \begin{pmatrix} u_{Rn} \\ v_{Rn} \\ f_R \end{pmatrix} \times R(\omega_L, \phi_L, \kappa_L) \begin{pmatrix} u_{Ln} \\ v_{Ln} \\ f_R \end{pmatrix} \right]$$

$$+ \begin{pmatrix} b_x \\ 0 \\ 0 \end{pmatrix}' \left[R(\omega_R, \phi_R, \kappa_R) \begin{pmatrix} u_{Rn} \\ v_{Rn} \\ f_R \end{pmatrix} \times \Delta R(\omega_L, \phi_L, \kappa_L) \begin{pmatrix} u_{Ln} \\ v_{Ln} \\ f_R \end{pmatrix} \right]$$

$$+ \begin{pmatrix} b_x \\ 0 \\ 0 \end{pmatrix}' \left[\Delta R(\omega_R, \phi_R, \kappa_R) \begin{pmatrix} u_{Rn} \\ v_{Rn} \\ f_R \end{pmatrix} \times R(\omega_L, \phi_L, \kappa_L) \begin{pmatrix} u_{Ln} \\ v_{Ln} \\ f_R \end{pmatrix} \right]$$

Recall that for any rotation matrix R, $\Delta R R' = S$, where S is a skew symmetric matrix. Hence $\Delta R = SR$, so that

$$R(\omega_R, \phi_R, \kappa_R) \begin{pmatrix} u_{Rn} \\ v_{Rn} \\ f_R \end{pmatrix} \times \Delta R(\omega_L, \phi_L, \kappa_L) \begin{pmatrix} u_{Ln} \\ v_{Ln} \\ f_R \end{pmatrix}$$

$$= R(\omega_R, \phi_R, \kappa_R) \begin{pmatrix} u_{Rn} \\ v_{Rn} \\ f_R \end{pmatrix} \times S_L R(\omega_L, \phi_L, \kappa_L) \begin{pmatrix} u_{Ln} \\ v_{Ln} \\ f_R \end{pmatrix}$$

Letting

$$\begin{pmatrix} p_{Rn} \\ q_{Rn} \\ s_{Rn} \end{pmatrix} = R(\omega_R, \phi_R, \kappa_R) \begin{pmatrix} u_{Rn} \\ v_{Rn} \\ f_R \end{pmatrix}$$

$$\begin{pmatrix} p_{Ln} \\ q_{Ln} \\ s_{Ln} \end{pmatrix} = L(\omega_L, \phi_L, \kappa_L) \begin{pmatrix} u_{Ln} \\ v_{Ln} \\ f_L \end{pmatrix}$$

$$S_L = \begin{pmatrix} 0 & \Delta\omega_{L3} & -\Delta\omega_{L2} \\ -\Delta\omega_{L3} & 0 & \Delta\omega_{L1} \\ \Delta\omega_{L2} & -\Delta\omega_{L1} & 0 \end{pmatrix},$$

then

$$\begin{pmatrix} b_x \\ 0 \\ 0 \end{pmatrix}' \left[R(\omega_R, \phi_R, \kappa_R) \begin{pmatrix} u_{Rn} \\ v_{Rn} \\ f_R \end{pmatrix} \times \Delta R(\omega_L, \phi_L, \kappa_L) \begin{pmatrix} u_{Ln} \\ v_{Ln} \\ f_L \end{pmatrix} \right]$$

$$= \begin{pmatrix} b_x \\ 0 \\ 0 \end{pmatrix}' \left\{ \begin{pmatrix} p_{Rn} \\ q_{Rn} \\ s_{Rn} \end{pmatrix} \times \left[\begin{pmatrix} \Delta\omega_{L1} \\ \Delta\omega_{L2} \\ \Delta\omega_{L3} \end{pmatrix} \times \begin{pmatrix} p_{Ln} \\ q_{Ln} \\ s_{Ln} \end{pmatrix} \right] \right\}$$

$$= b_x(q_{Rn}q_{Ln} + s_{Rn}s_{Ln} - q_{Rn}p_{Ln} - s_{Rn}p_{Ln}) \begin{pmatrix} \Delta\omega_{L1} \\ \Delta\omega_{L2} \\ \Delta\omega_{L3} \end{pmatrix}$$

Similarly,

$$\begin{pmatrix} b_x \\ 0 \\ 0 \end{pmatrix}' \left[\Delta R(\omega_R, \phi_R, \kappa_R) \begin{pmatrix} u_{Rn} \\ v_{Rn} \\ f_R \end{pmatrix} \times R(\omega_L, \phi_L, \kappa_L) \begin{pmatrix} u_{Ln} \\ v_{Ln} \\ f_L \end{pmatrix} \right]$$

$$= \begin{pmatrix} b_x \\ 0 \\ 0 \end{pmatrix}' \left[S_R R(\omega_R, \phi_R, \kappa_R) \begin{pmatrix} u_{Rn} \\ v_{Rn} \\ f_R \end{pmatrix} \times R(\omega_L, \phi_L, \kappa_L) \begin{pmatrix} u_{Ln} \\ v_{Ln} \\ f_L \end{pmatrix} \right]$$

$$= \begin{pmatrix} b_x \\ 0 \\ 0 \end{pmatrix}' \left\{ \left[-\begin{pmatrix} \Delta\omega_{R1} \\ \Delta\omega_{R2} \\ \Delta\omega_{R3} \end{pmatrix} \times \begin{pmatrix} p_{Rn} \\ q_{Rn} \\ s_{Rn} \end{pmatrix} \right] \times \begin{pmatrix} p_{Ln} \\ q_{Ln} \\ s_{Ln} \end{pmatrix} \right\}$$

$$= -\begin{pmatrix} b_x \\ 0 \\ 0 \end{pmatrix}' \left\{ \begin{pmatrix} p_{Ln} \\ q_{Ln} \\ s_{Ln} \end{pmatrix} \times \left[-\begin{pmatrix} \Delta\omega_{R1} \\ \Delta\omega_{R2} \\ \Delta\omega_{R3} \end{pmatrix} \times \begin{pmatrix} p_{Rn} \\ q_{Rn} \\ s_{Rn} \end{pmatrix} \right] \right\}$$

$$= bx(q_{Ln}q_{Rn} + s_{Ln}s_{Rn} - q_{Ln}p_{Rn} - s_{Ln}p_{Rn}) \begin{pmatrix} \Delta\omega_{R1} \\ \Delta\omega_{R2} \\ \Delta\omega_{R3} \end{pmatrix}$$

It is immediately apparent that $\Delta\omega_{L1}$ and $\Delta\omega_{R1}$ are negations of each other, so that at iteration t, where the current solution is $(\omega_R^t, \phi_R^t, \kappa_R^t, \omega_L^t, \phi_L^t, \kappa_L^t)$, the least-squares system may take the reduced form

$$\begin{pmatrix} -(q_{R1}^t q_{L1}^t + s_{R1}^t s_{L1}^t) & q_{R1}^t p_{L1}^t & s_{R1}^t p_{L1}^t & -q_{L1}^t p_{R1}^t & -s_{L1}^t p_{R1}^t \\ & & \vdots & & \\ -(q_{Rn}^t q_{Ln}^t + s_{Rn}^t s_{Ln}^t) & q_{Rn}^t p_{Ln}^t & s_{Rn}^t p_{Ln}^t & -q_{Ln}^t p_{Rn}^t & -s_{Ln}^t p_{Rn}^t \end{pmatrix} \begin{pmatrix} \Delta\omega_{L1} \\ \Delta\omega_{L2} \\ \Delta\omega_{L3} \\ \Delta\omega_{R2} \\ \Delta\omega_{R3} \end{pmatrix}$$

$$= \begin{pmatrix} q_{L1}^t s_{R1}^t - q_{R1}^t s_{L1}^t \\ \vdots \\ q_{Ln}^t s_{Rn}^t - q_{Rn}^t s_{Ln}^t \end{pmatrix}$$

where

$$\begin{pmatrix} p_{Rn}^t \\ q_{Rn}^t \\ s_{Rn}^t \end{pmatrix} = R(\omega_R^t, \phi_R^t, \kappa_R^t) \begin{pmatrix} u_{Rn} \\ v_{Rn} \\ f_R \end{pmatrix}$$

$$\begin{pmatrix} p_{Ln}^t \\ q_{Ln}^t \\ s_{Ln}^t \end{pmatrix} = R(\omega_L^t, \phi_L^t, \kappa_L^t) \begin{pmatrix} u_{Ln} \\ v_{Ln} \\ f_L \end{pmatrix}$$

Once a solution is determined for $(\Delta\omega_{L1}, \Delta\omega_{L2}, \Delta\omega_{L3}, \Delta\omega_{R2}, \Delta\omega_{R3})$, we can use Eq. (14.25) to obtain

$$\begin{pmatrix} \Delta\omega_L \\ \Delta\phi_L \\ \Delta\kappa_L \end{pmatrix} = \begin{pmatrix} \cos\kappa_L^t / \cos\phi_L^t & -\sin\kappa_L^t / \cos\phi_L^t & 0 \\ \sin\kappa_L^t & \cos\kappa_L^t & 0 \\ -\cos\kappa_L^t \tan\phi_L^t & \sin\kappa_L^t \tan\phi_L^t & 1 \end{pmatrix} \begin{pmatrix} \Delta\omega_{L1} \\ \Delta\omega_{L2} \\ \Delta\omega_{L3} \end{pmatrix}$$

$$\begin{pmatrix} \Delta\omega_R \\ \Delta\phi_R \\ \Delta\kappa_R \end{pmatrix} = \begin{pmatrix} \cos\kappa_R^t / \cos\phi_R^t & -\sin\kappa_R^t / \cos\phi_R^t & 0 \\ \sin\kappa_R^t & \cos\kappa_R^t & 0 \\ -\cos\kappa_R^t \tan\phi_R^t & \sin\kappa_R^t \tan\phi_R^t & -1 \end{pmatrix} \begin{pmatrix} -\Delta\omega_{L1} \\ \Delta\omega_{R2} \\ \Delta\omega_{R3} \end{pmatrix}$$

and the updated solution is determined from

$$\begin{pmatrix} \omega_R^{t+1} \\ \phi_R^{t+1} \\ \kappa_R^{t+1} \\ \omega_L^{t+1} \\ \phi_L^{t+1} \\ \kappa_L^{t+1} \end{pmatrix} = \begin{pmatrix} \omega_R^t \\ \phi_R^t \\ \kappa_R^t \\ \omega_L^t \\ \phi_L^t \\ \kappa_L^t \end{pmatrix} + \begin{pmatrix} \Delta\omega_R \\ \Delta\phi_R \\ \Delta\kappa_R \\ \Delta\omega_L \\ \Delta\phi_L \\ \Delta\kappa_L \end{pmatrix}$$

As in the absolute orientation problem, the quaternion representation for a rotation matrix can also be used here. Letting $a_R^t, b_R^t, c_R^t, d_R^t$ and $a_L^t, b_L^t, c_L^t, d_L^t$ be the quaternion parameters for the rotation of the right and left images, respectively, at iteration t we can use Eqs. (14.36) and (14.34) to obtain

$$\begin{pmatrix} \Delta a_L \\ \Delta b_L \\ \Delta c_L \end{pmatrix} = \frac{1}{2} \begin{pmatrix} d_L^t & -c_L^t & b_L^t \\ c_L^t & d_L^t & -a_L^t \\ -b_L^t & a_L^t & d_L^t \end{pmatrix} \begin{pmatrix} \Delta\omega_{L1} \\ \Delta\omega_{L2} \\ \Delta\omega_{L3} \end{pmatrix}$$

$$\Delta d_L = -\frac{a_L^t \Delta a_L + b_L^t \Delta b_L + c_L^t \Delta c_L}{d_L^t}$$

$$
\begin{pmatrix} \Delta a_R \\ \Delta b_R \\ \Delta c_R \end{pmatrix} = \frac{1}{2} \begin{pmatrix} d_R^t & -c_R^t & b_R^t \\ c_R^t & d_R^t & -a_R^t \\ -b_R^t & a_R^t & d_R^t \end{pmatrix} \begin{pmatrix} \Delta \omega_{R1} \\ \Delta \omega_{R2} \\ \Delta \omega_{R3} \end{pmatrix}
$$

$$
\Delta d_R = -\frac{a_R^t \Delta a_R + b_R^t \Delta b_R + c_R^t \Delta c_R}{d_R^t}
$$

The updated solution is then given by

$$
\begin{pmatrix} a_R^{t+1} \\ b_R^{t+1} \\ c_R^{t+1} \\ d_R^{t+1} \end{pmatrix} = k_R \left[\begin{pmatrix} a_R^t \\ b_R^t \\ c_R^t \\ d_R^t \end{pmatrix} + \begin{pmatrix} \Delta a_R \\ \Delta b_R \\ \Delta c_R \\ \Delta d_R \end{pmatrix} \right]
$$

where

$$
k_R = \frac{1}{\sqrt{(a_R^t + \Delta a_R)^2 + (b_R^t + \Delta b_R)^2 + (c_R^t + \Delta c_R)^2 + (d_R^t + \Delta d_R)^2}}
$$

and

$$
\begin{pmatrix} a_L^{t+1} \\ b_L^{t+1} \\ c_L^{t+1} \\ d_L^{t+1} \end{pmatrix} = k_L \left[\begin{pmatrix} a_L^t \\ b_L^t \\ c_L^t \\ d_L^t \end{pmatrix} + \begin{pmatrix} \Delta a_L \\ \Delta b_L \\ \Delta c_L \\ \Delta d_L \end{pmatrix} \right]
\tag{14.42}
$$

where

$$
k_L = \frac{1}{\sqrt{(a_L^t + \Delta a_L)^2 + (b_L^t + \Delta b_L)^2 + (c_L^t + \Delta c_L)^2 + (d_L^t + \Delta d_L)^2}}
$$

14.5   Interior Orientation

The *interior orientation*, or *inner orientation*, of a camera is specified by the camera constant f, the distance between the image plane and the camera lens; by the principal point (u_p, v_p), which is the intersection of the optic axis with the image plane in the measurement reference frame located on the image plane; and by the geometric distortion characteristics of the lens, which we assume here to be isotropic around the principal point. Several techniques for the determination of interior orientation have been published, including those of Abdel-Aziz and Karara (1974), Faig (1971), Dohler (1971), Mehrotra and Karara (1975), Karara and Faig (1972), Brown (1971, 1972), and Washer (1957, 1963).

To determine the inner orientation, one must actually solve the absolute orientation problem, with the additional parameters being those of the interior orientation.

The model we use is

$$(k_1 r_n^2 + k_2 r_n^4 + k_3 r_n^6)\begin{pmatrix} u_n - u_p \\ v_n - v_p \end{pmatrix} = \frac{f}{s_n}\begin{pmatrix} p_n \\ q_n \end{pmatrix}, \qquad n = 1, \ldots, N \qquad (14.43)$$

where

$$\begin{pmatrix} p_n(x_0, y_0, z_0, \omega, \phi, \kappa) \\ q_n(x_0, y_0, z_0, \omega, \phi, \kappa) \\ s_n(x_0, y_0, z_0, \omega, \phi, \kappa) \end{pmatrix} = R(\omega, \phi, \kappa)\begin{pmatrix} x_n - x_0 \\ y_n - y_0 \\ z_n - z_0 \end{pmatrix}$$

$$r_n^2(u_p, v_p) = (u_n - u_p)^2 + (v_n - v_p)^2$$

The unknowns are the parameters of the interior orientation $f, u_p, v_p, k_1, k_2, k_3$ and the parameters of the exterior orientation $x_0, y_0, z_0, \omega, \phi, \kappa$. To determine the values of the unknown parameters, we solve the nonlinear least-squares system Eq. (14.43) by iteratively linearizing it around the current approximate solution and solving the linearized system for the next approximate solution. The term $k_1 r_n^2 + k_2 r_n^4 + k_3 r_n^6$ accounts only for the radial geometric distortion. More complex distortion models are possible; for example, Brown (1966) discusses using both radial distortion and tangential distortions.

We linearize the left-hand side of Eq. (14.43) first.

$$\Big[(k_1 + \Delta k_1)r_n^2(u_p + \Delta u_p, v_p + \Delta v_p) + (k_2 + \Delta k_2)r_n^4(u_p + \Delta u_p, v_p + \Delta v_p)$$

$$+ (k_3 + \Delta k_3)r_n^6(u_p + \Delta u_p, v_p + \Delta v_p)\Big]\begin{bmatrix} u_n - (u_p + \Delta u_p) \\ v_n - (v_p + \Delta v_p) \end{bmatrix}$$

$$= [k_1 r_n^2(u_p, v_p) + k_2 r_n^4(u_p, v_p) + k_3 r_n^6(u_p, v_p)]\begin{pmatrix} u_n - u_p \\ v_n - v_p \end{pmatrix}$$

$$+ \left\{ [k_1 r_n^2(u_p, v_p) + k_2 r_n^4(u_p, v_p) + k_3 r_n^6(u_p, v_p)]\begin{pmatrix} -1 \\ 0 \end{pmatrix} \right.$$

$$+ [2k_1 r_n^2(u_p, v_p) + 4k_2 r_n^3(u_p, v_p) + 6k_3 r_n^5(u_p, v_p)]\begin{pmatrix} u_n - u_p \\ v_n - v_p \end{pmatrix} \right\} \Delta u_p$$

$$+ \left\{ [k_1 r_n^2(u_p, v_p) + k_2 r_n^4(u_p, v_p) + k_3 r_n^6(u_p, v_p)]\begin{pmatrix} 0 \\ -1 \end{pmatrix} \right.$$

$$+ [2k_1 r_n^2(u_p, v_p) + 4k_2 r_n^3(u_p, v_p) + 6k_3 r_n^5(u_p, v_p)]\begin{pmatrix} u_n - u_p \\ v_n - v_p \end{pmatrix} \right\} \Delta v_p$$

$$+ [r_n^2(u_p, v_p)\Delta k_1 + r_n^4(u_p, v_p)\Delta k_2 + r_n^6(u_p, v_p)\Delta k_3]\begin{pmatrix} u_n - u_p \\ v_n - v_p \end{pmatrix}$$

At iteration t, the left-hand side then has the following form:

$$= (k_1^t r_n^{t2} + k_2^t r_n^{t4} + k_3^t r_n^{t6}) \begin{pmatrix} u_n - u_p^t \\ v_n - v_p^t \end{pmatrix} + C_n^t \begin{pmatrix} \Delta u_p \\ \Delta v_p \\ \Delta k_1 \\ \Delta k_2 \\ \Delta k_3 \end{pmatrix}$$

where

$$C_n^t = \begin{pmatrix} c_{11n}^t & c_{12n}^t & c_{13n}^t & c_{14n}^t & c_{15n}^t \\ c_{21n}^t & c_{22n}^t & c_{23n}^t & c_{24n}^t & c_{25n}^t \end{pmatrix}$$

$$c_{11n}^t = -(k_1^t r_n^{t2} + k_2^t r_n^{t4} + k_3^t r_n^{t6}) + (u_n - u_p^t)(2k_1^t r_n^t + 4k_2^t r_n^{t3} + 6k_3^t r_n^{t5})$$

$$a_{12n}^t = (u_n - u_p^t)(2k_1^t r_n^t + 4k_2^t r_n^{t3} + 6k_3^t r_n^{t5})$$

$$a_{13n}^t = (u_n - u_p^t)r_n^{t2}$$

$$a_{14n}^t = (u_n - u_p^t)r_n^{t4}$$

$$a_{15n}^t = (u_n - u_p^t)r_n^{t6}$$

$$a_{21n}^t = (v_n - v_p^t)(2k_1^t r_n^t + 4k_2^t r_n^{t3} + 6k_3^t r_n^{t5})$$

$$a_{22n}^t = -(k_1^t r_n^{t2} + k_2^t r_n^{t4} + k_3^t r_n^{t6}) + (v_n - v_p^t)(2k_1^t r_n^t + 4k_2^t r_n^{t3} + 6k_3^t r_n^{t5})$$

$$a_{23n}^t = (v_n - v_p^t)r_n^{t2}$$

$$a_{24n}^t = (v_n - v_p^t)r_n^{t4}$$

$$a_{25n}^t = (v_n - v_p^t)r_n^{t6}$$

Now we linearize the right-hand side of Eq. (14.43). At iteration t, we have

$$(f^t + \Delta f) \frac{\begin{bmatrix} p_n(x_0^t + \Delta x_0, y_0^t + \Delta y_0, z_0^t + \Delta z_0, \omega^t + \Delta\omega, \phi^t + \Delta\phi, \kappa^t + \Delta\kappa) \\ q_n(x_0^t + \Delta x_0, y_0^t + \Delta y_0, z_0^t + \Delta z_0, \omega^t + \Delta\omega, \phi^t + \Delta\phi, \kappa^t + \Delta\kappa) \end{bmatrix}}{s_n(x_0^t + \Delta x_0, y_0^t + \Delta y_0, z_0^t + \Delta z_0, \omega^t + \Delta\omega, \phi^t + \Delta\phi, \kappa^t + \Delta\kappa)}$$

$$= f^t \frac{\begin{bmatrix} p_n(x_0^t, y_0^t, z_0^t, \omega^t, \phi^t, \kappa^t) \\ q_n(x_0^t, y_0^t, z_0^t, \omega^t, \phi^t, \kappa^t) \end{bmatrix}}{s_n(x_0^t, y_0^t, z_0^t, \omega^t, \phi^t, \kappa^t)} + \begin{pmatrix} \Delta f \\ \Delta x_0 \\ \Delta y_0 \\ \Delta z_0 \\ \Delta\omega \\ \Delta\phi \\ \Delta\kappa \end{pmatrix}$$

where

$$D_n^t = \begin{bmatrix} \dfrac{p_n(x_0^t, y_0^t, z_0^t, \omega^t, \phi^t, \kappa^t)}{s_n(x_0^t, y_0^t, z_0^t, \omega^t, \phi^t, \kappa^t)} & A_n^t B_n^t \\[2em] \dfrac{q_n(x_0^t, y_0^t, z_0^t, \omega^t, \phi^t, \kappa^t)}{s_n(x_0^t, y_0^t, z_0^t, \omega^t, \phi^t, \kappa^t)} & \end{bmatrix}$$

$$A_n^t = \frac{f^t}{s_n(x_0^t, y_0^t, z_0^t, \omega^t, \phi^t, \kappa^t)}$$

$$\times \begin{bmatrix} 1 & 0 & -p_n(x_0^t, y_0^t, z_0^t, \omega^t, \phi^t, \kappa^t)/s_n(x_0^t, y_0^t, z_0^t, \omega^t, \phi^t, \kappa^t) \\ 0 & 1 & -q_n(x_0^t, y_0^t, z_0^t, \omega^t, \phi^t, \kappa^t)/s_n(x_0^t, y_0^t, z_0^t, \omega^t, \phi^t, \kappa^t) \end{bmatrix}$$

and

$$B_n^t = \left[-R(\omega^t, \phi^t, \kappa^t)\ Q(\omega^t, \phi^t, \kappa^t, x_0^t, y_0^t, z_0^t) \right]$$

$$Q(\omega^t, \phi^t, \kappa^t, x_0^t, y_0^t, z_0^t) =$$

$$\left[\frac{\partial R}{\partial \omega}(\omega^t, \phi^t, \kappa^t) \begin{pmatrix} x_0 - x_0^t \\ y_0 - y_0^t \\ z_0 - z_0^t \end{pmatrix} \frac{\partial R}{\partial \phi}(\omega^t, \phi^t, \kappa^t) \begin{pmatrix} x_0 - x_0^t \\ y_0 - y_0^t \\ z_0 - z_0^t \end{pmatrix} \frac{\partial R}{\partial \kappa}(\omega^t, \phi^t, \kappa^t) \begin{pmatrix} x_0 - x_0^t \\ y_0 - y_0^t \\ z_0 - z_0^t \end{pmatrix} \right]$$

just as in Eq. (14.18).

Putting together the linearized left- and right-hand sides, we have

$$(k_1^t r_n^{t2} + k_2^t r_n^{t4} + k_3^t r_n^{t6}) \begin{pmatrix} u_n - u_p^t \\ v_n - v_p^t \end{pmatrix} + A_n^t \begin{pmatrix} \Delta u_p \\ \Delta v_p \\ \Delta k_1 \\ \Delta k_2 \\ \Delta k_3 \end{pmatrix} = \frac{f^t}{s_n^t} \begin{pmatrix} p_n^t \\ q_n^t \end{pmatrix} + D_n^t \begin{pmatrix} \Delta f \\ \Delta x_0 \\ \Delta y_0 \\ \Delta z_0 \\ \Delta \omega \\ \Delta \phi \\ \Delta \kappa \end{pmatrix}$$

Collecting all the adjustments together, we have

$$E_n^t \begin{pmatrix} \Delta u_p \\ \Delta v_p \\ \Delta k_1 \\ \Delta k_2 \\ \Delta k_3 \\ \Delta f \\ \Delta x_0 \\ \Delta y_0 \\ \Delta z_0 \\ \Delta \omega \\ \Delta \phi \\ \Delta \kappa \end{pmatrix} = (k_1^t r_n^{t2} + k_2^t r_n^{t4} + k_3^t r_n^{t6}) \begin{pmatrix} u_n - u_p^t \\ v_n - v_p^t \end{pmatrix} - \frac{f^t}{s_n^t} \begin{pmatrix} p_n^t \\ q_n^t \end{pmatrix} = h_n^t$$

where

$$E'_n = (-A'_n \; D'_n).$$

The total linearized least-squares system then becomes the following overconstrained linear system of $2N$ equations:

$$E' \begin{pmatrix} \Delta u_p \\ \Delta v_p \\ \Delta k_1 \\ \Delta k_2 \\ \Delta k_3 \\ \Delta f \\ \Delta x_0 \\ \Delta y_0 \\ \Delta z_0 \\ \Delta \omega \\ \Delta \phi \\ \Delta \kappa \end{pmatrix} = h'$$

where

$$E' = \begin{pmatrix} E'_1 \\ E'_2 \\ \vdots \\ E'_N \end{pmatrix} \quad \text{and } h' = \begin{pmatrix} h'_1 \\ h'_2 \\ \vdots \\ h'_N \end{pmatrix}$$

If we assume that the random perturbations associated with each row of the system are uncorrelated and have identical variance, the normal equation is

$$\begin{pmatrix} \Delta u_p \\ \Delta v_p \\ \Delta k_1 \\ \Delta k_2 \\ \Delta k_3 \\ \Delta f \\ \Delta x_0 \\ \Delta y_0 \\ \Delta z_0 \\ \Delta \omega \\ \Delta \phi \\ \Delta \kappa \end{pmatrix} = (E'^t E')^{-1} E'^t h'$$

Then the next iteration approximate solution is defined by

$$
\begin{pmatrix}
u_p^{t+1} \\
v_p^{t+1} \\
k_1^{t+1} \\
k_2^{t+1} \\
k_3^{t+1} \\
x_0^{t+1} \\
y_0^{t+1} \\
z_0^{t+1} \\
\omega^{t+1} \\
\phi^{t+1} \\
\kappa^{t+1}
\end{pmatrix}
=
\begin{pmatrix}
u_p^t \\
v_p^t \\
k_1^t \\
k_2^t \\
k_3^t \\
x_0^t \\
y_0^t \\
z_0^t \\
\omega^t \\
\phi^t \\
\kappa^t
\end{pmatrix}
+
\begin{pmatrix}
\Delta u_p \\
\Delta v_p \\
\Delta k_1 \\
\Delta k_2 \\
\Delta k_3 \\
\Delta x_0 \\
\Delta y_0 \\
\Delta z_0 \\
\Delta \omega \\
\Delta \phi \\
\Delta \kappa
\end{pmatrix}
$$

14.6   Stereo

Here we discuss the triangulation procedure on which is based the inference of the position of a 3D point (x, y, z) given to the perspective projection (u_L, v_L) on the left and (u_R, v_R) on the right image of a stereo pair. The triangulation procedure is a special case of forward section, which is the determination of a 3D point from the intersection of more than two rays. We assume that (u_L, v_L) and (u_R, v_R) are referenced to the principal point and that any lens geometric distortion has been compensated for. The triangulation procedure makes use of the parallax, which is the displacement in the perspective projection of a point caused by a translational change in the position of observation.

To illustrate the stereo triangulation procedure, we first consider the case in which the position of the left camera lens is at

$$
\begin{pmatrix}
-b_x/2 \\
0 \\
0
\end{pmatrix}
$$

and the position of the right camera lens is at

$$
\begin{pmatrix}
b_x/2 \\
0 \\
0
\end{pmatrix}.
$$

Assume that the image plane is at a distance f in front of each camera lens and that both cameras are oriented identically, with the x-axis of the camera reference frame oriented along the line defined by the position of the camera lenses. Let (x, y, z)

be a 3D point and let (u_L, v_L) and (u_R, v_R) be its perspective projection on the left and right images, respectively. Then

$$\begin{pmatrix} u_L \\ v_L \end{pmatrix} = \frac{f}{z} \begin{pmatrix} x + b_x/2 \\ y \end{pmatrix} \quad \text{and} \quad \begin{pmatrix} u_R \\ v_R \end{pmatrix} = \frac{f}{z} \begin{pmatrix} x - b_x/2 \\ y \end{pmatrix}$$

Note that in this situation $v_L = v_R$, so that the y-parallax is zero.

The solution for (x, y, z), given (u_L, v_L) and (u_R, v_R), can be obtained from the difference $u_L - u_R$, which is called the x-parallax. Now

$$u_L - u_R = \frac{f}{z} \left[x + \frac{b_x}{L} - \left(x - \frac{b_x}{L} \right) \right] = \frac{f}{z}(b_x)$$

Hence

$$z = \frac{f b_x}{u_L - u_R} \tag{14.44}$$

Once the depth z is determined, the (x, y) coordinates are easily determined from the perspective projection equations:

$$\begin{pmatrix} x \\ y \end{pmatrix} = \frac{z}{f} \begin{pmatrix} u_L \\ v_L \end{pmatrix} - \begin{pmatrix} b_x/2 \\ 0 \end{pmatrix} \tag{14.45}$$

$$= \frac{z}{f} \begin{pmatrix} u_R \\ v_R \end{pmatrix} + \begin{pmatrix} b_x/2 \\ 0 \end{pmatrix}$$

EXAMPLE 14.3

Suppose that the camera constant $f = 10$, the base $b_x = 4$, and the perspective projections are $(u_L, v_L) = (2, 3)$ and $(u_R, v_R) = (1.5, 3)$. Then by Eq. (14.44)

$$z = \frac{10 \times 4}{.5} = 80$$

and by Eq. (14.41)

$$\begin{pmatrix} x \\ y \end{pmatrix} = \frac{80}{10} \begin{pmatrix} 2 \\ 3 \end{pmatrix} - \begin{pmatrix} 2 \\ 0 \end{pmatrix} = \begin{pmatrix} 14 \\ 24 \end{pmatrix}$$

Hence

$$\begin{pmatrix} x \\ y \\ z \end{pmatrix} = \begin{pmatrix} 14 \\ 24 \\ 80 \end{pmatrix}$$

The relation $z = fb_x/(u_L - u_R)$ to determine the depth from the x-parallax is a classic relation that in real-world situations is actually close to being useless, for three simple reasons: (1) The observed perspective projections are subject to measurement errors, so that $v_L \neq v_R$ for corresponding points; (2) the camera reference frames for the left and right images may often have slightly different orientations; and (3) when there are two different cameras that take the left and right images, it is almost always the case that the camera constant $f_R \neq f_L$. This means that even if the relation $z = fb_x/(u_L - u_R)$ were generalized to include the effects of $f_R \neq f_L$ and to use u_L and u_R, where they now represent x-coordinates after rotating the measured left and right image frame coordinates to correct for the different orientation of the left and right camera reference frames, there would still be a problem because, owing to measurement and orientation error, $v_L \neq v_R$. This means that the ray

$$
L = \left\{ \begin{pmatrix} x \\ y \\ z \end{pmatrix} \middle| \begin{pmatrix} x \\ y \\ z \end{pmatrix} = \begin{pmatrix} b_x/2 \\ 0 \\ 0 \end{pmatrix} + \lambda_L \begin{pmatrix} u_L \\ v_L \\ f_L \end{pmatrix} \right\}
$$

associated with the left image point (u_L, v_L) and the ray

$$
R = \left\{ \begin{pmatrix} x \\ y \\ z \end{pmatrix} \middle| \begin{pmatrix} x \\ y \\ z \end{pmatrix} = \begin{pmatrix} -b_x/2 \\ 0 \\ 0 \end{pmatrix} + \lambda_R \begin{pmatrix} u_R \\ v_R \\ f_R \end{pmatrix} \right\}
$$

associated with the corresponding right image point (u_R, v_R) will not intersect in 3D space. In this case the best we can do is find that 3D point whose distance to L and R is minimized.

In the common reference frame that the relative orientation has already been computed for in the case of both images, let (x_L, y_L, z_L) be the position of the camera lens of the left image, and $R(\omega_L, \phi_L, \kappa_L)$ be the rotation that rotates the reference frame of the left image into the common reference frame. Similarly, let (x_R, y_R, z_R) be the position of the camera lens of the right image, and $R(\omega_R, \phi_R, \kappa_R)$ be the rotation that rotates the reference frame of the right image into the common reference frame. Let f_R and f_L be the camera constants associated with the right and left images, respectively. Then by Eq. (14.37), and provided there is no measurement noise, the point (x, y, z) must lie on the intersection of the two rays:

$$
\left\{ \begin{pmatrix} x \\ y \\ z \end{pmatrix} \middle| \begin{pmatrix} x \\ y \\ z \end{pmatrix} = \begin{pmatrix} x_L \\ y_L \\ z_L \end{pmatrix} + \lambda_L Q_L \begin{pmatrix} u_L \\ v_L \\ f_L \end{pmatrix} \right\}
$$

and

$$
\left\{ \begin{pmatrix} x \\ y \\ z \end{pmatrix} \middle| \begin{pmatrix} x \\ y \\ z \end{pmatrix} = \begin{pmatrix} x_R \\ y_R \\ z_R \end{pmatrix} + \lambda_R Q_R \begin{pmatrix} u_R \\ v_R \\ f_R \end{pmatrix} \right\}
$$

Since there is some measurement noise, we do not expect that a λ_L and λ_R exist such that

$$\begin{pmatrix} x_L \\ y_L \\ z_L \end{pmatrix} + \lambda_L Q_L \begin{pmatrix} u_L \\ v_L \\ f_L \end{pmatrix} = \begin{pmatrix} x_R \\ y_R \\ z_R \end{pmatrix} + \lambda_R Q_R \begin{pmatrix} u_R \\ v_R \\ f_R \end{pmatrix}$$

holds exactly. We do expect, however, that we can find a λ_L and λ_R such that ϵ^2 is minimized where

$$\epsilon^2 = \left\| \begin{pmatrix} x_L \\ y_L \\ z_L \end{pmatrix} + \lambda_L \begin{pmatrix} p_L \\ q_L \\ s_L \end{pmatrix} - \begin{pmatrix} x_R \\ y_R \\ z_R \end{pmatrix} + \lambda_R \begin{pmatrix} p_R \\ q_R \\ s_R \end{pmatrix} \right\|^2$$

and where

$$\begin{pmatrix} p_L \\ q_L \\ s_L \end{pmatrix} = Q_L \begin{pmatrix} u_L \\ v_L \\ f_L \end{pmatrix} \quad \text{and} \quad \begin{pmatrix} p_R \\ q_R \\ s_R \end{pmatrix} = Q_R \begin{pmatrix} u_R \\ v_R \\ f_R \end{pmatrix}$$

To determine the minimizing λ_L and λ_R, we take the partial derivatives of ϵ^2 with respect to λ_L and λ_R, set them to zero, and solve. This results in

$$\begin{pmatrix} \lambda_L \\ \lambda_R \end{pmatrix} = Ab$$

where

$$A = \begin{pmatrix} a_{11} & a_{12} \\ a_{21} & a_{22} \end{pmatrix}$$

$$a_{11} = k \begin{pmatrix} p_R \\ q_R \\ s_R \end{pmatrix}' \begin{pmatrix} p_R \\ q_R \\ s_R \end{pmatrix}$$

$$a_{12} = a_{21} = -k \begin{pmatrix} p_L \\ q_L \\ s_L \end{pmatrix}' \begin{pmatrix} p_R \\ q_R \\ s_R \end{pmatrix}$$

$$a_{22} = k \begin{pmatrix} p_L \\ q_L \\ s_L \end{pmatrix}' \begin{pmatrix} p_L \\ q_L \\ s_L \end{pmatrix}$$

$$b_1 = - \left[\begin{pmatrix} x_L \\ y_L \\ z_L \end{pmatrix} - \begin{pmatrix} x_R \\ y_R \\ z_R \end{pmatrix} \right]' \begin{pmatrix} p_L \\ q_L \\ s_L \end{pmatrix}$$

$$b_2 = - \left[\begin{pmatrix} x_L \\ y_L \\ z_L \end{pmatrix} - \begin{pmatrix} x_R \\ y_R \\ z_R \end{pmatrix} \right]' \begin{pmatrix} p_R \\ q_R \\ s_R \end{pmatrix}$$

$$k = \left[\begin{pmatrix} p_R \\ q_R \\ s_R \end{pmatrix}' \begin{pmatrix} p_R \\ q_R \\ s_R \end{pmatrix} \right] \left[\begin{pmatrix} p_L \\ q_L \\ s_L \end{pmatrix}' \begin{pmatrix} p_L \\ q_L \\ s_L \end{pmatrix} \right] - \left[\begin{pmatrix} p_L \\ q_L \\ s_L \end{pmatrix}' \begin{pmatrix} p_R \\ q_R \\ s_R \end{pmatrix} \right]^2$$

In the case when

$$\begin{pmatrix} x_R \\ y_R \\ z_R \end{pmatrix} - \begin{pmatrix} x_L \\ y_L \\ z_L \end{pmatrix} = \begin{pmatrix} bx \\ 0 \\ 0 \end{pmatrix} \quad b_1 = b_x p_L, b_2 = -b_x p_R$$

then

$$\lambda_L = \frac{-b_x p_L \begin{pmatrix} p_R \\ q_R \\ s_R \end{pmatrix}' \begin{pmatrix} p_R \\ q_R \\ s_R \end{pmatrix} + b_x p_R \begin{pmatrix} p_R \\ q_R \\ s_R \end{pmatrix}' \begin{pmatrix} p_L \\ q_L \\ s_L \end{pmatrix}}{k}$$

■ EXAMPLE 14.4

The position of the left lens is $\begin{pmatrix} 0 \\ 0 \\ 0 \end{pmatrix}$, and the position of the right lens is

$\begin{pmatrix} 2 \\ 0 \\ 0 \end{pmatrix}$. Hence $b_x = -2$. On the left image $\begin{pmatrix} p_L \\ q_L \\ s_L \end{pmatrix} = \frac{1}{6} \begin{pmatrix} 10 \\ 20 \\ 30 \end{pmatrix}$, and on the

right image $\begin{pmatrix} p_R \\ q_R \\ s_R \end{pmatrix} = \frac{1}{6} \begin{pmatrix} 8 \\ 20 \\ 30 \end{pmatrix}$.

$$\lambda_L = \frac{\frac{1}{36}(1364)(2 \times 10/6) + \frac{1}{30}(1380)(-2 \times 8/6)}{\left(\frac{1}{36} \times 1364\right)\left(\frac{1}{36} \times 1400\right) - \frac{1}{36^2} \times 1380} = 6$$

Then

$$\begin{pmatrix} x \\ y \\ z \end{pmatrix} = \begin{pmatrix} 0 \\ 0 \\ 0 \end{pmatrix} + 6\frac{1}{6} \begin{pmatrix} 10 \\ 20 \\ 30 \end{pmatrix} = \begin{pmatrix} 10 \\ 20 \\ 30 \end{pmatrix}$$

■

14.7   2D–2D Absolute Orientation

In the simple 2D absolute orientation problem, sometimes called the 2D–2D pose detection problem, we are given N 2D coordinate observations from the observed image: $x_1, ..., x_N$. These could correspond, for example, to the observed center position of all observed objects. We are also given the corresponding or matching N 2D coordinate vectors from the model: $y_1, ..., y_N$. In the usual inspection situation, establishing which observed vector corresponds to which model vector is simple because the object being observed is fixtured and its approximate position and orientation are known. The approximate rotational and translational relationship between the image coordinate system and the object coordinate system permits the matching to be done just by matching a rotated and translated image position to an object position. The match is established if the rotated image position is close enough to the object position.

In the ideal case, the simple 2D pose detection problem is to determine from the matched points a more precise estimate of a rotation matrix R and a translation t such that $y_n = Rx_n + t$, $n = 1, ..., N$. Since small observational errors are likely, the real problem must be posed as a minimization. Determine R and t that minimize the weighted sum of the residual errors ϵ^2:

$$\epsilon^2 = \sum_{n=1}^{N} w_n \|y_n - (Rx_n + t)\|^2 \tag{14.46}$$

The weights $w_n, n = 1, ..., N$, satisfy $w_n \geq 0$ and $\sum_{n=1}^{N} w_n = 1$. If there is no prior knowledge as to how the weights should be set, they can be defined to be equal: $w_n = 1/N$.

Upon expanding Eq. (14.46), we have

$$\epsilon^2 = \sum_{n=1}^{N} w_n \left[(y_n - t)'(y_n - t) - (y_n - t)'Rx_n - x_n'R'(y_n - t) + x_n'R'Rx_n \right]$$

Since R is a rotation matrix, it is orthonormal, so that $R^{-1} = R'$. Also, since $(y_n - t)'Rx_n$ is a scalar, it is equal to its transpose. Hence

$$\epsilon^2 = \sum_{n=1}^{N} w_n \left[(y_n - t)'(y_n - t) - 2(y_n - t)'Rx_n + x_n'x_n \right]$$

Taking the partial derivative of ϵ^2 with respect to the components of the translation t and setting the partial derivative to 0, we obtain

$$0 = \sum_{n=1}^{N} w_n \left[-2(y_n - t) + 2Rx_n \right]$$

Letting

$$\bar{x} = \sum_{n=1}^{N} w_n x_n \quad \text{and} \quad \bar{y} = \sum_{n=1}^{N} w_n y_n$$

immediately results in

$$\bar{y} = R\bar{x} + t$$

Substituting $\bar{y} - R\bar{x}$ for t in the expression for the residual error, we can do some simplifying:

$$\epsilon^2 = \sum_{n=1}^{N} w_n \Big\{ [y_n - (\bar{y} - R\bar{x})]'[y_n - (\bar{y} - R\bar{x})]$$

$$- 2[y_n - (\bar{y} - R\bar{x})]'Rx_n + x_n'x_n \Big\}$$

$$= \sum_{n=1}^{N} w_n \Big[(y_n - \bar{y})'(y_n - \bar{y}) + 2(y_n - \bar{y})'R\bar{x} + \bar{x}'R'R\bar{x}$$

$$- 2(y_n - \bar{y})'Rx_n - 2\bar{x}R'Rx_n + x_n'x_n \Big]$$

$$= \sum_{n=1}^{N} w_n \Big[(y_n - \bar{y})'(y_n - \bar{y})$$

$$- 2(y_n - \bar{y})'R(x_n - \bar{x}) + (x_n - \bar{x})'(x_n - \bar{x}) \Big] \qquad (14.47)$$

The counterclockwise rotation angle θ is related to the rotation matrix by

$$R = \begin{pmatrix} \cos\theta & -\sin\theta \\ \sin\theta & \cos\theta \end{pmatrix}$$

We want to take the partial derivative of ϵ^2 with respect to θ. Now we need a notation in which the two components of x_n and the two components of y_n can be written explicitly.

Letting

$$x_n = \begin{pmatrix} x_{1n} \\ x_{2n} \end{pmatrix}, \quad y_n = \begin{pmatrix} y_{1n} \\ y_{2n} \end{pmatrix}$$

$$\bar{x} = \begin{pmatrix} \bar{x}_1 \\ \bar{x}_2 \end{pmatrix}, \quad \bar{y} = \begin{pmatrix} \bar{y}_1 \\ \bar{y}_2 \end{pmatrix}$$

we obtain

$$(y_n - \bar{y})' R (x_n - \bar{x}) = (y_{n1} - \bar{y}_1)\cos\theta(x_{n1} - \bar{x}_1)$$

$$+ (y_{n1} - \bar{y}_1)(-\sin\theta)(x_{n2} - \bar{x}_2)$$

$$+ (y_{n2} - \bar{y}_2)\sin\theta(x_{n1} - \bar{x}_1)$$

$$+ (y_{n2} - \bar{y}_2)\cos\theta(x_{n2} - \bar{x}_2)$$

Setting to zero the partial derivative of ϵ^2 with respect to θ results in

$$0 = -2 \sum_{n=1}^{N} w_n \Big[(y_{n1} - \bar{y}_1)(-\sin\theta)(x_{n1} - \bar{x}_1)$$
$$+ (y_{n1} - \bar{y}_1)(-\cos\theta)(x_{n2} - \bar{x}_2)$$
$$+ (y_{n2} - \bar{y}_2)\cos\theta(x_{n1} - \bar{x}_1)$$
$$+ (y_{n2} - \bar{y}_2)(-\sin\theta)(x_{n2} - \bar{x}_2) \Big]$$

We let

$$A = \sum_{n=1}^{N} w_n [(y_{n1} - \bar{y}_1)(x_{n1} - \bar{x}_1) + (y_{n2} - \bar{y}_2)(x_{n2} - \bar{x}_2)]$$

$$B = \sum_{n=1}^{N} w_n [(y_{n1} - \bar{y}_1)(x_{n2} - \bar{x}_2) - (y_{n2} - \bar{y}_2)(x_{n1} - \bar{x}_1)]$$

Then

$$0 = A\sin\theta + B\cos\theta$$

Hence

$$\cos\theta = -A/\sqrt{A^2 + B^2} \qquad \text{and} \qquad \sin\theta = B/\sqrt{A^2 + B^2}$$

or

$$\cos\theta = A/\sqrt{A^2 + B^2} \qquad \text{and} \qquad \sin\theta = -B/\sqrt{A^2 + B^2}$$

The correct value for θ will, in general, be unique and will be the θ that minimizes ϵ^2. Thus the better of the two choices can always be easily determined by simply substituting each value for θ into the original expression for ϵ^2.

14.8   3D–3D Absolute Orientation

Suppose that we have determined the location of N 3D points $p_1,\ldots,p_N$ relative to the camera reference frame. Now suppose that in the object model reference frame these same N 3D points have location $q_1,\ldots,q_N$. To answer the question "Where is it?" in which "it" means the object, we must determine where the model reference frame is with respect to the camera reference frame. To answer the question "Where is it?" in which "it" means the camera, we must determine where the camera reference frame is with respect to the model reference frame. Hence we must determine a rotation matrix R and translation vector t that satisfy

$$p_n = Rq_n + t, \qquad n = 1,\ldots,N \tag{14.48}$$

The 3D–3D absolute orientation problem, sometimes called the 3D–3D pose estimation problem, is to infer R and t from $q_1, ..., q_N$ and $p_1, ..., p_N$. A generalization of this pose estimation problem is to assume an unknown scale. Then Eq. (14.48) takes the form

$$p_n = sRq_n + t, \qquad n = 1, \ldots, N \tag{14.49}$$

In what follows, we assume s is known to be 1.

To determine R and t, we set up a constrained least-squares problem. We will minimize $\sum_{n=1}^{N} w_n \| p_n - (Rq_n + t) \|^2$ subject to the constraint that R is a rotation matrix, that is, $R' = R^{-1}$. To be able to express these constraints using Lagrangian multipliers, we let

$$R = \begin{pmatrix} r_1' \\ r_2' \\ r_3' \end{pmatrix}$$

where each r_i is a 3×1 vector. The constraint $R' = R^{-1}$ then amounts to the six constraint equations

$$r_1' r_1 = 1$$
$$r_2' r_2 = 1$$
$$r_3' r_3 = 1$$
$$r_1' r_2 = 0$$
$$r_1' r_3 = 0$$
$$r_2' r_3 = 0$$

The least-squares problem with these constraints can be written as minimizing ϵ^2, where

$$\epsilon^2 = \sum_{n=1}^{N} \sum_{k=1}^{3} w_n (p_{nk} - r_k' q_n - t_k)^2 + \sum_{k=1}^{3} \lambda_k (r_k' r_k - 1) + 2\lambda_4 r_1' r_2 + 2\lambda_5 r_1' r_3 + 2\lambda_6 r_2' r_3$$

$$p_n = \begin{pmatrix} p_{n1} \\ p_{n2} \\ p_{n3} \end{pmatrix}, \qquad q_n = \begin{pmatrix} q_{n1} \\ q_{n2} \\ q_{n3} \end{pmatrix}, \qquad \text{and} \quad t = \begin{pmatrix} t_1 \\ t_2 \\ t_3 \end{pmatrix}$$

Taking the partial derivative of ϵ^2 with respect to t_n results in

$$\frac{\partial \epsilon^2}{\partial t_k} = \sum_{n=1}^{N} 2 w_n (p_{nk} - r_k' q_n - t_k) (-1), \quad k = 1, 2, 3$$

Setting these partials to zero results in

$$\sum_{n=1}^{N} w_n (p_n - Rq_n - t) = 0$$

By rearranging, we obtain

$$t = \bar{p} - R\bar{q} \qquad (14.50)$$

where

$$\bar{q} = \frac{\sum\limits_{n=1}^{N} w_n q_n}{\sum\limits_{n=1}^{N} w_n} \qquad \text{and} \qquad \bar{p} = \frac{\sum\limits_{n=1}^{N} w_n p_n}{\sum\limits_{n=1}^{N} w_n}$$

Thus once R is known, t is quickly determined. Substituting $\bar{p} - R\bar{q}$ for t in the definition of ϵ^2, results in

$$\epsilon^2 = \sum_{n=1}^{N} w_n \sum_{k=1}^{3} [p_{nk} - \bar{p}_n - r'_k(q_n - \bar{q})]^2 + \sum_{k=1}^{3} \lambda_k (r'_k r_k - 1)$$
$$+ 2\lambda_4 r'_1 r_2 + 2\lambda_5 r'_1 r_3 + 2\lambda_6 r'_2 r_3$$

where

$$\bar{p} = \begin{pmatrix} \bar{p}_1 \\ \bar{p}_2 \\ \bar{p}_3 \end{pmatrix}, \quad \bar{q} = \begin{pmatrix} \bar{q}_1 \\ \bar{q}_2 \\ \bar{q}_3 \end{pmatrix}$$

Now we take partial derivatives of ϵ^2 with respect to the components of each r_n. To write things more compactly, by $\frac{\partial \epsilon^2}{\partial r_n}$ we mean a 3×1 vector whose components are the partial derivative of ϵ^2 with respect to each of the components of r_n. Then

$$\frac{\partial \epsilon^2}{\partial r_1} = \sum_{n=1}^{N} 2w_n [p_{n1} - \bar{p}_1 - r'_1(q_n - \bar{q})](q_n - \bar{q})(-1) + 2\lambda_1 r_1 + 2\lambda_4 r_2 + 2\lambda_5 r_3$$

$$\frac{\partial \epsilon^2}{\partial r_2} = \sum_{n=1}^{N} 2w_n [p_{n2} - \bar{p}y_2 - r'_2(q_n - \bar{q})] \, (q_n - \bar{q}) \, (-1) + 2\lambda_2 r_2 + 2\lambda_4 r_1 + 2\lambda_6 r_3$$

$$\frac{\partial \epsilon^2}{\partial r_3} = \sum_{n=1}^{N} 2w_n [p_{n3} - \bar{p}_3 - r'_3(q_n - \bar{q})] \, (p_n - \bar{q}) \, (-1) + 2\lambda_3 r_3 + 2\lambda_5 r_1 + 2\lambda_6 r_2$$

Setting these partial derivatives to zero and rearranging, we obtain

$$\sum_{n=1}^{N} w_n (q_n - \bar{q})(q_n - \bar{q})' r_1 + \lambda_1 r_1 + \lambda_4 r_2 + \lambda_5 r_3 = \sum_{n=1}^{N} w_n (p_{n1} - \bar{p}_1)(q_n - \bar{q})$$

$$\sum_{n=1}^{N} w_n (q_n - \bar{q})(q_n - \bar{q})' r_2 + \lambda_4 r_1 + \lambda_2 r_2 + \lambda_6 r_3 = \sum_{n=1}^{N} w_n (p_{n2} - \bar{p}_2)(q_n - \bar{q})$$

$$\sum_{n=1}^{N} w_n (q_n - \bar{q})(q_n - \bar{q})' r_3 + \lambda_5 r_1 + \lambda_6 r_2 + \lambda_3 r_3 = \sum_{n=1}^{N} w_n (p_{n3} - \bar{p}_3)(q_n - \bar{q})$$

Let

$$A = \sum_{n=1}^{N} (q_n - \overline{q})(q_n - \overline{q})'$$

$$\Lambda = \begin{pmatrix} \lambda_1 & \lambda_4 & \lambda_5 \\ \lambda_4 & \lambda_2 & \lambda_6 \\ \lambda_5 & \lambda_6 & \lambda_3 \end{pmatrix}$$

and

$$B = (b_1 b_2 b_3) \quad \text{when} \quad b_k = \sum_{n=1}^{N} w_n (p_{nk} - \overline{p}_k)(q_n - \overline{q})$$

Then the above equation can be simply rewritten as

$$A R' + R'\Lambda = B$$

Multiplying both sides on the left by R, we have

$$R A R' + \Lambda = RB$$

Since $A = A'$, $(RAR')' = RAR'$. Since both RAR' and Λ are symmetric, the left-hand side must be symmetric. Hence the right-hand side is also symmetric. This means

$$R B = (RB)'$$

The solution for R now comes quickly. Let the singular-value decomposition of B be

$$B = UDV$$

where U and V are orthonormal and D is diagonal. Then

$$RUDV = (UDV)'R'$$
$$= V'DU'R'$$

By observation, a solution for R is immediately obtained as

$$R = V'U' \tag{14.51}$$

The matrix R is guaranteed to be orthonormal, but it is not guaranteed to have determinant $+1$. It might have determinant -1. Determinant -1 means that R incorporates a reflection in addition to the rotation. Such a result can happen if the observed points are few in number and are very noisy. In this case a more computationally expensive solution using either quaternions or a nonlinear optimization must be employed.

EXAMPLE 14.5

The vertices of the three 3D points in the world coordinate system are

$$
\begin{pmatrix} x_1 \\ y_1 \\ z_1 \end{pmatrix} = \begin{pmatrix} 0 \\ 0 \\ 6 \end{pmatrix}, \quad
\begin{pmatrix} x_2 \\ y_2 \\ z_2 \end{pmatrix} = \begin{pmatrix} 3 \\ 0 \\ 6 \end{pmatrix}, \quad
\begin{pmatrix} x_3 \\ y_3 \\ z_3 \end{pmatrix} = \begin{pmatrix} 0 \\ 4 \\ 6 \end{pmatrix}
$$

The vertices of these same 3D points in the sensor coordinate system are

$$
\begin{pmatrix} p_1 \\ q_1 \\ r_1 \end{pmatrix} = \begin{pmatrix} 0.96064 \\ -1.56383 \\ 11.17024 \end{pmatrix}, \quad
\begin{pmatrix} p_2 \\ q_2 \\ r_2 \end{pmatrix} = \begin{pmatrix} 3.84593 \\ -1.99419 \\ 11.87016 \end{pmatrix}, \quad
\begin{pmatrix} p_3 \\ q_3 \\ r_3 \end{pmatrix} = \begin{pmatrix} 1.47246 \\ 2.39025 \\ 11.49159 \end{pmatrix}
$$

Using Eq. (14.51) to solve for the rotation matrix, decomposing this matrix into the ω, ϕ, κ angular rotations, and using Eq. (14.50) to solve for the translation, we calculate

$$
t = \begin{pmatrix} -3.75434 \\ 0.52558 \\ -4.66637 \end{pmatrix} \quad \text{and} \quad
\begin{pmatrix} \omega \\ \phi \\ \kappa \end{pmatrix} = \begin{pmatrix} -0.082796 \\ 0.23547 \\ 0.14807 \end{pmatrix} \text{ radians}
$$

where the rotation matrix is given by Eq. (14.51).

Early solutions to the 3D–3D absolute orientation problem can be found in the photogrammetry literature beginning with Thompson (1958–59), Schut (1962–63), Tienstra (1969), and Pope (1970). Blais (1972) gives a solution to the problem in the case when there may be a scale factor or magnification different from 1. Sansò(1973) gives a solution to the problem by using quaternions. Huang (1987) and Haralick et al. (1987) discuss the singular value decomposition approach to the problem.

14.9   Robust M-Estimation

Least-squares techniques are ideal when the random data perturbations or measurement errors are Gaussian distributed. However, when a small fraction of the data have large non-Gaussian perturbations, least-squares techniques become techniques with least virtue. For with least-squares techniques, the perturbation of the answer caused by the perturbation of even a single data component grows linearly in an unbounded way with the data component perturbation.

This section briefly describes some robust techniques used in nonlinear regression problems. In particular, they can be used to robustly solve the equation that results from the linearization of the original nonlinear absolute orientation or relative orientation problem. We consider first the one parameter estimation problem.

Any estimate T_k defined by a minimization problem of the form

$$\min_{T_k} \sum_{i=1}^{n} \rho(x_i - T_k)$$

or by an implicit equation

$$\sum_{i=0}^{n} \psi(x_i - T_k) = 0$$

where ρ is an arbitrary function (called object function),

$$\psi(x - T_k) = \frac{\partial}{\partial T_k} \rho(x - T_k)$$

is called an M-estimate. This last equation can be written equivalently as

$$\sum_{i=0}^{n} w_i(x_i - T_k) = 0$$

where

$$w_i = \frac{\psi(x_i - T_k)}{x_i - T_k}, \quad i = 1,\ldots,n$$

This gives a formal representation of T_k as a weighted mean

$$T_k = \frac{\displaystyle\sum_{i=1}^{n} w_i x_i}{\displaystyle\sum_{i=1}^{n} w_i}$$

with weights depending on the sample (Huber, 1981). It is known that M-estimators minimize objective functions more general than the familiar sum of squared residuals associated with the sample mean. Among many forms of functions ρ and ψ proposed in the literature, Huber's and Tukey's form is discussed in this section. Huber derived the following robust ρ and ψ:

$$\rho(x) = \begin{cases} 0.5x^2, & \text{if } |x| \leq a; \\ a|x| - 0.5a^2, & \text{otherwise} \end{cases}$$

$$\psi(x) = \begin{cases} -a, & \text{if } x < -a; \\ x, & \text{if } |x| \leq a; \\ a, & \text{if } x > a \end{cases}$$

Tukey's ψ function (Hoaglin, Mosteller, and Tukey, 1983) can be expressed as

$$\psi(x) = \begin{cases} x\left(1 - \left(\frac{x}{a}\right)^2\right)^2, & \text{if } |x| \le a; \\ 0, & \text{if } |x| > a \end{cases}$$

where a is a tuning constant, 1.5 for Huber's and 6 for Tukey's.

The nonlinear regression problem can be formulated as follows: Let $f_i : E^m \to E$, $i = 1, \ldots, n$ be functions that map m-dimensional space into a real line. Let $\theta = (\theta_1, \theta_2, \ldots, \theta_m)' \in E^m$ be the m-dimensional unknown vector to be estimated. The solution to the set of n equations

$$f_i(\theta) = y_i, \quad i = 1, \ldots, n$$

that minimizes

$$\sum_{i=1}^{n} \rho\left(\frac{y_i - f_i(\theta)}{S}\right)$$

can be found in several different ways. To create a scale invariant version of the M-estimator, the robust estimate of scale such as the following is introduced:

$$S = \frac{\text{median}_i |y_i - f_i(\theta)|}{0.6745}$$

where 0.6745 is one-half the interquantile range of the Gaussian normal distribution $N(0, 1)$. Here we take the median of the nonzero deviations only, because with large m, too many residuals can equal zero (Hogg, 1979).

In robust estimation, the estimates are obtained only after an iterative process, because the estimates do not have closed forms. Two such iterative methods are presented here that can solve the minimization problem stated above (Huber, 1981).

14.9.1 Modified Residual Method

In the modified residual method, the residuals are modified by a proper ψ function before the least-squares problem is solved. The iterative procedure to determine θ is:

1. Choose an initial approximation θ^0.
2. Iterate. Given the estimation θ^k in step k, compute the solution in the $(k+1)$th step as follows:
3. Compute the modified residuals r_i^* for $i = 1, \ldots, n$.

$$r_i^* = \psi\left(\frac{r_i}{S^k}\right) S^k$$

where

$$r_i = y_i - f_i(\theta^k)$$

$$S^k = \frac{1}{.6745} \operatorname*{median}_{r_i \ne 0} |r_i|$$

4. Solve the least-squares problem $X\delta = r^*$, where $X = [x_{ij}]$ is the Jacobian matrix:

$$x_{ij} = \frac{\partial}{\partial\theta_j} f_i(\theta^k)$$

The solution for this equation can be found by using the standard least-squares method. If the singular-value decomposition of the matrix X is $X = U_1\Sigma_1 V'$, then the solution is $\hat{\delta} = V\Sigma_1^{-1}U_1'r^*$.

5. Set $\theta^{k+1} = \theta^k + \hat{\delta}$.

14.9.2 Modified Weights Method

Taking the derivative of the objective function ρ with respect to θ and setting it to zero, we get

$$\sum_i \psi\left(\frac{y_i - f_i(\theta)}{S}\right)\frac{\partial f_i(\theta)}{\partial\theta_j} = 0$$

In the standard weighted form,

$$\sum_i w_i r_i \frac{\partial f_i(\theta)}{\theta_j} = 0$$

where

$$w_i = \frac{\psi(\frac{r_i}{S})}{(\frac{r_i}{S})}$$

Therefore the iterative procedure to determine θ is:

1. Choose an initial approximation θ^0.
2. Iterate. Given θ^k at kth step, compute θ^{k+1} as follows:
3. Solve

$$PX\delta = Pr$$

where

$$P = \begin{pmatrix} \sqrt{w}_1 & & \\ & \ddots & \\ & & \sqrt{w}_N \end{pmatrix}$$

4. If $\hat{\delta}$ is the solution in step 3, then set

$$\theta^{k+1} = \theta^k + \hat{\delta}$$

14.9.3   Experimental Results

To illustrate the importance of robust solutions, we discuss experimental results obtained from several hundred thousand controlled experiments. This section describes how the controlled experiments are constructed and shows their results. Each result is presented as a graph where the sum of errors of the three rotation angles ω, ϕ, κ is plotted against various control parameters, such as the signal-to-noise ratio (SNR), the number of matched points, or the number of outliers, which will be defined later.

Data Set Generation

A set of 3D model points, $y_i = (y_{i1}, y_{i2}, y_{i3})', i = 1, \ldots, N$, is generated within a box defined by

$$y_{i1}, y_{i2}, y_{i3} \in [0, 10]$$

That is, the three coordinates are independent random variables, each uniformly distributed between 0 and 10. Next, three rotation angles are selected from an interval [20, 70], and the translation vector $t = (t_1, t_2, t_3)$ is also generated such that t_1 and t_2 are uniformly distributed within an interval [5, 15] and t_3 is within [20, 50]. With these transformation parameters, the 3D model points are rotated and translated in the 3D space, forming a set of 3D points $x_i, i = 1, \ldots, N$. At this stage, independent identically distributed Gaussian noise $N(0, \sigma)$ is added to all three coordinates of the transformed points x_i. To test the robustness of the algorithms, some fraction of the 3D points, x_i, are replaced with randomly generated 3D points, $z_i = (z_{i1}, z_{i2}, z_{i3})', i = 1, \ldots, M$. M is the number of the replaced 3D points and

$$z_{i1} = t_1 + v_{i1}$$
$$z_{i2} = t_2 + v_{i2}$$
$$z_{i3} = x_{i3}$$

where $v_{i1}, v_{i2}, i = 1, \ldots, M$ are independent random variables uniformly distributed within an interval [-5, 5]. These random points, z_i, are called outliers in our experiments. To get the matching set of 2D points, $x_i, i = 1, \ldots, N$, are perspectively projected onto the image plane. Given the 3D model points and the corresponding 2D points on the image plane, each algorithm is applied to find the three rotation angles and the translation vector.

One can notice from this description that there are three parameters we can control in each experiment: (1) the number of 3D model points N, (2) the standard deviation σ of the Gaussian noise, and (3) the number of outliers M. In the experimental result, we use SNR and the percentage of outliers PO in place of σ and M, respectively, where

$$SNR = 20 \log \frac{10}{\sigma} db$$

$$PO = \frac{M}{N} \times 100\%$$

Results

For each parameter setting, (N, SNR, PO), 1000 experiments are performed to get a reasonable estimate of the performance of the algorithm. Each experiment has three parts (E1, E2, and E3), as follows:

E1: Set $N = 20$. Estimate the sum of three rotation angle errors against SNR (20db to 80db in 10db step) for different PO (0% to 20% in 5% step).

E2: Set SNR $= 40$db. Estimate the sum of three rotation angle errors against PO (0% to 20% in 5% step) for different N (10 to 50 by steps of 10).

E3: Set PO $= 10$%. Estimate the sum of three rotation angle errors against SNR (20db to 80db in 10db step) for different N (10 to 50 by steps of 10).

For the linearized algorithm, the initial estimate of the three rotation angles are selected randomly within 15° of the true angles. The initial approximation of the translation vector is selected randomly within ±10 of the true translation vector. Figures 14.1 and 14.2 show the results of the least-squares adjustment algorithm of Section 14.3.1 and the robust M-estimate algorithm of Section 14.9.2, respectively. One more experiment is performed to compare the iterative least-squares algorithm with the robust iterative least-squares algorithm. Both algorithms are run with $N = 20$, PO $= 10$%, and SNR going from 20db to 40db in steps of 10db. Then the iterative least-squares algorithm is run with $N = 18$, PO $= 0$%, and SNR going from 20db to 40db in steps of 10db. This compares the efficiency of the robust technique against that of the nonrobust technique in the case when the latter uses only the nonoutlier points given to the robust technique. Figure 14.3 shows the result of this experiment.

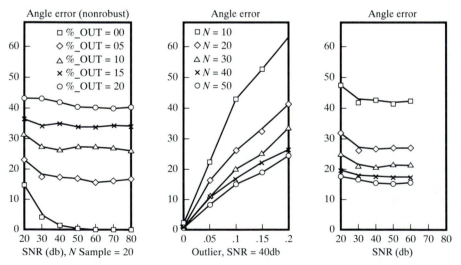

Figure 14.1 Performance characteristics of the least-squares adjustment algorithm.

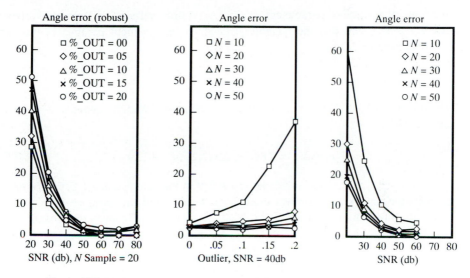

Figure 14.2 Performance characteristics of the robust M-estimate algorithm.

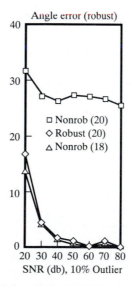

Figure 14.3 Efficiency of the robust technique operating on a data set of 20 points, 18 points having Gaussian noise and 2 outliers, against the nonrobust technique operating on a data set with 18 points having Gaussian noise.

14.10   Error Propagation

Suppose that input parameters $x_1,\ldots,x_N$ are subject to random errors $\Delta x_1,\ldots,\Delta x_N$ and that the quantity y depends on the input parameters through a known function f:

$$y = f(x_1,\ldots,x_N)$$

The quantities $x_1 + \Delta x_1,\ldots,x_N + \Delta x_N$ are observed, and the quantity $y + \Delta y$ is calculated by

$$y + \Delta y = f(x_1 + \Delta x_1,\ldots,x_N + \Delta x_N)$$

The errors $\Delta x_1,\ldots,\Delta x_N$ induce an error Δy in the calculated $y + \Delta y$. The error propagation analysis determines the expected value and the variance of $y + \Delta y$. The only information we have about $\Delta x_1,\ldots,\Delta x_N$ is their mean and covariance:

$$E\left[\begin{pmatrix} \Delta x_1 \\ \vdots \\ \Delta x_N \end{pmatrix}\right] = 0$$

$$E\left[\begin{pmatrix} \Delta x_1 \\ \vdots \\ \Delta x_N \end{pmatrix} (\Delta x_1,\ldots,\Delta x_N)\right] = \begin{pmatrix} \sigma_{11}\sigma_{12}\ldots\sigma_{1N} \\ \vdots \\ \sigma_{N1}\sigma_{N2}\ldots\sigma_{NN} \end{pmatrix}$$

We proceed by assuming that the random errors $\Delta x_1,\ldots,\Delta x_N$ are small and that it is therefore reasonable to approximate $f(x_1 + \Delta x_1,\ldots,x_N + \Delta x_N)$ by a second-order expansion of around $x_1,\ldots,x_N$:

$$y + \Delta y = f(x_1 + \Delta x_1,\ldots,x_N + \Delta x_N)$$

$$= f(x_1,\ldots,x_N) + \sum_{n=1}^{N} \frac{\partial f}{\partial x_n}(x_1,\ldots,x_N)\Delta x_n$$

$$+ \sum_{m=1}^{N}\sum_{n=1}^{N} \frac{\partial^2 f}{\partial x_m \partial x_n}(x_1,\ldots,x_N)\Delta x_m \Delta x_n$$

Hence by taking expectations on both sides, we have

$$E[y + \Delta y] = f(x_1,\ldots,x_N) + \sum_{n=1}^{N} \frac{\partial f}{\partial x_n}(x_1,\ldots,x_N)E[\Delta x_n]$$

$$\sum_{m=1}^{N}\sum_{n=1}^{N} \frac{\partial^2 f}{\partial x_m \partial x_n}(x_1,\ldots,x_N)E[\Delta x_m \Delta x_n]$$

Since $E[\Delta x_n] = 0$ and $E[\Delta x_m \Delta x_n] = \sigma_{mn}$, and since $y = f(x_1, \ldots, x_N)$,

$$E[y + \Delta y] = y + \sum_{m=1}^{N} \sum_{n=1}^{N} \frac{\partial^2 f}{\partial x_m \partial x_n}(x_1, \ldots, x_N)\sigma_{mn}$$

Since the values of $(x_1, \ldots, x_N)$ are not known, but the values $(x_1 + \Delta x_1, \ldots, x_N + \Delta x_N)$ are known, we make the approximation

$$\sum_{m=1}^{N} \sum_{n=1}^{N} \frac{\partial^2 f}{\partial x_m \partial x_n}(x_1, \ldots, x_N)\sigma_{mn} \approx \sum_{m=1}^{N} \sum_{n=1}^{N} \frac{\partial^2 f}{\partial x_m \partial x_n}(x_1 + \Delta x_1, \ldots, x_N + \Delta x_N)\sigma_{mn}$$

in order to evaluate the bias of the calculated variate $y + \Delta y$. An unbiased estimate of y is then given by $\tilde{y}$, where

$$\tilde{y} = f(x_1 + \Delta x_1, \ldots, x_N + \Delta x_N)$$
$$- \sum_{m=1}^{N} \sum_{n=1}^{N} \frac{\partial^2 f}{\partial x_m \partial x_n}(x_1 + \Delta x_1, \ldots, x_N + \Delta x_N)\sigma_{mn}$$

To calculate the variance of $y + \Delta y$, we assume that the second-order bias effect is sufficiently small so that to a first-order approximation $E[\Delta y] = 0$. In this case

$$V[y + \Delta y] = E[(\Delta y)^2]$$
$$= \sum_{m=1}^{N} \frac{\partial f}{\partial x_m}(x_1, \ldots, x_N) \sum_{n=1}^{N} \frac{\partial f}{\partial x_n}(x_1, \ldots, x_N)\sigma_{mn}$$

The variance is estimated by $\hat{\sigma}_y^2$,

$$\hat{\sigma}_y^2 = \sum_{m=1}^{N} \frac{\partial f}{\partial x_m}(x_1 + \Delta x_1, \ldots, x_N + \Delta x_N) \sum_{n=1}^{N} \frac{\partial f}{\partial x_n}(x_1 + \Delta x_1, \ldots, x_N + \Delta x_N)\sigma_{mn}$$

The calculated $y + \Delta y$ is not uncorrelated with an observed input $x_n + \Delta x_n$:

$$Cov(y + \Delta y, x_n + \Delta x_n) = E[\Delta y \Delta x_n]$$
$$= E\left[\sum_{m=1}^{N} \frac{\partial f}{\partial x_m}(x_1, \ldots, x_N)\Delta x_m \Delta x_n\right]$$
$$= E\left[\sum_{m=1}^{N} \frac{\partial f}{\partial x_m}(x_1, \ldots, x_N)\sigma_{mn}\right]$$

We therefore estimate the covariance by

$$E\left[\sum_{m=1}^{N} \frac{\partial f}{\partial x_m}(x_1 + \Delta x_1, \ldots, x_N + \Delta x_N)\sigma_{mn}\right]$$

14.10.1   Implicit Form

Error propagation can also be done in the implicit form. The known function f has the form

$$f(x_1, x_2, \ldots, x_N, y) = 0$$

The quantities $(x_1 + \Delta x_1, \ldots, x_N + \Delta x_N)$ are observed, and the quantity $y + \Delta y$ is determined to satisfy $f(x_1 + \Delta x_1, \ldots, x_N, y + \Delta y) = 0$. The errors $(\Delta x_1, \ldots, \Delta x_N)$ induce an error Δy in the calculated $y + \Delta y$. To determine the error, we proceed as before and linearize f around $f(x_1, x_2, \ldots, x_N, y)$:

$$f(x_1 + \Delta x_1, \ldots, x_N + \Delta x_N, y + \Delta y) = f(x_1, \ldots, x_n, y)$$

$$+ \sum_{n=1}^{N} \frac{\partial f}{\partial x_n}(x_1, \ldots, x_N, y) \Delta x_n$$

$$+ \frac{\partial f}{\partial y_n}(x_1, \ldots, x_N, y) \Delta y$$

Since $f(x_1 + \Delta x_1, \ldots, x_N + \Delta x_N, y + \Delta y) \approx 0$ and $f(x_1, x_2, \ldots, x_N, y) = 0$,

$$\Delta y = \frac{-\sum_{n=1}^{N} \frac{\partial f}{\partial x_n}(x_1, x_2, \ldots, x_N, y) \Delta x_n}{\frac{\partial f}{\partial y}(x_1, x_2, \ldots, x_N, y)}$$

Since $E[\Delta x_n] = 0$, it follows that $E[\Delta y] = 0$. The variance of interest is $V[y + \Delta y]$. Now $V[y + \Delta y] = V[\Delta y]$, and since $E[\Delta y] = 0$,

$$V[\Delta y] = E[(\Delta y)^2] = E\left[\frac{\left(-\sum_{n=1}^{N} \frac{\partial f}{\partial x_n}(x_1, x_2, \ldots, x_N, y) \Delta x_n \right)^2}{\frac{\partial f}{\partial y}(x_1, \ldots, x_N, y)^2} \right]$$

$$= \frac{\sum_{m=1}^{N} \sum_{n=1}^{N} \frac{\partial f}{\partial x_m}(x_1, \ldots, x_N, y) \frac{\partial f}{\partial x_n}(x_1, \ldots, x_N, y) \sigma_{mn}}{\left[\frac{\partial f}{\partial y}(x_1, \ldots, x_N, y) \right]^2} \tag{14.52}$$

As before, we estimate the variance of Δy by $\hat{\sigma}_y^2$, where

$$\hat{\sigma}_y^2 = \frac{\sum_{m=1}^{N} \sum_{n=1}^{N} \frac{\partial f}{\partial x_m}(x_1 + \Delta x_1, \ldots, x_N + \Delta x_N, y + \Delta y) \times \frac{\partial f}{\partial x_n}(x_1 + \Delta x_1, \ldots, x_N + \Delta x_N, y + \Delta y) \sigma_{mn}}{\frac{\partial f}{\partial y}(x_1 + \Delta x_1, \ldots, x_N + \Delta x_N, y + \Delta y)^2}$$

14.10.2 Implicit Form: General Case

Suppose that $x_1, \ldots, x_K$ are K $N \times 1$ vectors that represent true values of some quantities. Suppose $x_1 + \Delta x_1, \ldots, x_K + \Delta x_K$ are the K $N \times 1$ vectors that represent the noisy observed values of these quantities. The vectors $\Delta x_1, \ldots, \Delta x_K$ are then the random perturbations. Suppose that β is a $L \times 1$ vector representing the unknown true parameters. The noiseless model is

$$f(x_k, \beta) = 0, \quad k = 1, \ldots, K$$

With the noisy observations, the idealized model is

$$f(x_k + \Delta x_k, \beta + \Delta \beta) = 0, \quad k = 1, \ldots, K \tag{14.53}$$

This system of equations (14.53) is solved in the least-squares sense for $\beta + \Delta\beta$. To determine the expected value and covariance matrix for the solution $\beta + \Delta\beta$, we proceed by linearizing the equation:

$$f(x_k + \Delta x_k, \beta + \Delta\beta) = f(x_k, \beta) + \left[\frac{\partial f}{\partial x}(x_k, \beta) \right]' \Delta x_k + \left[\frac{\partial f}{\partial \beta}(x_k, \beta) \right]' \Delta\beta \tag{14.54}$$

Let g_k be the $N \times 1$ vector defined by

$$g_k = \frac{\partial f}{\partial x_k}(x_k, \beta)$$

Let h_k be the $L \times 1$ vector defined by

$$h_k = \frac{\partial f}{\partial \beta}(x_k, \beta)$$

and let H be the $K \times L$ matrix defined by

$$H = \begin{pmatrix} h_1' \\ \vdots \\ h_K' \end{pmatrix}$$

With $f(x_k + \Delta x_k, \beta + \Delta\beta) = f(x_k, \beta) = 0$, the system of equations (14.54) can be written

$$\begin{pmatrix} -g_1' \Delta x_1 \\ \vdots \\ -g_K' \Delta x_K \end{pmatrix} = H \Delta\beta$$

This system can be solved in the equal-weighted least-squares sense. There results

$$\Delta\beta = (H'H)^{-1} H' \begin{pmatrix} -g_1' \Delta x_1 \\ \vdots \\ -g_K' \Delta x_K \end{pmatrix}$$

From this it is clear that to the extent the linearization approximation is good, the expected value of $\Delta\beta$ is 0 when the expected value of Δx_k is 0, $k = 1,\ldots,K$.

To determine the covariance matrix of $\beta + \Delta\beta$, we examine

$$\Delta\beta\Delta\beta' = (H'H)^{-1}H' \begin{pmatrix} -g_1'\Delta x_1 \\ \vdots \\ -g_K'\Delta x_K \end{pmatrix} (\Delta x_1' g_1 \ldots \Delta x_K' g_K)\,[(H'H)^{-1}]'\,H$$

Since $E[\Delta x_k \Delta x_k'] = \sum_x$ and $E[\Delta x_k \Delta x_i] = 0$, $i \neq k$,

$$\sum_{\beta+\Delta\beta} = \sum_{\Delta\beta} = (H'H)^{-1}H' \begin{pmatrix} g_1'\sum_k g_1 & & \\ & \ddots & \\ & & g_K'\sum_k g_K \end{pmatrix} (H'H)^{-1}H \qquad (14.55)$$

Of course, the least-squares solution for $\beta+\Delta\beta$ does not guarantee $f(x_k+\Delta x_k, \beta+\Delta\beta) = 0$. Rather, what most likely happens is $f(x_k+\Delta x_k, \beta+\Delta\beta) = \xi_k$, where ξ_k is the kth residual error. In this case only if the residual errors are small enough—that is, if $g_k'\sum_k g_k >> \xi_k^2$—will Eq. (14.55) given for $\sum_{\beta+\Delta\beta}$ be accurate.

14.11   Summary

We have discussed the basic perspective projective or collinearity equation and have shown how to solve the exterior orientation, interior orientation, and relative orientation problems by using the standard and quaternion solutions. In doing so, we have shown how to take a nonlinear least-squares problem, linearize it, and solve the nonlinear problem by iteratively solving successive linearized least-squares problems.

We have discussed the stereo triangulation procedure, the 2D–2D and the 3D–3D pose estimation problems, and how to change the estimation problems into robust estimation problems. Finally, we have discussed the principles of error propagation with the explicit and implicit functional forms.

We have described all the techniques in this chapter from the framework of analytic photogrammetry. We have tried to provide sufficient detail so that people with a computer vision background can pick up the techniques and the technical jargon of analytic photogrammetry and thereby make the analytic photogrammetry literature accessible to vision people.

A considerable body of literature exists on analytic photogrammetric techniques. The journals in which much of this literature can be found are listed after the Exercises. The *Manual of Photogrammetry* is a good general reference, as are books such as Wolf (1983), second edition.

∎ Exercises

14.1. Suppose that $\omega = -0.088158$ radians, $\phi = 0.24665$ radians, and $\kappa = 0.150997$ radians. Determine the rotation matrix R given by

$$R = R(\kappa)R(\phi)R(\omega)$$

when $R(\kappa), R(\phi), R(\omega)$ are defined by Eq. (14.1).

14.2. Given the values of a rotation matrix R, determine a procedure to calculate the angles $\omega, \phi,$ and κ that satisfy

$$R = R(\kappa)R(\phi)R(\omega)$$

as defined in Eq. (14.1).

14.3. Write a program to determine the exterior orientation by using the standard solution technique of Section 14.3.1.

14.4. Write a program to determine the exterior orientation by using the quaternion technique of Section 14.3.3.

14.5. Three points in the world coordinate system are known:

$$\begin{pmatrix} x_1 \\ y_1 \\ z_1 \end{pmatrix} = \begin{pmatrix} 0 \\ 0 \\ 8 \end{pmatrix}, \quad \begin{pmatrix} x_2 \\ y_2 \\ z_2 \end{pmatrix} = \begin{pmatrix} 4 \\ 0 \\ 8 \end{pmatrix}, \quad \begin{pmatrix} x_3 \\ y_3 \\ z_3 \end{pmatrix} = \begin{pmatrix} 0 \\ 3 \\ 8 \end{pmatrix}$$

The observed perspective projections of these points are

$$\begin{pmatrix} u_1 \\ v_1 \end{pmatrix} = \begin{pmatrix} 0.0026 \\ 0.00955 \end{pmatrix}, \quad \begin{pmatrix} u_2 \\ v_2 \end{pmatrix} = \begin{pmatrix} 0.0229 \\ -0.012 \end{pmatrix}, \quad \begin{pmatrix} u_3 \\ v_3 \end{pmatrix} = \begin{pmatrix} 0.00472 \\ 0.00732 \end{pmatrix}$$

The distance between the image plane and the front of the camera is $f = 0.075$. With an initial estimate of 0 for $x_0, y_0, z_0, \omega, \phi, \kappa$, use one of the iterative techniques to solve for the exterior orientation and show that the translation and rotation are given by

$$\begin{pmatrix} x_0 \\ y_0 \\ z_0 \end{pmatrix} = \begin{pmatrix} -3.86712 \\ 0.470074 \\ -4.59090 \end{pmatrix}, \quad \begin{pmatrix} \omega \\ \phi \\ \kappa \end{pmatrix} = \begin{pmatrix} -0.088158 \\ 0.24665 \\ 0.150897 \end{pmatrix} \text{ radians}$$

14.6. Write a program to determine the relative orientation by using the standard technique of Section 14.4.1.

14.7. Write a program to determine the relative orientation by using the quaternion solution technique of Section 14.4.2.

14.8. In a stereo camera setup, there are six observed corresponding points in the left and right perspective projection images. They are given in the table below.

	1	2	3	n 4	5	6
u_{L_n}	0.00269	0.0267	0.00591	0.0779	-0.00801	-0.0540
v_{L_n}	0.0107	0.0066	0.0287	0.0405	-0.0327	-0.0327
u_{R_n}	-0.0235	0.0076	-0.0187	0.101	-0.0297	-0.0749
v_{R_n}	-0.0496	0.0455	0.0765	0.106	-0.0129	-0.00889

The distance between the center of perspectivity and the image plane is $f = 0.075$. The center of perspectivity for the left image is the origin. For the right image

$x_R = 5$. Use an initial estimate of 0 for $y_R, z_R, \omega_R, \phi_R, \kappa_R$ and show that the translation and rotation are given by

$$\begin{pmatrix} y_R \\ z_R \end{pmatrix} = \begin{pmatrix} -3.66073 \\ 3.61088 \end{pmatrix} \quad \text{and} \quad \begin{pmatrix} \omega_R \\ \phi_R \\ \kappa_R \end{pmatrix} = \begin{pmatrix} 0.049167 \\ -0.129778 \\ 0.076311 \end{pmatrix} \text{ radians}$$

■ Photometric Journals

Bildmessung und Luftbildwesen (BuL)
Organ der Deutschen Gesellschaft für Photogrammetrie
und Fernerkundung DGPF
Herbert Wichman Verlag
Postfach 210729
7500 Karlsruhe

Photogrammetria
Official Journal of the International Society for Photogrammetry
Elsevier Publishing Company
P.O. Box 330
ND 1000 AH Amsterdam

Photogrammetric Engineering
Journal of the American Society of Photogrammetry
105 Virginia Avenue
Fall Church, VA 22046

Photogrammetric Record
Published by the Photogrammetric Society, London
Department of Photogrammetry and Surveying
University College
Gower Street
London WC1E 6BT

Photogrammetrie
Bulletin tremestriel de la Sociétè belge de photogrammétrie
Boulevard Pachéco 34,
Bruxelles

Sociétè Française de Photogrammetrie
Bulletin, Institute Géographique National
2, avenue Pasteur
Saint-Mandé, F94

Nederlands Geodetisch Tijdschrift
Uitgave van de Nederlandse
Vereniging voor Geodesie
Centrale Directie Kadaster
Waltersingel 1,
Apeldoorn

Zeitschrift für Vermessungskunde (ZfV)
Organ des Deutschen Verins für Vermessungswesen
Konrad Wittwer-Verlag
Nordbahnhofstraße 16
7000 Stuttgart - 1

Der Vermessungsingenieur
Organ des Verbandes Deutscher Vermessungsingenieure
Druck- und Verlagshaus Chmierlorz
Thorwaldsenanlage 57
7500 Karlsruhe 21

Allgemeine Vermessungs Nachricten (AVN)
Ferdinand Dümmlers Verlag
Postfach 297
5300 Bonn 1

Vermessungstechnik
Geod-Kartographische Zeitschrift der Deutschen Demokratischen R.
für Wissenschaft und Praxis
VEB Verlag für Bauwesen
Französische Straße 13-14
Berlin 108

Österreichische Zeitscrhift für Vermessungswesen
Organ des Österreich. Vereins für Vermessungswesen
Verlagshaus M. Rohrer
A 2500 Baden bei Wein

Schweizerische Zeitschrift für Vermessungswesen
Organ des Schweiz. Vereins für Vermessungswesen
Fabag Druckerei AG
CH 8401 Winterthur

■ Bibliography

Abdel-Aziz, Y., and H. M. Karara, "Direct Linear Transformation from Comparative Co-ordinates into Object Space Coordinates in Close Range Photogrammetry," *Proceedings of the ASP Symposium on Close Range Photogrammetry,* Urbana, IL, 1974.

Arun, K. S., T. S. Huang, and S. D. Blostein, "Least Squares Fitting of Two 3D Point Sets" *IEEE Transactions on Pattern Analysis and Machine Intelligence*, Vol. PAMI-9, No. 5, Sept. 1987, pp. 698–700.

Bender, L. U., "Derivation of Parallax Equations," *Photogrammetric Engineering,* Vol. 33, 1967, pp. 1175–1179.

Blais, J. A. R., "Three-Dimensional Similarity," *Canadian Surveyor,* Vol. 1, 1972, pp. 71–76.

Boge, W. F., "Resection Using Iterative Least Squares," *Photogrammetric Engineering,* Vol. 31, 1965, pp. 701–714.

Bopp, H., and H. Krauss, "Ein Orientierungs- und Kalibrierungsverfahren für nichttopographische Anwendungen der Photogrammetrie," *ANV,* 1978, pp. 182–188.

Brown, D. C., "Decentering Distortion and the Definitive Calibration of Metric Cameras," *Photogrammetric Engineering,* Vol. 32, 1966, pp. 444–462.

——, "Close Range Camera Calibration," *Photogrammetric Engineering,* Vol. 37, 1971, pp. 855–866.

——, "Calibration of Close-Range Cameras," *12th Congress of the ISP,* Ottawa, 1972.

Campbell, J., "Zur Ausgleichung nach 'quasi-vermittelnden' Beobachtungen, *Zeitschrift für Vermessungswesen,* Vol. 11, 1969, pp. 441–445.

Caprile, B., and V. Torre, "Using Vanishing Points for Camera Calibration," *International Journal of Computer Vision,* Vol. 4, 1990, pp. 127–140.

Dohler, M., "Nahbildmessung mit Nicht-Mess Kammera," *Bildmessung und Luftbildwesen,* Vol. 29, 1971, pp. 17–76.

Dos Santos, F., "Generalization of Y-Parallax Differential Formula," *Photogrammetria,* Vol. 23, 1968, pp. 95–102.

Faig, W., "Calibration of Close Range Cameras," *Proceedings of the ASP Symposium on Close Range Photogrammetry,* Urbana, IL, 1971.

Finsterwalder, S., "Die Geometrischen Grundlagen der Photogrammetrie," *Jahresberichte der Deutschen Mathematiker-Vereinigung,* Vol. 6, 1897, pp. 1–41.

Finsterwalder, S., and W. Scheufele, "Das Rückwärtseinschneiden im Raum," *Sebastian Finsterwalder zum 75. Geburtstage,* Verlag Herbert Wichmann, Berlin, 1937, pp. 86–100.

Hallert, B., "Quality of Exterior Orientation," *Photogrammetric Engineering,* Vol. 32, 1966, pp. 464–475.

Haralick, R. M., C. Y. Lee, X. Zhuang, V. G. Vaidya, and M. B. Kim, "Pore Estimation From Corresponding Point Data," *IEEE Computer Society Workshop on Computer Vision*, Miami, FL, Nov. 30-Dec. 2, 1987, pp. 258–263.

Hinsken, L., "A Singularity-Free Algorithm for Spatial Orientation of Bundles," *International Archives of Photogrammetry and Remote Sensing,* Vol. 27, Kyoto, Japan, 1988.

Hoaglin, D. C., F Mosteller and J. W. Tukey, *Understanding Robust and Exploratory Data Analysis,* Wiley, New York, 1983, pp. 348–349.

Hofmann-Wellenhof, B., "Die gegenseitige Orientierung von zwei Strahlenbündeln bei Übereinstimmung, bei unbekannten Näherungswerten und durch ein nichtiteratives Verfahren," Deutsche Geodätische Kommission, Munich, 1979, pp. 5–67.

Hogg, R. V., "An Introduction to Robust Estimaton," in *Robustness in Statistics*, R.L. Launer and G.N. Wilkinson, (eds.), Academic Press, New York, 1979.

Horn, B. K. P., "Closed-Form Solution of Absolute Orientation Using Unit Quaternions," *Journal of the Optical Society of America A*, Vol. 4, 1987, pp. 629–642.

Horn, B. K. P., H. M. Hilden, and S. Negahdaripour, "Closed-Form Solution of Absolute Orientation Using Orthonormal Matrices," *Journal of the Optical Society of America A*, Vol. 5, 1988, pp. 1127–35.

Huber, P. J., *Robust Statistics*, Wiley, New York, 1981.

Karara, H. M., and W. Faig, "Interior Orientation in Close Range Photogrammetry: An Analysis of Alternative Approaches," *International Archives of Photogrammetry, Commission V*, ISP, 1972.

Killian, K., "Über das Rückwärtseinschneiden im Raum," *Österreichische Zeitschrift für Vermessungswesen*, Vol. 43, 1955, pp. 97–104, 171–179.

——, "Ebenes und räumliches Rckwärtseinschneiden eines Dreickes in Hinblick auf die Luftbildmessung," *Österreichische Zeitschrift für Vermessungswesen*, Vol. 54, 1966, pp. 181–192.

Kostka, R., and A. Reithofer, "Numerische Rekonstruktion reäumlicher Objekte aus Einzelbildern—Am Beispiel der Amateurbildauswertung von Unfallsituationen," *Österreichische Zeitschrift für Vermessungswesen*, Vol. 63, 1975, pp. 95–103.

Kruppa, E., "Zur Ermittlung eines Objektes aus zwei Perspektiven mit innere Orientierung," *Sitzungsberichte der Mathematisch-Naturwiss*, Klasse der Kaiserlichen Akademie der Wissenschaften (Wien), Abs. IIa, 122, 1913, pp. 1939–48.

Kubik, K., "Iterative Methoden zur Lösung des nichtlinearen Ausgleichsproblems," *Zeitschrift für Vermessungskunde*, Vol. 6, 1967, pp. 214–225.

Lehman, E. H., "Determining Exposure Point, Tilt, and Direction of Photograph from Three Known Ground Positions and Focal Length," *Photogrammetric Engineering*, Vol. 29, 1963, pp. 702–708.

Lenox, J. B., and J. R. Cuzzi, "Accurately Characterizing a Measured Change in Configuration," *American Society of Mechanical Engineers Publication*, Vol. 78-DET-50, 1978, pp. 1–7.

Mehrotra, R. C., and H. M. Karara, "A Computational Procedure and Software for Establishing a Stable Three-Dimensional Test Area for Close Range Application," *Proceedings of the ASP Symposium on Close Range Photogrammetry*, Champaign, IL, 1975.

Merritt, E., *Analytical Photogrammetry*, Pitman, New York, 1958.

——, "Explicit Three-Point Resection in Space," *Photogrammetric Engineering*, Vol. 15, 1949, pp. 649–665.

Müller, E., "Vorlesungen über Darstellende Geometrie," in *Die Linearen Abbildungen*, Vol. 1, E. Kruppa (ed.), Franz Deutike Verlag, Leipzig, 1923.

Müller, F., "Direkte (exakte) Lösung des einfachen Rückwärtseinschnei-dens im Raume," *Allgemeine Vermessungs-Nachrichten*, Vol. 50, 1925, pp. 249–256.

Pope, A., "An Advantageous Alternative Parametrization of Rotation for Analytic Photogrammetry," *Symposium on Computational Photogrammetry of the American Society of Photogrammetry*, Alexandria, VA, 1970.

Rinner, K., "Studie über eine rechnerische Lösung für die gegenseitige Orientierung photogrmmetrischer Aufnahmen," *Photogrammetria*, Vol. 5, 1942, pp. 41-54.

Rinner, K., "Analytische photogrammetrische Triangulation mit formtreuen Bündeln," *Über räumliche Drehungen,* Verlag der Bayerischen Akademie der Wissenschaften, Munich, 1957, pp. 5–40.

——, "Die Orientierung eines Richtungsbündels," *Österreichische Zeitschrift für Vermessungswesen,* Vol. 53, 1965, pp. 105–113.

Rosenfield, G. H., "The Problem of Exterior Orientation in Photogrammetry," *Photogrammetric Engineering,* Vol. 25, 1959, pp. 536–553.

Sansò, F., "An Exact Solution of the Roto-Translation Problem," *Photogrammetria,* Vol. 29, 1973, pp. 203–216.

Schermerhorn, W., "Introduction to the Theory of Error of Aerial Triangulation in Space," *Photogrammetria,* Vol. 3, 1940, pp. 138–147.

——, "Einleitung zur Fehertheorie der räumlichen Aerotriangulation," *Photogrammetria,* Vol. 4, 1941, pp. 28–45.

Schmid, H., "An Analytical Treatment of the Problem of Triangulation by Stereophotogrammetry," *Photogrammetria,* Vol. 13, 1956, pp. 67–77, 91–116.

Schmidt, H. H., and S. Heggli, "Räumliche Koordinatentransformation. Eine pseuo-lineare Formulierung als Annäherungslösung mit eune strenge Ausgleichung mit entsprecehndem Fortran-Programm," Institue für Geodäsie und Photogrammetrie der ETH Zürich, Mitteilungen, No. 23, 1978.

Schut, G. H., "An Analysis of Methods and Results in Analytical Aerial Triangulation," *Photogrammetria,* Vol. 14, 1957, pp. 16–32.

——, "Construction of Orthogonal Matrices and Their Application in Analytical Photogrammetry," *Photogrammetria,* Vol. 15, 1958–59, pp. 149–162.

——, "Remarks on the Theory of Analytical Aerial Triangulation," *Photogrammetria,* Vol. 16, 1959, pp. 57–66.

——, "On Exact Equations for the Computation of the Rotational Elements of Absolute Orientation," *Photogrammetria,* Vol. 17, 1962–63, pp. 34–37.

Smith, A. D. N., "The Explicit Solution of the Single Picture Resection Problem, with a Least Squares Adjustment to Redundant Control," *Photogrammetria,* Vol. 5, 1965, pp. 113–122.

Sobel, I., "On Calibrating Computer Controlled Cameras for Perceiving 3-D Scenes," *Artificial Intelligence,* Vol. 5, 1974, pp. 185–198.

Szczepanski, W., "Die Lösungsvorschläge für den räumlichen Rückwärts-einschnitt," Deutsche Geodätische Kommission, Reihe C: Dissertationen–Heft, No. 29, 1958, 1–144.

Thompson, E. H., "An Exact Linear Solution of the Problem of Absolute Orientation," *Photogrammetria,* Vol. 15, 1958–59, p. 163–179.

——, "A Rational Algebraic Formulation of the Problem of Relative Orientation," *Photogrammetric Record,* Vol. 3, 1959, pp. 152–159.

——, "The Projective Theory of Relative Orientation," *Photogrammetria,* Vol. 23, 1968, pp. 67–75.

——, "Space Resection without Interior Orientation," *Photogrammetric Record,* Vol. 7, 1971a, pp. 39–45.

——, "Space Resection: Failure Cases," *Photogrammetric Record,* Vol. 7, 1971b, pp. 201–204.

Tiestra J. M., "Calculation of Orthogonal Matrices," ITC Delft Series A, Vol. 48, 1969.

Van den Hout, C. M. A., and P. Stefanovič, "Efficient Analytical Relative Orientation," *Proceedings of the 13th Congress of the International Society for Photogrammetry,* Helsinki, 1976, pp. 1–22.

Van der Weele, A. J., "The Relative Orientation of Photographs of Mountainous Terrain," *Photogrammetria,* Vol. 16, 1959, pp. 161–169.

Van Roessel, "Estimating Lens Distortion with Orthogonal Polynomials," *Photogrammetric Engineering,* Vol. 36, 1970, pp. 584–588.

Veldpaus, F. E., H. J. Woltring, and L. J. M. G. Dortmans, "A Least-Squares Algorithm for the Equiform Transformation from Spatial Marker Co-ordinates," *Journal of Biomechanics,* Vol. 21, 1988, pp. 45–54.

Washer, F. E., "A Simplified Method of Locating the Point of Symmetry," *Photogrammetric Engineering,* Vol. 23, 1957, pp. 75–88.

———, "The Precise Evaluation of Linear Distortion," *Photogrammetric Engineering,* Vol. 29, 1963, pp. 327–332.

Wolf, H., *Ausgleichungsrechnung,* Ferd. Dümmler Verlag, Bonn, 1975.

Wolf, P. R., *Elements of Photogrammetry,* 2d ed., McGraw-Hill, New York, 1983.

Wunderlich, W., "Zur rechnerischen Durchführung des Vierpunktverfahrens," *Österreichische Zeitschrift für Vermessungswesen,* Vol. 45, 1957, pp. 9–13.

———, "Über die gefährlichen Örter bei zwei Achtpunktproblemen und einem Fünfpunktproblem," *Österreichische Zeitschrift für Vermessungswesen,* Vol. 64, 1977, pp. 119–128.

15 MOTION AND SURFACE STRUCTURE FROM TIME-VARYING IMAGE SEQUENCES*

15.1   Introducton

Motion analysis involves estimating the relative motion of objects with respect to each other and the camera, given two or more perspective projection images in a time sequence. A variety of real-world problems have motivated current motion analysis research. These include applications in industrial automation and inspection, robot assembly, autonomous vehicle navigation, biomedical engineering, remote sensing, and general three-dimensional-scene understanding. This chapter describes how information about an object's motion and surface structure can be recovered either from the observed optic flow or from point correspondences. We discuss both least-squares and robust techniques for the structure recovery problem.

15.2   The Fundamental Optic Flow Equation

We begin by discussing the relationship between the motion of a three-dimensional point and the corresponding motion of that point on the perspective projection image.

The point (x, y, z) on the moving rigid body has perspective projection (u, v) on the image plane, which we take to be a distance f in front of the camera lens, the origin. The motion of (x, y, z)—that is, its velocity—causes a motion of its perspective projection (u, v) on the image. We denote by $(\dot{u}, \dot{v})$ the velocity of the point (u, v). The optic flow image is the image in which pixel (u, v), the perspective projection of some point (x, y, z), contains its own motion $(\dot{u}, \dot{v})$. By the geometry of the perspective projection, we have

$$\begin{pmatrix} u \\ v \end{pmatrix} = \frac{f}{z} \begin{pmatrix} x \\ y \end{pmatrix} \tag{15.1}$$

---

* This chapter was written by **Xinhua Zhuang** and **Robert M. Haralick**.

To change Eq. (15.1) into the optic flow equation, we take time derivatives of both sides of Eq. (15.1):

$$\begin{pmatrix} \dot{u} \\ \dot{v} \end{pmatrix} = \frac{f}{z^2} \begin{pmatrix} z\dot{x} - x\dot{z} \\ z\dot{y} - y\dot{z} \end{pmatrix} \tag{15.2}$$

Solving Eq. (15.1) for $\begin{pmatrix} x \\ y \end{pmatrix}$ and substituting into Eq. (15.2) yields the fundamental optic flow equation:

$$\begin{pmatrix} \dot{u} \\ \dot{v} \end{pmatrix} = \frac{1}{z} \begin{pmatrix} f & 0 & -u \\ 0 & f & -v \end{pmatrix} \begin{pmatrix} \dot{x} \\ \dot{y} \\ \dot{z} \end{pmatrix} \tag{15.3}$$

The general solution of this equation is given by

$$\begin{pmatrix} \dot{x} \\ \dot{y} \\ \dot{z} \end{pmatrix} = \frac{z}{f} \begin{pmatrix} \dot{u} \\ \dot{v} \\ 0 \end{pmatrix} + \lambda \begin{pmatrix} u \\ v \\ f \end{pmatrix} \tag{15.4}$$

where λ is a free variable.

The first term of the solution equation (15.4) is the back-projected optic flow. It constitutes that particular solution to Eq. (15.3) in which all motion is in a plane parallel to the image plane. The second term is the general solution to the homogeneous equation. It indicates that any three-dimensional motion along the ray of sight is not captured in the optic flow.

15.2.1 Translational Motion

Suppose that we observe an N-point optic flow field $\{(u_n, v_n, \dot{u}_n, \dot{v}_n)\}_{n=1}^N$ of a corresponding set of unknown three-dimensional points $\{(x_n, y_n, z_n)\}_{n=1}^N$, all of which are assumed to be moving with the same but unknown velocity $(\dot{x}, \dot{y}, \dot{z})$. Then up to a multiplicative constant, the position of all the points and their common velocity can be determined from the optic flow field.

By Eq. (15.3) we have for each point

$$\begin{pmatrix} \dot{u}_n \\ \dot{v}_n \end{pmatrix} z_n + \begin{pmatrix} -f & 0 & u_n \\ 0 & -f & v_n \end{pmatrix} \begin{pmatrix} \dot{x}_n \\ \dot{y}_n \\ \dot{z}_n \end{pmatrix} = \begin{pmatrix} 0 \\ 0 \end{pmatrix}, \quad n = 1, \ldots, N$$

Putting all these equations into a single matrix equation, we obtain

$$A \begin{pmatrix} z_1 \\ z_2 \\ \vdots \\ z_N \\ \dot{x} \\ \dot{y} \\ \dot{z} \end{pmatrix} = 0 \tag{15.5}$$

where A is the $2N \times (N+3)$ matrix defined by

$$A = \begin{pmatrix} \dot{u}_1 & & & & -f & 0 & u_1 \\ \dot{v}_1 & & & & 0 & -f & v_1 \\ & \dot{u}_2 & & & -f & 0 & u_2 \\ & \dot{v}_2 & & & 0 & -f & v_2 \\ & & \ddots & & & & \vdots \\ & & & \dot{u}_N & -f & 0 & u_N \\ & & & \dot{v}_N & 0 & -f & v_N \end{pmatrix}$$

We would like to determine the solution to Eq. (15.5) that minimizes

$$(z_1, z_2, \ldots, z_N, \dot{x}, \dot{y}, \dot{z}) A'A \begin{pmatrix} z_1 \\ z_2 \\ \vdots \\ z_N \\ \dot{x} \\ \dot{y} \\ \dot{z} \end{pmatrix}$$

subject to the constraint that $z_1^2 + z_2^2 + \cdots + z_N^2 + \dot{x}^2 + \dot{y}^2 + \dot{z}^2 = k^2$, a nonzero constant. The solution is given by k times that eigenvector of $A'A$ having the smallest eigenvalue. $A'A$ has the following simple form:

$$A'A = \begin{pmatrix} \dot{u}_1^2 + \dot{v}_1^2 & & & & -f\dot{u}_1 & -f\dot{v}_1 & \dot{u}_1 u_1 + \dot{v}_1 v_1 \\ & \dot{u}_2^2 + \dot{v}_2^2 & & & -f\dot{u}_2 & -f\dot{v}_2 & \dot{u}_2 u_2 + \dot{v}_2 v_2 \\ & & \ddots & & & & \vdots \\ & & & \dot{u}_N^2 + \dot{v}_N^2 & -f\dot{u}_N & -f\dot{v}_N & \dot{u}_N u_N + \dot{v}_N v_N \\ -f\dot{u}_1 & -f\dot{u}_2 & & -f\dot{u}_N & Nf^2 & 0 & -f\sum_n u_n \\ -f\dot{v}_1 & -f\dot{v}_2 & & -f\dot{v}_N & 0 & Nf^2 & -f\sum_n v_n \\ \dot{u}_1 u_1 + \dot{v}_1 v_1 & \dot{u}_2 u_2 + \dot{v}_2 v_2 & & \dot{u}_N u_N + \dot{v}_N v_N & -f\sum_n u_n & -f\sum_n v_n & \sum_n u_n^2 + v_n^2 \end{pmatrix}$$

With a value for each depth z_n, the x- and y-coordinates can be determined directly from the perspective projection equations

$$\begin{pmatrix} x_n \\ y_n \end{pmatrix} = \frac{z_n}{f} \begin{pmatrix} u_n \\ v_n \end{pmatrix}, \quad n = 1, \ldots, N$$

A less calculation-intensive way to proceed is to solve for the velocity $(\dot{x}, \dot{y}, \dot{z})$ first and then solve for the depths $z_1, \ldots, z_N$. To do this, notice that from Eq. 15.4

$$\begin{pmatrix} \dot{x} \\ \dot{y} \\ \dot{z} \end{pmatrix}$$

is in the span of

$$\left\{ \begin{pmatrix} \dot{u} \\ \dot{v} \\ 0 \end{pmatrix}, \begin{pmatrix} u \\ v \\ f \end{pmatrix} \right\}$$

The vector

$$\begin{pmatrix} \dot{u} \\ \dot{v} \\ 0 \end{pmatrix} \times \begin{pmatrix} u \\ v \\ f \end{pmatrix} = \begin{pmatrix} \dot{v}f \\ -\dot{u}f \\ \dot{u}v - \dot{v}u \end{pmatrix}$$

is orthogonal to both

$$\begin{pmatrix} \dot{u} \\ \dot{v} \\ 0 \end{pmatrix} \quad \text{and} \quad \begin{pmatrix} u \\ v \\ f \end{pmatrix}$$

Hence it must be that

$$\begin{pmatrix} \dot{u} \\ \dot{v} \\ 0 \end{pmatrix} \times \begin{pmatrix} u \\ v \\ f \end{pmatrix}$$

is orthogonal to $\begin{pmatrix} \dot{x} \\ \dot{y} \\ \dot{z} \end{pmatrix}$. Therefore

$$\begin{pmatrix} \dot{v}f \\ -\dot{u}f \\ \dot{u}v - u\dot{v} \end{pmatrix}' \begin{pmatrix} \dot{x} \\ \dot{y} \\ \dot{z} \end{pmatrix} = 0$$

For N observed optic flow points, we want to solve the overconstrained system

$$A \begin{pmatrix} \dot{x} \\ \dot{y} \\ \dot{z} \end{pmatrix} = 0$$

subject to the constraint $\dot{x}^2 + \dot{y}^2 + \dot{z}^2 = k^2$, a positive constant, and where

$$A = \begin{pmatrix} \dot{v}_1 f & -\dot{u}_1 f & \dot{u}_1 v_1 - u_1 \dot{v}_1 \\ \vdots & & \\ \dot{v}_N f & -\dot{u}_N f & \dot{u}_N v_N - u_N \dot{v}_N \end{pmatrix}$$

The solution is given by that eigenvector of $A'A$ having the smallest eigenvalue. Here $A'A$ has the form

$$A'A = \begin{pmatrix} f^2 \sum_n \dot{v}_n^2 & -f^2 \sum_n \dot{u}_n \dot{v}_n & f \sum_n \dot{v}_n(\dot{u}_n v_n - u_n \dot{v}_n) \\ -f^2 \sum_n \dot{u}_n \dot{v}_n & f^2 \sum_n \dot{u}_n^2 & -f \sum_n \dot{u}_n(\dot{u}_n v_n - u_n \dot{v}_n) \\ f \sum_n \dot{v}_n(\dot{u}_n v_n - u_n \dot{v}_n) & -f \sum_n \dot{u}_n(\dot{u}_n v_n - u_n \dot{v}_n) & \sum_n (\dot{u}_n v_n - u_n \dot{v}_n)^2 \end{pmatrix}$$

Once $\begin{pmatrix} \dot{x} \\ \dot{y} \\ \dot{z} \end{pmatrix}$ is determined (up to a multiplicative constant), the depth $z_1, \ldots, z_N$

can be determined from Eq. (15.3). The z_n that minimizes $[z_n \dot{u}_n - (f\dot{x} - u_n \dot{z})]^2 +$
$[z_n \dot{v}_n - (f\dot{y} - v_n \dot{z})]^2$ is given by

$$z_n = \frac{\dot{u}_n(f\dot{x} - u_n \dot{z}) + \dot{v}_n(f\dot{y} - v_n \dot{z})}{\dot{u}_n^2 + \dot{v}_n^2}, \quad n = 1, \ldots, N$$

15.2.2 Focus of Expansion and Contraction

If a three-dimensional point (x, y, z) in translational motion results in a two-dimensional projected point (u, v) having no motion, then from Eq. (15.4)

$$\begin{pmatrix} \dot{x} \\ \dot{y} \\ \dot{z} \end{pmatrix} = \lambda \begin{pmatrix} u \\ v \\ f \end{pmatrix}$$

indicating that all the translational motion is in a direction along the ray of sight.

Suppose now that a set of three-dimensional points $\left\{ \begin{pmatrix} x_n \\ y_n \\ z_n \end{pmatrix} \right\}_{n=1}^{N}$ is in trans-

lational motion. Then from Eq. (15.4) it is apparent that

$$\frac{f}{\dot{z}_n} \begin{pmatrix} \dot{x}_n \\ \dot{y}_n \end{pmatrix} = \frac{z_n}{\dot{z}_n} \begin{pmatrix} \dot{u}_n \\ \dot{v}_n \end{pmatrix} + \begin{pmatrix} u_n \\ v_n \end{pmatrix} \tag{15.6}$$

Now considering that $z_n/\dot{z}_n$ is unknown, the geometric meaning of Eq. (15.6) is that

the unknown three-dimensional motion $f/\dot{z}_n \begin{pmatrix} \dot{x}_n \\ \dot{y}_n \end{pmatrix}$ must lie along a line passing

through $\begin{pmatrix} u_n \\ v_n \end{pmatrix}$ in a direction $\begin{pmatrix} \dot{u}_n \\ \dot{v}_n \end{pmatrix}$.

Then from Eq. (15.6)

$$\frac{f}{\dot{z}_n} \begin{pmatrix} \dot{x}_n \\ \dot{y}_n \end{pmatrix} = \begin{pmatrix} u_0 \\ v_0 \end{pmatrix} = \frac{z_n}{\dot{z}_n} \begin{pmatrix} \dot{u}_n \\ \dot{v}_n \end{pmatrix} + \begin{pmatrix} u_n \\ v_n \end{pmatrix}, \quad n = 1, \ldots, N \tag{15.7}$$

This means that the point $\begin{pmatrix} u_0 \\ v_0 \end{pmatrix}$ will be at the intersection of all the lines

$$\left\{ \begin{pmatrix} u \\ v \end{pmatrix} \middle| \begin{pmatrix} u \\ v \end{pmatrix} = \begin{pmatrix} u_n \\ v_n \end{pmatrix} + \lambda \begin{pmatrix} \dot{u}_n \\ \dot{v}_n \end{pmatrix} \right\}$$

Such a point is called the *focus of expansion* if the three-dimensional-point field is moving toward the camera, and the *focus of contraction* if the three-dimensional-point field is moving away from the camera. One way to determine the focus of expansion or contraction is to multiply Eq. (15.7) by $(\dot{v}_n - \dot{u}_n)$. This results in the set of equations

$$(\dot{v}_n - \dot{u}_n)\begin{pmatrix} u_0 \\ v_0 \end{pmatrix} = (\dot{v}_n - \dot{u}_n)\begin{pmatrix} u_n \\ v_n \end{pmatrix}, \quad n = 1, \dots, N$$

This overdetermined system of equations can be solved in the least-squares sense for (u_0, v_0). The solution can be written as

$$(u_0, v_0) = \begin{pmatrix} \sum_n \dot{u}_n^2 & \sum_n u_n v_n \\ \sum_n u_n v_n & \sum_n \dot{v}_n^2 \end{pmatrix} \begin{pmatrix} \sum_n \dot{v}_n(\dot{v}_n u_n - \dot{u}_n v_n) \\ -\sum_n \dot{u}_n(\dot{v}_n u_n - \dot{u}_n v_n) \end{pmatrix} \frac{1}{\sum_n \dot{u}_n^2 \sum_n \dot{v}_n^2 - (\sum_n u_n v_n)^2}$$

A least-squares solution for (u_0, v_0) minimizing the criterion

$$\epsilon^2 = \sum_n \left\| \begin{pmatrix} u_0 \\ v_0 \end{pmatrix} - \begin{pmatrix} u_n \\ v_n \end{pmatrix} - \lambda_n \begin{pmatrix} \dot{u}_n \\ \dot{v}_n \end{pmatrix} \right\|^2$$

can also be determined. The solution for it satisfies

$$\begin{pmatrix} u_0 \\ v_0 \end{pmatrix} = \left[I + \sum_n \begin{pmatrix} \dot{u}_n^2 & \dot{u}_n \dot{v}_n \\ \dot{u}_n \dot{v}_n & \dot{v}_n^2 \end{pmatrix} \frac{1}{\dot{u}_n^2 + \dot{v}_n^2} \right]^{-1}$$

$$\times \sum_n \left[I - \begin{pmatrix} \dot{u}_n^2 & \dot{u}_n \dot{v}_n \\ \dot{u}_n \dot{v}_n & \dot{v}_n^2 \end{pmatrix} \frac{1}{\dot{u}_n^2 + \dot{v}_n^2} \right] \begin{pmatrix} u_n \\ v_n \end{pmatrix} \quad (15.8)$$

15.2.3 Moving Line Segment

Suppose two three-dimensional points (x_1, y_1, z_1) and (x_2, y_2, z_2) are a known fixed distance apart and are moving with translational motion at a common velocity $(\dot{x}, \dot{y}, \dot{z})$. Let the corresponding optic flow for these points be $(u_1, v_1, \dot{u}_1, \dot{v}_1)$ and $(u_2, v_2, \dot{u}_2, \dot{v}_2)$, respectively. Then it is possible to determine the position, the orientation, and the velocity of the line segment.

From the perspective projection equations,

$$\begin{pmatrix} x_1 \\ y_1 \end{pmatrix} = \frac{z_1}{f} \begin{pmatrix} u_1 \\ v_1 \end{pmatrix}, \qquad \begin{pmatrix} x_2 \\ y_2 \end{pmatrix} = \frac{z_2}{f} \begin{pmatrix} u_2 \\ v_2 \end{pmatrix}$$

From the optic flow equation,

$$\begin{pmatrix} \dot{x} \\ \dot{y} \end{pmatrix} = \frac{z_1}{f} \begin{pmatrix} \dot{u}_1 \\ \dot{v}_1 \end{pmatrix} + \frac{\dot{z}}{f} \begin{pmatrix} u_1 \\ v_1 \end{pmatrix} = \frac{z_2}{f} \begin{pmatrix} \dot{u}_2 \\ \dot{v}_2 \end{pmatrix} + \frac{\dot{z}}{f} \begin{pmatrix} u_2 \\ v_2 \end{pmatrix} \quad (15.9)$$

From the known length of the line segment,

$$(x_2 - x_1)^2 + (y_2 - y_1)^2 + (z_2 - z_1)^2 = d^2 \quad (15.10)$$

The unknowns are $(x_1, y_1, z_1), (x_2, y_2, z_2)$, and $(\dot{x}, \dot{y}, \dot{z})$. There are nine unknowns and nine equations. The optic flow equation (15.9) permits us to obtain a least-squares solution for $\dot{z}$ in terms of z_1 and z_2. That is, from

$$\dot{z}\begin{pmatrix} u_2 - u_1 \\ v_2 - v_1 \end{pmatrix} = \begin{pmatrix} z_1\dot{u}_1 - z_2\dot{u}_2 \\ z_1\dot{v}_1 - z_2\dot{v}_2 \end{pmatrix}$$

we obtain

$$\dot{z} = \frac{(u_2 - u_1)(z_1\dot{u}_1 - z_2\dot{u}_2) + (v_2 - v_1)(z_1\dot{v}_1 - z_2\dot{v}_2)}{(u_2 - u_1)^2 + (v_2 - v_1)^2}$$

Substituting this back into the equation, we can solve for z_2 in terms of $z_1 : z_2 = kz_1$, where

$$k = \frac{\dot{u}_1(v_2 - v_1) - \dot{v}_2(u_2 - u_1)}{\dot{u}_2(v_2 - v_1) - \dot{v}_2(u_2 - u_1)}$$

Now we substitute the relations for (x_1, y_1, z_1) from the perspective projection equations into Eq. (15.10) to obtain

$$\left(\frac{kz_1u_2}{f} - \frac{z_1u_1}{f} \right)^2 + \left(\frac{kz_1v_2}{f} - \frac{z_1v_1}{f} \right)^2 + (kz_1 - z_1)^2 = d^2$$

Hence

$$z_1 = \frac{fd}{\sqrt{(ku_2 - u_1)^2 + (kv_2 - v_1)^2 + (k-1)^2}}$$

Having a value for z_1, we can obtain the value for z_2 from $z_2 = kz_1$. With z_1, z_2, and $\dot{z}$ determined, $\dot{x}$ and $\dot{y}$ are determined. And from the perspective projection equations, with z_1 and z_2 determined, x_1, y_1, x_2, and y_2 are determined.

15.2.4 Optic Flow Acceleration Invariant

Differentiating Eq. (15.3) with respect to time and solving for $(\ddot{x}, \ddot{y}, \ddot{z})$ results in

$$\begin{pmatrix} \ddot{x} \\ \ddot{y} \\ \ddot{z} \end{pmatrix} = \frac{z}{f}\begin{pmatrix} \ddot{u} \\ \ddot{v} \\ 0 \end{pmatrix} + \frac{2\dot{z}}{f}\begin{pmatrix} \dot{u} \\ \dot{v} \\ 0 \end{pmatrix} + \frac{\ddot{z}}{f}\begin{pmatrix} u \\ v \\ f \end{pmatrix} \tag{15.11}$$

If two three-dimensional points (x_1, y_1, z_1) and (x_2, y_2, z_2) move with the same velocity $(\dot{x}, \dot{y}, \dot{z})$ and acceleration $(\ddot{x}, \ddot{y}, \ddot{z})$, then by Eqs. (15.9) and (15.11) their optic flow $(u_1, v_1, \dot{u}_1, \dot{v}_1)$ and $(u_2, v_2, \dot{u}_2, \dot{v}_2)$ must satisfy

$$\frac{\dot{z}}{f}\begin{pmatrix} u_1 - u_2 \\ v_1 - v_2 \\ 0 \end{pmatrix} + \frac{z_1}{f}\begin{pmatrix} \dot{u}_1 \\ \dot{v}_1 \\ 0 \end{pmatrix} - \frac{z_2}{f}\begin{pmatrix} \dot{u}_2 \\ \dot{v}_2 \\ 0 \end{pmatrix} = 0$$

$$\frac{\ddot{z}}{f}\begin{pmatrix} u_1 - u_2 \\ v_1 - v_2 \\ 0 \end{pmatrix} + \frac{2\dot{z}}{f}\begin{pmatrix} \dot{u}_1 - \dot{u}_2 \\ \dot{v}_1 - \dot{v}_2 \\ 0 \end{pmatrix} + \frac{z_1}{f}\begin{pmatrix} \ddot{u}_1 \\ \ddot{v}_1 \\ 0 \end{pmatrix} - \frac{z_2}{f}\begin{pmatrix} \ddot{u}_2 \\ \ddot{v}_2 \\ 0 \end{pmatrix} = 0$$

Rewriting this system of equations, we obtain

$$
\begin{pmatrix}
\dot{u}_1 & -\dot{u}_2 & u_1 - u_2 & 0 \\
\dot{v}_1 & -\dot{v}_2 & v_1 - v_2 & 0 \\
\ddot{u}_1 & -\ddot{u}_2 & 2(\dot{u}_1 - \dot{u}_2) & u_1 - u_2 \\
\ddot{v}_1 & -\ddot{v}_2 & 2(\dot{v}_1 - \dot{v}_2) & v_1 - v_2
\end{pmatrix}
\begin{pmatrix}
z_1 \\
z_2 \\
\dot{z} \\
\ddot{z}
\end{pmatrix}
=
\begin{pmatrix}
0 \\
0 \\
0 \\
0
\end{pmatrix}
$$

For this homogeneous system to have any solution, the determinant must be zero. Hence we have the optic flow invariant for accelerated translational motion:

$$
\begin{vmatrix}
\dot{u}_1 & -\dot{u}_2 & u_1 - u_2 & 0 \\
\dot{v}_1 & -\dot{v}_2 & v_1 - v_2 & 0 \\
\ddot{u}_1 & -\ddot{u}_2 & 2(\dot{u}_1 - \dot{u}_2) & u_1 - u_2 \\
\ddot{v}_1 & -\ddot{v}_2 & 2(\dot{v}_1 - \dot{v}_2) & v_1 - v_2
\end{vmatrix}
= 0
$$

This constraint can be reexpressed as the orthogonality of the vectors:

$$
\begin{pmatrix}
(v_1 - v_2)^2 \\
(v_1 - v_2)(u_1 - u_2) \\
(u_1 - u_2)^2 \\
2(v_1 - v_2) \\
2(u_1 - u_2)
\end{pmatrix}
\quad \text{and} \quad
\begin{pmatrix}
\ddot{u}_1 \dot{u}_2 - \dot{u}_1 \ddot{u}_2 \\
\dot{u}_1 \ddot{v}_2 - \dot{u}_2 \ddot{v}_1 + \dot{v}_1 \ddot{u}_2 - \dot{v}_2 \ddot{u}_1 \\
\ddot{u}_1 \dot{v}_2 - \dot{v}_1 \ddot{v}_2 \\
(-\dot{v}_1 \ddot{u}_2 + \dot{v}_2 \ddot{u}_1)(u_1 - u_2) \\
(\dot{v}_1 \ddot{u}_2 - \dot{v}_2 \ddot{u}_1)(v_1 - v_2)
\end{pmatrix}
$$

15.3   Rigid-Body Motion

In rigid-body motion there is no relative motion of points in or on the rigid body with respect to one another. The points must always maintain a fixed position relative to one another. All the points move with the body as a whole. Hence the moving position of each point in or on the moving rigid body can be represented by the same rotational and translational transformation of the point from its initial position. This implies that there exists a rotation matrix $R(t)$ and translation vector $T(t)$ such that for all points p in or on the rigid body,

$$
p(t) = R(t)p(0) + T(t) \tag{15.12}
$$

Here $p(0)$ represents the initial position of the given point, $R(0) = I$, and $T(0) = 0$.

The motion of each point in or on the rigid body is described by its velocity vector, which is simply the time derivative of its position. Thus upon taking the time derivative of Eq. (15.12), we obtain

$$
\dot{p}(t) = \dot{R}(t)p(0) + \dot{T}(t) \tag{15.13}
$$

We can reexpress $\dot{p}$ of Eq. (15.13) in terms of its current position rather than its initial position by solving Eq. (15.12) for $p(0)$ and substituting $R^{-1}(t)[p(t) - T(t)]$ for $p(0)$ in Eq. (15.13). This yields

$$
\dot{p}(t) = \dot{R}(t)R^{-1}(t)p(t) + \dot{T}(t) - \dot{R}(t)R^{-1}(t)T(t) \tag{15.14}
$$

To simplify Eq. (15.14), we let

$$S(t) = \dot{R}(t)R^{-1}(t) = \dot{R}(t)R'(t) \qquad (15.15)$$

and

$$k(t) = \dot{T}(t) - S(t)T(t) \qquad (15.16)$$

to obtain

$$\dot{p}(t) = S(t)p(t) + k(t) \qquad (15.17)$$

The system of first-order ordinary linear differential equations represented by Eq. (15.17) can be uniquely solved for $p(t)$ in terms of the initial position $p(0)$:

$$p(t) = \exp\left(\int_0^t S(u)du\right) p(0) + \int_0^t \exp\left(\int_v^t S(u)du\right) [k(v)]dv \qquad (15.18)$$

Equation (15.18) states that the description of motion in Eq. (15.17) is equivalent, in the sense of having the identical information, to the description of point position in Eq. (15.12).

From Eq. (15.15)

$$\dot{R}(t) - S(t)R(t) = 0$$

which implies that

$$R(t) = \exp\left(\int_0^t S(u)du\right) \qquad (15.19)$$

since $R(0) = I$. From Eq. (15.16)

$$T(t) = \int_0^t \exp\left(\int_v^t S(u)du\right) [k(v)]dv \qquad (15.20)$$

since $T(0) = 0$. Hence the information in the motion description $[S(t), k(t)]$ is sufficient to recover the rotational and translational transformations as well as the position description. We will find it convenient to use the motion description $[S(t), k(t)]$ instead of the motion description $[R(t), T(t)]$, since $[S(t), k(t)]$ ties the current velocity to the current position.

Consider the meaning of Eq. (15.17). Reexpressing $k(t)$ in terms of $S(t)$ and $T(t)$, we have

$$\dot{p}(t) = S(t)[p(t) - T(t)] + \dot{T}(t) \qquad (15.21)$$

If there were no rotational component, $\dot{p}(t)$ would be exactly $\dot{T}(t)$. This suggests that $S(t)[p(t) - T(t)]$ must be tied to the velocity of the motion due to angular rotation. Now $p(t) - T(t)$ is simply the position of the point relative to the frame whose origin is at $T(t)$, and $S(t)[p(t) - T(t)]$ must therefore be the same as $\Omega(t) \times [p(t) - T(t)]$, where $\Omega(t)$ is the angular velocity of the rigid body.

We can understand this in a more formal way by first seeing that $S(t)$ is skew symmetric, $S(t) = -S(t)'$, and then relating the action of a skew symmetric matrix on a given vector with the cross-product of a vector whose components are the unique entries in the skew symmetric matrix and the given vector. Differentiating $I = R(t)R'(t)$ with respect to time yields

$$0 = R(t)\dot{R}'(t) + \dot{R}(t)R'(t)$$

$$= R(t)\dot{R}'(t) + S(t)$$

$$= \left[\dot{R}(t)R'(t)\right]' + S(t)$$

$$= S(t)' + S(t)$$

Hence $S(t)$ must have the form

$$S(t) = \begin{pmatrix} 0 & -w_z(t) & w_y(t) \\ w_z(t) & 0 & -w_x(t) \\ -w_y(t) & w_x(t) & 0 \end{pmatrix}$$

Now notice that

$$[S(t)p(t)]' = [-w_z(t)p_y(t) + w_y(t)p_z(t), w_z(t)p_x(t) - w_x(t)p_z(t),$$
$$- w_y(t)p_x(t) + w_x(t)p_y(t)]$$

where $p(t)' = [p_x(t), p_y(t), p_x(t)]$. This expression is exactly the same as the cross-product

$$\omega(t) \times p(t), \qquad \text{where} \qquad \omega(t)' = [w_x(t), w_y(t), w_z(t)]$$

Thus we can describe the rigid-body motion given by Eq. (15.21) by

$$\dot{p}(t) = \omega(t) \times p(t) + k(t) \qquad (15.22)$$

From the rigid-body-motion equation (15.22)

$$\begin{pmatrix} \dot{x} \\ \dot{y} \\ \dot{z} \end{pmatrix} = \omega \times \begin{pmatrix} x \\ y \\ z \end{pmatrix} + k \qquad (15.23)$$

and the perspective projection equation

$$\begin{pmatrix} x \\ y \\ z \end{pmatrix} = \frac{z}{f} \begin{pmatrix} u \\ v \\ f \end{pmatrix}$$

we can determine an expression for $\dot{z}$:

$$\dot{z} = \frac{z}{f}(w_x v - w_y u) + k_z$$

Substituting this expression for $\dot{z}$ back into Eq. (15.23), using the perspective projection equation, and eliminating the third component, which simply states $0 = 0$, we obtain, after some simplification,

$$
\begin{pmatrix} \dot{u} \\ \dot{v} \end{pmatrix} = \begin{pmatrix} \frac{f}{z} & 0 & -\frac{u}{z} & -\frac{uv}{f} & \frac{f^2+u^2}{f} & -v \\ 0 & \frac{f}{z} & -\frac{v}{z} & -\frac{f^2+v^2}{f} & \frac{uv}{f} & u \end{pmatrix} \begin{pmatrix} k_x \\ k_y \\ k_z \\ w_x \\ w_y \\ w_z \end{pmatrix}
\tag{15.24}
$$

The kernel of the 2×6 matrix of Eq. (15.24) is spanned by the orthogonal vectors

$$
\begin{pmatrix} u \\ v \\ f \\ 0 \\ 0 \\ 0 \end{pmatrix}
\begin{pmatrix} 0 \\ 0 \\ 0 \\ u \\ v \\ f \end{pmatrix}
\begin{pmatrix} uvz \\ -(u^2+f^2)z \\ fvz \\ -f^2 \\ 0 \\ uf \end{pmatrix}
\begin{pmatrix} f(u^2+v^2+f^2)z \\ 0 \\ -u(u^2+v^2+f^2)z \\ uvf \\ -(u^2+f^2)f \\ vf^2 \end{pmatrix}
$$

Any part of the motion induced by k_x, k_y, k_z, w_x, w_y, and w_z that is in the direction spanned by these four vectors will not be captured by the image optic flow.

The general solution of Eq. (15.24) can be written as the sum of a particular solution to the nonhomogeneous equation plus the general solution to the homogeneous equation. It is easy to verify that

$$
\frac{z}{f} \begin{pmatrix} \dot{u} \\ \dot{v} \\ 0 \\ 0 \\ 0 \\ 0 \end{pmatrix}
$$

is a particular solution to the nonhomogeneous equation. So the general solution of Eq. (15.24) is given by

$$
\begin{pmatrix} k_x \\ k_y \\ k_z \\ w_x \\ w_y \\ w_z \end{pmatrix} = \frac{z}{f} \begin{pmatrix} \dot{u} \\ \dot{v} \\ 0 \\ 0 \\ 0 \\ 0 \end{pmatrix} + \lambda_1 \begin{pmatrix} u \\ v \\ f \\ 0 \\ 0 \\ 0 \end{pmatrix} + \lambda_2 \begin{pmatrix} 0 \\ 0 \\ 0 \\ u \\ v \\ f \end{pmatrix}
$$

$$
+ \lambda_3 \begin{pmatrix} uvz \\ -(u^2+f^2)z \\ fvz \\ -f^2 \\ 0 \\ uf \end{pmatrix} + \lambda_4 \begin{pmatrix} f(u^2+v^2+f^2)z \\ 0 \\ -u(u^2+v^2+f^2)z \\ uvf \\ -(u^2+f^2)f \\ vf^2 \end{pmatrix}
$$

where $\lambda_1, \lambda_2, \lambda_3$, and λ_4 are free variables. The general solution indicates exactly how much we can tell about the unknown motion parameters k_x, k_y, k_z, w_x, w_y, and w_z from one measured point of optic flow.

Associated with rigid-body motion is a fundamental optic flow equation, just as in the case of translational motion. We can develop this optic flow equation in the following way. From the perspective projection equations,

$$\begin{pmatrix} u \\ v \\ f \end{pmatrix} = \frac{f}{z} \begin{pmatrix} x \\ y \\ z \end{pmatrix} \tag{15.25}$$

Taking derivatives of Eq. (15.25) with respect to time, we have

$$\begin{pmatrix} \dot{u} \\ \dot{v} \\ 0 \end{pmatrix} = \frac{f}{z^2} \left[z \begin{pmatrix} \dot{x} \\ \dot{y} \\ \dot{z} \end{pmatrix} - \dot{z} \begin{pmatrix} x \\ y \\ z \end{pmatrix} \right]$$

so that

$$z \begin{pmatrix} \dot{u} \\ \dot{v} \\ 0 \end{pmatrix} = f \begin{pmatrix} \dot{x} \\ \dot{y} \\ \dot{z} \end{pmatrix} - \frac{\dot{z}f}{z} \begin{pmatrix} x \\ y \\ z \end{pmatrix}$$

But for rigid-body motion,

$$\begin{pmatrix} \dot{x} \\ \dot{y} \\ \dot{z} \end{pmatrix} = \omega \times \begin{pmatrix} x \\ y \\ z \end{pmatrix} + k$$

Hence

$$z \begin{pmatrix} \dot{u} \\ \dot{v} \\ 0 \end{pmatrix} = f \left[\omega \times \begin{pmatrix} x \\ y \\ z \end{pmatrix} + k \right] - \frac{\dot{z}f}{z} \begin{pmatrix} x \\ y \\ z \end{pmatrix} \tag{15.26}$$

Now from the perspective projection equation,

$$\begin{pmatrix} x \\ y \\ z \end{pmatrix} = \frac{z}{f} \begin{pmatrix} u \\ v \\ f \end{pmatrix} \tag{15.27}$$

Upon substituting Eq. (15.27) into Eq. (15.26), we obtain

$$z \begin{pmatrix} \dot{u} \\ \dot{v} \\ 0 \end{pmatrix} = f \left[\omega \times \frac{z}{f} \begin{pmatrix} u \\ v \\ f \end{pmatrix} + k \right] - \dot{z} \begin{pmatrix} u \\ v \\ f \end{pmatrix}$$

After rearranging, we have

$$z \left[\begin{pmatrix} \dot{u} \\ \dot{v} \\ 0 \end{pmatrix} - \omega \times \begin{pmatrix} u \\ v \\ f \end{pmatrix} \right] = fk - \dot{z} \begin{pmatrix} u \\ v \\ f \end{pmatrix} \tag{15.28}$$

Notice that the vector $k \times \begin{pmatrix} u \\ v \\ f \end{pmatrix}$ is orthogonal to both k and $\begin{pmatrix} u \\ v \\ f \end{pmatrix}$. By multiplying both sides of Eq. (15.28) by $\left[k \times \begin{pmatrix} u \\ v \\ f \end{pmatrix} \right]'$ and dividing out z, we obtain the fundamental optic flow equation for rigid-body motion:

$$\left[k \times \begin{pmatrix} u \\ v \\ f \end{pmatrix} \right]' \left[\begin{pmatrix} \dot{u} \\ \dot{v} \\ 0 \end{pmatrix} - \omega \times \begin{pmatrix} u \\ v \\ f \end{pmatrix} \right] = 0 \qquad (15.29)$$

This equation is important because it does not involve depth. Writing out Eq. (15.29) and rearranging produces

$$k' \begin{pmatrix} -v^2 - f^2 & uv & uf \\ uv & -u^2 - f^2 & vf \\ uf & vf & -u^2 - v^2 \end{pmatrix} \omega + k' \begin{pmatrix} -f\dot{v} \\ f\dot{u} \\ -\dot{u}v + u\dot{v} \end{pmatrix} = 0$$

A least-squares solution for k and ω, given observed optic flow points $\{(u_n, v_n, \dot{u}_n, \dot{v}_n)\}_{n=1}^N$, would then minimize

$$\epsilon^2 = \sum_{n=1}^N (k' A_n \omega + k' b_n)^2$$

where

$$A_n = \begin{pmatrix} -v_n^2 - f^2 & u_n v_n & u_n f \\ u_n v_n & -u_n^2 - f^2 & v_n f \\ u_n f & v_n f & -u_n^2 - v_n^2 \end{pmatrix}$$

and

$$b_n = \begin{pmatrix} -f\dot{v}_n \\ f\dot{u}_n \\ -\dot{u}_n v_n + u_n \dot{v}_n \end{pmatrix}$$

The minimizing k, ω must satisfy

$$\begin{pmatrix} \frac{\partial \epsilon^2}{\partial k} \\ \frac{\partial \epsilon^2}{\partial \omega} \end{pmatrix} = 2 \sum_{n=1}^N (k' A_n \omega + k' b_n) \begin{pmatrix} A_n \omega + b_n \\ A_n k \end{pmatrix} = 0$$

which is a difficult nonlinear problem.

It can be quickly verified that

$$
\begin{pmatrix}
\dot{v}f^2 \\
-\dot{u}f^2 \\
(\dot{u}v - u\dot{v})f \\
\left[\dot{u}(v^2 + f^2) - uv\dot{v}\right]z \\
\left[\dot{v}(u^2 + f^2) - uv\dot{u}\right]z \\
-(u\dot{u} + v\dot{v})fz
\end{pmatrix}
$$

is orthogonal to

$$
\begin{pmatrix} \dot{u} \\ \dot{v} \\ 0 \\ 0 \\ 0 \\ 0 \end{pmatrix}
\begin{pmatrix} u \\ v \\ f \\ 0 \\ 0 \\ 0 \end{pmatrix}
\begin{pmatrix} 0 \\ 0 \\ 0 \\ u \\ v \\ f \end{pmatrix}
\begin{pmatrix} uvz \\ -(u^2 + f^2)z \\ fvz \\ -f^2 \\ 0 \\ uf \end{pmatrix}
\begin{pmatrix} f(u^2 + v^2 + f^2)z \\ 0 \\ -u(u^2 + v^2 + f^2)z \\ uvf \\ -(u^2 + f^2)f \\ vf^2 \end{pmatrix}
$$

Hence if there were no noise at each optic flow point $(u_n, v_n, \dot{u}_n, \dot{v}_n)$, one would have

$$
(q'_n \ r'_n z_n) \begin{pmatrix} k \\ \omega \end{pmatrix} = 0
$$

where

$$
q_n = \begin{pmatrix} \dot{v}_n f^2 \\ -\dot{u}_n f^2 \\ (\dot{u}_n v_n - u_n \dot{v}_n)f \end{pmatrix}
$$

$$
r_n = \begin{pmatrix} \dot{u}_n(v_n^2 + f^2) - u_n v_n \dot{v}_n \\ \dot{v}_n(u_n^2 + f^2) - u_n v_n \dot{u}_n \\ -(u_n \dot{u}_n + v_n \dot{v}_n)f \end{pmatrix}
$$

The least-squares solution for $k, \omega, z_1, \ldots, z_N$, given $\{(u_n, v_n, \dot{u}_n, \dot{v}_n)\}_{n=1}^N$, must minimize

$$
\epsilon^2 = \sum_{n=1}^N (k', \omega') \begin{pmatrix} q_n \\ r_n z_n \end{pmatrix} (q'_n, r'_n z_n) \begin{pmatrix} k \\ w \end{pmatrix}
$$

If $z_1, \ldots, z_N$ were known, then $\begin{pmatrix} k \\ \omega \end{pmatrix}$ would be that eigenvector of $\sum_{n=1}^N \begin{pmatrix} q_n \\ r_n z_n \end{pmatrix} (q'_n, r'_n z_n)$ having the smallest eigenvalue. If k and ω were known, then

$$
z_n = \frac{-\left[\omega' r_n q'_n k + (\omega' r_n q'_n k)'\right]}{2\omega' r_n r'_n \omega} \tag{15.30}
$$

One possible way to proceed is to assume initially that each z_n is the same nonzero constant greater than f and solve for k, ω. Then solve for $z_1, \ldots, z_N$ using Eq. (15.30) and iterate. In these iterations ϵ^2 will decrease in each iteration. This assures a solution that will at least be a local minimum of ϵ^2.

15.4   Linear Algorithms for Motion and Surface Structure from Optic Flow

We first concisely summarize existing results on planar patch motion recovery from optic flow velocities (Longuet-Higgins, 1984a; Negahdaripour and Horn, 1985). Then we give a simplified linear algorithm (Zhuang, Huang, and Haralick, 1988b) on curved general patch motion recovery from optic flow velocities.

15.4.1   The Planar Patch Case

Suppose a rigid planar patch is in motion in the half-space $z < 0$. Let $p(t)$ be an arbitrary object point on the planar patch at the time t, $p(t) = \left[x(t), y(t), z(t)\right]'$. Let $\left[u(t), v(t)\right]$ denote the central projective coordinates of $p(t)$ onto the image plane $z = f$ through the camera lens, which is located at the origin:

$$
\begin{cases}
u(t) = f\,\frac{x(t)}{z(t)} \\[2mm]
v(t) = f\,\frac{y(t)}{z(t)} \\[2mm]
p(t) = \frac{z(t)}{f}\left[u(t), v(t), f\right]'
\end{cases}
\tag{15.31}
$$

Let $[\dot{u}(t), \dot{v}(t)]$ denote the instantaneous velocity of the moving image point $[u(t), v(t)]$. We call $[u(t), v(t); \dot{u}(t), \dot{v}(t)]$ an optic flow image point.

The rigid planar patch motion is represented by a rigid-motion constraint:

$$
\dot{p}(t) = \omega(t) \times p(t) + k(t)
\tag{15.32}
$$

and a planar geometric constraint:

$$
n(t)'p(t) = 1
\tag{15.33}
$$

where $\omega(t)$ is the instantaneous rotational angular velocity, $k(t)$ is the instantaneous translational velocity, and the unit vector $n(t)$ is orthogonal to the moving planar patch. From Eqs. (15.32) and (15.33),

$$
\dot{p} = \omega \times p + kn'p
\tag{15.34}
$$

where for simplicity the time variable t has been omitted.

Let

$$
\omega = (\omega_1, \omega_2, \omega_3)',
$$

$$
\Omega = \begin{pmatrix} 0 & -\omega_3 & \omega_2 \\ \omega_3 & 0 & -\omega_1 \\ -\omega_2 & \omega_1 & 0 \end{pmatrix}
\tag{15.35}
$$

Then Eq. (15.34) could be written as

$$
\dot{p} = \Omega p + kn'p
$$
$$
= (\Omega + kn')p
\tag{15.36}
$$

Denote the 3×3 matrix $\Omega + kn'$ by W and its three row vectors by $w_1', w_2',$ and w_3'. The 3×3 matrix W is called the planar motion parameter matrix. Since Ω is skew symmetric, it is clear that $W + W' = kn' + nk'$, and that the three eigenvalues of $kn' + nk'$ are $k'n - \|k\| \cdot \|n\|$, 0, and $k'n + \|k\| \cdot \|n\|$.

Then Eq. (15.36) can be written as

$$\begin{pmatrix} \dot{x} \\ \dot{y} \\ \dot{z} \end{pmatrix} = W \begin{pmatrix} x \\ y \\ z \end{pmatrix} \tag{15.37}$$

From the perspective projection equations,

$$\begin{pmatrix} x \\ y \end{pmatrix} = \frac{z}{f} \begin{pmatrix} u \\ v \end{pmatrix} \tag{15.38}$$

Taking the time derivatives of these equations, we have

$$\begin{pmatrix} \dot{x} \\ \dot{y} \end{pmatrix} = \frac{1}{f} \begin{pmatrix} z\dot{u} + u\dot{z} \\ z\dot{v} + v\dot{z} \end{pmatrix} \tag{15.39}$$

Now we substitute Eqs. (15.38) and (15.39) into Eq. (15.37):

$$\begin{pmatrix} z\dot{u} + u\dot{z} \\ z\dot{v} + v\dot{z} \\ f\dot{z} \end{pmatrix} = z \begin{pmatrix} w_1' \\ w_2' \\ w_3' \end{pmatrix} \begin{pmatrix} u \\ v \\ f \end{pmatrix} \tag{15.40}$$

From the third row of Eq. (15.40),

$$\dot{z} = \frac{z}{f} w_3' \begin{pmatrix} u \\ v \\ f \end{pmatrix} \tag{15.41}$$

We substitute the expression for $\dot{z}$ of Eq. (15.41) into the first two rows of Eq. (15.40) to obtain the optical flow–planar motion equation

$$\left[\begin{pmatrix} w_1' \\ w_2' \end{pmatrix} - \frac{1}{f} \begin{pmatrix} u \\ v \end{pmatrix} w_3' \right] \begin{pmatrix} u \\ v \\ f \end{pmatrix} = \begin{pmatrix} \dot{u} \\ \dot{v} \end{pmatrix} \tag{15.42}$$

which is linear in the nine elements of W.

Let $f_n(W) = \left[\begin{pmatrix} w_1' \\ w_2' \end{pmatrix} - \frac{1}{f} \begin{pmatrix} u_n \\ v_n \end{pmatrix} w_3' \right] \begin{pmatrix} u_n \\ v_n \\ f \end{pmatrix}$, $n = 1 \ldots, N$. Then we have $2N$ linear

equations

$$f_n(W) = \begin{pmatrix} \dot{u}_n \\ \dot{v}_n \end{pmatrix}, \quad n = 1, \ldots, N \tag{15.43}$$

The optic flow–planar motion recovery requires first solving for the planar motion parameter matrix W and then finding $\omega, k,$ and n. The problem is thoroughly treated by Longuet-Higgins (1984a) and by Negahdaripour and Horn (1985). Their main results (with slight extensions) can be concisely summarized as follows:

1. The rank of the coefficient matrix of the set of N equations in Eq. (15.42), where $N \geq 4$, equals 8 if and only if there are four among the N image points—for instance $\{(u_i, v_i) : i = 1, 2, 3, 4\}$—such that no three of them are collinear. It is clear that this condition is independent of motion.

2. Under this condition, the general solution of the $2N$ homogeneous linear equations—that is, $f_n(W) = 0$, $n = 1, \ldots, N$—equals αI_3, where α is any real number and I_3 is the 3×3 identity matrix. Moreover, the motion parameter matrix W is uniquely determined by Eq. (15.43) and the constraint: The middle eigenvalue of $W + W'$ equals zero.

3. If we denote the three eigenvalues of $W + W'$ by $\lambda_1 \leq \lambda_2 \leq \lambda_3$ and the corresponding three orthonormal eigenvectors by q_1, q_2, q_3, where W has been uniquely determined and $\lambda_2 = 0$, then it holds that

$$\begin{aligned}
\lambda_1 < 0 < \lambda_3 \quad &\text{if} \quad k \times n \neq 0, \\
\lambda_1 = 0 < \lambda_3 \quad &\text{if} \quad k \neq 0, \; k \text{ and } n \text{ share the same direction,} \\
\lambda_1 < 0 = \lambda_3 \quad &\text{if} \quad k \neq 0, \; k \text{ and } -n \text{ share the same direction,} \\
\lambda_1 = 0 = \lambda_3 \quad &\text{if} \quad k = 0, \; n \text{ arbitrary.}
\end{aligned} \tag{15.44}$$

4. Using the geometric constraint $n'p = 1$ and the fact that the object is in motion in the half-space $z < 0$, we reach the following conclusion: If $\lambda_1 = 0 = \lambda_3$, then $k = 0$ and $\Omega = W$. Otherwise, either one of Eqs. (15.45) and (15.46) or one of Eqs. (15.47) and (15.48) must hold true:

$$\sqrt{\lambda_3} q_3' s_n < \sqrt{-\lambda_1} q_1' s_n, \qquad n = 1, \ldots, N \tag{15.45}$$

$$\sqrt{\lambda_3} q_3' s_n > \sqrt{-\lambda_1} q_1' s_n, \qquad n = 1, \ldots, N \tag{15.46}$$

$$\sqrt{\lambda_3} q_3' s_n < -\sqrt{-\lambda_1} q_1' s_n, \qquad n = 1, \ldots, N \tag{15.47}$$

$$\sqrt{\lambda_3} q_3' s_n > -\sqrt{-\lambda_1} q_1' s_n, \qquad n = 1, \ldots, N \tag{15.48}$$

where $s_n = (u_n, v_n, f)'$. Equation (15.45) implies

$$n = \frac{1}{2} \left(-\sqrt{-\lambda_1} q_1 + \sqrt{\lambda_3} q_3 \right), \quad \text{up to a positive factor } \alpha > 0 \tag{15.49}$$

$$k = \sqrt{-\lambda_1} q_1 + \sqrt{\lambda_3} q_3, \quad \text{up to a positive factor } \frac{1}{\alpha}$$

Equation (15.46) implies

$$n = \frac{1}{2} \left(\sqrt{-\lambda_1} q_1 - \sqrt{\lambda_3} q_3 \right), \quad \text{up to a positive factor } \alpha > 0 \tag{15.50}$$

$$k = -\sqrt{-\lambda_1} q_1 - \sqrt{\lambda_3} q_3, \quad \text{up to a positive factor } \frac{1}{\alpha}$$

And either Eq. (15.45) or Eq. (15.46) implies

$$\Omega = W - \frac{1}{2}(\lambda_1 q_1 q_1' + \sqrt{-\lambda_1 \lambda_3} q_1 q_3' - \sqrt{-\lambda_1 \lambda_3} q_3 q_1' + \lambda_3 q_3 q_3') \qquad (15.51)$$

Equation (15.47) implies

$$n = \frac{1}{2}\left(\sqrt{-\lambda_1} q_1 + \sqrt{\lambda_3} q_3\right), \text{ up to a positive factor } \alpha > 0$$

$$(15.52)$$

$$k = -\sqrt{-\lambda_1} q_1 + \sqrt{\lambda_3} q_3, \text{ up to a positive factor } \frac{1}{\alpha}$$

Equation (15.53) implies

$$n = \frac{1}{2}\left(-\sqrt{-\lambda_1} q_1 - \sqrt{\lambda_3} q_3\right), \text{ up to a positive factor } \alpha > 0$$

$$(15.53)$$

$$k = \sqrt{-\lambda_1} q_1 - \sqrt{\lambda_3} q_3, \text{ up to a positive factor } \frac{1}{\alpha}$$

And either Eq. (15.47) or Eq. (15.48) implies that

$$\Omega = W - \frac{1}{2}(\lambda_1 q_1 q_1' - \sqrt{-\lambda_1 \lambda_3}\ q_1 q_3' + \sqrt{-\lambda_1 \lambda_3}\ q_3 q_1' + \lambda_3 q_3 q_3') \qquad (15.54)$$

The ambiguity of planar motion recovery happens if and only if $\lambda_1 < 0 < \lambda_3$. One of Eqs. (15.49) and (15.50) and one of Eqs. (15.52) and (15.53) become true. In that case the two solutions denoted by $(n, k, \omega), (n^*, k^*, \omega^*)$, are related by

$$n^* \times k = 0, \quad k^* \times n = 0, \quad k^* n^{*'} = nk', \quad \omega^* = \omega + n \times k \qquad (15.55)$$

To resolve the ambiguity in this case, we need more information about the motion and the geometry.

15.4.2 General Case: Optic Flow–Motion Equation

Now we briefly set up the optic flow–motion equation that does not involve depth information and solve it by using a linear least-squares technique.

Let k^* be any nonzero vector that is collinear with k. Then from Eq. (15.28)

$$zk^* \times \left[\begin{pmatrix} \dot{u} \\ \dot{v} \\ 0 \end{pmatrix} - \Omega \times \begin{pmatrix} u \\ v \\ f \end{pmatrix} \right] = k^* \times \left[fk - \dot{z} \begin{pmatrix} u \\ v \\ f \end{pmatrix} \right] = -\dot{z}k^* \times \begin{pmatrix} u \\ v \\ f \end{pmatrix}$$

Since $z \neq 0$,

$$k^* \times \begin{pmatrix} u \\ v \\ f \end{pmatrix} \text{ and } k^* \times \left[\begin{pmatrix} \dot{u} \\ \dot{v} \\ 0 \end{pmatrix} - \Omega \times \begin{pmatrix} u \\ v \\ f \end{pmatrix} \right]$$

are collinear. But if a vector c is collinear with a cross-product $a \times b$, then $c'a = c'b = 0$. Hence

$$\left[k^* \times \begin{pmatrix} u \\ v \\ f \end{pmatrix} \right]' \left[\begin{pmatrix} \dot{u} \\ \dot{v} \\ 0 \end{pmatrix} - \Omega \times \begin{pmatrix} u \\ v \\ f \end{pmatrix} \right] = 0$$

or

$$\left[k^* \times \begin{pmatrix} u \\ v \\ f \end{pmatrix} \right]' \begin{pmatrix} \dot{u} \\ \dot{v} \\ 0 \end{pmatrix} = \left[k^* \times \begin{pmatrix} u \\ v \\ f \end{pmatrix} \right]' \left[\Omega \times \begin{pmatrix} u \\ v \\ f \end{pmatrix} \right] \qquad (15.56)$$

Let

$$k^* = \begin{pmatrix} k_1^* \\ k_2^* \\ k_3^* \end{pmatrix}, \quad \Omega = \begin{pmatrix} \omega_1 \\ \omega_2 \\ \omega_3 \end{pmatrix}$$

$$K^* = \begin{pmatrix} 0 & -k_3^* & k_2^* \\ k^* & 0 & -k_1^* \\ -k_2^* & k_1^* & 0 \end{pmatrix}, \quad \Omega = \begin{pmatrix} 0 & -\omega_3 & \omega_2 \\ \omega_3 & 0 & -\omega_1 \\ -\omega_2 & \omega_1 & 0 \end{pmatrix} \qquad (15.57)$$

Then

$$k^* \times (u,v,f)' = K^*(u,v,f)'$$

$$\omega \times (u,v,f)' = \Omega(u,v,f)' \qquad (15.58)$$

And Eq. (15.56) could be written as

$$(u,v,f)K^*\Omega(u,v,f)' + (\dot{u},\dot{v},0)'K^*(u,v,f) = 0 \qquad (15.59)$$

or

$$(u,v,f)\frac{K^*\Omega + \Omega K^*}{2}(u,v,f)' + (\dot{u},\dot{v},0)'K^*(u,v,f)' = 0 \qquad (15.60)$$

which does not involve the depth z and is called the optic flow–motion equation.

Next we show a necessary and sufficient condition called the rank assumption, under which a unique rotation ω and a nonzero vector k^*, which is collinear with k, can be determined simply from the optic flow–motion equation.

15.4.3   A Linear Algorithm for Solving Optic Flow–Motion Equations

Let L denote $(K^*\Omega + \Omega K^*/2)$. Then we have N linear homogeneous equations with unknowns L and K^* :

$$(u_i,v_i,f)L(u_i,v_i,f) + (\dot{u}_i,\dot{v}_i,o)K^*(u_i,v_i,f)' = 0, \quad i = 1,\ldots,N \qquad (15.61)$$

It is easy to see that Eq. (15.61) contains nine unknowns (six for L and three for K^*). It is also clear that being collinear with k, k^* equals either αk with α an arbitrary real number when $k \neq 0$ or an arbitrary 3×1 vector when $k = 0$. In other words, (L, K^*) will have one degree of freedom when $k \neq 0$ or three degrees of freedom when $k = 0$. Thus (L, K^*), which is defined by Ω and k^*, constitutes the general solution to the system of N linear homogeneous equations—that is, Eq. (15.61)—if it has a rank 8 when $k \neq 0$ or a rank 6 when $k = 0$. Let

$$
\begin{cases}
L = \begin{vmatrix} l_1 & l_4 & l_5 \\ l_4 & l_2 & l_6 \\ l_5 & l_6 & l_3 \end{vmatrix} \\[6pt]
h = (l_1, l_2, l_3, 2l_4, 2l_5, 2l_6, k_1^*, k_2^*, k_3^*)' \\[6pt]
a_i = (u_i^2, v_i^2, f^2, u_i v_i, u_i v_i, f \dot v_i, -f \dot u_i, \dot u_i v_i - \dot v_i u_i)
\end{cases}
\tag{15.62}
$$

Then Eq. (15.61) could be written in a more readable form:

$$
a_i h = 0, \quad i = 1, \ldots, N
\tag{15.63}
$$

This overconstrained least-squares problem can then be posed as the determination of h, $\|h\| = 1$, which minimizes

$$
h' \left(\sum_{i=1}^{N} a_i' a_i \right) h
\tag{15.64}
$$

That is, the solution to Eq. (15.63) coincides with the smallest eigenvalue vector of the 9×9 symmetric and nonnegative matrix $A = \sum_{i=1}^{N} a_i' a_i$ whose rank is the same as the $N \times 9$ coefficient matrix, that is, $\begin{pmatrix} a_1 \\ \vdots \\ a_N \end{pmatrix}_{N \times 9}$, of Eq. (15.63). Thus h, which is defined by Ω and k^* through L and K^*, coincides with the smallest eigenvalue vector of A if and only if the following rank assumption holds:

$$
\begin{cases}
\text{Rank}(A) = 8 & \text{when} \quad k \neq 0 \\
\text{Rank}(A) = 6 & \text{when} \quad k = 0
\end{cases}
\tag{15.65}
$$

To assure Eq. (15.65), we need at least eight points ($N \geq 8$) when $k \neq 0$ or six points ($N \geq 6$) when $k = 0$. More optic flow image points are preferable to smooth out any noise perturbations. From now on, we assume Eq. (15.65).

Once a nonzero smallest eigenvalue vector h of A is determined, L and $K^*(\neq 0)$ can then be determined. And the rotation ω could be uniquely determined through $L = (K^* \Omega + \Omega K^*/2)$. The latter is equivalent to

$$
\omega_1 k_1^* = \frac{l_1 - l_2 - l_3}{2}
\tag{15.66}
$$

$$
\omega_2 k_2^* = \frac{l_2 - l_3 - l_1}{2}
\tag{15.67}
$$

$$\omega_3 k_3^* = \frac{l_3 - l_1 - l_2}{2} \tag{15.68}$$

$$\omega_2 k_1^* + \omega_1 k_2^* = 2l_4 \tag{15.69}$$

$$\omega_1 k_3^* + \omega_3 k_1^* = 2l_5 \tag{15.70}$$

$$\omega_2 k_3^* + \omega_3 k_2^* = 2l_6 \tag{15.71}$$

When $k_1^* \neq 0$, Eqs. (15.66), (15.69), and (15.70) could be used to determine Ω. When $k_2^* \neq 0$, Eqs. (15.67), (15.69), and (15.71) could be used to determine Ω. And when $k_3^* \neq 0$, Eqs. (15.68), (15.70), and (15.71) could be used to determine Ω. For stability we recommend the following scheme to uniquely determine Ω:

Step 1: If $|k_1^*| \geq |k_2^*|, |k_3^*|$, then use Eqs. (15.66), (15.69), and (15.70).
Step 2: Otherwise, if $|k_2^*| \geq |k_3^*|$, then use Eqs. (15.67), (15.69), and (15.70).
Step 3: Otherwise, use Eqs. (15.68), (15.70), and (15.71).

Therefore, under the rank assumption equation (15.65), the optic flow-motion equation uniquely determines the instantaneous rotation Ω and a nonzero vector k^* that is collinear with k. As mentioned before, k^* equals αk when $k \neq 0$ or could be an arbitrary nonzero vector when $k = 0$.

Next we show how to simply determine the mode of motion (i.e., whether $k = 0$ or not), the direction of k, and the surface structure—that is, the relative depths—when $k \neq 0$.

15.4.4 Mode of Motion, Direction of Translation, and Surface Structure

From Eq. (15.32) we can see that

$$\dot{z} = z \begin{vmatrix} \omega_1 & \omega_2 \\ u & v \end{vmatrix} + k_3 \tag{15.72}$$

Using Eq. (15.72) to eliminate $\dot{z}$ in Eq. (15.28), we have

$$z \left\{ \begin{vmatrix} \omega_1 & \omega_2 \\ u & v \end{vmatrix} (u,v)' - \left(\begin{vmatrix} \omega_2 & \omega_3 \\ v & f \end{vmatrix} \begin{vmatrix} \omega_3 & \omega_1 \\ f & u \end{vmatrix} \right)' + (u,v)' \right\}$$

$$+ \{ k_3(u,v)' - (k_1, k_2)' \} = 0 \tag{15.73}$$

Let

$$\begin{cases} f_i(\omega) = \begin{vmatrix} \omega_1 & \omega_2 \\ u_i & v_i \end{vmatrix} (u_i, v_i)' - \left(\begin{vmatrix} \omega_2 & \omega_3 \\ v_i & f \end{vmatrix}, \begin{vmatrix} \omega_3 & \omega_1 \\ f & u_i \end{vmatrix} \right)' \\ g_i(k) = k_3(u_i, v_i)' - (k_1, k_2)' \end{cases} \tag{15.74}$$

Then we obtain N equations from Eq. (15.73) as follows:

$$z_i\{f_i(\omega) + (\dot{u}_i, \dot{v}_i)'\} + g_i(k) = 0, \quad i = 1, \ldots, N \tag{15.75}$$

Clearly $k = 0$ if and only if for at least two points $g_i(k) = 0$. Thus $k \neq 0$ if and only if for all points, except possibly for one, $f_i(\omega) + (u_i, v_i)' \neq 0$. This means that the mode of motion can be uniquely determined. When $k \neq 0$ is confirmed, the direction of translation and the surface structure can be simply determined as follows: From Eq. (15.75) we obtain

$$\frac{\{f_i(\omega) + (\dot{u}_i, \dot{v}_i)'\}z_i}{\|k\|} + g_i\left(\frac{k}{\|k\|}\right) = 0 \tag{15.76}$$

$$\frac{\|f_i(\omega) + (u_i, v_i)'\|^2 z_i}{\|k\|} + (f_i(\omega)' + (\dot{u}_i, \dot{v}_i))g_i\left(\frac{k}{\|k\|}\right) = 0 \tag{15.77}$$

It is clear from Eq. (15.77) that $[f_i(\omega)' + (\dot{u}_i, \dot{v}_i)]g_i(k/\|k\|)$ is positive except possibly for one point, where it is zero. It is also clear that $k/\|k\|$ equals $k^*/\|k^*\|$ or $-k^*/\|k^*\|$ (depending on whether the translation k has the same direction as k^* or $-k^*$), and correspondingly

$$[f_i(\omega)' + (\dot{u}_i, \dot{v}_i)]g_i\left(\frac{k}{\|k\|}\right) = \pm[f_i(\omega)' + (\dot{u}_i, \dot{v}_i)]g_i\left(\frac{k^*}{\|k^*\|}\right) \tag{15.78}$$

Thus combining Eqs. (15.77) and (15.78), we have

$$\frac{z_i}{\|k\|} = -\frac{\left|[f_i(\omega)' + (\dot{u}_i, \dot{v}_i)]g_i\left(\frac{k^*}{\|k^*\|}\right)\right|}{\|f_i(\omega) + (\dot{u}_i, \dot{v}_i)'\|^2} \tag{15.79}$$

Furthermore, the translation k has the same direction as k^* or $-k^*$ if and only if $\sum_i [f_i(\omega)' + (\dot{u}_i, \dot{v}_i)]g_i(k^*)$ is positive or negative.

15.4.5 Linear Optic Flow–Motion Algorithm and Simulation Results

As seen from the foregoing, there exists a linear algorithm to uniquely determine ω, the mode of motion, the direction of translation, and the relative depth map or surface structure $z_i/\|k\|$, $i = 1, \ldots, N$, when $k \neq 0$. Now we give that algorithm and some simulation results:

Step 1: Find the eigenvector h of A having the smallest eigenvalue. Set $k^* = (h_7, h_8, h_9)'$.

Step 2: If $(|k_1^*| \geq |k_2^*|, |k_3^*|)$, then set

$$\omega_1 = \frac{(h_1 - h_2 - h_3)}{2k_1^*}$$

$$\omega_2 = \frac{(h_4 - k_2^*\omega_1)}{k_1^*}$$

$$\omega_3 = \frac{(h_5 - k_3^*\omega_1)}{k_1^*}$$

Step 3: Otherwise, if $(|k_2^*| \geq |k_3^*|)$, then set

$$\omega_2 = \frac{(h_2 - h_3 - h_1)}{2k_2^*}$$

$$\omega_3 = \frac{(h_6 - k_3^* \omega_2)}{k_2^*}$$

$$\omega_1 = \frac{(h_4 - k_1^* \omega_2)}{k_2^*}$$

Step 4: Otherwise set

$$\omega_3 = \frac{(h_3 - h_1 - h_2)}{2k_3^*}$$

$$\omega_1 = \frac{(h_5 - k_1^* \omega_3)}{k_3^*}$$

$$\omega_2 = \frac{(h_6 - k_2^* \omega_3)}{k_3^*}$$

Step 5: If $\left(\sum_i ||f_i(\omega) + (\dot{u}_i, \dot{v}_i)'|| = 0 \right)$, then set $k = 0$.   Stop.

Step 6: Otherwise set

$$\frac{z_i}{||k||} = -\frac{\left| (f_i(\omega)' + (\dot{u}_i, \dot{v}_i))g_i(\frac{k^*}{||k^*||}) \right|}{||f_i(\omega) + (\dot{u}_i, \dot{v}_i)'||^2}$$

Step 7: The translation k has the same direction as k^* or $-k^*$ if $\sum_i (f_i(\omega)' + (\dot{u}_i, \dot{v}_i))g_i(k^*)$ is positive or negative. Stop.

The experiments needed to verify this algorithm should answer three questions: (1) What is the minimum number of points needed to compute motion and surface structure from accurate or noisy optic flow measurements in practice? (2) What is the likelihood that we will come across a set of optic flow image points that violate the rank assumption, that is, Eq. (15.65)? And (3) What is the accuracy of the estimated motion parameters from noisy optic flow measurements?

One way of testing the validity of the rank assumption is to generate many sets of data randomly and to compute the eigenvalues of A. If A has fewer than eight (six) nonzero eigenvalues when $k \neq 0(= 0)$, then the rank assumption is violated. This method, however, suffers from the defect that, since most eigenvalue-computing routines return inexact values even with accurate given data, we need a threshold around zero, below which we would like to say that the eigenvalues are zero or nearly zero. This obviously is a problem, since the rank assumption is not violated until more than one (three) eigenvalues are strictly zero when $k \neq 0(= 0)$.

Hence the approach used here is different. We have actually implemented the algorithm and tested it with many sets of randomly chosen data. Implicitly this tells

Table 15.1   The noiseless case

True Direction of Translation	Estimated Direction of Translation	True Rotation	Estimated Rotation	u	v	$\dot{u}$	$\dot{v}$
0.74	0.74	-1.59	-1.59	0.38e+00	-.90e+00	-.23e+02	-.57e+01
0.58	0.58	1.34	1.34	-.77e-01	-.19e+00	-.61e+01	-.28e+00
-0.34	-0.34	-4.41	-4.41	-.12e+01	0.42e+00	-.13e+01	-.54e+01
				-.27e+01	-.27e+01	0.23e+02	0.56e+02
				-.64e+00	-.50e+00	-.79e+01	-.11e+01
				-.41e+00	0.20e+01	-.13e+02	-.38e+02
				-.54e+00	-.75e+00	-.87e+01	0.91e+00
				-.26e+01	0.51e+00	0.20e+02	-.38e+02
0.74	0.74	-1.59	-1.59	0.32e+00	-.45e+00	-.65e+01	-.81e+01
0.58	0.58	1.34	1.34	0.30e+00	-.24e+00	-.53e+01	-.30e+01
-0.34	-0.34	-4.41	-4.41	-.72e+00	-.69e+00	-.51e+01	0.28e+01
				-.67e-01	0.40e+00	-.35e+01	-.44e+01
				-.17e+01	-.61e+00	0.77e-01	0.25e+01
				-.70e+00	-.62e+00	-.77e+01	0.27e+00
				0.50e+00	0.53e+00	-.26e+01	-.56e+01
				-.10e+00	-.65e+01	0.12e+03	0.28e+03
0.74	0.74	-1.59	-1.59	-.19e+00	-.35e+00	-.59e+01	-.13e+01
0.58	0.58	1.34	1.34	-.10e+00	0.42e+00	-.15e+01	-.25e+01
-0.34	-0.34	-4.41	-4.41	0.32e+00	-.34e+00	-.11e+02	-.57e+01
				0.14e+01	-.19e+01	-.43e+02	-.44e-01
				-.12e+00	0.75e-01	-.29e+01	0.25e+01
				-.23e+00	-.56e+00	-.79e+01	0.27e+00
				0.69e+00	0.64e+00	-.98e+01	-.56e+02
				0.58e+00	-.44e+00	-.20e+02	0.28e+02
0.74	0.74	-1.59	-1.59	-.85e+00	0.18e+01	-.38e+02	-.11e+03
0.58	0.58	1.34	1.34	-.26e+00	0.36e+00	-.66e+01	-.74e+01
-0.34	-0.34	-4.41	-4.41	0.35e+00	0.51e+00	-.19e+01	-.45e+01
				-.82e+00	-.59e-01	-.42e+01	-.24e+01
				-.20e+01	0.26e+01	0.83e+01	-.37e+02
				0.31e+00	0.40e+00	-.31e+01	-.49e+01
				0.11e+00	0.83e+00	-.14e+02	-.19e+02
				0.77e+00	-.20e+00	-.14e+02	-.97e+01
0.74	0.74	-1.59	-1.59	0.12e+01	0.46e+00	-.12e+02	-.14e+02
0.58	0.58	1.34	1.34	-.27e+00	0.45e+00	-.24e+01	-.34e+01
-0.34	-0.34	-4.41	-4.41	-.51e+00	0.95e+00	-.22e+01	-.70e+01
				-.68e+00	-.84e+00	-.62e+01	0.33e+01
				0.13e+01	-.34e+00	-.45e+02	-.23e+02
				-.89e-01	0.53e+00	-.13e+01	-.29e+01
				0.10e+01	0.16e+01	-.23e+02	-.33e+02
				0.89c+00	-.12e+00	-.12e+02	-.96e+01

Table 15.2   The noisy case: $|\dot{u} - \overline{\dot{u}}| + |\dot{v} - \overline{\dot{v}}| \leq 0.2$, where $(\dot{u}, \dot{v})$ is true optic flow velocity and $(\overline{\dot{u}}, \overline{\dot{v}})$ is erroneous optic flow velocity.

True Direction of Translation	Estimated Direction of Translation	True Rotation	Estimated Rotation	u	v	$\overline{u}$	$\overline{v}$
-0.69	-0.68	0.42	0.43	043.e+01	-.30e-01	0.95e+01	0.74e+01
-0.53	-0.54	-1.15	-1.16	0.13e+01	0.55e+00	0.24e+02	0.16e+02
0.49	0.49	0.22	0.17	-.13e+01	0.82e+00	-.19e+01	0.20e+02
				-.19e+00	-.13e+01	0.14e+02	-.40e+01
				0.24e+00	-.37e+00	0.51e+01	0.23e+01
				-.32e+00	0.30e+00	0.43e+01	0.68e+01
				-.96e-01	-.38e-01	0.35e+01	0.33e+01
				-.25e+00	-.49e+00	0.18e+02	0.93e+01
-0.69	-0.69	0.42	0.42	0.64e+00	0.45e+00	0.91e+01	0.76e+01
-0.53	-0.54	-1.15	-1.15	0.10e+00	-.44e+00	0.12e+02	0.52e+01
0.49	0.48	0.22	0.22	-.33e+00	0.31e+00	0.29e+01	0.50e+01
				0.45e+00	0.44e+00	0.70e+01	0.64e+01
				0.54e+00	0.87e+00	0.19e+02	0.20e+02
				-.51e+02	0.20e+02	-.21e+05	0.86e+04
				-.17e+02	-.16e+02	-.39e+04	-.36e+04
				0.68e+00	-.65e+00	0.11e+02	0.26e+01
-0.69	-0.67	0.42	0.43	-.11e+01	-.11e+00	0.96e-01	0.66e+01
-0.53	-0.54	-1.15	-1.16	0.10e+00	-.12e+01	0.16e+02	-.21e+01
0.49	0.49	0.22	0.16	0.10e+00	0.36e-01	-.69e+01	0.56e+01
				-.30e+00	0.54e+00	0.81e+01	0.13e+02
				-.25e+01	0.21e+01	0.18e+02	0.51e+02
				0.25e+01	-.20e+01	0.68e+02	-.12e+02
				0.31e+01	-.28e+01	0.34e+03	-.12e+03
				0.16e+00	0.17e+01	0.44e+02	0.80e+02
-0.69	-0.68	0.42	0.41	0.11e+01	0.11e+00	0.16e+02	0.87e+01
-0.53	-0.54	-1.15	-1.14	0.12e+00	-.47e+00	0.43e+01	0.18e+01
0.49	0.49	0.22	0.26	-.50e+00	0.30e+00	0.28e+01	0.61e+01
				-.58e+00	-.58e+00	0.20e+01	0.12e+01
				-.21e+02	-.80e+01	-.51e+04	-.18e+04
				-.82e+00	-.21e+01	0.93e+01	-.24e+02
				0.50e+00	0.55e+00	0.63e+01	0.61e+01
				-.21e+00	0.54e+00	0.42e+01	0.70e+01
-0.69	-0.67	0.42	0.43	-.049e+00	0.41e+00	0.17e+01	0.47e+01
-0.53	-0.53	-1.15	-1.16	-38e+00	-.20e+01	0.14e+02	-.16e+02
0.49	0.50	0.22	0.17	-.16e+00	0.80e+00	0.71e+01	0.12e+02
				0.24e+01	0.21e+01	0.60e+02	0.50e+02
				0.59e+00	0.43e+00	0.72e+01	0.61e+01
				0.86e+02	-.11e+03	0.61e+05	-.76e+05
				-.31e+02	-.31e+02	-.91e+04	-.93e+04
				0.94e+00	0.15e+01	0.29e+02	0.33e+02

us the likelihood of the occurrence of optic flow image-point sets that violate the rank assumption, since any time we encounter such a set of points, the algorithm will break down.

Several observations from our experiments follow: In spite of testing the algorithm with an extremely large number of sets of randomly generated points, we have not met a set of points that violates the rank assumption.

For the case when no noise occurs in the optic flow measurements, the algorithm is extremely accurate. Even when some limited noise appears, the algorithm works well, except that the mode of motion cannot be determined correctly. That is, the case of pure rotation (i.e., $k = 0$) cannot be recognized. It will affect how we determine the relative depth map and the direction of translation. Since the case of pure rotation admits an indefinite relative depth map and an indefinite direction of translation, the algorithm gives simply one out of an infinite set of consistent interpretations. Thus acceptance of the interpretation extracted by the algorithm could never cause a disaster as long as the noise is limited. For the sake of space, only a few experimental results are tabulated here. Table 15.1 lists five groups of noiseless optic flow data (last four columns), true direction of translation (the first column), true rotation (the third column), estimated direction of translation (the second column), and estimated rotation (the fourth column) by using the linear optic flow–motion algorithm given in Section 15.4.5. Table 15.2 treats in the same way five groups of noisy optic flow data.

15.5   The Two View–Linear Motion Algorithm

15.5.1   Planar Patch Motion Recovery from Two Perspective Views: A Brief Review

The planar patch motion estimation problem is important in practice. Earlier works on the subject include Tsai and Huang (1984) and Tsai, Huang, and Zhu (1982). A concise closed-form-solution algorithm will follow in the next section.

Two View–Planar Motion Equation

Suppose a rigid planar patch is in motion in the half-space $z < 0$ (Figure 15.1). Let $p_1 = (x_1, y_1, z_1)'$ represent an arbitrary object point on the patch before motion and let $p_2 = (x_2, y_2, z_2)'$ represent the same object point after motion.

Let (u_1, v_1) and (u_2, v_2) represent the central projective coordinates of p_1 and p_2 onto the image plane $z = f$ through the camera lens, that is, the origin:

$$u_1 = fx_1/z_1, \quad v_1 = fy_1/z_1$$

$$u_2 = fx_2/z_2, \quad v_2 = fy_2/z_2$$

The rigid-body-motion equation relates p_1 to p_2 as follows:

$$p_2 = R_0 p_1 + t_0 \tag{15.80}$$

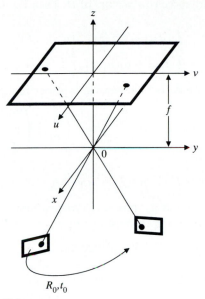

Figure 15.1 Imaging geometry for two view–planar motion.

and the planarity constrains p_1 by

$$n_0'p_1 = 1 \tag{15.81}$$

where

$R_0 =$ the 3×3 rotational matrix, $R_0'R_0 = I, |R_0| = 1$

$t_0 =$ the 3×1 translational vector

$n_0 =$ the 3×1 normal vector

Combining Eq. (15.80) with Eq. (15.81) produces the planar rigid-body-motion equation

$$p_2 = (R_0 + t_0n_0')p \tag{15.82}$$

Projecting the planar rigid-body motion onto the image plane $z = f$ produces

$$(u_2, v_2, f)' = \frac{z_1}{z_2}(R_0 + t_0n_0')(u_1, v_1, f)' \tag{15.83}$$

Let the planar rigid-motion parameter matrix be defined by

$$B = \begin{pmatrix} b_1 \\ b_2 \\ b_3 \end{pmatrix} = R_0 + t_0n_0' \tag{15.84}$$

where b_i, $i = 1, 2, 3$, are three row vectors of B. Then Eq. (15.83) could be written as

$$\begin{cases} (u_2, v_2)' = \frac{z_1}{z_2} \binom{b_1}{b_2} (u_1, v_1, f)' \\ f = \frac{z_1}{z_2} b_3 (u_1, v_1, f)' \end{cases}$$ (15.85)

From Eq. (15.85) we derive the two view–planar motion equation

$$b_3(u_1, v_1, f) \binom{u_2}{v_2} = f \binom{b_1}{b_2} \begin{pmatrix} u_1 \\ v_1 \\ f \end{pmatrix}$$ (15.86)

with the natural constraint

$$b_3(u_1, v_1, f)' > 0$$ (15.87)

Now the planar motion recovery problem involves first solving the planar rigid-motion parameter matrix B and then estimating R_0, t_0, and n_0.

Solving the Two View–Planar Motion Equation

The planar patch is said to be visible if a part of the patch with a nonzero area can be seen from the lens before and after motion. For a visible planar patch, Eq. (15.86) (considering the nine elements of B as unknown) has a rank 8. Thus any nonzero solution A to Eq. (15.86) relates to B by a unique scale factor k_0:

$$k_0 A = B = R_0 + t_0 n_0'$$ (15.88)

Usually the visibility condition might not be satisfied. However, we can prove that Eq. (15.86) has a rank 8 if and only if there exist four image-point correspondence pairs $(u_{i1}, v_{i1}) \leftrightarrow (u_{i2}, v_{i2})$, $i = 1, 2, 3, 4$, so that no three of both $\{(u_{i1}, v_{i1}) : i = 1, 2, 3, 4\}$ and $\{(u_{i2}, v_{i2}), i = 1, 2, 3, 4\}$ are collinear. We will assume this in the remaining part of this section.

Solving $k_0 A = R_0 + t_0 n_0'$

Let three eigenvalue vectors of $A'A$ be $\xi^2 \geq \eta^2 \geq \zeta^2$, $u_1, u_2, u_3 = u_1 \times u_2$, respectively. Then the following conditions hold:

1. The term k_0 is uniquely determined by $k_0^2 \eta^2 = 1$ and $k_0 a_3(u_1, v_1, f)' > 0$, where a_3 is the third-row vector of A.

(15.89)

2. If $\xi^2 = \eta^2 = \zeta^2$ and $k_0 \, det(A) > 0$, then $R_0 = k_0 A$ and $t_0 = 0$.
3. If $\xi^2 = \eta^2 = \zeta^2$ and $k_0 \, det(A) < 0$, then there are an infinite number of solutions for $\{R_0, t_0, n_0\}$:

$$R_0 = k_0 A [e_1, e_2, e_2 \times e_1][e_1, e_2, e_1 \times e_2]'$$
$$t_0 = 2R_0(e_2 \times e_1)$$
$$n_0 = e_1 \times e_2$$

where e_1, e_2 are two arbitrary unit vectors orthogonal to each other.

4. If $\xi^2 = \eta^2 > \zeta^2$, then there exists a unique solution for $\{R_0, t_0, n_0\}$:

$$R_0 u_i = k_0 A u_i, \ i = 1, 2$$
$$R_0 u_3 = (R_0 u_1) \times (R_0 u_2)$$
$$t_0 = (k_0 A - R_0) u_3$$
$$n_0 = u_3$$

After $R_0 u_i (i = 1, 2, 3)$ is determined, R_0 can be uniquely determined.

5. If $\xi^2 > \eta^2 = \zeta^2$, then there exists a unique solution for $\{R_0, t_0, n_0\}$:

$$R_0 u_i = k_0 A u_i, \ i = 2, 3$$
$$R_0 u_1 = (R_0 u_2) \times (R_0 u_3)$$
$$t_0 = (k_0 A - R_0) u_1$$
$$n_0 = u_1$$

6. If $\xi^2 > \eta^2 > \zeta^2$, then there are two solutions for $\{R_0, t_0, n_0\}$. Let

$$e_1 = \left(\pm \sqrt{1 - k_0^2 \zeta^2} \ u_1 - \sqrt{k_0^2 \xi^2 - 1} \ u_3 \right) / \sqrt{k_0^2 (\xi^2 - \zeta^2)}$$
$$e_2 = u_2$$
$$e_3 = e_1 \times e_2$$

Then

$$R_0 e_i = k_0 A e_1, \ i = 1, 2$$
$$R_0 e_3 = (R_0 e_1) \times (R_0 e_2)$$
$$t_0 = (k_0 A - R_0) e_3$$
$$n_0 = e_3$$

15.5.2 General Curved Patch Motion Recovery from Two Perspective Views: A Simplified Linear Algorithm

In the previous section we assumed that the observed surface part of a moving rigid body constitutes a planar patch. Now we discard this restriction and consider a general curved patch. Several equivalent linear algorithms (Longuet-Higgins, 1981; Tsai and Huang, 1984; Zhuang, Huang, and Haralick, 1986; and Zhuang, 1989) are devoted to this topic. The first three involve two possible decompositions of motion parameter matrix E (for definition, see Eq. 15.92 below). Zhuang (1989) showed that two decompositions could be avoided, and hence a simplification could be reached. The following explains that simplification and summarizes the corresponding algorithm.

Derivation of the Two View–Motion Equation

Recall that the three-dimensional coordinates of a point before and after motion are related by Eq. (15.80). Taking any nonzero vector T that is collinear with T_0 (i.e.,

$T \times T_0 = 0$) and taking its cross-product with both sides of Eq. (15.80), we obtain

$$\frac{z'}{z}T \times (u',v',f)' = T \times [R_0(u,v,f)'] \tag{15.90}$$

After taking the inner product of both sides of Eq. (15.90) with (u',v',f), we have

$$(u',v',f)(T \times R_0)(u,v,f)' = 0 \tag{15.91}$$

where $T \times R_0 = [T \times r_1, T \times r_2, T \times r_3]$; the terms r_1, r_2, and r_3 being the columns of R_0. Let the motion parameter matrix E be defined as

$$E = T \times R_0 \tag{15.92}$$

Then Eq. (15.91) states that for any image-point correspondence pair $(u,v) \leftrightarrow (u',v')$, the 3×3 matrix E satisfies the following two view–motion equation, which is linear and homogeneous in the nine elements of E:

$$(u',v',f)E(u,v,f)' = 0 \tag{15.93}$$

Denote the set of all observed image-point correspondence pairs $(u_i,v_i) \leftrightarrow (u_i',v_i')$, $i = 1,\ldots,N$, by P. Let

$$a_i = (u_i'u_i, u_i'v_i, u_i'f, v_i'u_i, v_i'v_i, v_i'f, fu_i, fv_i, f^2), \quad i = 1,\ldots,N$$

$$A = \begin{pmatrix} a_1 \\ \vdots \\ a_N \end{pmatrix}$$

$$E = \begin{pmatrix} h_1 & h_2 & h_3 \\ h_4 & h_5 & h_6 \\ h_7 & h_8 & h_9 \end{pmatrix}$$

$$h = (h_1 \ldots h_9)'$$

Then we can easily see that the linear equations with $[(u_i,v_i),(u_i',v_i')] \in P$ are equivalent to the following matrix linear equation for h:

$$Ah = 0 \tag{15.94}$$

Both Eqs. (15.93) and (15.94) will be called the two view–motion equation. We are to solve first for the intermediate unknowns h_i and then obtain the motion parameters from them. Any $T \times R_0$ with $T \times T_0 = 0$ satisfies both Eqs. (15.93) and (15.94) (the latter if $T \times R_0$ is rearranged as h); moreover, such a vector T, which is collinear with T_0, has one degree of freedom when $T_0 \neq 0$ and three degrees of freedom when $T_0 = 0$. Therefore the general solution of the two view–motion equation has at least one degree of freedom when $T_0 \neq 0$ and three degrees of freedom when $T_0 = 0$. In other words, the coefficient matrix A in Eq. (15.94) has a rank no larger than 8 when $T_0 \neq 0$ and no larger than 6 when $T_0 = 0$. If the rank of A equals 8, then the translation T_0 must be nonzero and the general solution must have one degree of freedom and hence coincides with $\alpha(T_0 \times R_0)$, where α is any real number. If the rank equals 6 and the translation T_0 is zero, then the general solution must have three degrees of freedom and hence coincides with $T \times R_0$, where T is any real vector.

Degeneracy and Surface Assumption

The two view–motion equation is called degenerate if the rank of A is less than 8 when $T_0 \neq 0$ or less than 6 when $T_0 = 0$. Thus when the two view–motion equation is not degenerate, any nonzero solution E can be decomposed into $T \times R_0$ with $T \times T_0 = 0$. In the following we always assume that the rank of A equals 8(6) when $T_0 \neq 0 (= 0)$.

The rank assumption has a simple geometric interpretation called the surface assumption. To simplify the interpretation, we assume that the object is stationary and the camera is moving. Let the origins of the camera's system be 0 and 0', respectively, before and after the motion. Then the rank assumption holds if and only if the three-dimensional points corresponding to P do not lie on a quadratic surface passing through 0 and 0' when $T_0 \neq 0$ or on a cone with its apex at 0 when $T_0 = 0$ (then 0 and 0' coincide). The first part of the surface assumption, which was originally derived by Longuet-Higgins (1984b) when assuming $T_0 \neq 0$, does not include the second part, that is, the case $T_0 = 0$.

To satisfy the surface assumption, it is apparent that at least six or eight point correspondence pairs are needed, depending on whether the translation T_0 is zero or not. In practice, more pairs are preferable to increase the probability that the surface assumption will be satisfied and to smooth out noise effects.

Decomposing E into $T \times R_0$ with $T \times T_0 = 0$

Under the surface assumption, any nonzero solution E of the two view–motion equation has a decomposition $T \times R_0$, with $T \times T_0 = 0$. Longuet-Higgins (1981) showed that, besides $T \times R_0$, E admits one and only one alternative decomposition $(-T) \times R_0'$, with R_0' being an orthonormal matrix of the first kind. Thus the surface assumption assures that any nonzero solution E of the two view–motion equation admits two and only two decompositions, one of which gives the true rotation R_0. When T appears in the decomposition giving the true rotation, then $(-T)$ will appear in the second decomposition, giving the false rotation, and vice versa. Thus in order to avoid the second, unnecessary decomposition, we need only find the right T involved in the first decomposition. To determine the right T, we first point out that

$$\text{Rank}(E) = 2$$
$$\|E\| = \sqrt{2}\|T\|$$
$$E'T = 0$$

That rank of E equals 2 is obvious, since $E = [T \times r_1, T \times r_2, T \times r_3]$, and its three columns span a two-dimensional plane. $||E|| = \sqrt{2}||T||$ is just an easy exercise. And $E'T = 0$ could also be shown easily:

$$E'T = \begin{pmatrix} < T \times r_1, T > \\ < T \times r_2, T > \\ < T \times r_3, T > \end{pmatrix} = 0$$

Thus the right T could be determined up to a sign by $E'T = 0$ and $||T|| = 1/\sqrt{2}||E||$. To determine the sign, we rewrite Eq. (15.90) as follows:

$$z'T \times (u', v', f)' = zT \times [R_0(u, v, f)'] = zE(u, v, f)' \tag{15.95}$$

which implies

$$\left(\sum_{i=1}^{n} < T \times (u_i', v_i', f)', E(u_i, v_i, f)' > \right) > 0$$

since $z_i < 0, z_i' < 0$, and, except for at most one image-point correspondence pair, $T \times (u_i', v_i', f)' \neq 0$. As a result, the right T is uniquely determined by

$$\begin{cases} E'T = 0 \\ ||T|| = \frac{1}{\sqrt{2}}||E|| \\ \sum_{i=f}^{n} [< T \times (u_i', v_i', f)', E(u_i, v_i, f)' >] > 0 \end{cases} \tag{15.96}$$

Once the right T is determined, the true rotation R_0 could be uniquely determined through $E = T \times R_0$ as follows (cf. Longuet-Higgins, 1981, for example):

$$R_0 = [W_2 \times W_3, W_3 \times W_1, W_1 \times W_2] - T \times E \tag{15.97}$$

with $W_1 = T \times E_1, W_2 = T \times E_2$, and $W_3 = T \times E_3$, and $E_i(i = 1, 2, 3)$ representing the ith column of E.

15.5.3 Determining Translational Orientation

Since T is collinear with T_0, T_0 should have the same orientation as T or $(-T)$. Taking a cross-product with both sides of Eq. (15.80) by (u', v', f), we obtain

$$z(u', v', f)' \times [R_0(u, v, f)'] + (u', v', f)' \times T_0 = 0 \tag{15.98}$$

which implies that, since $z < 0$, T_0 has the same orientation as T or $(-T)$ if and only if $(u', v', f)' \times [R_0(u, v, f)']$ has the same orientation as $(u', v', f)' \times T$ or $[-(u', v', f)' \times T]$, and in turn if and only if $\sum_{i=1}^{n} \{< (u_i', v_i', f)' \times [R_0(u_i, v_i, f)'], (u_i', v_i', f)' \times T >\} \geq 0$ or ≤ 0.

15.5.4 Determining Mode of Motion and Relative Depths

The mode of motion indicates whether or not the translation is zero. It is easy to show that $T_0 = 0$ if and only if $(u', v', f)'$ is always collinear with $R_0(u, v, f)'$, and in turn if and only if $\sum_{i=1}^{n} \|(u'_i, v'_i, f)' \times [R_0(u_i, v_i, f)']\| = 0$.

When $T_0 \neq 0$, we could determine the relative depths $z/\|T_0\|, z'/\|T_0\|$ as follows:

$$\begin{cases} z/\|T_0\| = -\|T \times (u', v', f)'\|/\|(u', v', f)' \times [R_0(u, v, f)']\| \\ z'/\|T_0\| = -\|T \times [R_0(u, v, f)']\|/\|(u', v', f)' \times [R_0(u, v, f)']\| \end{cases} \tag{15.99}$$

15.5.5 A Simplified Two View–Motion Linear Algorithm

Step 1: Use a least-squares approach to solve Eq. (15.93).
Step 2: Use a least-squares approach to solve $E'T = 0$ in Eq. (15.96).
Step 3: Determine R_0 by Eq. (15.97).
Step 4: Determine the translational orientation as in Section 15.5.3.
Step 5: Determine the relative depths by Eq. (15.99).

■ **EXAMPLE 15.1**

$$T_0 = (0,0,0), \qquad \begin{bmatrix} 1/\sqrt{2} & 1/\sqrt{2} & 0 \\ -1/\sqrt{2} & 1/\sqrt{2} & 0 \\ 0 & 0 & 1 \end{bmatrix}$$

Six points in image plane $z = 1$ before motion:

$$(0.63, -0.93)$$
$$(2.09, 0.10)$$
$$(0.53, 1.43)$$
$$(1.85, 1.83)$$
$$(1.29, 0.41)$$
$$(-1.32, -0.12)$$

Six points in image plane $z = 1$ after motion:

$$(-0.21, -1.10)$$
$$(1.54, -1.41)$$
$$(1.39, 0.63)$$
$$(2.60, -0.01)$$
$$(1.20, -0.62)$$
$$(-1.01, 0.85)$$

Computed E, R, T :

$$E = \begin{bmatrix} 0.27 & -0.27 & -0.02 \\ 0.27 & 0.27 & -0.59 \\ -0.41 & 0.43 & -0.00 \end{bmatrix}$$

$$R = \begin{bmatrix} 0.71 & 0.71 & 0.00 \\ -0.71 & 0.71 & -0.00 \\ 0.00 & -0.00 & 1.00 \end{bmatrix}$$

$$T = \begin{bmatrix} 0.59 \\ -0.02 \\ 0.39 \end{bmatrix}$$

Thus the algorithm gives the correct rotation, $R_0 = R$.

■

EXAMPLE 15.2

$$T_0 = (0, 0, -1)^{\pm}, \qquad R_0 = \begin{bmatrix} 1/\sqrt{2} & 1/\sqrt{2} & 0 \\ -1/\sqrt{2} & 1/\sqrt{2} & 0 \\ 0 & 0 & 1 \end{bmatrix}$$

Eight points in image plane $z = 1$ before motion:

$$(-0.04, 0.96)$$
$$(-0.09, -1.22)$$
$$(-0.67, 0.91)$$
$$(1.17, 1.29)$$
$$(1.10, 0.65)$$
$$(-0.13, -0.98)$$
$$(-1.13, -1.19)$$
$$(1.03, -0.37)$$

Eight points in image plane $z = 1$ after motion:

$$(0.41, 0.44)$$
$$(-0.60, -0.52)$$
$$(0.1, 0.67)$$
$$(1.07, 0.06)$$
$$(0.62, -0.16)$$
$$(-0.45, -0.35)$$
$$(-0.89, -0.02)$$
$$(0.29, -0.62)$$

Computed E, R, T :

$$E = \begin{bmatrix} -0.50 & 0.50 & 0.00 \\ -0.50 & -0.50 & -0.00 \\ -0.00 & 0.00 & 0.00 \end{bmatrix}$$

$$R = \begin{bmatrix} -0.71 & 0.71 & 0.00 \\ -0.71 & 0.71 & 0.00 \\ -0.00 & 0.00 & 1.00 \end{bmatrix}$$

$$T = \begin{bmatrix} -0.00 \\ 0.00 \\ 0.71 \end{bmatrix}$$

Thus the algorithm gives the correct rotation, $R_0 = R$, and the correct translational orientation, $T_0/\|T_0\| = -T/\|T\|$. The latter is determined as in Section 15.5.3.

15.5.6 Discussion and Summary

For the case when no noise appears in the image-point correspondence pairs measurements, the algorithmn is extremely accurate. When a small noise appears in the measurements, it works well except that the mode of motion cannot be determined correctly. That is, the case of pure rotation (i.e., $T_0 = 0$) cannot be recognized. It will affect how one determines the relative depth map and the translational orientation. Since the case of pure rotation admits an indefinite relative depth map and an indefinite translational orientation, the algorithm gives just one of the infinite consistent interpretations. Thus acceptance of the interpretation extracted by the algorithm could never cause a disaster as long as the noise is limited.

15.6 Linear Algorithm for Motion and Structure from Three Orthographic Views

In his influential work on motion analysis, Ullman (1979) showed that for the orthographic case four-point correspondences over three views are sufficient to determine the motion and structure of the four-point rigid configuration. The problem has been reexamined by Aloimonos and Brown (1986), Zhuang, Huang, and Haralick (1988a), and Huang and Lee (1989). Here we show a simple linear procedure to solve Ullman's three-views–motion problem.

15.6.1 Problem Formulation

We assume that the image plane is stationary and that three orthographic views at the time t_1, t_2, and t_3, respectively, are taken of a rigid object moving in the three-dimensional object space. By processing the three views, we intend to determine the motion and structure of the three-dimensional object.

Let (x, y, z) be the object space coordinates, and (u, v) the image space coordinates. The u- and v-axes coincide with the x- and y-axes (in particular, the origins of the x–y–z coordinate system and the u–v system coincide). Let

$$\begin{aligned}
(x, y, z) &= \text{object-space coordinates of a point } P \text{ on the rigid object at } t_1, \\
(x', y', z') &= \text{object-space coordinates of the same point } P \text{ at } t_2, \\
(x'', y'', z'') &= \text{object-space coordinates of } P \text{ at } t_3, \\
(u, v) &= \text{image-space coordinates of } P \text{ at } t_1, \\
(u', v') &= \text{image-space coordinates of } P \text{ at } t_2, \\
(u'', v'') &= \text{image-space coordinates of } P \text{ at } t_3.
\end{aligned}$$

Then

$$(x', y', z')' = R(x, y, z)' + T_r \tag{15.100}$$
$$(x'', y'', z'')' = S(x, y, z)' + T_s \tag{15.101}$$

where $R = (r_{ij})_{3 \times 3}$, and $S = (s_{ij})_{3 \times 3}$ are a rotation matrix, and $T_r = (t_{ri})_{3 \times 1}$ and $T_s = (t_{si})_{3 \times 1}$ are a 3×1 translation vector. The problem we are trying to solve is: Given four image-point correspondences

$$(u_i, v_i) \leftrightarrow (u_i', v_i') \leftrightarrow (u_i'', v_i''), \quad i = 1, 2, 3, 4 \tag{15.102}$$

determine $(R, T_r), (S, T_s)$, and (x_i, y_i, z_i), $i = 1, 2, 3, 4$. Note that with orthographic projections,

$$\begin{aligned}
(X, Y) &= (x, y) \\
(X', Y') &= (x', y') \\
(X'', Y'') &= (x'', y'')
\end{aligned} \tag{15.103}$$

and therefore it is obvious that t_{r3}, t_{s3} can never be determined. We can hope to determine the depths of the object points only to within an additive constant. What we are trying to determine, then, are R, S, t_{ri}, t_{si}, $i = 1, 2, z_i - z_1$, $i = 2, 3, 4$. We can decompose the rigid-body motion from t_1 to t_2 (t_3) as a rotation $R(S)$ around the point (x_1, y_1, z_1) followed by a translation $T_r + (x_1', y_1', z_1')'$ $[T_s + (x_1'', y_1'', z_1'')']$. To determine R, S, $z_i - z_1$, $i = 2, 3, 4$, we can then use the following equations:

$$(u_i' - u_1', v_i' - v_1')' = \overline{R}(u_i - u_1, v_i - v_1, z_i - z_1)' \tag{15.104}$$
$$(u_i'' - u_1'', v_i'' - v_1'')' = \overline{S}(u_i - u_1, v_i - v_1, z_i - z_1)', \quad i = 2, 3, 4 \tag{15.105}$$

where $\overline{R} = (r_{ij})_{2 \times 3}, \overline{S} = (s_{ij})_{2 \times 3}$.

Once $R, S, z_i - z_1$, $i = 2, 3, 4$, are determined, both $z_i' - z_1'$ and $z_i'' - z_1''(i = 2, 3, 4)$ can be uniquely determined, and moreover both t_{ri} and $t_{si}(i = 1, 2)$ can be determined as a function of z_1.

15.6.2 Determining $r_{33}, s_{33}, (r_{13}, r_{23}, r_{31}, r_{32}), (s_{13}, s_{23}, s_{31}, s_{32})$

Both r_{33} and s_{33}, and $(r_{13}, r_{23}, r_{31}, r_{32})$ and $(s_{13}, s_{23}, s_{31}, s_{32})$ up to a sign, can be uniquely determined from three orthographic views. Let

$$a_1 = (u_2 - u_1, u_3 - u_1, u_4 - u_1)$$
$$a_2 = (v_2 - v_1, v_3 - v_1, v_4 - v_1)$$
$$a_3 = (z_2 - z_1, z_3 - z_1, z_4 - z_1)$$
$$b_1 = (u_2' - u_1', u_3' - u_1', u_4' - u_1')$$
$$b_2 = (v_2' - v_1', v_3' - v_1', v_4' - v_1')$$
$$c_1 = (u_2'' - u_1'', u_3'' - u_1'', u_4'' - u_1'')$$
$$c_2 = (v_2'' - v_1'', v_3'' - v_1'', v_4'' - v_1'')$$
$$A = \begin{pmatrix} a_1 \\ a_2 \\ a_3 \end{pmatrix}$$
$$\overline{A} = \begin{pmatrix} a_1 \\ a_2 \end{pmatrix}, \overline{B} = \begin{pmatrix} b_1 \\ b_2 \end{pmatrix}, \overline{C} = \begin{pmatrix} c_1 \\ c_2 \end{pmatrix}$$

Then $\overline{A}, \overline{B}$, and $\overline{C}$ are known and Eq. (15.105) can be written as

$$\overline{B} = \overline{R}A \qquad (15.106)$$
$$\overline{C} = \overline{S}A \qquad (15.107)$$

We further assume that

$$\text{Rank}\ (A) = 3 \qquad (15.108)$$

Now we try to determine R, S, and a_3. If

$$\text{Rank}\ \left(\frac{\overline{A}}{\overline{B}} \right) < 3$$

that is, $(= 2)$, then $r_{13} = r_{23} = 0$ because of Eq. (15.108). Moreover, r_{11}, r_{12}, r_{21}, and r_{22} are uniquely determined by

$$\begin{pmatrix} r_{11} & r_{12} \\ r_{21} & r_{22} \end{pmatrix} = (\overline{BA'})(\overline{AA'})^{-1} \qquad (15.109)$$

Thus it is obvious that $r_{31} = r_{32} = 0$, and $r_{33} = 1$ or -1, depending on whether $\begin{vmatrix} r_{11} & r_{12} \\ r_{21} & r_{22} \end{vmatrix} = 1$ or not. In summary, when rank $\left(\frac{\overline{A}}{\overline{B}} \right)$ is less than 3, the rotation matrix R can be uniquely determined even though Eq. (15.104) does not contain any information about a_3.

If both rank $\left(\dfrac{\overline{A}}{B}\right)$ and rank $\left(\dfrac{\overline{A}}{C}\right)$ are less than 3, then both R and S can be uniquely determined. However, in this case the row vector a_3 cannot be recovered.

If rank $\left(\dfrac{\overline{A}}{B}\right) < 3$ and rank $\left(\dfrac{\overline{A}}{C}\right) = 3$, then R is uniquely determined, and both S and a_3 have infinite solutions (cf. Section 15.6.3). Similarly, when rank $\left(\dfrac{\overline{A}}{B}\right) = 3$ and rank $\left(\dfrac{\overline{A}}{C}\right) < 3$, then S is uniquely determined, and both R and a_3 have infinite solutions. Therefore the case that really interests us is

$$\text{Rank}\left(\dfrac{\overline{A}}{B}\right) = 3 \tag{15.110}$$

$$\text{Rank}\left(\dfrac{\overline{A}}{C}\right) = 3 \tag{15.111}$$

In the following we show how $r_{33}, S_{33}, (r_{13}, r_{23}, r_{31}, r_{32})$ up to a sign and $(S_{13}, S_{23}, S_{31}, S_{32})$ up to a sign, can be uniquely determined under the assumption equations (15.110) and (15.111).

Using the assumption equation (15.108), Eqs. (15.106) and (15.107), and the following identities:

$$(r_{32}, -r_{31}, 0) = (-r_{23}, r_{13})\overline{R} \tag{15.112}$$

$$(S_{32}, -S_{31}, 0) = (-S_{23}, S_{13})\overline{S} \tag{15.113}$$

we get

$$(r_{32}, -r_{31}, 0)A = (-r_{23}, r_{13})\overline{B} \tag{15.114}$$

$$(S_{32}, -S_{31}, 0)A = (-S_{23}, S_{13})\overline{C} \tag{15.115}$$

or equivalently

$$(r_{32}, -r_{31})\overline{A} = (-r_{23}, r_{13})\overline{B} \tag{15.116}$$

$$(S_{32}, -S_{31})\overline{A} = (-S_{23}, S_{13})\overline{C} \tag{15.117}$$

or still equivalently

$$[\overline{A}', \overline{B}'](r_{32}, -r_{31}, r_{23}, -r_{13})' = 0 \tag{15.118}$$

$$[\overline{A}', \overline{C}'](S_{32}, -S_{31}, S_{23}, -S_{13})' = 0 \tag{15.119}$$

Because of the assumption equations (15.110) and (15.111), both $(r_{13}, r_{23}, r_{31}, r_{32})$ up to a scale factor α and $(S_{13}, S_{23}, S_{31}, S_{32})$ up to a scale factor β are uniquely determined by Eqs. (15.118) and (15.119). That is, there exist four 1×2 row vectors $u, v, \overline{u}, \overline{v}$ such that

$$(r_{13}, r_{23}) = \alpha u$$
$$(r_{31}, r_{32}) = \alpha v$$

$$\tag{15.120}$$

$$(S_{13}, S_{23}) = \beta \overline{u}$$
$$(S_{31}, S_{32}) = \beta \overline{u}$$

Because of $r_{13}^2 + r_{23}^2 = r_{31}^2 + r_{32}^2 \leq 1$ and $r_{13}^2 + r_{23}^2 > 0$ (the latter is implied by assumption equation 15.110), we confirm that

$$\alpha \neq 0 \text{ and } ||u|| = ||v|| > 0 \tag{15.121}$$

Similarly, using assumption equation (15.111), we confirm that

$$\beta \neq 0 \text{ and } ||\bar{u}|| = ||\bar{v}|| > 0 \tag{15.122}$$

Without loss of generality, in the proceeding discussion we assume

$$||u|| = ||v|| = ||\bar{u}|| = ||\bar{v}|| = 1 \tag{15.123}$$

Using Eqs. (15.106), (15.107), and (15.120), we can further derive

$$
\begin{aligned}
u\bar{B} &= u\bar{R}A \\
&= u \begin{pmatrix} r_{11} & r_{12} \\ r_{21} & r_{22} \end{pmatrix} \bar{A} + u \begin{pmatrix} r_{13} \\ r_{23} \end{pmatrix} a_3 \\
&= \frac{1}{\alpha}(r_{13}, r_{23}) \begin{pmatrix} r_{11} & r_{12} \\ r_{21} & r_{22} \end{pmatrix} \bar{A} + \alpha a_3 \\
&= \frac{1}{\alpha}(r_{11}r_{13} + r_{21}r_{23}, r_{12}r_{13} + r_{22}r_{23})\bar{A} + \alpha a_3 \\
&= -\frac{1}{\alpha}r_{33}(r_{31}, r_{32})\bar{A} + \alpha a_3 \\
&= -r_{33}v\bar{A} + \alpha a_3
\end{aligned}
\tag{15.124}
$$

and

$$\bar{u}\bar{C} = -s_{33}\bar{v}\bar{A} + \beta a_3 \tag{15.125}$$

Eliminating a_3 from Eqs. (15.124) and (15.125) leads to

$$\beta[u\bar{B} + r_{33}(v\bar{A})] = \alpha[\bar{u}\bar{C} + s_{33}(\bar{v}\bar{A})] \tag{15.126}$$

or, in a more readable form,

$$[(u\bar{B})', (v\bar{A})', (\bar{v}\bar{A})'](\beta/\alpha, r_{33}\beta/\alpha, -s_{33})' = (\bar{u}\bar{C})' \tag{15.127}$$

If the 3×3 coefficient matrix of Eq. (15.127) has a rank 3, then $\beta/\alpha, r_{33}$, and s_{33} are all uniquely determined, as easily seen. Moreover, α and β are uniquely determined up to a sign as follows:

$$
\begin{aligned}
\alpha &= \pm\sqrt{1 - r_{33}^2} \\
\beta &= \alpha(\beta/\alpha) \\
&= \pm\sqrt{1 - r_{33}^2}(\beta/\alpha)
\end{aligned}
$$

As a result, both $(r_{13}, r_{23}, r_{31}, r_{32})$ up to a sign and $(s_{13}, s_{23}, s_{31}, s_{32})$ up to a sign are uniquely determined:

$$(r_{13}, r_{23}, r_{31}, r_{32}) = \pm\sqrt{1 - r_{33}^2}(u, v) \tag{15.128}$$

$$(s_{13}, s_{23}, s_{31}, s_{32}) = \pm\sqrt{1 - r_{33}^2}(\beta/\alpha)(\bar{u}, \bar{v}) \tag{15.129}$$

$$\begin{pmatrix} I_2 & \\ & -1 \end{pmatrix} R \left[\begin{pmatrix} I_2 & \\ & -1 \end{pmatrix} S \right]$$

15.6.3 Solving a Unique Orthonormal Matrix R

Now we show how to determine a unique orthonormal matrix R, that is, $R'R = I_3$, given $(r_{13}, r_{23}, r_{31}, r_{32}, r_{33})$, obeying $0 < r_{13}^2 + r_{23}^2 = r_{31}^2 + r_{32}^2 = 1 - r_{33}^2 \leq 1$. Thus there exist two candidates for the rotation matrix $R(S)$, since $(r_{13}, r_{23}, r_{31}, r_{32})(s_{13}, s_{23}, s_{31}, s_{32})$ is uniquely determined only up to a sign, even though $r_{33}(s_{33})$ is uniquely determined. The right one can be determined by the conditions $det(R) = 1[det(S) = 1]$. The other one is a reflection matrix and equals

Using the identity equation (15.112) again and noticing

$$(r_{13}, r_{23})\overline{R} = (-r_{33}r_{31}, -r_{33}r_{32}, r_{13}^2 + r_{23}^2), \tag{15.130}$$

we obtain

$$\begin{pmatrix} -r_{23} & r_{13} \\ r_{13} & r_{23} \end{pmatrix} \overline{R} = \begin{pmatrix} r_{32} & -r_{31} & 0 \\ -r_{33}r_{31} & -r_{33}r_{32} & r_{13}^2 + r_{23}^2 \end{pmatrix} \tag{15.131}$$

which leads to

$$\overline{R} = \frac{1}{r_{13}^2 + r_{23}^2} \begin{pmatrix} -r_{23} & r_{13} \\ r_{13} & r_{23} \end{pmatrix} \begin{pmatrix} r_{32} & -r_{31} & 0 \\ -r_{33}r_{31} & -r_{33}r_{32} & r_{13}^2 + r_{23}^2 \end{pmatrix} \tag{15.132}$$

$$R = \begin{pmatrix} \overline{R} \\ r_3 \end{pmatrix} \tag{15.133}$$

where $r_3 = (r_{31}, r_{32}, r_{33})$.

We can also directly verify that the matrix R computed by Eqs. (15.132) and (15.133) is orthonormal, that is, $R'R = I_3$.

15.6.4 Linear Algorithm to Uniquely Solve R, S, a_3

Now we are ready to give the algorithm to determine R, S, and a_3 under the assumption equations (15.110) and (15.111):

Step 1: Solve four 1×2 unit row vectors $u, v, \bar{u}, \bar{v}$ by Eqs. (15.118) and (15.119).

Step 2: Solve $\beta/\alpha, r_{33}, s_{33}$ by Eq. (15.127).

Step 3: Determine $r_{13}, r_{23}, r_{31}, r_{32}$ and $s_{13}, s_{23}, s_{31}, s_{32}$ by Eqs. (15.128) and (15.129).

Step 4: Determine R by Eqs. (15.132) and (15.133) and the condition $det(R) = 1$. Similarly, determine S.

Step 5: Determine a_3 by

$$a_3 = \frac{1}{r_{13}^2 + r_{23}^2} [(r_{13}, r_{23})\overline{B} + r_{33}(r_{31}, r_{32})\overline{A}] \tag{15.134}$$

15.6.5 Summary

Given two orthographic views, one cannot finitely determine the motion and structure of a rigid object, no matter how many point correspondences are used, as shown by Huang. From the argument in Section 15.6.2, we saw that both (r_{13}, r_{23}) and (r_{31}, r_{32}) could be determined only up to an arbitrary scale factor $\alpha, 0 < |\alpha| \leq 1$, from two orthographic views. For each $\alpha, 0 < |\alpha| \leq 1$, Eqs. (15.119), (15.132), and (15.133) determine two orthonormal matrices when letting $r_{33} = \pm\sqrt{1 - \alpha^2}$. One of them is a rotation matrix. Thus the number of solutions is uncountably infinite as α varies.

Given three orthographic views, one can uniquely determine motion and structure when assuming Eqs. (15.120) and (15.121) and

$$\text{Rank}\left([(u\overline{B})', (v\overline{A})', (\overline{v}\overline{A})']\right) = 3 \tag{15.135}$$

Clearly, the assumptions of Eqs. (15.110) and (15.111) imply the assumption equation (15.108). However, it is not clear whether assumption equation (15.135) is always assured by Eqs. (15.110) and (15.111). If not, then Eqs. (15.110), (15.111), and (15.135) constitute a set of necessary and sufficient conditions to uniquely determine the motion and structure of a rigid object from three orthographic views, as we see from the argument in Section 15.6.2.

15.7 Developing a Highly Robust Estimator for General Regression

15.7.1 Inability of the Classical Robust M-Estimator to Render High Robustness

As is well known, the classical robust estimator, such as the M-, L-, or R-estimator (Huber, 1981), possesses the following properties:

1. It has a reasonably good (optimal or nearly optimal) efficiency at the assumed noise distribution.

2. It is robust in the sense that the degradation in performance caused by a small number of outliers is relatively small.

3. Somewhat larger deviations from the assumed distribution do not cause a catastrophe.

The MF-estimator to be explored and developed here will represent a new brand of robust estimator that possesses properties 1 and 2 as well as the following property, which is much stronger than property 3:

4. The degradation in performance caused by somewhat larger deviations from the assumed distribution is still relatively small.

Before we construct a highly robust estimator, we first need to say a few words about the minimax approach, which was widely used in developing the classical robust M-estimator.

Assume that the true underlying shape F lies in some neighborhood P_ϵ of the assumed standard normal distribution Φ, where $P_\epsilon(\Phi) = \{F \mid F = (1 - \epsilon)\Phi + \epsilon H, H \in M\}$, with M representing the set of unknown foreign distributions; that the observations are independent with common distribution $F(x - \theta)$; and that the location parameter θ is to be estimated.

The minimax approach to robustly estimating the location parameter is based on minimizing its maximum asymptotic variance for all possible distributions $F \in P_\epsilon$. Suppose that F_0 attains maximum asymptotic variance for location in the set P_ϵ; then the corresponding probability density function f_0 has following form (Huber, 1981):

$$f_0(x) = \frac{1 - \epsilon}{\sqrt{2\pi}} \exp(\frac{-x^2}{2}), \qquad \text{for } |x| \leq a$$

$$= \frac{1 - \epsilon}{\sqrt{2\pi}} \exp(\frac{a^2}{2} - a|x|), \qquad \text{for } |x| > a \qquad (15.136)$$

with the robust control parameter a and the outlier proportion parameter ϵ connected through

$$\frac{2\phi(a)}{a} - 2\Phi(-a) = \frac{\epsilon}{1 - \epsilon} \qquad (15.137)$$

where $\phi = \Phi'$ is the standard normal density. Moreover, the asymptotically efficient maximum likelihood estimate of location for F_0 (called the M-estimate) in fact has been proved to possess certain minimax properties in P_ϵ (Huber, 1981).

Clearly, for the M-estimate of location, the robust distribution function F_0, instead of the standard normal distribution Φ, is exclusively used to model each possible distribution function F in the neighborhood $P_\epsilon(\Phi)$. As we know, the maximum likelihood estimate of location obtained by using the standard normal distribution Φ leads to the least-squares estimate.

The reason why the function f_0 (or F_0) has a reasonably good efficiency at the assumed standard normal distribution Φ is that, with a high probability, the residual $|x - \theta|$ will be less than or equal to a; in other words, the M-estimator will do mostly the same job as the least-squares estimator does.

The reason why the function f_0 (or F_0) is robust when only a small number of outliers exist is that, with a probability of about or less than $1 - \epsilon$, the residual $|x - \theta|$ will be less than or equal to a; for this major part of residuals the M-estimator behaves much like the least-squares estimator as demanded, and with a probability of about or perhaps less than ϵ, the residual $|x - \theta|$ will be larger than a. To gain robustness, this minor part of the residuals is also taken care of by the M-estimator. To see the above point clearly, we should point out that the unknown outlier process is modeled as the standard normal process by the least-squares estimator, that is, $h(x) = \phi(x)$, or as the process whose probability density $h(x)$ equals zero as $|x| \leq a$ or $\frac{1-\epsilon}{\epsilon\sqrt{2\pi}}[\exp(\frac{a^2}{2} - a|x|) - \exp(-\frac{x^2}{2})]$ as $|x| > a$ by the M-estimator. The important fact that the magnitude of an outlier residual is more likely to be larger than the magnitude of a nonoutlier residual is fully neglected by the least-squares estimator, which produces a nonrobust estimate. The same fact is more or less reflected in the construction of the M-estimator. In fact, the possibility that an outlier with a residual magnitude will be less than or equal to a is completely excluded by the M-estimator. From the previous expression for $h(x)$, it is easy to verify that the probability density function of outliers has only two modes, peaks approximately around $\pm(a + \sqrt{a^2 + 8}/2)$, and flattens as residual magnitudes go beyond $\pm(a + \sqrt{a^2 + 8}/2)$.

The reason why the function f_0 cannot render high robustness is that the robust control parameter a is tied to the outlier proportion parameter ϵ and approaches zero as ϵ tends to 1. That means when the outlier proportion ϵ becomes larger, all nonoutlier residuals ought to be smaller to comply with the M-estimator. If the nonoutlier residuals do not become correspondingly smaller, some or even all of them will be treated as outliers by the M-estimator; the contained information useful for location estimation will thus be lost, often resulting in bad estimates. The situation becomes even worse when a limited sample size is used in the estimation process, since the M-estimator simply cannot recognize enough nonoutliers for a reasonably good estimate.

It seems that if we model each possible distribution function F in the neighborhood $P_\epsilon(\Phi)$ by using a single fixed robust distribution function, a high robustness will never be possible. In our development, we attempt to fit the most likely values to each unknown foreign probability density function at the observed data individually instead of completely modeling each unknown foreign probability density function. Specifically, we will use more heuristic reasoning rather than purely mathematical reasoning. In fact, we will combine the Bayes statistical decision rule with a number of deeply explored heuristic considerations and turn the general robust regression problem into a model fitting problem that is not only more flexible but also more tractable.

15.7.2 Partially Modeling Log Likelihood Function by Using Heuristics

Assume that p unknown parameters $\theta_1, \cdots, \theta_p$, shortened as the vector θ, are to be estimated from N observations $y_1, \cdots, y_N$, each of which is an m-dimensional vector. Assume that $f_k : R_p \to R_m, k = 1, \cdots, N$. Let r_k be the residual between y_k and $f_k(\theta)$, that is, $r_k = y_k - f_k(\theta)$. Furthermore, assume that each single observation y_k with probability $1 - \epsilon$ is a "good" one, that is, not an outlier, and with probability ϵ, a "bad" one, that is, an outlier, where $0 \le \epsilon \le 1$. In the former case the residual r_k is Gaussian distributed with zero mean and an unknown covariance matrix $\sigma^2 I_m$; in the latter case it obeys an unknown foreign distribution. All r_k's are independent and identically distributed with the common probability density function f, namely,

$$f(r_k) = \frac{1 - \epsilon}{(\sqrt{2\pi}\sigma)^m} \exp\left(-\frac{\|r_k\|^2}{2\sigma^2}\right) + \epsilon h(r_k) \tag{15.138}$$

where h is unknown. Thus the log likelihood function of observing $y_1, \cdots, y_N$ conditioned on $\theta_1, \cdots, \theta_p, \sigma, \epsilon, h(r_1), \cdots, h(r_N)$ is expressed by Q as follows:

$$Q = \log P[y_1, \cdots, y_N \mid \theta_1, \cdots, \theta_p, \sigma, \epsilon, h(r_1), \cdots, h(r_N)]$$

$$= \log \prod_k P[y_k \mid \theta_1, \cdots, \theta_p, \sigma, \epsilon, h(r_k)]$$

$$= \sum_k \log P[y_k \mid \theta_1, \cdots, \theta_p, \sigma, \epsilon, h(r_k)]$$

$$= \sum_k \log f(r_k) \tag{15.139}$$

which, combined with Eq. (15.138), constitutes the first model assumption. To be successful in gaining high robustness, we need to further explore possible heuristics implicit in the regression problem instead of hurrying to maximize the log likelihood function Q.

According to Eq. (15.138), the probability of the observation y_k being a nonoutlier conditioned on $\theta, \sigma, \epsilon, h(r_k)$ is given by λ_k:

$$\lambda_k = \frac{1 - \epsilon}{(\sqrt{2\pi}\sigma)^m} \exp\left(-\frac{\|r_k\|^2}{2\sigma^2}\right) / f(r_k) \tag{15.140}$$

Using the Bayes statistical decision rule, we can classify the observation y_k as a nonoutlier if $\lambda_k > 0.5$ or as an outlier otherwise. Let G denote the indices of "good" observations and B the indices of "bad" observations, where

$$G = \{k : \lambda_k > 0.5\}$$
$$B = \{k : \lambda_k \le 0.5\} \tag{15.141}$$

The second model assumption consists of the following heuristic condition:

$$\frac{\#G}{N} = 1 - \epsilon$$

$$\frac{\#B}{N} = \epsilon \tag{15.142}$$

where $\#$ represents "the number of."

To obtain a reliable estimation, a minimum of "good" observations are required. The minimum number, denoted as L, is very problem dependent and can be experimentally or theoretically determined. The results will no longer be reliable when $\#G$ drops below L. Thus for the third model assumption we use the following heuristic condition:

$$\#G \geq L \tag{15.143}$$

The fourth model assumption states that all $h(r_k)$ could be taken as equal, namely,

$$h(r_k) = \delta, \quad k = 1, \cdots, N \tag{15.144}$$

This assumption comes from the heuristic consideration that for two different observations the one with a smaller residual magnitude should more likely be a nonoutlier than the other one. Thus, the partition $\{G, B\}$ ought to have the following property:

$$\max\{\|r_k\| : k \in G\} < \min\{\|r_k\| : k \in B\} \tag{15.145}$$

Let

$$g = \min\left\{\frac{1 - \epsilon}{(\sqrt{2\pi}\sigma)^m} \exp\left(-\frac{\|r_k\|^2}{2\sigma^2}\right) : k \in G\right\}$$

$$b = \max\left\{\frac{1 - \epsilon}{(\sqrt{2\pi}\sigma)^m} \exp\left(-\frac{\|r_k\|^2}{2\sigma^2}\right) : k \in B\right\} \tag{15.146}$$

Then because of Eq. (15.145), it follows that

$$g > b \tag{15.147}$$

which indicates that the partition $\{G, B\}$ can be equally well defined by

$$G = \left\{k : \frac{1 - \epsilon}{(\sqrt{2\pi}\sigma)^m} \exp\left(\frac{-\|r_k\|^2}{2\sigma^2}\right) > \epsilon\delta\right\}$$

$$B = \left\{k : \frac{1 - \epsilon}{(\sqrt{2\pi}\sigma)^m} \exp\left(\frac{-\|r_k\|^2}{2\sigma^2}\right) \leq \epsilon\delta\right\} \tag{15.148}$$

where δ can be an arbitrary number in the interval $\frac{1}{\epsilon}[b,g]$. This means that the choice of δ can be quite broad. This observation will play an important role in our algorithmic development.

Let the fitting error of detected nonoutliers be defined by $\max\{\|r_k\| : k \in G\}$, and let the distance of detected outliers from detected nonoutliers be defined by $\min\{\|r_k\| : k \in B\} - \max\{\|r_k\| : k \in G\}$. For each set of parameters $\{\theta_1, \cdots, \theta_p, \sigma, \epsilon, \delta\}$, a partition $\{G, B\}$ of $\{1, \cdots, N\}$ can be generated by Eq. (15.141). It is reasonable to search for a parameter set having the least possible fitting error and the largest possible distance. This requirement will amount to minimizing the following cost function:

$$C(G,B) = \frac{\#G}{N} \max_G \|r_k\| + \frac{\#B}{N} \frac{1}{\min_B \|r_k\| - \max_G \|r_k\|} \tag{15.149}$$

The minimal cost requirement constitutes the last heuristic condition, namely,

$$C(G,B) = \min \tag{15.150}$$

Now the general robust regression problem can be stated as finding the parameter set $\{\theta_1, \cdots, \theta_p, \sigma, \epsilon, \delta\}$ by maximizing the log likelihood function of the model that is partially modeled by basic assumptions consisting of Eqs. (15.138), (15.139), (15.141) to (15.144), and (15.150). That is,

$$\max Q(\theta_1, \cdots, \theta_p, \sigma, \epsilon, \delta) = \sum \log \left\{ \frac{1-\epsilon}{(\sqrt{2\pi}\sigma)^m} \exp\left(-\frac{\|r_k\|^2}{2\sigma^2}\right) + \epsilon\delta \right\}$$

subject to

$$\#G \geq L$$
$$\#G = (1-\epsilon)N$$
$$C(G,B) = \min \tag{15.151}$$

where condition equations (15.138), (15.139), and (15.144) have been included in the model log likelihood function Q, and where L is a problem-dependent number to be experimentally or theoretically determined.

Combining the Bayes statistical decision rule with a number of heuristic considerations has clearly turned the general robust regression problem into a more appropriate model-fitting problem. The algorithm so developed is called the MF-estimator.

15.7.3 Discussion

The M-, L-, and R-estimators are all residual based; the MF-estimator is no exception. Assume that $m = 1$ (i.e., one-dimensional case) and that the true parameters are $\theta_1, \cdots, \theta_p$, forming N true residuals, $r_k = y_k - f_k(\theta_1, \cdots, \theta_p), k = 1, \cdots, N$.

Suppose those N residuals can be distinctively divided into two groups, the nonoutlier set G and the outlier set B, so that each residual in G is small in magnitude and much smaller than each residual in B. To use the M-estimator effectively, the following three conditions must be satisfied:

$$\#G \geq (1 - \epsilon)N \quad \text{or} \quad \#B \leq \epsilon N,$$
$$|r_k| \leq a, \quad r_k \text{ distributed around zero if } k \in G,$$
$$|r_k| > a, \quad r_k \text{ distributed around } \pm \frac{a + \sqrt{a + 8}}{2} \text{ or beyond if } k \in B.$$

Suppose we add more outliers to B but leave G alone, so that the two sets, G and newly formed B, can still be distinctively divided as before. Then the second condition will not hold any more because the number of outliers becoming larger leads to $a \to 0$. Comparatively, adding more qualified outliers will not influence the effective use of the MF-estimator. The following two conditions are necessary for a good use of the M-, L-, and R-estimators and are also sufficient for a good use of the MF-estimator:

1. A large enough number of qualified nonoutliers to permit a satisfactory nonlinear least-squares fitting;

2. Qualified outliers fairly far from nonoutliers based on the residual consideration.

Unlike the M-estimator, which uses a single probability density function to model all possible unknown outlier probability densities, the MF-estimator assumes only that all $h(r_k)$ are equally valued. It does not assume which value they should take nor what the whole shape of $h(r_k)$ should be. From residual-based considerations, the assumption of all $h(r_k)$ being equally valued is reasonable, especially when only a limited sample size is allowable, as is the case in many computer vision problems.

From a practical point of view, however, it is hard to guarantee that observed or processed data have a large enough distance between outliers and nonoutliers. Therefore in order for the proposed MF-estimator to be practically useful, this problem must be solved. In other words, the second condition stated above must be removed from a practical use of the proposed MF-estimator. This and other issues will be discussed in Section 15.13.

15.7.4 MF-Estimator

To solve the model-fitting problem, that is, Eq. (15.151), what we really need is to maximize Q w.r.t. $\theta, \sigma, \epsilon, \delta$ subject to $\#G = (1 - \epsilon)N$. As we will see, it is quite simple to include condition equation (15.143), that is, $\#G \geq L$, in the MF-estimator. As for condition equation (15.150), that is, $C(G, B) = \min$, it will be reached through only a few trials for the initial value of δ, because a whole interval exists instead of a single value for a good, workable choice of δ, as we stated before.

To maximize Q subject to $\#G = (1 - \epsilon)N$, we basically follow the gradient-ascent rule. The partial derivatives of Q w.r.t. $\theta_1, \cdots, \theta_p, \sigma, \epsilon, \delta$ are as follows:

$$\frac{\partial Q}{\partial \theta_i} = \frac{1}{\sigma^2} \sum_k \lambda_k r_k' \frac{\partial f_k}{\partial \theta_i}, \quad i = 1, \cdots, p \tag{15.152}$$

$$\frac{\partial Q}{\partial \sigma} = \frac{1}{\sigma} \left\{ \frac{1}{\sigma^2} \sum_k \|r_k\|^2 \lambda_k - m \sum_k \lambda_k \right\} \tag{15.153}$$

$$\frac{\partial Q}{\partial \epsilon} = \frac{N - \lambda}{\epsilon} - \frac{\lambda}{1 - \epsilon}, \quad \text{where} \quad \lambda = \sum_k \lambda_k \tag{15.154}$$

$$\frac{\partial Q}{\partial \delta} = \frac{N - \lambda}{\delta} \tag{15.155}$$

We then define the location, scale, and distribution steps to determine $\theta_1, \cdots, \theta_p, \sigma, \epsilon, \delta$, respectively, from the nth to the $(n+1)$th iterative step, as follows:

Location Step: Because of Eq. (15.152), to assure $\sum_i \frac{\partial Q}{\partial \theta_i} \Delta \theta_i \geq 0$ it is necessary

to have $\sum_k \lambda_k r_k' \left(\sum_i \frac{\partial f_k}{\partial \theta_i} \Delta \theta_i \right) \geq 0$, which is satisfied if

$$\lambda_k r_k = (\frac{\partial f_k}{\partial \theta_1}, \cdots, \frac{\partial f_k}{\partial \theta_p}) \Delta \theta, \quad k = 1, \cdots, N \tag{15.156}$$

Let

$$X = \begin{pmatrix} \frac{\partial f_1}{\partial \theta_1}, & \cdots, & \frac{\partial f_1}{\partial \theta_p} \\ \vdots & \ddots & \vdots \\ \frac{\partial f_N}{\partial \theta_1}, & \cdots, & \frac{\partial f_N}{\partial \theta_p} \end{pmatrix} \tag{15.157}$$

If the singular-value decomposition of X is

$$X = UWV' \tag{15.158}$$

(where $U'U = I_p$, $V'V = VV' = I_p$, and $W = \text{diag}[\sigma_1^2, \cdots, \sigma_p^2]$), then the least-squares solution to (19) is given by

$$\Delta \theta = VW^{-1}U'(\lambda_1 r_1', \cdots, \lambda_N r_N')' \tag{15.159}$$

Scale Step: Because of Eq. (15.153), we simply define the value of σ at the $(n+1)$th iterative step as follows:

$$\sigma^2 = \frac{1}{m} \sum_k \frac{\lambda_k}{\lambda} \|r_k\|^2 \tag{15.160}$$

Distribution Step: First, we consider how to change meaningfully the outlier proportion, ϵ. This is quite simple. The change of ϵ from the nth step to the $(n+1)$th step should be made so that $\frac{\partial Q}{\partial \epsilon} \geq 0$ and $0 < \epsilon + \Delta\epsilon < 1$. We need only set

$$
\Delta\epsilon = \begin{cases}
-\alpha_1\epsilon & \text{if } \dfrac{1}{N}\left(\dfrac{N-\lambda}{\epsilon} - \dfrac{\lambda}{1-\epsilon}\right) < -\alpha_2 \\
\alpha_1(1-\epsilon) & \text{if } \dfrac{1}{N}\left(\dfrac{N-\lambda}{\epsilon} - \dfrac{\lambda}{1-\epsilon}\right) > \alpha_2 \\
0 & \text{otherwise}
\end{cases}
\tag{15.161}
$$

because of Eq. (15.154), where $0 < \alpha_1, \alpha_2 < 1$.

To change δ meaningfully, however, we must not only take care of $\frac{\partial Q}{\partial \epsilon}\Delta\epsilon + \frac{\partial Q}{\partial \delta}\Delta\delta \geq 0$ and $\delta + \Delta\delta > 0$ but also reinforce the heuristic condition, $\#G = (1-\epsilon)N$ (see Eq. 15.151). As pointed out before, for each set of parameter trial values, a partition $\{G, B\}$ defined by Eq. (15.141) can be calculated. Suppose $\#G > (1-\epsilon)N$ with the current trial value of ϵ and calculated G. That means that the outliers are estimated lower than needed. To correct the imbalance, we need only increase $\epsilon\delta$. If the change of ϵ has already been determined as positive, we leave δ alone; otherwise we need to increase δ. Similarly, when $\#G < (1-\epsilon)N$ and $\Delta\epsilon > 0$, we need to decrease δ. Combining these ideas with the requirements that $\frac{\partial Q}{\partial \epsilon}\Delta\epsilon + \frac{\partial Q}{\partial \delta}\Delta\delta \geq 0$ and $\delta + \Delta\delta > 0$, we set

$$
\Delta\delta = \begin{cases}
\tau\delta & \text{if } \dfrac{\#G}{(1-\epsilon)N} > 1 + \alpha_3 \text{ and } \Delta\epsilon < 0 \\
-\tau\delta & \text{if } \dfrac{\#G}{(1-\epsilon)N} < 1 - \alpha_3 \text{ and } \Delta\epsilon > 0
\end{cases}
\tag{15.162}
$$

with

$$
\tau = \min\left\{\alpha_4, \frac{\alpha_5}{N-\lambda}\left(\frac{N-\lambda}{\epsilon} - \frac{\lambda}{1-\epsilon}\right)\Delta\epsilon\right\}, \quad 0 < \alpha_3, \alpha_4, \alpha_5 < 1
$$

All $\lambda_k, \lambda, r_k, X$, and so on, in Eqs. (15.157) to (15.162) are calculated by using the trial values for $\theta, \sigma, \epsilon, \delta$ at the nth iterative step.

Up to now, three of the five heuristic conditions have been included in the algorithmic development. These are Eqs. (15.139) to (15.142) and Eq. (15.144). It is time to include condition equations (15.143) and (15.150) in our algorithmic development. We can simply summarize the MF-estimator as follows:

Step 1: Choose an initial approximation: $\theta^0, \sigma^0, \epsilon^0$, and δ^0.

Step 2: Iterate. Given the estimation $\theta^n, \sigma^n, \epsilon^n$, and δ^n at the nth step, compute the change $\Delta\theta, \Delta\sigma, \Delta\epsilon$, and $\Delta\delta$ by using Eqs. (15.159) to (15.163).

Step 3: After convergence, compute the corresponding partition and cost, that is, $C(G, B)$.

Step 4: If $\#G \geq L$ and $C(G, B) < \zeta$, where ζ is a prespecified small-cost bound, then go to step 7.

Step 5: If $\#G < L$, then the appropriate initial value of δ that leads to the most preferable partition $\{G, B\}$ with $\#G \geq L$ and $C(G, B) = \min$ cannot go beyond the interval $(0, \delta^0)$ and thus will be found by applying the golden bisection technique to the interval $(0, \delta^0)$ while following steps 2 and 3. Go to step 7.

Step 6: $\delta^0 \longleftarrow 2\delta^0$, and go back to step 2.

Step 7: Reestimate the parameters $\theta_1, \cdots, \theta_p$ by iteratively applying the nonlinear least-squares method to the good observation data, that is, G, to increase the estimation precision. Stop.

15.8   Optic Flow–Instantaneous Rigid-Motion Segmentation and Estimation

In this section we first formulate the optic flow–single rigid-motion estimation problem such that it becomes a general regression problem. Then we turn the optic flow–multiple rigid-motion segmentation and estimation problem into a number of successive optic flow–single rigid-motion estimation problems.

15.8.1   Single Rigid Motion

Suppose a rigid body is in motion in the half-space $z < 0$. Let $p(t)$ be the position vector of an object point at the time t, $p(t) = [x(t), y(t), z(t)]'$. Let $[X(t), Y(t)]$ denote the central projective coordinates of p(t) onto the image plane $z = f$:

$$X(t) = f \frac{x(t)}{z(t)}$$

$$Y(t) = f \frac{y(t)}{z(t)}$$

$$p(t) = \frac{z(t)}{f} [X(t), Y(t), f]' \tag{15.163}$$

Let $[u(t), v(t)]$ denote the noisy observation of the instantaneous velocity of the moving image point $[X(t), Y(t)]$. We call each element $\{[X(t), Y(t)], [u(t), v(t)]\}$ a noisy optic flow image point. Suppose we are given N optic flow image points $\{[X_k(t), Y_k(t)], [u_k(t), v_k(t)]\}$, $k = 1, \cdots, N$. The instantaneous representation of the rigid motion is described by

$$\dot{p}(t) = \omega(t) \times p(t) + k(t) \tag{15.164}$$

where $\omega(t) = [\omega_1(t), \omega_2(t), \omega_3(t)]'$ represents the instantaneous rotational angular velocity of the rigid motion and $k(t) = [k_1(t), k_2(t), k_3(t)]'$ the instantaneous translational velocity.

Differentiating Eq. (15.163), could express $\dot{p}$ in Eq. (15.164) as

$$\dot{p} = \frac{\dot{z}}{f}(X, Y, f)' + \frac{z}{f}(u, v, 0)' \tag{15.165}$$

where for simplicity the time variable t has been omitted. When we combine

Eqs. (15.164) and (15.165), it follows that

$$\frac{\dot{z}}{f}(X,Y,f)' + \frac{z}{f}(u,v,0)' = \frac{z}{f}\omega \times (X,Y,f)' + k \qquad (15.166)$$

from which we could eliminate $\dot{z}$ and express the ideal instantaneous velocity of each moving image point (X_k, Y_k) as

$$\mu_k = \left(\begin{vmatrix} \omega_2 & \omega_3 \\ Y_k & f \end{vmatrix}, \begin{vmatrix} \omega_3 & \omega_1 \\ f & X_k \end{vmatrix}\right)'$$
$$- \frac{1}{f}\begin{vmatrix} \omega_1 & \omega_2 \\ X_k & Y_k \end{vmatrix}(X_k, Y_k)' + \frac{1}{z_k}[f(k_1, k_2)' - k_3(X_k, Y_k)'] \qquad (15.167)$$

An observation $(u_k, v_k)'$ is a noisy instance of μ_k, as stated before. Let $r_k = (u_k, v_k)' - \mu_k$ denote the residual between them. The optic flow–single instantaneous-motion estimation problem is to infer ω, k, and $z_1, \cdots, z_N$ from $[(X_1, Y_1), (u_1, v_1)], \cdots, [(X_N, Y_N), (u_N, v_N)]$.

Assume each single optic flow velocity observation (u_k, v_k) at (X_k, Y_k) with probability $1 - \epsilon$ is a "good" one, that is, not an outlier; with probability ϵ, a "bad" one, that is, an outlier, where $0 \leq \epsilon \leq 1$. In the former case the residual r_k is Gaussian distributed with zero mean and an unknown covariance matrix $\sigma^2 I_2$. In the latter case the residual r_k obeys an unknown foreign distribution. Further, assume that the r_k's are independent and identically distributed. Thus the optic flow–single-motion estimation problem can be stated in terms of finding the parameter set $\omega, k, z_1, \cdots, z_N, \sigma, \epsilon, \delta$ by solving the model-fitting problem (15.151) with $m = 2$ and $L = 4$. The latter condition, that is, $L = 4$, is experimentally determined and seems to be much weaker than the eight-point requirement for the linear optic flow–motion algorithm (see Zhuang, Huang, and Haralick, 1988b). The same requirement was observed earlier for the linear two-view–motion algorithm (see Tsai and Huang, 1984).

We can also pose the problem in the following way: How many good point correspondences are really needed in order to apply the nonlinear least-squares estimator? Let L denote that number. Then there are $L + 6$ unknowns, that is, three rotational angles, three translational components, and L depths, and only $2L$ observed coordinates. As seen from Eq. (15.167), we can determine only three translational components and L depths up to a scale factor. Normalizing translational components and depths by dividing each by the total sum of depths, we reduce the total unknowns from $L+6$ to $L+5$. If $L = 5$, the number of observed coordinates is equal to the number of unknowns. Since we determined experimentally that $L = 4$ is the smallest L which gives a solution, there must be some reason behind it. One explanation, not necessarily satisfactory is that besides the $2L$ observed coordinates, there are L inequality constraints: Each normalized depth should be positive.

15.8.2 Multiple Rigid Motions

Suppose multiple rigid bodies are in motion in the half-space $z < 0$. Suppose that N observed noisy optic flow image points $[(X_1, Y_1), (u_1, v_1)], \cdots, [(X_N, Y_N), (u_N, v_N)]$

come from those multiple instantaneous rigid motions. The optic flow–multiple rigid-motion estimation problem is to infer multiple rigid-motion parameters and depths from those observations. As we explained earlier, we can handle the problem if the proposed MF-estimator turns out to be highly robust.

15.9   Experimental Protocol

To prove that the MF-estimator is highly robust, we can conduct experiments with the simplest location estimation, simulated optic flow–single rigid-motion estimation, and simulated optic flow–multiple rigid-motion segmentation and estimation. We choose the simplest location estimation by using both the M-estimator and the MF-estimator in order to compare the performance of the proposed MF-estimator with the M-estimator. We choose simulated optic flow–single rigid-motion estimation and simulated optic flow–multiple rigid-motion segmentation and estimation in order to show the strength of the proposed MF-estimator in solving significant applications under the condition that the outlier and nonoutlier residuals can be distinctively divided. As already pointed out, this condition must be removed in order for the MF-estimator to be practically useful. This constitutes our further research (see Section 15.13).

15.9.1   Simplest Location Estimation

Suppose there are N one-dimensional observations, $x_1, \cdots, x_N$, on a location parameter θ. The observations are independent, with common density function $f(x - \theta)$, which is modeled by

$$f(x) = \frac{1 - \epsilon}{\sqrt{2\pi}} \exp(-\frac{|x|^2}{2}) + \epsilon h(x)$$

where $h(x)$ is unknown. To estimate the location θ, we use both the M-estimator and the MF-estimator.

The location parameter θ is randomly selected. The $N - M$ random numbers from $N(0, 1)$ are generated and added to θ to form $N - M$ good observations about θ. The remaining M observations are outliers. They are randomly generated, for instance, from a uniform distribution, under the condition that each of them stays a certain distance from θ. In our experiments the outlier residual magnitudes are larger than 2.5. All nonoutlier residual magnitudes do not exceed 2 with the probability 95%. Thus the outlier and nonoutlier residuals are "distinctively" divided.

The experimental results with hundreds of thousands of trials convincingly show that the MF-estimator performs much better than the M-estimator. For a sample size $N = 10, 20, 30$, the largest possible outlier proportion that can ever be reached is $\epsilon = 0.5, 0.5, 0.5$, respectively, by the M-estimator, or $\epsilon = 0.8, 0.9, 0.9\dot{3}$, respectively, by the MF-estimator. For the MF-estimator, only two good points or observations are needed to get an accurate location estimate, no matter how many outliers occur. The superior performance shown by the MF-estimator in the location estimation case will also be proved in the optic flow–rigid-motion estimation case.

15.9.2 Optic Flow–Rigid-Motion Segmentation and Estimation

To measure the performance of the MF-estimator for optic flow–single (or multiple) rigid motion (or motions), several hundred controlled experiments were conducted.

A Single Rigid-Motion Case

Data Set Generation: A set of three-dimensional points, $p_k = (x_k, y_k, z_k)'$, $k = 1, \cdots, N$, is generated within a box defined by $x_k, y_k \in [20, 30], z_k \in [15, 55]$ at the time t. That is, each x- and y-component is uniformly distributed between 20 and 30; each z-component, between 15 and 55. These three-dimensional points are perspectively projected on the image plane to generate the image points (X_k, Y_k), $k = 1, \cdots, N$, which are assumed known in our experiments. Next both the three instantaneous rotational angles and the instantaneous translational components are randomly selected—for instance, from a uniform distribution. With these parameters the three-dimensional velocities $\dot{p}_k$, $k = 1, \cdots, N$, are formed and then perspectively projected on the image plane $z = f$ to compose the optic flow μ_k, $k = 1, \cdots, N$, which represents the instantaneous velocity of the moving image point (X_k, Y_k), $k = 1, \cdots, N$. At this stage independent identically distributed Gaussian noise $N(0, \sigma^2)$ is added to all components of some optic flow velocities μ_k. The formed residuals are less than 2 with the probability 95%. To test the robustness of the proposed MF-estimator, we randomly change the remaining optic flow velocities such that the formed residuals are larger than 2.5σ and thus become outliers. In this manner we finally obtain the noisy optic flow velocities (u_k, v_k), $k = 1, \cdots, N$. Given the two-dimensional optic flow image points $[(X_k, Y_k), (u_k, v_k)]$: $k = 1, \cdots, N$, the MF-estimator is applied to find the three rotational angles, the translational vector, and the depths of some three-dimensional points p_k that produce nonoutlier optic flow velocities.

There are three parameters to be controlled: the sample size N, the outlier number M, and the standard deviation σ of the Gaussian noise. In the experimental results we use $\epsilon = M/N$ and SNR instead of M and σ, where $\text{SNR} = 20 \log \left\{ \max \sqrt{(u_k^2 + v_k^2)} / \sigma \right\}$ db. The initial approximation of the three rotational angles is selected randomly within $\pm 15°$ of the true angles. The initial approximation of the three translational components is selected randomly within ± 10 of the true translational components. The initial approximation of the depths is selected randomly within ± 5 of the true depths. The MF-estimator requires the solution of a nonlinear equation. Therefore, like the solution to most non-linear equations it requires a good initial guess in order to work.

Results

For each parameter setting $(N, \epsilon, \text{SNR})$, hundreds of experiments must be conducted to get a reasonable evaluation of the performance of the proposed MF-estimator. Three different sets of experiments follow:

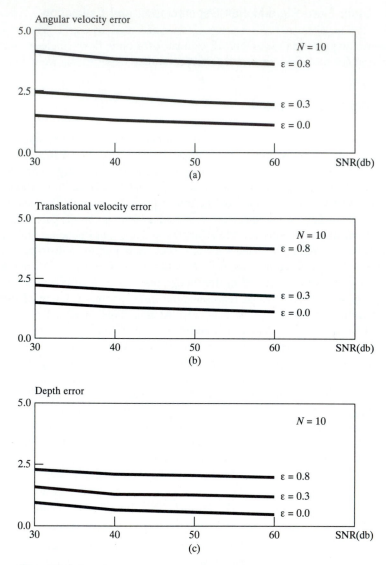

Figure 15.2 Experimental results of Op1. The average rotational angular velocity error, the average translational velocity error, and the average depth error are depicted as curves in (a), (b), and (c), respectively.

Op1: Set $N = 10$. Estimate the average rotational angular velocity error, the average translational velocity error, and the average depth error against SNR (40db to 70db in 10db step) for different $\epsilon(0.0$ to 0.6 in steps of 0.1).

Op2: Set $\epsilon = 0.3$. Estimate the average rotational angular velocity error, the average translational velocity error, and the average depth error against SNR (40db to 70db in 10db step) for different $N(6$ to 20 in steps of 5).

Op3: Set SNR $= 60$db. Estimate the average rotational angular velocity error, the

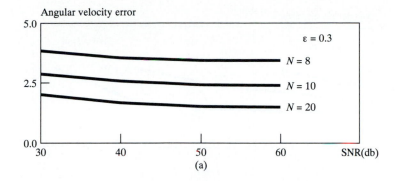

(a)

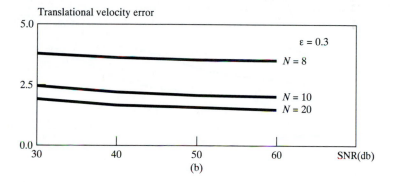

(b)

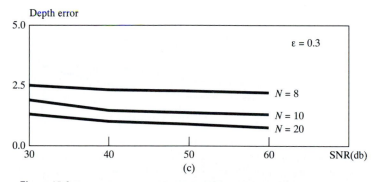

(c)

Figure 15.3 Experimental results of Op2. The average rotational angular velocity error, the average translational velocity error, and the average depth error are depicted as curves in (a), (b), and (c), respectively.

average translational velocity error, and the average depth error against ϵ (0.0 to 0.8 in steps of 0.1) for different N(5 to 20 in steps of 5).

The results shown in Figs. 15.2 to 15.4 are promising in that the proposed MF-estimator almost reaches the highest possible robustness. Only four good optic flow image points are needed to obtain an accurate estimate of motion and depth

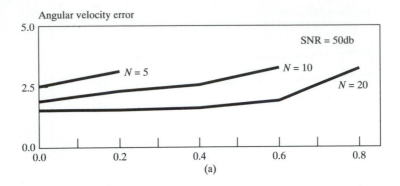

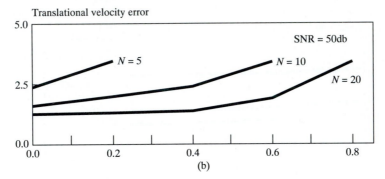

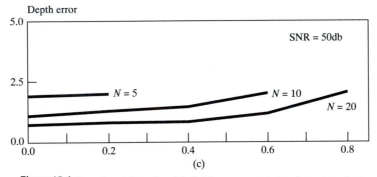

Figure 15.4 Experimental results of Op3. The average rotational angular velocity error, the average translational velocity error, and the average depth error are depicted as curves in (a), (b), and (c), respectively.

parameters, no matter how many outliers occur. (In current experiments the sample size $N \leq 20$.)

Besides the figures, we also provide Tables 15.3 to 15.6, which list four sets of data consisting of the optic flow image points, with true, initial, and estimated instantaneous motion parameters derived by using the MF algorithm of Sections. 15.7.2 and 15.7.4.

Table 15.3   Example of optic flow–single rigid-motion estimation; sample $= 10$, $\varepsilon = 0.00$,
SNR $= 50.0$.

	True Value	Initial Value	Error	Estimated Value	Error
ω_1	15.0429	0.0925	-14.9504	13.0896	-1.9444
ω_2	20.3644	11.5670	-8.7974	22.3527	1.9883
ω_3	25.5522	35.8781	10.3260	22.6679	-2.8843
κ_1	44.4788	34.8713	-9.6075	42.6005	-1.8783
κ_2	39.8518	38.9193	-0.9379	40.1535	0.3016
κ_3	31.4791	34.2375	2.7584	33.0393	1.5602
σ	0.0008	0.7498	0.7940	0.0156	0.0148
ε	0.0000	0.3000	0.3000	0.0664	0.0664

Mean angular error $= 2.272$, mean translational error $= 1.247$, mean depth error
$= 1.030$.

Optic Flow Data

Index	(X,Y)	True Optic Flow	Noisy Optic Flow
1	(1.2662, 1.7580)	(-0.1703, -0.6328)	(-0.1702, -0.6327)
2	(0.8496, 0.9530)	(0.7150, 0.5772)	(0.7152, 0.5772)
3	(0.6941, 0.5347)	(0.7239, 0.6462)	(0.7239, 0.6475)
4	(1.1375, 1.1260)	(0.3335, 0.5484)	(0.3327, 0.5497)
5	(0.9295, 1.2583)	(0.4644, 0.1626)	(0.4653, 0.1626)
6	(1.2279, 1.3862)	(0.1683, 0.1633)	(0.1705, 0.1628)
7	(0.9583, 0.8343)	(0.6870, 0.8237)	(0.6869, 0.8232)
8	(0.5265, 0.6700)	(0.6981, 0.4070)	(0.6980, 0.4063)
9	(0.7598, 0.6901)	(0.7860, 0.7291)	(0.7863, 0.7291)
10	(1.4346, 1.1770)	(0.0851, 0.7519)	(0.0857, 0.7503)

Multiple Rigid-Motion Case

The high robustness exhibited by the MF-estimator for optic flow–single rigid motion is significant in segmenting and estimating multiple instantaneous rigid motions.
As stated before, relative to one rigid motion, the other rigid motions and outliers
can all be considered together as outliers. Thus the MF-estimator can be used to
segment and estimate the multiple instantaneous rigid motions one after another
if each instantaneous rigid motion has at least four good optic flow image points
when the sample size $N \leq 20$. In this respect the classical robust M-estimator
with low robustness is useless. Many experiments with the M-estimator indicate

Table 15.4   Second example of optic flow–single rigid-motion estimation; sample $= 10$, $\varepsilon = 0.20$, SNR $= 50.0$.

	True Value	Initial Value	Error	Estimated Value	Error
ω_1	15.0429	0.0925	-14.9504	13.4621	-1.6168
ω_2	20.3644	11.5670	-8.7974	22.3745	2.0100
ω_3	25.5522	35.8781	10.3260	24.9423	-0.6099
κ_1	44.4788	34.8713	-9.6075	44.5221	0.0432
κ_2	39.8518	38.9193	-0.9379	40.2651	0.4133
κ_3	31.4791	34.2375	2.7584	33.6314	2.1523
σ	0.0008	0.9202	0.9194	0.0464	0.0456
ε	0.0000	0.3000	0.1000	0.2014	0.0014

Mean angular error $= 1.412$, mean translational error $= 0.870$, mean depth error $= 0.695$.

Optic Flow Data

Index	(X,Y)	True Optic Flow	Noisy Optic Flow
1	(1.2662, 1.7580)	(-0.1703, -0.6328)	(-0.1702, -0.6327)
2	(0.8496, 0.9530)	(0.7150, 0.5772)	(0.7152, 0.5772)
3	(0.6941, 0.5347)	(0.7239, 0.6462)	(0.7239, 0.6475)
4	(1.1375, 1.1260)	(0.3335, 0.5484)	(0.3327, 0.5497)
5	(0.9295, 1.2583)	(0.4644, 0.1626)	(0.4653, 0.1626)
6	(1.2279, 1.3862)	(0.1683, 0.1633)	(0.1705, 0.1628)
7	(0.9583, 0.8343)	(0.6870, 0.8237)	(0.6869, 0.8232)
8	(0.5265, 0.6700)	(0.6981, 0.4070)	(0.6980, 0.4063)
9	(0.7598, 0.6901)	(0.7860, 0.7291)	(3.7134, 3.6593)
10	(1.4346, 1.1770)	(0.0851, 0.7519)	(2.3602, 3.1379)

that it can reach a robustness of only $\epsilon \leq 0.2$ or so (see Haralick et al., 1989). Figures 15.5 to 15.7 show the results of applying the proposed MF-estimator in segmenting and estimating two instantaneous rigid motions. When the sample size—that is, N—increases, the question of whether or not the minimum number of good points required by a single rigid motion estimation—that is, L—remains the same or increases needs to be studied further.

Table 15.5   Third example of optic flow–single rigid-motion estimation; sample = 10, $\varepsilon = 0.40$, SNR = 50.0.

	True Value	Initial Value	Error	Estimated Value	Error
ω_1	15.0429	0.0925	-14.9504	15.2761	0.2332
ω_2	20.3644	11.5670	-8.7974	21.7553	1.3909
ω_3	25.5522	35.8781	10.3260	10.0419	-15.5103
κ_1	44.4788	34.8713	-9.6075	37.9511	-6.5277
κ_2	39.8518	38.9193	-0.9379	48.3581	8.5063
κ_3	31.4791	34.2375	2.7584	32.5550	1.0759
σ	0.0008	1.2346	1.2338	0.0538	0.0530
ε	0.4000	0.3000	-0.1000	0.4014	0.0014

Mean angular error = 5.711, mean translational error = 5.370, mean depth error = 1.860.

Optic Flow Data

Index	(X,Y)	True Optic Flow	Noisy Optic Flow
1	(1.2662, 1.7580)	(-0.1703, -0.6328)	(-0.1702, -0.6327)
2	(0.8496, 0.9530)	(0.7150, 0.5772)	(0.7152, 0.5772)
3	(0.6941, 0.5347)	(0.7239, 0.6462)	(0.7239, 0.6475)
4	(1.1375, 1.1260)	(0.3335, 0.5484)	(0.3327, 0.5497)
5	(0.9295, 1.2583)	(0.4644, 0.1626)	(0.4653, 0.1626)
6	(1.2279, 1.3862)	(0.1683, 0.1633)	(0.1705, 0.1628)
7	(0.9583, 0.8343)	(0.6870, 0.8237)	(3.6144, 3.7539)
8	(0.5265, 0.6700)	(0.6981, 0.4070)	(2.9732, 2.7930)
9	(0.7598, 0.6901)	(0.7860, 0.7291)	(3.0366, 3.0553)
10	(1.4346, 1.1770)	(0.0851, 0.7519)	(2.8640, 3.1051)

15.10   Motion and Surface Structure from Line Correspondences

Several formulations for computing motion and surface structure from image-line correspondences currently exist (Mitiche et al., 1986; Liu and Huang, 1986; Faugeras et al., 1987; Spetsakis and Aloimonos, 1987). They are based on different constraints, involving different equations in different unknowns.

The following discussion concerns only the general rigid motion of straight-line structures. The technique we use is close to that of Liu and Huang (1986).

Table 15.6   Fourth example of optic flow–single rigid-motion estimation; sample $= 10$, $\varepsilon = 0.60$, SNR $= 50.0$.

	True Value	Initial Value	Error	Estimated Value	Error
ω_1	15.0429	0.0925	-14.9504	16.1118	1.0689
ω_2	20.3644	11.5670	-8.7974	25.3727	5.0083
ω_3	25.5522	35.8781	10.3260	22.9863	-2.5659
κ_1	44.4788	34.8695	-9.6093	37.8397	-6.6391
$\cdot\kappa_2$	39.8518	38.6904	-1.1615	38.8358	-1.0160
κ_3	31.4791	26.2979	-5.1813	29.6518	-1.8273
σ	0.0008	2.4498	2.4490	0.0559	0.0551
ε	0.6000	0.3000	-0.3000	0.6033	0.0033

Mean angular error $= 2.881$, mean translational error $= 3.161$, mean depth error $= 3.538$.

Optic Flow Data

Index	(X,Y)	True Optic Flow	Noisy Optic Flow
1	(1.2662, 1.7580)	(-0.1703, -0.6328)	(-0.1702, -0.6327)
2	(0.8496, 0.9530)	(0.7150, 0.5772)	(0.7152, 0.5772)
3	(0.6941, 0.5347)	(0.7239, 0.6462)	(0.7239, 0.6475)
4	(1.1375, 1.1260)	(0.3335, 0.5484)	(0.3327, 0.5497)
5	(0.9295, 1.2583)	(0.4644, 0.1626)	(3.3918, 3.0928)
6	(1.2279, 1.3862)	(0.1683, 0.1633)	(2.4434, 2.5493)
7	(0.9583, 0.8343)	(0.6870, 0.8237)	(2.9376, 3.1499)
8	(0.5265, 0.6700)	(0.6981, 0.4070)	(3.4769, 2.7603)
9	(0.7598, 0.6901)	(0.7860, 0.7291)	(2.9389, 2.8429)
10	(1.4346, 1.1770)	(0.0851, 0.7519)	(2.3072, 3.5143)

15.10.1 Problem Formulation

A viewing system is modeled as in Fig. 15.8. A line l in three-dimensional space has projection L on the image plane $z = f$. Line L is in known plane Π, which we refer to subsequently as the projective plane of l.

Consider a set of lines in three-dimensional space, that is, $\{l_1,\ldots,l_K\}$, which is moved onto a set of lines $\{l'_1,\ldots,l'_K\}$ by a rigid motion (R',T') at time t' or onto a set of lines $\{l''_1,\ldots,l''_K\}$ by a rigid motion (R'',T''), respectively. The projections

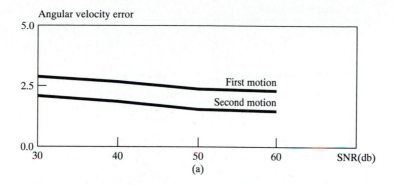

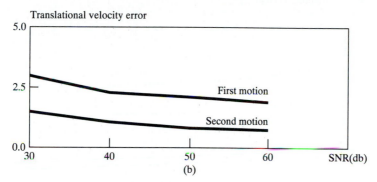

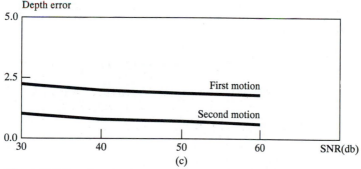

Figure 15.5 Experimental results of using five well-matched point pairs for each motion using three outliers in segmenting and estimating two rigid motions. The average rotational angular velocity error, the average translational velocity error, and the average depth error are depicted as curves in (a), (b), and (c), respectively.

of $\{l_1, \ldots, l_K\}$, $\{l_1', \ldots, l_K'\}$, or $\{l_1'', \ldots, l_K''\}$ on the image plane are $\{L_1, \ldots, L_K\}$, $\{L_1', \ldots, L_K'\}$, or $\{L_1'', \ldots, L_K''\}$, respectively. Let their respective projective planes be $\{\Pi_1, \ldots, \Pi_K\}$, $\{\Pi_1', \ldots, \Pi_K'\}$, or $\{\Pi_1'', \ldots, \Pi_K''\}$, respectively. Now, given K triples of line correspondences in three views, that is,

$$L_i \leftrightarrow L_i' \leftrightarrow L_i'', \quad i = 1, \ldots, K \tag{15.168}$$

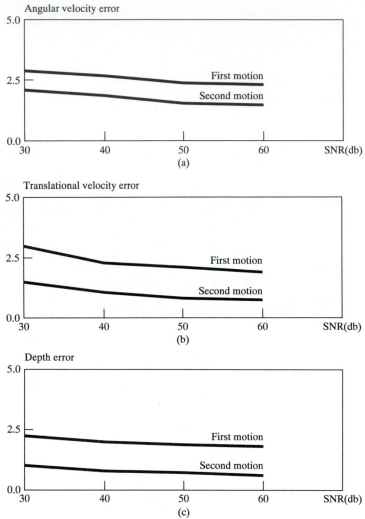

Figure 15.6 Experimental results of using five well-matched point pairs for each motion using no outliers in segmenting and estimating two rigid motions. The average rotational angular velocity error, the average translational velocity error, and the average depth error are depicted as curves in (a), (b), and (c), respectively.

the motion and surface structure from line correspondence problems is to infer R', R'', T', T'' and l_i, $i = 1, \ldots, K$.

15.10.2   Solving Rotation Matrices $\mathbf{R'}, \mathbf{R''}$ and Translations $\mathbf{T'}, \mathbf{T''}$

Let n_i, n_i', n_i'' denote the normals of the ith projective planes Π_i, Π_i', and Π_i'' at three different times, respectively. The three planes are determined by the camera lens,

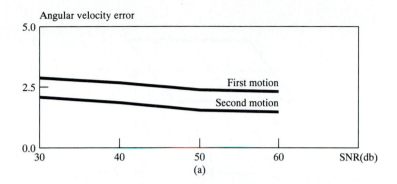

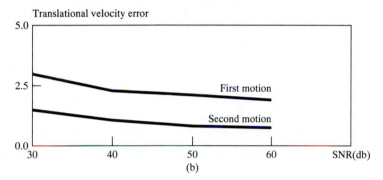

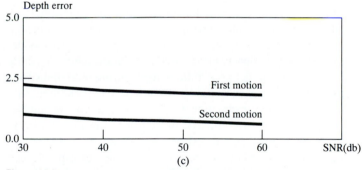

Figure 15.7 Experimental results of using five well-matched point pairs for each motion using five outliers in segmenting and estimating two rigid motions. The average rotational angular velocity error, the average translational velocity error, and the average depth error are depicted as curves in (a), (b), and (c), respectively.

that is, the origin, and by the ith image lines L_i, L'_i, and L''_i at three different times, respectively. Suppose that p represents any point on the line l_i. Then

$$n'_i p = 0$$
$$n'^{'}_i (R'p + T') = 0$$
$$n'^{''}_i (R''p + T'') = 0$$

$$(15.169)$$

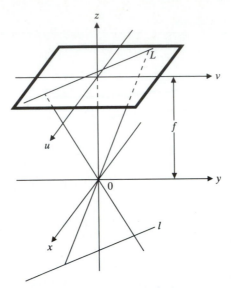

Figure 15.8 Cartesian reference system—central projection.

The first two equations define a unique three-dimensional line solution if and only if

$$\text{Rank} \begin{bmatrix} n_i^t \\ n_i^{'t} R' \end{bmatrix} = 2 \tag{15.170}$$

The reason that two views of any number of lines are not sufficient to recover structure and motion from line correspondences can be explained in terms of Eq. (15.170). For two views of any number of image-line correspondences, there exist an infinite number of choices for R', so that all those conditions in Eq. (15.170) are satisfied. Hence we can get an infinite number of interpretations for rotation R' from those image line correspondences in two views. Since condition Eq. (15.170) is independent of any choice of translation T', there exist an infinite number of choices for T' too. By Eq. (15.170) and the first two equations in Eq. (15.169), an infinite number of three-dimensional line recoveries can be determined correspondingly.

For three views of a certain number of image-line correspondences, the story is different. The three equations in (15.169) define a unique three-dimensional line solution if and only if

$$\text{Rank} \begin{bmatrix} n_i^t \\ n_i^{'t} R' \\ n_i^{''t} R'' \end{bmatrix} = 2 \tag{15.171}$$

and

$$\text{Rank} \begin{bmatrix} n_i^t & 0 \\ n_i^{'t} R' & \left(n_i^{'t} T' \right) \\ n_i^{''t} R'' & \left(n_i^{''t} T'' \right) \end{bmatrix} = 2 \tag{15.172}$$

To get some insight on how many triples of line correspondences over three views are really needed, we assume for the time being that both rotation matrices

R' and R'' are known and that condition equation (15.171) holds for each triple of line correspondences over three views. Denote three columns of the matrix in Eq. (15.171) by a_i, b_i, and c_i. Then all three vectors a_i, b_i, c_i are known. For each index i, two of three are not collinear. Denote these two by u_i, v_i. Then condition equation (15.172) leads to the following condition for T' and T'' :

$$|u_i, v_i, w_i| = 0 \qquad (15.173)$$

where

$$w_i = (0, n_i'^t T', n_i''^t T'')^t$$

Expanding Eq. (15.173), we obtain a linear homogeneous equation of T', T'' :

$$\begin{vmatrix} u_{i1} & v_{i1} \\ u_{i3} & v_{i3} \end{vmatrix} n_i'^t T' = \begin{vmatrix} u_{i1} & v_{i1} \\ u_{i2} & v_{i2} \end{vmatrix} n_i''^t T'' \qquad (15.174)$$

which implies that both T' and T'' up to a scale factor can be uniquely determined in general if five triples of line correspondences over three views are given. This result was first obtained by Liu and Huang (1986).

Now we return to the general case. Condition equations (15.171) and (15.172) lead to

$$|a_i, b_i, c_i| = 0 \qquad (15.175)$$
$$|a_i, b_i, w_i| = 0 \qquad (15.176)$$
$$|b_i, c_i, w_i| = 0 \qquad (15.177)$$
$$|a_i, c_i, w_i| = 0 \qquad (15.178)$$

The first equation and one of the last three constitute two independent nonlinear equations for twelve unknowns: θ', ϕ', ψ', for R'; θ'', ϕ'', ψ'', for R''; three components for T'; and another three components for T''. At least six triples of image-line correspondences over three views are needed in order to have the same number of equations.

15.10.3 Solving Three-Dimensional Line Structure

Once rotation matrices R', R'' and translations T', T'' are solved, each three-dimensional line l_i can be determined by Eq. (15.169).

15.11   Multiple Rigid Motions from Two Perspective Views

The estimation of three-dimensional motion parameters of a rigid body or multiple rigid bodies is important in motion analysis. Its applications include scene analysis, motion prediction, robotic vision, and on-line dynamic industrial processing. Thompson (1959) developed the nonlinear equations and gave a solution procedure that determines a rotation matrix guaranteed to be orthonormal. Roach and Aggarwal (1980) developed a nonlinear algorithm and dealt with noisy data. The linear motion estimation algorithm was developed by Longuet-Higgins (1981), extended by Tsai and Huang (1984), unified by Zhuang, Huang, and Haralick (1986), and simplified by Zhuang (1989). The linear algorithm has the advantage of being simpler and faster than the nonlinear ones. Furthermore it always can find the unique solution except in a degenerate case. The linear algorithm, however, works well only when there is limited noise and no corresponding point-matching errors. An incorrect match makes a point in the first view correspond to an incorrect point in the second view. As argued in Fischler and Bolles (1981) and Haralick (1986), the estimators used by machine vision procedures must be robust. Good results of robustly estimating a single rigid motion by using the classical M-estimator are reported in Haralick et al. (1989). As shown, however, the allowable contamination—that is, the outlier proportion ϵ—is only about 0.2. With such a low robustness, it is generally impossible for the M-estimator to be used for segmenting and estimating multiple rigid motions.

The multiple-motion segmentation and estimation problem with point data arises from two views taken of the same objects, each of which can be thought of as having undergone an unknown rigid-body motion from the first view to the second view. The image-point correspondences are obtained by a matching procedure. The motion segmentation and estimation problem with corresponding point data begins with such a corresponding point data set. Its solution is a procedure which uses that data set to segment and estimate one rigid motion from the others. Segmenting and estimating multiple rigid motions constitutes a tough but essential problem in image time-sequence motion analysis.

15.11.1 Problem Statement

The basic geometry of the problem is sketched in Fig. 15.9. The origin of the coordinate system coincides with the camera lens, and a rigid body is in motion in the half-space $z < 0$. Let p be the position vector of an object point before motion, $p = (x, y, z)'$. Let u denote the central perspective projection of p onto the image plane $z = f$. Let q represent the same object point after motion and v its noisy central perspective projection. It is well known in kinematics that

$$q = Rp + T \tag{15.179}$$

where R is the 3×3 rotation matrix and T is the 3×1 translation vector. In terms of Euler angles θ, ϕ, and ψ, the three rows of R can be represented as follows:

$$r_1' = (\cos \psi \cos \theta, \quad \sin \psi \cos \theta, \quad -\sin \theta)$$
$$r_2' = (-\sin \psi \cos \phi + \cos \psi \sin \phi \sin \theta, \; \cos \psi \cos \phi + \sin \psi \sin \phi \sin \theta, \; \sin \psi \cos \theta)$$
$$r_3' = (\sin \psi \sin \phi + \cos \psi \cos \phi \sin \theta, \; -\cos \psi \sin \phi + \sin \psi \cos \phi \sin \theta, \; \cos \phi \cos \theta)$$

Suppose we are given N corresponding image-point pairs: u_k, v_k; $k = 1, \cdots, N$. Let μ_k denote the ideal matching point of u_k on the image plane, namely,

$$\mu_k = \frac{\frac{z_k}{f}(r_1, r_2)'(u_k', f)' + (t_1, t_2)'}{\frac{z_k}{f} r_3'(u_k', f)' + t_3} \tag{15.180}$$

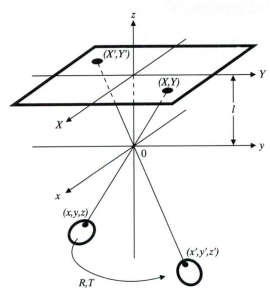

Figure 15.9 Imaging geometry for two-view–motion.

An observation v_k is a noisy instance of μ_k, as stated before. Let $r_k = v_k - \mu_k$ denote the residual between them. The image-point correspondences–single rigid-motion estimation problem is to infer R, T, and $z_1, \cdots, z_N$ from $u_1, \cdots, u_N$ and $v_1, \cdots, v_N$.

Assume that each single observation v_k with probability $1 - \epsilon$ is a "good" one, that is, not an outlier, and with probability ϵ a "bad" one, that is, an outlier, where $0 \le \epsilon \le 1$. In the former case the residual r_k is Gaussian distributed with zero mean, and an unknown covariance matrix $\sigma^2 I_2$, in the latter the residual r_k obeys an unknown foreign distribution. Evidently we could say that the r_k's are independent and identically distributed.

Now the image-point correspondences–single rigid-motion estimation problem can be stated as finding the parameter set θ, ϕ, ψ, T, z_k's, σ, ϵ, δ by solving model-fitting equation (15.151) with $m = 2$ and $L = 4$. The latter condition, that is, $L = 4$, is experimentally determined and seems to be much weaker than the eight-point requirement for the linear two view–motion algorithm.

We can also pose the problem as follows: How many good point correspondences are needed in order to apply the nonlinear least-squares estimator? Let L denote that number. Then there are $L + 6$ unknowns—that is, three rotational angles, three translational components, and L depths—and only $2L$ observed coordinates. As seen from Eq. (15.180), we can determine only three translational components and L depths up to a scale factor. Normalizing translational components and depths by dividing each by the total sum of depths, we reduce the total unknowns from $L + 6$ to $L + 5$. If $L = 5$, the number of observed coordinates is equal to the number of unknowns. The fact that $L = 4$ produces answers experimentally needs an explanation which we at this time do not have.

15.11.2   Simulated Experiments

To measure the performance of the MF-estimator, several hundred thousand controlled experiments were conducted.

A Single Rigid-Motion Case

Data Set Generation: A set of three-dimensional points, $p_k = (x_k, y_k, z_k)'$, $k = 1, \cdots, N$, is generated within a box defined by $x_k, y_k, z_k \in [20, 30]$; that is, each component is uniformly distributed between 20 and 30. These three-dimensional points are perspectively projected on the image plane to generate the image points u_k, which are assumed known in our experiments. Next, three rotational angles are randomly selected—for instance, from a uniform distribution—and the translational components are also randomly selected—for instance, within an interval $[20, 50]$. Having these parameters, the three-dimensional points are rotated and translated in the three-dimensional space forming the three-dimensional points q_k, $k = 1, \cdots, N$, which are then perspectively projected on the image plane to form the image points μ_k. At this stage independent identically distributed Gaussian noise $N(0, \sigma^2)$ is added to all components in some of the image points μ_k. To test the robustness of the MF-estimator, the remaining image points μ_k are randomly moved in the image plane to form outliers. In this manner we finally obtain all image points v_k in the second view. Given the two-dimensional image points u_k, $k = 1, \cdots, N$, and the corresponding two-dimensional image points v_k, $k = 1, \cdots, N$, the MF-estimator is applied to find the three rotational angles, the normalized translation vector, and the normalized depths of some three-dimensional points p_k that produce nonoutlier image points v_k in the second view under the rigid motion.

There are three parameters to be controlled: the sample size N, the outlier number M, and the standard deviation σ of the Gaussian noise. In the experimental results we use the proportion of outliers ϵ and the signal-to-noise ratio SNR instead of M and σ, where $\epsilon = M/N$ and $\text{SNR} = 20\log(\max \|v_k\|/\sigma)db$. The initial approximation of the three rotational angles is selected randomly within $\pm15°$ of the true angles. The initial approximation of the three translational components is selected randomly within ±10 of the true translational components. The initial approximation of the depths is selected randomly within ±5 of the true depths. Since the MF-estimator is a nonlinear approach it requires a good initial guess in order to work. The step sizes chosen for α_i are $\alpha_1 = 0.01$, $\alpha_2 = 0.01$, $\alpha_3 = 0.1$, $\alpha_4 = 0.05$, $\alpha_5 = 0.05$, and $\alpha_6 = 0.5$.

Results

For each parameter setting $(N, \epsilon, \text{SNR})$, hundreds of experiments must be conducted to get a reasonable evaluation of the performance of the MF-estimator. Three different sets of experiments follow:

Ex1: Set $N = 10$. Estimate the average three-rotational-angle error, the average three-translational-component error and the average depth error against SNR (40db to 70db in 10db step) for different ϵ (0.0 to 0.6 in steps of 0.1).

Ex2: Set $\epsilon = 0.2$. Estimate the average three-rotational-angle error, the average three-translational-component error, and the average depth error against SNR (40db to 70db in 10db step) for different N (5 to 30 in steps of 5).

Ex3: Set SNR $= 60$db. Estimate the average three-rotational-angle error, the average three-translational-component error, and the average depth error against ϵ (0.0 to 0.8 in steps of 0.1) for different N (5 to 30 in steps of 5).

The results shown in Figs. 15.10 to 15.12 convincingly prove that the MF-estimator almost reaches the highest possible robustness. Only 4 or occasionally 5 well-matched image-point pairs are needed to get an accurate estimate of motion and depth parameters, no matter how many outliers occur. (In our experiments the sample size $N \leq 30$.) Comparatively, 8, 16, or 24 well-matched image-point pairs are needed in the M-estimator when $N = 10, 20, 30$, respectively, as reported in Haralick et al. (1989).

Tables 15.7 to 15.10 list four sets of data consisting of corresponding image-point pairs, with true, initial, and estimated motion parameters derived by using the MF algorithm of Sections 15.7.2 and 15.7.4.

Multiple Rigid-Motion Case

The high robustness exhibited by the MF-estimator is significant in segmenting and estimating multiple rigid motions. Relative to one rigid motion, the other rigid motions and outliers can all be considered together as outliers. Thus the MF-estimator can be used to segment and estimate the multiple rigid motions one after another if each rigid motion owns four or five well-matched point pairs when the sample size $N \leq 30$. In this respect the M-estimator with low robustness is useless. Figure 15.13 shows the results of applying the MF-estimator in segmenting and estimating two rigid motions. When the sample size increases, the question of whether or not the minimum number of good image-point correspondences required by a single rigid-motion estimation remains the same or increases needs to be studied further.

15.12   Rigid Motion from Three Orthographic Views

Estimating motion and surface structure from multiple views (perspective projections or orthographic projections) is of course important to the field of computer vision. We have shown how to compute multiple rigid motions from two perspective views by using the MF-estimator. Both translation and depths can be determined only up to a scale factor. When the object or objects are relatively far from the camera, those normalized translational components and depths have little use in determining the true translation and depths. The errors in translation and depths will surely be magnified inappropriately by scaling. Huang (1989) emphasizes that in this situation it is better to use orthographic views as an approximation to the imaging processes.

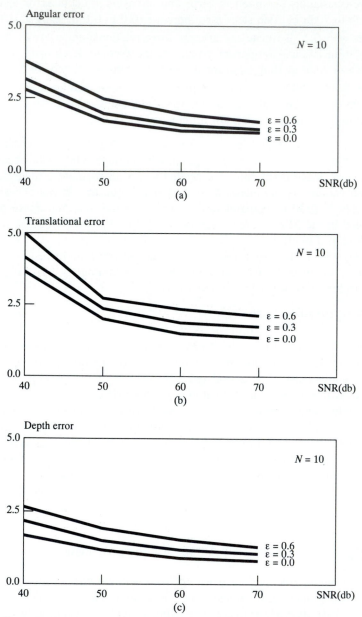

Figure 15.10 Experimental results of E1. The average rotational angular velocity error, the average translational velocity error, and the average depth error are depicted as curves in (a), (b), and (c), respectively.

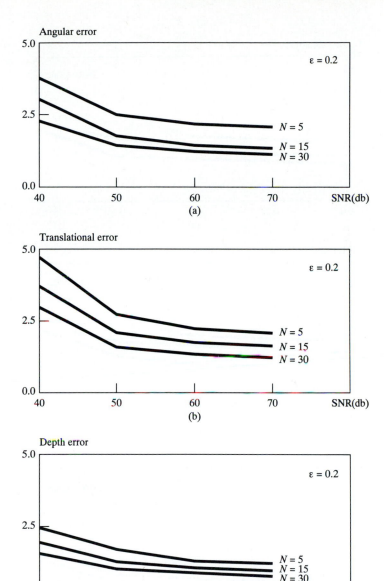

Figure 15.11 Experimental results of E2.

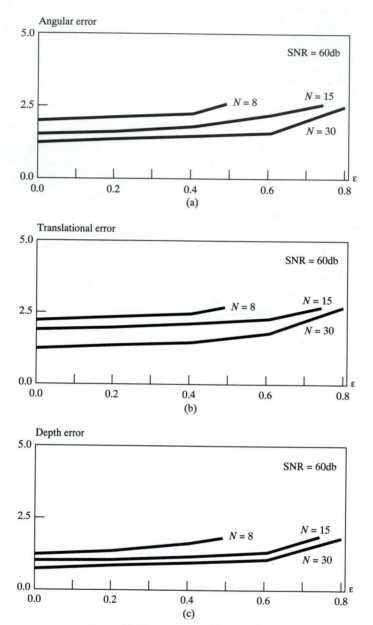

Figure 15.12 Experimental results of E3.

Table 15.7   Example of two-view–single rigid-motion estimation; sample $= 10$, $\varepsilon = 0.00$, SNR $= 50.0$.

	True Value	Initial Value	Error	Estimated Value	Error
θ	32.9484	18.9797	-13.9687	33.8716	0.8693
ϕ	38.5453	32.4590	-6.0863	39.0121	0.4668
ψ	38.1637	27.3706	-10.7931	41.2477	3.0840
t_1	20.1650	10.1873	-9.9776	21.0536	0.8887
t_2	40.6195	33.4134	-7.2060	42.2353	1.6158
t_3	47.4349	46.6808	-0.7541	46.7992	-0.6357
σ	0.0031	0.0489	0.0458	0.0030	-0.0001
ε	0.0000	0.3000	0.3000	0.0664	0.0664

Mean angular error $= 1.473$, mean translational error $= 1.047$, mean depth error $= 1.759$.

Two-View Data

Index	True First View	True Second View	Noisy Second View
1	(0.9912, 1.1367)	(0.4907, 0.9411)	(0.4910, 0.9414)
2	(1.1552, 1.1795)	(0.5166, 0.9152)	(0.5174, 0.9152)
3	(1.1676, 1.1169)	(0.5041, 0.8923)	(0.5041, 0.8972)
4	(0.8979, 0.8733)	(0.4366, 0.8930)	(0.4335, 0.8981)
5	(0.8393, 0.7826)	(0.4098, 0.8767)	(0.4133, 0.8769)
6	(0.8156, 1.0327)	(0.4628, 0.9807)	(0.4713, 0.9789)
7	(0.9647, 1.2215)	(0.5117, 0.9852)	(0.5114, 0.9835)
8	(1.0274, 1.3137)	(0.5264, 0.9851)	(0.5259, 0.9824)
9	(1.0324, 0.9241)	(0.4623, 0.8706)	(0.4635, 0.8704)
10	(1.1692, 1.1067)	(0.5080, 0.8913)	(0.5104, 0.8852)

Ullman (1979) showed that for the orthographic case four-point correspondences over three views are sufficient to determine the motion and structure of the four-point rigid configuration. However, the conditions of uniqueness and convergence in his nonlinear method are not clear. Aloimonos and Brown (1986), Zhuang, Huang, and Haralick (1988a), and Huang and Lee (1989) have reexamined the problem. Zhuang, Huang, and Haralick (1988a) showed a simple algorithm and gave the necessary and sufficient conditions for a unique solution.

Here we show a robust version of the algorithm that, as a viable and practical means for computing structure and motion, may have potentially wide application,

Table 15.8   Second example of two-view–single rigid-motion estimation; sample $=10$, $\varepsilon = 0.20$, SNR $= 50.0$.

	True Value	Initial Value	Error	Estimated Value	Error
θ	32.9484	18.9797	-13.9687	33.7409	0.7926
ϕ	38.5453	32.4590	-6.0863	38.6155	0.0702
ψ	38.1637	27.3706	-10.7931	39.3268	1.1631
t_1	20.1650	10.1873	-9.9776	20.8243	0.6593
t_2	40.6195	33.4134	-7.2060	41.1412	0.5218
t_3	47.4349	46.6808	-0.7541	46.8536	-0.5814
σ	0.0031	0.0695	0.0664	0.0029	-0.0002
ε	0.2000	0.3000	0.1000	0.2007	0.0007

Mean angular error $= 0.675$, mean translational error $= 0.587$, mean depth error $= 1.515$.

Two-View Data

Index	True First View	True Second View	Noisy Second View
1	(0.9912, 1.1367)	(0.4907, 0.9411)	(0.4910, 0.9414)
2	(1.1552, 1.1795)	(0.5166, 0.9152)	(0.5174, 0.9152)
3	(1.1676, 1.1169)	(0.5041, 0.8923)	(0.5041, 0.8972)
4	(0.8979, 0.8733)	(0.4366, 0.8930)	(0.4335, 0.8981)
5	(0.8393, 0.7826)	(0.4098, 0.8767)	(0.4133, 0.8769)
6	(0.8156, 1.0327)	(0.4628, 0.9807)	(0.4713, 0.9789)
7	(0.9647, 1.2215)	(0.5117, 0.9852)	(0.5114, 0.9835)
8	(1.0274, 1.3137)	(0.5264, 0.9851)	(0.5259, 9824)
9	(1.0324, 0.9241)	(0.4623, 0.8706)	(1.8897, 2.3008)
10	(1.1692, 1.1067)	(0.5080, 0.8913)	(1.2831, 1.7773)

since in many practical situations orthographic views give a good approximation of the imaging processes.

15.12.1   Problem Formulation and Algorithm

We assume that the image plane is stationary and that three orthographic views at times t_1, t_2, and t_3, respectively, are taken of a rigid object moving in the three-dimensional object space. By processing the three views, we intend to determine the motion and structure of the three-dimensional object.

Table 15.9   Third example of two-view–single rigid-motion estimation; sample = 10, ε = 0.40, SNR = 50.0.

	True Value	Initial Value	Error	Estimated Value	Error
θ	32.9484	18.9797	-13.9687	33.3735	0.4251
ϕ	38.5453	32.4590	-6.0863	40.0652	1.5199
ψ	38.1637	27.3706	-10.7931	40.2500	2.0863
t_1	20.1650	10.1873	-9.9776	21.59388	1.4289
t_2	40.6195	33.4134	-7.2060	43.1687	2.5493
t_3	47.4349	46.6808	-0.7541	49.8716	2.4367
σ	0.0031	0.0966	0.0935	0.0033	0.0002
ε	0.4000	0.3000	-0.1000	0.3998	-0.0002

Mean angular error = 1.344, mean translational error = 2.138, mean depth error = 1.760.

Two-View Data

Index	True First View	True Second View	Noisy Second View
1	(0.9912, 1.1367)	(0.4907, 0.9411)	(0.4910, 0.9414)
2	(1.1552, 1.1795)	(0.5166, 0.9152)	(0.5174, 0.9152)
3	(1.1676, 1.1169)	(0.5041, 0.8923)	(0.5041, 0.8972)
4	(0.8979, 0.8733)	(0.4366, 0.8930)	(0.4335, 0.8981)
5	(0.8393, 0.7826)	(0.4098, 0.8767)	(0.4133, 0.8769)
6	(0.8156, 1.0327)	(0.4628, 0.9807)	(0.4713, 0.9789)
7	(0.9647, 1.2215)	(0.5117, 0.9852)	(1.9391, 2.4154)
8	(1.0274, 1.3137)	(0.5264, 0.9851)	(1.3015, 1.8712)
9	(1.0324, 0.9241)	(0.4623, 0.8706)	(1.2129, 1.6968)
10	(1.1692, 1.1067)	(0.5080, 0.8913)	(1.7869, 1.7445)

Let (x, y, z) be the object space coordinates and (X, Y) the image space coordinates. The X- and Y-axes coincide with the x- and y-axes (in particular, the origins of the x-y-z coordinate system and the X-Y coordinate system coincide). Let

(x, y, z) = object-space coordinates of a point on the rigid object at t_1,

(x', y', z') = object-space coordinates of the same point at t_2,

(x'', y'', z'') = object-space coordinates of the point at t_3,

(X, Y) = image-space coordinates of the point at t_1,

(X', Y') = image-space coordinates of the point at t_2,

(X'', Y'') = image-space coordinates of the point at t_3.

Table 15.10   Fourth example of two-view–single rigid-motion estimation; sample = 10, $\varepsilon = 0.60$, SNR = 50.0

	True Value	Initial Value	Error	Estimated Value	Error
θ	32.9484	18.9797	-13.9687	34.9244	1.9761
ϕ	38.5453	32.4590	-6.0863	35.9650	-2.5803
ψ	38.1637	27.3706	-10.7931	34.1910	-3.9727
t_1	20.1650	10.1873	-9.9776	23.70958	3.5446
t_2	40.6195	33.4134	-7.2060	43.4554	2.8359
t_3	47.4349	46.6808	-0.7541	51.3569	3.9220
σ	0.0031	0.9667	0.9636	0.0019	0.0012
ε	0.6000	0.3000	-0.3000	0.6013	0.0013

Mean angular error = 2.843, mean translational error = 3.434, mean depth error = 1.701.

Two-View Data

Index	True First View	True Second View	Noisy Second View
1	(0.9912, 1.1367)	(0.4907, 0.9411)	(0.4910, 0.9414)
2	(1.1552, 1.1795)	(0.5166, 0.9152)	(0.5174, 0.9152)
3	(1.1676, 1.1169)	(0.5041, 0.8923)	(0.5041, 0.8972)
4	(0.8979, 0.8733)	(0.4366, 0.8930)	(0.4335, 0.8981)
5	(0.8393, 0.7826)	(0.4098, 0.8767)	(1.8372, 2.3069)
6	(0.8156, 1.0327)	(0.4628, 0.9807)	(1.2378, 1.8667)
7	(0.9647, 1.2215)	(0.5117, 0.9852)	(1.2623, 1.8115)
8	(1.0274, 1.3137)	(0.5264, 0.9851)	(1.8052, 1.8384)
9	(1.0324, 0.9241)	(0.4623, 0.8706)	(1.1152, 1.4844)
10	(1.1692, 1.1067)	(0.5080, 0.8913)	(1.2301, 2.1536)

Thus

$$(x', y', z')^t = R'(x, y, z)^t + T' \tag{15.181}$$

$$(x'', y'', z'')^t = R''(x, y, z)^t + T'' \tag{15.182}$$

where R' and R'' are rotation matrices, $R' \sim \theta', \phi', \psi'$, and $R'' \sim \theta'', \phi'', \psi''$, and where T' and T'' are translation vectors. The superscript t represents transposition.

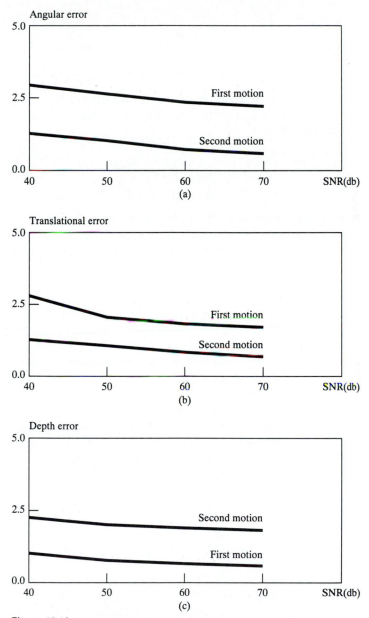

Figure 15.13 Experimental results of having five well-matched point pairs for each motion and having no outliers in segmenting and estimating two rigid motions. The average rotational angular velocity error, the average translational velocity error, and the average depth error are depicted as curves in (a), (b), and (c), respectively.

Note that with orthographic projections,

$$(X,Y) = (x,y)$$
$$(X',Y') = (x',y')$$
$$(X'',Y'') = (x'',y'') \tag{15.183}$$

Let $\overline{R}', \overline{R}''$ denote a 2×3 matrix consisting of the first two rows of R' or R'', respectively, and let $\overline{T}', \overline{T}''$ denote a 2×1 vector consisting of the first two components of T' or T'', respectively. Then

$$(x',y')^t = \overline{R}'(x,y,z)^t + \overline{T}'$$
$$(x'',y'')^t = \overline{R}''(x,y,z)^t + \overline{T}'' \tag{15.184}$$

Let v', v'' denote a noisy observation of $(x',y')^t$ and $(x'',y'')^t$, respectively. Suppose we are given N corresponding image triples: $u_k = (x_k, y_k)^t, v'_k, v''_k; \ k = 1, \cdots, N$. An observation v'_k or v''_k is a noisy instance of $u'_k = (x'_k, y'_k)^t$ or $u''_k = (x''_k, y''_k)^t$, respectively, as stated before. Let $r'_k = v'_k - u'_k$, $r''_k = v''_k - u''_k$ denote corresponding residuals. The three-orthographic-view–single rigid-motion estimation problem is to infer $\theta', \phi', \psi', \theta'', \phi'', \psi'', \overline{T}', \overline{T}''$, and $z_1, \cdots, z_N$ from $u_k, v'_k, v''_k; k = 1, \cdots, N$.

Assume that each single observation v'_k or v''_k with probability $1 - \epsilon'$ or $1 - \epsilon''$ is a "good" one, that is, not an outlier, and with probability ϵ' or ϵ'', a "bad" one, that is, an outlier, where $0 \le \epsilon', \epsilon'' \le 1$. In the former case the residual r'_k or r''_k is Gaussian distributed with zero mean and an unknown covariance matrix $\sigma'^2 I_2$ or $\sigma''^2 I_2$; in the latter case each obeys an unknown foreign distribution. Evidently we could say that both r'_k and r''_k are independent and identically distributed, and the N residuals in the second view are independent of the N residuals in the third view.

Now the three-orthographic-view–single rigid-motion estimation problem can be stated as finding the parameter set $\theta', \phi', \psi', \theta'', \phi'', \psi'', \overline{T}', \overline{T}'', z_1, \cdots, z_N, \sigma', \sigma'', \epsilon', \epsilon'', \delta', \delta''$ by solving the following model-fitting problem:

$$\max \sum_{k=1}^{N} \log \left\{ \frac{1 - \epsilon'}{(\sqrt{2\pi}\sigma')^2} \exp\left(-\frac{\|r'_k\|^2}{2\sigma'^2}\right) + \epsilon'\delta' \right\}$$
$$+ \sum_{k=1}^{N} \log \left\{ \frac{1 - \epsilon''}{(\sqrt{2\pi}\sigma'')^2} \exp\left(-\frac{\|r''_k\|^2}{2\sigma''^2}\right) + \epsilon''\delta'' \right\}$$

$$\begin{aligned}
\text{subject} \quad \text{to} \quad &\#G' \ge L' \\
&\#G' = (1 - \epsilon')N \\
&\min C(G', B') \\
&\#G'' \ge L'' \\
&\#G'' = (1 - \epsilon'')N \\
&\min C(G'', B'') \tag{15.185}
\end{aligned}$$

where G' and G'' denote the indices of those residuals r'_k and r''_k, respectively, whose probability of being a nonoutlier is larger than 0.5, and where B' and B'' denote the indices of the remaining residuals r'_k and r''_k, respectively.

Let λ'_k and λ''_k denote the probability of r'_k and r''_k being nonoutliers, respectively, namely,

$$\lambda'_k = \frac{\frac{1-\epsilon'}{2\pi\sigma'^2} \exp\left\{-\frac{\|r'_k\|^2}{2\sigma'^2}\right\}}{\frac{1-\epsilon'}{2\pi\sigma'^2} \exp\left\{-\frac{\|r'_k\|^2}{2\sigma'^2}\right\} + \epsilon'\delta'}$$

$$\lambda''_k = \frac{\frac{1-\epsilon''}{2\pi\sigma''^2} \exp\left\{-\frac{\|r''_k\|^2}{2\sigma''^2}\right\}}{\frac{1-\epsilon''}{2\pi\sigma''^2} \exp\left\{-\frac{\|r''_k\|^2}{2\sigma''^2}\right\} + \epsilon''\delta''}$$

(15.186)

Let λ' or λ'' denote the sum $\sum \lambda'_k$ or $\sum \lambda''_k$, respectively. Let

$$X' = \begin{pmatrix} \frac{\partial u'_1}{\partial \theta'} & \frac{\partial u'_1}{\partial \phi'} & \frac{\partial u'_1}{\partial \psi'} & \frac{\partial u'_1}{\partial t'_1} & \frac{\partial u'_1}{\partial t'_2} \\ \vdots & \vdots & \vdots & \vdots & \vdots \\ \frac{\partial u'_N}{\partial \theta'} & \frac{\partial u'_N}{\partial \phi'} & \frac{\partial u'_N}{\partial \psi'} & \frac{\partial u'_N}{\partial t'_1} & \frac{\partial u'_N}{\partial t'_2} \end{pmatrix}$$

$$X'' = \begin{pmatrix} \frac{\partial u''_1}{\partial \theta''} & \frac{\partial u''_1}{\partial \phi''} & \frac{\partial u''_1}{\partial \psi''} & \frac{\partial u''_1}{\partial t''_1} & \frac{\partial u''_1}{\partial t''_2} \\ \vdots & \vdots & \vdots & \vdots & \vdots \\ \frac{\partial u''_N}{\partial \theta''} & \frac{\partial u''_N}{\partial \phi''} & \frac{\partial u''_N}{\partial \psi''} & \frac{\partial u''_N}{\partial t''_1} & \frac{\partial u''_N}{\partial t''_2} \end{pmatrix}$$

Let the singular-value decomposition for X' or X'' be $X' = U'W'V''$ or $X'' = U''W''V'''$, respectively. Then, using the MF-estimator with some modification, we can work out the three-view–motion algorithm. In the following we only show how to make a meaningful change from the nth iteration to the $(n + 1)$th iteration for the location, scale, and distribution parameters.

Location Step:

$$\begin{pmatrix} \Delta\theta' \\ \Delta\phi' \\ \Delta\psi' \\ \Delta t'_1 \\ \Delta t'_2 \end{pmatrix} = V'W'^{-1}U'' \begin{pmatrix} \lambda'_1 r'_1 \\ \vdots \\ \lambda'_N r'_N \end{pmatrix}$$

$$\begin{pmatrix} \Delta\theta'' \\ \Delta\phi'' \\ \Delta\psi'' \\ \Delta t''_1 \\ \Delta t''_2 \end{pmatrix} = V''W''^{-1}U''' \begin{pmatrix} \lambda''_1 r''_1 \\ \vdots \\ \lambda''_N r''_N \end{pmatrix}$$

Let

$$Z_k = \begin{pmatrix} \frac{\partial u'_k}{\partial z_k} \\ \frac{\partial u''_k}{\partial z_k} \end{pmatrix}$$

and

$$\hat{z}_k = z_k + q\|Z_k\|^{-2}Z'_k \begin{pmatrix} \lambda'_k r'_k \\ \lambda''_k r''_k \end{pmatrix}$$

where $0 < q < 1$. We define

$$\Delta z_k = \begin{cases} z_{min} - z_k & \text{if } \hat{z}_k < z_{min} \\ z_{max} - z_k & \text{if } \hat{z}_k \geq z_{max} \\ \hat{z}_k - z_k & \text{otherwise} \end{cases}$$

where z_{min}, z_{max} are the prespecified lower and upper bounds of depths, respectively.

Scale Step:

$$\sigma' = \frac{1}{2} \sum_k \frac{\lambda'_k}{\lambda'} \|r'_k\|^2$$

$$\sigma'' = \frac{1}{2} \sum_k \frac{\lambda''_k}{\lambda''} \|r''_k\|^2$$

Distribution Step:

$$\Delta\epsilon' = \begin{cases} -\alpha_1\epsilon' & \text{if} & \frac{1}{N}\left(\frac{N-\lambda'}{\epsilon'} - \frac{\lambda'}{1-\epsilon'}\right) < -\alpha_2 \\ \alpha_1(1-\epsilon') & \text{if} & \frac{1}{N}\left(\frac{N-\lambda'}{\epsilon'} - \frac{\lambda'}{1-\epsilon'}\right) > \alpha_2 \\ 0 & \text{otherwise} \end{cases}$$

$$\Delta\epsilon'' = \begin{cases} -\alpha_1\epsilon'' & \text{if} & \frac{1}{N}\left(\frac{N-\lambda''}{\epsilon''} - \frac{\lambda''}{1-\epsilon''}\right) < -\alpha_2 \\ \alpha_1(1-\epsilon'') & \text{if} & \frac{1}{N}\left(\frac{N-\lambda''}{\epsilon''} - \frac{\lambda''}{1-\epsilon''}\right) > \alpha_2 \\ 0 & \text{otherwise} \end{cases}$$

$$\Delta\delta' = \begin{cases} \tau'\delta' & \text{if} & \frac{\#G'}{(1-\epsilon')N} > 1 + \alpha_3 & \text{and} & \Delta\epsilon' < 0 \\ -\tau'\delta' & \text{if} & \frac{\#G'}{(1-\epsilon')N} < 1 - \alpha_3 & \text{and} & \Delta\epsilon' > 0 \end{cases}$$

$$\Delta\delta'' = \begin{cases} \tau''\delta'' & \text{if} & \frac{\#G''}{(1-\epsilon'')N} > 1 + \alpha_3 & \text{and} & \Delta\epsilon'' < 0 \\ -\tau''\delta'' & \text{if} & \frac{\#G''}{(1-\epsilon'')N} < 1 - \alpha_3 & \text{and} & \Delta\epsilon'' > 0 \end{cases}$$

15.12.2   Simulated Experiments

Data Set Generation: A set of three-dimensional points, $p_k = (x_k, y_k, z_k)^t, k = 1, \cdots, N$, is generated within a box defined by $x_k, y_k \in [20, 30]$, $z_k \in [20, 40]$; that is, each x- and y-component is uniformly distributed between 20 and 30, and each z-component, between 20 and 40. These three-dimensional points are orthographically projected on the image plane to generate the image points $u_k = (x_k, y_k)^t$, which constitute the first view. Next, two sets of three rotational angles are randomly selected—for instance, from a uniform distribution—and two sets of three translational components from [20, 50] are selected for both the second and the third views. Having these parameters, the three-dimensional points p_k are rotated and translated in the three-dimensional space forming the three-dimensional points $p'_k, k = 1, \cdots, N$ for the second view and $p''_k, k = 1, \cdots, N$ for the third view, which are then orthographically projected on the image plane to form the image points u'_k and u''_k, $k = 1, \cdots, N$, for the second and third views. At this stage independently distributed Gaussian noise $N(0, \sigma'^2)$ or $N(0, \sigma''^2)$ is added to all components of some image points u'_k or u''_k, respectively. To test the robustness of the algorithm proposed in the previous subsubsection, the remaining image points u'_k or u''_k are randomly moved in the image plane to form outliers. Finally, we obtain the image points v'_k in the second view and v''_k in the third view. The algorithm is applied to find two sets of three rotational angles, two sets of the first two translational components, and depths of some three-dimensional points that produce nonoutlier image points in both the second and the third views under the rigid motion.

There are five parameters to be controlled: the sample size N, the outlier number M' in the second view and M'' in the third view, and the standard deviations σ' and σ'' of the Gaussian noises. In the experimental results we use $\epsilon' = M'/N$, $\epsilon'' = M''/N$, and SNR', SNR'' instead of $M', M'', \sigma', \sigma''$, where SNR' $= 20\log(\max\|v'_k\|)/\sigma'$db and SNR'' $= 20\log(\max\|v''_k\|)/\sigma''$db. Let ϵ denote $\max\{\epsilon', \epsilon''\}$. To simplify the experiments, we let SNR' $=$ SNR'' $=$ SNR and assume that the indices of outliers in the third view are the indices of some outliers in the second view when $\epsilon' \geq \epsilon''$; and we assume that the indices of outliers in the second view are the indices of some outliers in the third view when $\epsilon'' \geq \epsilon'$. Now there are only three parameters to be controlled: N, ϵ, and SNR.

Results

For each parameter setting $(N, \epsilon, \text{SNR})$, hundreds of experiments must be conducted to get a reasonable evaluation of the performance of the algorithm. Three different sets of experiments follow:

Ex1: Set $N = 10$. Estimate the average six-rotational-angle error, the average four-translational-component error, and the average depth error against SNR (40db to 70db in 10db step) for different ϵ (0.0 to 0.5 in 0.1 step).

Ex2: Set $\epsilon = 0.2$. Estimate the average rotational-angle error, the average four-translational-component error, and the average depth error against SNR (40db to 70db in 10db step) for different N (5 to 20 in 5 step).

Ex3: Set SNR = 60db. Estimate the average rotational-angle error, the average four-translational-component error, and the average depth error against ϵ (0.0 to 0.8 in 0.1 step) for different N (5 to 30 in 5 step).

The initial approximation of the six rotational angles is selected randomly within $\pm 15°$ of the true angles. The initial approximation of the four translational components is selected randomly within ± 10 of the true translation components. The initial approximation of the depths is selected randomly within ± 5 of the true depths. The results are shown in Figs. 15.13 to 15.17.

15.12.3 Further Research on the MF-Estimator

At least two major problems remain to be solved for the MF-estimator to be practically useful. One is what we call the distance problem, and the other is the requirement for a good initial approximation.

To solve the distance problem, we should divide the whole set of residuals into three sets: the good one, the bad one, and the fuzzy one. The good one will contribute to the least-squares parameter estimation; the bad one will be far enough from the good one (that is why it is called bad) so as not to prevent the good one from contributing to the least-squares parameter estimation; the fuzzy one neither is far enough from the good one nor has small enough residuals. The effect of this part of the data should be kept limited so it cannot interfere too badly in the least squares estimation process.

To indirectly satisfy the requirement for a good initial approximation, we must realize that the problem is actually related to global optimization. It is common to be trapped in local maxima (or local minima) while following the gradient-ascent (or gradient-descent) rule. To be initial-value independent or to avoid being trapped in local extrema, we must not strictly observe the gradient-ascent (or gradient-descent) rule for maximization (or minimization). Rather we must apply a stochastic search rule such as the one used in the Boltzmann machine to find the global minimum of energy function. To directly satisfy the requirement for a good initial approximation, we must use our knowledge of the parameter space to be searched in order to search it efficiently.

15.13 Literature Review

The following is a brief survey of approaches to the computation of optic flow or image-point correspondences and the inference of each object's motion and surface structure.

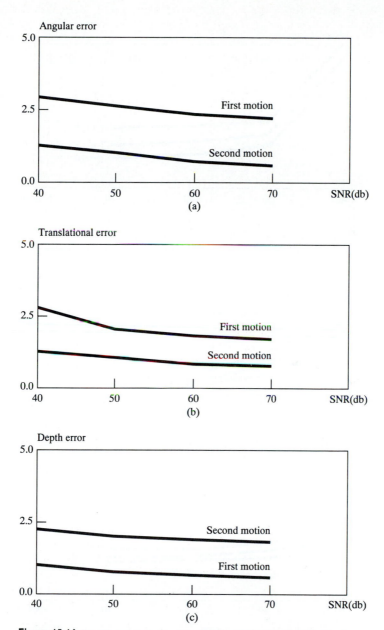

Figure 15.14 Experimental results of having five well-matched point pairs for each motion and having five outliers in segmenting and estimating two rigid motions. The average rotational angular velocity error, the average translational velocity error, and the average depth error are depicted as curves in (a), (b), and (c), respectively.

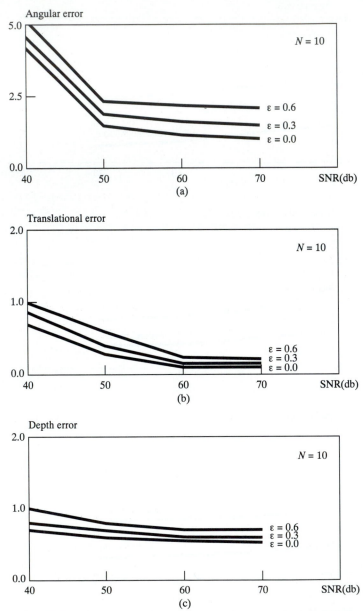

Figure 15.15 Experimental results of Ex1. The average rotational angular veloc-ity error, the average translational velocity error, and the average depth errors are depicted as curves in (a), (b), and (c), respectively.

15.13.1   Inferring Motion and Surface Structure

As we have seen, methods for inferring three-dimensional information from optic flow or image-point correspondences can be roughly divided into three classifications:

1. Use of individual sets of feature points
2. Use of local optic flow information about a single point
3. Use of the entire optic flow field

All methods except a few robust ones, to various degrees, rely on an accurate calculation of the optic flow or image-point correspondences as their input; such optic flow fields or image point correspondences, however, are not yet widely available. In the following we concentrate on the first classification, that is, inferring motion and surface structure by using feature-point correspondences. As suggested earlier, good optic flow velocities or image-point correspondences can be obtained with a high probability only at prominent feature points.

Few of the studies on inferring single rigid motion and surface structure from feature-point correspondences systematically discuss the effect of noise. Thompson (1959) developed the nonlinear equations and gave a solution procedure that determined a rotation matrix guaranteed to be orthonormal. His technique was to linearize the nonlinear equations and iterate. Roach and Aggarwal (1980) developed a nonlinear algorithm and dealt with noisy data. The linear motion estimation algorithm was developed by Longuet-Higgins (1981), extended by Tsai and Huang (1984), unified by Zhuang, Huang, and Haralick (1986), and simplified by Zhuang (1989). The feature-point correspondences come from two views taken at two different times. When the interval between the two times becomes infinitesimal, the point correspondence information takes the form of optic flow velocities. The related linear algorithm was developed by Zhuang, Huang, and Haralick (1988b). Linear algorithms are simpler and faster than the nonlinear ones. Furthermore, the linear algorithm can always find a unique solution except in the degenerate cases. The linear approach works well when there is limited noise and no corresponding matching errors. The minimum number of point correspondences needed varies according to whether the set of equations is linear or nonlinear. Usually the nonlinear technique requires fewer points than the linear one. However, the nonlinear approach does require a good initial guess in order to work. Nagel (1981), Prazdny (1981b), and Lawton (1982, 1983) deal with the initial estimation problem as a search through motion parameter space.

Despite all the results obtained over the years, almost none of these inference techniques have been successfully applied to feature-point correspondences calculated from real imagery. Jerian and Jain (1984) have attempted to use Tsai and Huang's approach with an optic flow field obtained by using Nagel's taxicab scene, but the results have been poor. Roach and Aggarwal (1980) have observed that their algorithm requires as many as 10 significant digits of accuracy in the input in order to work. Fang and Huang (1982, 1984b) have applied their technique successfully by using the well-defined corner points of a moving robot arm.

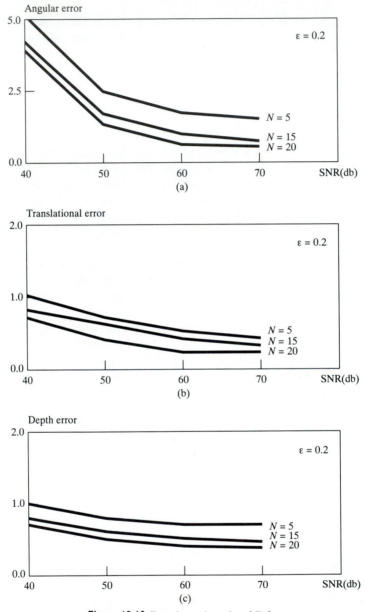

Figure 15.16 Experimental results of Ex2.

The reason why all these algorithms are not practical is simply that none takes seriously the nature of errors in optic flow or image-point correspondences. This problem, however, can be remedied by using the classic robust M-estimator if the proportion of outliers is not high (cf. Haralick et al., 1989).

Very little has been published on inferring multiple rigid motions from optic

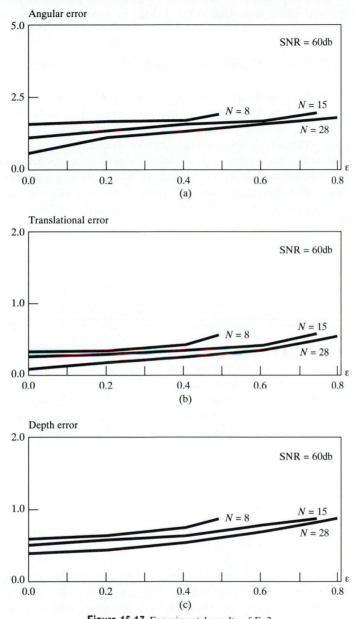

Figure 15.17 Experimental results of Ex3.

flow or image-point correspondences (but see Fennema and Thompson, 1979; Adiv, 1985; and Murray and Buxton, 1987). Adiv's method first segments the continuous optic flow into regions that approximate planar surfaces. A least-squares approach is used to estimate motion and surface structure for each planar surface. Adjacent surfaces with consistent motion and surface structure parameters are assumed to belong to the same three-dimensional object. None of these techniques takes outliers into consideration, which they need to do in order to work in the real world.

15.13.2 Computing Optic Flow or Image-Point Correspondences

Computing optic flow or image-point correspondences usually involves detecting prominent image features over time and matching them. This is called the *correspondence problem*. A basic assumption generally used in optic flow computation is that the image changes enough from time to time so that these changes can be detected, but not so quickly so that correspondence cannot be established. One might think that this assumption can be easily realized by simply adjusting the sampling rate accordingly. However, this ignores aliasing effects that might occur. Another problem is that in areas where optic flow contains abundant information—for example, at occlusion boundaries, specular points (points where the plane tangent to the surface is frontal with respect to the observer), or near a focus of expansion—it is usually quite difficult for any existing methods to calculate optic flow precisely. Other error sources include noise and digitization effects in image formation. These are usually considered insignificant and are ignored. Many methods for computing optic flow first compute normal velocity. The normal velocity of a point on a moving contour is simply the two-dimensional velocity of that point with respect to the contour. Normal velocities are computed because of the aperture problem (Marr and Ullman, 1981). To calculate the true optic flow velocities from a normal velocity field, one has to use additional constraints, such as the smoothness constraint of Horn and Schunck (1981) and Hildreth (1982, 1983).

Several matching and differential methods exist for computing optic flow or image-point correspondences. Matching methods include template, token, and image matching, and they work for both optic flow and image-point correspondences. Differential methods include difference and gradient techniques, and they work only for optic flow.

Template matching involves finding iconic models of moving objects in one frame and matching them against the next frame. Various correlation methods (Aggarwal, Davis, and Martin, 1981) are used to accomplish the matching. The main advantage of template matching is its computational efficiency. However, the iconic models must be known a priori so that they can be tracked. This scene-dependent knowledge is not widely available.

Token matching (Barnard and Thompson, 1980) establishes correspondences between image features in adjacent images in a scene. These features, often called tokens, are usually pixels that are distinguishable from neighboring pixels, for example, corner points. Tokens could also be other image features, such as regions or object segments. Token matching relies on good token finders.

Image matching is similar to token matching, only here small image patches from one image are matched to find a corresponding image patch on a second image. Techniques for image matching are discussed in Chapter 16.

Finding reliable prominent points in an image is computationally expensive, even though the computation is local and parallel. Besides, their precise locations are often difficult to obtain. The result is that optic flow or image-point correspondences cannot usually be found precisely. Matching methods also fail in computing optic flow when significant changes occur between adjacent images.

Difference methods look for significant intensity changes, pixel by pixel or region by region, in adjacent images of a scene. This is done by subtracting one image from another and thresholding the result. One problem with difference methods is that objects whose interiors are homogeneous produce no pixel-by-pixel differences. No optic flow information can be produced in such regions. Problems also occur when nonhomogeneous low-contrast objects move across a nonhomogeneous background. To date, difference methods have not been used with much success in the calculation of optic flow.

Gradient methods (Kearney, Thompson, and Boley, 1982) use the relationships between the spatial and temporal gradients of image intensity to compute optic flow. The motion-constraint equation (Horn and Schunck, 1981) specifies that the magnitude of irradiance change over time at some point (x, y) must equal the magnitude of irradiance change over space at some fixed time t. That is,

$$\frac{\partial I}{\partial x}\frac{\delta x}{\delta I} + \frac{\partial I}{\partial x}\frac{\delta y}{\delta t} + \frac{\partial I}{\partial t} = 0 \tag{15.187}$$

where I is the image intensity at (x, y) at time t and δx and δy are the x- and y-displacements of point (x, y) between t and $t + \delta t$. The motion-constraint equation for a point defines a line in flow field; the true optic flow velocity of the point must lie somewhere on this line. Because of aperture problems, we can retrieve only the component of the optic flow velocity normal to a moving intensity distribution (such as that exhibited by the image intensity profile of a contour).

The terms $\delta x / \delta t$ and $\delta y / \delta t$ are simply the components of the optic flow velocity at point (x, y) at time t. The normal optic flow velocity is simply the projection of $(\delta x / \delta t, \delta y / \delta t)$ along the direction $(\partial I / \partial x, \partial I / \partial y)$. Since the displacements (or disparities) are often treated as equivalents, the motion-constraint equation is often written as

$$\frac{\partial I}{\partial x}\delta x + \frac{\partial I}{\partial y}\delta y + \frac{\partial I}{\partial t}\delta t = 0 \tag{15.188}$$

This equation is not valid at image intensity discontinuities, because in those places the brightness patterns will not be constant. It actually assumes that the intensity of a pixel does not change significantly over a short time interval; hence the scene illumination must be fairly constant and the motions between adjacent images in a scene must be small. Thus the reflectance properties of objects must be well behaved; specular points on a surface will not be analyzable. With the exception of Nagel (1981, 1983), gradient methods generally assume that local image intensity gradients are linear—which is not generally true. Schunck (1986) has shown that the motion-constraint equation is valid only for pure translation under orthographic projection. It is not valid for rotations under orthographic projection or for any motion under perspective projection. The motion-constraint equation contains two unknowns $\delta x / \delta t$ and $\delta y / \delta t$. Thus an additional constraint is needed to solve for the two unknowns; these assumptions, which are usually related to the global structure of the flow field, are not generally true either.

The main attribute of gradient methods is that they are based on very simple local computation that may be performed in parallel at a low-level processing. As we have seen, they are also based on many unrealistic assumptions and do not work well when applied to real imagery (Jain, 1984; Jerian and Jain, 1988). Although gradient methods tend to produce dense flow fields, the optic flow so produced is not accurate.

■ Exercises

15.1. Show that, in rigid-body motion, if $k = 0$, then the general solution for w is given by

$$w = \begin{bmatrix} uv\dot{u} - \dot{v}(u^2 + f^2) \\ \dot{u}(v^2 + f^2) - uv\dot{v} \end{bmatrix} \frac{1}{f(u^2 + v^2 + f^2)} + \lambda \begin{pmatrix} u \\ v \\ f \end{pmatrix}$$

and that

$$w' \begin{bmatrix} \dot{u}(v^2 + f^2) - uv\dot{v} \\ (u^2 + f^2) - uv\dot{u} \\ -f(\dot{u}u + \dot{v}v) \end{bmatrix} = 0$$

15.2. If the optic flow $\{(u_n, v_n, \dot{u}_n, \dot{v}_n)\}_{n=1}^N$ is given, show that a least-squares solution for w is given by that eigenvector of $\sum_{n=1}^{N} s_n s_n'$ having the smallest eigenvalue when

$$s_n = \begin{bmatrix} \dot{u}_n(v_n^2 + f^2) - u_n v_n \dot{v}_n \\ \dot{v}_n(v_n^2 + f^2) - u_n v_n \dot{u}_n \\ -f(\dot{u}_n u_n + \dot{v}_n v_n) \end{bmatrix}$$

15.3. a. Show that, in the general solution for w, when $k = 0$ there results

$$\begin{pmatrix} w_x \\ w_y \end{pmatrix} = w_z \begin{pmatrix} u \\ v \end{pmatrix} + \frac{1}{f(u^2 + v^2 + f^2)} \begin{bmatrix} uv & -(u^2 + f^2) \\ v^2 + f^2 & -uv \end{bmatrix} \begin{pmatrix} \dot{u} \\ \dot{v} \end{pmatrix}$$

b. Determine the solution to the least-squares problem minimizing

$$\epsilon^2 = \sum_{n=1}^{N} \left\| \begin{pmatrix} w_x \\ w_y \end{pmatrix} - w \begin{pmatrix} u \\ v \end{pmatrix} - \frac{1}{f(u^2 + v^2 + f^2)} \begin{bmatrix} uv & -(u^2 + f^2) \\ v^2 + f^2 & -uv \end{bmatrix} \begin{pmatrix} \dot{u} \\ \dot{v} \end{pmatrix} \right\|$$

15.4. Show that

$$\begin{pmatrix} u \\ v \\ f \\ 0 \\ 0 \\ 0 \end{pmatrix} \begin{pmatrix} 0 \\ 0 \\ 0 \\ u \\ v \\ f \end{pmatrix} \begin{pmatrix} 0 \\ -(u^2 + f^2 + v^2)z \\ 0 \\ -f^2 \\ 0 \\ uf \end{pmatrix} \begin{pmatrix} f(u^2 + v^2 + f^2)z \\ 0 \\ -u(u^2 + v^2 + f^2)z \\ 0 \\ -f(u^2 + v^2 + f^2) \end{pmatrix}$$

is a basis for the kernel of

$$\begin{pmatrix} \frac{f}{z} & 0 & -\frac{u}{z} & -\frac{uv}{f} & \frac{u^2+f^2}{f} & -v \\ 0 & \frac{f}{z} & -\frac{v}{z} & -\frac{v^2+f^2}{f} & \frac{uv}{f} & u \end{pmatrix}$$

15.5. Show that the kernel of

$$\begin{pmatrix} \frac{f}{z} & 0 & -\frac{u}{z} & -\frac{uv}{f} & \frac{f^2+u^2}{f} & -v \\ 0 & \frac{f}{z} & -\frac{v}{z} & -\frac{f^2+v^2}{f} & \frac{uv}{f} & u \end{pmatrix}$$

is spanned by the vector

$$e_1 = \begin{pmatrix} u \\ v \\ f \\ 0 \\ 0 \\ 0 \end{pmatrix}, e_2 = \begin{pmatrix} 0 \\ 0 \\ 0 \\ u \\ v \\ f \end{pmatrix}, e_3 = \begin{pmatrix} 0 \\ -z(u^2+v^2+f^2) \\ 0 \\ -f^2 \\ 0 \\ uf \end{pmatrix}, e_4 = \begin{pmatrix} \frac{-z(u^2+v^2+f^2)(u^2+f^2)}{f^2} \\ \frac{-uv(u^2+v^2+f^2)z}{f^2} \\ 0 \\ -uv \\ (u^2+f^2) \\ fv \end{pmatrix}$$

Hence the general solution to Eq. (15.24) can be written as

$$\begin{pmatrix} k_x \\ k_y \\ k_z \\ w_x \\ w_y \\ w_z \end{pmatrix} = \sum_{i=1}^{4} \lambda_i e_i + \frac{z}{f} \begin{pmatrix} \dot{u} \\ \dot{v} \\ 0 \\ 0 \\ 0 \\ 0 \end{pmatrix}$$

15.6. From the general solution given in Exercise 15.5,

$$\begin{pmatrix} k_z \\ w_x \\ w_y \\ w_z \end{pmatrix} = \lambda_1 \begin{pmatrix} f \\ 0 \\ 0 \\ 0 \end{pmatrix} + \lambda_2 \begin{pmatrix} f \\ 0 \\ 0 \\ 0 \end{pmatrix} + \lambda_3 \begin{pmatrix} 0 \\ -f^2 \\ 0 \\ uf \end{pmatrix} + \lambda_4 \begin{pmatrix} 0 \\ -uv \\ u^2+f^2 \\ fv \end{pmatrix}$$

Use this relation and the general solution to find a relation between (k_x, k_y) and (k_z, w_x, w_y, w_z).

15.7. Suppose that an oriented line segment of known length is in translational motion $(\dot{x}, \dot{y}, \dot{z})$. Let one end of the line segment be (x_1, y_1, z_1) and the other end be (x_2, y_2, z_2). Suppose that the known orientation is (m_1, m_2, m_3). Then

$$\begin{pmatrix} x_2 \\ y_2 \\ z_2 \end{pmatrix} = \begin{pmatrix} x_1 \\ y_1 \\ z_1 \end{pmatrix} + \rho \begin{pmatrix} m_1 \\ m_2 \\ m_3 \end{pmatrix}$$

Let the optic flow for the points (x_1, y_1, z_1) be $(u_1, v_1, \dot{u}_1, \dot{v}_1)$ and $(u_2, v_2, \dot{u}_2, \dot{v}_2)$. Assume that $(u_2 - u_1)(\dot{v}_2 - \dot{v}_1) \neq (\dot{u}_2 - \dot{u}_1)(v_2 - v_1)$. Determine the position of each endpoint and the velocity $(\dot{x}, \dot{y}, \dot{z})$.

15.8. Verify that

$$p(t) = \exp\left(\int_0^t S(u)du\right) p(0) + \int_0^t \exp\left(\int_v^t S(u)du\right) k(v)dv$$

solves the linear differential equations (15.17), that is, $\dot{p}(t) = S(t)p(t) + k(t)$.

15.9.   Prove that

$$(q'_n, r'_n z_n) \binom{k}{\omega} = 0$$

where

$$q_n = \begin{pmatrix} \dot{v}_n f^2 \\ -\dot{u}_n f^2 \\ (\dot{u}_n v_n - u_n \dot{v}_n)f \end{pmatrix}$$

$$r_n = \begin{pmatrix} \dot{u}_n(v_n^2 + f^2) - u_n v_n \dot{v}_n \\ \dot{v}_n(u_n^2 + f^2) - u_n v_n \dot{u}_n \\ -(u_n \dot{u}_n + v_n \dot{v}_n)f \end{pmatrix}$$

15.10.   Prove that the three eigenvalues of $kn' + nk'$ are $k'n - ||k|| \cdot ||n||$, 0, and $k'n + ||k|| \cdot ||n||$, where k, n refer to Eqs. (15.34) and (15.36) in Section 15.4.1.

15.11.   Writing Eq. (15.42) as follows:

$$\begin{bmatrix} u & v & f & 0 & 0 & 0 - u/f(u,v,f) \\ 0 & 0 & 0 & u & v & f - v/f(u,v,f) \end{bmatrix} \begin{pmatrix} w_1 \\ w_2 \\ w_3 \end{pmatrix} = \begin{pmatrix} \dot{u} \\ \dot{v} \end{pmatrix}$$

prove that the rank of the coefficient matrix of a set of linear equations in Eq. (15.42), with $N \geq 4$, equals 8 if and only if there are four among the N image points such that no three of them are collinear.

Exercises 15.12 to 15.19 represent a step-by-step solution to determining planar motion and structure from optic flow.

15.12.   Prove that the general solution of the corresponding $2N$ homogeneous equations in Eq. (15.42) equals αI_3 when the rank of the coefficient matrix equals 8.

15.13.   Prove that the motion parameter matrix W is uniquely determined by Eq. (15.42) and the constraint, that is, the middle eigenvalue of $W + W'$, equals zero when the rank equals 8.

15.14.   When the motion parameter matrix W is uniquely determined by Eq. (15.42), W must equal $\Omega + kn'$, and $W + W'$ equals $kn' + nk'$ correspondingly. Prove that the three eigenvalues of $kn' + nk'$ are $k'n - ||k|| \cdot ||n||$, 0, and $k'n + ||k|| \cdot ||n||$. Let $\lambda_1 = k'n - ||k|| \cdot ||n||$, $\lambda_2 = 0$, $\lambda_3 = k'n + ||k|| \cdot ||n||$. Prove that

$$\lambda_1 < 0 < \lambda_3 \quad \text{if } k \times n \neq 0$$
$$\lambda_1 = 0 < \lambda_3 \quad \text{if } k \neq 0, k \text{ and } n \text{ share the same direction}$$
$$\lambda_1 < 0 = \lambda_3 \quad \text{if } k \neq 0, k \text{ and } -n \text{ share the same direction}$$
$$\lambda_1 = 0 = \lambda_3 \quad \text{if } k = 0$$

15.15.   Explain that when $W = \Omega + n'$ and $\Omega + \Omega' = 0$, then $\lambda_1 = 0 = \lambda_3$ implies $k = 0$ and $W = \Omega$, where $\lambda_1, 0, \lambda_3$ are the eigenvalues of $W + W' = kn' + nk'$.

15.16.   Let $\lambda_1, 0$, and λ_3 be three eigenvalues of the symmetric matrix $kn' + nk'$. Let q_1, q_2, and q_3 be three corresponding orthonormal eigenvectors. Using $\lambda_i q_i = (n'q_i)k +$

$(k'q_i)n$, $i = 1,2,3$, prove that

$$\lambda_i = 2(n'q_i)(k'q_i), \quad i = 1,3$$

$$\lambda_i(n'q_i) = (n'q_i)(n'k) + (k'q_i)\|n\|^2, \quad i = 1,2,3$$

$$\lambda_i(k'q_i) = (n'q_i)\|k\|^2 + (k'q_i)(k'n), \quad i = 1,2,3$$

$$(n'q_i)^2\|k\|^2 = (k'q_i)^2\|n\|^2, \quad i = 1,2,3$$

$$(n'q_i)^4\|k\|^2 = \frac{\lambda_i^2}{4}\|n\|^2, \quad i = 1,2,3$$

$$(n'q_i)^2\|k\| = \frac{|\lambda_i|}{2} \cdot \|n\|, \quad i = 1,2,3$$

$$(n'q_i)\sqrt{\|k\|} = \pm\sqrt{\frac{|\lambda_i|}{2}}\sqrt{\|n\|}, \quad i = 1,2,3$$

$$(k'q_i)\sqrt{\|n\|} = \pm \operatorname{sign}(\lambda_i)\sqrt{\frac{|\lambda_i|}{2}}\sqrt{\|k\|}, \quad i = 1,2,3$$

15.17. Using the results from the previous exercise, prove that there are four choices for (n,k) :

    a. $n = \dfrac{1}{2}\left(-\sqrt{-\lambda_1}\, q_1 + \sqrt{\lambda_3}\, q_3\right)$ up to a scale factor $\alpha > 0$

         $k = \sqrt{-\lambda_1}\, q_1 + \sqrt{\lambda_3}\, q_3$ up to $\frac{1}{\alpha}$

    b. $n = \dfrac{1}{2}\left(\sqrt{-\lambda_1}\, q_1 - \sqrt{\lambda_3}\, q_3\right)$ up to $\alpha > 0$

         $k = -\sqrt{-\lambda_1}q_1 - \sqrt{\lambda_3}\, q_3$ up to $\frac{1}{\alpha}$

    c. $n = \dfrac{1}{2}\left(\sqrt{-\lambda_1}\, q_1 + \sqrt{\lambda_3}\, q_3\right)$ up to $\alpha > 0$

         $k = -\sqrt{-\lambda_1}\, q_1 + \sqrt{\lambda_3}\, q_3$ up to $\frac{1}{\alpha}$

    d. $n = \dfrac{1}{2}\left(-\sqrt{-\lambda_1}\, q_1 - \sqrt{\lambda_3}\, q_3\right)$ up to $\alpha > 0$

         $k = \sqrt{-\lambda_1}\, q_1 - \sqrt{\lambda_3}\, q_3$ up to $\frac{1}{\alpha}$

15.18. Using the constraints $n'p = 1$ and $z < 0$, prove (1) if (a) gives the genuine solution for (n,k), then $\sqrt{\lambda_3}q'_3s_n < \sqrt{-\lambda_1}q'_1s_n$, $n = 1,\ldots,N$; (2) if (b) gives the genuine solution for (n,k), then $\sqrt{\lambda_3}\,q'_3s_n > \sqrt{-\lambda_1}\,q'_1s_n$, $n = 1,\ldots,N$; (3) if (c) gives the genuine solution for (n,k), then $\sqrt{\lambda_3}\,q'_3s_n < -\sqrt{-\lambda_1}\,q'_1s_n$, $n = 1,\ldots,N$; and (4) if (d) gives the genuine solution for (n,k), then $\sqrt{\lambda_3}\,q'_3s_n > -\sqrt{-\lambda_1}\,q'_1s_n$, $n = 1,\ldots,N$ where $s_n = (u_n, v_n, f)'$. Thus both (a) and (b) and both (c) and (d) cannot be true simultaneously.

15.19. Prove that there are two choices for (n,k) if and only if $\lambda_1 < 0 < \lambda_3$ and both one of (a,b) and one of (c,d) become true.

15.20. Prove that the two solutions denoted by $(n,k,\omega),(n^*,k^*,\omega^*)$ are related by $n^* \times k = 0$, $k^* \times n = 0$, $k^*n^{*'} = nk'$, $\omega^* = \omega + n \times k$.

15.21. Prove that (a) or (b) implies the same ω, and that (c) or (d) implies the same ω.

Exercises 15.22 to 15.29 represent a step-by-step solution to determining planar motion and structure from two views.

15.22. While establishing Eq. (15.86) in Section 15.5.1, we use any nonzero vector k^* that is collinear with k. Why? Can we simply use k itself?

15.23. Prove that Eq. (15.86) has a rank 8 if and only if there exist four image-point correspondence pairs $(u_{i1}, v_{i1}) \leftrightarrow (u_{i2}, v_{i2})$, $i = 1, 2, 3, 4$ so that no three of both $\{(u_{i1}, v_{i1}) : i = 1, 2, 3, 4\}$ and $\{(u_{i2}, v_{i2}) : i = 1, 2, 3, 4\}$ are collinear.

15.24. If Eq. (15.86) has a rank 8, then any nonzero solution A to Eq. (15.86) relates to $R_0 + t_0 n_0'$ by a unique scale factor k_0, $k_0 A = R_0 + t_0 n_0'$. Prove it.

15.25. Let three eigenvalue vectors of $A'A$ be $\xi^2 \geq \eta^2 \geq \zeta^2$, $u_1, u_2, u_3 = u_1 \times u_2$, respectively. Prove that k_0 is uniquely determined by $k_0^2 \eta^2 = 1$ and $k_0 a_3 (u_1, v_1, f)' > 0$, where a_3 is the third-row vector of A, where A is a non-zero solution to Eq. (15.86), and $k_0 A = R_0 + t_0 n_0'$.

15.26. Prove that if $\xi^2 = \eta^2 = \zeta^2$, then $k_0 A$ is orthonormal, where A is a nonzero solution to Eq. (15.86), and $k_0 A = R_0 + t_0 n_0'$.

15.27. Prove that if $\xi^2 = \eta^2 = \zeta^2$ and $k_0 \det(A) > 0$, then $k_0 A$ is the orthonormal matrix of the first kind and $R_0 = k_0 A$, $t_0 = 0$, where A is a nonzero solution to Eq. (15.86), and $k_0 A = R_0 + t_0 n_0'$.

15.28. Prove that if $\xi^2 = \eta^2 = \zeta^2$ and $k_0 \det(A) < 0$, then there are an infinite number of solutions for $\{R_0, t_0, n_0\}$:

$$R_0 = k_0 A(e_1, e_2, e_2 \times e_1)(e_1, e_2, e_1 \times e_2)'$$

$$t_0 = 2R_0(e_2 \times e_1)$$

$$n_0 = e_1 \times e_2$$

where e_1, e_2 are two arbitrary unit vectors orthogonal to each other.

15.29. Prove that the middle eigenvector u_2 is orthogonal to both n_0 and $R_0' t_0$.

15.30. Let $e_1 = u_2 \times n_0$, $e_2 = u_2$, $e_3 = n_0$. Prove that

$$R_0 e_i = k_0 A e_i, \quad i = 1, 2$$

$$R_0 e_3 = (k_0 A e_1) \times (k_0 A e_2)$$

$$t_0 = (k_0 A - R_0) e_3$$

15.31. Let $e_1 = \alpha u_1 + \beta u_3$. Prove that

$$\alpha(n_0' u_1) + \beta(n_0' u_3) = 0$$

$$\alpha(k_0^2 \xi^2 - 1)u_1 + \beta(k_0^2 \zeta^2 - 1)u_3 = n_0[\alpha t_0'(k_0 A u_1) + \beta t_0'(k_0 A u_3)]$$

$$\alpha(k_0^2 \xi^2 - 1) = (n_0' u_1)[\alpha t_0'(k_0 A u_1) + \beta t_0'(k_0 A u_3)]$$

$$\beta(k_0^2 \zeta^2 - 1) = (n_0' u_3)[\alpha t_0'(k_0 A u_1) + \beta t_0'(k_0 A u_3)]$$

$$\alpha(k_0^2 \xi^2 - 1)(n_0' u_3) = \beta(k_0^2 \zeta^2 - 1)(n_0' u_1)$$

$$(k_0^2 \xi^2 - 1)\alpha^2 + (k_0^2 \zeta^2 - 1)\beta^2 = 0$$

$$k_0^2(\xi^2 - \zeta^2)\alpha^2 = 1 - k_0^2 \zeta^2$$

$$\alpha = \pm \sqrt{\frac{1 - k_0^2 \zeta^2}{k_0^2(\xi^2 - \zeta^2)}}, \quad \text{if } \xi^2 > \zeta^2$$

$$\beta = \pm\sqrt{1 - \alpha^2}$$

$$e_1 = \frac{\left(\pm\sqrt{1 - k_0^2\xi^2}\ u_1 - \sqrt{k_0^2\xi^2 - 1}\ u_3\right)}{\sqrt{k_0^2(\xi^2 - \zeta^2)}}$$

up to a sign, if $\xi^2 > \zeta^2$.

15.32. Assume that $E = T \times R_0$, with R_0 being orthonormal. Prove that $R_0 = [W_2 \times W_3, W_3 \times W_1, W_1 \times W_2] - T \times E$, where $W_i = T \times E_i$, E_i is the ith column of E.

15.33. Given two orthographic views, why can't we finitely determine the motion and structure of a rigid object?

15.34. Why can't a classical robust estimator, such as the M-estimator, render a high robustness?

■ Bibliography

Adiv, G., "Determining Three-Dimensional Motion and Structure from Optical Flow Generated by Several Moving Objects," *IEEE Transactions on Pattern Analysis and Machine Intelligence,* Vol. PAMI-7, 1985, pp. 384–401.

Aggarwal, J. K., "Three-Dimensional Description of Objects and Dynamic Scene Analysis," *Digital Image Analysis,* S. Levioldi, (ed.), Pitman Publishing, Inc., 1983, pp. 29–46.

Aggarwal, J. K., L. S. Davis, and W. N. Martin, "Correspondence Processes in Dynamic Scene Analysis," *Proceedings of the IEEE,* Vol. 69, 1981, pp. 562–572.

Aggarwal, J. K., and M. J. Magee, "Determining Motion Parameters Using Intensity Guided Range Sensing," *Pattern Recognition,* Vol. 19, 1986, pp. 169–180.

Aggarwal, J. K., and W. N. Martin, "Analyzing Dynamic Scenes Containing Multiple Moving Objects," *Springer Series in Information Sciences,* T. S. Huang (ed.), Vol. 5, Image Sequence Analysis, 1981, pp. 355–380.

Aggarwal, J. K., and Y. F. Wang, "Structure and Motion Computation from Point or Line Correspondences in Images," *Advances in Image Processing and Pattern Recognition,* V. Cappillinni and R. Marconi, (eds.), Elsevier, 1986, pp. 171–178.

Akita, K., "Image Sequence Analysis of Real World Human Motion," *Pattern Recognition,* Vol. 17, 1984, pp. 73–83.

Aloimonos, J., and C. M. Brown, "Perception of Structure from Motion," *Proceedings of the IEEE Conference on Computer Vision and Pattern Recognition,* Miami Beach, FL, 1986, pp. 510–517.

Asada, M., and S. Tsuji, "Representation of Three-Dimensional Motion in Dynamic Scenes," *Computer Vision, Graphics, and Image Processing,* Vol. 21, 1983, pp. 118–144.

Asada, M., M. Yachida, and S. Tsuji, "Representation of Motions in Time-Varying Imagery," *Proceedings of the Sixth International Conference on Pattern Recognition,* Munich, 1982, pp. 306–309.

——, "Analysis of Three-Dimensional Motions in Blocks World," *Pattern Recognition,* Vol. 17, 1984, pp. 57–71.

Ayala, I. L., et al., "Moving-Target Tracking Using Symbolic Registration," *IEEE Transactions on Pattern Analysis and Machine Intelligence,* Vol. PAMI-4, 1982, pp. 515–520.

Baker, H. H., "Surface Reconstruction from Image Sequences," *Proceedings of the Second International Conference on Computer Vision,* Tampa, FL, 1988.

Baker, H. H., R. C. Bolles, and D. H. Marimont, "Determining Scene Structure from a Moving Camera: Epipolar-Plane Image Analysis," *SRI International,* Menlo Park, CA, 1986.

Barnard, S. T., and W. B. Thompson, "Disparity Analysis of Images," *IEEE Transactions on Pattern Analysis and Machine Intelligence,* Vol. PAMI-2, 1980, pp. 333–340.

Barron, J. L., A. D. Jepson, and J. K. Tsotsos, "The Feasibility of Motion and Structure Computations," *Proceedings of the Second International Conference on Computer Vision,* Tampa, FL, 1988.

Bazakos, M. E., D. P. Panda, and D. Duncan, "Passive Ranging in Outdoor Environment," *IEEE International Conference on Robotics and Automation,* St. Louis, MO, 1985, pp. 836–842.

Berger, B. J., and E. A. Parrish, Jr., "An Iterative Implementation of the TSVIP Video Tracking Algorithm," *IEEE Proceedings of the 14th Annual Southeastern Symposium on System Theory,* Blacksburg, VA, 1982, pp. 256–261.

Bernd, N., "Exploiting Image Formation Knowledge for Motion Analysis," *IEEE Transactions on Pattern Analysis and Machine Intelligence,* Vol. PAMI-2, 1980, pp. 550–554.

Brandt, A. V., and W. Tengler, "Obtaining Smoothed Optical Flow of Fields by Modified Block Matching," *Proceedings of the Fifth Scandinavian Conference on Image Analysis,* Stockholm, 1987, pp. 523–529.

Bridwell, N. J., and T. S. Huang, "A Discrete Spatial Representation for Lateral Motion Stereo," *Computer Vision, Graphics, and Image Processing,* Vol. 21, 1983, pp. 33–57.

Broida, T. J., and R. Chellappa, "Estimation of Object Motion Parameters from Noisy Images," *IEEE Transactions on Pattern Analysis and Machine Intelligence,* Vol. PAMI-8, 1986, pp. 90–99.

Bruss, A. R., and B. K. P. Horn, "Passive Navigation," *Computer Vision, Graphics, and Image Processing,* Vol. 21, 1983, pp. 3–20.

Burkhardt, H., and N. Diehl, "Simultaneous Estimation of Rotation and Translation in Image Sequences," in *Signal Processing III: Theories and Applications,* I. T. Young et al. (eds.), Elsevier, North-Holland, 1986, pp. 821–824.

Chen, H. H., "Motion and Depth from Binocular Orthographic Views," *Proceedings of the Second International Conference on Computer Vision,* Tampa, FL, 1988.

Chen, H. H., and T. S. Huang, "Using Motion from Orthographic Projections to Prune 3D Point Matches," AT&T Bell Laboratories, Holmdel, NJ, 1989.

Cowart, A. E., W. E. Snyder, and W. H. Ruedger, "The Detection of Unresolved Targets Using the Hough Transform," *Computer Vision, Graphics, and Image Processing,* Vol. 21, 1983, pp. 222–238.

Cyganski, D., and J. A. Orr, "Object Identification and Orientation Estimation from Point Set Tensors," *Proceedings of the Seventh International Conference on Pattern Recognition,* Montreal, 1984.

——, "Applications of Tensor Theory to Object Recognition and Orientation Determination," *IEEE Transactions on Pattern Analysis and Machine Intelligence,* Vol. PAMI-2, 1985, pp. 662–673.

Davis, L. S., Z. Wu, and H. Sun, "Contour-Based Motion Estimation," *Computer Vision, Graphics, and Image Processing,* Vol. 23, 1983, pp. 313–326.

Dengler, J., and M. Schmidt, "The Dynamic Pyramid—A Model for Motion Analysis with Controlled Continuity," *International Journal of Pattern Recognition and Artificial Intelligence,* Vol. 2, 1988, pp. 275–286.

D'Haeyer, J., "Determining Motion of Image Curves from Local Pattern Changes," *Computer Vision, Graphics, and Image Processing,* Vol. 34, 1986, pp. 166–188.

Diehl, N., and H. Burkhardt, "Planar Motion Estimation with a Fast Converging Algorithm," *Proceedings of the Eighth International Conference on Pattern Recognition,* Paris, 1986.

Dreschler, L. S., and H. H. Nagel, "On the Selection of Critical Points and Local Curvature Extrema of Region Boundaries for Interframe Matching," *Proceedings of the Sixth International Conference on Pattern Recognition,* Munich, 1982a.

———, "Volumetric Model and 3D Trajectory of a Moving Car Derived from Monocular of TV-Frame Sequences of a Street Scene," *Computer Graphics and Image Processing,* Vol. 20, 1982b, pp. 199–228.

Fan, T. I., and K. S. Fu, "A Syntactic Approach to Time-Varying Image Analysis," *Computer Graphics and Image Processing,* Vol. 11, 1979, pp. 138–149.

Fang, J. Q., and T. S. Huang, "A Corner-Finding Algorithm for Image Analysis and Registration," *Proceedings of the Second National Conference on Artificial Intelligence,* Vol. AAAI-82, 1982, pp. 46–49, Pittsburgh, PA.

———, "Solving Three-Dimensional Small-Rotation Motion Equations: Uniqueness, Algorithms, and Numerical Results," *Computer Vision, Graphics, and Image Processing,* Vol. 26, 1984a, pp. 183–206.

———, "Some Experiments on Estimating the 3D Motion Parameters of a Rigid Body from Two Consecutive Image Frames," *IEEE Transactions on Pattern Analysis and Machine Intelligence,* Vol. PAMI-6, 1984b, pp. 545–554.

Fennema, C. L., and W. B. Thompson, "Velocity Determination in Scenes Containing Several Moving Objects," *Computer Graphics and Image Processing,* Vol. 9, 1979, pp. 301–315.

Ferrie, F. P., M. D. Levine, and S. W. Zucker, "Cell Tracking: A Modeling and Minimization Approach," *IEEE Transactions on Pattern Analysis and Machine Intelligence,* Vol. PAMI-4, 1982, pp. 277–290.

Fischler, M., and R. Bolles, "Random Sample Consensus: A Paradigm for Model Fitting with Applications to Image Analysis and Automated Cartography," *Communications of the ACM,* Vol. 24, 1981, pp. 281–395.

Fitzpatrick, J. M., and M. R. Leuze, "A Class of One-to-One Two-Dimensional Transformations," *Computer Vision, Graphics, and Image Processing,* Vol. 39, 1987, pp. 369–382.

Fu, K. S., and T. I. Fan, "Tree Translation and Its Application to a Time-Varying Image Analysis Problem," *IEEE Transactions on Systems, Man, and Cybernetics,* Vol. SMC-12, 1982, pp. 856–867.

Fukui, I., "Difference and Indefinite Sum of Walsh Function," Mechanical Engineering Laboratory, Systems Science Department, Namiki 1-2, Sakura-mura, Niihari-gun, Ibaraki 305, Japan.

Gerbrands, J. J., F. Booman, and J. H. C. Reiber, "Computer Analysis of Moving Radiopaque Markers from X-Ray Cinefilms," *Computer Graphics and Image Processing,* Vol. 11, 1979, pp. 35–48.

Haralick, R. M., "Computer Vision Theory: The Lack Thereof," *Computer Vision, Graphics, and Image Processing,* Vol. 36, 1986, pp. 372–386.

Haralick, R. M., et al., "Pose Estimation from Corresponding Point Data," *IEEE Transactions on Systems, Man, and Cybernetics,* Vol. SMC-19, 1989, pp. 1426–1446.

Harris, C. G., and J. M. Pike, "3D Positional Integration from Image Sequences," *Image and Vision Computing,* Vol. 6, 1988, pp. 87–90.

Hayden, C. H., R. C. Gonzalez, and A. Ploysongsang, "A Temporal Edge-Based Image Segmentor," *Pattern Recognition,* Vol. 20, 1987, pp. 281–290.

Hildreth, E. C., "The Integration of Motion Information along Contours," *Proceedings of the IEEE Workshop on Computer Vision Representation and Control,* Rindge, NH, 1982, pp. 83–91.

Horn, B. K. P., and B. G. Schunck, "Determining Optical Flow," *Artificial Intelligence,* Vol. 17, 1981, pp. 185–203.

Hsu, Y. Z., H. H. Nagel, and G. Rekers, "New Likelihood Test Methods for Change Detection in Image Sequences," *Computer Vision, Graphics, and Image Processing,* Vol. 26, 1984, pp. 73–106.

——, "Motion and Structure from Two Orthographic Projections," *Proceedings of the Fifth Scandinavian Conference on Image Analysis,* Stockholm, Vol. 1, 1987, pp. 499–504.

——, "Motion Estimation from Stereo Sequences," in *Machine Vision from Inspection and Measurement,* H. Freeman (ed.), Academic Press, New York, 1989.

Huang, T. S., and J. Q. Fang, "Estimating 3D Motion Parameters: Some Experimental Results," *Proceedings of SPIE, Intelligent Robots: Third International Conference on Robot Vision and Sensor Controls,* Cambridge, MA, 1983.

Huang, T. S., and C. H. Lee, "Motion and Structure from Orthographic Views," *IEEE Transactions on Pattern Analysis and Machine Intelligence,* 1989, Vol. PAMI-11, pp. 536–540.

Huber, P., *Robust Statistics,* Wiley, New York, 1981.

Jacobus, C. J., R. T. Chien, and J. M. Selander, "Motion Detection and Analysis of Matching Graphs of Intermediate-Level Primitives," *IEEE Transactions on Pattern Analysis and Machine Intelligence,* Vol. PAMI-2, 1980, pp. 495–510.

Jain, R., "Dynamic Scene Analysis Using Pixel-Based Processes," *Computer,* Vol. 14, 1981a, pp. 12–18.

——, "Extraction of Motion Information from Peripheral Processes," *IEEE Transactions on Pattern Analysis and Machine Intelligence,* Vol. PAMI-3, 1981b, pp. 489–503.

——, "Direct Computation of the Focus of Expansion," *IEEE Transactions on Pattern Analysis and Machine Intelligence,* Vol. PAMI-5, 1983a, pp. 58–64.

——, "Segmentation of Frame Sequence Obtained by a Moving Observer," *Research Publication GMR-4247,* Computer Science Department, General Motors Research Laboratories, Warren, MI, 1983b.

Jain, R., W. N. Martin, and J. K. Aggarwal, "Segmentation Through the Detection of Changes Due to Motion," *Computer Graphics and Image Processing,* Vol. 11, 1979, pp. 13–34.

Jain, R., and H. H. Nagel, "On the Analysis of Accumulative Difference Pictures from Image Sequences of Real World Scenes," *IEEE Transactions on Pattern Analysis and Machine Intelligence,* Vol. PAMI-1, 1979, pp. 206–214.

——, "Dynamic Scene Analysis," in *Progress in Pattern Recognition,* T. Kanade and A. Rosenfeld (eds.), 1984, pp. 125–168, North Holland, Amsterdam.

Jain, R., S. L. Bartlett, and N. O'Brien, "Motion Stereo Using Ego-Motion Complex Logarithmic Mapping," *IEEE Transactions on Pattern Analysis and Machine Intelligence,* Vol. PAMI-9, 1987, pp. 356–369.

Jain, R., and N. O'Brien, "Ego-Motion Complex Logarithm Mapping," *Intelligent Robots and Computer Vision,* SPIE Vol. 1984, pp. 16–23.

Jayaramamurthy, S. N., and R. Jain, "An Approach to the Segmentation of Textured Dynamic Scenes," *Computer Vision, Graphics, and Image Processing,* Vol. 21, 1983, pp. 239–261.

Jenkin, M., and J. K. Tsotsos, "Applying Temporal Constraints to the Dynamic Stereo Problem," *Computer Vision, Graphics, and Image Processing,* Vol. 33, 1986, pp. 16–32.

Jerian, C., and R. Jain, "Determining Motion Parameters for Scenes with Translation and Rotation," *IEEE Transactions on Pattern Analysis and Machine Intelligence,* Vol. PAMI-6, 1984, pp. 523–530.

——, "Polynomial Methods for Structure from Motion," *Proceedings of the Second International Conference on Computer Vision,* Tampa, FL, 1988, pp. 197–206.

Kahn, P., "Local Determination of a Moving Contrast Edge," *IEEE Transactions on Pattern Analysis and Machine Intelligence,* Vol. PAMI-7, 1985, pp. 402–409.

Kearney, J. K., W. B. Thompson, and D. L. Boley, "Gradient Based Estimation of Disparity," *Proceedings of the Conference on Pattern Recognition and Image Processing,* Las Vegas, Nev., 1982, pp. 246–251.

Kim, Y. C., and J. K. Aggarwal, "Determining Object Motion in a Sequence of Stereo Images," *IEEE Journal of Robotics and Automation,* Vol. RA-3, 1987, pp. 599–614.

Kittler, J., and J. Illingworth, "Minimum Error Thresholding," *Pattern Recognition,* Vol. 19, 1986, pp. 41–47.

Koenderink, J. J., and A. J. van Doorn, "Optic Flow," *Vision Research,* Vol. 26, 1986, pp. 161–180.

——, "Facts on Optic Flow," *Biological Cybernetics,* Vol. 56, 1987, pp. 247–254.

Kories, R., and G. Zimmermann, "Motion Detection in Image Sequences: An Evaluation of Feature Detectors," *Proceedings of the Seventh International Conference on Pattern Recognition,* Montreal, 1984.

Kwangyoen, W., L. S. Davis, and P. Thrift, "Motion Estimation Based on Multiple Local Constraints and Nonlinear Smoothing," *Pattern Recognition,* Vol. 16, 1983, pp. 563–570.

Lawton, D. T., "Motion Analysis via Local Translational Processing," *Proceedings of the IEEE Workshop on Computer Vision Representation and Control,* Rindge, NH, 1982, pp. 59–72.

——, "Processing Translational Motion Sequences," *Computer Vision, Graphics, and Image Processing,* Vol. 22, 1983, pp. 116–144.

Legters, G. R., and T. Y. Young, "A Mathematical Model for Computer Image Tracking," *IEEE Transactions on Pattern Analysis and Machine Intelligence,* Vol. PAMI-4, 1982, pp. 583–594.

Levine, M. D., P. B. Noble, and Y. M. Youssef, "Understanding Blood Cell Motion," *Computer Vision, Graphics, and Image Processing,* Vol. 21, 1983, pp. 58–84.

Liu, Y., and T. S. Huang, "Estimation of Rigid Body Motion Using Straight Line Correspondences," *Computer Vision, Graphics, and Image Processing,* Vol. 43, 1988, pp. 37–52.

Longuet-Higgins, H. C., "A Computer Algorithm for Reconstructing a Scene from Two Projections," *Nature,* Vol. 293, 1981, pp. 133–135.

——, "The Visual Ambiguity of a Moving Plane," *Proceedings of the Royal Society of London,* Series B, Vol. 223, 1984a, pp. 165–175.

Longuet-Higgins, H. C., "The Reconstruction of a Scene from Two Projections-Configurations That Defeat the 8-point Algorithm," *Proceedings of the First Conference on Artificial Intelligence Applications,* Denver, 1984b, pp. 395–397.

Marr, D., and S. Ullman, "Directional Selectivity and Its Use in Early Visual Processing," *Proceedings of the Royal Society of London,* Series B, Vol. 211, 1981, pp. 151–180.

Martin, W. N., and J. K. Aggarwal, "Volumetric Descriptions from Dynamic Scenes," *Pattern Recognition Letters 1,* Elsevier, North-Holland, 1982, pp. 107–113.

——, "Volumetric Descriptions of Objects from Multiple Views," *IEEE Transactions on Pattern Analysis and Machine Intelligence,* Vol. PAMI-5, 1983, pp. 150–158.

Murray, D. W., and B. F. Buxton, "Scene Segmentation from Visual Motion Using Global Optimization," *IEEE Transactions on Pattern Analysis and Machine Intelligence,* Vol. PAMI-9, 1987, pp. 161–180.

Nagel, H. H., "Representation of Moving Rigid Objects Based on Visual Observations," *IEEE Computer,* Vol. 14, 1981, pp. 29–39.

——, "Displacement Vectors Derived from Second-Order Intensity Variations in Image Sequences," *Computer Vision, Graphics, and Image Processing,* Vol. 21, 1983, pp. 85–117.

Neumann, B., "Exploiting Image Formation Knowledge for Motion Analysis," *IEEE Transactions on Pattern Analysis and Machine Intelligence,* Vol. PAMI-2, 1980, pp. 550–553.

——, "Optical Flow," *Computer Graphics,* Vol. 18, 1984, pp. 17–19.

Prazdny, K., "A Simple Method for Recovering Relative Depth Map in the Case of a Translating Sensor," *Proceedings of the Seventh International Joint Conference on Artificial Intelligence,* Vancouver, BC, Canada, 1981a, pp. 698–699.

——, "Determining the Instantaneous Direction of Motion from Optical Flow Generated by a Curvilinearly Moving Observer," *Computer Graphics and Image Processing,* Vol. 17, 1981b, pp. 238–248.

——, "On the Information in Optical Flows," *Computer Vision, Graphics, and Image Processing,* Vol. 22, 1983, pp. 239–259.

Roach, J. W., and J. K. Aggarwal, "Determining the Movement of Objects from a Sequence of Images," *IEEE Transactions on Pattern Analysis and Machine Intelligence,* Vol. PAMI-2, 1980, pp. 554–562.

Rosenfeld, A., L. S. Davis, and A. M. Waxman, "Vision for Autonomous Navigation," *Proceedings of the First Conference on Artificial Intelligence Applications,* Denver, 1984, pp. 140–141.

Schunck, B. G., "The Image Flow Constraint Equation," *Computer Vision, Graphics, and Image Processing,* Vol. 35, 1986, pp. 20–46.

——, "Image Flow Segmentation and Estimation by Constraint Line Clustering," *IEEE Transactions on Pattern Analysis and Machine Intelligence,* Vol. 11, 1989, pp. 1010–27.

Thompson, E. H., "A Rational Algebraic Formulation of the Problem of Relative Orientation," *Photogrammetric Record,* Vol. 3, 1959, pp. 152–159.

Tsai, R. Y., "Multiframe Image Point Matching and 3D Surface Reconstruction," *IEEE Transactions on Pattern Analysis and Machine Intelligence,* Vol. PAMI-5, 1983, pp. 159–174.

Tsai, R., and T. S. Huang, "Uniqueness and Estimations of Motion Parameter of Rigid Objects with Curved Surfaces," Technical Report R-921, UIC-ENG-81-2352, Coordinated Science Laboratory, University of Illinois, Urbana, August, 1981.

Tsai, R. Y., and T. S. Huang, "Uniqueness and Estimation of Three-Dimensional Motion Parameters of Rigid Objects with Curved Surfaces," *IEEE Transactions on Pattern Analysis and Machine Intelligence,* Vol. PAMI-6, 1984, pp. 13–26.

Tsai, R. Y., T. S. Huang, and W. L. Zhu, " Estimating Three-Dimensional Motion Parameters of a Rigid Planar Patch, II: Singular Value Decomposition," *IEEE Transactions on Acoustics, Speech, and Signal Processing,* Vol. ASSP-30, 1982, pp. 525–534.

Tsotsos, J. K., et al., "A Framework for Visual Motion Understanding," *IEEE Transactions on Pattern Analysis and Machine Intelligence,* Vol. PAMI-2, 1980, pp. 563–573.

Tsuji, S., M. Osada, and M. Yachida, "Tracking and Segmentation of Moving Objects in Dynamic Line Images," *IEEE Transactions on Pattern Analysis and Machine Intelligence,* Vol. PAMI-2, 1980, pp. 516–522.

Tsuji, S., Y. Tagi, and M. Asada, "Dynamic Scene Analysis for a Mobile Robot in a Man-Made Environment," *Proceedings of the IEEE International Conference on Robotics and Automation,* St. Louis, MO, 1985.

Tsuji, S., and M. Yachida, "Efficient Analysis of Dynamic Images Using Plan," *Proceedings of the US-JAPAN Seminar,* Tokyo, 1978.

Tsukiyama, T., and Y. Shirai, "Detection of the Movements of Persons from a Sparse Sequence of TV Images," *Pattern Recognition,* Vol. 18, 1985, pp. 207–213.

Turner, L. F., "A System for the Automatic Recognition of Moving Patterns," *IEEE Transactions on Information Theory,* Vol. IT-12, 1966, pp. 195–205.

Ullman, S., *The Interpretations of Visual Motion,* MIT Press, Cambridge, MA, 1979.

——, "Analysis of Visual Motion by Biological and Computer Systems," *Computer,* Vol. 14, 1981, pp. 57–69.

——, "Analysis of Visual Motion by Biological and Computer Systems," in *Readings in Computer Vision: Issues, Problems, Principles and Paradigms,* M. A. Fischler and O. Firschein (eds.), Morgan Kaufmann, Los Altos, CA, 1987, pp. 132–144.

Vemuri, B. C., K. R. Diller, and J. K. Aggarwal, "A Model for Characterizing the Motion of the Solid-Liquid Interface in Freezing Solutions," *Pattern Recognition,* Vol. 17, 1984, pp. 313–319.

Vemuri, B., et al., "Image Analysis of Solid-Liquid Interface Morphology in Freezing Solutions," *Pattern Recognition,* Vol. 16, 1983, pp. 51–61.

Waxman, A. M., B. Kamgar-Parsi, and M. Subbarao, "Closed-Form Solutions to Image Flow Equations for 3D Structure and Motion," *International Journal of Computer Vision,* Vol. 1, 1987, pp. 239–258.

Webb, J. A., and J. K. Aggarwal, "Visually Interpreting the Motion of Objects in Space," *IEEE Computer,* Vol. 14, August, 1981, pp. 40–46.

——, "Structure from Motion of Rigid and Jointed Objects," *Artificial Intelligence,* Vol. 19, 1982, pp. 107–130.

——, "Shape and Correspondence," *Computer Vision, Graphics, and Image Processing,* Vol. 21, 1983, pp. 145–160.

Weng, J., T. S. Huang, and N. Ahuja, "Error Analysis of Motion Parameter Estimation from Image Sequences," *Proceedings of the International Conference on Computer Vision,* London, 1987a, pp. 703–707.

——, "3D Motion Estimation, Understanding, and Prediction from Noisy Image Sequences," *IEEE Transactions on Pattern Analysis and Machine Intelligence,* Vol. PAMI-9, 1987b, pp. 370–389.

——, "Motion Parameter Determination from Two Perspective Views: Error Analysis," Coordinated Science Laboratory, University of Illinois, Urbana, 1987c.

Westphal, H., and H. H. Nagel, "Toward the Derivation of Three-Dimensional Descriptions from Image Sequences for Nonconvex Moving Objects," *Computer Vision, Graphics, and Image Processing,* Vol. 34, 1986, pp. 302–320.

Woodham, R. J., "Analyzing Images of Curved Surfaces," *Artificial Intelligence,* Vol. 17, 1981, pp. 117–140.

Yachida, M., "Determining Velocity Map by 3D Iterative Estimation," *Proceedings of the Seventh International Joint Conference on Artificial Intelligence,* Vancouver, BC, 1981.

——, "Determining Velocity Maps by Spatio-Temporal Neighborhoods from Image Sequences," *Computer Vision, Graphics, and Image Processing,* Vol. 21, 1983, pp. 262–279.

Yachida, M., M. Asada, and S. Tsuji, "Automatic Analysis of Moving Images," *IEEE Transactions on Pattern Analysis and Machine Intelligence,* Vol. PAMI-3, 1981, pp. 12–19.

——, "Automatic Motion Analysis System of Moving Objects from the Records of Natural Processes," *Proceedings of the Fourth International Joint Conference on Pattern Recognition,* pp. 726–730.

Yachida, M., M. Ikeda, and S. Tsuji, "A Plan-Guided Analysis of Cineangiograms for Measurement of Dynamic Behavior of Heart Wall," *IEEE Transactions on Pattern Analysis and Machine Intelligence,* Vol. PAMI-2, 1980, pp. 537–543.

Yalamanchili, S., W. M. Martin, and J. K. Aggarwal, "Extraction of Moving Object Descriptions via Differencing," *Computer Graphics and Image Processing,* Vol. 18, 1982, pp. 188–201.

Yasumoto, Y. and G. Medioni, "Robust Estimation of Three-Dimensional Motion Parameters from a Sequence of Image Frames Using Regularization," *IEEE Transactions on Pattern Analysis and Machine Intelligence,* Vol. PAMI-8, 1986, pp. 464–471.

Yen, B. L., and T. S. Huang, "Determining 3D Motion and Structure of a Rigid Body Using the Spherical Projection," *Computer Vision, Graphics, and Image Processing,* Vol. 21, 1983, pp. 21–32.

——, "Determining 3D Motion/Structure of a Rigid Body over Two Frames Using Correspondences of Straight Lines Lying on Parallel Planes," *Proceedings of the Seventh International Conference on Pattern Recognition,* Montreal, Canada, 1984.

Zhuang, X., "A Simplification to Linear Two View Motion Algorithms," *Computer Vision, Graphics, and Image Processing,* Vol. 46, 1989, pp. 175–178.

Zhuang, X., and R. M. Haralick, "Rigid Body Motion and the Optic Flow Image," *Proceedings of the First Conference on Artificial Intelligence Applications,* Denver, 1984, pp. 366–375.

Zhuang, X., T. S. Huang, and R. M. Haralick, "Two-View Motion Analysis: A Unified Algorithm," *Journal of the Optical Society of America,* Vol. 3, 1986, pp. 1492–1500.

——, "A Simple Procedure to Solve Motion and Structure from Three Orthographic Views," *IEEE Journal of Robotics and Automation,* Vol. 4, 1988a, pp. 236–239.

——, "A Simplified Linear Optic Flow-Motion Algorithm," *Computer Vision, Graphics, and Image Processing,* Vol. 42, 1988b, pp. 334–344.

Zisserman, A., P. Giblin, and A. Blake, "The Information Available to a Moving Observer from Specularities," *Image and Vision Computing,* Vol. 7, 1989, pp. 38–42.

Zucker, S. W., "The Fox and the Forest: Towards a Type 1/Type 2 Constraint for Early Optical Flow," in *Representation and Perception,* N. Badler and J. Tsotsos (eds.), Elsevier, North Holland, 1986.

16 IMAGE MATCHING

16.1.1 Image Matching and Object Reconstruction

Many computer vision tasks require the analysis of two or more images. Time-varying sequences for recognizing parts on a conveyor belt based on their three-dimensional shape or for the visual inspection of the geometry of manufactured parts; for the medical diagnosis of beating hearts; for monitoring land use; for deriving topographic maps from satellite or aerial imagery; can only be accomplished if, at least, pairs of related images are available. Other examples include the analysis of slices of computer tomography images. In principle, the inherent goal of these tasks is object reconstruction, that is, the determination of an object's pose or shape.

The situation we predominantly want to discuss here is symbolically sketched in Fig. 16.1(a). A concrete example is given in Fig. 16.1(b). The object is mapped into the images I^1 and I^2, and possibly further images I^k, via the transformations T_0^1 and T_0^2, and possibly additional transformations T_0^k. These transformations describe all aspects of the imaging process, namely, illumination, reflectance, sensing, and the like. They essentially depend on the geometry of the object (form, pose), of the illumination sources (point, line, or area source), of the sensing devices (interior orientation), and so on.

The mapping can be summarized as

$$I^k = T_0^k(p_G, p_R, p_P, p_A, \ldots)$$

This chapter was written by Wolfgang Förstner.

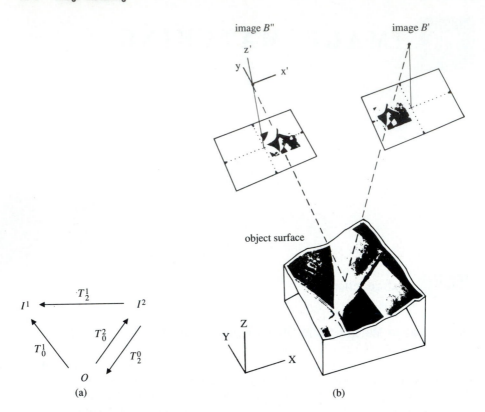

Figure 16.1 General setup of object reconstruction and image matching. (a) The object is mapped to I^1 and I^2 via the transformations T_0^1 and T_0^2. If enough images I^k are available, the reconstruction of the imaging process is possible in principle. If the reflectance properties of the surface O can be derived from one image, the compound transformation $T_2^1 = T_0^1 \cdot (T_0^2)^{-1}$ contains only geometric parameters that may be derived by using image-matching techniques. (b) The setup for two perspective images (from Wrobel, 1987). The reconstruction of the surface is possible either by a direct solution of the parameters describing the surface's geometry and reflectances or by forward intersection after having established correspondences between homologue image features and interpolation.

where $p_G, p_R, p_P, p_A, \ldots$ represent generally unknown parameters specifying the geometry and the reflectance properties of the object, the pose of the camera k, and the atmospheric response. Nearly all object reconstruction tasks, such as those mentioned above, can be modeled this way, possibly after including dynamic models for describing the change of the object over time. If enough images are taken, one in principle can recover the parameters from the observed intensities. Otherwise further regularization has to be applied by putting additional constraints onto the parameters. For analyzing perspective images, the most general case that has been realized assumes that the atmospheric parameters are known, the orientation parameters are restricted to six per image, and the reflectance and surface functions are described in a finite-element fashion (Helava, 1988; Ebner et al., 1987; Wrobel, 1987).

The general model assumes that the relation between object points or features and image points or features is established. This correspondence between object and image can be found automatically if constraints hidden in the transformation can be made explicit—for example, if the object's geometry is restricted to a polyhedron and if the surface reflectance obeys some regularities. In many cases these constraints and regularities are tight enough to enable one to find homologous features in the images—that is, to establish correspondences between image features—without explicitly referring to the object's properties. In the extreme case, only two images are sufficient to recover the object's form. It is this special case of image matching that we want to discuss in this chapter.

Formally, we want to assume that the transformations T_o^k are one to one, except perhaps up to a few parameters. Thus we assume that we can derive the reflectance parameters p_R from one image, given all other parameters. This implicitly excludes transparent surfaces or occlusions. The image I^2 can be predicted from I^1 by using the compound transformation

$$T_2^1 = T_0^1 \cdot T_2^0 = T_0^1 \cdot (T_0^2)^{-1}(p_G, p_R, p_{P1}, p_{P2}, p_A, \ldots)$$

The geometric part now contains the inverse perspective of image I^2, thus including the surface undulations, and also the perspective of image I^1. Only part of the pose parameters, p_{P1} and p_{P2}, can be derived, namely, those that do not depend on the external coordinate system of the object. The radiometric part is eliminated or at least restricted to a few parameters p_R if the reflectance does not depend strongly on the shape or the pose of the object, as in, for example, diffuse reflectance. In this case the difference between I^1 and I^2 is governed by the form of the object.

Under these restrictions the task of image analysis in all these applications is to establish correspondence between the images or to bring the images into registration. For perspective images the techniques of analytic photogrammetry (see Chapter 14) may be used to determine pose parameters or three-dimensional coordinates of object points.

16.1.2   The Principle of Image Matching

Image matching starts from two digital or digitized images or image patches I' and I''. Their size may vary from 5×5 pixels for tracking points in image sequences to $10,000 \times 10,000$ pixels for registering satellite images. Let the points P' and P'' of images I' and I'' have coordinates (r', c') and (r'', c'') and intensities g' and g'', respectively. Assume that if (r', c') and (r'', c'') are corresponding points, their coordinates can be related by

$$(r', c') = T_G(r'', c''; p_G) \tag{16.1}$$

where T_G is a specified geometric spatial-mapping function reflecting the knowledge about the geometric relation between the images, and p_G is a set or a vector of unknown parameters.

The intensities on one image can be related to those on the other by

$$g' = T_I(g''; p_I) \tag{16.2}$$

where the intensity-mapping function T_I contains the knowledge about the intensity relation between the images, again with the vector p_I being unknown. This leads to the complete model of *image matching*

$$g'(r',c') = T_I\{g''[T_G(r'',c''; p_G)]; p_I\} \tag{16.3}$$

The mapping functions T_G and T_I may be deterministic, stochastic, and/or piecewise continuous.

For optical or range images, the correspondence of points P' and P'' means that P' and P'' relate to the same point on the object. An *arbitrary* pair of points (P', P'') may have two states: P' and P'' are corresponding or they are not corresponding. As shown in Fig. 16.2, the problem of image matching or correspondence now consists of two parts:

1. Finding all corresponding points;
2. Determining the parameters p_G and p_I of the mapping functions T_G and T_I.

The solutions of parts 1 and 2 are not fully equivalent: The solution of part 1 usually does not imply a solution of part 2, as no explicit or implicit determination of the intensity mapping is available. But the solution of part 1 can be derived from the solution of part 2 if the spatial mapping is applied to all positions. In practice, however, this is neither feasible nor necessary, as the mapping function for physical reasons can generally be assumed at least piecewise smooth; thus only a limited number of corresponding points are sufficient to reconstruct the mapping function.

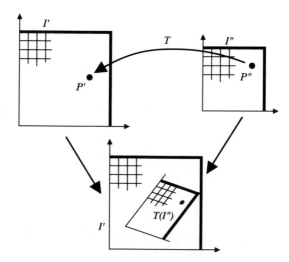

Figure 16.2 Principle of image matching and correspondence: (1) Find corresponding points P' and P'', and (2) determine mapping function T.

This implicitly assumes that the sampling theorem holds for the representation of the geometric spatial mapping.

Most approaches use only a limited number of points and explicitly or implicitly use mapping functions of the type represented in Eqs. (16.1) and (16.2). The solution is always based on the intensity functions g' and g'' or on the attributes a' and a'' of the points

$$P' = P'(r', c'; a') \quad \text{and} \quad P'' = P''(r'', c''; a'')$$

derived from the intensity functions in a neighborhood around (r', c') and (r'', c'').

The solution then is achieved in a three-step procedure:

1. Select appropriate *image features* in one or in both images, possibly by using an interest operator to derive a list of interesting points or edges in one image and another list for the second image. The selection of image features in only one image may be done using the intensity function as the describing attribute. The list of *all* (feasible) positions in the second image must then be searched by an area- or intensity-based image-matching procedure.

2. Find corresponding feature or point pairs (P'_i, P''_j), with P'_i from the first list and P''_j from the second list, that fulfill the criteria of *similarity* and *consistency*. Similarity is based on the attributes a'_i and a''_j of the features or points and thus on the properties of the intensity function, whereas consistency is based on the degree to which the spatial-mapping function is fullfilled. Both similarity and consistency usually are used in an *algorithm* to find an optimal solution, which obviously requires a relative weighting between similarity and consistency.

3. For stereo images, one may have to interpolate the parallaxes $(r'' - r', c'' - c')$ between the selected feature or point pairs in order to obtain a dense parallax field. This interpolation may be based on the *spatial-mapping function*. However, quite different interpolation schemes may also be used, especially if T_G is only given implicitly, for example, by the algorithm to optimize the consistency.

Practically all methods for image matching follow these steps but use significantly different image features, similarity or consistency measures, algorithms, and mapping functions. Some methods are based on the structure of the image content, and thus on a relational description of the images (Price, 1985; Boyer and Kak, 1986, 1988). These methods will be discussed in the following two chapters in detail.

16.1.3 Image Matching Procedures

Table 16.1, which is by no means exhaustive, summarizes the main properties of some matching algorithms. The applications lie in image sequence analysis (TV scenes, robotics, tracking, visual navigation) and in surface reconstruction from stereo images. The approach of Stockman, Kopstein, and Bennett (1982) and Stockman (1987) has also been applied to object location in industrial environments.

Most *interest operators* either try to use no knowledge about the scene and select edges, blobs, or statistically defined points or they rely on special properties

Table 16.1   Properties of some correspondence algorithms for image matching.

Author	Year	Features and Attributes of Interest Operator	Similarity Measure	Mapping Function and Interpolation	Algorithm	Application
Hannah	1974 1989	Variance Moravec	Correlation	NP /–	Local search H	ST
Barnard and Thompson	1980	Moravec	Intensity difference	NP/–	DR	ST
Dreschler	1981	Corners Blobs	Intensity difference class Interest value	NP /–	DR	IS
Baker and Binford	1982	Edges	Sign Strength	NP/ Linear	DP	ST
Grimson	1981 1985	Edges	Direction sign	NP/ PW, smooth	NN	ST
Stockman Kopstein, and Bennett Stockman	1982 1987	Abstract edges	Class	Similarity transfor- mation	Clustering	Registration object location
Benard	1983	edges blobs	grad. + intens. diff., direction	NP linear	DP	ST

Source: Adapted from Förstner, 1986a

Note: NP=non-parametric, PW=piecewise, DR=discrete relaxation, R=relaxation, IS=image sequence, DP=dynamic program, NN=nearest neighbor, H=hierarchical, ST=stereo, and ACF=autocorrelation function.

of the scene, assuming the ability to find corners or junctions of edges. In all cases the attributes of the features are derived from local neighborhoods of the points.

The applied *similarity* measures reflect the assumed intensity-mapping function. In the most simple case of identity mapping $g' = T_I(g'') = g''$, intensity differences serve to measure similarity. If the product moment correlation coefficient is used to measure similarity, T_I implicitly is assumed to be linear: $g' = ag'' + b$. If, however, the intensities are assumed to vary regionwise, properties of edges may be used to measure similarity.

A similar link between *consistency measure* or *interpolation* method and surface type can be established in the case of stereo images in normal orientation

Table 16.1 — *Continued*

Author	Year	Features and Attributes of Interest Operator	Similarity Measure	Mapping Function and Interpolation	Algorithm	Application
Nagel and Enkelmann	1983	Corners	Intensity difference	PW smooth	R	IS
	1986		Interest value	dto.		
Förstner	1984 1986	Roundness and curvature of ACF corners Blobs	SNR Interest value Distinctness	Affine dto.	R	ST
Zimmermann and Kories	1984	Blobs	Class	NP/–	NN	IS
Faugeras Ayache and Faverjon	1985/ 1987	Edges	Orientation sign	NP /–	Hypothesis Verify technique	ST IS
Ohta and Kanade	1985	Edges	Intensity difference	NP	DP	ST
Kanade	1987		Interest value	Linear		
Barnard	1986	Intensity values	Intensity constraint	PW/ dto.	Stochastic R H	ST
Witkin, Terzopoulos, and Kass	1987	—	Correlation	PW smooth	H multigrid	general signal

(Fig. 16.3). Constant or piecewise constant T_G (a) corresponds to surfaces or surface patches parallel to the image planes; linear T_G (b) corresponds to locally tilted planar surface patches; and smooth or piecewise smooth T_G or interpolation function (c) corresponds to smooth or piecewise smooth surfaces. Observe that, owing to occlusions (d), smooth surfaces may lead to discontinuities in the mapping.

The main difference between the methods is how an *algorithmic* solution for the consistency of the match is achieved. Most methods are iterative. *Relaxation schemes* seem to be the most flexible for a solution of the highly complex optimization problem, as no parametric form of the mapping function is required, though a parametric form may be used to advantage. In order to speed up convergence, hierarchical or scale space techniques, possibly together with multigrid methods, are

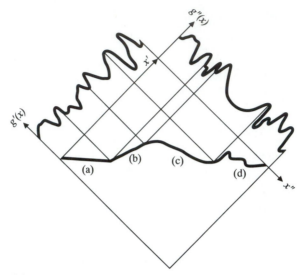

Figure 16.3 One-dimensional surface profile observed from two line cameras by using orthogonal projection: (a) horizontal, (b) sloped, (c) smooth, and (d) piecewise smooth with occlusions.

applied. These approaches guarantee a solution nearly in linear time. *Dynamic programming* also needs no explicit geometric model but optimizes a cost function that takes the similarity of the features into account. In image sequence analysis, consistency can be achieved by selecting the *nearest neighbor* to a point, taking care that the mapping is one to one. These methods can handle occlusions, specifically if the mapping function is assumed to be piecewise smooth. An interesting noniterative approach is the search for a *cluster in parameter space* of the transformation.

The scale space technique of Witkin, Terzopoulos, and Kass (1987) conceptually can handle more than two images, a feature they demonstrated with one-dimensional signals. Table 16.1 does not include the recent approaches by Ebner et al. (1987), Helava (1988), and Wrobel (1987), who independently proposed surface reconstruction schemes with an explicit, finite element approach. They assume piecewise smooth object surfaces. Conceptually they can handle multiple images and then recover possibly locally varying parameters of a reflectance model.

Though quite a variety of approaches exist, certain trends can be observed:

- Most algorithms contain a finite element description that includes discontinuities, which seems to be sufficient for matching two images. If more than two images are available, image matching will loose its power, as truly three-dimensional representations of the object to be recovered have to be used.

- Both intensity-based and feature-based methods are in practical use and may be combined. The result of the feature-based method usually provides the basis for the intensity-based procedure.

Table 16.2   Precision of image matching and edge detection (all figures in pixels).

Author	Year	Empirical Findings	Computer Simulations	Theoretical Values	Applications/Remarks
Sharp, Christensen, and Gilman	1965	1	—	—	Digital terrain models
Bernstein	1973 (cf. 1983)	0.1	—	—	Registration
Klaasman	1975	—	—	0.05	Edge detection
Cafforio and Rocca	1976	0.1	—	—	TV image sequences
McGillem and Svedlov	1976	—	—	0.5/SNR	Registration
Hill	1980	—	0.02–0.1	—	Binary images Target location
Huang/Hsu	1981	—	0.02–0.1	—	Parallax estimation
Förstner	1982	—	—	0.01–0.1	Target location
Thurgood and Mikhail	1982	—	0.02–0.1	—	Target location
Ackermann and Pertl	1983	0.1–0.2	—	—	Parallax estimation
Ho	1983	—	—	0.02–0.2	Binary images Target location
Grün/Baltsavias	1985	0.05–0.1	—	—	Parallax estimation
Vosselman	1986	0.02–0.03	—	—	Target location

Source: Adapted from Förstner, 1984.

- Hierarchical or scale space methods are commonly used to solve the problem of initial approximate values, which nearly all methods require.

Unfortunately, most approaches lack self-diagnosis, and little information on the achievable accuracy is provided by the matching technique itself. Self-diagnosis, however, is indispensable for real applications in order to automatically detect failures of the system and to be certain that the system produces meaningful answers.

Table 16.2 summarizes reported results on the precision of intensity-based matching algorithms and edge detectors in terms of standard deviations. The results are given separately for empirical findings or estimations, for computer simulations, and for values derived theoretically. There is a tendency toward subpixel accuracy, namely, 0.1 pixel or better. Such results were reported by Bernstein (1973) and Cafforio and Rocca (1976). They seem to be realistically obtainable under production conditions and come close to the corresponding results from computer simulations and theoretical studies.

This chapter shows how and under which conditions such high accuracies can be achieved with both intensity-based and feature-based methods. Our development of the algorithms will provide tools for self-diagnosis that can be used first to design matching procedures properly and then to check the performance at run time. The underlying concepts utilize classical least-squares estimation techniques and error propagation, allowing one to derive the precision of the final results in terms of the intensity signal, the noise, and the particular algorithm.

Section 16.2 discusses intensity-based matching of one-dimensional signals, Section 16.3 generalizes the results to two-dimensions. The selection of distinct points in Section 16.4 is discussed. Our approach is based on the expected matching accuracy. It also provides an interpretation of the selected points being independent of intensity-based matching, namely as selection of corners and centers of circular symmetric features. Section 16.5 presents a feature-based matching algorithm using a maximum likelihood type estimation for the geometric transformation. Section 16.6 provides the necessary tools for deriving three-dimensional coordinates from points matched in stereo pairs. Empirical results from an implemented system close the chapter.

16.2   Intensity-Based Matching of One-Dimensional Signals

Intensity-based matching techniques directly refer to the model equation (16.3) and aim at estimating and evaluating the parameters p_G and p_I. To give insight into the principles, we first develop methods for matching one-dimensional signals. As even this task is demanding when taking the statistical properties of the data into account, we restrict this discussion to the case in which one of the signals is assumed to be perfectly known. This situation is relevant to an object location procedure. The model then can be written as

$$g(x) = T_I\{f[T_G(y; p_g)]; p_I\} + n(x)$$

where g', g'', r', and r'' have been replaced by g, f, x, and y, respectively, and the observational noise component $n(x)$ is stated explicitly. We first assume that $f(y)$ is given by sampled data and a fitting or interpolation scheme that allows one to use a derived or estimated continuous $f(y)$ from which the first derivatives can be obtained analytically. This is just like the facet model (Chapter 8). Because of the highly nonlinear character of the estimation problem, we always assume that approximate values $p_I^{(0)}$ and $p_G^{(0)}$ are known from some prior information or prediction scheme. This initial approximation permits us to replace the nonlinear problem by a linear substitute problem whose solution then gives rise to better approximations. In this way we naturally arrive at an iterative solution to the nonlinear problem. We therefore always assume that second-order effects are sufficiently small specifically

(a) when interpolating f or its derivative, (b) when neglecting the random nature of f if it is derived from real data, or (c) when deriving variances for the estimated parameters. A variety of controlled tests have shown that the approximations related to these assumptions are fully acceptable in most applications.

16.2.1 The Principle of Differential Matching

The simplest kind of matching is that of two one-dimensional signals derived from two views of the same scene. For now, one could think of these signals as being image rows.

   To simplify the explanation of the estimation procedure involved in the matching, we assume that there are no intensity changes due to viewing direction and that

$$y(x_i) = x_i - \hat{u}(x_i)$$

where $\hat{u}(x_i)$ is the unknown deformation at x_i to produce the corresponding point in the object. The *nonlinear model* can thus be expressed as

$$g(x_i) = f[x_i - \hat{u}(x_i)] + n(x_i), \qquad i = 1, \ldots, m$$

which is valid for all m observed values $g(x_i)$, $i = 1, \ldots, m$, and where $n(x_i)$ is the additive noise at position x_i, which we assume to be independent and identically distributed with mean zero and variance σ_n^2. The nonlinear model makes the assumed observation process explicit. An example for the functions f and g is given in Fig. 16.4. We now assume that we already have a function approximation u_0 of the

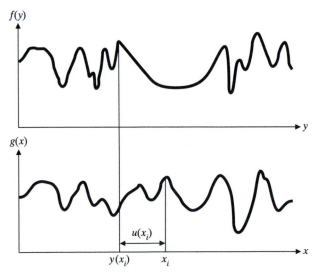

Figure 16.4 Assumed observation process. The observed function $g(x)$ is a noisy version of the deformed function $f(y)$. The deformation $u(x) = x - y(x)$ refers to the coordinates of the observed function $g(x)$ and is positive in the example point x_i, as the point x_i corresponding to $y(x_i)$ is to the right of $y(x_i)$.

function $\hat{u}$. For example, this approximation can be considered given because of the highly constrained geometry inherent in many factory-floor vision tasks. With approximate values $u_0(x_i)$ given, we can write

$$\hat{u}(x_i) = u_0(x_i) + \widehat{\Delta u}(x_i)$$

where the $\widehat{\Delta u}(x_i)$ is the unknown correction for the unknown value $\hat{u}(x_i)$. Thus the nonlinear model can be written as

$$g_i = f(x_i - \hat{u}_i) + n_i$$
$$= f(x_i - u_{0i} - \widehat{\Delta u}_i) + n_i, \qquad i = 1, \ldots, m$$

with the abbreviations $g_i = g(x_i)$, $\hat{u}_i = \hat{u}(x_i)$, $n_i = n(x_i)$ $u_{0i} = u_0(x_i)$, and $\widehat{\Delta u}_i = \widehat{\Delta u}(x_i)$.

Now we linearize f around the point $x_i - u_{0i}$ to obtain

$$g_i = f(x_i - u_{0i}) - f'(x_i - u_{0i})\widehat{\Delta u}_i + \frac{1}{2}f''(x_i - u_{0i} - \xi\widehat{\Delta u}_i)(\widehat{\Delta u}_i)^2 + n_i$$

for some $\xi \in [0, 1]$, and where $f'(y) = df/dy$ and $f''(y) = d^2f/dy^2$.

In the following we assume that f' does not vanish and that the second-order term is negligible. With the differences

$$\Delta g_i = \Delta g(x_i) = g(x_i) - f(x_i - u_{0i})$$

and the derivative

$$f_i' = f'(x_i - u_{0i})$$

we then obtain the *linearized model* in which Δg_i and f_i' are known and $\widehat{\Delta u}_i$ is unknown:

$$\Delta g_i = -f_i'\,\widehat{\Delta u}_i + n_i, \qquad i = 1, \ldots, m \qquad (16.4)$$

or explicitly

$$g(x_i) - f[x_i - u_0(x_i)] = -\frac{df(y)}{dy}\Big|_{y=x_i-u_0(x_i)} \widehat{\Delta u}(x_i) + n(x_i), \qquad i = 1, \ldots, m$$

We can easily determine the random variable $\widehat{\Delta u}_i$ assuming $f_i' \neq 0$:

$$\widehat{\Delta u}_i = -\frac{\Delta g_i}{f_i'} \qquad (16.5)$$

or explicitly

$$\widehat{\Delta u}_i = -\frac{g(x_i) - f[x_i - u_0(x_i)]}{\frac{df(y)}{dy}\big|_{y=x_i-u_0(x_i)}}$$

Thus

$$\hat{u}_i = u_{0i} + \widehat{\Delta u}_i$$

The estimated local deformation—that is, the shift $\hat{u}(x_i)$—therefore simply uses the difference of the functions at the corresponding position $x_i - u_0(x_i)$ and x_i of f and g and the slope f' (Fig. 16.5). This procedure is also termed the *differential approach* when estimating the optical flow (cf. Huang, 1981). As the variance $\sigma^2_{\Delta g_i}$

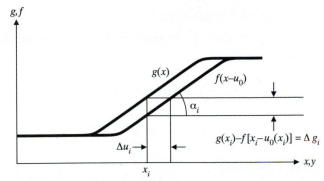

Figure 16.5 Principle of the applied differential approach to estimate the local deformation $\Delta u(x_i)$. The functions g and f are assumed to be locally approximable by a linear function. Then one is able to derive the shift $\Delta u_i = -\Delta g_i / f_i'$ from the difference Δg_i of f and g at the corresponding positions x_i and $x_i - u_0(x_i)$, which depends on the approximate deformation $u_0(x_i)$ and on the known slope $f_i' = \tan \alpha$.

of Δg_i is equal to the noise variance σ_n^2, we obtain for the standard deviation $\sigma_{\hat{u}_i}$ of the estimate $\hat{u}_i = u_0(x_i) + \widehat{\Delta u_i}$

$$\sigma_{\hat{u}_i} = \frac{\sigma_n}{f_i'} \tag{16.6}$$

The precision of the estimate obviously depends on the slope f_i' of the given function and is high at edges with high slope, which is to be expected. No estimate for u is available if $f_i' = 0$, and for small f_i' the weight, $w_i = f_i'^2 / \sigma_n^2$, of this estimate is low.

As there is no redundancy in this estimate, it is unreliable in the sense that errors in the approximate values $u_0(x_i)$ and the model—especially wrong assumptions relative to the invariance of lighting—fully influence the estimate. Therefore additional constraints are necessary to improve the result. Specifically, knowledge about the structure of $u(x)$ can allow the use of larger windows. In the following section we discuss deformation functions $u(x)$ with increasing complexity and add parameters for brightness and contrast differences.

16.2.2 Estimating an Unknown Shift

We assume here that $\hat{u}(x)$ is only a uniform shift, $\hat{u}(x) = \hat{u}$, so that $y(x_i) = x_i - \hat{u}$. Further, we assume that u_0 is an initial approximation for u, so that we can write $\hat{u} = u_0 + \widehat{\Delta u}$ for all x. Then by linearizing f around $x_i - u_0$ for each i, there results

$$f[x_i - (u_0 + \widehat{\Delta u})] = f(x_i - u_0) - f'(x_i - u_0)\widehat{\Delta u}$$

The linearized model can then be expressed as

$$g(x_i) = f(x_i - u_0) - f'(x_i - u_0)\widehat{\Delta u} + n_i, \quad i = 1, \ldots, m$$

or in short

$$\Delta g_i = -f_i' \widehat{\Delta u} + n_i, \qquad i = 1, \ldots, m \tag{16.7}$$

holding for all m observed values g_i. Minimizing $\Omega = \sum_{i=1}^{m} n_i^2$ by choosing the appropriate value of $\widehat{\Delta u}$ yields

$$\frac{\partial \Omega}{\partial \widehat{\Delta u}} = \frac{\partial}{\partial \widehat{\Delta u}} \sum_{i=1}^{m} (f_i' \Delta g_i + f_i' \widehat{\Delta u})^2$$

$$= 2 \sum_{i=1}^{m} f_i \Delta g_i + 2 \left(\sum f_i'^2 \right) \widehat{\Delta u} = 0$$

from which follows the estimate

$$\widehat{\Delta u} = -\frac{\sum_{i=1}^{m} f_i' \, \Delta g_i}{\sum_{i=1}^{m} f_i'^2} = -\frac{\sum_{i=1}^{m} [g(x_i) - f(x_i - u_0)] f'(x_i)}{\sum_{i=1}^{m} [f'(x_i)]^2} \tag{16.8}$$

To get information about the precision of the estimate $\hat{u}$, we write the estimate as

$$\hat{u} = u_o + \sum_{i=1}^{m} a_i \Delta g_i = u_o + a^T \Delta g$$

where

$$a_i = -f_i' / \sum f_i'^2$$

$$a = \begin{pmatrix} a_1 \\ a_2 \\ \vdots \\ a_m \end{pmatrix}$$

$$\Delta g = \begin{pmatrix} \Delta g_1 \\ \Delta g_2 \\ \vdots \\ \Delta g_m \end{pmatrix}$$

Letting $\sigma_{\hat{u}}^2 = E[(\hat{u} - E(\hat{u})]^2$, we obtain

$$\sigma_{\hat{u}}^2 = E\{[u_o + a^T \, \Delta g - E(u_o + a^T \, \Delta g)][u_o + \Delta g^T a - E(u_o + \Delta g^T a)]\}$$

$$= E\{a^T [\Delta g - E(\Delta g)][\Delta g^T - E(\Delta g^T)]a\}$$

$$= a^T E\{[\Delta g - E(\Delta g)][\Delta g^T - E(\Delta g^T)]\}a$$

Since the covariance matrices $\Sigma_{\Delta g \Delta g}$ and Σ_{nn} are equal to $Diag(\sigma_{n_i}^2) = \sigma_n^2 \cdot I$,

$$\sigma_{\hat{u}}^2 = a^T \, \Sigma_{nn} a = a^T Diag(\sigma_{n_i}^2) a$$

$$= \left(\sum_{i=1}^{m} a_i^2 \right) \sigma_n^2$$

As

$$\sum_{i=1}^{m} a_i^2 = \frac{\sum_{i=1}^{m} f_i'^2}{(\sum_{i=1}^{m} f_i'^2)^2} = \frac{1}{\sum_{i=1}^{m} f_i'^2}$$

we finally obtain for the variance of $\hat{u}$

$$\sigma_{\hat{u}}^2 = \frac{\sigma_n^2}{\sum_{i=1}^{m} f_i'^2} \tag{16.9}$$

If we define the *mean squared gradient* of f_i' by

$$\sigma_{f'}^2 \doteq \frac{\sum_{i=1}^{m} f_i'^2}{m} \tag{16.10}$$

we can represent the standard deviation for the estimated shift $\hat{u}$ by

$$\sigma_{\hat{u}} = \frac{\sigma_n}{\sqrt{m}\,\sigma_{f'}} \tag{16.11}$$

In an intuitive manner the standard direction depends on the number m of the observed values, the noise variance σ_n^2, and the texture or edge busyness measured by the mean squared gradient of the object.

We would obtain the same result for the estimate $\hat{u}$ and its variance by taking the weighted average of the individual $\widehat{\Delta u}_i$ from Eq. (16.5). For weights we use

$$w_i = \frac{f_i'^2}{\sum_{j=1}^{m} f_j'^2} \tag{16.12}$$

which sum to 1 and are inversely proportional to the variances of the $\widehat{\Delta u}$ that are given by Eq. (16.6). That is,

$$\widehat{\Delta u} = \frac{\sum_{i=1}^{m} \widehat{\Delta u}_i w_i}{\sum w_i}$$

$$= -\frac{\sum_{i=1}^{m} \frac{\Delta g_i}{f_i'} f_i'^2}{\sum_{i=1}^{m} f_i'^2} \tag{16.13}$$

The estimate obviously is independent of the assumed variance σ_n^2 of the noise.

As the common denominator $\sum f_i'^2$ in the weights of Eq. (16.12) cancels in the estimates equation (16.13), one can simply use $w_i = f_i'^2$ as weights. Thus only those parts where the slopes f_i' are nonzero contribute to the estimate, which accords with intuition. The selection of edges is often based on the relative maxima of f'^2; this can be interpreted as a selection of places that locally contribute the most to the estimate of the signal's geometry. The interest operator presented in Section 16.4 is a two-dimensional version of this idea.

If m is large enough, the noise variance can be estimated from the residuals of the least-squares fit:

$$\hat{\sigma}_n^2 = \frac{1}{m-1} \sum_{i=1}^{m} \left[g(x_i) - f(x_i - \hat{u}) \right]^2 \tag{16.14}$$

One can show (Koch, 1987) that the precision of the estimated noise standard deviation $\hat{\sigma}_n$ is

$$\hat{\sigma}_{\hat{\sigma}_n} = \frac{1}{\sqrt{2(m - m_p)}} \hat{\sigma}_n \tag{16.15}$$

with the number $m_p = 1$ of the unknown parameters, if the simple model of Eq. (16.7) really holds. Thus at least 72 samples are necessary to yield an accuracy of better than 10% in $\hat{\sigma}_n$.

EXAMPLE 16.1

Let an edge model $f(y_i)$ and the observed edge $g(x_i)$ be given:

i	1	2	3	4	5	6	7	8	9	10	11	12	13	14
f_i	10	10	10	10	20	30	40	40	40	40	40	40	40	40
g_i	—	—	—	14	13	14	26	37	42	41	42	—	—	—
Δg_i	—	—	—	4	-7	-16	-14	-3	2	1	2	—	—	—
f_i'	0	0	0	5	10	10	5	0	0	0	0	0	0	0
n_i	—	—	—	4	3	-4.8	-2.8	-1.8	2	1	2	—	—	—

The pixel spacing is Δx, the gradients $f_i' = (f_{i+1} - f_{i-1})/2[gr/\Delta x]$, and the approximate value $u_0 = 0[\Delta x]$. Omitting the first and last three observations, we obtain the estimate

$$\hat{u} = -\frac{\sum f_i' g_i}{\sum f_i'^2} = -\frac{5 \cdot 4 + 10 \cdot (-7) + 10 \cdot (-16) + 5 \cdot (-14)[gr^2/\Delta x]}{25 + 100 + 100 + 25[gr^2/\Delta x^2]}$$

$$= -\frac{-280[gr^2/\Delta x]}{250[gr^2/\Delta x^2]} = 1.118[\Delta x]$$

The theoretical precision of this estimate, when assuming $\sigma_n = 1[gr]$, is

$$\sigma_{\hat{u}} = \frac{\sigma_n}{\sqrt{\sum f_i'^2}} = \frac{1[gr]}{\sqrt{250[gr^2/\Delta x^2]}} \approx 0.06[\Delta x]$$

and thus less than one-tenth of a pixel. Only observations 4, 5, 6, and 7 have nonzero weights and hence contribute $\hat{u}$. The residuals $n(x_i)$ are comparatively large. The estimated noise standard deviation according to Eq. (16.14) is $\sigma_n \approx \sqrt{1/(8-1) \cdot 45.32[gr^2]} \approx 3.1[gr]$. This is significantly larger than $1[gr]$. There are obvious differences in brightness, contrast, and possibly also in scale.

16.2.3 Estimating Unknown Shift and Scale

We now augment our model by assuming that the transformation contains a parameter compensating for unknown scale:

$$y_i = s\,(x_i - x_0) - \hat{u} = s\,\bar{x}_i - \hat{u}$$

Note our use of a fixed reference point x_0, which we will later choose for our convenience. From here on we work in the reduced coordinates

$$\bar{x}_i = x_i - x_0$$

The linear model reads as

$$g(x_i) = f(s\,\bar{x}_i - \hat{u}) + n(x_i), \quad i = 1, \ldots, m$$

Using approximate values s_0 and u_0 for the unknown parameters s and $\hat{u}$ and linearizing around $s_0\bar{x}_i - u_0$ yields

$$g(x_i) = f(s_0\bar{x}_i - u_0) - f'(s_0\bar{x}_i - u_0)\,\widehat{\Delta u}$$

$$+ f'(s_0\bar{x}_i - u_0)\,\bar{x}_i\,\widehat{\Delta s} + n(x_i)$$

This result leads to the linearized model

$$\Delta g_i = -f'_i\widehat{\Delta u} + f'_i\bar{x}_i\widehat{\Delta s} + n(x_i), \quad i = 1, \ldots, m$$

again holding for all m observed values, and where $\widehat{\Delta s}$ satisfies $s = s_0 + \widehat{\Delta s}$.

Minimizing $\Omega = \sum n^2(x_i)$ with respect to the two unknown parameters, we set $\partial\Omega/\partial\widehat{\Delta u} = 0$ and $\partial\Omega/\partial\widehat{\Delta s} = 0$ to yield

$$\frac{\partial\Omega}{\partial\widehat{\Delta u}} = \frac{\partial\Omega}{\partial\widehat{\Delta u}}\sum_{i=1}^{m}(\Delta g_i + f'_i\,\widehat{\Delta u} - f'_i\bar{x}_i\,\widehat{\Delta s})^2$$

$$= 2\sum_{i=1}^{m} f'_i(\Delta g_i + f'_i\,\widehat{\Delta u} - f'_i\bar{x}_i\,\widehat{\Delta s}) = 0$$

and analogously

$$\frac{\partial\Omega}{\partial\widehat{\Delta s}} = -2\sum_{i=1}^{m} f'_i x_i(\Delta g_i + f'_i\,\widehat{\Delta u} - f'_i\bar{x}_i\,\widehat{\Delta s}) = 0$$

After some rearranging we can obtain the matrix equation

$$\begin{pmatrix} \sum f'^2_i & -\sum f'^2_i\bar{x}_i \\ -\sum f'^2_i\bar{x}_i & \sum f'^2_i\bar{x}^2_i \end{pmatrix} \begin{pmatrix} \widehat{\Delta u} \\ \widehat{\Delta s} \end{pmatrix} = \begin{pmatrix} -\sum f'_i\,\Delta g_i \\ \sum f'_i\bar{x}_i\,\Delta g_i \end{pmatrix}$$

where the sums are to be taken over all i, $i = 1, \ldots, m$. The solution of this 2×2 normal equation system yields the least-squares estimates for the unknown corrections $\widehat{\Delta u}$ to u_0 and $\widehat{\Delta s}$ to s_0:

$$\hat{u} = u_0 + \widehat{\Delta u}, \quad \hat{s} = s_0 + \widehat{\Delta s}$$

The variance of the noise can again be derived from the residuals

$$\hat{\sigma}_n^2 = \frac{1}{m-2} \sum (\Delta g_i + f_i' \widehat{\Delta u} - f_i' \bar{x}_i \widehat{\Delta s})^2$$

In order to analyze the estimates in detail, we shift the coordinate system such that the off-diagonal term in the normal equation becomes zero. The shift x_0 then has to be chosen so that $\sum f_i'^2 \bar{x}_i = 0$ or

$$x_0 = \frac{\sum_{i=1}^{m} f_i'^2 x_i}{\sum f_i'^2} \quad \text{with } f_i' = \frac{df(y)}{dy}\bigg|_{y=s_0(x_i-x_0)-u_0} \tag{16.16}$$

This is the *weighted center of gravity of the object,* where the weights are again the squares of the gradient of f evaluated at the corresponding positions $s_0(x_i - x_0) - u_0$.

With this choice of x_0, the estimates $\widehat{\Delta u}$ and $\widehat{\Delta s}$ are statistically independent. The estimate here for $\widehat{\Delta u}$ is identical to the estimate without scale parameter. Thus if one is free to choose a point in the image to be transformed into the object by using $y_i' = \hat{s}(x_i - x_0) - \hat{u}$, one may use the weighted center of gravity and solve for the shift only. Also, the precision of this shift is independent of the scale.

In general, a point x_i in the image corresponds to the *predicted point* $\hat{y}_i = \hat{s}(x_i - x_0) - \hat{u}$ in the object. The variance of the predicted point $\hat{y}_i$ is given by

$$\sigma_{\hat{y}_i}^2 = \sigma_n^2 \left(\frac{1}{\sum f_i'^2} + \frac{(x_i - x_0)^2}{\sum f_i'^2 \bar{x}_i^2} \right) \tag{16.17}$$

Thus the variance of points, transformed from the image into the object, increases with increasing distance from the weighted center of gravity. The weighted center of gravity x_0 from Eq. (16.16) is the point that, when transformed into the object, has the smallest variance.

We will generalize this result in the two-dimensional case and use it as a basis for the derivation of an interest operator in Section 16.4.

■ EXAMPLE 16.2

With the same data as in Example 16.1, we want to determine the shift u and a geometric scale s. We assume $u_0 = 1$, approximating the first estimate, and $s_0 = 1$. The 2×2 normal equation is

$$\begin{pmatrix} 250[gr^2/\Delta x^2] & -1625[gr^2/\Delta x] \\ -1625[gr^2/\Delta x] & 10725[gr^2] \end{pmatrix} \begin{pmatrix} \widehat{\Delta u} \\ \widehat{\Delta s} \end{pmatrix} = \begin{pmatrix} 100[gr^2/\Delta x] \\ -685[gr^2] \end{pmatrix}$$

which yields the estimates

$$\hat{u} = u_0 + \widehat{\Delta u} = 1 - 1.002[\Delta x] = -0.002[\Delta x]$$

and

$$\hat{s} = s_0 + \widehat{\Delta s} = 1 - 0.216[1] = 0.784$$

With the assumed noise standard deviation $\sigma_n = 1[gr]$ and the inverse $Q = N^{-1}$,

$$\begin{pmatrix} 0.2640[\Delta x^2/gr^2] & 0.0400[\Delta x/gr^2] \\ 0.0400[\Delta x/gr^2] & 0.006156[1/gr^2] \end{pmatrix}$$

we obtain the theoretical standard deviations according to

$$\sigma_{\hat{u}} = \sigma_n \sqrt{Q_{11}} = 1[gr] \cdot \sqrt{0.2640[\Delta x/gr^2]} \approx 0.52[\Delta x]$$

$$\sigma_{\hat{s}} = \sigma_n \sqrt{Q_{22}} = 1[gr] \cdot \sqrt{0.006156[1/gr^2]} \approx 0.08$$

The shift is significantly less accurate than in Example 16.1 without the scale parameter. The geometric scale can be determined to approximately 8%. The reason for the increased standard deviation of $\hat{u}$ is that $\hat{u}$ now refers to the origin of the coordinate system alone, which (cf. the table in Example 16.1) lies to the left of the used interval.

The weighted center of gravity of f is

$$x_{0f} = \frac{\sum f_i'^2 x_i}{\sum f_i'^2} = \frac{5 \cdot 4 + 10 \cdot 5 + 10 \cdot 6 + 5 \cdot 7[gr^2/\Delta x]}{250[gr^2/\Delta x^2]} = 5.5[\Delta x]$$

In a similar manner we obtain

$$x_{0g} = 7.12[\Delta x]$$

If we now transform x_{0g}, we get the predicted point $y_{0g} = \hat{s} \cdot x_{0g} - \hat{u} = 0.784 \cdot 7.12[\Delta x] - (-0.002[\Delta x]) = 5.59[\Delta x]$, which is very close to the weighted center of gravity $5.5[\Delta x]$ of f. Its precision is identical to the precision of the shift in Example 16.1, namely, $\sigma_{y_{0g}} = 0.06[\Delta x]$, as it is independent on the scale! This also can be proved by error propagation using $y_{0g} = (-1 \; x_{0g})(\hat{u} \; \hat{s})^T$. Thus

$$\sigma_{y_{0g}}^2 = \sigma_n^2(-1 \; x_{0g})Q(-1 \; x_{0g})^T$$

16.2.4 Compensation for Brightness and Contrast

If object and image have different brightness and contrast, we have to compensate for this difference. Without any geometric scale parameter, the nonlinear model then reads

$$g(x_i) = af(x_i - u) + b + n(x) \tag{16.18}$$

where a represents change in contrast and b change in brightness.

If we start from an approximate value a_0 for a and u_0 for u, we obtain the linearized model in a manner similar to that just described. We let $\Delta g_i = \Delta g(x_i) = g(x_i) - a_0 f(x_i - u_0)$ and obtain

$$\Delta g_i = -a_0 f_i' \widehat{\Delta u} + f_i \widehat{\Delta a} + \hat{b} + n_i, \qquad i = 1, \ldots, m$$

(We do not need an approximate value for b, as the model equation 16.18 is linear in b.)

The normal equation system for the least-squares solution is then given by

$$
\underbrace{\begin{pmatrix} a_0^2 \sum f_i'^2 & -a_0 \sum f_i' f_i & -a_0 \sum f_i' \\ -a_0 \sum f_i' f_i & \sum f_i^2 & \sum f_i \\ -a_0 \sum f_i' & \sum f_i & \sum 1 \end{pmatrix}}_{Q^{-1}} \begin{pmatrix} \widehat{\Delta u} \\ \widehat{\Delta a} \\ \hat{b} \end{pmatrix} = \begin{pmatrix} -a_0 \sum f_i' \Delta g_i \\ \sum f_i \Delta g \\ \sum \Delta g_i \end{pmatrix}
$$

$$(16.19)$$

where

$$
f_i' = \frac{df(y)}{dy}\bigg|_{y=x_i-u_0}
$$

and $f_i = f(x_i - u_0)$, and again the sums are to be taken over all i.

The estimates thus are

$$
\hat{u} = u_o + \widehat{\Delta u}, \quad \hat{a} = a_0 + \widehat{\Delta a}, \text{ and } \hat{b}
$$

The precision of the estimates can be obtained by using the inverse $Q = (q_{ij})$ of Q^{-1} of the 3×3 normal system in Eq. (16.19), that is

$$
\sigma_{\hat{u}} = \sigma_n \sqrt{q_{11}}
$$
$$
\sigma_{\hat{a}} = \sigma_n \sqrt{q_{22}}
$$
$$
\sigma_{\hat{b}} = \sigma_n \sqrt{q_{33}}
$$

The noise variance may be estimated from

$$
\hat{\sigma}_n^2 = \frac{1}{m-3} \sum_{i=1}^{m} \left[g(x_i) - \hat{a} f(x_i - \hat{u}) - \hat{b} \right]^2 \qquad (16.20)
$$

EXAMPLE 16.3

With the same data as in Example 16.1 we want to determine the shift and two radiometric parameters for brightness and contrast. We obtain the 3×3 normal equation system (16.19) (cited without dimensions):

$$
\begin{pmatrix} 250 & -750 & -30 \\ -750 & 7900 & 230 \\ -30 & 230 & 8 \end{pmatrix} \begin{pmatrix} \hat{u} \\ \hat{a} \\ \hat{b} \end{pmatrix} = \begin{pmatrix} 100 \\ -90 \\ -1 \end{pmatrix}
$$

The estimates are therefore

$$
\hat{u} = 1.796[\Delta x], \quad \hat{a} = 0.883, \quad \hat{b} = 6.226[gr]
$$

With the inverse normal matrix $Q = N^{-1}$

$$
\begin{pmatrix} 0.007877 & -0.000686 & 0.04909 \\ -0.000686 & 0.000836 & -0.02662 \\ 0.04909 & -0.02662 & 1.074 \end{pmatrix}
$$

and the estimated noise standard deviation from Eq. (16.20)

$$\sigma_n = \sqrt{20.48[gr^2]/(8-3)} \approx 2.0[gr]$$

we obtain the empirical standard deviations for the estimates

$$\hat{\sigma}_{\hat{u}} \approx 2.0\sqrt{0.007877} = 0.18[\Delta x]$$

and similarly

$$\hat{\sigma}_{\hat{a}} \approx 0.059, \text{ and } \hat{\sigma}_{\hat{b}} = 2.1[gr]$$

The fit between model and data obviously is significantly better than in the two previous examples as the estimated noise. The standard deviation is reduced to only 2[gr], which is due to the better modeling of the radiometry.

16.2.5 Estimating Smooth Deformations

If the transformation $y = T_G(x) = y(x)$ (cf. Section 16.2.1) cannot be approximated by a linear function, or if one wants to use larger windows but the transformation is still smooth, it may be represented by

$$y_i = y(x_i) = x_i - u(x_i)$$

with some smoothness constraints on $u(x)$. These could refer to the first, second, or higher derivatives of y or any function $D_u = D(u', u'', \ldots)$ of the derivatives of $u(x_i)$. Thus we could require

$$E\left[D_{u_j}(u)\right] = 0, \qquad V\left[D_{u_j}(u)\right] = \sigma_{D_j}^2, \quad j = 1, \ldots, k \qquad (16.21)$$

holding for all j from 1 through k. D_{u_j}, for example, stands for a linear combination

$$D_{u_j} = a_1 \, u_j'[u(x)] + a_2 \, u_j''[u(x)] + \cdots$$

of the derivatives of u evaluated at x_j.

We now want to estimate the $u_i = u(x_i)$ by using the observed values

$$g(x_i) = f\left[x_i - u(x_i)\right] + n(x_i) \qquad (16.22)$$

and the a priori knowledge about the smoothness of $u(x)$ in a Bayesian manner. We can maximize the a posteriori probability

$$p(u \mid g, D_u) = \frac{p(g \mid u) \, p(D_u \mid u)}{\int \int \, p(g \mid u) \, p(D_u \mid u) \, dg dD_u}$$

Assuming Gaussian distribution for both $g_i = g(x_i)$ and $D_{u_j}(u)$ with

$$p(g_i \mid u) = \frac{1}{\sqrt{2\pi\sigma_{n_i}^2}} \exp\left[-\frac{1}{2}\left(\frac{g_i - f(y_i)}{\sigma_{n_i}}\right)^2\right]$$

$$p(D_{u_j} \mid u) = \frac{1}{\sqrt{2\pi\sigma_{D_j}^2}} \exp\left[-\frac{1}{2}\left(\frac{D_{u_j}}{\sigma_{D_j}}\right)^2\right]$$

and the independence of g_i and D_{u_j}, we obtain the expression

$$\Omega_1(u) = -\log p(u \mid g, D_u)$$

$$= \frac{1}{2} \sum_i \left(\frac{g_i - f(x_i - u_i)}{\sigma_{n_i}} \right)^2 + \frac{1}{2} \sum_j \left(\frac{D_{u_j}(u(x))}{\sigma_{D_j}} \right)^2 + \text{const.}$$

to be minimized with respect to u_i.

Now we treat the case in which the smoothness is measured by the second derivative $u''(x_j)$ of $u(x)$. The variances of n_i and of $u''(x_j)$ are assumed to be constant; thus $\sigma_{n_i}^2 = \sigma_n^2$ and $\sigma_{u''_j}^2 = \sigma_c^2$ (c stands for curvature). Then we have to minimize

$$\Omega(u) = 2[\Omega_1(u) - \text{const.}] = \frac{1}{\sigma_n^2} \sum_i [g_i - f(x_i - u_i)]^2 + \frac{1}{\sigma_c^2} \sum_j \{u''_j[u(x)]\}^2$$

(16.23)

The first term in Eq. (16.23) measures the similarity between the given and the observed functions; the second term measures the smoothness. This is equivalent to using, in addition to Eq. (16.21), the fictitious second-derivative curvature-related observations

$$c(x_j) = u''(x_j) + v(x_j), \qquad j = 1, \ldots, k \tag{16.24}$$

with $c(x_j) = 0$ and $\sigma_{v_j}^2 = \sigma_c^2$, and to determine the $u(x_i)$ by a weighted LS technique with the weights being the inverse of the variances. The random variations in the second derivative are represented by $v(x_j)$. If we now represent the second derivative $u''(x_j)$ by

$$u''(x_j) = u(x_{j-1}) - 2u(x_j) + u(x_{j+1})$$

with approximate values $u_0(x_j) = u_{0j}$, and thus

$$u(x_j) = u(x_j) + \Delta u_0(x_j)$$
$$= u_{0j-1} - 2u_{0j} + u_{0j+1} + (\Delta u_{j-1} - 2\Delta u_j + \Delta u_{j+1})$$

then we arrive at the linearized model

$$\Delta g_i = -f'_i \Delta u_i + n_i, i = 1, \ldots, m \tag{16.25}$$
$$-c_{0j} = \Delta u_{j-1} - 2\Delta u_j + \Delta u_{j+1} + v_j, j = 2, \ldots, m-1 \tag{16.26}$$

with

$$\Delta g_i = g(x_i) - f[x_i - u_0(x_i)]$$
$$f'_i = \frac{df(y)}{dy}\bigg|_{y = x_i - u_0(x_i)}$$
$$c_{0j} = u_{0j-1} - 2u_{0j} + u_{0j+1}$$
$$v_j = v(x_j)$$

Equation (16.25) represents the observation process and Eq. (16.26) represents the smoothness constraint.

The model can be conveniently written in matrix notation

$$\Delta g = -A_1\,\Delta u + n$$
$$-c = A_2\,\Delta u + v \tag{16.27}$$

with the $m \times m$ matrix A_1 containing the derivatives f_i' on the main diagonal

$$A_1 \underset{m \times m}{=} \text{Diag}\,(f_i') = \begin{pmatrix} f_1' & & & \\ & \ddots & & \\ & & f_i' & \\ & & & \ddots & \\ & & & & f_m' \end{pmatrix}$$

and the $(m-2) \times m$ matrix A_2 consisting of $m-2$ rows with $(1-2\ 1)$ around the main diagonal

$$\underset{(m-2)\times m}{A_2} = \begin{pmatrix} 1 & -2 & 1 & & & & \\ & 1 & -2 & 1 & & & \\ & & 1 & -2 & 1 & & \\ & & & \ddots & & & \\ & & & 1 & -2 & 1 & \\ & & & & \ddots & & \\ & & & & 1 & -2 & 1 \end{pmatrix}$$

Minimizing

$$\Omega(u) = w_n \cdot n^T n + w_c \cdot v^T v \tag{16.28}$$

with respect to u yields the normal equation system

$$[w_n A_1^T A_1 + w_c\,A_2^T A_2]\,\Delta u = [w_n A_1^T \Delta g - w_c A_2^T c] \tag{16.29}$$

with the weights $w_n = 1/\sigma_n^2$ and $w_c = 1/\sigma_c^2$. The normal equation matrix N is a pentadiagonal matrix

$$w_n \cdot \text{Diag}\,(f_i'^2) + w_c \cdot \begin{pmatrix} 1 & -2 & 1 & & & & & \\ -2 & 5 & -4 & 1 & & & & \\ 1 & -4 & 6 & -4 & 1 & & & \\ & 1 & -4 & 6 & -4 & 1 & & \\ & & & \ddots & & & & \\ & & & 1 & -4 & 6 & -4 & 1 \\ & & & & 1 & -4 & 5 & -2 \\ & & & & & 1 & -2 & 1 \end{pmatrix}$$

and the right-hand sides are given by

$$
\begin{pmatrix}
w_n & \cdot & f_1' & \cdot & \Delta g_i & - & w_c & \cdot & & & c_{01} \\
w_n & \cdot & f_2' & \cdot & \Delta g_2 & - & w_c & \cdot & & (-2c_{01} & +c_{02}) \\
w_n & \cdot & f_3' & \cdot & \Delta g_3 & - & w_c & \cdot & (c_{01} & -2c_{02} & +c_{03}) \\
\vdots & & \vdots & & \vdots & & \vdots & & \vdots & \vdots & \vdots \\
w_n & \cdot & f_{m-2}' & \cdot & \Delta g_{m-2} & - & w_c & \cdot & (c_{0m-2} & -2c_{0m-2} & +c_{0m}) \\
w_n & \cdot & f_{m-1}' & \cdot & \Delta g_{m-1} & - & w_c & \cdot & (c_{0m-1} & -2c_{0m}) \\
w_n & \cdot & f_m' & \cdot & \Delta g_m & - & w_c & \cdot & c_{0m} & &
\end{pmatrix}
$$

This system can be solved by factorization in $O(m)$ time, leading to estimates $\widehat{\Delta u} = (\widehat{\Delta u_i})$ and thus to $\hat{u}_i = u_{0i} + \widehat{\Delta u_i}$ or $\hat{u} = u_0 + \widehat{\Delta u}$.

Often the noise variances σ_n^2 and σ_c^2 are not known. Without proof (Förstner, 1985) we give an estimate for both variances, which in an iterative manner could be used to obtain optimal estimates for both the u_i and the variances σ_n^2 and σ_c^2. The estimation requires the diagonal terms q_{ii} of the inverse $Q = N^{-1}$ of the normal equation matrix. They can be calculated by using a standard routine from a library for solving sparse equation systems.

Including the case of individual variances $\sigma_{n_i}^2$ and $\sigma_{c_j}^2$, the estimates are given by

$$
\hat{\sigma}_{n_i}^2 = \frac{1}{r_n} \left(\sum_{i=1}^{m} \left(g(x_i) - f\left[x_i - \hat{u}(x_i)\right] \right)^2 / \sigma_{n_i}^2 \right) \cdot \sigma_{n_i}^2
$$

and

$$
\hat{\sigma}_{c_j}^2 = \frac{1}{r_c} \left(\sum_{j=2}^{m-1} \left[\hat{u}(x_{j-1}) - 2\hat{u}(x_j) + \hat{u}(x_{j+1}) \right]^2 / \sigma_{c_j}^2 \right) \cdot \sigma_{c_j}^2
$$

with

$$
Q = N^{-1}
$$

$$
q_{ii} = (Q)_{ii}
$$

$$
u_n = \sum_{i=1}^{m} \{ f'[x_i - \hat{u}(x_i)] \}^2 \cdot q_{ii} / \sigma_{n_i}^2
$$

$$
r_n = m - u_n
$$

$$
r_c = (m - 2) - r_n
$$

which simplify for $\sigma_{n_i} = \sigma_n$ and $\sigma_{c_j} = \sigma_c$. These estimates assume that no outliers in the observations $g(x_i)$ and no discontinuities are present. The case including outliers and discontinuities may use robust estimation techniques, iteratively weighting down the individual observations depending on the residuals n_i and v_j.

The relative precision of the standard deviations $\hat{\sigma}_n$ and $\hat{\sigma}_c$ is approximately $1/\sqrt{2r_n}$ and $1/\sqrt{2r_c}$, thus leading to reliable results only for sufficiently long profiles. The technique may easily be generalized to two-dimensional functions.

16.2.6 Iterations and Resampling

As $f(x)$ is highly nonlinear, the estimates in general are only improved approximate values. Especially if the initial approximations are crude, the estimates have to be further refined if one wants to exploit the precision inherent in the data. This leads to an iterative estimation scheme.

We obtain the estimates after the νth iteration by, for example,

$$\hat{a}^{(\nu)} = \hat{a}^{(\nu-1)} + \widehat{\Delta a}^{(\nu)}$$
$$\hat{u}^{(\nu)} = \hat{u}^{(\nu-1)} + \widehat{\Delta u}^{(\nu)} \tag{16.30}$$

The constant shift b of the intensity values can be estimated directly, thus we take $b^{(0)} = 0$ for all iterations.

As the positions $y_i = x_i - u(x_i)$ are noninteger values in general, we have to interpolate. Specifically we need

$$f[x_i - u_0(x_i)]$$

and

$$\left. \frac{df(y)}{dy} \right|_{y=x_i-u_0(x_i)}$$

For choosing a proper interpolation, we have to require that the optimization function $\Omega(u)$ be smooth or, equivalently, that the elements in the normal equations change continuously with changing approximate values $u_0(x_i)$. Otherwise small changes of $u_0(x_i)$ may lead to large changes in the estimates and hence to unstable results.

Therefore $f(y)$ and $f'(y)$ at least have to be interpolated linearly:

$$f(x_i - u) = (1 - u/\Delta x)f_i + \frac{u}{\Delta x}f_{i-1} \tag{16.31}$$

$$f'(x_i - u) = (1 - u/\Delta x)f'_i + \frac{u}{\Delta x}f'_{i-1}, \qquad 0 \le u < \Delta x \tag{16.32}$$

where Δx is the spacing of the x_i, $0 \le u \le \Delta x$, and, for example, with the first derivative $f'(x_i)$ from a second-order facet model

$$f'_i = \frac{1}{2\Delta x}(f_{i+1} - f_{i-1}) \tag{16.33}$$

Here we use the abbreviations f_i for $f(x_i)$ and f'_i for $f'_i(x_i)$. Obviously $f(x_i - u)$ and $f'(x_i - u)$ do *not* refer to the same smooth function $f(x)$, as $f'(x_i - u)$ refers to a second-order interpolation scheme of $f(x)$. If we require that $f(x_i - u)$ and $f'(x_i - u)$ refer to the same smooth function, we have to interpolate $f(x_i - u)$ with a second-order function, for example,

$$f(x_i + u) = f_i + f'_i \cdot u + \frac{1}{2}f''_i \cdot u^2, \ |u| < \frac{1}{2}\Delta x$$

with $f'(x_i)$ as in Eq. (16.33) and

$$f''_i = \frac{2}{\Delta x^2}[f_{i+1} - 2f_i + f_{i-1}]$$

As the slope at $x_i + \Delta x/2$ is then $2(f_{i+1} - f_i)/\Delta x$, and thus twice the first differences, one could instead use the *smoothed* function

$$\tilde{f}(x_i + u) = \overline{f}_i + f'_i \cdot u + \frac{1}{2}f''_i \cdot u^2, \quad |u| < \frac{1}{2}\Delta x$$

with

$$\overline{f}_i = \frac{1}{8}(f_{i-1} + 6f_i + f_{i-1})$$

$$f'_i = \frac{1}{2\Delta x}(f_{i+1} - f_{i-1})$$

$$f''_i = \frac{1}{\Delta x^2}(f_{i+1} - 2f_i + f_{i-1})$$

Both functions $f(x_i + u)$ and $\tilde{f}(x_i + u)$ pass through the points $\frac{1}{2}(f_{i+1} + f_i)$ and $\frac{1}{2}(f_{i-1} + f_i)$ and have a common slope with the neighboring interpolation elements.

■ EXAMPLE 16.4

a. With the same data as in Example 16.1, we now perform a second iteration for the shift estimate only. The main data are summarized in the table:

i	4	5	6	7	8	9	10	11
y	2.88	3.88	4.88	5.88	6.88	7.88	8.88	9.88
g	14	13	14	26	37	42	41	42
f	10	10	18.81	28.80	38.80	40	40	40
Δg	4.00	3.00	-4.81	-2.80	-1.80	2.00	1.00	2.00
f'	0	4.4	9.4	10	5.6	0.6	0	0
n	4.00	3.00	-1.93	0.07	1.08	2.00	1.00	2.00

Linear interpolation has been applied to both f and f' (cf. Eqs. 16.31 and 16.32); for example, $f(6 - u^{(1)}) = f(6 - 1.118) = f(5 - 0.118) = (1 - 0.118) \cdot f_5 + 0.118 \cdot f_4 = 0.882 \cdot 20 + 0.118 \cdot 10 = 18.81$. The normal equations are

$$\left(\sum_i f'^2_i \right) \cdot \widehat{\Delta u} = \sum_i f'_i g_i \equiv 239.4 \cdot \widehat{\Delta u} = 68.725$$

leading to the correction $\Delta u^{(2)} = 0.288[\Delta x]$ and to a better estimate $u^{(2)} = u^{(1)} + \Delta u^{(2)} = 1.406[\Delta x]$ for the shift. Observe the second iteration leading to a change (0.288) in the estimate that is significantly larger than the internal precision (0.06, cf. Example 16.1) of the estimate.

b. In the case of the model of Example 16.3 with the additional radiometric parameters, a second iteration would yield the shift with its standard deviation

$$\hat{u} = 1.61[\Delta x], \quad \sigma_{\hat{u}} = 0.15[\Delta x]$$

and the estimated noise standard deviation

$$\sigma_n = 1.43[gr]$$

We will compare this result with the one obtained by cross-correlation in Examples 16.5 and 16.6.

■

16.2.7 Matching of Two Observed Profiles

We have assumed that $f(x)$ is known. Now we want to extend the matching procedure to the case in which *both* profiles are corrupted by noise. We will show how we can reduce the procedure to those methods developed so far.

We start with two noisy profiles

$$g_1(x_i) = f(x_i) + n_1(x_i)$$

$$g_2(x_i) = f(y_i) + n_2(x_i)$$

and the geometric model

$$y_i = x_i - u(x_i)$$

We could have applied a symmetric model distributing $u(x_i)$ equally on both signals (cf. Horn, 1987, sec. 13.9). This would have led to the same results. We again assume that the noise components are independent and white with standard deviations σ_{n_1} and σ_{n_2}. The first profile $g_1(x_i)$ is just the observed function $f(x)$, and $g_2(x_i)$ is the observed and deformed function $f(x)$. With approximate values $u_0(x_i)$, thus $u(x_i) = u_0(x_i) + \Delta u(x_i)$, and again neglecting higher-order terms, we can write

$$g_1[x_i - u_0(x_i)] = f[x_i - u_0(x_i)] + n_1[x_i - u_0(x_i)]$$

and linearize

$$g_2(x_i) = f[x_i - u_0(x_i)] - \frac{df(y)}{dy}\bigg|_{y=x_i-u_0(x_i)} \cdot \Delta u(x_i) + n_2(x_i)$$

Using the abbreviations

$$\Delta g_i = g_2(x_i) - g_1[x_i - u_0(x_i)]$$

$$f'_i = \frac{df(y)}{dy}\bigg|_{y=x_i-u_0(x_i)}$$

and

$$\overline{n}_i = n_2(x_i) - n_1[x_i - u_0(x_i)]$$

we obtain the linearized model

$$\Delta g_i = -f'_i \cdot \Delta u(x_i) + \overline{n}_i$$

in full correspondence to Eq. (16.4).

To use this linearized model, we must take into account that the f_i' have to be estimated from the given data. This could be achieved by restoring $g_1(x)$ and $g_2(x)$, yielding $\hat{g}_1(x)$ and $\hat{g}_2(x)$, which are actually estimates for $f(x)$, and by taking the average of the first derivatives $\hat{g}_1'(x)$ and $\hat{g}_2'(x)$ at the corresponding position. Thus, for example,

$$\hat{f}_i' = \frac{1}{2}\{\hat{g}_1'[x_i - u_0(x_i)] + \hat{g}_2'(x_i)\}$$

The restoration of g_1 and g_2 can be based on any of the noise cleaning techniques discussed in Chapter 7. The interpolation, however, again has to take into account the requirements on the smoothness of the optimization function, as discussed earlier. This noise-cleaning step for obtaining reliable estimates for the first derivatives is necessary to achieve consistent results.

However, the determination of the function differences $\Delta g_i = g_2(x_i) - g_1[x_i - u_0(x_i)]$ must be based on the original data, as only then does the error model of white noise hold. The interpolation, which may be needed to obtain $g_1[x_i - u_0(x_i)]$, introduces only negligible correlations (≤ 0.5) between neighboring Δg_i's. If the degree of this smoothing is kept low, we can assume that $\bar{n}(x_i) = n_2(x_i) - n_1[x_i - u_0(x_i)]$ are still white but with variance $\sigma_{\bar{n}}^2 = \sigma_{n_1}^2 + \sigma_{n_2}^2$. For $\sigma_{n_1} = \sigma_{n_2}$ we thus obtain $\sigma_{\bar{n}}^2 = 2\sigma_n^2$. For a constant-shift model $\hat{u}(x_i) = \hat{u}$,

$$\hat{\sigma}_{\bar{n}}^2 = \frac{1}{m-1} \sum_{i-1}^{m} \left[g_2(x_i) - g_1\left(x_i - \hat{u}\right) \right]^2$$

is now an estimate for the variance of the noise difference between the two profiles and, in the case of $\sigma_{n_1} = \sigma_{n_2}$, can be used to obtain an estimate

$$\sigma_n = \frac{\sigma_{\bar{n}}}{\sqrt{2}}$$

for the noise standard deviation and thus for the observational precision.

16.2.8 Relations to Cross-Correlation Techniques

The earliest applications of intensity-based image matching in remote sensing used cross-correlation techniques. The model assumed simply a shift between two corresponding image sections. To compensate for different brightness and contrast, a linear transformation in the intensity values was assumed, though not always explicitly stated.

We want to discuss the relations between this classical technique and the differential approach, as they give insight into the similarities and differences of both methods. Though only one-dimensional images are treated, the result can be directly transferred to more dimensions. The model reads as

$$g_1(x_i) = f(x_i) + n_1(x_i)$$
$$g_2(x_i) = f(x_i - u) + n_2(x_i)$$

or

$$g_2(x_i) = g_1(x_i - u) + \bar{n}(x_i)$$

with

$$\bar{n}(x_i) = n_2(x_i) - n_1(x_i)$$

The principle of estimating u is to search for the maximal correlation coefficient ρ_{12} of g_1 and g_2 :

$$\max_u \frac{Cov\left[g_1(x_i - u), g_2(x_i)\right]}{\sqrt{V[g_1(x_i - u)] \cdot V[g_2(x_i)]}} = \frac{\sigma_{g_1 g_2}(u)}{\sigma_{g_1}(u)\sigma_{g_2}} = \rho_{12}(u) \rightarrow \hat{u} \tag{16.34}$$

with

$$\sigma_{g_1 g_2}(u) = \frac{1}{m-1}\left[\sum_{i=1}^{m} g_1(x_i - u)g_2(x_i) - \frac{1}{m}\sum_{i=1}^{m} g_1(x_i - u)\sum_{i=1}^{m} g_2(x_i)\right]$$

$$\sigma_{g_1}^2(u) = \frac{1}{m-1}\left[\sum_{i=1}^{m} g_1^2(x_i - u) - \frac{1}{m}\left(\sum_{i=1}^{m} g_1(x_i - u)\right)^2\right]$$

$$\sigma_{g_2}^2 = \frac{1}{m-1}\left[\sum_{i=1}^{m} g_2^2(x_i) - \frac{1}{m}\left(\sum_{i=1}^{m} g_2(x_i)\right)^2\right] \tag{16.35}$$

As the empirical mean and variances are taken into account, different brightness and contrast are immediately compensated for. Thus cross-correlation corresponds to least-squares matching, using the model of Section 16.2.4. Observe that the cross-correlation term here is the product moment cross-correlation coefficient from statistics, whereas the cross-correlation term as often used in electrical engineering refers to the mean product of two signals and is not normalized with respect to the mean and the variance.

The search for the estimate $\hat{u}$ in general leads to an integer position. The rounding error is dominant in most applications; therefore an interpolation of the correlation function is useful. Let the integer position of the maximum of $\rho(u)$ be u_0 and the two neighboring positions be u_- and u_+, and let the corresponding values of $\rho(u)$ be ρ_0, ρ_- and ρ_+. Then using the quadratic interpolation of $\rho(u)$ leads to the following estimate $\hat{u}$:

$$\hat{u} = u_0 - \frac{\rho'(u_0)}{\rho''(u_0)} = u_0 - \frac{\frac{1}{2\Delta x}(\rho_+ - \rho_-)}{\frac{1}{\Delta x^2}(\rho_+ - 2\rho_0 + \rho_-)} = u_0 - \frac{\rho_+ - \rho_-}{2(\rho_+ - 2\rho_0 + \rho_-)} \cdot \Delta x \tag{16.36}$$

where Δx is the spacing between the u_i.

Thus subpixel accuracy is achievable. Not only was this initially claimed by Bernstein (1973), but it was consistently proved empirically and theoretically by many researchers. The correlation coefficient was used as an acceptance criterion for the problem of setting a proper threshold to reject bad matches. Also, the weakness of the correlation coefficient leads to the use of additional measures of performance, such as the "drop" or "slope" of the correlation function (Helava, 1976; Panton, 1978). This was used to prevent matches with a very flat correlation function from passing.

■ **EXAMPLE 16.5**

With the data from Example 16.1, we want to estimate the shift by using cross-correlation and to compare it with the results of the model, adding further parameters for the differences in brightness and contrast. We therefore take the same set of observed values $g^{(4)}$ to $g^{(11)}$ and correlate them with windows of size 8 for f, thus setting $g_1 = g$ and $g_2 = f$ in Eq. (16.35). With the variances $\sigma_g^2 = \sigma_{g_2}^2 = 179.98[gr^2]$ and $\sigma_f^2(u) = \sigma_{g_1}^2(u)$, the covariances $\sigma_{g_1g_2}^2(u) = \sigma_{fg}(u)$, and the correlation coefficients $\rho_{12}(u)$ in the table

u	$\sigma_{g_1}^2(u)$	$\sigma_{g_1g_2}(u)$	$\rho_{12}(u)$
3	183.9	160.5	0.8823
2	200.0	187.9	0.9901
1	183.9	175.2	0.9628
0	135.7	128.2	0.8294
−1	55.4	64.1	0.6423
−2	12.5	20.9	0.4405

we obtain the optimal integer position $u_0 = 2$. The subpixel estimate is $\hat{u} = 2 - (.9628 - 0.8823)/(0.8823 - 2 \cdot 0.9901 + 0.9628)/2 = 2 - 0.30 = 1.70[\Delta x]$ (Eq. 16.36), which is close to the estimate $1.61[\Delta x]$ in the second iteration of Example 16.4.

■

We now relate these measures to those developed in the context of the differential approach. First let us assume that the two images are not shifted or that the optimal shift has been applied to g_2. Then assuming

$$g_1(x) = f(x) + n_1(x)$$
$$g_2(x) = a[f(x) + n_2(x)] + b$$

where n_1 and n_2 are independent white noise with variance $\sigma_{n_1} = \sigma_{n_2} = \sigma_n$ and f is stochastic with variance σ_f^2 and independent from n_1 and n_2, we obtain the variances

$$\sigma_{g_1}^2 = \sigma_f^2 + \sigma_n^2$$
$$\sigma_{g_2}^2 = a^2(\sigma_f^2 + \sigma_n^2)$$

and the covariance

$$\sigma_{g_1g_2} = a \cdot \sigma_f^2$$

and therefore the correlation coefficient ρ_{12} (cf. Ballard and Brown, 1982)

$$\rho_{12} = \frac{\sigma_{g_1g_2}}{\sigma_{g_1}\sigma_{g_2}} = \frac{\sigma_f^2}{\sigma_f^2 + \sigma_n^2} \tag{16.37}$$

or, with the signal-to-noise ratio,

$$SNR^2 = \frac{\sigma_f^2}{\sigma_n^2}$$

$$\rho_{12} = \frac{SNR^2}{SNR^2 + 1} \tag{16.38}$$

or

$$SNR^2 = \frac{\rho_{12}}{1 - \rho_{12}} \tag{16.39}$$

Thus the maximum achievable correlation coefficient is limited by the signal-to-noise ratio. On the other hand, we can derive the signal-to-noise ratio from the empirical correlation coefficient. Moreover, if we know the variance $\sigma_{g_1}^2$ of the observed signal, we can derive the noise variance

$$\sigma_n^2 = \sigma_{g_1}^2 (1 - \rho_{12}) = \sigma_f^2 \frac{1 - \rho_{12}}{\rho_{12}} \tag{16.40}$$

As the correlation coefficient is identical to the cosine of the angle γ between the vectors g_1 and g_2, we have three equivalent measures for evaluating the intensity of two signals. Examples are:

$$\begin{array}{lll}
SNR = 10 & \rho = 0.99 & \gamma = 8° \\
SNR = 3.4 & \rho = \sqrt{2}/2 & \gamma = 45° \\
SNR = 1 & \rho = 0.5 & \gamma = 60°
\end{array} \tag{16.41}$$

Values of ρ smaller than 0.5 thus correspond to a signal-to-noise ratio less than 1. Reasonable critical values for ρ_{12} lie in the range between 0.5 and 0.7.

If we use Eqs. (16.37) and (16.39), we can rewrite the variance of the estimated shift from Eq. (16.11):

$$\sigma_{\hat{u}}^2 = \frac{1}{m} \cdot \frac{1}{SNR^2} \cdot \frac{\sigma_f^2}{\sigma_{f'}^2} = \frac{1}{m} \cdot \frac{1 - \rho_{12}}{\rho_{12}} \cdot \frac{\sigma_f^2}{\sigma_{f'}^2} \tag{16.42}$$

It now becomes obvious that the correlation coefficient gives only partial information about the precision of a match: the number of points used—thus the window size and the ratio $\sigma_f^2/\sigma_{f'}^2$—is not used.

This ratio $\sigma_{f'}^2/\sigma_f^2$ relating the "edge busyness" $\sigma_{f'}^2$ to the variance of the signal can be shown to be proportional to the effective bandwidth of the signal, and, owing to the moment theorem, equal to the (negative) *curvature* of the autocorrelation function (ACF) of $f(x)$ (Papoulis, 1965; Ryan, Gray, and Hunt, 1980). Thus two profiles with the same length and the same variance may lead to significantly different matching precisions (Fig. 16.6). The standard deviation $\sigma_x(B)$ obtainable with profile B is four times higher than the standard deviation $\sigma_x(A)$ with profile $A : \sigma_x(B) = 4 \cdot \sigma_x(A)$.

Using the second derivative $\rho''(u) = (\rho_+ - 2\rho_0 + \rho_-)/\Delta x^2$ of the empirical correlation function $\rho(u)$ (cf. Eq. 16.34), we obtain an expression for the variance

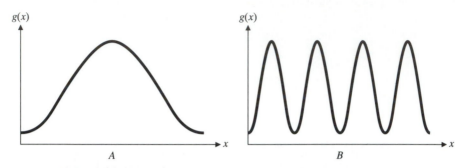

Figure 16.6 Two profiles of common contrast but different sharpness. If one assumes a constant noise variance, matching with profile A leads to four times smaller a standard deviation of the estimated shift than does matching with profile B.

of estimated shift

$$\sigma_x^2 = \frac{1}{m} \cdot \frac{1 - \rho_0}{\rho_0} \cdot \frac{\Delta x^2}{-\rho_+ + 2\rho_0 - \rho_-} \tag{16.43}$$

with the pixel spacing Δx. This may be used for evaluating the result of correlation techniques and, of course, holds if the geometric model of a pure shift is approximately valid and the position $\hat{u}$ has been estimated by some interpolation of the cross-correlation function (cf. Eq. 16.36).

EXAMPLE 16.6

The result of the correlation in Example 16.5 can now be evaluated. The standard deviation of the estimated shift according to Eq. (16.43) results in $\sigma_{\hat{u}}^2 = 1/8 \cdot (1 - 0.9901)/0.9901/(-0.8823 + 2 \cdot 0.9901 - 0.9628) = 0.009251$, and thus $\sigma_{\hat{u}} = 0.096[\Delta x]$, which seems a bit too optimistic. The estimated noise standard deviation, according to Eq. (16.40), $\sigma_n = 1.41[gr]$, is very close to the one $(1.43[gr])$ obtained in the second iteration of Example 16.4. The difference probably results from the weak estimate of the curvature of the cross-correlation function from only three points, which likely is biased toward too high a curvature (Berman, 1989).

6.3   Intensity-Based Matching of Two-Dimensional Signals

The differential techniques for intensity-based matching can be directly transferred to the case of two-dimensional functions $f(r,c)$ and $g(r,c)$.

16.3.1   The Principle and the Relation to Optical Flow

In the most general case of matching an observed intensity function to a given one, we start from the nonlinear model

$$g(r_i, c_i) = f(p_i, q_i) + n(r_i, c_i) \qquad i = 1, \ldots, m \tag{16.44}$$

with

$$\begin{pmatrix} p \\ q \end{pmatrix}_i = \begin{pmatrix} r \\ c \end{pmatrix}_i - \begin{pmatrix} \hat{u}(r,c) \\ \hat{v}(r,c) \end{pmatrix}_i \tag{16.45}$$

again assuming that the object is given at a grid, together with an appropriate interpolation scheme, and that the noise is white with variance σ_n^2.

If we assume that approximate values $u_0(r_i, c_i)$ of the unknown *deformation* $[\hat{u}(r,c), \hat{v}(r,c)]$ are known, we obtain the linearized model

$$\Delta g_i = -f_{r,i} \cdot \widehat{\Delta u}_i - f_{c,i} \cdot \widehat{\Delta v}_i + n_i \tag{16.46}$$

with

$$\Delta g_i = g(r_i, c_i) - f\left[r_i - u_0(r_i, c_i), c_i - v_0(r_i, c_i)\right]$$

$$f_{r_i} = \left. \frac{\partial f(p, q)}{\partial p} \right|_{p = r_i - u_0(r_i, c_i), q = c_i - v_0(r_i, c_i)}$$

$$f_{c_i} = \left. \frac{\partial f(p, q)}{\partial q} \right|_{p = r_i - u_0(r_i, c_i), q = c_i - v_0(r_i, c_i)}$$

$$\widehat{\Delta u}_i = \hat{u}(r_i, c_i) - u_0(r_i, c_i)$$

$$\widehat{\Delta v}_i = \hat{v}(r_i, c_i) - v_0(r_i, c_i)$$

$$n_i = n(r_i, c_i)$$

Here we want to relate Eq. (16.46) to the optical flow equation used in motion analysis.

If one treats f as a time-varying intensity field $f(r, c, t)$ with the velocity field $[u(r,c), v(r,c)]$, the nonlinear model equation (16.44) can be written as

$$f\left[r + u(r,c) \cdot dt, \ c + v(r,c) \cdot dt, \ t + dt\right] = f(r, c, t)$$

with the the noise term omitted, thus referring to the expected value of g. If we now linearize, for the sake of simplicity at $u = v = 0$, we obtain

$$f(r,c,t) + \frac{\partial f(r,c,t)}{\partial r} \cdot u(r,c) \cdot dt + \frac{\partial f(r,c,t)}{\partial c} \cdot v(r,c) \cdot dt + \frac{\partial f(r,c,t)}{\partial t} dt = f(r,c,t)$$

Setting $\nabla f = (f_r, f_c)^T$ the spatial gradient of f, $V = (u, v)^T$ the velocity field, and $\dot{f} = \partial f / \partial t$ the temporal change of f, we obtain

$$\nabla f \cdot V + \dot{f} = 0$$

the *optical flow equation*.

The equation simply states that the intensity *of corresponding points* is time invariant, and changes in f are due only to V. This relation approximately holds for image sequences but neither at occlusions nor if the time spacing Δt is large. While occlusions cannot explicitly be handled with the model equation (16.44), changes in illumination or reflectance may be taken into account by additional parameters.

Now Eq. (16.46) does not allow us to estimate $\Delta u(r_i, c_i)$ and $\Delta v(r_i, v_i)$ simultaneously, as only one observation—namely, Δg_i—is available. This so-called *aperture problem* can be solved only by adding constraints on the deformation field $V(r,c) = [u(r,c), \; v(r,c)]$. Several stages of constraints are of practical use here:

1. $V(r,c)$ is constant. This is the classical assumption used also for cross-correlation (Barnea and Silverman, 1972).

2. $V(r,c)$ is a linear function in r,c. This would be a first-order approximation of V, which for smooth deformations can be used for local estimates. The first attempts to exceed a shift were made by Schalhoff and McVey (1979) and Huang (1981) (cf. the review given by Förstner, 1984).

3. $V(r,c)$ is smooth. This assumption allows one to model real situations, as in satellite imagery, medical imagery, or reconstruction of smooth surfaces.

4. $V(r,c)$ most generally is best modeled to be piecewise smooth. Then occlusions may also be handled, though no explicit reference is made to the three-dimensional structure in the model.

5. If image matching is used to evaluate stereo images; if the relative orientation of the cameras, including their interior orientation, is known; and if the rows of both images are parallel to the baseline, that is, the line through the two projection centers, then $u(r,c) \equiv 0$ for all r,c. Thus only $v(r,c)$ is to be estimated, which reduces the problem to profile matching, discussed in the previous section. Now, however, the matching can be based on two-dimensional windows, thereby stabilizing the estimates (cf. Section 16.6.3).

We want to treat only the first two cases—those with unknown contrast shift and unknown linear transformation—but will include parameters for differences in brightness or contrast. In addition, we discuss two special cases in order to motivate the interest operator to select distinct points. The more general problem of estimating a complete displacement field was treated in Chapter 15.

16.3.2   Estimating Constant-Shift Parameters

The simplest model for matching is

$$p = x - u, \qquad q = y - v$$

This assumes solely a shift (u,v) between the two windows of concern.

The linearized model then reads as

$$\Delta g_i = -f_{r_i} \cdot \widehat{\Delta u} - f_{c_i} \cdot \widehat{\Delta v} + n_i, \quad i = 1, \ldots, m \tag{16.47}$$

with

$$f_{r_i} = \frac{\partial f(p,q)}{\partial r}\bigg|_{\substack{r=r_i-u_0 \\ c=c_i-v_0}}$$

$$f_{c_i} = \frac{\partial f(p,q)}{\partial c}\bigg|_{\substack{r=r_i-u_0 \\ c=c_i-v_0}}$$

$$\widehat{\Delta u} = \hat{u} - u_0,$$

$$\widehat{\Delta v} = \hat{v} - v_0,$$

$$\Delta g_i = g(r_i,c_i) - f(r_i - u_0, c_i - v_0)$$

$$u_i = u(r_i,c_i)$$

for all m pixels (r_i,c_i) in the window. In most cases the points (r_i,c_i) form a grid with $m = m_r m_c$ points, but Eq. (16.47) also can be used if the points do *not* form a grid.

The normal equations are then given by

$$N\hat{y} = h \tag{16.48}$$

with the 2×2 normal equation matrix N

$$N = \begin{pmatrix} \sum\limits_{i=1}^{m} f_{r_i}^2 & \sum\limits_{i=1}^{m} f_{r_i} f_{c_i} \\ \sum\limits_{i=1}^{m} f_{r_i} f_{c_i} & \sum\limits_{i=1}^{m} f_{c_i}^2 \end{pmatrix} = \begin{pmatrix} N_{11} & N_{12} \\ N_{21} & N_{21} \end{pmatrix} \tag{16.49}$$

and the right-hand sides

$$h = -\begin{pmatrix} \sum\limits_{i=1}^{m} f_{r_i} \cdot \Delta g_i \\ \sum\limits_{i=1}^{m} f_{c_i} \cdot \Delta g_i \end{pmatrix} \tag{16.50}$$

and $\hat{y} = \begin{pmatrix} \widehat{\Delta u} \\ \widehat{\Delta v} \end{pmatrix}$, which can be resolved for $\hat{y}$, for example, by using the weight coefficient matrix of the unknown parameters $\hat{y}$

$$Q = N^{-1} = \frac{1}{N_{11}N_{22} - N_{12}^2} \begin{pmatrix} N_{22} & -N_{12} \\ -N_{12} & N_{11} \end{pmatrix}$$

leading to the estimates $\hat{u} = u_0 + \widehat{\Delta u}$ and $\hat{v} = v_0 + \widehat{\Delta v}$. We also obtain an estimate for the unknown variance factor, identical here to the noise variance

$$\sigma_n^2 = \frac{1}{m-2} \cdot \sum\limits_{i=1}^{m} [g(r_i,c_i) - f(r_i - \hat{u}, c_i - \hat{v})]^2$$

This can then be used to obtain the covariance matrix $\Sigma_{\hat{y}\hat{y}}$ of the unknown shift $\hat{u}$ and $\hat{v}$

$$\Sigma_{\hat{y}\hat{y}} = \hat{\sigma}_n^2 \cdot Q = \hat{\sigma}_n^2 \cdot \begin{pmatrix} \sum_{i=1}^{m} f_{r_i}^2 & \sum_{i=1}^{m} f_{r_i} f_{c_i} \\ \sum_{i=1}^{m} f_{r_i} f_{c_i} & \sum_{i=1}^{m} f_{c_i}^2 \end{pmatrix}^{-1} \tag{16.51}$$

from which follow the standard deviations

$$\hat{\sigma}_{\hat{u}} = \hat{\sigma}_n \cdot \sqrt{Q_{11}} = \hat{\sigma}_n \cdot \sqrt{\frac{N_{22}}{N_{11}N_{22} - N_{12}^2}}$$

and

$$\hat{\sigma}_{\hat{v}} = \hat{\sigma}_n \cdot \sqrt{Q_{22}} = \hat{\sigma}_n \cdot \sqrt{\frac{N_{11}}{N_{11}N_{22} - N_{12}^2}}$$

The estimates are correlated with correlation coefficient

$$\rho_{\hat{u}\hat{v}} = \frac{-N_{12}}{\sqrt{N_{11}N_{22}}}$$

which could be used to further evaluate the estimates.

If we use estimates

$$\sigma_{fr}^2 \doteq \frac{1}{m} \sum_{i=1}^{m} f_{r_i}^2 = \frac{N_{11}}{m}$$

$$\sigma_{fc}^2 \doteq \frac{1}{m} \sum_{i=1}^{m} f_{c_i}^2 = \frac{N_{22}}{m}$$

and

$$\sigma_{frfc}^2 \doteq \frac{1}{m} \sum_{i=1}^{m} f_{r_i} f_{c_i} = \frac{N_{12}}{m}$$

for the *local squared gradient,* or

$$\nabla \widehat{f \nabla f^t} = \begin{pmatrix} \sigma_{fr}^2 & \sigma_{frfc} \\ \sigma_{fcfr} & \sigma_{fc}^2 \end{pmatrix} \tag{16.52}$$

we can rewrite Eq. (16.51) as

$$\Sigma_{\hat{y}\hat{y}} = \frac{\hat{\sigma}_n^2}{m} \cdot \begin{pmatrix} \sigma_{fr}^2 & \sigma_{frfc} \\ \sigma_{fcfr} & \sigma_{fc}^2 \end{pmatrix}^{-1} \tag{16.53}$$

Thus in addition to the estimates $\hat{u}$ and $\hat{v}$, we obtain a measure for the (internal) precision of the estimates. This precision intuitively depends on the following factors:

1. *The noise variance* σ_n^2. If we have good a priori knowledge about σ_n^2, we can introduce it instead of the estimated value $\hat{\sigma}_n^2$. As the model Eq. (16.47) will in many cases be oversimplified, we will not generally use $\hat{\sigma}_n^2$.

2. *The number m of used pixels*. Thus the standard deviation decreases linearly with the width of a square window. This holds only as long as new information is collected when increasing the window, and thus only as long as the average gradient in the window remains constant. Otherwise Eq. (16.51) demonstrates no advantage in using larger windows, as the sums in brackets remain practically constant.

3. *The average squared gradient in the window* $(\widehat{\nabla f \nabla f}^T)$. This indicator of the edge busyness is decisive for the precision of the match. We can show that it measures the curvature of the autocovariance (cross-correlation) function of $f(r,c)$, specifically the negative Hessian of the autocorrelation function of $f(r,c)$, if we assume f has zero mean and is thus a direct generalization of the relations to cross-correlation techniques discussed in Section 16.2.8.

The advantage of using Eqs. (16.51) and (16.53) is that, if we assume the noise variance is constant over an entire image, they enable us to determine in advance those places where we can expect high precision— that is, before we actually perform the matching procedure. This is because the equations depend only on the content of *one* window, and not on the actual observations, which would require two windows. Even if both windows are distorted by noise, $f(r,c)$ may be replaced by an adequate estimate from only one of the two windows, also in this case allowing us to check in advance the precision of the match.

Figure 16.7 shows 10 small windows from which the interior 16×16 pixels are used to evaluate the expected precision of a match. The confidence ellipses shown below the template assume a noise standard deviation of $\sigma_n = 5$ gray values. With 99% probability the true. shift will lie in the area depicted by the ellipse around the estimated shift. The pixel size is $\Delta x = 20\mu m$. The smallest confidence ellipses have major axes of less than $1\mu m$ and thus less than $1/20$ of a pixel. The largest confidence ellipse is obtained at the edge point, indicating that one cannot expect good accuracy for the position along the edge. The other confidence ellipses reflect reasonably well the image content with respect to the expected precision of matching. Thus the covariance matrix, proportional to the inverse of the normal equation matrix, is an ideal measure with which to evaluate the *distinctness* of small windows. As it is a statistical measure, essentially derived by averaging the quadratic gradient, no interpretation of the image content within the window is performed; therefore corners, circles, and random texture are treated simply from the standpoint of distinctness. The interest operator, discussed in Section 16.4, was actually motivated by searching for windows that guarantee good matching accuracy.

16.3.3 Estimating Linear Transformations

In the most important case of matching two small windows, we assume the model

$$g_1(r_i,c_i) = f(r_i,c_i) + n_1(r_i,c_i)$$

$$g_2(r_i,c_i) = h\left[f(p_i,q_i)\right] + n_2(r_i,c_i)$$

$$\begin{pmatrix} p \\ q \end{pmatrix}_i = \begin{pmatrix} a_1 & a_2 \\ a_4 & a_5 \end{pmatrix} \cdot \begin{pmatrix} r \\ c \end{pmatrix}_i + \begin{pmatrix} a_3 \\ a_6 \end{pmatrix} \tag{16.54}$$

$$h(f) = a_7 \cdot f + a_8 \tag{16.55}$$

where the noise n_1 and n_2 in both observed images g_1 and g_2 is white with standard deviations σ_{n_1} and σ_{n_2}, and the eight parameters a_k, $k = 1,\ldots,8$, are unknown. If we assume that the approximate values a_{0k}, $k = 1,\ldots,8$, are known, we obtain the linearized model

$$\begin{aligned} \Delta g_i =\ & f_{r_i} \cdot r_i \cdot \widehat{da_1} \\ &+ f_{r_i} \cdot c_i \cdot \widehat{da_2} \\ &+ f_{r_i} \qquad \widehat{da_3} \\ &+ f_{c_i} \cdot r_i \cdot \widehat{da_4} \\ &+ f_{c_i} \cdot c_i \cdot \widehat{da_5} \\ &+ f_{c_i} \qquad \widehat{da_6} \\ &+ f_i \qquad \widehat{da_7} \\ &+ 1 \qquad \widehat{da_8} + \overline{n}_i \end{aligned}$$

with

$$\widehat{da_k} = \hat{a}_k - a_{0k}, \qquad k = 1,\ldots,8$$
$$\Delta g_i = g_2(r_i, c_i) - g_1(p_{0i}, q_{0i})$$
$$\overline{n}_i = n_2(r_i, c_i) - n_1(p_{0i}, q_{0i})$$

and estimates for f_{r_i}, f_{c_i}, and f_i, for example,

$$f_{r_i} = \frac{\partial \hat{g}_1(r_i, c_i)}{\partial r}$$
$$f_{c_i} = \frac{\partial \hat{g}_1(r_i, c_i)}{\partial c}$$
$$f_i = \hat{g}_1(r_i, c_i)$$

with

$$\begin{pmatrix} p \\ q \end{pmatrix}_{0i} = \begin{pmatrix} a_{01} & a_{02} \\ a_{04} & a_{05} \end{pmatrix} \cdot \begin{pmatrix} r \\ c \end{pmatrix}_i + \begin{pmatrix} a_{03} \\ a_{06} \end{pmatrix}$$

where $\hat{g}_1$—that is, the smoothed version of g_1 (cf. Section 13.2.7)—is restored. Again at least linear interpolation is required for g_1, f_{r_i}, and f_{c_i} in order to avoid unstable solutions.†

---

†An averaging of g_1 and g_2 and of $\partial g_1 / \partial p$ and $\partial g_2 / \partial r$, and the like, needs to take the geometric and radiometric transformations into account.

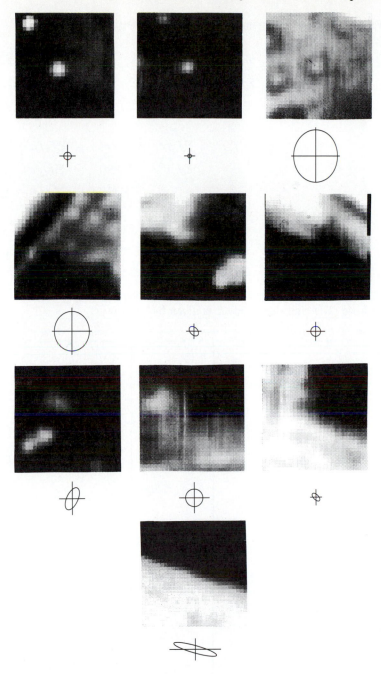

Figure 16.7 Precision of matching to be expected for 10 small windows. Although 32×32 pixels are shown, only the interior 16×16 pixels are assumed to be used. The noise standard deviation is assumed to be $\sigma_n = 5[gr]$. The 99% confidence ellipses given in μm refer to a pixel size of 20μm. Observe the extremely high precision to be expected and the elongated ellipse at the edge.

We then have the following normal equation matrix N:

$$
\begin{pmatrix}
\Sigma f_r^2 r^2 & \Sigma f_r^2 rc & \Sigma f_r^2 r & | & \Sigma f_r f_c r^2 & \Sigma f_r f_c rc & \Sigma f_r f_c r & | & \Sigma f_r fr & \Sigma f_r r \\
 & \Sigma f_r^2 c^2 & \Sigma f_r^2 c & | & \Sigma f_r f_c cr & \Sigma f_r f_c c^2 & \Sigma f_r f_c c & | & \Sigma f_r fc & \Sigma f_r c \\
 & & \Sigma f_r^2 & | & \Sigma f_r f_c r & \Sigma f_r f_c c & \Sigma f_r f_c & | & \Sigma f_r f & \Sigma f_r \\
- & - & - & | & - & - & - & | & - & - \\
 & & & | & \Sigma f_c^2 r^2 & \Sigma f_c^2 rc & \Sigma f_c^2 r & | & \Sigma f_c f_r r & \Sigma f_c r \\
 & & & | & & \Sigma f_c^2 c^2 & \Sigma f_c^2 c & | & \Sigma f_c fc & \Sigma f_c c \\
 & & & | & & & \Sigma f_c^2 & | & \Sigma f_c f & \Sigma f_c \\
- & - & - & | & - & - & - & | & - & - \\
 & & & | & & & & | & \Sigma f^2 & \Sigma f \\
 & & & | & & & & | & & \Sigma 1
\end{pmatrix}
$$

$$(16.56)$$

omitting the index i for convenience and solving the sums over all pixels (r_i, c_i) in the windows. The right-hand side h is given by

$$
h = \begin{pmatrix}
\sum_{i=1}^{m} f_{r_i} r_i \Delta g_i \\[2mm]
\sum_{i=1}^{m} f_{r_i} c_i \Delta g_i \\[2mm]
\sum_{i=1}^{m} f_{r_i} \Delta g_i \\[2mm]
- \; - \; - \\[2mm]
\sum_{i=1}^{m} f_{c_i} r_i \Delta g_i \\[2mm]
\sum_{i=1}^{m} f_{c_i} c_i \Delta g_i \\[2mm]
\sum_{i=1}^{m} f_{c_i} \Delta g_i \\[2mm]
- \; - \; - \\[2mm]
\sum_{i=1}^{m} f_i \Delta g_i \\[2mm]
\sum_{i=1}^{m} 1 \Delta g_i
\end{pmatrix}
\qquad (16.57)
$$

The normal equation system $N\hat{y} = h$ for the eight unknown parameters $\hat{y} = (\widehat{\Delta a_k})$ yields the six corrections Δu_k, $k = 1, \ldots, 6$, to the approximate values of the geometric transformation and the correction $\widehat{\Delta a_7}$ and $\widehat{\Delta a_8}$ for the radiometric parameter; thus

$$\hat{a}_k = a_{0k} + \widehat{\Delta a_k}, \qquad k = 1, \ldots, 8$$

which may be used as new approximate values in a further iteration.

The main effort consists in building the normal equation matrix. If g_2 is not very noisy, simply a restored version $\hat{g}_1$ or $\hat{g}_2$ could be used for determining the f_i, f_{r_i}, and f_{c_i}, which then need not be updated during an iteration sequence. Only the right-hand sides h of Eq. (16.57) have to be recomputed in each iteration. This speeds up the computation and increases the radius of convergence (Burkhardt and Moll, 1979).

The estimated noise variance

$$\hat{\sigma}_n^2 = \frac{1}{m-8} \cdot \sum_{i=1}^{m} [\hat{a}_7 g_1(\hat{p}_i, \hat{q}_i) + \hat{a}_8 - g_2(\hat{r}_i, \hat{c}_i)]^2 \qquad (16.58)$$

with

$$\begin{pmatrix} \hat{p}_i \\ \hat{q}_i \end{pmatrix} = \begin{pmatrix} \hat{a}_1 & \hat{a}_2 \\ \hat{a}_4 & \hat{a}_5 \end{pmatrix} \begin{pmatrix} r_i \\ c_i \end{pmatrix} + \begin{pmatrix} \hat{a}_3 \\ \hat{a}_6 \end{pmatrix}$$

now is a reliable estimate in many cases, especially if windows that are not too large or not too small are used (e.g., between 9×9 and 31×31), and if the distortions between the two images are smooth enough to be modeled by a linear transformation. The two parameters compensating for brightness and contrast are sufficient but also necessary in most practical cases.

The inverse Q of the normal equation matrix may be used to yield the covariance matrix and from there the standard deviation $\sigma_{\hat{a}_k}$ of the estimate

$$\hat{\sigma}_{\hat{a}_k} = \hat{\sigma}_n \cdot \sqrt{Q_{kk}}$$

thus also specifying the precision of the estimated scale, rotation, and shears.

The small windows, however, often do not contain enough detail to enable one to determine all eight parameters. Especially the scales (a_1, a_5, and a_7) and the shears (a_2 and a_4) are frequently not estimable. Therefore a priori knowledge about the transformation may be introduced in a Bayesian manner by using additional observations (possibly fictitious ones)

$$da_k = \widehat{da_k} + n_{a_k}, \qquad w_{a_k} = \frac{\sigma_n^2}{\sigma_{a_k}^2}, \qquad k = 1, \ldots, 8$$

with individual weights depending on the quality, specifically the standard deviations σ_{a_k} of the corrections to the a priori values da_k. This leads to the modified and stabilized normal equation system

$$[N + \text{Diag}(w_{a_k})] \cdot \hat{y} = h$$

The right-hand sides h, because of the corrections assumed to be $da_k = 0$, remain unchanged. The following standard deviations can be recommended:

Scales, shears	$\sigma_{a_k} = 0.1 - 1$	$k = 1,2,4,5,7$
Geometric shifts	$\sigma_{a_k} = 1 - 10$ [pixel]	$k = 3,6$
Radiometric shift	$\sigma_{a_k} = 10 - 100$ [gray value]	$k = 8$

The noise standard deviation σ_n has to be estimated or guessed. The result is not too sensitive against errors of a factor 2 in these assumed standard deviations.

16.3.4 Invariant Points

Much as in Section 16.2.3 (Eq. 16.16), we now want to find points within the window that, when transferred into the other window, are invariant with respect to changes or errors in scale or rotation. To simplify the derivation, we restrict the analysis to three parameters.

The first model, including two shifts and scale, yields the symmetric normal equation matrix

$$N = \begin{pmatrix} N_{11} & N_{12} & N_{13} \\ N_{21} & N_{22} & N_{23} \\ N_{31} & N_{32} & N_{33} \end{pmatrix} = \begin{pmatrix} \sum_i f_{r_i}^2 & \sum_i f_{r_i} f_{c_i} & \sum_i f_{r_i}(f_{r_i}\bar{r}_i + f_{c_i}\bar{c}_i) \\ & \sum_i f_{c_i}^2 & \sum_i f_{c_i}(f_{r_i}\bar{r} + f_{c_i}\bar{c}_i) \\ & & \sum_i (f_{r_i}\bar{r}_i + f_{c_i}\bar{c}_i)^2 \end{pmatrix} \quad (16.59)$$

using $\bar{r}_i = r_i - r_0$ and $\bar{c}_i = c_i - c_0$ with the unknown reference point (r_0, c_0). If we require the estimates $\hat{u}$ and $\hat{v}$ of the two shifts u and v to be independent on the estimated scale, then we have to determine r_0 and c_0 from

$$N_{13} = 0 = \sum f_{r_i}[f_{r_i}(r_i - r_0) + f_{c_i}(c_i - c_0)]$$
$$N_{23} = 0 = \sum f_{c_i}[f_{r_i}(r_i - r_0) + f_{c_i}(c_i - c_0)]$$

leading to the 2×2 system of equations

$$\begin{pmatrix} \sum_i f_{r_i}^2 & \sum_i f_{r_i} f_{c_i} \\ \sum_i f_{r_i} f_{c_i} & \sum_i f_{c_i}^2 \end{pmatrix} \begin{pmatrix} r_0 \\ c_0 \end{pmatrix} = \begin{pmatrix} \sum_i (f_{r_i}^2 r_i + f_{r_i} f_{c_i} c_i) \\ \sum_i (f_{r_i} f_{c_i} r_i + f_{c_i}^2 c_i) \end{pmatrix} \quad (16.60)$$

for (r_0, c_0). This point has several important properties:

- If only the shift parameters $(\hat{r}, \hat{c})$ would have been determined, but the contents of the windows of concern are different in scale, then the shift should be applied to (r_0, c_0) in $g(r, c)$, as the transferred point $(r_0 - \hat{r}, c_0 - \hat{c})$ then would not be biased owing to scale differences.

- The transferred point $(r_0 - \hat{r}, c_0 - \hat{c})$ has the minimum variance (cf. Eq. 16.17), provided the model with three parameters is adequate.

- The invariance of (r_0, c_0) with respect to scale differences especially holds for *corner points* (cf. Fig. 16.8a, b), for example, points where several (≥ 2) edges meet. This is because (in the noiseless case) all gradients $\nabla f_i = (f_{r_i}, f_{c_i})$ are orthogonal to $(r_i - r_0, c_i - c_0)$, leading to the condition $f_{r_i}\bar{r}_i + f_{c_i}\bar{c}_i = 0$ and thus to $N_{13} = N_{23} = 0$. This indicates that the estimates for the shifts and the scale are uncorrelated. But then also $N_{33} = 0$ (cf. Eq. 16.59), from which it is clear that the scale cannot be determined. The condition mentioned above was also used by Negahdaripour and Horn (1989)

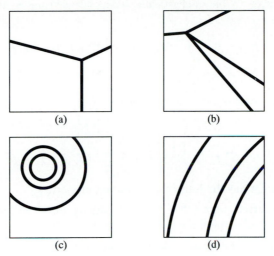

Figure 16.8 Two pairs of images where a scale difference (a,b) and a rotation (c,d) cannot be determined or can be only weakly determined. The points (r_0,c_0) and (r'_0,c'_0) can be determined directly from eqs.(16.60) and (16.61). Estimating the shifts (r,c) without additional geometric parameters for scale or rotation (e.g., by two-dimensional cross-correlation) and applying the shifts to these points leads to unbiased transferred points.

to estimate the focus of expansion, leading to the same normal equation system, Eq. (16.60).

The same reasoning can be followed to determine the point $(r_o^\bullet, c_0^\bullet)$ in the window that, when transferred to the other window, is invariant with respect to differences in *rotation*. If we want to estimate an unknown rotation in addition to unknown shifts $(\hat{r}, \hat{c})$, the nonlinear model

$$g(r_i, c_i) = f(\cos\phi \cdot r_i + \sin\phi \cdot c_i - u, -\sin\phi \cdot r_i + \cos\phi \cdot c_i - v) + n(r_i, c_i)$$

with $\phi \approx 0$ in linearized form, reads as

$$g(r_i, c_i) = f(r_i + \phi c_i - u, -\phi r_i + c_i - v) + n(r_i, c_i)$$

This leads to the 3×3 normal equation matrix

$$N = \begin{pmatrix} N_{11} & N_{12} & N_{13} \\ N_{21} & N_{22} & N_{23} \\ N_{31} & N_{32} & N_{33} \end{pmatrix}$$

$$= \begin{pmatrix} \sum_i f_{r_i}^2 & -\sum_i f_{r_i} f_{c_i} & -\sum_i f_{r_i}(f_{r_i}\overline{c}_i - f_{c_i}\overline{r}_i) \\ & \sum_i f_{c_i}^2 & \sum_i f_{c_i}(f_{r_i}\overline{c} - f_{c_i}\overline{r}_i) \\ & & \sum_i (f_{r_i}\overline{c}_i - f_{c_i}\overline{r}_i)^2 \end{pmatrix}$$

For the shifts (r,c) to be stochastically independent from ϕ, the conditions for $(r_0^\bullet, c_0^\bullet)$ now are

$$N_{13} = 0 = \sum f_{r_i}\left[f_{r_i}(c_i - c_0^\bullet) - f_{c_i}(r_i - r_0^\bullet)\right]$$

$$N_{23} = 0 = \sum f_{c_i}\left[f_{r_i}(c_i - c_0^\bullet) - f_{c_i}(r_i - r_0^\bullet)\right]$$

Therefore $(r_0^\bullet, c_0^\bullet)$ in this case can be estimated from the 2×2 equation system:

$$\begin{pmatrix} \sum_i f_{c_i}^2 & -\sum_i f_{r_i}f_{c_i} \\ -\sum_i f_{r_i}f_{c_i} & \sum_i f_{r_i}^2 \end{pmatrix} \begin{pmatrix} r_0^\bullet \\ c_0^\bullet \end{pmatrix} = \begin{pmatrix} \sum_i (f_{c_i}^2 \cdot r_i - f_{r_i}f_{c_i} \cdot c_i) \\ \sum_i (-f_{r_i}f_{c_i} \cdot r_i + f_{r_i}^2 \cdot c_i) \end{pmatrix} \tag{16.61}$$

Also, this point has the following important properties:

- The transferred point $(r_0^\bullet - \hat{r}, c_0^\bullet - \hat{c})$ among all others has the highest precision, provided the three-parameter model holds.

- If one wishes to determine the shift parameters (r,c) from the two windows without taking a rotation into account, the transferred point $(r_0^\bullet - \hat{r}, c_0^\bullet - \hat{c})$ is unbiased, that is, invariant to possible rotations between the two windows around $(r_0^\bullet, c_0^\bullet)$ and $(r_0^\bullet - \hat{r}, c_0^\bullet - \hat{c})$, respectively.

- The invariance of $(r_0^\bullet, c_0^\bullet)$ with respect to rotation differences especially holds for the centers of figures, which are circularly symmetric with respect to $(r_0^\bullet, c_0^\bullet)$ (cf. Fig. 16.8c, d). The gradients are expected to be parallel to (r_i, c_i), leading to the condition $f_{r_i}\bar{c}_i - f_{c_i}\bar{r}_i = 0$. Therefore not only does $N_{13} = N_{23} = 0$, proving the independence of the shift estimates and the rotation, but also $N_{33} = 0$, showing that the rotation is not determinable.

We will use these relations and the similarity of the normal equation matrices (16.49), (16.60), and (16.61) as a basis for the interest operator discussed next.

16.4   An Interest Operator

16.4.1   Introduction

Image matching as well as general image analysis may require one to find interesting points in the image. *Interesting* here has several meanings, depending on the context:

1. *Distinctness.* Points should be distinct, that is, distinguishable from immediate neighbors. This definition especially excludes points sitting on the same edge. Distinct points may be corners, blobs, highly textured places, and so on. One way to measure distinctness is to compare the intensity function within a window at a point with the intensity function of all surrounding windows. One

could, for example, use the correlation coefficient. If the maximum of the correlation coefficients of the point with its neighbors is small, then the point is dissimilar to all neighbors and is thus a distinct point. This condition is identical to requiring the autocorrelation function at a point to be peaked or to show a high curvature in all directions.

2. *Invariance.* The position as well as the selection of the interesting point should be invariant with respect to the expected geometric and radiometric distortions, which may include robustness with regard to gross or unexpected errors. Invariance and distinctness obviously are the main properties that interesting points should have, as they influence all subsequent steps in the analysis.

3. *Stability.* The position as well as the selection should be invariant with respect to viewing excluding "virtual interestingness." This is to ensure that interesting points in the image correspond to interesting points in the object. For example, corner points of polyhedra can be assumed stable, while T-junctions usually are unstable, as they almost always result from occlusions. Stability thus is decisive for image-matching three-dimensional reconstruction as well as for general image analysis tasks.

4. *Uniqueness.* Whereas distinctness aims at local separability, uniqueness aims at global, that is, imagewide, separability. This is to avoid locally distinct but repetitive features or points that confuse or at least slow down many matching procedures. Unique points thus may significantly increase the reliability of the results of matching and analysis procedures. Uniqueness is probably the notion closest to *interestingness* and the reason to use the term *interest operator* for procedures extracting such points.

5. *Interpretability.* While the previous notions specifying interestingness can be used for both matching and image analysis, we may in addition require the extracted points to have a meaning with respect to image interpretation. Such points then may be corners, junctions of lines, centers of circles, rings, and so on.

Chapter 7 discussed several interest operators for detecting corners. Some, like Moravec's operator (Moravec, 1980; Thorpe, 1983) do not locate corners precisely enough; others rely on a geometric description of the intensity surface and thus are not able to handle corners with more than two edges meeting. All techniques for finding corners of polyhedra with three or even more edges meeting require extracting edges before grouping them in order to obtain corner points. Techniques for finding the centers of circles follow the same type of procedure.

Here we discuss an interest operator that practically fulfills all the criteria mentioned above and contains the extraction procedures as a special case (Paderes, Mikhail, and Förstner, 1984; Förstner, 1986a; Förstner and Gülch, 1987). It follows a three-step procedure:

1. *Selection of optimal windows.* The selection is based on the average gradient magnitude within a window of prespecified size. Searching for local maxima, while suppressing windows on edges, guarantees (local) distinctness. The measure used is, also invariant with respect to rotation.

2. *Classification of the image function within the selected windows.* The classification distinguishes between types of singular points such as corners, rings, and spirals, on one hand, and isotropic texture, on the other. Excluding spirals here, a classification of corners, rings, and general texture, based on a statistical test, is available.

3. *Estimation of the optimal point within the window as the classification.* The estimation is precise for corners and for the centers of circular symmetric features or spirals.

The interest operator has several salient features:

1. The selection of the windows is *optimal* with respect to the following tasks:

   - Find windows that guarantee optimal precision for matching (cf. Section 16.4.4);
   - Find corners;
   - Find centers of circular symmetric features;
   - Find centers of logarithmic spirals (possibly).

2. The selection of the windows is the same for all these tasks and needs no a priori knowledge of the number of edges meeting at a corner or the number of rings.

3. The corner points are *invariant* with respect to *rotations* of a polyhedron in three-dimensions around that corner. In addition, the operator is scale invariant at corners. This is probably the most important property.

4. The *decision* with respect to corner, circular symmetric features, and texture can be based on an *F-test*.

5. The estimation of the optimal point within the window can be represented as a least-squares fit that allows a *rigorous evaluation* of its precision. Specifically, one can derive a covariance matrix for the located point, which may be used in further steps of a geometric analysis.

6. The procedure for finding the optimal windows as well as for locating the optimal points within the windows allows performance in parallel and thus enables real-time feature extraction—in the extreme, down to a few lines' delay during the scanning of the image.

We will discuss the three steps in detail, omitting the operator's property of locating centers of spirals. As the window selection and the distinction are based on the point location within the windows, we treat these estimation procedures first.

16.4.2 Estimating Corner Points

Let us first assume that an $m_r \times m_c$ window is known to contain a corner point $p_0 = (r_0, c_0)'$. The aim is to obtain an estimate $\hat{p}_0 = (\hat{r}_0, \hat{c}_0)'$ for the corner point. As we do not know how many edges intersect at p_0, we treat the edge elements individually, each being representative of a straight line passing through $p_i = (r_i, c_i)'$, $i = 1, \ldots, m$ (cf. Fig. 16.9 a). We assume that the estimated point $\hat{p}_i$ is the point closest to all straight lines crossed by the edge elements in the window, taking individual uncertainty into account.

Let $p_i' = (r_i, c_i)$ be a pixel within the window, with

$$e_i' = e(r_i, c_i) = \nabla f_i' = (f_{r_i}, f_{c_i}) = \left[f_{r_i}(r_i, c_i), f_{c_i}(r_i, c_i) \right]$$

its gradient, and

$$\frac{\nabla f_i'}{|\nabla f_i|} = (\cos \phi_i, \sin \phi_i) = \left[\cos \phi(r_i, c_i), \sin \phi(r_i, c_i) \right]$$

its unit vector in the direction of the gradient. The straight line passing through $p_i = (r_i, c_i)$ parallel to the edge direction is given by

$$(p - p_i)' e_i = 0$$

with $p = (r, c)'$.

We now make two assumptions that allow a noniterative solution for determining the unknown corner point $p_0 = (r_0, c_0)'$:

- The directions ϕ_i of the edge elements, that is, of the lines orthogonal to ∇f_i passing through (r_i, c_i), are error free.
- All uncertainty is taken care of by the distances $l_i = p_i' e_i$ which we assume to be the original observations.

The weights of the edge elements are assumed to be $w_i = ||\nabla f_i||^2$, as if the edge position across had been determined in an image-matching approach using

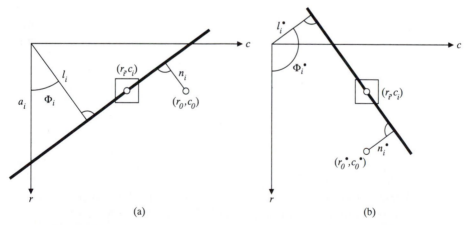

Figure 16.9 Models for estimating corners (a) and centers of circular symmetric features (b). In (a) the edge element through (r_i, c_i) is represented by a straight line (l_i, ϕ_i). The optimal point (r_0, c_0), a corner, is the one that minimizes the weighted sum of the squared distances n_i. In (b) the slope element through (r_i, c_i) is represented by the straight line (l_i', ϕ_i'). The optimal point, the center of the circular feature, minimizes the weighted sum of the distances n_i'.

a model of the edge (cf. Eq. 16.6). Moreover, the individual edge elements are assumed to be independent. Then the linear model reads as

$$l(r_i, c_i) = \cos \phi(r_i, c_i) \cdot \hat{r}_0 + \sin \phi(r_i, c_i) \cdot \hat{c}_0 + n(r_i, c_i)$$

or, with $l_i = l(r_i, c_i)$, $\phi_i = \phi(r_i, c_i)$, and $n_i = n(r_i, c_i)$,

$$l_i = \cos \phi_i \cdot \hat{r}_0 + \sin \phi_i \cdot \hat{c}_0 + n_i, \quad i = 1, \ldots, m \tag{16.62}$$

which is assumed to hold for all $m = m_r \times m_c$ pixels (r_i, c_i). The weight of l_i is assumed to be

$$w_i = ||\nabla f_i||^2 = f_r^2(r_i, c_i) + f_c^2(r_i, c_i)$$

Minimizing $\Omega(r_0, c_0) = \sum_{i=1}^{m} n_i^2 \cdot w_i$ with respect to $\hat{r}_0$ and $\hat{c}_0$ yields

$$\frac{1}{2} \frac{\partial \Omega(\hat{r}_0, \hat{c}_0)}{\partial \hat{r}_0} = \sum_{i=1}^{m} \cos \phi_i \cdot (l_i - \cos \phi_i \cdot \hat{r}_0 - \sin \phi_i \cdot \hat{c}_0) \cdot w_i = 0$$

$$\frac{1}{2} \frac{\partial \Omega(\hat{r}_0, \hat{c}_0)}{\partial \hat{c}_0} = \sum_{i=1}^{m} \sin \phi_i \cdot (l_i - \cos \phi_i \cdot \hat{r}_0 - \sin \phi_i \cdot \hat{c}_0) \cdot w_i = 0$$

which leads to the normal equation system

$$\begin{pmatrix} \sum_{i=1}^{m} w_i \cos^2 \phi_i & \sum_{i=1}^{m} w_i \cos \phi_i \cdot \sin \phi_i \\ \sum_{i=1}^{m} w_i \cos \phi_i \cdot \sin \phi_i & \sum_{i=1}^{m} w_i \sin^2 \phi \end{pmatrix} \begin{pmatrix} \hat{r}_0 \\ \hat{c}_0 \end{pmatrix}$$

$$= \begin{pmatrix} \sum_{i=1}^{m} l_i w_i \cos \phi_i \\ \sum_{i=1}^{m} l_i w_i \sin \phi_i \end{pmatrix}$$

Substituting $l_i = r_i \cos \phi_i + c_i \sin \phi_i$ and using $f_{r_i} = f_r(r_i, c_i) = ||\nabla f_i|| \cos \phi_i$ and $f_{c_i} = f_c(r_i, c_i) = ||\nabla f_i|| \sin \phi_i$, we arrive at

$$\begin{pmatrix} \sum_{i=1}^{m} f_{r_i}^2 & \sum_{i=1}^{m} f_{r_i} f_{c_i} \\ \sum_{i=1}^{m} f_{r_i} f_{c_i} & \sum_{i=1}^{m} f_{c_i}^2 \end{pmatrix} \begin{pmatrix} \hat{r}_0 \\ \hat{c}_0 \end{pmatrix} = \begin{pmatrix} \sum_{i=1}^{m} (f_{r_i}^2 r_i + f_{r_i} f_{c_i} c_i) \\ \sum_{i=1}^{m} (f_{r_i} f_{c_i} r_i + f_{c_i}^2 c_i) \end{pmatrix} \tag{16.63}$$

which is identical to Eq. (16.60). This was to be expected, as the intensity function at an ideal corner is invariant with respect to scale when taking the corner point as center of the origin.

We can also write Eq. (16.63) in the form

$$\left(\sum_{i=1}^{m} W_i \right) \hat{p}_0 = \sum_{i=1}^{m} (W_i \cdot p_i) \tag{16.64}$$

with the singular-weights matrices

$$W_i = \nabla f_i \nabla f_i' = ||\nabla f_i||^2 \cdot e_i \, e_i' = ||\nabla f_i||^2 \cdot \begin{pmatrix} \cos^2 \phi_i & \cos \phi_i \sin \phi_i \\ \cos \phi_i \sin \phi_i & \sin^2 \phi_i \end{pmatrix} \quad (16.65)$$

This shows $\hat{p}_0$ to be the *weighted center of gravity* of all points p_i with *the squared gradient* $\nabla f_i \nabla f_i'$ *as weight matrix,* thus using the model

$$p_i = \hat{p}_0 + n_i \quad \text{or} \quad \begin{pmatrix} r_i \\ c_i \end{pmatrix} = I_2 \cdot \begin{pmatrix} \hat{r}_0 \\ \hat{c}_0 \end{pmatrix} + \begin{pmatrix} n_{r_i} \\ n_{c_i} \end{pmatrix} \quad (16.66)$$

We will use this model in the following because of its simplicity.

We can arrive at a third interpretation of the normal equation system (16.63). If we divide the model equation (16.62) by $\cos \phi_i$ and substitute $s_i = \tan \phi_i$ and $\bar{n}_i = n_i / \cos \phi_i$, we obtain for the observation $a_i = l_i / \cos \phi_i$

$$a_i = \hat{r}_0 + s_i \cdot \hat{c}_0 + \bar{n}_i \quad (16.67)$$

which is the intercept of the straight line through p_i with the r-axis (see Fig. 16.9). If we now take (a_i, s_i) as the representation of the edge element in Hough space with intercept and slope as parameters, we can interpret the model equation (16.67) to be a straight-line fit in Hough space (cf. Chapter 7) with unknown parameters $(\hat{r}_0, \hat{c}_0)$, which in the image domain correspond to the intersection point of the edge elements. If we take the proper weights for the observations a_i, namely, $||\nabla f_i||^2 \cos \phi_i^2$, we arrive at the same normal equation system as before Eq. (16.63).

We can easily transfer this reasoning to the case in which the window is supposed to contain a circular symmetric feature, such as a circle or a set of rings. The idea is to use the slope elements, that is, the straight lines going through the points $p_i' = (r_i, c_i)$ and having the direction of the gradient ∇f_i. If the window contains a circular symmetric feature, then these lines intersect at the center $p_0^\bullet = (r_0^\bullet, c_0^\bullet)$. With the unit vector $e_i^\bullet$ being orthogonal to $e_i, e_i^\bullet = (-\sin \phi_i, \cos \phi_i)$, the equation for these straight lines reads as

$$(p - p_i)' \cdot \tilde{e}_i = 0$$

With the same reasoning as above, thus fixing the orientations ϕ_i and taking the distance $l_i^\bullet = l_i^\bullet(r_i, c_i) = p_i' e_i^\bullet = -r_i \sin \phi_i + c_0 \cdot \cos \phi_i$ as the original observation, we obtain the linear model

$$l_i^\bullet = -\sin \phi_i \cdot \hat{r}_0^\bullet + \cos \phi_i \cdot \hat{c}_0^\bullet + n_i^\bullet, \quad i = 1, \ldots, m \quad (16.68)$$

which is valid for all pixels in the window and assumes the weight to be $w_i = ||\nabla f_i||^2$. The resulting normal equation system for the center $(\hat{r}_0^\bullet, \hat{c}_0^\bullet)$ of the circular symmetric feature reads as

$$\begin{pmatrix} \sum_{i=1}^m w_i \sin^2 \phi_i & \sum_{i=1}^m w_i \cos \phi_i \sin \phi_i \\ \sum_{i=1}^m w_i \cos \phi_i \sin \phi_i & \sum_{i=1}^m w_i \cos^2 \phi_i \end{pmatrix} \begin{pmatrix} \hat{r}_0^\bullet \\ \hat{c}_0^\bullet \end{pmatrix} = \begin{pmatrix} -\sum_{i=1}^m w_i l_i^\bullet \sin \phi_i \\ \sum_{i=1}^m w_i l_i^\bullet \cos \phi_i \end{pmatrix} \quad (16.69)$$

which can also be written as

$$
\begin{pmatrix} \sum\limits_{i=1}^{m} f_{c_i}^2 & -\sum\limits_{i=1}^{m} f_{r_i} f_{c_i} \\ -\sum\limits_{i=1}^{m} f_{r_i} f_{c_i} & \sum\limits_{i=1}^{m} f_{r_i}^2 \end{pmatrix} \begin{pmatrix} \hat{r}_0' \\ \hat{c}_0' \end{pmatrix} = \begin{pmatrix} \sum\limits_{i=1}^{m} (f_{c_i}^2 r_i - f_{r_i} f_{c_i} c_i) \\ \sum\limits_{i=1}^{m} (-f_{r_i} f_{c_i} r_i + f_{r_i}^2 c_i) \end{pmatrix} \tag{16.70}
$$

This equation system is identical to Eq. (16.61) for the point that is invariant with respect to rotation during intensity-based matching.

Equation (16.70) can be written as

$$
\left(\sum_{i=1}^{m} W_i^\bullet \right) \cdot p_0^\bullet = \sum_{i=1}^{m} (W_i^\bullet \cdot p_i)
$$

Now, however, with the singular-weight matrix

$$
W_i^\bullet = \|\nabla f_i\|^2 e_i^\bullet \cdot e_i^\bullet = \|\nabla f_i\|^2 \begin{pmatrix} \sin^2 \phi_i & -\cos \phi_i \sin \phi_i \\ -\cos \phi_i \sin \phi_i & \cos^2 \phi_i \end{pmatrix} \tag{16.71}
$$

the center of circular symmetric features can also be interpreted as the weighted center of gravity. Thus the functional models equation (16.66) for estimating corners and centers of circular symmetric features is the same, whereas the weights W_i and $W_i^\bullet$ are different.

Here also the third interpretation of the estimation procedure using the Hough space can be applied: The determination of the intersection of all slope elements corresponds to the determination of the parameter $(\hat{r}_0^\bullet, \hat{c}_0^\bullet)$ of a straight line in Hough space. This may be achieved by dividing Eq. (16.68) by $-\sin \phi_i$, as $a_i^\bullet = -l_i^\bullet / \sin \phi_i$ is the intercept of the line passing through $p_i(r_i, c_i)$ and having the direction of the gradient.

The estimation procedures provided in this discussion may be used independently in selecting the windows and in determining the window content. But with the least-squares models for estimating the distinct points, we obtain statistical means for automatically classifying the window content and searching for those positions of the window within the image that generate the best local estimate of the image features.

16.4.3   Evaluation and Classification of Selected Windows

We first want to evaluate the estimates $\hat{p}_0$ and $\hat{p}_0'$ for corners and centers of circular symmetric features. In both cases we need an estimate $\hat{\sigma}_n^2$ for the noise variance σ_n^2 and the inverse Q of the 2×2 normal equation matrix N.

We immediately obtain the weighted sums Ω and Ω' of the squared residuals:

$$
\Omega = \sum_{i=1}^{m} (r_i - \hat{r}_0, c_i - \hat{c}_0) \cdot W_i \cdot (r_i - \hat{r}_0, c_i - \hat{c}_0)' = \sum_{i=1}^{m} n_i^2 w_i
$$

$$
\Omega^\bullet = \sum_{i=1}^{m} (r_i - \hat{r}_0^\bullet, c_i - \hat{c}_0^\bullet)' \cdot W_i^\bullet \cdot (r_i - \hat{r}_0^\bullet, c_i - \hat{c}_0^\bullet) = \sum_{i=1}^{m} n_i^{\bullet 2} w_i^\bullet
$$

from which follows the estimate for the noise variance

$$\hat{\sigma}_n^2 = \frac{\Omega}{m-2} \text{ and } \hat{\sigma}_{n^\bullet}^2 = \frac{\Omega^\bullet}{m-2} \quad (16.72)$$

Thus we obtain the covariance matrices

$$\Sigma_{pp} = D\begin{pmatrix} \hat{r}_0 \\ \hat{c}_0 \end{pmatrix} = \hat{\sigma}_n^2 \cdot \begin{pmatrix} \sum f_{r_i}^2 & \sum f_{r_i} f_{c_i} \\ \sum f_{r_i} f_{c_i} & \sum f_{c_i}^2 \end{pmatrix}^{-1} \quad (16.73)$$

and

$$\Sigma_{p^\bullet p^\bullet} = D\begin{pmatrix} \hat{r}_0^\bullet \\ \hat{c}_0^\bullet \end{pmatrix} = \hat{\sigma}_{n^\bullet}^2 \cdot \begin{pmatrix} \sum f_{c_i}^2 & -\sum f_{r_i} f_{c_i} \\ -\sum f_{r_i} f_{c_i} & \sum f_{r_i}^2 \end{pmatrix}^{-1}$$

for the two estimated points $\hat{p}$ and $\hat{p}^\bullet$.

■ EXAMPLE 16.7

Figure 16.10 shows 18 small windows between 5×5 and 9×9 pixels in size. The edge and slope elements sit between the pixels as the Roberts gradient is taken for determining the first derivatives. The hypothesis about the image content is indicated by a small cross for corners and a circle for circular symmetric features. The true points with 99% probability lie within the shown confidence ellipses. The estimation procedure obviously can handle corners with multiple edges coming together. Extrapolation is performed, and

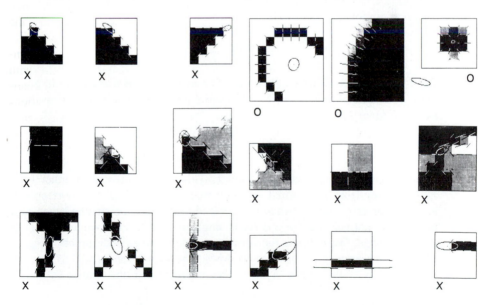

Figure 16.10 Weighted centers of gravity for 18 windows with 99%-confidence ellipses. The assumed model is indicated by the cross (corners) and the circle (circular symmetric figures). Observe the high precision and the extrapolation capacity of the estimation procedure.

the intersection point of the edge lines can be estimated. The example of a simple straight line passing through the window is solved by introducing the center $P_c = (r_1, c_1)$ of the window as a priori information in a Bayesian manner, thus adding the observations

$$\begin{pmatrix} r_1 \\ c_1 \end{pmatrix} = \begin{pmatrix} \hat{r}_0 \\ \hat{c}_0 \end{pmatrix} + \begin{pmatrix} n_{r_1} \\ n_{c_1} \end{pmatrix}$$

with a very low weight $(\sigma_1^2 \cdot I_2)$ to the other observed points p_i.

The covariance matrices Σ_{pp} and $\Sigma_{p\bullet p\bullet}$ can be represented by three values, which we will use later for the window selection step:

1. The average precision of the point or the weight w of the point, assuming $\sigma_n^2 = 1$

$$w = \frac{1}{\frac{1}{2} tr \Sigma} = \frac{1}{\frac{1}{2} tr N^{-1}} = \frac{det\ N}{\frac{1}{2} tr\ N} \qquad (16.74)$$

with N being the 2×2 matrix in Eqs. (16.60) or (16.61). If the eigenvalues of N are equal, then

$$w = \frac{1}{2} tr\ N \qquad (16.75)$$

2. The direction of the major axis of the confidence ellipse

$$\phi = \frac{1}{2} \arctan \frac{-2N_{12}}{N_{11} - N_{22}}$$

3. The form of the confidence ellipse, which may be derived from the ratio λ_1/λ_2 of the two eigenvalues of N or Σ or (to avoid the calculation of the eigenvalues) from the *form factor*

$$q = 1 - \left(\frac{\lambda_1 - \lambda_2}{\lambda_1 + \lambda_2}\right)^2 = \frac{4 \det N}{tr^2 N} = \frac{4 \det \Sigma}{tr^2 \Sigma} \qquad (16.76)$$

which lies in the range between 0 and 1. This can be determined either from Σ directly or from the normal equation matrix N. Here $q = 1$ corresponds to a circular error ellipse, whereas $q = 0$ indicates that the smaller eigenvalue is zero, the normal equation matrix is singular, and thus the window contains a straight edge. We will use the form parameter q to exclude points lying on edges when searching for optimal windows. The term q can be said to measure the circularity of the ellipse or the degree of isotropy of the texture within the window.

As the normal equation matrix for the corner is the same as for matching two windows, assuming a shift only (cf. Eq. 16.47), the confidence ellipses of the matching examples have the same form and orientation as those one would have obtained for the estimated corner point in these windows (cf. Fig. 16.7).

The classification of the window content aims at deciding whether

H_0 :   the window contains an isotropic texture;

H_A :   the window contains a corner;   (16.77)

$H_{A\bullet}$:   the window contains a circularly symmetric feature.

We therefore assume q to be large enough to exclude windows containing straight edges.

Owing to the orthogonality of the models, one can use the test statistic

$$T = \frac{\Omega}{\Omega_\bullet} \sim F_{m-2,m-2} \qquad (16.78)$$

which is F-distributed with $m - 2$ and $m - 2$ degrees of freedom, where m is the number of edge or slope elements. With two critical values k_1 and $k_2 = 1/k_1$, we can classify the window content:

$$T > k_1 \rightarrow \text{circular symmetric feature}$$
$$T < k_2 \rightarrow \text{corner}$$
$$\text{Else} \rightarrow \text{isotropic texture.}$$

Thus the test checks which of the models in (16.77) hold or not.

The parameterization of a spiral would permit estimation of the parameter, including the corner and the circular symmetric windows as special cases (cf. Bigün, 1990).

EXAMPLE 16.8

For the binary window

$$
f = \overset{\downarrow}{r} \quad
\begin{matrix}
& \rightarrow c & & & \\
0 & 0 & 0 & 0 & 0 \\
1 & 1 & 1 & 1 & 0 \\
1 & 1 & 1 & 0 & 0 \\
1 & 1 & 0 & 0 & 0 \\
1 & 0 & 0 & 0 & 0
\end{matrix}
$$

(cf. Fig. 16.10 #1) we obtain the derivatives f_r, f_c, and the product $f_r \cdot f_c$.

$$
f_r = \begin{array}{c|cccc}
 & 1 & 2 & 3 & 4 \\
\hline
1 & 2 & 2 & 2 & 1 \\
2 & 0 & 0 & -1 & -1 \\
3 & 0 & -1 & -1 & 0 \\
4 & -1 & -1 & 0 & 0
\end{array}
\qquad
f_c = \begin{array}{c|cccc}
 & 1 & 2 & 3 & 4 \\
\hline
1 & 0 & 0 & 0 & -1 \\
2 & 0 & 0 & -1 & -1 \\
3 & 0 & -1 & -1 & 0 \\
4 & -1 & -1 & 0 & 0
\end{array}
$$

$$f_r \cdot f_c = \begin{array}{c|cccc} & 1 & 2 & 3 & 4 \\ \hline 1 & 0 & 0 & 0 & -1 \\ 2 & 0 & 0 & 1 & 1 \\ 3 & 0 & 1 & 1 & 0 \\ 4 & 1 & 1 & 0 & 0 \end{array}$$

Similarly, gradients $f_r(r,c) = f(r + \frac{1}{2}, c - \frac{1}{2}) + f(r + \frac{1}{2}, c + \frac{1}{2}) - f(r - \frac{1}{2}, c - \frac{1}{2}) - f(r - \frac{1}{2}, c + \frac{1}{2})$, and $f_c(r,c)$ are used. The coordinate system is chosen such that the gradients refer to integer positions. The normal equation system (16.60) thus reads as

$$\begin{pmatrix} 19 & 5 \\ 5 & 7 \end{pmatrix} \begin{pmatrix} \hat{r}_0 \\ \hat{c}_0 \end{pmatrix} = \begin{pmatrix} (13 \cdot 1 + 2 \cdot 2 + 2 \cdot 3 + 2 \cdot 4) + (1 \cdot 1 + 2 \cdot 2 + 2 \cdot 3 + 0 \cdot 4) \\ (-1 \cdot 1 + 2 \cdot 2 + 2 \cdot 3 + 2 \cdot 4) + (1 \cdot 1 + 2 \cdot 2 + 2 \cdot 3 + 2 \cdot 4) \end{pmatrix}$$
$$= \begin{pmatrix} 42 \\ 36 \end{pmatrix}$$

This leads to estimates $\hat{r}_0 = 114/108 = 1.056, \hat{c}_0 = 474/108 = 4.389$. The weight sum of the squared residuals is

$$\Omega = 5/12 = 0.4166$$

In case one would assume that the window represents a circular symmetric feature,

$$\Omega^{\bullet} = 77/12 = 6.4166$$

Therefore the test statistic Eq. (16.78) is obtained as

$$T = \Omega/\Omega^{\bullet} = 5/77 \approx 0.065$$

which is significantly smaller than 1, indicating that the window represents a corner.

16.4.4 Selection of Optimal Windows

We can now describe the goal of the window selection scheme precisely: The interest operator should find distinct points—points that are discernible from immediate neighbors. We therefore search for local optima of the expected precision for estimating the location of points within a window or for matching the window with one or another image.

Recall that the weight w of the estimated points depends on the noise variance σ_n^2 and the signal content of the window, specifically, the average squared gradient. If one assumes constant σ_n^2, the search need be based only on the average squared gradient N or its inverse. The decisive values—namely, the form factor q and the traces $tr N$ and $tr N^{-1}$—are identical for the three tasks:

- Area based matching,
- Corner estimation,
- Estimation of the center of a circular symmetric feature.

As for large q, and thus nearly isotropic window texture $trN \approx 4trN^{-1}$, we base the selection of the optimal window on $w_i = \frac{1}{2}trN_i$ and q_i and require the following:

1. The confidence ellipse should be round. This is to ensure that selected windows do not contain a straight edge or strongly oriented texture, but rather that the point determination is equally precise in all directions. This leads to the condition

$$q_i > q_{min}$$

2. The confidence ellipse should be smaller than those obtained from neighboring windows and should not exceed a certain size. This ensures maximum local separability or distinctness and good accuracy of the point determination and leads to the conditions

$$w_i > w_{min}$$

and

$$w_i \geq w_l, \text{ for all } l \in \{ \text{ neighborhood of } i\}$$

The procedure for finding the centers of optimal windows is therefore the following:
Input parameters:

- A noise-cleaned image $f(r,c)$
- A gradient operator
- The window size for the operator (m_r, m_c)
- A neighborhood size for finding local extrema
- A threshold q_{min} for q
- A threshold w_{min} for w

1. Determine two derivative images $f_r(r,c)$ and $f_c(r,c)$ by using the gradient operator.
2. Determine the three images containing $f_r^2(r,c), f_r(r,c) \cdot f_c(r,c)$ and $f_c^2(r,c)$.
3. Convolve the three images with a box filter of size $m_r \times m_c$ yielding the three images $N_{11}(r,c), N_{12}(r,c)$, and $N_{22}(r,c)$, which represent the elements of the normal equation matrix for all positions in the image.
4. Determine the images $q(r,c)$ and $w(r,c)$ from Eqs. (16.76) and (16.75).
5. Threshold $w(r,c)$ leading to $w^*(r,c)$:

$$w^*(r,c) = \begin{cases} w(r,c) & \text{if } w(r,c) > w_{min} \text{ and } q(r,c) > q_{min} \\ 0 & \text{else} \end{cases}$$

6. Suppress the nonmaxima of $w^*(r,c)$, setting all w^* to zero where there is no relative maximum of w^* within the prespecified window.

Remarks:

a. Any noise-cleaning procedure that preserves edges and corners may be used. If noise is moderate, a linear filter, preferably Gaussian, is sufficient. In this case the noise cleaning and the gradient determination may be combined by using a larger window for the gradient operator. If the images are already cleaned, even Roberts gradient works sufficiently well.

b. The window size has to be chosen according to the task and the image content. If the selected points are used for matching highly textured images, window sizes of 5×5 already give reliable results. If corners of polyhedra or other interpretable features are to be detected, the window size should be chosen as large as possible so that two neighboring points on an average do not fall into the same window.

c. The neighborhood for the nonmaximum suppression also has to be chosen according to the task and the image content. A large neighborhood suppresses more points but also may eliminate good points sitting near stronger ones. Taking the window and neighborhood sizes to be equal has been proved a reasonable choice in many applications, such as stereo, image sequence analysis, and general image analysis.

d. The threshold q_{min} for the form parameter q due to Eq. (16.76) can be based on a critical value for the ratio $\lambda_1/\lambda_2 (\lambda_1 > \lambda_2)$ of the eigenvalues of N or Σ. If the window contains a straight edge, the larger eigenvalue equals $m \cdot \sigma_g^2$, that is, is proportional to the variance of the gradient across the edge, whereas the smaller eigenvalue equals $m \cdot \sigma_{n'}^2$, that is, is proportional to the variance of the noise gradient. A requirement that the window significantly contain no edge of $\lambda_1/\lambda_2 = 4$, say, using Eq. (16.76), leads to a threshold of $q_{min} = 0.64$. It corresponds to requiring the ratio of the two semiaxes of the error ellipse to be smaller than 2, or the two sides of a corner to meet at an acute angle larger than approximately $53° \approx 2 \arctan(1/2)$.

e. A similar reasoning leads to a choice for the threshold w_{min}. The weight should be significant with respect to noise in the input image (possibly already restored). The expectation of w_i is $m \cdot \sigma_{n'}^2$, where m is the number of edge elements used for determining w. The noise gradient variance $\sigma_{n'}^2$ may be estimated from

$$\sigma_{n'}^2 = \frac{t_{f'} \cdot M}{\#(f'^2 < t_{f'})}$$

where $t_{f'}$ is a threshold for the gradients $||\nabla f||^2 = f'^2$; where $\#(f'^2 < t_{f'})$ is the number of gradients of the image that are smaller than the chosen threshold; and where M is the total number of the edge elements of the image used for this estimate. The threshold $t_{f'}$ should be approximately three to five times an initial estimate of the noise gradient variance $\sigma_{n'}^2$, which could be obtained from a histogram of the f'^2. The estimate is based on the assumption that the noise in the image is white and Gaussian, and therefore the gradient squares $f_i'^2$ at flat areas are χ_2^2-distributed $[\exp(-x), x > 0]$

and that edges influence only the right-hand side of the (cumulative) histogram (theoretically: 1-exp(-x)), which — in contrast to Vorhees' and Poggio's approach (1987) — does not necessarily have to be built up explicitly, except for an initial guess. The threshold w_{min} then can be chosen to be a critical value for the noise gradient, that is, $km\sigma_n^2$, with $k = 10$, say, guaranteeing that only significant windows are selected. Though because of the discreteness of the intensity values and possibly the additional smoothing, the initial guess σ_n^2, may be 0, t_f, should be chosen > 0.

f. It may happen that different selected windows refer to the same distinct point. One representative may be selected *after* the class and the *accurate position* of the points have been determined and their equivalence checked; the one with the highest weight may then be chosen. The reference of multiple windows to the same point partially results from using a box filter for determining the elements of the normal equation system in step 3 and may be reduced, not eliminated, by using a Gaussian, a triangular, or an equivalent filter with a clear maximum at the center of the filter mask.

16.4.5 Uniqueness of Selected Points

The uniqueness of the selected points can be based on their similarity derived from the describing attributes. In our case this would be the intensity function in a window *around the optimal position* of the selected points, as the selected windows in certain cases may sit arbitrarily. Those points that have no features or attributes in common with other points should obtain the highest uniqueness measure.

We want to present a uniqueness measure that is based on the total correlation of a point with all others. It is based on the assumption that the attributes, here the intensity values, are Gaussian distributed (Förstner, 1988). We need the correlation matrix

$$R = (\rho_{ij})$$

between all selected points, with $\rho_{ii} = 1$ and ρ_{ij}, for example, from Eq. (16.34), replacing x_i by (r_i, c_i) and setting $u = 0$, that is, assuming no shift. Large correlation coefficients indicate high similarity. Then the total correlation ρ_i of point P_i and all others P_l can be derived from

$$\rho_i^2 = r_i'(R_{ii})^{-1}r_i = 1 - \frac{1}{(R^{-1})_{ii}} \tag{16.79}$$

where r_i is the vector containing all correlation coefficients between P_i and the P_l, R_{ii} is the submatrix of R after canceling the ith column and the ith row, and $(R^{-1})_{ii}$ is the ith diagonal element of the inverse of R. The total correlation obviously is a weighted mean over all correlations contained in r_i. For only two points the total correlation reduces to the normal correlation.

We now may define the uniqueness of point P_i by

$$u_i = \frac{1 - \rho_i}{\rho_i} \tag{16.80}$$

This is analogous to the signal-to-noise ratio in Eq. (16.40) but takes into account the inverse reasoning here: High correlations lead to low total uniqueness, whereas low total correlations lead to high uniqueness.†

The practical calculation has to take into account that R may be singular, especially if the number of points is larger than the number of attributes. Then one can calculate the uniqueness by using the correlation coefficient from

$$\rho_i^2 = 1 - \frac{1}{[(R + \delta \cdot I)^{-1}]_{ii}} + k \tag{16.81}$$

$$k = \frac{\delta(n + \delta)}{n - 1 + \delta}$$

with δ a small number, for example, 0.001, and n the number of points involved. The correction term k is motivated by assuming R to consist only of 1's, thus assuming the worst case, namely, that all points are completely similar. Then, owing to the $\delta \cdot I$-term modifying R, the correlation shows its largest value, namely, $1 - k$, not 1 as it should be.

Remarks:

1. The uniqueness measure based on the correlation coefficient of the surrounding intensity function derived from Eq. (16.34) is not invariant with respect to rotation, scale, or other distortions. Using the correlation coefficient from Eq. (16.37) derived from the least-squares fit between the intensity functions may provide this invariance, however, since it is practical only for moderate distortions.

2. This definition of uniqueness can be applied to all types of attributes as long as a correlation matrix can be derived, for example, based on a metric $d^2(P_i, P_j)$ between the points and using a correlation function, for example, $1/[1 + (d/d_0)^2]$, guaranteeing R to be at least positive semidefinite. This opens the door to measuring uniqueness based on attributes that are invariant with respect to a wide class of transformations, especially also symbolic attributes.

3. Both distinctness and uniqueness are based on the same similarity measure, the correlation coefficient. If we follow the reasoning that leads to Eq. (16.39), high distinctness corresponds to high curvature of the autocorrelation function, thus low correlation between a point and its immediate neighbors, whereas high uniqueness goes along with low correlation with the other selected points.

4. For large numbers n of points, the calculation of the inverse, being of order $O(n^3)$, may be prohibitive. This can be circumvented by using the maximum

---

†A different definition could apply an information-theoretic view and use the mutual information $H_i = -1/2\log(1 - \rho_i^2)$ (Papoulis, 1984, Eq. 75–98) of the P_i and all P_1, leading to the definition $u_i' = 1/H_i$. (Förstner, 1988). u_i is a monotonic function of u_i'. We prefer u_i here, as we want to use it as a factor for modifying the weight $1/\sigma_x^2$ of correspondences derived from the expected precision, where $\text{SNR}^2 = \rho/(1 - \rho)$ is used (cf. below).

square correlation $\rho_i^2 = \max_j(\rho_{ij}^2)$ of P_i with all other P_j to determine the uniqueness (cf. Eqs. 16.80 and 16.81).

EXAMPLE 16.9

Figure 16.11 shows two images with the selected windows classified according to Section 16.4.3. The estimated points obviously sit at the right positions, when one takes the discreteness of the presentation into account. Observe that in the upper right of Fig. 16.11(b) even an extrapolation of the edges was

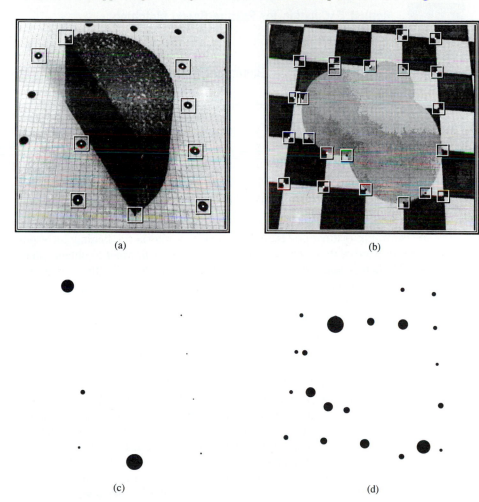

(a)

(b)

(c)

(d)

Figure 16.11 Two images (a) and (b) with selected windows and optimally estimated points after classification. Parts (c) and (d) show the uniqueness measure $(\rho/1 - \rho)$ represented by the area of the circles. Observe that the repetitive features show low uniqueness.

achieved. Also the uniqueness measures, represented by the area of the circles in (c) and (d), correspond to intuition. The images were prefiltered with a 7×7 binomial filter. The window sizes were 17×17 and 21×21, respectively; Roberts gradient and the threshold $q_{min} = 0.75$ were used.

16.5   Robust Estimation for Feature-Based Matching

16.5.1   The Principle of Feature-Based Matching

Recall the three steps of feature-based matching (cf. Section 16.1.2):

1. Selecting features by using some interest operator or some other feature extraction scheme;
2. Finding correspondences between the features by using some similarity and consistency measure;
3. Interpolating between the parallaxes by using a spatial-mapping function.

After having discussed criteria for selecting appropriate features and one possibility for extracting distinct points, we now develop criteria for finding correspondences and for interpolating between the matched features in order to obtain a dense parallax field. Steps 2 and 3 are often designed separately. Specifically, Step 2, finding corresponding features, often does not utilize the same information that is later used for Step 3, interpolating the parallax field. Consequently the strength of the model, namely, the spatial-mapping function, is not exploited, thereby leading to a suboptimal solution of the correspondence step, unless a refinement goes along with the interpolation. Therefore integrating steps 2 and 3 may lead to optimal procedures. Intensity-based matching is an example, the features there being just the pixels themselves. Optical flow using differential techniques—the scale space filtering technique of Witkin, Terzopoulos, and Kass, (1987) or the maximum likelihood technique proposed by Cernusch-Frias, et al. (1989)—falls into the same category. Also, techniques using dynamic programming (Baker and Binford, 1982; Ohta and Kanade, 1985) to obtain global optima while matching the features (line segments) integrate correspondence and interpolation, though this is not made explicit.

For finding an optimal solution, the similarity of the features and the consistency of the correspondence have to be weighted properly. In the following we will discuss a measure for similarity that can be used directly in an estimation process for the parameters of the mapping function serving as a model for consistency. The structure of the setup is closely related to the one by Barnard and Thompson (1980). The individual steps, however, are replaced in order to arrive at an algorithm that tracks the uncertainty from the basic observations, the intensities, to the final result, the spatial-mapping function, and then allows one to evaluate the individual steps and their effect on the result:

1. The similarity between the extracted features leads to a preliminary list of correspondences, including their weights. We restrict the discusssion to feature points, as described in the previous section.
2. A hypothesis for the mapping function is found by using a robust estimation procedure, similar to relaxation techniques, and by enforcing the one-to-one correspondence between the image features.
3. The final parameters of the mapping function are achieved by using a maximum likelihood estimate, allowing a rigorous evaluation of the match.

Though the spatial-mapping function may be quite general, for example, a piecewise-smooth deformation field, we treat only the case of an affine transformation between the images for simplicity. The generalization is straightforward.

16.5.2 The Similarity Measure

We want to determine the weight

$$w_{12} = \frac{1}{\hat{\sigma}^2_{p_{12}}}$$

of the parallax p_{12} in row and column directions by using the average variance

$$\hat{\sigma}^2_{p_{12}} = \frac{1}{2}\left(\hat{\sigma}^2_{r_{12}} + \hat{\sigma}^2_{c_{12}}\right)$$

of the points 1 and 2 in the left and the right image of concern. In order to arrive at a simple expression, we assume that the selected points have isotropic texture ($q = 1$, cf. Eq. 16.76), assuming derivatives f_r and f_c to be uncorrelated (covariance $= 0$), and that the points have equal variance. Then, using Eq. (16.73), we obtain

$$\hat{\sigma}^2_{p_{12}} = \frac{1}{2}\hat{\sigma}^2_n\left(\frac{1}{\sum f_r^2} + \frac{1}{\sum f_c^2}\right)$$

and with $\sum f_r^2 = \sum f_c^2$ (because of $q = 1$), we have

$$\hat{\sigma}^2_{p_{12}} = \hat{\sigma}^2_n \frac{1}{\frac{1}{2}\sum f_r^2 + f_c^2}$$

With the weights $w_i = \frac{1}{2}trN_i = \frac{1}{2}(\sum f_r^2 + \sum f_c^2)_i, i = 1,2$ of the two extracted points, and with $w = \sqrt{w_1 \cdot w_2}$ as an average, this yields

$$\hat{\sigma}^2_{p_{12}} = \frac{\hat{\sigma}^2_n}{\sqrt{w_1 \cdot w_2}}$$

Thus with some arbitrary variance factor σ_0^2, the weight w_{12} of the parallax can be determined from

$$w_{12} = \frac{\sqrt{w_1 \cdot w_2}}{\hat{\sigma}_n^2} \cdot \sigma_0^2 \qquad (16.82)$$

If we now use Eq. (16.40) and replace σ_f^2 by $\sigma_{f_1} \cdot \sigma_{f_2}$, we finally obtain

$$w_{12} = m \cdot \frac{\rho_{12}}{1 - \rho_{12}} \cdot \frac{\sigma_o^2}{\sigma_{f_1} \cdot \sigma_{f_2}} \cdot \sqrt{w_1 \cdot w_2} = m \cdot \frac{\sigma_o^2}{\sigma_n^2} \cdot \sqrt{w_1 \cdot w_2} \qquad (16.83)$$

This weight of the parallax can be determined by using:
- The correlation coefficient ρ_{12} of the two windows around the points based on Eq. (16.34), with the shift being zero;
- The empirical standard deviation σ_f of the intensities $f_1(r,c)$ and $f_2(r,c)$ in the two windows;
- The weights of the two windows, essentially being the variances of the gradients;
- The variance factor σ_0^2, usually taken to be 1.

Remarks

1. As the aim is to measure the similarity of the selected points that have been determined to subpixel values, the windows used for measuring the similarity have to refer to these estimated points. Thus they are usually placed on noninteger positions, which requires resampling. Rounding to full integer positions introduces errors of $1/\sqrt{12} \approx 0.3$ pixel spacing.

2. The weights w_i and the variances $\sigma_{f_i}^2$ can be computed separately and possibly in parallel for both images, the complexity being proportional to the total number of points selected and the window size. The main effort is to compute the correlation coefficients ρ_{12}, which in principle have to be determined over all pairings of points for the left and the right images. Heuristics, discussed later, may reduce this effort.

3. As the weight mainly depends on the correlation coefficient, it is only invariant with respect to geometric shifts and to linear radiometric transformations. As shown by Svedlow, McGillem, and Anuta (1976), geometric deviations below $20°$ or 30% do not have much influence on the correlation coefficient, thereby making this value useful for a wide class of applications. Strong rotations or occlusions, of course, cannot be handled with this similarity measure.

4. The three parts of the weight can be interpreted as similarity $[\rho_{12}/(1 - \rho_{12})]$, strength $(\sigma_{f_1} \cdot \sigma_{f_2})$, and distinctness $(\sqrt{w_1 \cdot w_2})$ and therefore can be replaced by any other measure having these properties. This especially holds for the correlation coefficient ρ_{12} or equivalently—including the strength—the empirical noise variance σ_n^2. The noise variance could also be derived from an intensity-based match of the two windows, then possibly leading to invariance with respect to more complex geometric differences. Using a different measure ρ with the properties of a correlation coefficient—for example, one derived from a distance d

and an appropriate correlation function, such as $\rho(d) = 1/(1+d^2)$—allows one to arrive at similarity measures that are invariant with respect to a much wider class of transformations than simply a geometric shift and a linear radiometric tranformation; measures of symbolic attributes of the points might also be included (cf. Exercise 16.19).

16.5.3 Heuristics for Selecting Candidate Pairs

The first step after selecting image features is to find *candidate pairs* of features. This is to reduce the algorithmic complexity in the final match. Here all types of a priori knowledge may be included and all types of strategies may be applied. The only requirement is that these heuristics and strategies be conservative, that is, they should eliminate only truly wrong correspondences. Some of these heuristics and strategies follow (cf. Ballard and Brown, 1982):

1. The *expected parallax* may be used to exclude feature pairs that are unlikely. This expectation may result from a *scale space* approach using an image *pyramid,* where the result of one level serves as approximation for the level below. The model of the object or, the movement of the object or the camera, may lead to weak or even strong constraints on the parallax field. When one knows the focus of expansion in an image sequence with constant velocity and no rotation, or when one knows the relative orientation of the two cameras, then the corresponding points have to sit on straight lines, the epipolar lines (cf. below).

2. These heuristics may be coupled with the requirement that the similarity of the features be above a certain threshold. One possibility is to require a minimum correlation coefficient, such as $\rho_{12} \geq 0.5$, which, according to Eq. (16.41), corresponds to requiring SNR ≥ 1, a reasonable threshold. Or one may require that $\sigma_p = \sigma_o/\sqrt{w_{12}}$ be better than some standard deviation, such as 2 pixels.

3. Finally, the uniqueness (cf. Section 16.4.5) of the selected points may also be used, for example, to reduce the number of candidate pairs. This can be done separately for both images by canceling all points that are too similar to one or several points within the same image. A criterion could be that the total correlation coefficient (e.g. Eq. 16.79) must be larger than a threshold to keep the point an interesting one. Or the weight w_{12} of the correspondence could be modified by including the uniqueness leading to

$$w_{12} = \frac{\sigma_0^2}{\sigma_n^2} \sqrt{w_1 \cdot w_2} \cdot \sqrt{u_1 \cdot u_2} \tag{16.84}$$

(cf. Eq. 16.82) thus increasing the weight with the average uniqueness of the two points in concern. (This is why we used the definition from Eq. 16.80 and not the one based on the mutual information.)

Other heuristics may be used too. The problem with all of them is that up to now no immediate evaluation of their effect on the result has been possible, specifically with respect to the savings in the algorithmic complexity of the total procedure.

The result of this selection step is a list $\{(r',c'),(r'',c''),w\}$ of candidate pairs with their coordinates in the right and left images and the weight of the correspondence as additional attribute.

16.5.4 Robust Estimation for Determining the Spatial-Mapping Function

The preliminary correspondences now have to be compared to see whether they are consistent with some model. The most general model applied here could be a local smoothness constraint that should hold almost everywhere (Barnard and Thompson, 1980; Terzopoulos, 1986b). The application of a finite-element description of the spatial-mapping function would be particulary appropriate. In order to demonstrate the principle, we restrict our discussion to the same model as the one we used with intensity-based matching, namely, a linear geometric transformation between the images. The model can be expressed as the parallaxes $p = r'' - r'$ and $q = c'' - c'$ by Eq. 16.54.

The model is assumed to hold for the correct correspondences. Assuming that the random errors are small, we can treat r' and c' to be fixed, which is an acceptable approximation and sufficient in most cases. The parallaxes p_k and q_k of the corresponding point a are supposed to have the same weight.

The normal equations for the least-squares estimates have a special structure that can be exploited to reduce the numerical effort, namely,

$$\begin{bmatrix} \sum r_k'^2 w_k & \sum r_k' c_k' w_k & \sum r_k' w_k \\ \sum r_k' c_k' w_k & \sum c_k'^2 w_k & \sum c_k' w_k \\ \sum r_k' w_k & \sum c_k' w_k & \sum w_k \end{bmatrix} \begin{bmatrix} \hat{a}_1 & \hat{a}_4 \\ \hat{a}_2 & \hat{a}_5 \\ \hat{a}_3 & \hat{a}_6 \end{bmatrix} = \begin{bmatrix} \sum p_k r_k' w_k & \sum q_k r_k' w_k \\ \sum p_k c_k' w_k & \sum q_k c_k' w_k \\ \sum p_k w_k & \sum q_k w_k \end{bmatrix} Q^{-1}$$

(16.85)

with the residuals

$$n_k = \begin{pmatrix} n_p \\ n_q \end{pmatrix}_k = \begin{pmatrix} \hat{a}_1 & \hat{a}_2 \\ \hat{a}_4 & \hat{a}_5 \end{pmatrix} \begin{pmatrix} r' \\ c' \end{pmatrix}_k + \begin{pmatrix} \hat{a}_3 \\ \hat{a}_6 \end{pmatrix} - \begin{pmatrix} p \\ q \end{pmatrix}_k \qquad k = 1,\ldots,l \quad (16.86)$$

We may obtain an estimate for the variance factor

$$\hat{\sigma}_o^2 = \frac{1}{2l - 6} \sum_k n_k' n_k w_k \tag{16.87}$$

which, together with the inverse $Q = N^{-1}$ of the normal equation matrix N, can be used to determine the estimated standard deviation of the unknowns:

$$\hat{\sigma}_{\hat{a}_1} = \hat{\sigma}_{\hat{a}_4} = \hat{\sigma}_o \sqrt{Q_{11}}$$
$$\hat{\sigma}_{\hat{a}_2} = \hat{\sigma}_{\hat{a}_5} = \hat{\sigma}_o \sqrt{Q_{22}} \tag{16.88}$$
$$\hat{\sigma}_{\hat{a}_3} = \hat{\sigma}_{\hat{a}_6} = \hat{\sigma}_o \sqrt{Q_{33}}$$

With the contribution h_k of each observation p_k or q_k to the determination of the six unknowns

$$h_k = (r_k' \ c_k' \ 1) \cdot Q^{-1} \cdot (r_k' \ c_k' \ 1)' \cdot w_k \tag{16.89}$$

we obtain (cf. Huber, 1981; Förstner, 1987) the standard deviation of the residuals

$$\hat{\sigma}_{\hat{n}_p} = \hat{\sigma}_{\hat{n}_q} = \hat{\sigma}_o \sqrt{(1 - h_k)/w_k}$$

which are used for the evaluation of the correspondences.

This estimation procedure leads to correct results only if the true correspondences are known. Actually we have no information on which of the elements in the list of preliminary correspondences is correct. But we can now define precisely what we mean by consistency:

1. The true correspondences should fullfill the geometric model. The wrong correspondences therefore can be treated as *outliers* with respect to the geometric model.
2. The true correspondences should form a one-to-one mapping.

We will exploit the requirement on the geometric consistency first. One obvious possibility is to assume either

$$A: \quad n_k \sim N(o, \sigma_k^2 I_2) \quad \text{with probability } 1 - \epsilon$$

or

$$B: \quad n_k \sim U[-L, +L] * U[-L, +L] \quad \text{with probability } \epsilon$$

Case A states that the parallaxes follow a normal distribution with zero mean and a small standard deviation; thus the two points concerned are corresponding ones. Case B states that the parallaxes come from a broad distribution, namely, a uniform distribution with a large L; thus the two points are not corresponding. The probability density function of the standardized true residuals $t = n_k/\sigma_{n_p}$ or $t = n_q/\sigma_{n_q}$ can then be stated as

$$f(x) = (1 - \epsilon) \cdot \varphi(x) + \epsilon \cdot H(x)$$

which is a narrow density function with heavy tails. This gives rise to a robust estimation. The maximum-likelihood estimate can be approximated by the sum of

$$\sum_k \rho \left(\frac{n_k}{\sigma_k} \right) \rightarrow \min$$

with ρ being a nonconvex function. (Here ρ should not be confused with the correlation coefficient used above.) With the method of modified residuals (cf. Huber, 1981), the solution can be found by iteratively weighting down the residuals, with the weight function $w(x) = \rho'(x)/x$ being of the form

$$w(x) = e^{-x^2/2}$$

The idea is that large residuals, indicating wrong correspondences, are weighted down in the course of iterations.

Only for convex minimum functions $\rho(x)$ is convergence guaranteed if the problem is linear (for a more precise statement of these conditions, cf. Huber, 1981).

Therefore the first few iterations should be performed with a convex function $\rho(x)$, which leads to a weight function

$$w(x) = \frac{1}{\sqrt{1 + x^2}}$$

Including the iterative adaptation of the variance factor, the total procedure is as follows:

1. Choose initial values for the parameters $a_j^{(0)}, j = 1, \ldots, 6$, say $(1, 0, 0, 1, 0, 0)$, and the weights $w_k^{(0)}$ (cf. the discussion above). Set the iteration number $\nu = 0$.

2. Solve the weighted least-squares problem according to Eqs. (16.85) to (16.89), which specifically leads to new estimates $\hat{a}_j^{(\nu+1)}, n_k^{(\nu)}, \hat{\sigma}_o^{2(\nu)}$, and $h_k^{(\nu)}$.

3. Update the weights according to

$$w_k^{(\nu+1)} = w_k^{(0)} \cdot w \left(\frac{|n_k^{(\nu)}| \cdot \sqrt{w_k^{(0)}}}{c \cdot \hat{\sigma}_o^{(\nu)} \cdots \sqrt{1 - h_k^{(\nu)}}} \right) \qquad (16.90)$$

Use the weight function $w(x) = 1/\sqrt{1+x^2}$ during the first few (3) iterations; use the weight function $w(x) = e^{-x^2/2}$ for the last two or three iterations; c should be in the range between 1.5 and 2.5.

4. If the change of $\hat{a}_j$ is less than a certain percentage, say 0.1, of their standard deviation, then stop; otherwise, increase ν by 1 and and go to 2.

The results of this robust estimation are parameters $\hat{a}_j$, which in general are close to the true ones and thus are not influenced by the wrong correspondences. In addition, we get weights that—owing to the fast falloff of the exponential function— are practically zero for clearly wrong matches and are significantly less than the original weights. Therefore all correspondences with weights $w_k^{(\nu)}$ that are significantly less than the original weight $w_k^{(0)}$—for example, $< 0.1 w_k^{(0)}$—are rejected and excluded from the following steps.

Though the remaining correspondences now are consistent with the geometric model they need not be unique, as nearby points or image features that are similar to the same points or image features in the other image might still remain. The deviations from the geometric model now intuitively can serve as a criterion for finding a one-to-one mapping: The list of correspondences keeps those which have the smaller—possibly weighted—residual. With these a final least-squares fit for determining the parameters of the geometric model is performed. In this estimation process one must use the original weights, not taking into account the weighting function or the possibly used uniqueness factors. This is necessary if we want to evaluate the final result by using statistical techniques. The *robust estimation* thus is used only *for finding a good hypothesis* for the match. An example is given in Fig. 16.12.

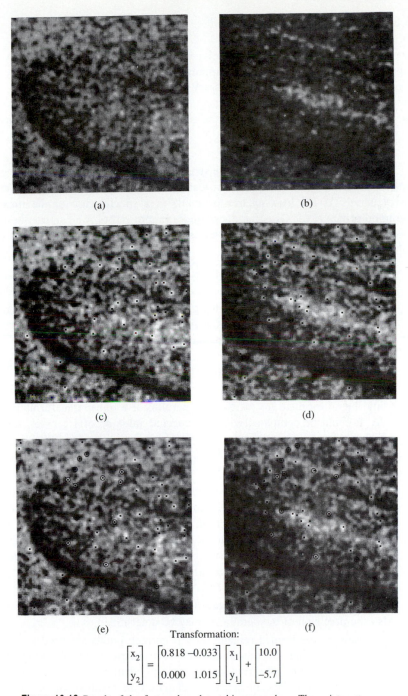

$$\begin{bmatrix} x_2 \\ y_2 \end{bmatrix} = \begin{bmatrix} 0.818 & -0.033 \\ 0.000 & 1.015 \end{bmatrix} \begin{bmatrix} x_1 \\ y_1 \end{bmatrix} + \begin{bmatrix} 10.0 \\ -5.7 \end{bmatrix}$$

Figure 16.12 Result of the feature-based matching procedure. The points automatically selected in the original images (a) and (b) are shown in (c) and (d). The final correspondences are indicated by circles in (e) and (f). Correspondences with residuals > 2 pixels are indicated by crosses.

16.5.5   Evaluating the Final Result

The result of the previous step, namely, finding the most likely parameters of the mapping function or a one-to-one mapping of the extracted features, finally has to be evaluated. This evaluation is necessary

- to be sure the solution is correct;
- to have quantitative measures for the quality of the parameters and the parallaxes to be used in the following steps.

These requirements motivated the strict separation between hypothesis generation (robust estimation) and hypothesis testing after final parameter estimation using least squares: As the least-squares principle is a special case of the maximum-likelihood principle for Gaussian-distributed random errors, we can use the derived density functions of the estimation processes and apply classical hypothesis tests to evaluate the result:

1. A global check whether data and model are consistent uses Ω, the sum of the squared residuals where the sum is taken over all accepted correspondences.

   The ratio Ω/σ_o^2 distributed with χ_{2m-6}^2 degrees of freedom. The denominator is the variance factor used for determining the weights. In case one were to use *unit weights* in the last estimation step, the ratio

   $$\hat{\sigma}_0^2 = \frac{\Omega}{(2m-6)}$$

   would be an estimate for the average parallax if a linear transformation really held. It could be compared with an expected value σ_0^2 of 1/2 pixel, say, using a Fisher test:

   $$\frac{\hat{\sigma}_0^2}{\sigma_0^2} \approx F_{2m-6,\infty}$$

   If the global test is rejected, one may conclude that the model is oversimplified or the data are much worse than expected. Without further testing, no hint is given on the cause of the rejection. If one is not certain about the a priori value σ_n, one may assume σ_n to be the result of a previous estimation step and assume a limited number $m' < \infty$ of degrees of freedom in the Fisher distribution. This leads to larger critical values $F_{2m-6,m',\alpha} > F_{2m-6,\infty,\alpha}$, and thus to a more conservative test (cf. Spiegelhalter, 1985).

2. The precision of the estimated parameters can be determined in the usual way. Also the precision of the parallaxes may be determined as $\sigma_{p_k} = \sigma_{\hat{q}_k} = \sigma_0 \sqrt{h_k}$.

3. The result can be termed reliable only if enough points are used to determine the spatial-mapping function and if they are distributed well enough that the result is not sensitive to errors in the correspondences. When leaving one correspondence out of the estimation, the effect on the result should be small. This maximum influence on one of the parameters a_i is bounded by

   $$\Delta_k a_i \leq \frac{1}{1-h_k} \sqrt{n_k' n_k \cdot h_k \cdot q_{ii}}$$

with h_k from Eq. (16.89), n_k from Eq. (16.86), and q_{ii} again being the diagonal element of the inverse $Q = N^{-1}$ (for the derivation, see Förstner, 1987), with $1 - h_k$ being identical to the redundancy number r_k.

4. The previous measures refer only to the extracted features, not to the original image. A final check could measure the correlation coefficient (similar to Eq. 16.35) after one has rectified the left image. This projection of one image into the other is the decisive check on the correctness of the matching, as all available information is exploited. Of course a wrong match due to repetitive patterns cannot be detected this way, a limitation common to all tests relying only on the image content. Such errors can be detected only if new information that has not been used in the image-matching step becomes available.

We finally have arrived at describing the match of the two images both by a set of corresponding image features, namely, distinct points, and by a mapping function, enabling one to determine the deformation field at each position.

16.6   Structure from Stereo by Using Correspondence

Now we deal with a special application of image matching. We discuss procedures for recovering the three-dimensional structure of objects from image pairs. This forms a link to Chapter 14 on analytic photogrammetry, where the basic relations were worked out. We assume that the interior orientation of the cameras is determined by some calibration procedure. It could be based on a set of targeted control points given in object space. These targets in the image could be located by the intensity-based matching procedure discussed in Section 16.3. We also assume that at least the relative orientation of the two cameras is known. It could be determined from a set of (≥ 5) matched image points. For reasons of stability at least *five groups of points* should be used (Förstner, 1987), which in general requires no great additional effort. If only the relative orientation is known, we can derive three-dimensional coordinates of the object in a local coordinate system that has an arbitrary origin, orientation, and scale. If in addition some (≥ 3) given points have been identified in the images, we can determine the exterior orientation of the cameras.

16.6.1   Epipolar Geometry

Image matching can be tremendously simplified if the relative orientation is known, as the two-dimensional search space is reduced to a one-dimensional one by the so-called epipolar geometry inherent in the oriented image pair. Figure 16.13 shows the general setup of two cameras. The projection centers O' and O'' form the baseline of length b; the principal points H' and H'' are assumed to be the origin of the two image coordinate systems (u', v') and (u'', v''), derived from the pixel coordinates (r', c') and (r'', c'') by using the interior orientation. The object point $P(x, y, z)$ then is mapped into $P'(u', v')$ and $P''(u'', v'')$ in the image planes μ'

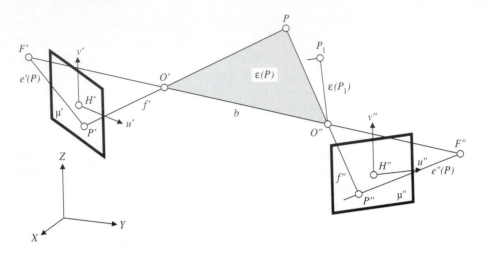

Figure 16.13 Epipolar geometry of a general image pair: image planes f' and f'', projection centers O' and O'', principal points u' and u'', image coordinate systems (u',v') and (u'',v''), baseline b, epipolar plane $\epsilon(P)$, epipolar lines $e'(P)$ and $e''(P)$, and epipoles F' and F''.

and μ''. Because of the geometric model of the perspective projection—specifically the collinearity condition—the five points P,O',O'',P', and P'' lie in one plane, the so-called *epipolar plane* $\epsilon(P)$ associated with P. The intersection lines of $\epsilon(P)$ with μ' and μ'' result in the two *epipolar lines* $e'(P)$ and $e''(P)$ associated with P. For points P_1 not sitting in the same epipolar plane, we obtain different pairs of epipolar lines. All epipolar planes form a pencil of planes passing through the baseline $b = (O'O'')$. The epipolar lines intersect in the *epipoles* F' and F'', which are the intersection of the baseline b intersects with the image planes μ' and μ'', respectively. Thus in general epipolar lines are not parallel.

The main advantage of these geometric relationships is that the epipolar plane $\epsilon(P)$ is defined by P',O', and O''. Thus when only one image point is given, the epipolar line $e''(P)$ is fixed and P'' must sit on this line. Therefore search is necessary in only one dimension. *The epipolar line constraint is the strongest constraint in image matching and should be used as soon as available.* Specifically it is independent of the shape of the object.

Remark:

The epipolar line constraint holds as long as the lens distortions are not too large to be negligible or, more generally, as long as the perspective projection can be modeled by a projective projection, where straight lines in object space map into straight lines in image space. This includes the affine distortion caused by different pixel distances in row and column directions. For high-precision applications the lens distortion (part of the interior orientation) has to be determined and taken into account to correct the observed image (pixel) coordinates in order to exploit the epipolar constraint as a crisp geometric condition. If one does not perform a

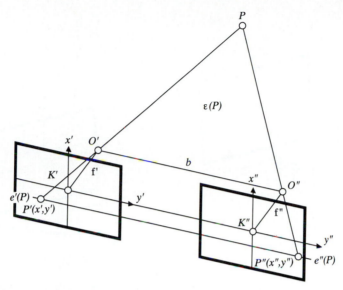

Figure 16.14 Epipolar geometry of a normal image pair: image plane; projection centers O' and O''; principal points K' and K''; image coordinate system (x', y') and (x'', y''); baseline b; x' is parallel to x'', which is parallel to b; focal lengths $f', f''(< 0)$; epipolar plane $\epsilon(P)$; epipolar lines $e'(P)$ and $e''(P)$ identical.

reduction of the observed pixel coordinates for lens distortion, the point P'' corresponding to P' still has to sit on a line that, however, is not straight anymore. In the following we want to neglect possible deviations from the ideal geometry.

If the images are in "normal" position, the determination of epipolar lines reduces to triviality (cf. Fig. 16.14): If the image planes μ' and μ'' are identical and parallel to b, the epipolar lines e' and e'' are parallel to b. Thus the search space for P'', given $P'(x', y')$, is the line $y = y'$.

16.6.2 Generation of Normal Images

It may be useful to rectify evaluation image pairs such that the geometry of a normal image pair can be exploited both for correspondence and for determining the three-dimensional coordinates of object points. Though this advantage cannot be exploited when three or more images are used, it has great computational advantages in the special case of two images, as this rectification has to be performed only once. The method for rectifying the images given below is independent of the representation used for the given pose of the two cameras. It uses the fact that any perspective projection is a projective projection. The principal idea is to replace the images in planes μ' and μ'' by images in plane ν, while keeping the geometry of the bundle of rays spanned by the points in μ' and μ'' and the projection centers O' and O''. Figure 16.15 shows the planes μ' and μ'', the new plane ν, and the images $P', P'', \underline{P}'$,

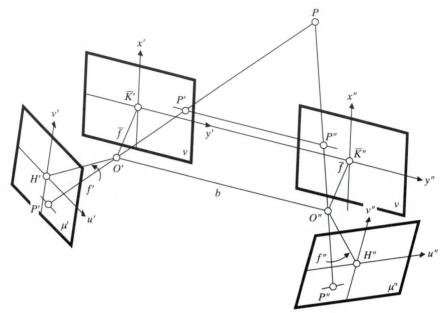

Figure 16.15 Relation between general image pair and normal image pair: common projection centers O' and O'', common baseline; common focal length f; and collinear, normal images in positive position (focal length >0).

and $\underline{P}''$ of P. The procedure has to guarantee that P' maps into $\underline{P}$. Four points in each image are sufficient. The procedure follows:

1. Choose a plane ν parallel to b. This fixes the focal length $\underline{f}$ of the normal images. Choose the new image coordinate systems (x', y') and (x'', y'') with origins $\overline{K}'$ and $\overline{K}''$, where x' is parallel to x'', which is parallel to b, and where $\overline{K}'$ and $\overline{K}''$ are the principal points of the new images and $\overline{K}'O'$ is parallel to $\overline{K}''O''$, which is parallel to ν. Choose a common pixel spacing in the normal images.

2. For each of the images:
   a. Choose four points well distributed over the new image, for example, forming a square. The coordinates are (x_i, y_i), $i = 1, \ldots, 4$, measured in pixels of the normal image.
   b. Project the four points into the original image, using the known pose of the cameras. This yields four coordinates (u_i, v_i), $i = 1, \ldots, 4$.
   c. Solve the equation system

   $$(g\, x_i + h\, y_i + 1) \cdot u_i = a\, x_i + b\, y_i + c$$
   $$(g\, x_i + h\, y_i + 1) \cdot v_i = d\, x_i + e\, y_i + f$$

   which is linear in the parameters a to h of the projective transformation

   $$(u, v) = T(x, y; a, \ldots, h)$$

   d. Rectify the original image. Each pixel (x_j, y_j) in the normal image determines its position (u_j, v_j) in the original image and derives its intensity value by resampling (linear or higher order).

For this discussion we now assume such normal images to be available.

16.6.3 Specializing the Image-Matching Procedures

In the preceding sections we assumed that no geometric relation between the two images was known. Moreover, we approximated the nonlinear mapping between the images by a linear transformation with six parameters (cf. Eqs. 16.54 and 16.55). We now know that the y-coordinates y' and y'' of corresponding points are equal. Thus the complete model reads as

$$x' = a_1 x'' + a_2 y'' + a_3$$
$$y' = y''$$
$$g' = h_1 g'' + h_2$$

We need only five parameters. In an intuitive manner they correspond to depth (a_3), to slope along the epipolar line (a_1), to slope along the epipolar line (a_1), and to brightness (h_2) and contrast (h_1) (cf. Fig. 16.15).

When using the intensity-based matching technique in Section 16.3, only the rows and columns 4 to 6 in Eqs. (16.56) and (16.57) have to be canceled. This is because the parameters a_4, a_5, and a_7 do not have to be determined, owing to the epipolar constraint, and are thus eliminated from the estimation process. The normal equations (16.56) and (16.57) therefore are reduced to a 5×5 linear equation system. The number of degrees of freedom reduces to $m - 5$, changing the ratio $1/(m - 8)$ in Eq. (16.58) to $1/(m - 5)$.

The feature-based matching algorithm also uses the geometric model from above, but the selection of distinct points has to be modified: We are interested only in the x-parallaxes, as the y-parallaxes are zero. Instead of searching for local maxima of $\sum f_r^2 + f_c^2$, we now only need to search for local maxima of $\sum_r \sum_c f_c^2$.

These local maxima yield the centers of optimal windows whose size should be a minimum of three rows and five columns, which proves to be sufficient in good imagery. If accuracy greater than $1/2$ pixel is necessary, larger windows should be used. The optimal point within the window is the weighted center of gravity from Eq. (16.16), but now the sum is evaluated over the entire window, which leads to more stable results.

Remark:

Maximizing f_x^2 is equivalent to searching for the zeros of $2 \cdot f_x \cdot f_{xx}$, and if $f \neq 0$, it is equivalent to determining the zero crossings of the second derivative. But since $f_x \neq 0$, the second derivative $2 \cdot f_{xx}^2 + 2 \cdot f_x \cdot f_{xxx}$ of f_x^2 reduces to $2 \cdot f_x \cdot f_{xxx}$, which is negative because the maxima of f_x^2 are searched for. Thus spurious zero crossings of the second derivative are automatically excluded. Therefore the interest operator, reduced to one dimension, is equivalent to searching edges across the epipolar lines.

16.6.4 Precision of Three-Dimensional Points from Image Points

If we know the coordinates (x', y') and (x'', y'') of corresponding points in the normal images, the three-dimensional coordinates can easily be determined, as shown in Chapter 14. With the object-space coordinate system defined there, we have

$$\begin{pmatrix} x \\ y \\ z \end{pmatrix} = \frac{b}{x'' - x'} \cdot \begin{pmatrix} x' \\ y' \\ f \end{pmatrix} + \begin{pmatrix} -b/2 \\ 0 \\ 0 \end{pmatrix}$$

with the base length b and the focal length f.

Following the line of thought of this chapter, we finally want to determine the precision of the three-dimensional coordinates. Instead of going to the general case with arbitrary orientation of the cameras, we want to restrict the discussion to the coordinates of this "normal model" of the object. If the right base b is chosen, the model is in the same scale as the object.

We first want to give the precision of the z-coordinate. With the parallax $p = x'' - x'$, we have

$$z = \frac{b \cdot f}{p}$$

or $p \cdot z = b \cdot f$, which results in $z\, dp + p\, dz = f\, db + b\, df$, or if $p, f,$ and b are given, we have

$$\frac{dz}{z} = \frac{db}{b} + \frac{df}{f} - \frac{dp}{p}$$

Therefore we obtain the simple relation for the relative precision σ_z/z :

$$\left(\frac{\sigma_z}{z} \right)^2 = \left(\frac{\sigma_b}{b} \right)^2 + \left(\frac{\sigma_f}{f} \right)^2 + \left(\frac{\sigma_p}{p} \right)^2$$

EXAMPLE 16.10

Given

$$f = 50\text{mm}, \qquad\qquad \sigma_f = 0.1\text{mm}$$
$$b = 200\text{mm}, \qquad\qquad \sigma_b = 0.1\text{mm}$$
$$z = 2\text{ m} = 2000\text{mm}$$
$$p = fb/z = 5\text{mm} \qquad\quad \sigma_p = 0.005\text{mm}$$

we have

$$\left(\frac{\sigma_z}{z} \right)^2 = \left(\frac{1}{500} \right)^2 + \left(\frac{1}{2000} \right)^2 + \left(\frac{1}{1000} \right)^2$$
$$= \frac{16 + 1 + 4}{2000^2} = \frac{21}{2000^2}$$

thus a relative precision of $\sqrt{21}/2000 \approx 1/400$, and so $\sigma_z = \sqrt{21}\text{mm} = 4.5$ mm. Obviously the worst relative precision counts, which in this case is the focal length with 1/500 relative precision.

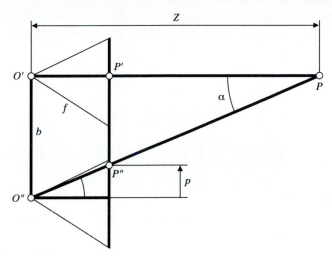

Figure 16.16 Assumed relation between parallax p and parallactic angle α.

In the case of perfect calibration and orientation, the relative precision of z reduces to

$$\frac{\sigma_z}{z} = \frac{\sigma_p}{p}$$

an extremely simple relation to remember. Using $z = fb/p$, we also obtain

$$\frac{\sigma_z}{z} = \frac{z}{f} \cdot \frac{\sigma_p}{b}$$

which relates the relative precision σ/z of z to the relative distance z/f, and the relative precision σ_p/b of the parallax, measured in units of the baseline. The relative precision thus decreases linearly with distance of p.

Finally, we can use the *parallactic angle* $\alpha \approx \arctan(p/f)$ at point P (cf. Fig. 16.16) to relate the relative distance precision to the angular error $\sigma_\alpha = \sigma_p/f$:

$$\frac{\sigma_z}{z} = \frac{z}{b} \cdot \frac{\sigma_p}{f} = \cot \alpha \cdot \sigma_\alpha$$

or for small angles α,

$$\frac{\sigma_z}{z} \approx \frac{\sigma_\alpha}{\alpha}$$

Thus for angles not much larger than σ_α, the relative precision of z also is very low. This situation may easily occur in navigation applications, when the time interval between successive frames is short.

■ **EXAMPLE 16.11**

The precision of z under the conditions mentioned above, but with f and b error free, is

$$\sigma_z = z \cdot \frac{\sigma_p}{p} = 2\text{mm}$$

which obviously would be more than a factor 2 too optimistic if the orientation parameters (f and b) were uncertain. We can also derive the precision of depth differences $h = z_2 - z_1$ of two points of an object, which is given by

$$h = z_2 - z_1 = f \, b \cdot \left(\frac{1}{p_2} - \frac{1}{p_1} \right)$$

We obtain for $z_1 \approx z_2 \approx z$, thus $h/z << 1$:

$$\left(\frac{\sigma_h}{h} \right)^2 = \left(\frac{\sigma_f}{f} \right)^2 + \left(\frac{\sigma_b}{b} \right)^2 + 2 \left(\frac{z}{h} \right)^2 \cdot \left(\frac{\sigma_p}{p} \right)^2$$

■ **EXAMPLE 16.12**

With the same assumptions as above, namely,

$$\sigma_f/f = 1/500,$$

$$\sigma_b/f = 1/2000,$$

$$\sigma_p/p = 1/1000$$

$$h = z/5 = 400\text{mm}$$

we obtain

$$\left(\frac{\sigma_h}{h} \right)^2 = \left(\frac{1}{500} \right)^2 + \left(\frac{1}{2000} \right)^2 + 2 \cdot 25 \cdot \left(\frac{1}{1000} \right)^2$$

$$= \frac{16 + 1 + 2 \cdot 25 \cdot 4}{2000^2} = \frac{217}{2000^2}$$

and thus

$$\sigma_h = \sqrt{217}/2000 \cdot 400\text{mm} \approx 3\text{mm}$$

This is *less* than $\sigma_z = 4.5$mm from above.

Observe that the effect of the errors in f and b are negligible here, $\sqrt{17}/2000^2$ against $\sqrt{200}/2000^2$. The reason why the depth differences are much more precise is that possible errors in f and b cancel when taking the difference $z_2 - z_1$.

This result can be generalized for the case when all parameters of the pose are uncertain. The precision of the absolute depth depends strongly on the precision of the relative depth, or depth differences depend mainly on the parallax precision. Thus

$$\frac{\sigma_h}{h} = \sqrt{2} \cdot \frac{\sigma_p}{p}$$

Nearly all other effects cancel. These results can easily be applied:

1. If the form of an object has to be determined, only relative depth is required. Thus most orientation errors cancel, and the precision of the form is determined by the accuracy of the parallaxes.

2. The same holds if one reference point in object space is given and the pose of the object has to be determined with respect to this (close) reference point.

3. However, if the pose of the camera has to be determined relative to some object, as in navigation, *all* orientation errors influence the overall accuracy. Thus all orientation parameters have to be determined with great care.

These are rules of thumb; a proper planning should apply a rigorous analysis.

Finally, we give the accuracy of the x- and y-coordinates of the point in object space:

$$x = z \frac{x'}{f} - \frac{b}{2}$$

$$y = \frac{z \cdot y'}{f}$$

Under the assumption $\sigma_f = \sigma_b = 0$,

$$\left(\frac{\sigma_x}{x}\right)^2 = \left(\frac{\sigma_z}{z}\right)^2 + \left(\frac{\sigma_{x'}}{x'}\right)^2$$

and

$$\left(\frac{\sigma_y}{y}\right)^2 = \left(\frac{\sigma_z}{z}\right)^2 + \left(\frac{\sigma_{y'}}{y'}\right)^2$$

■ EXAMPLE 16.13

Figure 16.17(a) shows a stereo image pair taken with a photogrammetric stereo camera. The object is a model of an engine crank case. The task is to determine the object's surface along profiles in prespecified planes in order to provide the input for a CAD system, where the model is then processed further. The accuracy requirements are 0.4mm tolerance. In order to cover the complete surface, six stereo image pairs were taken. The orientation of the camera is determined on the basis of precisely measured targeted control points, partly visible in Fig. 16.17.

The baseline of the stereo camera is $b = 0.8$m; the distance to the object is approximately 1.5m; the focal length f is 0.1m. Figure 16.17(b) shows a stereo image pair of the same object now illuminated with a texture projector to permit measurements over the complete surface.

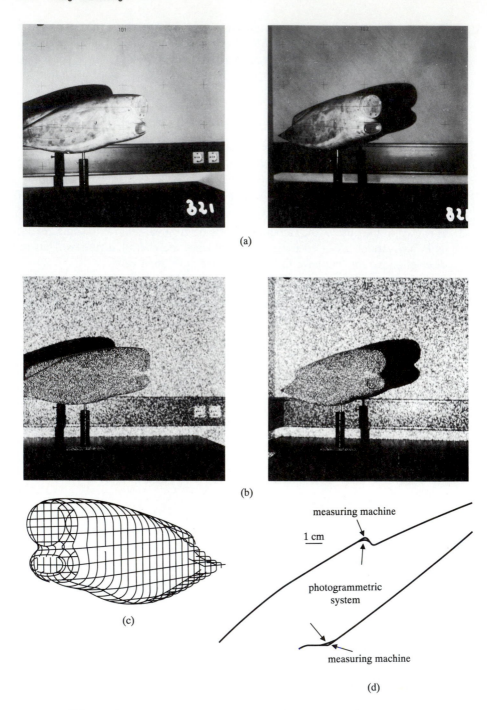

Figure 16.17 (a) stereo pair; (b) illustrated by texture projects (c) three-dimensional (d) profile compression.

The images are digitized with a pixel size of 20μm. Interactively given an initial match, the profiles are automatically measured by using the differential approach discussed in Section 16.3 with the specific model of Section 16.6.3 (cf. Schewe and Förstner, 1986; Schewe, 1989). The result of the measured profiles of four of the six stereo image pairs is shown in Fig. 16.17(c). Figure 16.17(d) shows two pairs of profiles, one measured with the photogrammetric system including the automatic matching procedure, and the other measured with a classical mechanical three-dimensional measuring machine. The deviations mainly are below 0.2μm. Only at the edges of the surface does the photogrammetric system produce a certain amount of blending, which is due to the limitation of the model.

Exercises

16.1. Image matching, in contrast to object reconstruction, should be used only if a geometric transformation between the images is an appropriate representation of the model or if certain invariant low-level features can be found in the images of concern. Give examples for stereo tasks in which image matching is appropriate: (1) for the complete image content; (2) for 80% of the image content; and (3) for less than 50% of the image content shown in both images.

16.2. You want to determine the distance between two holes of an object by using a digital camera with an accuracy of 1:10000. Is this possible? Describe the mensuration design, the size of the camera used (in pixels), and the technique used for locating the hole (refer to Table 16.2).

16.3. Give practical examples for the different types of object surfaces discussed in Fig. 16.2.

16.4. (Reading study.) Study the papers of Barnard and Thompson (1980), Stockman (1987), Ohta and Kanade (1985), and Witkin, Terzopoulos, and Kass (1987): (1) What type of surfaces do they refer to when applied to image matching? (2) How do they achieve consistent results? (3) How far do the algorithms provide means for evaluating the final result?

16.5. (Computer study.) Use Example 16.1 and apply it to the models "shift and scale," "shift + brightness + contrast," "shift + brightness," and all four parameters. Repeat the experiment with: (a) two ideal, noise-free ramp edges; (b) ramp edges with artificial noise ($\sigma_n = 2$); and (c) a noisy box with a shift that is not an integer. Generate the data using noninteger values for the geometric and radiometric transformation parameters. First use real numbers for the gray values, then use gray values rounded to integers. Compare the results with respect to the actual differences in the transformation parameters, the theoretical standard deviations of the estimates, assuming $\sigma_n = 2$, and the empirical standard deviation of the estimated parameters.

16.6. Name conditions such that the intensities g' and g'' at the corresponding points of
two images can be related reasonably well by

a. $g'(r',c') = g''(r'',c'')$

or

b. $g'(r',c') = ag''(r'',c'') + b$

Name typical situations in which these simple models do not hold (refer to Chapter 12 on illumination).

16.7. Name conditions such that extracted edges or edge points refer to the same object
point when: (a) the camera is moving; and (b) the lighting conditions change. Name
conditions such that extracted edges or edge points do *not* refer to the same object
point when: (a) the position of the camera is changing; (b) the position of the light
source is changing; and (c) the pose of the object is changing.

16.8. Under which conditions are two images enough for reconstructing the surface visible
in both images? Under which conditions may a third image be of advantage? In
which geometric relation should the three cameras be? Refer to Sections 16.3–16.6
and explain where the inability of matching possibly shows up. What objective
criteria could be used to detect situations in which no image matching is possible
with two and three cameras, respectively? (*Hint:* Analyze the matching ability in
dependency on the type of texture, on the orientation of edges, and on the complexity
of the object's surface.)

16.9. The ratio $\underline{T} = \Omega/\sigma^2 = (m-1) \cdot \hat{\sigma}_n^2/\sigma_n^2$ (cf. Eq. 16.14) is χ^2-distributed with $m-1$
degrees of freedom. The expectation and variance of $\underline{T}$ is given by

$$E(\underline{T}) = m-1, \quad V(\underline{T}) = \sigma_{T^2} = 2(m-1).$$

Prove Eq. (16.15). (*Hint:* If two stochastic variables are related by $\underline{s} = \sqrt{\underline{t}}$, then
their standard deviations are approximately related by $\sigma \approx \frac{1}{2}\sigma_t\sqrt{E(\underline{t})}$. Apply this
approximation to $\underline{t} = \hat{\sigma}_n^2$ after having derived the variance of $\hat{\sigma}_n^2 = i\underline{T} \cdot \sigma_n^2/(m-1)$.
Stochastic variables are underscored.)

16.10. Estimate the weighted center of gravity of: (a) a smoothed step edge

$$g_a(x) \begin{cases} a/c \cdot x & 0 \le x \le c \\ a & x > c \\ 0 & x < 0 \end{cases}$$

$$(c \ne 0)$$

(b) a smoothed box

$$g_b(x) = g_a(x+d) - g_a(x-e)$$

(c) a skew triangle

$$g_c(x) = \begin{cases} \frac{c}{a}x + c & -a \le x \le 0 \\ a & x > c \\ 0 & \text{otherwise} \end{cases}$$

$$(a \ne 0, b \ne 0)$$

with the window w not sitting symmetric to the signal and using integrals instead of
sums $s : x_o = \int_w x \cdot f''^{12}(x)dx / \int_w f''^{12}(x)dx$ (see Fig. 16.18). Compare the center
of gravity to the center of the window. Vary the window size and its position,

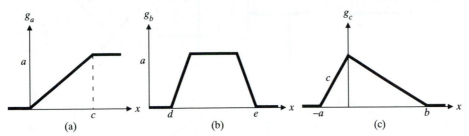

Figure 16.18 Images for Exercise 16.10.

that is, the lower and upper bounds of the integrals. Repeat the comparison with sampled data.

16.11. Prove the precision of the weighted center of gravity to be identical to Eq. (16.9).

16.12. (a) Apply a second iteration to Example 16.1 (cf. Exercise 16.2). Evaluate the improvement in position with respect to the standard deviation of the shifts. Repeat the experiment for two profiles that differ by more than two pixels in position. Does the second (or a third) iteration significantly improve the result? (b) Assume two ramp edges of width d and arbitrary height being shifted by an amount $> d$ with respect to each other. Explain why no solution for the shift can be obtained when using the differential approach. Where does this show up in the practical solution (use simulated data)? Compare the difference of the position of the weighted centers of gravity with the generated shift between the ramps. Comment on the results of (a) and (b) with respect to the rate and the range of convergency of the differential approach and its limitations and on the range of convergency of the feature-based matching techniques.

16.13. Use robust estimation for the shift of two profiles, one having one or two outliers. How does the robust procedure react (a) if the outliers are at a flat portion of the signal? (b) if the outliers are at an edge?

16.14. Generalize the subpixel estimate using correlation and its precision to two dimensions. (a) Show that the position can be obtained from

$$(\hat{r}, \hat{c})' = (r_0, c_0)' - [H\rho|_{(r_0,c_0)}]^{-1} \cdot \nabla\rho|_{(r_0,c_0)}$$

where $H\rho$ and $\nabla\rho$ are the Hessian and the gradient of $\rho(r,c)$

$$H\rho = \begin{pmatrix} \rho_{rr} & \rho_{rc} \\ \rho_{cr} & \rho_{cc} \end{pmatrix} \qquad \nabla\rho = \begin{pmatrix} \rho_r \\ \rho_c \end{pmatrix}$$

evaluated at the integer position (r_0, c_0), being the shift where the correlation coefficient is maximum. The gradient can be determined by using the Sobel operator, and the elements of the Hessian by using a similar convolution kernel. (b) Show that the covariance matrix of the estimated shift can be determined from

$$D\begin{pmatrix} \hat{r} \\ \hat{c} \end{pmatrix} = \frac{1}{m} \cdot \frac{1 - \rho_{12}}{\rho_{12}} \cdot [-H\rho|_{(r_0,c_0)}]^{-1} \cdot \Delta x^2$$

16.15. How can the optical flow equation be generalized: (a) in the case of different brightness? (b) in the case of different brightness and contrast? (*Hint:* Refer to model Eq. 16.16 or use an appropriate function of the image densities.)

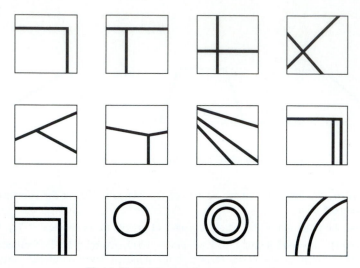

Figure 16.19 Windows for Exercise 16.17.

16.16. Specialize the estimate for a constant shift in two dimensions (Eq. 16.48) for binary images. Use logical operations. (*Hint:* $f_r(r,c)$ and $f_c(r,c)$ can only have the values $-1, 0$, and $+1$. The positive part f_{r+} of f_r can be determined from $f_{r+} = \{[f(r,c) \text{ xor } f(r+1,c)] \text{ and } f(r+1,c)\}$, while the negative part is $f_{r-} = \{[f(r,c) \text{ xor } f(r+1,c)] \text{ and } f(r,c)\}$, assuming 0=false and 1=true. Thus, for example, the sum $S = \sum_{rc} f_r(r,c)$ can be split into $S = S_+ - S_-$, with $S_+ = \sum_{rc} f_{r+}$ and $S_- = \sum_{rc} f_{r-}$.)

16.17. Apply the interest operator to the images in Fig. 16.19. (a) Compare the center of the optimal window with the estimated point. (b) Vary the window size used and describe the effect on the optimal position of the window and on the estimated point within the window.

16.18. Show that the correlation matrix of the three windows in Fig. 16.20 is

$$R = \begin{pmatrix} 1.000 & 0.033 & 0.199 \\ 0.033 & 1.000 & 0.260 \\ 0.199 & 0.260 & 1.000 \end{pmatrix}$$

and the uniqueness is

$$u_1 = 4.077, \; u_2 = 2.868, \; u_3 = 2.122$$

16.19. Apply the uniqueness measure to the symbol sequences "south," "north," and "august." The Levenshtein distances are

$$D = \begin{pmatrix} 0 & 2 & 5 \\ 2 & 0 & 6 \\ 5 & 6 & 0 \end{pmatrix}$$

Use the correlation function $\rho_{ij} = 1/[1+(d_{ij}/l_{max})^2]$, l_{max} being the maximal length of either symbol sequence, to show that the uniqueness of the three sequences is

$$u_1 = 0.136, \; u_2 = 0.161, \; u_3 = 0.697$$

 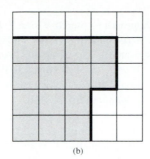 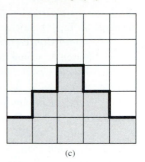

(a)   (b)   (c)

Figure 16.20 Three windows for Exercise 16.18.

16.20. Refer to Chapter 14 (analytical photogrammetry) and assume two cameras in the normal case. The focal length is $f = 35mm$, the base length is $b = 1m$, the pixel size is $20\mu m$, and the distance to the object varies between 9 and 10m. The standard deviation of the parallaxes is 0.1 pixels. (a) How accurately can you determine the length l of the object parallel to the basis? (b) Determine the standard deviation of the distance of a point relative to the cameras. (c) How accurately can the depth difference $d_1 - d_2$ of two points be determined? Assume $h = d_1 - d_2 = 1m$. What relative accuracy σ_h/h can you achieve?

■ Bibliography

Ackermann, F., and A. Pertl, "Zu ordnung Leleiner Bildflächen durch digitale Korrelation zur Verhnüpfung verschiedener oder verschniedenartiger Bilder in Anwerdingsbereich," *Photogrammetrie und Fernerkindung,* DFG-Abschlupbericht, 1983.

Akey, M. L., and R. Mitchell, "Detection and Subpixel Location of Objects in Digitized Aerial Images," *Proceedings of the Seventh Conference on Pattern Recognition,* Montreal, 1984, pp. 411–414.

Alliney, S., and C. Morandi, "Digital Image Registration Using Projections," *IEEE Transactions on Pattern Analysis and Machine Intelligence,* Vol. PAMI-8, 1986, pp. 222–233.

Altmann, J., and H. J. P. Reitbock, "A Fast Correlation Method for Scale-and Translation-Invariant Pattern Recognition," *IEEE Transactions on Pattern Analysis and Machine Intelligence,* Vol. PAMI-6, 1984, pp. 46–57.

Ayache, N., and B. Faverjon, "A Fast Stereovision Matcher Based on Prediction and Recursive Verification of Hypothesis," *Proceedings of the Third Workshop on Computer Vision,* Bellaire, MI, 1985, pp. 27–37.

———, "Efficient Registration of Stereo Images by Matching Graph Descriptions of Edge Segments," *International Journal on Computer Vision,* Vol. 1, 1987, pp. 107–131.

Baker, H. H., and T. O. Binford, "Depth from Edges and Intensity Based Stereo," *Proceedings of the International Joint Conference on Artificial Intelligence,* Vancouver, 1982, pp. 631–636.

Ballard, D. H., and C. M. Brown, *Computer Vision,* Prentice-Hall, Englewood Cliffs, NJ, 1982.

Barnard, S. T., "A Stochastic Approach to Stereo Vision," *Proceedings of the Fifth National Conference on Artificial Intelligence,* Philadelphia, 1986, pp. 676-690.

——, "Stereo Matching by Hierarchical, Microcanonical Annealing," *Proceedings of the Image Understanding Workshop,* Los Angeles, 1987, pp. 792-797.

Barnard, S. T., and W. B. Thompson, "Disparity Analysis of Images," *IEEE Transactions on Pattern Analysis and Machine Intelligence,* Vol. PAMI-2, 1980, pp. 333-340.

Barnea, D. I., and H. F. Silverman, "A Class of Algorithm for Fast Digital Image Registration," *IEEE Transactions on Computers,* Vol. C-21, 1972, pp. 179-186.

Benard, M., "Automatic Stereophotogrammetry: A Method Based on Feature Detection and Dynamic Programming," *Proceedings of the Specialist Workshop on Pattern Recognition in Photogrammetry,* Graz, Austria, 1983.

Berenstein, C. A., L. N. Kanal, D. Lavine and E. C. Olson, "A Geometric Approach to Subpixel Registration Accuracy," *Computer Vision, Graphics, and Image Processing,* Vol. 40, 1987, pp. 334-360.

Berman, M., "Large Sample Bias in Least Squares Estimators of a Circular Arc and Its Radius," *Computer Vision, Graphics, and Image Processing,* Vol. 45, 1989, pp. 126-128.

Bernstein, R., "Scene Correction (Precision Processing) of ERTS Sensor Data Using Digital Image Processing Techniques," *Proceedings of the Third ERTS Symposium,* Vol. 1-A, NASA SP-351, 1973.

——, "Image Geometry and Rectification," *Manual of Remote Sensing,* 2d ed., American Society Photogrammetrics, Little Falls Church, Va, 1983, chap. 21.

Bigün, J., "A Structure Feature for Some Image Processing Applications Based on Spiral Functions," *Computer Vision, Graphics, and Image Processing,* Vol. 51, 1990, pp. 166-194.

Bigün, J., and G. H. Granlund, "Optimal Orientation Detection of Linear Symmetry," *Proceedings of the First International Conference on Computer Vision,* London, 1987, pp. 433-438.

Blostein, S. T., and T. S. Huang, "Quantization Errors in Stereo Triangulation," *Proceedings of the First International Conference on Computer Vision,* London, 1987, pp. 325-334.

Bolles, R.C., L.H. Quam, M.A. Fischler, and H.C. Wolf, "Automatic Determination of Image-to-Database Correspondences," *Proceedings of the Sixth International Joint Conference on Artificial Intelligence*, Tokyo, 1979, pp. 73-78.

Borgefors, G., "Hierarchical Chamfer Matching: A Parametric Edge Matching Algorithm," *IEEE Transactions on Pattern Analysis and Machine Intelligence,* Vol. 10, 1988, pp. 849-865.

Bouthemy, P., and A. Benveniste, "Modeling of Atmospheric Disturbances in Meteorological Pictures," *IEEE Transactions on Pattern Analysis and Machine Intelligence,* Vol. PAMI-6, 1984, pp. 587-600.

Boyer, K. L., and A. C. Kak, "Symbolic Stereo from Structural Descriptions," School of Electrical Engineering, Purdue University, West Lafayette, IN, TR-EE 86-12, 1986.

——, "Structural Stereopsis for 3-D Vision," *IEEE Transactions on Pattern Analysis and Machine Intelligence,* Vol. PAMI-10, 1988, pp. 144-166.

Burkhardt, H., and H. Moll, "A Modified Newton-Raphson Search for the Model-Adaptive Identification of Delays," *Proceedings of the IFAC Symposium on the Identification and System Parameter Estimation,* Darmstadt, W. Germany, 1979, pp. 1279-86.

Burr, D. J., "A Dynamic Model for Image Registration," *Computer Graphics and Image Processing,* Vol. 15, 1981, pp. 102–112.

Cafforio, C., and F. Rocca, "Methods for Measuring Small Displacements of Television Images," *IEEE Transaction on Information Theory,* Vol. 22, 1976, pp. 573–579.

Castro, E. de, and C. Morandi, "Registration and Rotated Images Using Finite Fourier Transforms," *IEEE Transactions on Pattern Matching and Machine Intelligence,* Vol. PAMI-9, 1987, pp. 700–703.

Cernuschi-Frias, B., et al., "Toward a Model-Based Bayesian Theory for Estimating and Recognizing Parameterized 3-D Objects Using Two or More Images Taken from Different Positions," *IEEE Transactions on Pattern Analysis and Machine Intelligence,* Vol. PAMI-11, 1989, pp. 1028–52.

Dreschler, L., "Ermittlung markanter Punkte auf den Bildern bewegter Objekte und Berechnung einer 3D-Beschreibung auf dieser Grundlage," *Dissertation Fachbereich Informatik,* Universität Hamburg, 1981.

Ebner, H., "Berücksichtigung der lokalen Geländeform bei der Höheninter polation mit finiten Elementen," *Bildmessung und Luftbildwesen,* Vol. 51, 1983, pp. 3–9.

Ebner, H., et al., "Integration von Bildzuordnung und Objektrekonstruktion innerhalb der Digitalen Photogrammetrie," *Bildmessung und Lufbildwesen,* Vol. 55, 1987, pp. 194–203.

Förstner, W., "On the Geometric Precision of Digital Correlation," *International Archives of Photogrammetry and Remote Sensing,* Vol. 24-III, Helsinki, 1982, pp. 176–189.

——, "Reliability and Discernability of Extended Gauss-Markov Models," *Deutsche Geodätische Kommision,* München A 98, Munich, 1983, pp. 79–103.

——, "Quality Assessment of Object Location and Point Transfer Using Digital Image Correlation Techniques," *International Archives of Photogrammetry and Remote Sensing,* Vol. 25-A3a, Rio de Janeiro, 1984, pp. 197–219.

——, "Determination of the Additive Noise Variance in Observed Autoregressive Processes Using Variance Component Estimation Techniques," *Statistics and Decisions,* Suppl. Issue No. 2, 1985, pp. 263–274.

——, "A Feature Based Correspondence Algorithm for Image Matching," *International Archives of Photogrammetry and Remote Sensing,* Vol. 26-3/3, Rovaniemi, 1986a, pp. 150–166.

——, "On Automatic Measurement of Digital Surface Models," *Schriftenreihe des Instituts für Photogrammetrie der Universität Stuttgart 11,* Stuttgart, 1986b, pp. 69–90.

——, "Reliability Analysis of Parameter Estimation in Linear Models with Applications to Mensuration Problems in Computer Vision," *Computer Vision, Graphics, and Image Processing,* Vol. 40, 1987, pp. 273–310.

——, "Statistische Verfahren für die automatische Bildanalyse und ihre Bewertung bei der Objekterkennung und- vermessung," Habilitationsschrift, Stuttgart, 1988.

Förstner, W., and E. Gülch, "A Fast Operator for Detection and Precise Location of Distinct Points, Corners, and Centres of Circular Features," *Proceedings of the Intercommission Conference on Fast Processing of Photogrammetric Data,* Interlaken, Switzerland, 1987, pp. 281-305.

Gale, W.A., *Artificial Intelligence and Statistics,* Addison-Wesley, Reading, MA, 1985.

Giri, N.C., *Multivariate Statistical Inference,* Academic Press, New York, 1977.

Goshtasby, A., "Piecewise Cubic Mapping Functions for Image Registration," *Pattern Recognition,* Vol. 20, 1987, pp. 525–533.

——, "Image Registration by Local Approximation Methods," *Image and Vision Computing,* Vol. 6, 1988, pp. 255–261.

Grimson, W. E. L., *From Images to Surfaces: A Computational Study on the Human Early Visual System,* MIT Press, Cambridge, MA, 1981.

——, "Computational Experiments with a Feature Based Stereo Algorithm," *IEEE Transactions on Pattern Analysis and Machine Intelligence,* Vol. PAMI-7, 1985, pp. 17–34.

Grün, A., and E. Baltsavias, "Adaptive Least Squares Correlation with Geometrical Constraints," *Proceedings of the SPIE,* Vol. 595, Cannes, 1985.

Haggren, H., "Photogrammetric Prototype System for Real-Time Engineering Applications," *Optics in Engineering Measurement,* SPIE Proceedings, Vol. 599, 1985, pp. 330–335.

Hannah, M. J., "Computer Matching of Areas in Stereo Images," Ph.D. Diss., Stanford University, Stanford, CA, Report STAN-CS-74-483, 1974.

——, "A System for Digital Stereo Matching," *Phototgrammetric Engineering & Remote Sensing,* 1989.

Haralick, R. M., "Statistical and Structural Approach to Texture," *Proceedings of the IEEE,* Vol. 67, 1979, pp. 786–804.

Helava, U. V., "Digital Correlation in Photogrammetric Instruments," *International Archives of Photogrammetry and Remote Sensing,* Vol. 23-2, Helsinki, 1976.

——, "Object-Space Least-Squares Correlation," *Phototgrammetric Engineering & Remote Sensing,* Vol. 54, 1988, pp. 711–714.

Herbin, M., et al., "Automated Registration of Dissimilar Images: Application to Medical Imagery," *Computer Vision, Graphics, and Image Processing,* Vol. 47, 1989, pp. 77–88.

Herman, M., and T. Kanade, "The 3D Mosaic Scene Understanding System," in Pentland, 1986, pp. 322–358.

Hill, J. W., "Dimensional Measurement for Quantized Images," *SRI Project 4391,* Stanford Research Institute, Menlo Park, 1980.

Ho, C. S., "Precision of Digital Vision Systems," *IEEE Transactions on Pattern Analysis and Machine Intelligence,* Vol. PAMI-5, 1983, pp. 593–601.

Horn, B. K. P., *Robot Vision,* McGraw-Hill, New York, 1987.

Horn, B. K. P., and B. L. Bachman, "Using Synthetic Images to Register Real Images with Surface Models," *Communications of the ACM,* Vol. 21, 1978, pp. 914–924.

Huang, T. S. (ed.), *Image Sequence Analysis,* Springer, New York, 1981.

Huber, P., *Robust Statistics,* Wiley, New York, 1981.

Kass, M., and A. Witkin, "Analysing Oriented Patterns," *Computer Vision, Graphics, and Image Processing,* Vol. 37, 1987, pp. 362–385.

Klaasman, H., "Some Aspects on the Accuracy of the Approximated Position of a Straight Line on a Square Grid," *Computer Graphics and Image Processing,* Vol. 4, 1975, pp. 225–235.

Koch, K. R., *Parameterschätzung und Hypothesentests,* in linearen Modellen, Dümmler, 1987.

Kories, R. R., "Bildzuordnungsverfahren für die Auswertung von Bildfolgen," *Schriftenreihe des Instituts für Photogrammetrie der Universität Stuttgart,* Vol. 11, 1986, pp. 157–168.

Lam, K. P., "Position Determination Using Generalized Multidirectional Gradient Codes," *Computer Vision, Graphics, and Image Processing,* Vol. 28, 1984, pp. 228–239.

Lavine, D., B. A. Lambird, and L. N. Kanae, "Recognition of Spatial Point Patterns," *Pattern Recognition,* Vol. 16, 1983, pp. 289–295.

Li, X., and R. C. Dubes, "The First Stage in Two-Stage Template Matching," *IEEE Transactions on Pattern Analysis and Machine Intelligence,* Vol. PAMI-7, 1985, pp. 700–707.

Lindenberger, J., "Consideration of Observation Errors when Modelling Digital Terrain Profiles," *Proceedings of the Workshop on Progress in Digital Terrain Modelling of WG III/3 of the International Society of Photogrammetry and Remote Sensing,* Lyngby, Denmark, 1987, pp. 227-238.

Longuet-Higgins, M. S., "The Statistical Analysis of a Random Moving Surface," *Philosophical Transactions of the Royal Society of London,* Ser. A 249, 1957, pp. 321–387.

Maître, H., and Y. Wu, "Improving Dynamic Programming to Solve Image Registration," *Pattern Recognition,* Vol. 20, 1987, pp. 443–462.

Markarian, H., et al., "Digital Correction for High-Resolution Images," *Photogrammetric Engineering,* Vol. 39, 1973, pp. 1311–20.

——, *Implementation of Digital Techniques for Correcting High Resolution Images,* American Institute of Aeronautics and Astronautics, Report No. A72–10454, 1972.

McClure, D. E., "Image Models in Pattern Theory," *Image Modeling*, Rosenfeld (ed.), Academic Press, New York, 1981, pp. 259–276.

McGillem, C. D., and M. Svedlov, "Image Restoration Error Variance as a Measure of Overlay Quality," *IEEE Transactions on Geoscience Electronics,* Vol. 14, 1976, pp. 44–49.

——, "Optimum Filter for Minimizing of Image Registration Error Variance," *IEEE Transactions on Geoscience Electronics,* Vol. 15, 1977, pp. 257–259.

Medioni, G., and R. Nevatia, "Matching Images Using Linear Features," *IEEE Transactions on Pattern Analysis and Machine Intelligence,* Vol. PAMI-6, 1984, pp. 675–685.

Merickel, M., and M. McCarthy, "Registration of Contours for 3-D Reconstruction," in *Proceedings of the Seventh Annual Conference of the IEEE Engineering in Medicine and Biology Society,* M. Merickel and M. McCarthy (eds.), Vol. 1, 1985, pp. 616–620.

Mikhail, E. M., and F. Ackermann, *Observations and Least Squares,* Dun-Donelly, New York, 1976.

Milgram, D. L., "Computer Methods for Creating Photomosaics," *IEEE Transactions on Computers, Correspondence,* 1975, pp. 1113-19.

Moravec, H., *Obstacle Avoidance and Navigation in the Real World by a Seeing Robot Rover,* Technical Report CMU-Ri-TR3, Carnegie-Mellon University, Pittsburgh, 1980.

Munteau, C., "Evaluation of the Sequential Similarity Detection Algorithm Applied to Binary Images," *Pattern Recognition,* Vol. 13, 1981, pp. 167–175.

Nagel, H.-H., "Displacement Vectors Derived from Second Order Intensity Variations in Image Sequences," *Computer Vision, Graphics, and Image Processing,* Vol. 21, 1981, pp. 85–117.

Nagel, H.-H., and W. Enkelmann, "Iterative Estimation of Displacement Vector Fields from TV-Frame Sequences," *Proceedings of the Second European Signal Processing Conference,* Erlangen, Germany, 1983, pp. 299–302.

——, "An Investigation of Smoothness Constraints for the Estimation of Displacement Vector Fields from Image Sequences," *IEEE Transactions on Pattern Analysis and Machine Intelligence,* Vol. PAMI-8, 1986, pp. 565-593.

Negahdaripour, S., and B. K. P. Horn, "A Direct Method for Locating the Focus of Expansion," *Computer Vision, Graphics, and Image Processing,* Vol. 46, 1989, pp. 303–326.

Ogawa, H., "Labeled Point Pattern Matching by Delaunay Triangulation and Maximal Cliques," *Pattern Recognition,* Vol. 19, 1986, pp. 35–40.

Ohta, Y., and T. Kanade, "Stereo by Intra- and Inter-Scanline Search Using Dynamic Programming," *IEEE Transactions on Pattern Analysis and Machine Intelligence,* Vol. PAMI-7, 1985, pp. 139–154.

Ohta, Y., K. Takano, and K. Ikeda, "A Highspeed Matching System Based on Dynamic Programming," *Proceedings of the First International Conference on Computer Vision,* London, 1987, pp. 335–342.

Paderes, F. C., E. M. Mikhail, and W. Förstner, "Rectification of Single and Multiple Frames of Satellite Scanner Imagery Using Points and Edges as Control," *Proceedings of the NASA Symposium on Mathematical Pattern Recognition and Image Analysis,* Houston, 1984.

Pagano, M., "Estimation of Models of Autoregressive Signal Plus White Noise," *Annals of Statistics,* Vol. 2. 1984, pp. 99-108.

Panton, D. J., "A Flexible Approach to Digital Stereo Mapping," *Proceedings of the DTM Symposium, American Society of Photogrammetry,* St. Louis, 1978, pp. 32–60.

Papoulis, A., *Probability, Random Variables, and Stochastic Processes,* McGraw-Hill, New York, 1965; 2d ed., 1984.

Park, S. K., and R. A. Schowengerdt, "Image Reconstruction by Parametric Cubic Convolution," *Computer Vision, Graphics, and Image Processing,* Vol. 23, 1983, pp. 258–272.

Pentland, A. P. (ed.), *From Pixels to Predicates,* Ablex, Norwood, NJ, 1986.

Pereira, J. A. G., and N. D. A. Mascarenhar, "Digital Image Registration by Sequential Analysis," *Computers and Graphics,* Vol. 8, 1984, pp. 247–253.

Pertl, A., "Digital Image Correlation with the Analytical Plotter Planicomp C100," *International Archives of Photogrammetry and Remote Sensing,* Vol. 25-A3a, Rio de Janeiro, 1984.

Price, K. E., "Relaxation Matching Techniques—A Comparison," *IEEE Transactions on Pattern Analysis and Machine Intelligence,* Vol. PAMI-7, 1985, pp. 617–623.

Rice, T. A., and L. H. Jamieson, "Scaling and Rotational Registration," in *Computing Structures and Image Processing,* M. J. B Duff et al. (eds.), Academic Press, San Diego, 1985.

Rosenfeld, A., (ed.), *Image Modeling,* Academic Press, New York, 1981.

Ryan, T. W., R. T. Gray, and B. R. Hunt, "Prediction of Correlation Errors in Stereo Pair Images," *Optical Engineering,* Vol. 19, 1980.

Sadjadi, F. A., "Performance Evaluation of Correlations of Digital Images Using Different Separability Measures," *IEEE Transactions on Pattern Analysis and Machine Intelligence,* Vol. PAMI-4, 1982, pp. 436–441.

Schachter, B., "Long Crested Wave Models," *Image Modeling*, Rosenfeld (ed.), Academic Press, New York, 1981, pp. 327–342.

Schalroff, R. J., and E.S. McVey, "Algorithms Development for Real Time Automatic Video Tracking Systems," *Proceedings of the Third International Computers and Applications Conference,* Chicago, 1979, pp. 504–511.

Schewe, H., "Automatische photogrammetrische Erfassung von Industrie oberflächen," *IDENT/Vision,* Eindelfingen, 1989.

Schewe, H., and W. Förstner, "The Program PALM for the Automatic Line and Surface Measurement Using Image Matching Techniques," *International Archives of Photogrammetry and Remote Sensing,* Vol. 26-3/2, Rovaniemi, 1986, pp. 608–622.

Schulte, S., "Modellierung von Beobachtungsreihen durch ein erweitertes Autoregressives Modell," *Deutsche Geodätische Kommision,* München C 327, Munich, 1987.

Shapiro, L. G., and R. M. Haralick, "A Metric for Comparing Relational Descriptions," *IEEE Transactions on Pattern Analysis and Machine Intelligence,* Vol. PAMI-7, 1985, pp. 90-94.

Sharp, J. V., R. L. Christensen, and W. L. Gilman, "Automatic Map Compilation Using Digital Techniques," *Photogrammetric Engineering,* Vol. 31, 1965, pp. 223–239.

Smith, G. B., and H. C. Wolf, "Image-to-Image Correspondence: Linear-Structure Matching," *Proceedings of the NASA Symposium on Mathematical Pattern Recognition and Image Analysis,* Houston, 1984.

Spiegelhalter, D. J., "A Statistical View of Uncertainty in Expert Systems," in *Artificial Intelligence and Statistics,* W. A. Gale (ed.), Addison-Wesley, Reading, MA, 1985.

Steiner, D., "Digital Geometric Picture Correction Using a Piecewise Zero-Order Transformation," *Remote Sensing of Environment,* Vol. 3, 1974, pp. 261–283.

Stockman, G. C., "Object Recognition and Localization via Pose Clustering," *Computer Vision, Graphics, and Image Processing,* Vol. 40, 1987, pp. 361–387.

Stockman, G. C., S. Kopstein, and S. Bennett, "Matching Images to Models for Image Registration and Object Location via Clustering," *IEEE Transactions on Pattern Analysis and Machine Intelligence,* Vol. PAMI-4, 1982, pp. 229–241.

Svedlov, M., C. McGillem, and P. Anuta, "Analytical and Experimental Design and Analysis of an Optimal Processor for Image Registration," LARS Inf. Note 090776, Purdue University, West Lafayette, IN, 1976.

Terzopoulos, D., "Image Analysis Using Multigrid Relaxation Methods," *IEEE Transactions on Pattern Analysis and Machine Intelligence,* Vol. PAMI-8, 1986a, pp. 129–139.

—— "Regularization of Inverse Visual Problems Involving Discontinuities," *IEEE Transactions on Pattern Analysis and Machine Intelligence,* Vol. PAMI-8, No. 2, 1986, pp. 129–139.

—— "The Computation of Visible-Surface Representations," *IEEE Transactions on Pattern Analysis and Machine Intelligence,* Vol. PAMI-10, 1988, pp. 417–438.

Thorpe, C. E., *An Analysis of Interest Operators for FiDO,* Technical Report CMU-Ri-TR-83-19, Carnegie-Mellon University, Pittsburgh, 1983.

Thurgood, J. D., and E. M. Mikhail, "Subpixel Mensuration of Photogrammetric Targets in Digital Images," School of Civil Engineering, Purdue, University, West Lafayette, IN, CH-PH-82-2, 1982.

Tian, Q., and M. N. Huhns, "Algorithms for Subpixel Registration," *Computer Vision, Graphics, and Image Processing,* Vol. 35, 1986, pp. 220–233.

Venot, A., J. F. Lebruchec, and J. C. Roucayrol, "A New Class of Similarity Measures for Robust Image Registration," *Computer Vision, Graphics, and Image Processing,* Vol. 28, 1984, pp. 176–184.

Voorhees, H., and T. Poggio, "Detecting Blobs as Textons in Natural Images," *Proceedings of the Image Understanding Workshop,* Los Angeles, 1987, pp. 892–899.

Vosselman, G., "An Investigation into the Precision of a Digital Camera," Engineering Thesis, TH Delft, Department of Geodesy, Delft, 1986.

Vosselman, G., and W. Förstner, "The Precision of a Digital Camera," *International Archives of Photogrammetry and Remote Sensing,* 27-B1, Kyoto, 1988, pp. 148–157.

Wang, C. Y., et al., "Some Experiments in Relaxation Image Matching Using Corner Features," *Pattern Recognition,* Vol. 16, 1983, pp. 167–182.

Witkin, A., D. Terzopoulos, and M. Kass, "Signal Matching through Scale Space," *International Journal on Computer Vision,* Vol.1, 1987, pp. 133-144.

Wrobel, B., "A New Approach to Computer Stereo Vision and to Digital Photogrammetry," *Proceedings of the Intercommunication Conference on Fast Processing of Photographs,* Interlaken, Switzerland, 1987, pp. 231-258.

Zimmermann, G., and R. Kories, "Eine Familie von Bildmerkmalen für die Bewegungsbestimmung in Bildfolgen," *Informatik-Fachberichte,* Vol. 87, Springer, 1984, pp. 147–153.

CHAPTER

17

THE CONSISTENT-
LABELING PROBLEM

17.1   Introduction

In all phases of machine vision, and particularly in high-level vision, there is a standard class of problems that must be solved. In this class of problems we are given a set of objects and a set of possible names or labels for those objects. We are also given a set of constraints that limits the possible labels for each object. Sometimes the constraints are unary; in this case certain features of an object limit the allowable labels for that object. For example, a long, thin region of an airport scene may be a runway or taxiway, but not a passenger terminal. Often the constraints are binary; this means that a particular label for one object limits the possible set of labels for a related object. For example, if region A is a chair seat and region B is connected to and beneath region A, then region B may be a chair leg, but it is unlikely to be a telephone. Binary constraints were first used in computer vision by Barrow and Popplestone (1971). In general, the constraints may be N-ary for arbitrary positive integer N. In the general case, the set of labels allowed for sets of N mutually constrained objects is limited. The goal is to find a label for each object, so that the set of labels satisfies the given constraints. We formalize the problem as follows (Haralick and Shapiro, 1979, 1980).

An N-ary *consistent-labeling problem* (CLP) is a 4-tuple CLP $= (U, L, T, R)$. The first component U of the 4-tuple is a set of M *units* $U = \{1, \ldots, M\}$, which are the objects to be labeled. The second component L is the set of possible *labels*. The third component T is called the *unit-constraint relation*. T is an N-ary relation over the set U of units. If an N-tuple $(u_1, u_2, \ldots, u_N)$ belongs to T, then we say that units $u_1, u_2, \ldots, u_N$ mutually constrain one another. Groups of regions in an image that are adjacent or groups of parallel line segments can potentially mutually constrain one another. Finally, the fourth component R is called the *unit-label constraint relation*. R is an N-ary relation over the set $U \times L$ of unit-label pairs. If an N-tuple $[(u_1, l_1), (u_2, l_2), \ldots, (u_N, l_N)]$ belongs to R, then the units $u_1, u_2, \ldots, u_N$

may be assigned the corresponding labels $l_1, l_2, \ldots, l_N$. Thus the elements of R are allowable labelings of ordered size-N subsets of the unit set U.

The only groups of units that are constrained are the N-tuples of T, the unit-constraint relation. A *labeling* of a subset $\hat{U} = \{u_1, u_2, \ldots, u_n\}$ of U is a mapping $f : \hat{U} \to L$ from $\hat{U}$ to L. A labeling f of a subset $\hat{U}$ of the units is *consistent* if whenever $u_1, u_2, \ldots,$ and u_N are in $\hat{U}$ and the N-tuple $(u_1, u_2, \ldots, u_N)$ is in T, then $[(u_1, f(u_1)), (u_2, f(u_2)), \ldots, (u_N, f(u_N))]$ is in R. The goal of the consistent-labeling problem is to find one or all consistent-labelings of the unit set U.

17.2   Examples of Consistent-Labeling Problems

Consistent-labeling problems arise in computer vision, in artificial intelligence, and in science and engineering in general. In this section we show how several different problems fit the consistent-labeling formalism and thus can be solved by similar techniques. We begin with two puzzles and then discuss some low- and high-level vision problems that can be formulated as consistent-labeling problems.

17.2.1   The N-Queens Problem

In the N-queens problem, we are given an $N \times N$ chessboard and N queens. The queens must be placed on the chessboard in such a way that no queen can capture any other queen. This means that no two queens may be in the same row, same column, or same diagonal of the chessboard. The N-queens problem can be modeled as a consistent-labeling problem in which the unit set $U = \{1, 2, 3, \ldots, N\}$ is the set of rows on the chessboard and the label set $L = \{1, 2, 3, \ldots, N\}$ is the set of columns. Since there will be exactly one queen per row, a labeling will specify the column on which a queen is placed in each row. Every pair of rows will constrain one another, so the unit-constraint relation T will be the set $T = \{(u_i, u_j) \mid u_i, u_j \in U \text{ and } u_i \neq u_j\}$. The unit-label constraint relation $R = \{[(u_i, l_i), (u_j, l_j)] \mid (u_i, u_j) \in T, l_i \neq l_j, l_i, l_j \in L, \mid u_i - u_j \mid \neq \mid l_i - l_j \mid\}$ includes those pairs of unit-label pairs that represent two squares on the chessboard on which two queens can stand without capturing each other. For example, the pair $[(1, 1), (2, 4)]$ is in R, since a queen on row 1, column 1, does not affect a queen on row 2, column 4. But the pairs $[(1, 1), (2, 2)]$ and $[(1, 1), (3, 3)]$ are not in R, because the queens can capture each other diagonally. A consistent labeling will solve the N-queens problem by telling us, for each row, the column in which to place a queen, so that the constraints are satisfied and no queens can be captured.

17.2.2   The Latin-Square Puzzle

The Latin-square puzzle is an $n \times n$ matrix with n^2 objects that must be arranged on the matrix, one per square. We will consider a 4×4 puzzle for ease of illustration. In this case each object is one of four colors $C = \{red, blue, green, yellow\}$ and has one of four shapes $S = \{circle, square, triangle, octagon\}$. The problem is to arrange the objects such that each row, each column, and each of the two main diagonals of the matrix contains exactly one object of each color and exactly one object of each shape.

One way of modeling the Latin-square puzzle is to let the 16 squares of the matrix be the set of units $U = \{1, 2, \ldots, 16\}$. Then the labels are the objects to be placed on the squares, such as red square, blue triangle, yellow circle, blue circle, and so on. We can represent L as the Cartesian product set $L = C \times S$. The constraints are along the rows, columns, and main diagonals, so we can model T as a quaternary constraint; $T = \{(u_1, u_2, u_3, u_4) \mid u_1, u_2, u_3,$ and u_4 all lie in the same row, column, or diagonal$\}$. The unit-label constraint relation R would then consist of quadruples of unit-label pairs of the form $\{[u_1, (c_1, s_1)], [u_2, (c_2, s_2)], [u_3, (c_3, s_3)], [u_4, (c_4, s_4)]\}$, where (u_1, u_2, u_3, u_4) is in T, $(c_i, s_i) \in L$ for $i = 1, \ldots, 4$, and when $i \neq j$, $c_i \neq c_j$ and $s_i \neq s_j$.

17.2.3 The Edge-Orientation Problem

The edge-orientation problem arises in low-level vision. A local edge operator has been applied to an image and has determined, for each pixel, the strength of an edge passing through it in each of eight possible directions. Because of image noise, the output of this edge operator is also noisy. In order to produce cleaner edges, we wish to combine the output of the edge operator with the knowledge that most meaningful edges in the real world are highly continuous with low curvature. That is, in a sequence of edge pixels, the maximum angle by which any small edge segment can bend with respect to its predecessor or successor edge segment is limited to some maximum bending angle. The maximum bending angle between two adjacent pixels x and x' is a function of the spatial relationship between the two pixels. For example, if pixel x' is in the northeast relationship to pixel x, and pixel x has an edge direction pointing upward, we might restrict the allowable directions of pixel x' to those between 45° and 90° counterclockwise from the horizontal. However, if pixel x' is in the northwest relationship to x, we would instead restrict the allowable directions of x' to those between 90° and 135°. For simplicity, assume that there is a predicate $\Delta(x, l, x', l')$ that returns *true* if l and l' are compatible labels for pixels x and x' and *false* otherwise. We wish to use this information in order to assign an edge orientation or the value *none* to each pixel.

In the consistent-labeling model of this problem, U would be the set of pixels of the image, and L would be the set of possible edge orientations, including the special value *none*. Suppose that $E(x)$ is the set of possible edge orientations of pixel x, based on the results of the local edge operator. Let $Nbd(x)$ be the set of neighboring pixels to pixel x. Only edge orientations of pixels in the neighborhood of a given pixel can constrain the edge orientation of the given pixel. Thus $T = \{(x, x') \mid x' \in Nbd(x)\}$. Pairs of neighboring pixels have compatible labels if the difference between their edge orientations satisfies the Δ predicate or if at least one is labeled *none*. So the unit-label constraint relation $R = \{[(x, l), (x', l')] \mid (x, x') \in$

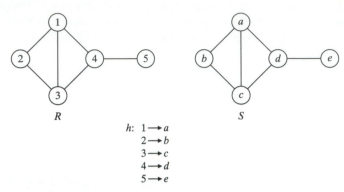

$$h:\ 1 \longrightarrow a$$
$$2 \longrightarrow b$$
$$3 \longrightarrow c$$
$$4 \longrightarrow d$$
$$5 \longrightarrow e$$

Figure 17.1 Graph isomorphism from graph R to graph S.

$T,\ l \in E(x),\ l' \in E(x'),$ and $\neg(l = none\ or\ l' = none) \Rightarrow \Delta(x,l,x',l')\}.$
Robinson (1977) used this idea to clean an edge image having oriented edges.

17.2.4 The Subgraph-Isomorphism Problem

A *graph* G is a pair (V,E), where V is a set of vertices and E is a nonreflexive, symmetric binary relation over V. Graphs are used in many areas of engineering and computer science to represent problems involving binary relationships between objects. It is often necessary to determine whether two graphs representing two different entities are identical, except for the labels of the vertices, indicating that the two objects have the same structure. We say that a graph $G = (V,E)$ is *isomorphic* to a graph $G' = (V',E')$ if there is a one-one, onto mapping f from V to V', satisfying that $(u,v) \in E \Leftrightarrow [f(u),f(v)] \in E'$. The mapping f is called a *graph isomorphism*. Figure 17.1 illustrates the concept of graph isomorphism. It is also common to want to know whether a given graph is identical to a *subgraph* of a larger graph. A graph $G = (V,E)$ is isomorphic to a subgraph of a second graph $G' = (V',E')$ if there is a one-one mapping f from V to V', satisfying that $(u,v) \in E \Rightarrow [f(u),f(v)] \in E'$.

The subgraph-isomorphism problem translates directly to a consistent-labeling problem. The unit-set U is the set of vertices V of G, the label-set L is the set of vertices V' of G', and the unit-constraint relation T is just the edge set E of G. The unit-label constraint relation R is defined by $R = \{[(u,u'),(v,v')] \mid (u,v) \in E,\ (u',v') \in E',$ and $u \neq v \Rightarrow u' \neq v'\}$. The graph-isomorphism problem translates into a dual consistent-labeling problem: Both the mapping $f : V \rightarrow V'$ and its inverse must be consistent labelings.

17.2.5 The Relational-Homomorphism Problem

The subgraph-isomorphism problem is a special case of a more general problem that we call the *relational-homomorphism problem*. There are two differences between

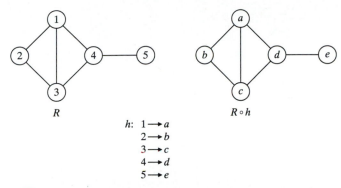

h: $1 \longrightarrow a$
$2 \longrightarrow b$
$3 \longrightarrow c$
$4 \longrightarrow d$
$5 \longrightarrow e$

Figure 17.2 Composition of a binary relation R with a mapping h.

the two. First, the relational-homomorphism problem is defined on N-ary relations, instead of just binary. Second, a homomorphism is also a structure-preserving mapping, but it is not as strict as an isomorphism in that the mapping does not have to be one-one. The relational-homomorphism problem can be defined as follows:

Let A and B be two sets. Let $T \subseteq A^N$ be an N-ary relation over set A. Let $f : A \rightarrow B$ be a function that maps elements of set A into set B. We define the *composition* of T with f, written $T \circ f$, by

$$T \circ f = \{(b_1, \cdots, b_N) \in B \mid \text{there exists}$$
$$(a_1, \cdots, a_N) \in T \text{ with } f(a_i) = b_i, i = 1, \cdots, N\}.$$

Figure 17.2 illustrates the composition of a binary relation with a mapping.

Let $S \subseteq B^N$ be a second N-ary relation. A *relational homomorphism* from T to S is a mapping $f : A \rightarrow B$ that satisfies $T \circ f \subseteq S$. That is, when a relational homomorphism is applied to each component of an N-tuple of T, the result is an N-tuple of S. Figure 17.3 illustrates the concept of a relational homomorphism. The problem of finding relational homomorphisms is sometimes called the problem of relational matching.

A relational homomorphism maps the elements of A to a subset of the elements of B having all the same interrelationships that the original elements of A had. In computer vision it is common for set A to represent the primitive parts of some object or object model and for set B to represent the primitive parts of some other object or description extracted from an image. If A is a much smaller set than B, then finding a one-one relational homomorphism is equivalent to finding a copy of a small object as part of a larger object. Finding a chair in an office scene is an example of such a task. If A and B are about the same size, then finding a relational homomorphism is equivalent to determining that the two objects are similar. A *relational monomorphism* is a relational homomorphism that is one-one. Such a function maps each primitive in A to a unique primitive in B. A monomorphism indicates a stronger match than a homomorphism.

Finally, a *relational isomorphism* f from an N-ary relation T to an N-ary relation S is a one-one relational homomorphism from T to S, and f^{-1} is a relational

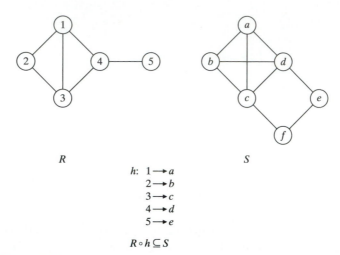

R

S

h: $1 \longrightarrow a$
$2 \longrightarrow b$
$3 \longrightarrow c$
$4 \longrightarrow d$
$5 \longrightarrow e$

$R \circ h \subseteq S$

Figure 17.3 Relational homomorphism h from binary relation R to binary relation S.

homomorphism from S to T. In this case, A and B have the same number of elements, each primitive in A maps to a unique primitive in B, and each primitive in A is mapped to by some primitive of B. Also, every tuple in T has a corresponding tuple in S, and vice versa. A relational isomorphism is the strongest kind of match: a symmetric match. A graph isomorphism is a binary-relational isomorphism.

The relational-homomorphism problem fits the consistent-labeling model in much the same way that the graph-isomorphism problem did. A is the set of units and B is the set of labels. The unit-constraint relation is simply the relation T of the relational homomorphism problem. The unit-label relation R is given by $R = \{[(u_1,l_1),\dots,(u_N,l_N)] \mid (u_1,\dots,u_N) \in T \text{ and } (l_1,\dots,l_N) \in S\}$. A consistent labeling that is a solution to this problem is a relational homomorphism from A to B.

The relational-homomorphism problem, as just defined, deals with objects, but not their attributes, and with relationships among objects, but not the attributes of those relationships. If we are trying to find a mapping from a model of an object to a description extracted from an image, it is likely that each primitive should match to a primitive of a similar shape and that some numeric value or values will be associated with each relationship, such as the angle between a pair of connecting lines. Thus instead of dealing only with two sets A and B and ordinary N-ary relations $T \subseteq A^N$ and $S \subseteq B^N$, we would like to have a property vector $P(x)$ associated with each primitive a of A and b of B, and a property vector $Q(t)$ associated with each tuple t of T and each tuple s of S. For these attributed primitives and attributed relations, a relational homomorphism would be a mapping $f : A \rightarrow B$ satisfying

1. $P(a) = P[f(a)]$ for each $a \in A$;
2. For each $t \in T$, $t \circ f \in S$ and $Q(t) = Q[f(t)]$.

This attributed relational-homomorphism problem is also a consistent-labeling problem with a limited set of labels for each unit. In this case the unit-label constraint relation is further constrained by the property values of the tuples.

17.3   Search Procedures for Consistent Labeling

Suppose we are given a consistent-labeling problem $CLP = (U, L, T, R)$. We wish to find the set of all consistent labelings $f : U \rightarrow L$ that satisfy the constraints specified by T and R. Of course, there may be no consistent labelings, in which case the algorithm should return the empty set. The general method for solving this problem is a *backtracking tree search*. Such a tree search algorithm can be improved upon by adding nonheuristic methods for pruning the tree. These methods may include a procedure called *forward checking* and one called *discrete relaxation* to help speed up the tree search. We will start by discussing the basic backtracking tree search and then cover several of its important variants. Freuder (1978) discusses a breadth-first variant of this kind of search.

17.3.1   The Backtracking Tree Search

A backtracking tree search begins with the first unit of U. This unit can potentially match each label in set L. Each of these potential assignments is a node at level 1 of the tree. The algorithm selects one of these nodes, makes the assignment, selects the second unit of U, and begins to construct the children of the first node, which are nodes that map the second unit of U to each possible label of L. At this level some of the nodes may be ruled out because they violate the constraints. The process continues to level $|U|$ of the tree. The paths from the root node to any successful nodes at level $|U|$ are the consistent labelings. Figure 17.4 illustrates a simple

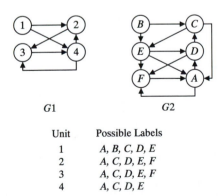

Unit	Possible Labels
1	A, B, C, D, E
2	A, C, D, E, F
3	A, C, D, E, F
4	A, C, D, E

Figure 17.4 Simple digraph-matching problem: to find a subgraph of graph $G2$ that is isomorphic to graph $G1$. The possible labels for each unit were chosen on the basis that indegree$[f(u)] \geq$ indegree(u) and outdegree$[f(u)] \geq$ outdegree(u) for all units u.

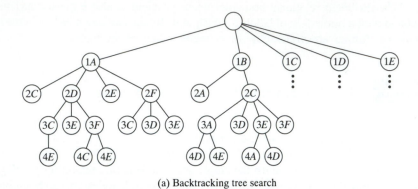

(a) Backtracking tree search

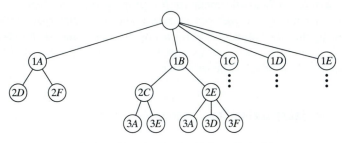

(b) Backtracking tree search with forward checking

Figure 17.5 Portion of the tree search for solving the graph-matching problem of Fig. 17.4 by using backtracking alone (a) and backtracking with forward checking (b). No solution is found in the portion shown, since the one correct mapping begins with $(1,E)$.

digraph-matching problem, and Fig. 17.5 a shows a portion of the backtracking tree search attempting to find a solution. The algorithm for a backtracking tree search is given below. The inputs to the algorithm are U, the set of units; L, the set of labels; T, the unit-constraint relation; R, the unit-label constraint relation; and f, the partial labeling accumulated so far. On the initial call to the procedure, f should be the empty set. Each instantiation of the recursive procedure adds a unit-label pair to the labeling. When a complete consistent labeling is found, it is printed by the procedure.

procedure treesearch (U, L, f, T, R);
$u := \text{first}(U)$;
for each $l \in L$ **do**
    $f' := f \cup \{(u,l)\}$;
    $OK := \text{true}$;
    for each N-tuple $(u_1, \ldots, u_N)$ in T containing component u
        and whose other components are all in domain(f) **do**
        if $((u_1, f'(u_1)), \ldots, (u_N, f'(u_N)))$ is not in R
        then begin $OK := \text{false}$; break **end**;
    end for;

```
        if OK then
          begin
            U' = remainder(U);
            if isempty(U')
            then output(f')
            else treesearch(U',L,f',T,R);
          end
      end for
    end treesearch;
```

17.3.2 Backtracking with Forward Checking

The backtracking tree search has exponential time complexity. Although there are no known polynomial algorithms in the general case, there are a number of algorithms that can cut down search time by reducing the size of the tree that is searched. *Forward checking* (Haralick and Elliott, 1980) is one such method. It is based on the idea that once a unit-label pair (u,l) is instantiated at a node in the tree, the constraints imposed by the relations cause instantiation of some future unit-label pairs (u',l') to become impossible. Suppose that (u,l) is instantiated high in the tree and that the subtree beneath that node contains nodes with first components $u_1, u_2, \ldots, u_n, u'$. Although (u',l') is impossible for any instantiations of $u_1, u_2, \ldots, u_n$, it will be tried by the backtracking tree search in every path that reaches its level in the tree.

Figure 17.6(a) illustrates this problem with a segment of a backtracking tree search for the 6-queens problem. In this segment the labels 1, 3, 5, and 6 for unit 5 appear twice at level 5 of the tree. But the instantiation of the pair (1,1) at level 1 rules out labels 1 and 5 for unit 5, and the instantiation of the pair (2,3) at level 2 rules out labels 3 and 6 for unit 5. In the plain backtracking tree search, this is not detected until unit 5 is instantiated at level 5 in the tree. Thus redundant testing is performed to detect a condition twice at level 5 that could have been predicted earlier in the tree.

The principle of forward checking is to rule out a pair (u',l') that is incompatible with a pair (u,l) at the time that (u,l) is instantiated, and to keep a record of that information. The data structure used to store the information is called a *future-error table (FTAB)*. There is one future-error table for each level of recursion in the tree search. Each table is a matrix having one row for each element of U and one column for each element of L. For any uninstantiated or *future unit* $u' \in U$ and potential label $l' \in L$, $FTAB(u',l') = 1$ if it is still possible to instantiate (u',l'), given the history of instantiations already made. $FTAB(u',l') = 0$ if (u',l') has already been ruled out owing to some previous assignment. When a pair (u,l) is instantiated by the backtracking tree search, the forward-checking procedure is called to examine all pairs (u',l') of future units and their remaining possible labels. For each pair (u',l') that is incompatible with the assignment of (u,l), forward checking sets $FTAB(u',l')$ to 0. Thus, after forward checking is completed, all future unit-label pairs (u',l') that are incompatible with the current assignment (u,l) or any

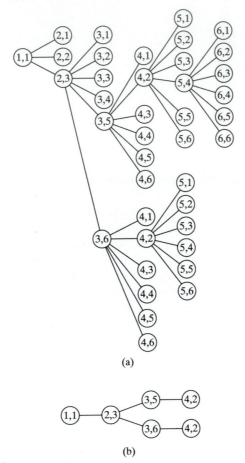

(a)

(b)

Figure 17.6 Segment of the tree search for solving the 6-queens problem by using backtracking alone (a) and backtracking with forward checking (b).

previous assignment have their positions in *FTAB* set to zero. If for any future unit u', $FTAB(u',l')$ becomes 0 for all labels $l' \in L$, then instantiation of (u,l) fails immediately, because there is no label for unit u' that is compatible with the current and previous assignments. Figure 17.6 (b) shows how the search tree generated by the backtracking tree search is reduced by forward checking in the 6-queens problem example, where 39 nodes have been reduced to 6. Figure 17.5 (b) shows the tree generated by the forward-checking tree search for the same portion of the graph-matching problem as that shown in Fig. 17.5 (a). Here the 28 nodes shown have been reduced to 14. Figure 17.7 shows the status of the future-error table during the forward-checking tree search of Fig. 17.5 (b).

The forward-checking tree search procedure is given below. Input parameters U, L, f, T, and R are the same as for the backtracking tree search. The additional input parameter *FTAB* is the future-error table, which we have implemented as

1A

	A	B	C	D	E	F
2	X	X	0	1	0	1
3	X	X	1	1	1	1
4	X	X	0	1	0	X

2D

	A	B	C	D	E	F
3	X	X	1	X	0	1
4	X	X	0	X	X	X

2F

	A	B	C	D	E	F
3	X	X	0	0	0	X	Fail
4	X	X	0	1	0	X

1B

	A	B	C	D	E	F
2	0	X	1	0	1	0
3	1	X	1	1	1	1
4	0	X	1	0	1	X

2C

	A	B	C	D	E	F
3	1	X	X	0	1	0
4	0	X	X	1	0	X

3A

	A	B	C	D	E	F
4	0	X	X	0	0	X	Fail

3E

	A	B	C	D	E	F
4	0	X	X	0	0	X	Fail

Figure 17.7 Status of the future-error table during the backtracking tree search with forward checking.

2E

	A	B	C	D	E	F
3	1	X	0	1	X	1
4	0	X	1	0	X	X

3A

	A	B	C	D	E	F	
4	0	X	0	0	X	X	Fail

3D

	A	B	C	D	E	F	
4	0	X	0	0	X	X	Fail

3F

	A	B	C	D	E	F
4	0	X	0	0	X	X

Figure 17.7 *Continued.*

a two-dimensional array. The initial future-error table should be a $| U | \times | L |$ array with all array elements set to 1. Before each recursive call, the procedure copies the current future-error table $FTAB$ to a temporary table $NEWFTAB$, updates the temporary table by using forward checking, and passes the temporary table to the next instantiation. This essentially creates a stack of future-error tables. In languages in which lists are the primary data structures, the future-error table could be implemented as a list of lists or a vector of lists.

procedure forward_checking_treesearch $(U, L, f, FTAB, T, R)$;
$u := $ first(U);
for each $l \in L$ **do**
  if $FTAB(u, l) == 1$ **then**

```
    begin
      f' := f ∪ {(u,l)};
      U' := remainder(U);
      if isempty(U')
      then output(f')
      else
        begin
          NEWFTAB := copy(FTAB);
          OK := forward_check(NEWFTAB,u,l,U',L,T,R,f');
          if OK then forward_checking_treesearch (U',L,f',NEWFTAB,T,R)
        end
      end
  end for
end forward_checking_treesearch;


function forward_check(FTAB,u,l,future_units,L, T, R, f');
for each u' ∈ future_units do
  forward_check := false;
  for each l' ∈ L with FTAB(u',l') == 1 do
    if compatible(u,l,u',l',T,R,f')
    then forward_check := true
    else FTAB(u',l') := 0
  end for
  if forward_check == false then break
end for
end forward_check;
```

For binary relations the utility function *compatible*, which determines whether an instantiation of (u',l') is possible given instantiation (u,l), is very simple. Units u and u' constrain one another only when either (u,u') or (u',u) is in T. Thus the algorithm for a function compatible for binary relations is as follows:

```
function b_compatible(u,l,u',l',T,R,f');
if (u,u') ∈ T and not [(u,l),(u',l')] ∈ R or
   (u',u) ∈ T and not [(u',l'),(u,l)] ∈ R
then b_compatible := false
else b_compatible := true
end b_compatible;
```

Note that for binary functions, the last argument f' to function b_compatible is not used, but it is included here for consistency.

For N-ary relations, $N > 2$, those N-tuples of T where u and u' are among the components and all other components are already instantiated must be examined.

The code for N-ary relations follows:

```
function compatible(u,l,u',l',T,R,f');
f" := f' ∪ {(u',l')};
compatible := true;
for each t = (u₁,...,uₙ) ∈ T containing u and u' whose other components
    are in domain (f") do
    if [(u₁,f"(u₁)),...,(uₙ,f"(uₙ))] is not in R
    then begin compatible := false; break end
end for
end compatible;
```

The binary procedure is very fast, since its time complexity is constant. The general procedure, if implemented as stated here, would have to examine each N-tuple of T. For a software implementation, it would be desirable to design the data structures for T, R, and f so that only the appropriate N-tuples are tested. A specialized hardware implementation could offer even more flexibility.

17.3.3   Backtracking with Discrete Relaxation

The forward-checking algorithm prunes the search tree of nodes representing unit-label pairs that have been ruled out, on the basis of incompatibility with some instantiated unit-label pair of the partial mapping constructed so far. We can go one step further and consider incompatibility between pairs of future unit-label pairs. Suppose unit u is a future (uninstantiated) unit, and label l is a possible label for u. If there is no possible label l' for some other future unit u' that is compatible with label l for unit u, then the pair (u,l) can be ruled out. The iterative procedure that employs this principle is called *discrete relaxation*. Discrete relaxation, which was first described by Ullmann (1965, 1966) in a character recognition application, is a polynomial complexity procedure that may be applied iteratively to a consistent-labeling problem to reduce the possible labels for each unit prior to a tree search, or it may be applied at each node of the tree. In the first case, for very tightly-constrained problems, the tree search can be greatly reduced or eliminated altogether (Waltz, 1970). When a problem is not so tightly constrained, the instantiation of the unit-label pairs at each node of the tree will constrain it further. In this case, applying a discrete relaxation procedure at each node of the tree may be more useful. We present here a discrete relaxation procedure called the *psi* operator (Haralick and Shapiro, 1979) that makes use of the same data structure (the future-error table) employed by the forward-checking algorithm. The *psi* operator is given in the context of a backtracking tree search with forward checking. It can also be executed alone, prior to the search; in this case all units are the future units.

```
procedure look_ahead_treesearch(U,L,f,FTAB,T,R);
u := first(U);
for each l ∈ L do
    if FTAB(u,l) == 1 then
```

```
begin
    f' := f ∪ {(u,l)};
    U' := remainder(U);
    if isempty(U')
    then output(f')
    else
        begin
            NEWFTAB := copy(FTAB);
            OK := forward_check(NEWFTAB,u,l,U',L,T,R,f');
            if OK then OK := psi(NEWFTAB,U',L,T,R,f');
            if OK then look_ahead_treesearch(U',L,f',NEWFTAB,T,R)
        end
    end
end for
end look_ahead_treesearch;


function psi(FTAB,future_units,L,T,R,f');
for each u ∈ future_units do
    psi := false;
    for each l ∈ L with FTAB(u,l) == 1 do
        for each u' > u ∈ future_units do
            OKP := false;
            for each l' ∈ L with FTAB(u',l') == 1 do
                if compatible(u,l,u',l',T,R,f')
                then begin OKP := true; break end
            end for;
            if OKP == false then break
        end for;
        if OKP then psi := true else FTAB(u,l) := 0
    end for
    if psi == false then break
end for
end psi;
```

Figure 17.8 shows how the *psi* operator works on the tree search at the node
(1,A). In this case the forward-checking tree search investigated several nodes under (1,A) before failing. The *psi* operator, however, determined that (1,A) is not
possible, because there is no label for unit 2 left that is compatible with the remaining labels for units 3 and 4. Thus the look-ahead tree search does not investigate
any nodes under (1,A). In general, the *psi* operator prunes the tree more than does
the forward-checking operator. However, it also has to do much more work at each
node, and many problems take longer to execute when using *psi* than when using
forward checking. The *psi* operator has been employed in a variety of applications, such as subgraph isomorphism (Ullmann, 1976; McGregor, 1979), network
consistency (Mackworth, 1977; Freuder, 1978), and pattern matching (Ullmann,
1979).

1A

	A	B	C	D	E	F
2	X	X	0	1	0	1
3	X	X	1	1	1	1
4	X	X	0	1	0	X

FTAB after
forward checking and
before *psi*.

```
psi: u ← 2
   psi ← false
      ℓ ← D
         u′ ← 3
         OKP ← false
            ℓ′ ← yes    OKP ← true
         u′ ← 4
         OKP ← false
            ℓ′ ← D    No
      OKP is false
      FTAB(2,0) ← 0
      ℓ ← F
         u′ ← 3
         OKP ← false
            ℓ′ ← C    No
            ℓ′ ← D    No
            ℓ′ ← E    No
            ℓ′ ← F    No
      OKP is false
      FTAB(2,F) ← 0
      psi is still false
      no label for 2 will work
```

Figure 17.8 Execution of the *psi* operator at the node (1,A) of the tree.

17.3.4 Ordering the Units

All the algorithms presented so far have assumed the units were arranged in some given order and have used that exact ordering of the units to perform the search. The ith unit was instantiated at level i in the tree. This is simple to program, but it is not necessarily the optimal strategy. Suppose at some level of the tree, unit i has n_i labels and unit j has n_j labels that have not been ruled out. If $n_j < n_i$, then, if we assume all other factors are equal, the procedure is likely to fail and backtrack sooner under unit j than under unit i. The sooner a failure occurs in a tree search, the larger is the subtree that gets pruned and the faster the procedure executes. Thus it is better to choose the unit that has the fewest labels left as the next unit, at each stage of the tree search. It is easy to modify the algorithms to keep track of how many labels remain for each unit and which unit has the least number of labels left. The tree search then selects this unit as the next one to try.

17.3.5   Complexity

The consistent-labeling problem is an NP-complete problem. Thus any algorithm to solve the general problem on a sequential processor has exponential complexity. In the worst case, where $R = (U \times L)^N$, every nonleaf node will have exactly $| L |$ successors. Thus there will be one (root) node at level 0, $| L |$ nodes at level 1, $| L |^2$ nodes at level 2, ..., and $| L |^M$ nodes at level M, for a total of $(| L |^{M+1} -1)/(| L | -1)$ nodes. In most real problems the constraints cause some backtracking. While the full tree is not generated, the backtracking tree search still has exponential complexity. Forward checking and look-ahead both can drastically reduce the number of nodes searched but do not change the overall complexity of the general problem. It is more interesting to consider the actual number of consistency checks, which is directly related to the execution time of the algorithm, either in particular applications or for an abstract model that tells us the probability of a given consistency check succeeding or failing.

Haralick and Elliot (1978) analyzed the case of binary consistent-labeling problems in which the probability that a consistency check succeeds is independent of the pair of units or labels involved and independent of whatever labels may already have been assigned to instantiated units. With this assumption, it was determined, both analytically and experimentally, that of the three variants presented above, the forward-checking tree search performed best (least number of consistency checks and in practice smallest execution time), the tree search with look-ahead was next, and the plain backtracking tree search was the worst, as expected. The look-ahead tree search investigated significantly fewer nodes in the tree than the forward-checking tree search, but did so much extra work at each node that the performance was worse. Figure 17.9 illustrates the relationship between number of units and consistency checks in the N-Queens problem (where number of labels equals number of units) for the three variants. Choosing the units by always selecting the one with fewest remaining labels reduced the number of consistency checks in both forward-checking and look-ahead tree searches (plain backtracking cannot use this strategy, since it does not remove labels from consideration).

17.3.6   The Inexact Consistent-Labeling Problem

The consistent-labeling problem requires all the constraints of the problem to be satisfied by any solution. When we try to apply the techniques for solving consistent-labeling problems to data extracted from real images, it becomes apparent very soon that we are expecting too much. Suppose that we are trying to find a copy of a model, expressed as a line drawing, in a line-segment data structure extracted from a real image. Although a human may claim to "see" all the lines from the model in the original image, the extracted structure will generally have lines that are missing, partially missing, extra, and distorted. Thus the tree search to find a mapping from the lines of the line drawing to the segments extracted from the image is likely to fail if it insists on a label for each unit and every constraint being satisfied. For such real-life problems, we need to formulate a consistent-labeling model that

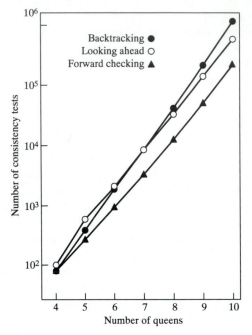

Figure 17.9 Relationship between number of units and consistency checks in the
N-queens problem for the backtracking tree search, the forward-checking tree
search, and the look-ahead tree search.

allows some error to be taken into account. We call the resultant model the *inexact
consistent-labeling problem* (Shapiro and Haralick, 1980, 1981).

An inexact consistent-labeling problem is a 6-tuple ICLP $= (U,L,T,R,w,\epsilon)$.
U, L, T, and R are the same as in a normal or exact consistent-labeling problem. The function w is a mapping from $T \times L^N$ to the interval [0,1]. In general, $w(u_1,\ldots,u_N,l_1,\ldots,l_N)$ is the error that would result from associating units
$(u_1,\ldots,u_N)$ with labels $(l_1,\ldots,l_N)$. The function w will usually depend on
the application problem. One useful way of defining w is to assign a weight
$weight(t)$ to each tuple $t \in T$. The weights model the concept that some relationships are more important than others. Then $w(u_1,\ldots,u_N,l_1,\ldots,l_N)$ is defined
as $weight[(u_1,\ldots,u_N)]$ for any set of labels $(l_1,\ldots,l_N)$. When the application does
not dictate an obvious set of weights, each tuple may be assigned the weight $1/\,|\,T\,|$.

An *inexact consistent labeling* is a mapping $f : U \to L$ where the sum of the
errors incurred by f on all N-tuples of units that constrain one another is less than
a given threshold ϵ. Thus in the above notation, the mapping f must satisfy:

$$\sum_{(u_1,\ldots,u_N)\in T} w[u_1,\ldots,u_N,f(u_1),\ldots,f(u_N)] \leq \epsilon$$

All of the tree-search procedures given above can be extended to the inexact
consistent-labeling framework. We will demonstrate the extension for the case of

the forward-checking tree search. Besides knowing about the weighting function w and the error threshold ϵ, the inexact forward-checking tree search deals with three kinds of errors: past, current, and future. The variable *past_error* is an input to the tree-search procedure, representing the error of the partial mapping that has been constructed so far. The tree search should initially be called with a value of 0 for *past_error*, and the value of the variable is never allowed to exceed the error threshold ϵ. As unit-label pairs are added to the mapping and some constraints are not satisfied by the resultant function, *past_error* will increase.

The variable *current_error* is a local variable to the tree-search procedure representing the error associated with the current pair (u,l). This is the error that the addition of the pair (u,l) to the partial mapping would add to the error already associated with the partial mapping. At the time of considering a pair (u,l), the tree search does not have to compute its *current_error*. It has been gradually computed all along by the update procedure, which is called every time a pair is added to the mapping. The value of *current_error* for a pair (u,l) is found in the future-error table at position $FTAB(u,l)$. Thus, in the inexact forward-checking tree search, the future-error table contains real numbers instead of 0s and 1s. The table is initialized to all zero (no errors), and the value for a pair (u',l') increases whenever a unit-label pair (u,l) that is inconsistent with (u',l') is added to the mapping being constructed. Furthermore, $FTAB$ is augmented by an extra vector array $MINERR$. $MINERR(u)$ is used to store the minimum error of all the labels for a given unit. It is also initialized to all zero.

Finally, the third variable *future_error* represents the possible error that can be incurred by the instantiation of future (not yet instantiated) units. Since the future-error table associates an accumulated error with each future unit and possible label based on the compatibility between that future unit-label pair and the partial mapping, *future_error* can be estimated by the sum over all future units u of $MINERR(u)$, the minimum error for any label of u. This sum is guaranteed to be not greater than the real future error, which has to take into account not only the error caused by future units interacting with past units, but also the error caused by future units interacting with future units. The forward-checking algorithm does not investigate the latter type of error; that is left to look-ahead procedures such as *psi*. The variable *future_error* is also set to zero for the initial call to the tree-search procedure.

The following procedures—inexact_forward_checking_treesearch, inexact_forward_check, and inexact_compatible—are modeled after the forward_checking_treesearch, forward_check, and compatible procedures given in the previous section. They use the concepts of the real-valued future error table ($FTAB$ augmented by $MINERR$) and past error, current error, and future error to find inexact consistent labelings given the unit set U, the label set L, the unit-constraint relation T, the unit-label constraint relation R, the weighting function w, and the error threshold ϵ.

function inexact_forward_checking_treesearch($U,L,f,FTAB,MINERR,T,R,$
$w,past\_error,future\_error,epsilon$);
$u := \text{first}(U)$;

```
for each l ∈ L do
  current_error := FTAB(u,l);
  if past_errror + current_error + future_error − MINERR(u) ≤ epsilon
  then
    begin
      f' := f ∪ {(u,l)};
      U' := remainder(U);
      if isempty(U')
      then output(f')
      else
        begin
          NEWFTAB := copy(FTAB);
          new_future_error := inexact_forward_check(NEWFTAB,
            MINERR,u,l,U',L,T,R,w,past_error,f');
          if new_future_error + past_error + current_error ≤ epsilon
          then inexact_forward_checking_treesearch(U',L,f',NEWFTAB,
            MINERR,T,R,w,past_error + current_error,
            new_future_error,epsilon)
        end
    end
end for
end inexact_forward_checking_treesearch;




function inexact_forward_check(FTAB,MINERR,u,l,future_units,L,T,
    R,w,past_error,f');
inexact_forward_check := 0;
for each u' ∈ future_units do
  smallest_error := 9999999;
    for each l' ∈ L with past_error + inexact_forward_check + FTAB(u',l')
        ≤ epsilon do
    error := inexact_compatible(u,l,u',l',T,R,w,f');
    FTAB(u',l') := FTAB(u',l') + error;
    if FTAB(u',l') < smallest_error
    then smallest_error := FTAB(u',l')
  end for
  MINERR(u) := smallest_error;
  inexact_forward_check := inexact_forward_check + smallest_error;
  if inexact_forward_check+past_error > epsilon then break
end for
end inexact_forward_check;

function inexact_compatible(u,l,u',l',T,R,w,f');
f'' := f' ∪ {(u',l')};
inexact_compatible := 0;
```

1B

	A	B	C	D	E	F	
2	$\frac{1}{6}$	X	0	$\frac{1}{6}$	0	$\frac{1}{6}$	Past error = 0
3	0	X	0	0	0	0	Current error = 0
4	$\frac{1}{6}$	X	0	$\frac{1}{6}$	0	$\frac{1}{6}$	Future error = 0

2A

	A	B	C	D	E	F	
3	X	X	$\frac{1}{6}$	0	$\frac{1}{6}$	0	Past error = 0
							Current error = $\frac{1}{6}$
4	X	X	0	$\frac{2}{6}$	0	$\frac{1}{6}$	Future error = 0

Figure 17.10 Part of the inexact forward checking procedure for the digraph-matching problem of Fig. 17.4, assuming equal weights of 1/6 on each arc of digraph G1.

for each $t = (u_1, \ldots, u_N) \in T$ containing u and u'
    whose other components are in domain(f'') **do**
    if $[(u_1, f''(u_1)), \ldots, (u_N, f''(u_N))]$ is not in R
    then inexact_compatible := inexact_compatible$+w(u_1, \ldots, u_N, f''(u_1), \ldots,$
        $f''(u_N))$
end for
end inexact_compatible;

Suppose that we add a weight of 1/6 to each of the arcs of digraph G1 in Fig. 17.4 and want to find a subgraph of digraph G2 that is "almost isomorphic" to G1. If we define "almost isomorphic" as not missing more than two arcs, we can set ϵ at 2/6 and solve the corresponding inexact consistent-labeling problem. Figure 17.10 illustrates the forward-checking procedure for a small portion of the tree search under node (1,B) to solve this problem. Notice that the inexact version of the tree search cannot eliminate labels as rapidly as the exact version. Label D for unit 4 is not eliminated when (1,B) is first instantiated. Instead it merely accumulates an error of 1/6. It is not until (1,B) and (2,A) have been instantiated that (4,D) is ruled out, since the sum of its error of 2/6 and the error already incurred by $f = \{(1, B), (2, A)\}$ of 1/6 adds up to 3/6, which is greater than ϵ.

17.4   Continuous Relaxation

In exact consistent-labeling procedures, a label l for a unit u is either possible or impossible at any stage of the tree search. As soon as a unit-label pair (u, l) is found to be incompatible with some already instantiated pair, the label l is marked as illegal for unit u. In the inexact consistent-labeling procedures, a label l for a unit u starts out with zero error attached. Gradually, as the pair (u,l) is found to be incompatible with other pairs, the error associated with (u,l) increases. If the error of (u,l) plus the error of the partial mapping plus the estimated future error exceeds some threshold, then again label l is considered illegal for unit u. This property of calling a label either possible or impossible in the preceding algorithms makes them *discrete* algorithms. In contrast, we can associate with each unit-label pair (u,l) a real number representing the probability or certainty that unit u can be assigned label l. In this case the corresponding algorithms are called *continuous*. In this section we look at a labeling algorithm called *continuous relaxation* for symmetric binary relations.

A continuous-relaxation labeling problem is a 6-tuple CRLP $=$ $(U,L,T,$ $P,C,R)$. As before, U is a set of units, L is a set of labels for those units, and $T \subseteq U^2$ is the binary unit-constraint relation. L is usually given as the union over all units i of L_i, the set of allowable labels for unit i. Suppose that $|U| = n$. Then P is a set of n functions $P = \{p_1 \ldots, p_n\}$, where $p_i(l)$ is the a priori probability that label l is valid for unit i. C is a set of n^2 compatibility coefficients $C = \{C_{ij}\}, i = 1, \ldots, n; j = 1, \ldots, n$. C_{ij} can be thought of as the influence that unit j has on the labels of unit i. Thus if we view the unit-constraint relation T as a graph, we can view C_{ij} as a weight on the edge between unit i and unit j. Finally, R is a set of n^2 functions $R = \{r_{ij}\}, i = 1, \ldots, n; j = 1, \ldots, n$, where $r_{ij}(l,l')$ is the compatibility of label l for unit i with label l' for unit j. This is also a generalization of the R in the discrete algorithms. In the discrete case, $r_{ij}(l,l')$ can be 1, meaning that $[(i,l),(j,l')]$ is allowed, or 0, meaning that the combination is incompatible. The solution of a continuous-relaxation labeling problem, like that of a consistent-labeling problem, is a mapping $f : U \to L$ that assigns a label to each unit. Unlike the discrete case, there is no external definition stating what conditions such a mapping f must satisfy. Instead, the definition of f is implicit in the procedure that produces it. This procedure is known as *continuous relaxation* and was first described in Rosenfeld, Hummel, and Zucker (1976).

As discrete relaxation algorithms like forward checking and look-ahead iterate to remove possible labels from the label set L_i of a unit i, continuous relaxation iterates to update the probabilities associated with each unit-label pair. The initial probabilities are defined by the set P of functions defining a priori probabilities. The algorithm starts with these initial probabilities at step 0. Thus we define the probabilities at step 0 by

$$p_i^0(l) = p_i(l)$$

for each unit i and label l. At each iteration k of the relaxation, a new set of probabilities $\{p_i^k(l)\}$ is computed from the previous set and the compatibility information. In order to define $p_i^k(l)$, we first introduce a piece of it, $q_i^k(l)$ defined by

$$q_i^k(l) = \sum_{j \in T(i)} C_{ij}[\sum_{l' \in L_j} r_{ij}(l,l')p_j^k(l')]$$

where $T(i) = \{j | (i,j) \in T\}$. The function $q_i^k(l)$ represents the influence that the current probabilities associated with labels of other units constrained by unit i have on the label of unit i. With this formulation, the formula for updating the p_i^k's can be written as

$$p_i^{k+1}(l) = \frac{p_i^k(l)[1 + q_i^k(l)]}{\sum\limits_{l' \in L_i} p_i^k(l')[1 + q_i^k(l')]}$$

The numerator of the expression allows us to add to the current probability $p_i^k(l)$ a term that is the product $p_i^k(l)q_i^k(l)$ of the current probability and the opinions of other related units, based on the current probabilities of their own possible labels. The denominator normalizes the expression by summing over all possible labels for unit i. Applications of and extensions to the basic relaxation idea may be found in the references. We limit our remaining discussion to the probabilistic formalization.

Zucker and Mohammed (1978) suggested rewriting and modifying the relation in such a way that the compatibility coefficients had the meaning of conditional probabilities. They rewrote $r_{ij}(l,l')$ as $P_{ij}(l|l')$, with the interpretation that $P_{ij}(l \mid l')$ is the conditional probability that unit i takes the label l, given that unit j takes the label l'. Of course it is required that $P_{ij}(l \mid l') \geq 0$ and

$$\sum_l P_{ij}(l \mid l') = 1$$

The modified probabilistic relaxation equation takes the form

$$P(l_i, t+1) = \frac{P(l_i, t) \prod\limits_{j \in L_i} \sum\limits_{y_j} P_{ij}(l_i \mid y_j)P(y_j, t)}{\sum\limits_{z_i} P(z_i, t) \prod\limits_{j \in L_i} \sum\limits_{y_j} P_{ij}(z_i \mid y_j)P(y_j, t)}$$

This form is certainly more suggestive of a probability interpretation for $P(l_i, t)$. Whatever the interpretation might be, however, it is not immediately apparent from the relaxation equation.

In either of these forms, the relaxation iteration was considered useful regardless of the number of times it had been previously iterated. The reason is that after each iteration the weights or probabilities were thought to be more consistent with prior expectations of neighboring label dependencies. Thus the question of whether the iterations converge arose naturally, as did the question about the meaning of the fixed point. Aspects of these questions were answered by Zucker, Krishnamurthy, and Haar (1978) and Haralick (1983).

Peleg (1980) attempted to give a more solid meaning to the question about what the probabilities were in a modified Zucker relation. He suggested that the Zucker compatibility coefficients r_{ij} should take the form

$$r_{ij}(l_i, y_j) = \frac{P_{ij}(l_i, y_j)}{P(l_i)P(y_j)}$$

and that the probabilities $P(l_i, t)$ were really just estimates that the unit i took the label l given that the previous estimate for this probability was $P(l_i, t-1)$. The justification given for this interpretation was based in part on a probability derivation with some conditional independence assumptions followed by some approximations. Kirby (1980) also gives an analysis of the Peleg relaxation equation; the product rule equation he gives is similar to the one we develop here, but the interpretation is different.

In our notation the relaxation equation that Peleg gives is

$$P(l_i,\, t+1) = \sum_{j \in L_i} c_{ij} \frac{P(l_i,t) \sum_{l_j} P(l_j \mid t) r_{ij}(l_i,l_j)}{\sum_{z_i} P(z_i,t) \sum_{y_j} P(y_j,t) r_{ij}(z_i,y_j)}$$

where the c_{ij} are weights that are nonnegative and that, for each index i, sum on the j index to 1.

Haralick (1983) uses a derivation that has some similarities to the Peleg derivation and shows that, with some appropriate conditional independence assumptions, each succeeding iteration of the relaxation equation produces a conditional probability that a unit takes a label, given the context that is one neighborhood-width larger than the unit's context at the previous iteration. With this meaning, the question about convergence is irrelevant. The iterations can continue until the entire context is taken into account. Further iterations than this will no longer create interpretations of conditional probability of the label given the entire context. This meaning also explains a behavior sometimes noticed in relaxation experiments: The emerging probabilities sometimes appear to be getting better for the first few iterations, after which they appear to get worse. In these instances what is happening is that the required conditional independence assumptions are getting further and further away from a good modeling of reality. The error introduced eventually overtakes the benefit produced by the larger context, and the probabilities get worse.

Our interpretation for $P(q_i, k)$ states that it is the conditional probability that unit i takes label q_i given the kth level context. Furthermore, the context at each iteration grows by an entire neighborhood width surrounding the previous level context.

To make these remarks precise, we will have to change the notation so that the context is explicitly written. Context means the units and their corresponding measurements, where the units come from some general neighborhood. Initially a measurement is made of each unit. We denote by d_i the measurement made of unit i. This is its immediate context. The neighborhood context for unit i is the measurement d_i plus all the measurements of units in the neighborhood of unit i. The next larger context for unit i is measurement d_i plus all the measurements of units in the neighborhood of unit i plus all the measurements of units in the neighborhood of the neighbors of unit i. The global context consists of all the units $1, \ldots, I$.

We denote by $Z_i(t)$ the units in the tth-level context for unit i; and by L_i, the set of neighbors for unit i. Then $Z_i(1) = \{i\}$. The units in the successive-level context can be defined iteratively by $Z_i(t+1) = \{j \mid \text{for some } k \in Z_j(t), \ j \in L_k\}$.

The purpose of the probabilistic relaxation is to compute, for each unit i and label q, the conditional probability $P(q_i \mid d_1, \ldots, d_I)$, where it is understood that a subscript n on a label or measurement designates that the label or measurement is for unit n. Thus $P(q_2)$ designates a generally different probability value from $P(q_3)$, even if $q_2 = q_3$. For a more complete notation, we would write $P_2(q_2)$ for $P(q_2)$. We use the shorter notation to avoid writing unnecessarily complex expressions.

We will need to write conditional probabilities, such as $P(q_i \mid d_1, \ldots, d_I)$, but only where the condition is on measurements for some arbitrary subset S of units whose names are not explicitly known. We denote this kind of conditional probability by $P(q_i \mid d_k : k \in S)$. Thus if $S = \{1, 3, 6, 7\}$, we write $P(q_i \mid d_k : k \in S)$ for $P(q_i \mid d_1, d_3, d_6, d_7)$. Likewise, if $T = \{2, 3, 4\}$, we write $P(q_n : n \in T \mid d_k : k \in S)$ for $P(q_2, q_3, q_4 \mid d_1, d_3, d_6, d_7)$.

In this notation the relaxation begins with $P[q_i \mid d_k : k \in Z_i(1)]$ and terminates with the probabilities $P(q_i \mid d_k : k \in \{1, \ldots, I\})$. Letting

$$r_{nm}(q_n, q_m) = \frac{P(q_n, q_m)}{P(q_n)P(q_m)}$$

we have the interpretation for the Peleg relaxation equation above:

$$P[q_i \mid d_k : k \in Z_i(t+1)] =$$

$$\frac{P[q_i \mid d_k : k \in Z_i(t)] \prod_{j \in T(i)} \sum_{q_j} P[q_k \mid d_k : k \in Z_j(t)] r_{ij}(q_i, q_j)}{\sum_{t_i} P[t_i \mid d_k : k \in Z_i(t)] \prod_{j \in T(i)} \sum_{q_j} P[q_j \mid d_k : d \in Z_j(t)] r_{ij}(t_i, q_j)}$$

17.5   Vision Applications

We have defined and stated algorithms for the consistent-labeling problem, the inexact consistent-labeling problem, and the continuous-relaxation problem. We now discuss several vision applications using these concepts.

Line Labeling with Discrete Relaxation

Discrete relaxation was first used by Waltz (1972) to label line drawings of polyhedral objects. Waltz worked with reasonably complex blocks world scenes with shadows and cracks. We will limit our discussion here to simple blocks world scenes of planar polyhedra with no shadows or cracks and trihedral vertices (3D vertices where exactly three surfaces come together). Huffman (1971) and Clowes (1971) showed that under these conditions, only four kinds of line segment junctions could be found in perfect line drawings extracted from the scenes. Figure 17.11 shows a typical line drawing, and Figure 17.12 illustrates the four types of junctions, which are called L, FORK, ARROW, and T junctions, and were originally defined by Guzman (1970).

The lines in a two-dimensional line drawing of one or more three-dimensional objects of this class can each be assigned one of four labels that indicate something about the three-dimensional structure of the objects and the viewpoint from which

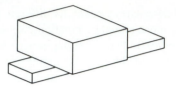

Figure 17.11 Line drawing of simple polyhedral objects.

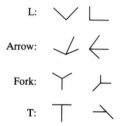

Figure 17.12 The four kinds of junctions found in line drawings of planar polyhedral objects restricted to having no shadows or cracks and only trihedral three-dimensional vertices.

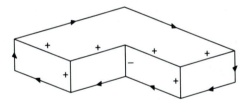

Figure 17.13 Line labeling of a two-dimensional line drawing of a three-dimensional object.

the object is being viewed. Each line segment is either an interior line segment of some object or a boundary line segment that separates one object from another or from the background. Interior line segments have two possible labels: " + " meaning convex and " − " meaning concave. Boundary line segments are labeled by an arrow indicating the direction in which one would have to walk along that boundary in order to keep a visible surface of the object on the right. Thus each boundary segment gets an arrow in one direction or the other; the corresponding labels are specified by > and <. Figure 17.13 illustrates the line labeling of a simple object.

In order to label a line drawing in this fashion, the labeler has to be familiar with the three-dimensional world. Huffman and Clowes deduced that out of the 208 combinatorially possible ways to label junctions from trihedral vertices, only 18 were physically possible. These are shown in Fig. 17.14. The constraints provided

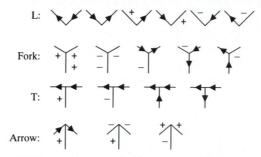

Figure 17.14 Labelings of the 18 physically possible junctions.

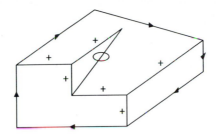

Figure 17.15 An impossible object. There is no valid label for the line marked with a circle.

by these limited physical possibilities allow us to express the line-labeling problem as a consistent-labeling problem. The set of units U is the set of line segments of the line drawing. The set of labels L is the label set $\{+, -, >, <\}$. Only lines that meet at a junction constrain one another, so $T = \{(u_i, u_j) \mid u_i \text{ and } u_j \text{ meet at a junction}\}$. Finally, the unit-label constraint relation R is defined according to the legal junction labelings by $R = \{[(u_i, l), (u_j, l')] \mid u_i \text{ and } u_j \text{ meet at a junction}\}$, and it is physically possible for line l_i to have label l and line l_j to have label l' at that type of junction.

A backtracking tree search or one using forward checking or look-ahead can solve this problem for a given line drawing. If the line drawing represents a possible object of the given object class, a consistent labeling will be found. No labeling will be found for impossible objects. Figure 17.15 illustrates this case with a line drawing of an object that is illegal because the top surface cannot be planar. The labeling procedure will be able to label all the lines consistently, except the one marked with a circle. That line must be labeled with a minus sign to satisfy the constraints of its arrow junction, but would require a boundary label to satisfy the constraints of its L junction. Thus no consistent labeling can be found.

Waltz used a form of discrete relaxation similar to the *psi* operator to solve his problems. He found that in most cases the relaxation removed so many labels that it determined the consistent labeling without performing any tree search at all.

Inexact Matching in Two-Dimensional Shape Recognition

Two-dimensional shape matching is a recognition task that arises in many different machine vision applications. Character recognition and sheet-metal-part recognition are two major examples. One of the early shape-matching systems (Shapiro, 1981) is a good example of an inexact consistent-labeling problem. The input shapes are polygons extracted from a digital image and represented by a sequence of boundary points. The boundary points are converted into shape primitives of two types: simple, near-convex pieces of the shape and intrusions into the boundary of the shape. The conversion is achieved through a graph-theoretic clustering procedure described in Shapiro and Haralick (1979). Figure 17.16 (a) shows a typical input shape, and Fig. 17.16 (b) shows the primitives constructed. Notice that the primitives are allowed to overlap. Primitive 1, the vertical stroke, overlaps parts of all three of the other simple part primitives.

Shapes are represented by a relational model consisting of two ternary relations: R_i, the *intrusion relation*, and R_p, the *protrusion relation*. Intuitively the intrusion relation consists of triples of the form (simple part 1, intrusion, simple part 2), where the two simple parts touch or nearly touch and constitute part of the boundary of the intrusion. The protrusion relation consists of triples of the form (intrusion 1, simple part, intrusion 2), where the simple part protrudes between the two intrusions. Figure 17.16 (c) shows the R_i and R_p relations for the shape of Fig. 17.16 (a). Both the R_i and the R_p relations are symmetric.

The decomposition shown in Fig. 17.16 is of a perfect prototype letter E that can be used as a model. Other hand-printed letter E's may decompose differently.

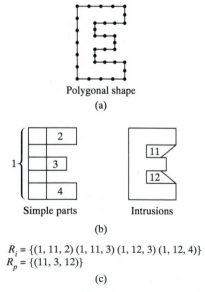

Polygonal shape
(a)

Simple parts   Intrusions

(b)

$R_i = \{(1, 11, 2)\,(1, 11, 3)\,(1, 12, 3)\,(1, 12, 4)\}$
$R_p = \{(11, 3, 12)\}$

(c)

Figure 17.16 Decomposition of a polygonal shape (a) into primitives called simple parts and intrusions (b), and its relational model (c).

In particular, there will often be extra primitives and extra or different relationships among primitives. Sometimes a primitive of the model will be missing entirely in the unknown, extracted shape. These characteristics of the problem make it suitable for the inexact consistent-labeling formalism. The units are the primitives of the model, and the labels are the primitives of the unknown shape plus the special label NIL for those model primitives that find no match in the unknown shape. The unit-constraint relation T is the union of R_i and R_p for the model shape. (This works because the intersection of R_i and R_p is always empty.) If we let T' be the union of R_i and R_p for the unknown shape, then the unit-constraint relation R is given by $R = \{[(u_i,l_i),(u_j,l_j),(u_k,l_k)] \mid (u_i,u_j,u_k) \in T, (l_i,l_j,l_k) \in T'$, type $(u_i) =$ type (l_i), type $(u_j) =$ type (l_j), and type $(u_k) =$ type (l_k), where type $=$ simple part or intrusion$\}$. The weights for inexact matching are assigned to the triples of the model. As discussed previously, the triples can be equally weighted or more important relationships can be weighted more heavily than less important relationships. An inexact matching tree-search procedure with forward checking or discrete relaxation can then be used to solve the shape-matching problem.

17.5.1 Image Matching Using Continuous Relaxation

A good example of a system using continuous relaxation is the image-matching system of Price (1985). The units to be matched are regions of a scene. The matching can be from a model to an image or from an image to another image. For purposes of explanation we assume the matching is from a model to an image. The goal of the matching procedure is to determine which regions in the image correspond to the given region in the model. The system employs the concept that initial strong matches or *islands of confidence* can provide the context needed for finding correspondences for less well-defined objects.

The operation of the matching system is similar to the relaxation procedure we have described. First, initial likelihoods are computed for each feasible unit-label pair. Then the relaxation scheme is applied iteratively until the likelihood of at least one unit-label pair is higher than a threshold. Then the unit-label pairs with the high likelihoods are used to compute new likelihoods for all other feasible unit-label pairs. The relaxation scheme is applied again until the likelihood is reached of another pair being higher than the threshold. Iterations continue until no more assignments can be made.

Each primitive of the model and each primitive of the image is described by an m-dimensional feature vector. The features include parameters such as region size, intensity, location, texture, and shape measures. Relationships among primitives such as adjacency, relative position, and distance form the relational component of the model or image description. The initial compatibility of a unit-label pair (u,n) is computed according to the formula

$$R(u,n) = \sum_{k=1}^{m} |V_{uk} - V_{nk}| W_k S_k$$

where V_{uk} is the kth feature of unit u, V_{nk} is the kth feature of label n, W_k is a normalization weight for feature k, and S_k is the task-dependent strength of feature k. The formula $R(u, n)$ is then converted to the range $(0, 1)$ by applying the function $f(u, n)$ defined by

$$f(u, n) = \frac{a}{R(u, n) + a}$$

where a is a constant that controls the steepness of the function. The function $f(u, n)$ corresponds to $p_u^o(n)$ in the continuous-relaxation formalism we discussed and is used to compute initial probabilities.

The compatibility of a particular unit-label pair (u_i, n_k) with the current possible assignments at all related units is given by the formula

$$Q_i(n_k) = \frac{1}{|N(u_i)|} \sum_{u_j \in N(u_i)} \sum_{n_l \in W_j} c(u_i, n_k, u_j, n_l) p_j(n_l) + \alpha f(u_i, n_k) p_i(n_k)$$

where $N(u_i)$ is the neighborhood or set of units related to unit u_i; W_j is the set of likely labels for unit u_j; $p_i(n_k)$ is the current probability of assigning label n_k to unit u_i; α is a factor between 0 and 1 that adjusts the relative importance of features and relations; and $c(u_i, n_k, u_j, n_l)$ computes the relational compatibility of unit u_i having label n_k with unit u_j having label n_l.

The updating formula is also a variant of the continuous-relaxation updating rule given in the previous section. Let $\overline{P}_i^{(n)}$ be the vector of probabilities associated with each possible label for unit u_i at step n. The updating rule is given by

$$\overline{P}_i^{(n+1)} = \overline{P}_i^{(n)} + \rho_n P_i \{\overline{g}_i^{(n)}\}$$

where ρ_n is a positive step size to control convergence speed, P_i is a linear projection operator that forces $\overline{P}^{(n+1)}$ to be a vector of probabilities, and $\overline{g}_i^{(n)}$ is a gradient function vector whose components are given by

$$g_i(n_k) = -Q_i(n_k) - p_i(n_k) \alpha f(u_i, n_k)$$
$$- \sum_{u_j | u_i \in N(u_j)} \frac{1}{|N(u_j)|} \sum_{n_l \in W_j} c(u_j, n_l, u_i, n_k) p_j(n_l)$$

The intuitive purpose of the gradient function is to give the direction of greatest change in the criteria. The updating function then takes a step in this direction.

Price compared four different relaxation updating schemes for the same general matching problem and found that this gradient-based optimization approach gave the best results.

17.6   Characterizing Binary-Relation Homomorphisms

Because relational matching is such a basic instance of consistent labeling and because it often occurs in the context of binary relations, in this section we describe a characterization of binary-relation homomorphisms (Haralick, 1978). We begin with the definition of binary relations. The natural generalization to labeled relations may be found in Haralick and Kartus (1978).

Definition 17.1: Let $R \subseteq A \times A$, $S \subseteq B \times B$, and $H \subseteq A \times B$. Here H is a *homomorphism* of R into S if and only if

1. H is defined everywhere on A (for every $a \in A$, there exists a $b \in B$ such that $(a,b) \in H$);
2. H is single-valued on B, that is, $(a,b) \in H$ and $(a,b') \in H$ imply $b = b'$;
3. $R \circ H \subseteq S$.

The relation $R \circ H$ is called the homomorphic image of R under H.

This definition of homomorphism is the same as that in Harary (1969). The specific problem that we are interested in is, given relations R and S, determine all the homomorphisms of R into S.

17.6.1   The Winnowing Process

Suppose that the relation $H \subseteq A \times B$ is a homomorphism of R into S; then what can we find out about the relation among H, R, and S? Certainly if H is a homomorphism from R to S, if H associates b with a, and if R associates a' with a, then all the elements that H associates with a' must be a subset of all the elements that S associates with b (Fig. 17.17). Likewise if H is a homomorphism from R to S, if H associates b with a, and if R associates a with a', then all the elements that H associates with a' must be a subset of all the elements that S^{-1} associates with b (Fig. 17.18).

As a direct consequence of this fact, if H is a homomorphism of R to S and a belongs to the domain of R, then whatever H associates with a must be contained in the domain of S. Likewise, if H is a homomorphism of R to S and a belongs to the range of R, then whatever H associates with a must be contained in the range of S.

Figure 17.19 shows two simple binary relations and their corresponding digraphs, which we use as an example for the rest of the section. For these relations we have Domain $S = \{a,b,c,d\}$ and Range $S = \{b,c,d,e\}$. Since 1 is in the domain of R, any homomorphism can only associate 1 with some of the elements in $\{a,b,c,d\}$. Since 2 and 5 are in the range of R, any homomorphism must associate 2 or 5 with some of the elements in $\{b,c,d,e\}$. Since 3 and 4 are both in the range and domain of R, any homomorphism must associate 3 or 4 with some of the elements in $\{a,b,c,d\} \cap \{b,c,d,e\} = \{b,c,d\}$. Thus any homomorphism H of R to S must be a subset of the relation T shown in Fig. 17.20.

This idea of determining more-restricted relations that must contain any homomorphism can be generalized. Suppose it is known that T is a relation that contains a homomorphism H, but H is not known. The relation T can be used to match pairs

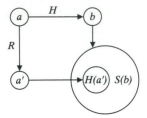

Figure 17.17 How functions that are homomorphisms are constrained.

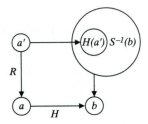

Figure 17.18 How functions that are homomorphisms are constrained.

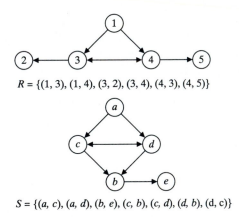

$R = \{(1, 3), (1, 4), (3, 2), (3, 4), (4, 3), (4, 5)\}$

$S = \{(a, c), (a, d), (b, e), (c, b), (c, d), (d, b), (d, c)\}$

Figure 17.19 Digraphs for the example relations R and S.

from the relation R to the relation S: the pair (a_1, a_2) of R is matched to the pair (b_1, b_2) of S only if $(a_1, b_1) \in T$ and $(a_2, b_2) \in T$. This set of matched pairs may be reduced in size under the following conditions. If the pair (a_1, b_1) is in T, if some pair of R has a component of value a_1, and if this pair of R cannot be matched by T to some pair in S having a corresponding component value of b_1, then the pair (a_1, b_1) can be removed from T, and T will still contain all the homomorphisms it originally contained.

We call this process of iteratively taking inconsistent pairs out of T the winnowing process, and it bears a close relationship to the Waltz filtering process (Waltz

$$
\begin{array}{c|l}
T & \\
\hline
1 & a,b,c,d \\
2 & b,c,d,e \\
3 & b,c,d \\
4 & b,c,d \\
5 & b,c,d,e
\end{array}
$$

Figure 17.20 Consequences of tracing the constraints. If H is a homomorphism of relation R to relation S (shown in Fig. 17.19), then H must be a subset of the relation T.

1970, 1972). Proposition 17.1 gives a formal statement of the winnowing process and proves that after winnowing the inconsistent associations out of a given relation, the new relation will contain all the homomorphisms it originally contained. As an immediate consequence of Proposition 17.1, its corollary states that homomorphisms are fixed points of the winnowing process. This suggests that the winnowing process is a natural one to consider for determining homomorphisms.

Before stating and proving Proposition 17.1, its corollary, or other propositions, we will need some convenient notational conventions. Let $R \subseteq A_1 \times A_2$. We define the following sets related to R :

$$
\Delta_n R = \{a_n \in A_n |\ \text{for some}\ (a_1,a_2) \in A_1 \times A_2, (a_1,a_2) \in R\}, n = 1\ \text{or}\ 2
$$

$$
R_n(a) = \{(a_1,a_2) \in R | a_n = a\}, n = 1\ \text{or}\ 2
$$

Proposition 17.1: The winnowing process. Let $R \subseteq A \times A$, $S \subseteq B \times B$, $H \subseteq T \subseteq A \times B$. Define one iteration of the winnowing process by

$$
G = \{(a_1,a_2,b_1,b_2) \in R \times S | (a_1,b_1) \in T\ \text{and}\ (a_2,b_2) \in T\}
$$

$$
Q = \left\{(a,b) \in T | b \in \bigcap_{n=1}^{2} \bigcap_{(a_1,a_2) \in R_n(a)} \Delta_n G(a_1,a_2)\right\}
$$

If H is defined everywhere and $R \circ H \subseteq S$, then $H \subseteq Q \subseteq T$.

Proof:

Let $(a,b) \in H$. To show $(a,b) \in Q$, we need only show

$$
b \in \bigcap_{n=1}^{2} \bigcap_{(a_1,a_2) \in R_n(a)} \Delta_n G(a_1,a_2)
$$

If for any $n, R_n(a) = \emptyset$, we have

$$\bigcap_{(a_1, a_2) \in R_n(a)} \Delta_n G(a_1, a_2) = B$$

Since $H \subseteq T, (a, b) \in H$ implies $(a, b) \in T$, so that $(a, b) \in Q$, and this case causes no problem.

Let n be any index such that $R_n(a) \neq \phi$. We will show

$$b \in \bigcap_{(a_1, a_2) \in R_n(a)} \Delta_n G(a_1, a_2)$$

The argument for $n = 1$ or $n = 2$ is similar, so without loss of generality, suppose $n = 1$.

Let $(a_1, a_2) \in R_1(a)$. By definition of $R_1(a)$, we must have $a_1 = a$. Now for convenience we let $b_1 = b$, so that $(a_1, b_1) = (a_1, b) \in H$. Since H is defined everywhere, there exists b_2 such that $(a_2, b_2) \in H$. Now $(a_1, a_2) \in R$, $(a_1, b_1) \in H$, and $(a_2, b_2) \in H$ imply $(b_1, b_2) \in R \circ H$. But by supposition $R \circ H \subseteq S$, so that $(b_1, b_2) \in R \circ H$ implies $(b_1, b_2) \in S$. And $(a_1, a_2) \in R, (b_1, b_2) \in S, (a_1, b_1) \in H, (a_2, b_2) \in H$, and $H \subseteq T$ imply $(a_1, a_2, b_1, b_2) \in G$. Hence $b = b_1 \in \Delta_1 G(a_1, a_2)$ and $(a, b) \in Q$.

Corollary 17.1: Homomorphisms are fixed points of the winnowing process.
Suppose $R \circ H \subseteq S$ and H is defined everywhere. Define

$$G = \{(a_1, a_2, b_1, b_2) \in R \times S | (a_1, b_1) \in H \text{ and } (a_2, b_2) \in H\}$$

$$Q = \left\{ (a, b) \in H | b \in \bigcap_{n=1}^{2} \bigcap_{(a_1, a_2) \in R_n(a)} \Delta_n G(a_1, a_2) \right\}$$

Then $H = Q$.

Proof:

It is obvious from the definition of Q that $Q \subseteq H$. From the proposition, $H \subseteq Q$; hence $H = Q$. And so the homomorphism H must be a fixed point of the winnowing process.

If the winnowing takes place successively, first constructing a relation T_2 from a given T_1 and then a T_3 from a given T_2, and so on, it is obvious that if $H \subseteq T_1$, then $H \subseteq T_n$ for each n, and for finite-sized relations there must be an N such that $T = T_n$ for all $n \geq N$. In other words, the process converges to a fixed point in a number M of steps that must be smaller than $\#T_1$.

Using the T in Fig. 17.20 as T_1, Fig. 17.21 shows how certain associations of the pairs of R with the pairs of S can be eliminated from consideration. As a result, a new relation T_2 can be constructed that is contained in T_1, and T_2 contains all the homomorphisms T_1 contains. Figure 17.22 shows what happens if the winnowing

	ac	ad	be	cb	cd	db	dc
13			/				
14			/				
32	/	/					
34	/	/	/				
43	/	/	/				
45	/	/					

T_2	Allowed Possibilities
1	a,c,d
2	b,c,d,e
3	c,d
4	c,d
5	b,c,d,e

Figure 17.21 Illustration of Proposition 17.2. Since $R \circ H \subseteq S$ implies $H(3) \subseteq \{b,c,d\}$ (see Fig. 17.17), all possible relationships that allow 3 to be mapped to a or e are illegal. Hence $(1,3)$ cannot be mapped by H to b,e; $(3,2)$ cannot be mapped to (a,c) or (a,d); and $(3,4)$ cannot be mapped by H to (b,e). Since $H(4) \subseteq \{b,c,d\}$, $(1,4)$ and $(3,4)$ cannot be mapped by H to (a,c) or (a,d). All the illegal possibilities are marked out as per Proposition 17.2. Allowed possibilities are tabulated on the right.

	ac	ad	be	cb	cd	db	dc
13			/	/		/	
14			/	/		/	
32	/	/	/				
34	/	/	/	/		/	
43	/	/	/	/		/	
45	/	/	/				

T_3	Allowed Possibilities
1	a,c,d
2	b,c,d
3	c,d
4	c,d
5	b,c,d

Figure 17.22 Illustration of Proposition 17.2. Since $R \circ H \subseteq S$ implies $H(3) \subseteq \{c,d\}$ (see Fig. 17.18), $(1,3)$ or $(4,3)$ cannot be mapped to (c,b) or (d,b), and $(3,2)$ or $(3,4)$ cannot be mapped to (b,e). Since $H(4) \subseteq \{c,d\}$, $(1,4)$ or $(3,4)$ cannot be mapped to (c,b) or (d,b), and $(4,5)$ cannot be mapped to (b,e). The disallowed mappings are marked out. As per Proposition 17.2. Allowed possibilities are tabulated on the right.

process begins by using the T_2 of Fig. 17.21 to construct the smaller relation T_3 shown in Fig. 17.22. Further iterations cannot reduce T_3 any more.

Unfortunately the resulting relation T_3 is not single valued, and although it is a fixed point of the winnowing process, it does not necessarily have the composition property: $R \circ T_3 \subseteq S$. Hence not all fixed points of the winnowing process are homomorphisms. Proposition 17.2 tells us which fixed points of the winnowing process are guaranteed to have the composition property: Any single-valued relation f that is a fixed point of the winnowing process is guaranteed to have the composition property $R \circ f \subseteq S$.

Proposition 17.2: Let $R \subseteq A \times A$, $S \subseteq B \times B$. Let $f \subseteq A \times B$ be single-valued. Define

$$G = \{(a_1,a_2,b_1,b_2) \in R \times S \,|\, (a_1,b_1) \in f, (a_2,b_2) \in f\}$$

Suppose

$$f = \left\{ (a,b) \in A \times B \,\middle|\, b \in \bigcap_{n=1}^{2} \bigcap_{(a_1,a_2) \in R_n(a)} \Delta_n G(a_1,a_2) \right\}$$

Then $R \circ f \subseteq S$.

Proof:

Let $(b_1,b_2) \in R \circ f$. Then for some $(a_1,a_2) \in R$, $(a_1,b_1) \in f$ and $(a_2,b_2) \in f$. Now $(a_1,b_1) \in f$ implies

$$b_1 \in \bigcap_{n=1}^{2} \bigcap_{(\alpha_1,\alpha_2) \in R_n(a_1)} \Delta_n G(\alpha_1,\alpha_2)$$

Since $(a_1,a_2) \in R_1(a_1)$, $b_1 \in \Delta_1 G(a_1,a_2)$.

Obviously $G(a_1,a_2) \neq \emptyset$, so let $(\beta_1,\beta_2) \in G(a_1,a_2)$. Now $(a_1,a_2,\beta_1,\beta_2) \in G$ implies $(\beta_1,\beta_2) \in S$, and $(a_1,b_1) \in f$ and $(a_1,b_1) \in f$ imply $b_1 = \beta_1$. Likewise $(a_2,\beta_2) \in f$ and $(a_2,b_2) \in f$ imply $b_2 = \beta_2$. Then $(b_1,b_2) = (\beta_1,\beta_2) \in S$, so that $R \circ f \subseteq S$.

Thus our search for homomorphisms is now limited to trying to determine all the single-valued, everywhere-defined fixed points of the winnowing process. It would be nice if we could take two or more fixed points of the winnowing process, somehow combine them in an appropriate way, and then have the resulting relation also be a fixed point of the process. With such a mechanism, it might be possible to generate the single-valued fixed points of the winnowing process from some small set of easily determined fixed points of the process.

There is a natural place to look for the easily determined fixed points of the winnowing process. Suppose we begin the process with a relation that allows everything in A to be paired with everything in B, except that the element $a_1 \in A$ is allowed to be paired only with the element $b_1 \in B$. The successive winnowing process then will determine a fixed point, called the basis relation $T^{a_1 b_1}$, that contains all homomorphisms that map a_1 to b_1. Figure 17.23 illustrates the results of successive winnowing where the element 1 is constrained to map to a.

The intersection of $T^{a_1 b_1}$ with another basis relation $T^{a_2 b_2}$ will contain all homomorphisms mapping a_1 to b_1 and a_2 to b_2.

Intersections of basis relations, where the intersections stay defined everywhere, become candidates for homomorphisms. Proposition 17.3 states that if f is any everywhere-defined relation from A to B and has the representation

$$f = \bigcap_{(a,b) \in f} T^{ab}$$

where T^{ab} is any fixed point of the winnowing process that maps a to only b, then f is a fixed point of the winnowing process.

Proposition 17.3: Let $R \subseteq A \times A$, $S \subseteq B \times B$, and $f \subseteq A \times B$. For each $(a,b) \in A \times B$, let $T^{ab} \subseteq A \times B$ satisfy the following conditions:

	ac	ad	be	cb	cd	db	dc
13			/	/	/	/	/
14			/	/	/	/	/
32							
34							
43							
45							

T_1^{1a}	
1	a
2	a,b,c,d,e
3	a,b,c,d,e
4	a,b,c,d,e
5	a,b,c,d,e

	ac	ad	be	cb	cd	db	dc
13			/	/	/	/	/
14			/	/	/	/	/
32	/	/	/				
34	/	/	/	/		/	
43	/	/	/	/		/	
45	/	/	/				

T_2^{1a}	
1	a
2	b,c,d
3	c,d
4	c,d
5	b,c,d

	ac	ad	be	cb	cd	db	dc
13			/	/	/	/	/
14			/	/	/	/	/
32	/	/	/				
34	/	/	/	/		/	
43	/	/	/	/		/	
45	/	/	/				

T_3^{1a}	
1	a
2	b,c,d
3	c,d
4	c,d
5	b,c,d

Figure 17.23 Successive eliminations of mapping possibilities by the construction of T^{1a} by successive winnowing.

1. $(a,\beta) \in T^{ab}$ implies $b = \beta$.

2. $T^{ab} = \left\{ (\alpha,\beta) \in A \times B \,\middle|\, \beta \in \bigcap_{n=1}^{2} \bigcap_{(a_1,a_2) \in R_n(\alpha)} \Delta_n G^{ab}(a_1,a_2) \right\}$ where

$$G^{ab} = \{(a_1,a_2,b_1,b_2) \in R \times S \,|\, (a_1,b_1) \in T^{ab} \text{ and } (a_2,b_2) \in T^{ab}\}$$

If f is defined everywhere and

$$f = \bigcap_{(a,b) \in f} T^{ab}$$

then f is a fixed point of the winnowing process; that is, if

$$G = \{(a_1, a_2, b_1, b_2) \in R \times S \,|\, (a_1, b_1) \in f \text{ and } (a_2, b_2) \in f\}$$

and

$$f' = \left\{ (\alpha, \beta) \in f \,|\, \beta = \bigcap_{n=1}^{2} \bigcap_{(a_1, a_2) \in R_n(\alpha)} \Delta_n G(a_1, a_2) \right\}$$

then

$$f' = f$$

Proof:

By Proposition 17.1, $f' \subseteq f$, so all we need show is that $f \subseteq f'$. Let $(\alpha, \beta) \in f$. We want to show that

$$\beta \in \bigcap_{n=1}^{2} \bigcap_{(a_1, a_2) \in R_n(\alpha)} \Delta_n G(a_1, a_2)$$

Either $R_n(\alpha) = \emptyset, n = 1, 2$, or not. If $R_n(\alpha) = \emptyset, n = 1, 2$, then

$$\bigcap_{n=1}^{2} \bigcap_{(a_1, a_2) \in R_n(\alpha)} \Delta_n G(a_1, a_2) = B$$

so that the assertion is trivially true.

If $R_n(\alpha) \neq \emptyset$ for some n, then $R_1(\alpha) \neq \emptyset$ or $R_2(\alpha) \neq \emptyset$. Suppose $R_1(\alpha) \neq \emptyset$. Then let $(a_1, a_2) \in R_1(\alpha)$. Since f is defined everywhere, there exists a γ such that $(a_2, \gamma) \in f$. But

$$f = \bigcap_{(a,b) \in f} T^{ab}$$

so that surely $(\alpha, \beta) \in T^{a_2 \gamma}$. Hence by definition of $T^{a_2 \gamma}$,

$$\beta \in \bigcap_{n=1}^{2} \bigcap_{(\alpha_1, \alpha_2) \in R_n(\alpha)} \Delta_n G^{a_2 \gamma}(\alpha_1, \alpha_2)$$

But $(a_1, a_2) \in R_1(\alpha)$, so that we must have $\beta \in \Delta_1 G^{a_2 \gamma}(a_1, a_2)$. Hence there exists a δ such that $(\beta, \delta) \in G^{a_2 \gamma}(a_1, a_2)$. Thus $(\beta, \delta) \in S$, $(a_1, \beta) \in T^{a_2 \gamma}$, and $(a_2, \delta) \in T^{a_2 \gamma}$. But $(a_2, \delta) \in T^{a_2 \gamma}$ implies $\delta = \gamma$, so we have $(\beta, \gamma) = (\beta, \delta) \in S$.

Now $(\beta, \gamma) \in S$, $(a_1, \beta) \in f$, $(a_2, \gamma) \in f$, and $(a_1, a_2) \in R$ imply $(\beta, \gamma) \in G(a_1, a_2)$, so that

$$\beta \in \Delta_1 G(a_1, a_2)$$

Since this is true for each $(a_1, a_2) \in R_1(\alpha)$,

$$\beta \in \bigcap_{(a_1, a_2) \in R_n(\alpha)} \Delta_1 G(a_1, a_2)$$

If $R_2(\alpha) \neq \emptyset$, a similar argument shows

$$\beta \in \bigcap_{(a_1,a_2) \in R_2(\alpha)} \Delta_2 G(a_1, a_2)$$

Hence

$$\beta \in \bigcap_{n=1}^{2} \bigcap_{(a_1,a_2) \in R_n(\alpha)} \Delta_n G(a_1, a_2)$$

17.6.2   Binary-Relation Homomorphism Characterization

From the preceding proposition it is only a short way to a characterization of the binary-relation homomorphism, for if T^{ab} is in fact not just any fixed point of the winnowing process that allows a to be mapped only to b, but instead is the basis relation determined by the winnowing process that begins with $\{(a,b)\} \cup (A - \{a\}) \times B$, then we should certainly have

$$H \subseteq \bigcap_{(a,b) \in H} T^{ab}$$

for any homomorphism H, from Proposition 17.1; and since a homomorphism is defined everywhere, the form

$$\bigcap_{(a,b) \in H} T^{ab}$$

must be single valued, so that

$$H \supseteq \bigcap_{(a,b) \in H} T^{ab}$$

This gives us

$$H = \bigcap_{(a,b) \in H} T^{ab}$$

for any homomorphism H.

Conversely, if H is any everywhere-defined binary relation having the representation

$$H = \bigcap_{(a,b) \in H} T^{ab}$$

then Proposition 17.3 states that H is a fixed point of the winnowing process, since H is defined everywhere. And since $(a,b') \in T^{ab}$ implies $b' = b$, H defined everywhere implies that

$$\bigcap_{(a,b) \in H} T^{ab}$$

is single valued. By Proposition 17.2,

$$\bigcap_{(a,b)\in H} T^{ab}$$

is single valued. It is also a fixed point of the winnowing process, which implies that

$$\bigcap_{(a,b)\in H} T^{ab}$$

has the composition property

$$R \circ \bigcap_{(a,b)\in H} T^{ab} \subseteq S$$

And since

$$\bigcap_{(a,b)\in H} T^{ab}$$

is defined everywhere, is single valued, and has the composition property, it must be a homomorphism. This leads us to the binary-relation homomorphism characterization theorem.

Theorem 17.1: Binary-relation homomorphism characterization. Let $R \subseteq A \times A, S \subseteq B \times B$, and $H \subseteq A \times B$. For each $(a,b) \in A \times B$ let

$$T_1^{ab} = \{(a,b)\} \cup (A - \{a\}) \times B$$

Define $T_2^{ab}, \ldots, T_n^{ab}, \ldots$ iteratively by

$$T_{n+1}^{ab} = \left\{ (\alpha,\beta) \in T_n^{ab} \Big| \beta = \bigcap_{n=1}^{2} \bigcap_{(a_1,a_2)\in R_n(\alpha)} \Delta_n G^{ab}(a_1,a_2) \right\}$$

where

$$G^{ab} = \{(a_1,a_2,b_1,b_2) \in R \times S | (a_1,b_1) \in T_n^{ab} \text{ and } (a_2,b_2) \in T_n^{ab}\}$$

Suppose for all $n \geq N$ and for all $(a,b) \in A \times B, T^{ab} = T_n^{ab}$. Then H is a homomorphism of R into S if and only if

1. $\bigcap_{(a,b)\in H} T^{ab}$ is defined everywhere;
2. $H = \bigcap_{(a,b)\in H} T^{ab}$

Proof:

Suppose H is a homomorphism of R into S. Let $(\alpha, \beta) \in H$. Then by Proposition 17.1, $H \subseteq T^{\alpha\beta}$. Since this is true for each $(\alpha, \beta) \in H$,

$$H \subseteq \bigcap_{(\alpha,\beta)\in H} T^{\alpha\beta}$$

Now let

$$(a, b) \in \bigcap_{(\alpha,\beta)\in H} T^{\alpha\beta}$$

Since H is defined everywhere, there exists a γ such that $(a, \gamma) \in H$. Then

$$(a, b) \in \bigcap_{(\alpha,\beta)\in H} T^{\alpha\beta}$$

implies $(a, b) \in T^{\alpha\gamma}$. But by construction $T^{\alpha\gamma}$, $(a, b) \in T^{\alpha\gamma}$ implies $b = \gamma$. Hence $(a, b) = (\alpha, \gamma) \in H$, so that

$$\bigcap_{(\alpha,\gamma)\in H} T^{\alpha\beta} \subseteq H$$

Now

$$H \subseteq \bigcap_{(\alpha,\gamma)\in H} T^{\alpha\beta} \quad \text{and} \quad \bigcap_{(\alpha,\gamma)\in H} T^{\alpha\beta} \subseteq H$$

imply

$$\bigcap_{(\alpha,\gamma)\in H} T^{\alpha\beta} = H$$

And since H is a homomorphism, it is defined everywhere, so that

$$\bigcap_{(\alpha,\gamma)\in H} T^{\alpha\beta}$$

is defined everywhere.

Suppose

$$H = \bigcap_{(\alpha,\gamma)\in H} T^{\alpha\beta}$$

is defined everywhere. By Proposition 17.3, H is a fixed point of the winnowing process. Also H is single valued, for if $(a, b_1) \in H$ and $(a, b_2) \in H$, then $(a, b_1) \in T^{ab_2}$, which implies $b_1 = b_2$. By Proposition 17.2, any single-valued relation H that is a fixed point of the winnowing process has the composition property $R \circ H \subseteq S$. Now by definition of homomorphism, H is a homomorphism of R into S.

17.6.3 Depth-First Search for Binary-Relation Homomorphisms

The binary-relation characterization theorem allows all homomorphisms of R into S to be found by a depth-first search in the following manner. Suppose we are looking for homomorphisms that map the element $1 \in A$ to the element $a \in B$. We can determine by the successive winnowing process the basis relation T^{1a} that must contain all such homomorphisms. Now, T^{1a} may have other elements of A that are uniquely mapped to elements of B. If so, we can determine the basis relation for these pairs and take the intersection of all of them with T^{1a}. The resulting intersection must contain any homomorphism that maps 1 to a. If the resulting intersection relation has additional elements that are uniquely mapped, more intersections can be taken. When the resulting intersection has no more additional elements that are uniquely mapped, then one of three cases exists:

1. The intersection is not defined everywhere, in which case no homomorphism mapping 1 to a exists.

2. The intersection relation f is defined everywhere and is single valued, in which case

$$f = \bigcap_{(a,b) \in f} T^{ab}$$

so that f is a homomorphism.

3. The intersection relation is defined everywhere and is not single valued, in which case a choice must be made in a branch of the depth-first search.

The choice is to map to a unique element of B one of those elements of A having possible multiple associations with the elements of B. In this last case, once such a choice is made, the corresponding basis relation must be intersected with the previously intersected relations. This brings us back to the point of looking for additional uniquely mapped pairs. From here the search iterates until each branch of the tree terminates in one of the first two cases.

Figure 17.24 lists all the basis relations that are defined everywhere for our example relation R and S. Figure 17.25 illustrates the tree determined by a depth first search. The tree shows all the eight possible relational homomorphisms. Figure 17.26 shows the eight homomorphisms and their corresponding homomorphic images.

	T^{1a}	T^{2b}	T^{2c}	T^{2d}	T^{3c}	T^{3d}	T^{4c}	T^{4d}	T^{5b}	T^{5c}	T^{5d}
1	a	acd	a	a	a	a	a	a	acd	a	a
2	bcd	b	c	d	bd	bc	bc	bd	bcd	bd	bc
3	cd	cd	d	c	c	d	d	c	cd	c	d
4	cd	cd	c	d	d	c	c	d	cd	d	c
5	bcd	bcd	bd	bc	bc	bd	bd	bc	b	c	d

Figure 17.24 Tabulation of basis relations that are defined everywhere.

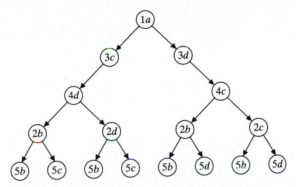

Figure 17.25 Tree determined by a depth-first search. The tree shows all eight possible relational homomorphisms that map 1 to A.

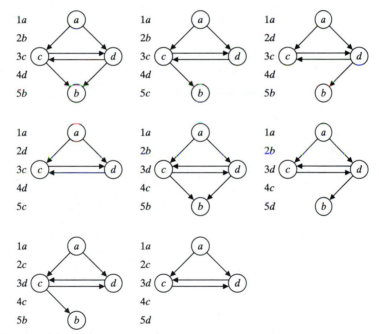

Figure 17.26 The eight homomorphisms and their corresponding homomorphic images.

A simple specialization of this iterative process allows the determination of whether one relation is isomorphic to a part of another. In the depth-first search, terminate any branch when two elements from A map to the same element from B. This guarantees that the resulting intersection relation will be one-to-one. Since we assumed the relations R and S to be finite, one-to-one homomorphisms are isomorphisms.

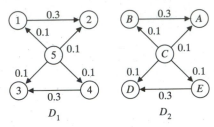

Figure 17.27 The weighted digraphs D_1 and D_2 of Exercise 17.6.

■ Exercises

17.1. Find a consistent labeling in the following consistent-labeling problem.

$$U = \{1,2,3,4\}$$
$$L = \{a,b,c,d\}$$
$$T = \{(2,1),(2,4),(3,2),(3,4)\}$$
$$R = \{[(2,a),(1,b)],[(2,a),(1,c)],[(2,d),(1,a)],[(2,d),(1,c)],$$
$$[(2,c),(1,b)],[(3,a),(2,b)],[(3,a),(2,c)],[(3,d),(2,a)],$$
$$[(3,d),(2,c)],[(3,c),(2,b)],[(2,a),(4,b)],[(2,a),(4,c)],$$
$$[(2,d),(4,a)],[(2,d),(4,c)],[(2,c),(4,b)],[(3,a),(4,b)],$$
$$[(3,a),(4,c)],[(3,d),(4,a)],[(3,d),(4,c)],[(3,c),(4,b)]\}$$

17.2. A view-class model of a three-dimensional object consists of a set of two-dimensional view classes of the object. Each view class can be represented by a set of line segments that appear in the views associated with that class plus a ternary relation W_V that contains triples of line segments (l,m,n), where l,m,n form a U-shaped figure. A view v that appears in an image also consists of a set of line segments plus a ternary relation W_v defined in the same manner as W. The view-class matching problem is to match v to V by finding a mapping from the line segments of v to those of V that satisfies $(l,m,n) \in W_v \Rightarrow [f(l),f(m),f(n)] \in W_V$. Express this problem formally as a consistent-labeling problem.

17.3. Trace and compare the results of (a) the backtracking tree search and (b) the backtracking tree search with forward checking on the consistent-labeling problem defined in Exercise 17.1.

17.4. Trace and compare the results of the backtracking tree search with (a) forward checking and (b) discrete relaxation on the consistent-labeling problem defined in Exercise 17.1.

17.5. Modify the backtracking tree search with forward checking to choose the unit that has the fewest labels left, as described in Section 17.3.4.

17.6. Given weighted digraph D_1 and digraph D_2 shown in Fig. 17.27, show how the inexact forward-checking tree search works to find a mapping from $\{1,2,3,4,5\}$ to $\{A,B,C,D,E\}$ with error less than 0.25. Use the definition that $w(u_i,u_j,l_i,l_j) = $ **weight**(u_i,u_j).

17.7. Translate the problem of Exercise 17.6 into a continuous relaxation labeling problem by using the weights as compatibility coefficients and defining $r_{ij}(l,l')$ as 0 if $(i,j) \in D_1$ and $(l,l') \notin D_2$, and as 1 otherwise. The a priori probabilities $p_1(l)$ should be set to 1 if node i of D_1 has the same indegree and outdegree as node I of D_2, to 0.75 if either the indegree or the outdegree (but not both) differs by 1, and to 0.5 otherwise.

17.8. a. Show how a backtracking tree search can be used to find a labeling of the line drawing of Fig. 17.3 with labels $\{+, -, <, >\}$.

b. Compare to a backtracking tree search with forward checking.

c. Compare to a backtracking tree search with discrete relaxation.

d. Try applying discrete relaxation to the problem with no tree search and iterate until a fixed point is reached.

17.9. Give an algorithm that carries out the depth-first search for binary-relation homomorphisms described in Section 17.6.3.

■ Bibliography

Barrow, H. G., and R. J. Popplestone, "Relational Descriptions in Picture Processing," in *Machine Intelligence 6,* B. Meltzer (ed.), American Elsevier, New York, 1971, pp. 377–396.

Clowes, M., "On Seeing Things," *Artificial Intelligence,* Vol. 2, pp. 79–116, 1971.

Danker, A. J., and A. Rosenfeld, "Blob Detection by Relaxation," *IEEE Transactions on Pattern Analysis and Machine Intelligence,* Vol. PAMI-3, 1981, pp. 79–92.

Eklundh, J-O., and A. Rosenfeld, "Some Relaxation Experiments Using Triples of Pixels," *IEEE Transactions on Systems, Man, and Cybernetics,* Vol. SMC-10, 1980, pp. 150–153.

Elfving, T., and J-O. Eklundh, "Some Properties of Stochastic Labeling Procedures," *Computer Vision, Graphics, and Image Processing,* Vol. 20, 1982, pp. 158–170.

Faugeras, O. D., "Scene Labeling: An Optimization Approach," *Pattern Recognition,* Vol. 12, 1980, pp. 339–347.

——, "Decomposition and Decentralization Techniques in Relaxation Labeling," *Computer Vision, Graphics, and Image Processing,* Vol. 16, 1981(a), pp. 341–355.

——, "Improving Consistency and Reducing Ambiguity in Stochastic Labeling: An Optimization Approach," *IEEE Transactions on Pattern Analysis and Machine Intelligence,* Vol. PAMI-3, 1981(b), pp. 412–423.

——, "Semantic Description of Aerial Images Using Stochastic Labelling," *IEEE Transactions on Pattern Analysis and Machine Intelligence,* Vol. PAMI-3, 1981(c), pp. 633–642.

——, "Relaxation Labeling and Evidence Gathering," *Proceedings of the Sixth International Conference on Pattern Recognition,* Munich, 1982, pp. 405–412.

Fekete, G., J-O. Eklundh, and A. Rosenfeld, "Relaxation: Evaluation and Applications," *IEEE Transactions on Pattern Analysis and Machine Intelligence,* Vol. PAMI-3, 1981, pp. 459–468.

Freuder, E. C., "Synthesizing Constraint Expressions," *Communications of the ACM,* Vol. 21, 1978, pp. 958–966.

Guzman, A., "Decomposition of a Visual Scene into Three-Dimensional Bodies," in *Automatic Interpretation and Classification of Images,* A. Graselli (ed.), Academic Press, New York, 1969, pp. 243–276.

Haralick, R. M., "The Characterization of Binary Relation Homomorphisms," *International Journal of General Systems,* Vol. 4, 1978, pp. 113–121.

——, "An Interpretation of Probabilistic Relaxation," *Computer Vision, Graphics, and Image Processing,* Vol. 22, 1983, pp. 388–395.

——, "An Interpretation for Probabilistic Relaxation," *Computer Vision, Graphics, and Image Processing,* Vol. 22, 1985, pp. 388–395.

Haralick, R. M., and G. Elliott, "Increasing Tree Search Efficiency for Constraint Satisfaction Problems," *Proceedings of the Sixth International Joint Conference on Artificial Intelligence,* Tokyo, 1979, pp.263–313.

——, "Increasing True Search Efficiency for Constraint Satisfaction Problems," *Artificial Intelligence,* Vol. 14, 1980, pp. 263–313.

Haralick, R. M., and J. Kartus, "Arrangements, Homomorphisms, and Discrete Relaxation," *IEEE Transactions on Systems, Man, and Cybernetics,* Vol. SMC-8, 1978, pp. 600–612.

Haralick, R. M., and L. G. Shapiro, "The Consistent Labeling Problem I," *IEEE Transactions on Pattern Analysis and Machine Intelligence,* Vol. PAMI-1, 1979, pp. 173–184.

——, "The Consistent Labeling Problem II," *IEEE Transactions on Pattern Analysis and Machine Intelligence,* Vol. PAMI-2, 1980, pp. 193–203.

Harary, F., *Graph Theory,* Addison-Wesley, Reading, MA, 1969.

Huffman, D., "Impossible Objects as Nonsense Sentences," in *Machine Intelligence,* B. Meltz and D. Michie (eds.), Edinburgh University Press, Edinburgh, 1971, Vol. 5, pp. 295–323.

Hummel, R., and S. Zucker, "On the Foundations of Relaxation Labeling Processes," *IEEE Transactions on Pattern Analysis and Machine Intelligence*, Vol. PAMI-5, 1983, pp. 267–287.

Kirby, R. L, "A Product Rule Relaxation Method," *Computer Vision, Graphics, and Image Processing,* Vol. 13, 1980, pp. 158–189.

Kitchen, L., and A. Rosenfeld, "Discrete Relaxation for Matching Relational Structures," *IEEE Transactions on Systems, Man, and Cybernetics,* Vol. SMC-9, 1979, pp. 869–874.

——, "Scene Analysis Using Region-Based Constraint Filtering," *Pattern Recognition,* Vol. 17, 1984, pp. 189–203.

Kittler, J., and J. Föglein, "On Compatibility and Support Functions in Probabilistic Relaxation," *Computer Vision, Graphics, and Image Processing,* Vol. 34, 1986, pp. 257–267.

Kuschel, S. A., and C. V. Page, "Augmented Relaxation Labeling and Dynamic Relaxation Labeling," *IEEE Transactions on Pattern Analysis and Machine Intelligence,* Vol. PAMI-4, 1982, pp. 676–682.

Lev, A., S. W. Zucker, and A. Rosenfeld, "Iterative Enhancement of Noisy Images," *IEESMC,* Vol. SMC-7, 1977, pp. 435–442.

Lloyd, S. A., "Relaxation Labelling: A Theoretical Analysis and Some New Algorithms," *Proceedings of the Third International Scandinavian Conference on Image Analysis,* Copenhagen, 1983, pp. 128–133.

Mackworth, A., "Consistency in Network Relations," *Artificial Intelligence,* Vol. 8, 1977, pp. 99–118.

McGregor, J. J., "Relational Consistency Algorithms and Their Application in Finding Subgraph and Graph Isomorphisms," *Information Sciences,* Vol. 18, 1979, pp. 362–383.

Mohammed, J. L., R. A. Hummel, and S. W. Zucker, "A Gradient Projection Algorithm for Relaxation Methods," *IEEE Transactions on Pattern Analysis and Machine Intelligence,* Vol. PAMI-5, 1983, pp. 330–332.

Nagin, P. A., A. R. Hanson, and E. M. Riseman, "Variations in Relaxation Labeling Techniques," *Computer Vision, Graphics, and Image Processing,* Vol. 17, 1981, pp. 33–51.

———, "Studies in Global and Local Histogram–Guided Relaxation Algorithms," *IEEE Transactions on Pattern Analysis and Machine Intelligence,* Vol. PAMI-4, 1982, pp. 263–276.

O'Leary, D. P., and S. Peleg, "Analysis of Relaxation Processes: The Two-Node Two-Label Case," *IEEE Transactions on Systems, Man, and Cybernetics,* Vol. SMC-13, 1983, pp. 618–623.

Peleg, S., "Ambiguity Reduction in Handwriting with Ambiguous Segmentation and Uncertain Interpretation," *Computer Vision, Graphics, and Image Processing,* Vol. 10, 1979, pp. 235–245.

———, "A New Probabilistic Relaxation Scheme," *IEEE Transactions on Pattern Analysis and Machine Intelligence,* Vol. PAMI-2, 1980, pp. 362–369.

Price, K. E. "Relaxation Matching Techniques—A Comparison," *IEEE Transactions on Pattern Analysis and Machine Intelligence,* Vol. PAMI-7, 1985, pp. 617–623.

Robinson, G., "Edge Detection by Compass Gradient Masks," *Computer Graphics and Image Processing,* Vol. 6, 1977, pp. 492–501.

Rosenfeld, A., "Iterative Methods in Image Analysis," *Pattern Recognition,* Vol. 10, 1978, pp. 181–187.

Rosenfeld, A., R. Hummel, and S. Zucker, "Scene Labeling by Relaxation Operations," *IEEE Transactions on Systems, Man, and Cybernetics*, Vol. SMC-6, 1976, pp. 420–453.

Rutkowski, W., S. Peleg, and A. Rosenfeld, "Shape Segmentation Using Relaxation," *IEEE Transactions on Pattern Analysis and Machine Intelligence,* Vol. PAMI-3, 1981, pp. 368–375.

Schachter, B. J., et al., "An Application of Relaxation Methods to Edge Reinforcement," *IEEE Transactions on Systems, Man, and Cybernetics,* Vol. SMC-7, 1977, pp. 813–816.

Shapiro, L. G., "A Structural Model of Shape", *IEEE Transactions on Pattern Analysis and Machine Intelligence*, Vol. PAMI-2, 1980, pp. 111–126.

Shapiro, L. G., and R. M. Haralick, "Decomposition of Two-Dimensional Shapes by Graph-Theoretic Clustering," *IEEE Transactions on Pattern Analysis and Machine Intelligence*, Vol. PAMI-1, 1979, pp. 10–20.

———, "Algorithms for Inexact Matching," *Proceedings of the Fifth International Conference on Pattern Recognition,* Miami, 1980, pp. 202–207.

———, "Structural Descriptions and Inexact Matching," *IEEE Transactions on Pattern Analysis and Machine Intelligence,* Vol. PAMI-3, 1981, pp. 504–519.

Thathachar, M. A. L, and P. S. Sastry, "Relaxation Labeling with Learning Automata," *IEEE Transactions on Pattern Analysis and Machine Intelligence,* Vol. PAMI-8, 1986, pp. 256–268.

Ullman, S., "Relaxation and Constrained Optimization by Local Processes," *Computer Graphics and Image Processing,* Vol. 10, 1979, pp. 115–125.

Ullmann, J. R., "Parallel Recognition of Idealised Line Characters," *Kybernetik,* Vol. 2, 1965, pp. 221–226.

Ullman, J. R., "Associating Parts of Patterns," *Information and Control,* Vol. 9, 1966, pp. 583–601.

———, "An Algorithm for Subgraph Homomorphism," *Journal of the ACM,* Vol. 23, 1976, pp. 31–42.

———, "The Pattern Matching Problem," *Institute of Physics Conference Series,* No. 44, 1979, pp. 50–66.

Waltz, D. L., "Understanding Line Drawings of Scenes with Shadows," in *The Psychology of Computer Vision,* P. Winston (ed.), McGraw-Hill, New York, 1970, pp. 19–92.

———, "Generating Semantic Descriptions from Drawings of Scenes with Shadows," MIT Technical Report A1271, Nov. 1972.

Yamamoto, H., "A Method of Deriving Compatibility Coefficients for Relaxation Operators," *Computer Vision, Graphics, and Image Processing,* Vol. 10, 1979, pp. 256–271.

Zucker, S. W., "Low-Level Vision, Consistency, and Continuous Relaxation," *Proceedings of the Second National Conference of the Canadian Society for Computational Studies of Intelligence,* Toronto, 1978, pp. 107–116.

Zucker, S. W., R. A. Hummel, and A. Rosenfeld, "An Application of Relaxation Labeling to Line and Curve Enhancement," *IEEE Transactions on Computers,* Vol. C-26, 1977, pp. 394–403.

Zucker, S. W., E. V. Krishnamurthy, and R. L. Haar, "Relaxation Processes for Scene Labeling: Convergence, Speed, and Stability," *IEEE Transactions on Systems, Man, and Cybernetics,* Vol. SMC-8, 1978, pp. 41–48.

Zucker, S. W., and J. L. Mohammed, "Analysis of Probabilistic Relaxation Labeling Processes," *Proceedings of the IEEE Computer Society Conference on Pattern Recognition and Image Processing,* Chicago, 1978, pp. 307–312.

18 OBJECT MODELS AND MATCHING

18.1   Introduction

Object recognition is one of the most important aspects of computer vision. In order to recognize and identify objects, the vision system must have one or more stored models of the objects that may appear in the universe it deals with. This chapter discusses the object models used by vision systems and the matching procedures used for recognizing objects. The topics covered are two-dimensional models, three-dimensional models, general methods for matching, and model organization. Actual vision systems are described in the next chapter.

18.2   Two-Dimensional Object Representation

Two-dimensional object representation and recognition by computer began in the 1960s and has been an active area of research ever since. Surveys can be found in Pavlidis (1978a, 1978b). Two-dimensional shape analysis is useful in a number of applications of machine vision, including medical image analysis, aerial image analysis, and manufacturing. The method used for shape recognition often depends on the particular representation selected. Thus we begin by looking at the various representations that have been used. They fall loosely into five classes: representation by global features, by local features, by boundary description, by skeleton, and by two-dimensional parts.

18.2.1 Global Feature Representation

A two-dimensional object can be thought of as a binary image. The pixels of the object have value 1, and the pixels outside the object have value 0. Because of this relationship, it is natural to represent shapes by using some of the same features we used to represent binary images (Chapter 3). Commonly used features for two-dimensional shape representation include area, perimeter, moments, circularity, and elongation. Some of the earliest shape recognition work utilized moments and Fourier descriptors.

Shape Recognition by Moments

Let f be a binary image function, and let $S = \{(x, y) \mid f(x, y) = 1\}$ represent a two-dimensional shape. For each pair of nonnegative integers (j, k), the digital (j, k)th moment of S is given by

$$M_{jk}(S) = \sum_{(x,y) \in S} x^j y^k$$

$M_{00}(S)$ is then just $\#S$. *Moment invariants* are functions of the digital moments that are invariant under certain shape transformations. For two-dimensional shape recognition, we would like to have quantities that are moment invariants under translation, rotation, scaling, and some kinds of skewing.

The *center of gravity* $(\bar{x}, \bar{y})$ of S can be expressed in terms of some moments:

$$\bar{x} = \frac{M_{10}(S)}{M_{00}(S)}$$

$$\bar{y} = \frac{M_{01}(S)}{M_{00}(S)}$$

Using the center of gravity, we can define the *central (j, k)th moment of S* by

$$\mu_{jk} = \sum_{(x,y) \in S} (x - \bar{x})^j (y - \bar{y})^k$$

The central moments are translation invariant, since if $S^* = \{(x^*, y^*) \mid x^* = x + a, \ y^* = y + b, (x, y) \in S\}$, then

$$\bar{x}(S^*) = \frac{\displaystyle\sum_{(x^*, y^*) \in S^*} x^*}{M_{00}(S)}$$

$$= \frac{\displaystyle\sum_{(x,y) \in S} (x + a)}{M_{00}(S)}$$

$$= \bar{x}(S) + a$$

$$\bar{y}(S^*) = \bar{y}(S) + b$$

and

$$\mu_{jk}(S^*) = \sum_{(x^*,y^*)\in S^*} [x^* - \bar{x}(S^*)]^j [y^* - \bar{y}(S^*)]^k$$

$$= \sum_{(x,y)\in S} \{x + a - [\bar{x}(S) + a]\}^j (y + b - [\bar{y}(S) + b])^k$$

$$= \sum_{(x,y)\in S} (x - \bar{x})^j (y - \bar{y})^k$$

$$= \mu_{jk}(S)$$

The standard deviation can be expressed in terms of moments:

$$\sigma_x = \sqrt{\frac{\mu_{20}}{M_{00}}}$$

$$\sigma_y = \sqrt{\frac{\mu_{02}}{M_{00}}}$$

Alt (1962) normalized the coordinates by their respective standard deviation to obtain the normalized coordinates

$$x' = \frac{(x - \bar{x})}{\sigma_x}$$

$$y' = \frac{(y - \bar{y})}{\sigma_y}$$

Thus the mean values of x' and y' are both 0, and the variances are both 1. Normalizing by area, Alt obtained the normalized moments defined by

$$m_{jk} = \frac{\sum (x')^j (y')^k}{M_{00}}$$

These moments are invariant under translation, scale, and, in general, affine transformations of the form $x^* = ax + b$, $y^* = cy + d$ (referred to as "stretching" and "squeezing" transformations), since if $S^* = \{(x^*,y^*) \mid x^* = ax + b, \ y^* = cy + d, \ (x,y) \in S\}$, then we have

$$m_{jk}(S^*) = \frac{\displaystyle\sum_{(x^*,y^*)\in S^*} \left(\frac{x^* - \bar{x}(S^*)}{\sigma_x(S^*)}\right)^j \left(\frac{y^* - \bar{y}(S^*)}{\sigma_y(S^*)}\right)^k}{M_{00}(S^*)}$$

$$= \frac{\displaystyle\sum_{(x,y)\in S} \frac{a^j [x - \bar{x}(S)]^j}{a^j (\sigma_x)^j (S)} \frac{c^k [y - \bar{y}(S)]^k}{c^k (\sigma_y)^k (S)}}{M_{00}(S)}$$

$$= m_{jk}(S)$$

The details of the proof that $(1)(m_{00})(S) = (m_{00})(S^*)$; (2) $\bar{x}(S^*) = a\bar{x}(S) + b$, $\bar{y}(S^*) = c\bar{y}(S) + d$; and (3) (σ_x) $(S^*) = a(\sigma_x)(S)$, $(\sigma_y)(S^*) = c(\sigma_y)(S)$ are left for an exercise.

Alt's moments were, by design, not rotation invariant, since he wanted the characters 6 and 9 to be distinct. Hu (1962) defined the *normalized central moments* of S to be

$$\eta_{jk} = \frac{\mu_{jk}}{\mu_{00}^\gamma}, \quad \gamma = \frac{j+k}{2} + 1$$

From these normalized moments he defined seven functions that are rotation invariant:

$$\phi(1) = \eta_{20} + \eta_{02}$$
$$\phi(2) = (\eta_{20} - \eta_{02})^2 + 4\eta_{11}^2$$
$$\phi(3) = (\eta_{30} - 3\eta_{12})^2 + (3\eta_{21} - \eta_{03})^2$$
$$\phi(4) = (\eta_{30} + \eta_{12})^2 + (\eta_{21} + \eta_{03})^2$$
$$\phi(5) = (\eta_{30} - 3\eta_{12})(\eta_{30} + \eta_{12})[(\eta_{30} + \eta_{12})^2 - 3(\eta_{21} + \eta_{03})^2]$$
$$+ (3\eta_{21} - \eta_{03})(\eta_{21} + \eta_{03})[3(\eta_{30} + \eta_{12})^2 - (\eta_{21} + \eta_{03})^2]$$
$$\phi(6) = (\eta_{20} - \eta_{02})[(\eta_{30} + \eta_{12})^2 - (\eta_{21} + \eta_{03})^2]$$
$$+ 4\eta_{11}(\eta_{30} + \eta_{12})(\eta_{21} + \eta_{03})$$
$$\phi(7) = (3\eta_{21} - \eta_{03})(\eta_{30} + \eta_{12})[(\eta_{30} + \eta_{12})^2 - 3(\eta_{21} + \eta_{03})^2]$$
$$- (\eta_{30} - 3\eta_{12})(\eta_{21} + \eta_{03})[3(\eta_{30} + \eta_{12})^2 - (\eta_{21} + \eta_{03})^2]$$

The last function, $\phi(7)$, is also skew invariant.

Maitra (1979) generalized the moment invariants to take into account change in illumination and contrast. Hsia (1981) noted that since the dynamic range of the normalized moments can be quite large, it is more convenient to use $\log |\phi(i)|$. When the sign of $\phi(i)$ is important, $Sgn[\phi(i)]\log|\phi(i)|$ would be appropriate to use. Teague (1980) and Khotanzad and Hong (1988) discussed Zernike moments that are also rotation invariant.

Moment invariants are useful in two-dimensional object recognition, can be implemented in real time (Anderson, 1985), have been used in aircraft identification (Dudan, Breeding, and McGhee, 1977), and have even been extended to three-dimensional object recognition (Reeves et al., 1988; Reeves and Taylor, 1989; Sadjadi, 1984). They have also been used to determine orientation of three-dimensional structures in medical images (Faber and Stokely, 1988). However, they are not sufficient for distinguishing all shapes, they can be very sensitive to noise, and their values drastically change with occlusion. These are problems typical of all global features. Bamieh and De Figueiredo (1986) supplement moments with an attributed graph for object recognition. Teh and Chin (1988) give a good review of moment methods.

Shape Recognition with Fourier Descriptors

Fourier descriptors provide another means for extracting global features from two-dimensional shapes. Rather than characterizing the entire area of the shape, Fourier descriptors usually are defined to characterize the boundary. The main idea is to represent the boundary as a function of one variable $\phi(t)$, expand $\phi(t)$ in its Fourier series, and use the coefficients of the series as Fourier descriptors (FDs). A finite number of these FDs can be used to describe the shape.

There have been several different suggestions for defining ϕ and constructing the FDs (e.g., Zahn and Roskies, 1972; Granlund, 1972; Richard and Hemami, 1974; Persoon and Fu, 1977; Crimmins, 1982; Dekking and Van Otterloo, 1986). We follow Persoon and Fu in their modification of Granlund's FD definition for curves represented by polygons, the most common representation in computer vision.

Let γ be a clockwise-oriented, simple closed curve represented by the parameterized function

$$Z(l) = [x(l), y(l)], \quad 0 \le l \le L$$

where l is the arc length along γ. A point moving along the curve generates the complex function

$$u(l) = x(l) + jy(l)$$

which is a periodic function with period L. The Fourier series expansion of $u(l)$ is given by

$$u(l) = \sum_{-\infty}^{\infty} a_n e^{jn(2\pi/L)l}$$

The FDs are the coefficients $\{a_n\}$ defined by

$$a_n = \frac{1}{L} \int_0^L u(l) e^{-j(2\pi/L)nl} \, dl$$

Persoon and Fu assume that the two-dimensional shape to be described is represented as a sequence of m points $< V_0, V_1, \ldots, V_m = V_0 >$. From the sequence of points, they define a sequence of unit vectors

$$b_k = \frac{V_{k+1} - V_k}{|V_{k+1} - V_k|}$$

and a sequence of cumulative differences

$$l_k = \sum_{i=1}^{k} |V_i - V_{i-1}|, \quad k > 0$$

$$l_0 = 0$$

The FDs are then defined by

$$a_n = \frac{1}{L\left(\frac{n2\pi}{L}\right)^2} \sum_{k=1}^{m} (b_{k-1} - b_k)e^{-jn(2\pi/L)l_k}$$

Using these FDs, Persoon and Fu defined a distance measure to be used in shape comparison. Suppose α and β are two curves to be compared and that $\{a_n\}$ is the sequence of FDs of α, $\{b_n\}$ is the sequence of FDs of β, and M is the number of harmonics used. Then the distance measure is given by

$$d(\alpha, \beta) = \left[\sum_{n=-M}^{M} |a_n - b_n|^2 \right]^{\frac{1}{2}}$$

In order to compare an unknown curve α to a model curve β, they developed a numeric procedure that solves for the scale, rotation, and starting point that minimizes $d(\alpha, \beta)$. The resultant distance is a measure of the similarity between α and β.

Profitt (1982) discusses a normalization technique that can be used in conjunction with FD representation. Chellappa and Bagdazin (1984), using an autoregressive model, obtain estimates of the variances of the Fourier coefficients. Lin and Chellappa (1987) give a procedure for estimating the Fourier coefficients under the constraint of a known value for perimeter^2/area. This improves the estimate even when some part of the true boundary has been occluded. Stracklee and Nagelkerke (1983) note that when the shape is represented as the tangent angle of the boundary as a function of arc length, then truncating the Fourier coefficient representation can produce a representation in which the reconstructed boundary does not close on itself. They give a procedure for the estimation of the Fourier coefficient representation of the tangent angle function that guarantees the reconstructed boundary will close on itself.

18.2.2 Local Feature Representation

A two-dimensional object can also be characterized by its local features, their attributes, and their interrelationships. The most commonly used local features in industrial-part recognition are holes and corners. Holes can be detected by a connected component procedure followed by boundary tracing or, if the shapes of the holes are known in advance, through the operations of binary mathematical morphology (Chapter 5). Corner detection can be performed on a binary image or on a gray tone image (Chapter 7).

Local features must be organized into some type of structure for matching. The most common type of structure is a graph whose nodes represent local features and their properties or measurements and whose edges represent relationships among the features. For example, if the features are all corners, then the angle at which the lines meet is a feature property, and each corner can be related to its two adjacent neighbors around the external and internal boundaries of the object. When holes are included, their areas and shapes are obvious properties to use. However, there is not one spatial relationship that is the obvious one to use for describing the relationships among holes or among holes and corners. Distance from hole to hole or from hole to corner is one possibility. If the object will not be occluded, then the centroid can be used as a focal point and the positions of all other features expressed in relation to the position of the centroid. This approach is described later in this chapter.

A feature-based system for recognition of industrial parts that uses both holes and corners was developed by Bolles and Cain (1982). Their local-feature-focus method finds one key feature, the focus feature, in an image and uses it to predict a few nearby features to look for. It uses graph matching to find the largest cluster of image features matching a cluster of object features near the focus feature. Once such a cluster is found, a hypothesis verification procedure adds more features and also checks the boundary of the object. The features used in their example system were regions that had properties of intensity (black or white), area-to-axis ratio, and corners in which the size of the included angle was the measured property.

The system's knowledge of the most important features to look for in each object and of a cluster of other features that should stand in certain relationships to the focus feature was generated automatically from training images. The training program identified similar local features in different objects, computed symmetries, marked structurally equivalent features that could not be distinguished locally, built feature-centered descriptions, selected nearby features, and ranked the focus features according to the size of the graph that would have to be matched.

18.2.3 Boundary Representation

Boundary representation is the most common representation for two-dimensional objects. There are three main ways to represent the boundary of an object: (1) as a sequence of points, (2) by its chain code (Freeman, 1974), and (3) as a sequence of line segments.

The Boundary as a Sequence of Points

Generally, the points of the boundary come from some kind of border-following or edge-tracking algorithm performed on a digital image. The result of such an operation is a list of pixel coordinates. The list can be maintained as a whole, converted into one of the other two main boundary representations, or processed to produce a smaller list of *interest points*. Interest points are points on the boundary that have some special property that makes them useful in a given matching algorithm. The affine-invariant matching algorithm, defined later in this chapter, requires a set of interest points that are described as being sharp convexities or deep concavities of the boundary of the shape.

One method of extracting these interest points from the original sequence of boundary points of the curve is the curve-partitioning algorithm described in Phillips and Rosenfeld (1987). Given a point P on the curve and a fixed arc length k, there is a set of chords that have arc length k and span the part of the curve containing P. Let $d(P,C)$ be the perpendicular distance from a point P to a chord C whose span includes P, and let $M(P,C)$ be the maximum distance from P to all such chords. P is a partition point of the curve if the value of $M(P,C)$ is a local maximum (for the given k) and also exceeds a threshold $t(k)$.

This method finds points of high curvature along the boundary. It can be modified to select a point P that is the median point in a sequence of points $< P_1, \ldots, P_n >$ for which $M(P_i, C), i = 1, \ldots, n$, are all local maxima. In this way it detects not only very sharp corners but also points of high curvature along the boundary that are part of a section of approximately constant curvature.

Freeman (1978b) discusses extracting interest points that are discontinuities in curvature, points of inflection, curvature maxima, points of intersection, and points of tangency. Langridge (1982) discusses a technique for locating initial points that are discontinuities in the curve.

The Chain Code Representation

Although an ordered list of the coordinates of boundary pixels is a sufficient representation for any boundary, it may be too fine a representation for many applications. Freeman (1961, 1974) developed a representation called chain encoding that can be used at any level of quantization and that saves space required for the row and column coordinates. A boundary (or any curve) to be encoded is first quantized by placing over it a square grid whose side length determines the resolution of the encoding. Figure 18.1(a) shows a curved boundary overlaid by a grid. The marked points are the grid intersections that are closest to the curve and are to be used in the chain encoding of the curve. (The precise definition of the grid intersect quantization is given in Freeman, 1974.) Figure 18.1(b) shows the line segments called *links* that will be used to approximate the curve.

Each link can be horizontal, vertical, or diagonal. Figure 18.1(c) shows an encoding scheme whereby each of the eight possible directions a link may have is assigned an integer between 0 and 7. The curve is encoded as a string of digits each between 0 and 7 representing each of the links associated with the quantization points of the curve. A *chain* or *chain encoding* is written in the form

$$A = a_1 a_2 \ldots a_n$$

or

$$A = \overset{n}{\underset{i=1}{C}} a_i$$

Figure 18.1(d) gives the chain encoding of the curve of Fig. 18.1(a).

Not only do chain codes provide a convenient, space-saving representation for storing boundaries, but they also provide a representation that can be directly used by many algorithms for processing boundaries. Freeman (1974) discusses algorithms for such operations as the inverse of a chain, the width and height of a chain, the first and second moments about the x-axis, the residue of a chain, the rotation of a chain, and others. Corner finding is another operation that can be performed on the chain encoding of a boundary. Kuhl and Giardina (1982) give a direct procedure for the determination of the Fourier coefficients from chain codes. Cederberg (1979)

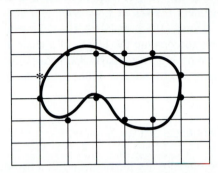

(a) Original curve with grid intersect quantization points.
The asterisk (*) marks the starting point of the curve.

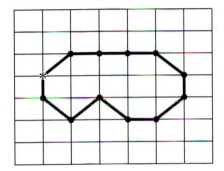

(b) Links associated with the quantization points of (a).

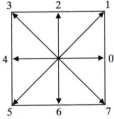

100076543532

(c) Encoding scheme.   (d) Chain code for the curve of (a).

Figure 18.1 Chain encoding of a boundary curve.

and Schmidt (1985) discuss techniques for converting raster scan image data to a chain code representation.

The length of a chain code having n chains can be simply estimated as n. However, this estimate is not very accurate. Freeman (1970) first suggested something better: Letting n_e be the number of even chain codes, n_o the number of odd chain codes, and n_c the number of corners, Freeman suggested $L = n_e + \sqrt{2}n_o$. An unbiased estimate of perimeter length L is given by $L = .948(n_e + \sqrt{2}n_o)$ and is accurate to about 2.6% (Groen and Verbeek, 1978). An even more accurate estimate, given by Vossepoel and Smeulders (1982), is $L = .98n_e + 1.406n_o - .091n_c$, which is accurate to about 6%. Dorst (1985) and Dorst and Smeulders (1986, 1987) discuss more accurate ways to compute the length of a chain-encoded curve. Several generalizations have been developed to make chain codes more efficient or accurate (Freeman, 1978b; Minami and Shinohara, 1986; Sriraman, Kpolowitz, and Mohan, 1989).

The Boundary as a Sequence of Line Segments

The third common representation for the boundary of a two-dimensional shape is as a sequence of line segments. Although any sequence of points can be thought of as a sequence of line segments, this representation is generally used after the original sequence of boundary points has been segmented into a set of line segments representing near-linear portions of the boundary. Pavlidis' split-and-merge algorithm (Pavlidis and Horowitz, 1974) is one possible way to achieve such segmentation. Fitting line segments to the clusters of adjacent collinear points detected by a Hough transform (see Chapter 11) or grouped together by a line-finding procedure such as the Burns line detector (Burns, Hanson, and Riseman, 1986) is another possibility.

Once the sequence of line segments has been computed by some method, it can be converted into a model of the shape that can be used in shape recognition or other matching tasks. A model for representing and matching sequences of line segments was given by Davis (1979). Davis represented a line segment sequence by the sequence of junction points $< X_i, Y_i, \alpha_i >$ where a pair of lines meet at coordinate location (X_i, Y_i) with angle magnitude α_i. Given a sequence $O = O_1, O_2, \ldots, O_n$ of junction points representing the boundary of a model object O and a similar sequence $T = T_1, T_2, \ldots T_m$ representing the boundary of a test object T, the goal is to find an association $F : \{1, 2, \ldots, m\} \Rightarrow \{1, 2, \ldots, n\} \cup \{missing\}$ that satisfies $i < j \Rightarrow F(i) < F(j)$ or either $F(i) = missing$ or $F(j) = missing$.

Davis used constraints on both sides (line segments) and angles to define what is meant by a best mapping for this problem. Let $M(i, j)$ be a local evaluation function that measures the goodness of the match of junction i of T to junction j of O, based on the difference between the angles α_i and α_j. Let $S_{ij}(i', j')$ be a measure of the consistency of mapping junction i to junction i' and junction j to junction j', based on the difference between the segment lengths of $T_i T_j$ and $O_{i'} O_{j'}$. The cost of a mapping F is given by

$$C(F) = \sum_{i=1}^{m} M[i, F(i)] + \sum_{i=1}^{m} \sum_{j=1}^{m} S_{ij}[F(i), F(j)] + P(m_T) + P(m_O)$$

where P is a penalty function for missing angles.

The problem of finding the mapping F that minimizes $C(F)$ is an inexact consistent-labeling problem, as discussed in the last chapter. Davis solved the problem with a discrete relaxation procedure that was able to prune the inconsistent correspondences between junctions and between line segments from the search space.

The relational-distance measure defined later in this chapter is a general matching formalism for this type of problem.

18.2.4 Skeleton Representation

Although the boundary of a two-dimensional object gives full information on the shape of the object, this may not be the most suitable information for matching. Particularly for shapes that can be thought of as a union of long, sometimes thin parts called *strokes*, the essence of the shape can be described as a sequence of line segments that capture the linearity of the strokes. Figure 18.2 shows a set of characters and the line segments that can be used to characterize them. Blum (1973) and Blum and Nagel (1978) defined the *symmetric axis transform* of a two-dimensional object as the set of maximal circular disks that fit inside the object. The object can be represented by its *symmetric axis* (the locus of the centers of these maximal disks) plus the set of distances of these centers to the boundary of the object. This was discussed from the point of view of mathematical morphology in Chapter 5. Figure 18.3 gives the symmetric axis of the characters in Fig. 18.2 one other two-dimensional object. The symmetric axis is one example of a *skeleton* description of a two-dimensional object.

The symmetric axis is not always completely representative of the strokes of an object. Notice that the symmetric axis of a rectangle, rather than being a single line, consists of five line segments. This property and the fact that the symmetric axis is extremely sensitive to noise make it difficult to use in matching. To solve this problem, Brady and Asada (1984) developed a separate definition for skeleton

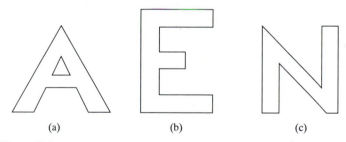

(a)　　　　(b)　　　　(c)

Figure 18.2 Line segments that characterize the strokes of a set of characters.

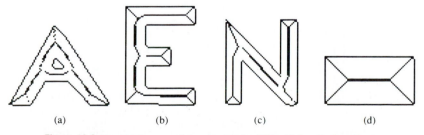

(a)　　　(b)　　　(c)　　　(d)

Figure 18.3 Symmetric axes of the characters of Fig. 18.2 and other shapes.

called the *axis of smoothed local symmetries*. A *local symmetry* is the midpoint
P of a line segment BA joining a pair of points A and B on the boundary of a shape
such that the angle α between BA and the outward normal n_A at A is the same as
the angle between BA and the inward normal n_B at B, as illustrated in Fig. 18.4.
The loci of local symmetries that are maximal with respect to forming a smooth
curve are called *axes* or *spines*. The *cover* of an axis is the portion of the shape
subtended by the axis. If the cover of an axis is properly contained in the cover
of a second axis, the first axis is *subsumed* by the second. Figure 18.5 illustrates
the axes of symmetry of a rectangle. The short diagonal axes are subsumed by
the horizontal and vertical axes and can be either deleted or relegated to a lower
place in a hierarchical description of the shape (see Chapter 19 for hierarchical
representations). Figure 18.6 shows the smoothed local symmetries (from Brady
and Asada) of other objects.

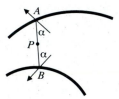

Figure 18.4 A point P that is a local symmetry with respect to boundary points
A and B.

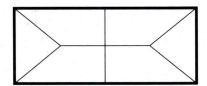

Figure 18.5 Symmetric axes of local symmetry of a rectangle.

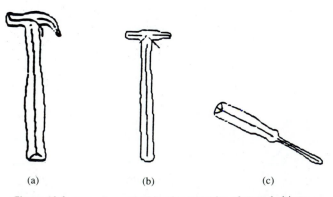

(a)   (b)   (c)

Figure 18.6 Axes of smoothed local symmetries of several objects.

18.2.5 Two-Dimensional Part Representation

The last representation for two-dimensional shapes is a decomposition of the entire area of the shape into smaller two-dimensional parts. The parts, their attributes, and their interrelationships can then be used to form a structural description of the shape. A possible structural description for two-dimensional shapes was given in Chapter 17. A relational-distance approach to matching is described later in this chapter. The main question left is, how do we define the parts of a shape?

Pavlidis tackled the shape-decomposition problem with several different approaches. One early approach was the decomposition of shapes into *primary convex subsets* (Pavlidis, 1968). The primary convex subsets and the nuclei (regions where primary convex subsets overlap) form the nodes of a labeled graph representing the original shape (Pavlidis, 1972). Figure 18.7 shows an example of the letter *E* decomposed into primary convex subsets and nuclei. A related approach was the decomposition into convex parts, T-shaped parts, and spirals by Feng and Pavlidis (1975). Although the second method relaxed the constraints of the first, it did not address the problem of noisy and deformed shapes.

Shapiro and Haralick (1979) designed a method to decompose shapes into *near-convex* pieces. The near-convexity allows noisy or somewhat distorted instances of perfect shapes to have the same or nearly the same decompositions as the perfect shapes themselves. This method calculates a binary relation that captures important information about the shape of a two-dimensional object represented as a sequence of points. The binary relation is constructed as follows:

Given two points P_1 and P_2 on the boundary of the object, we can construct a straight line from P_1 to P_2. If the line lies completely interior to the boundary of the object, then the ordered pair (P_1, P_2) belongs to the relation L_I. The L_I relation, sometimes called the *visibility relation*, can be computed in $O(n^2)$ time (Lee, 1983). Once the relation has been computed, a graph-theoretic clustering

Figure 18.7 Decomposition of a shape into primary convex subsets and nuclei. The nuclei are the shaded areas of overlap.

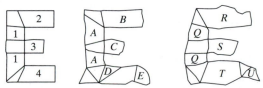

Figure 18.8 Decomposition of three similar shapes into near-convex pieces. The pieces are allowed to overlap.

algorithm that was given in Shapiro and Haralick (1979) determines the clusters of the relation. The points in each cluster define the near-convex parts of the object. Figure 18.8 illustrates the decompositions of several imperfect letter E's from our structural-shape-matching paper (Shapiro, 1980). The structural descriptions and inexact matching of these shapes were described in the previous chapter.

Phillips (1985) decomposes a shape into compact pieces by using the shape skeleton and the distance values associated with it.

18.3   Three-Dimensional Object Representations

Three-dimensional object recognition presents an even greater challenge than two-dimensional shape recognition. Perhaps because of this, a wide variety of different representations for three-dimensional objects have been tried. In this section we describe some of the most common representations.

Surveys and other information representing three-dimensional objects can be found in Baer, Eastman, and Henrion (1979); Requicha (1980, 1983); Aggarwal et al. (1981); Barnhill (1983); Cohen (1983); Gordon (1983); Hanna, Abel, and Greenberg (1983); Requicha and Voelcker (1983); Koparkar and Mudur (1984); Besl and Jain (1985); Chin and Dyer (1986); Lee and Jea (1987); and Xie and Calvert (1988).

18.3.1   Local Features Representation

The concept of local features that was used to describe two-dimensional shapes can be readily extended to three-dimensional objects. There are several major differences between the two-dimensional and three-dimensional matching procedures. First, the data tend to be different. Whereas some systems continue to use standard two-dimensional gray tone images of three-dimensional objects, many systems are using range data obtained from laser range finders, light striping, photometric stereo, or other techniques that can determine depth. Thus these systems start with the depth information and from it try to infer the surfaces, edges, corners, holes, and other features of the object. Second, with or without depth data, only a portion of the object is usually viewed. Thus the matching is more difficult than for two-dimensional objects with no self-occlusion properties. Two of the main techniques for matching three-dimensional object models—the viewpoint consistency constraint and view-class matching—are described later in this chapter. As far as extensibility of the local-feature-focus technique described earlier for two-dimensional objects, Horaud and Bolles (1986) developed a three-dimensional version of the local-feature-focus method, using range data for input and edges for features.

18.3.2   Wire Frame Representation

Since many of the early three-dimensional vision systems worked with polyhedral objects, edges have been the main local feature used for recognition or pose estimation. A three-dimensional object model that consists of only the edges of the object is called a *wire frame* model. The very first computer vision system (Roberts, 1965) used wire frame models of polyhedral objects. The models were matched to line drawings extracted from gray tone images of the scene. Matching was based on the topological arrangement of the line segments. This system also follows the paradigm of viewpoint consistency that is described later in this chapter.

18.3.3 Surface-Edge-Vertex Representation

One of the most general and most commonly used representations for three-dimensional object models and for three-dimensional data is the *surface-edge-vertex* representation. The representation is a data structure containing a list of the surfaces of the object, a list of the edges of the object, a list of the vertices of the object, and usually the topological relationships that specify the surfaces on either side of an edge and the vertices on either end of a line segment. The surface-edge-vertex representation is used heavily in the VISIONS (Visual Integration by Semantic Interpretation of Natural Scenes) system (Hanson and Riseman, 1978) at the University of Massachusetts. Hierarchical, relational models are discussed as a form of knowledge in the next chapter; here we describe a complete surface-edge-vertex model used in much of our own work and, in particular, in the PREMIO (Prediction in Matching Images to Objects) system (Camps, Shapiro, and Haralick, 1989).

The PREMIO three-dimensional object model is a hierarchical, relational model (see Chapter 19) with five levels: world, object, face/edge/vertex, surface/boundary, arc/two-dimensional piece, and a one-dimensional piece. The world level at the top of the hierarchy is concerned with the arrangement of the different objects in the world. The object level is concerned with the arrangement of the different faces, edges, and vertices that form the objects. The face level describes a face in terms of its surfaces and its boundaries. The surface level specifies the elemental pieces that form those surfaces, and the two-dimensional piece level describes these pieces and specifies arcs that form the boundaries. Finally, the one-dimensional piece level describes the elemental pieces that form the arcs.

The PREMIO object model is a topological object model because it represents not only the geometry of the objects but also the relations among their faces, edges, and vertices. This information is redundant in the sense that it can be derived from the geometric information, but it is included to speed up the system. The object model is an instance of a general spatial data structure (Shapiro and Haralick, 1980) that can be used to represent any spatial information or relational data and is defined as follows:

An *atom* is a unit of data that will not be further broken down. Examples of atoms are a matrix of real numbers and a string of characters. An *attribute-value table* is a set of pairs: $A/V = \{(a,v)\}$, where a is an attribute and v is the value associated with a. Both a and v can be atoms or more complex structures. A *spatial data structure D* is a set $D = \{R_1, R_2, \ldots, R_k\}$ of k relations. Each relation R_i has dimension n_i and a sequence of domain sets $S_{i1}, S_{i2}, \ldots, S_{in_i}$ such that $R_i \subseteq S_{i1} \times S_{i2} \times \cdots \times S_{in_i}$. The elements of the domain sets may be atoms or spatial data structures.

The spatial data structure is the building block of the PREMIO model and represents a geometric entity. Each spatial data structure has one distinguished binary relation, the attribute-value table, containing the global attributes of the entity that the structure represents. A geometric entity might be formed from parts or be related to other instances of entities of lower levels. Each instance of an entity of a lower level consists of a pointer to the lower-level entity and a transformation that is used to transform the physical points of the lower entity, expressed in its own coordinate system, to the coordinate system being used at the current level. A list of these parts or entity instances forms a unary relation, and the interrelationships among them form n-ary relations.

The basic geometric entities used in the PREMIO model are the world; the objects in the world; the object faces, edges, and vertices; the face boundaries; the face surfaces; the two-dimensional pieces that form the face surfaces; the boundary arcs; and the one-dimensional pieces that form these arcs. Thus there are ten different spatial data structure (SDS) types: WORLD SDS, OBJECT SDS, EDGE SDS, VERTEX SDS, FACE SDS, BOUNDARY SDS, SURFACE SDS, 2D-PIECE SDS, ARC SDS, and 1D-PIECE SDS. A diagram of the complete model is shown in Figure 18.9.

At the world level, the world as seen by PREMIO is represented as a set of instances of any number of objects. At this level there is only one type of spatial data structure, the WORLD SDS, consisting of two relations: a WORLD A/V relation and an OBJECTS relation. The WORLD A/V relation contains the bounding box that contains the entire world. The OBJECTS relation is an unary relation listing the object instances in the world and giving for each one an object name or identifier, a type indicating the kind of object, a pointer to a spatial data structure describing the object at the next level of the hierarchy, and a pointer to a matrix representing a transformation by which the points in the object can be transformed into the world coordinates system.

At the object level, objects are considered as sets of faces that intersect along edges, which in turn intersect at vertices. Objects are represented by OBJECT SDS's. An OBJECT SDS consists of four relations: an OBJECT A/V relation, a FACES relation, an EDGES relation, and a VERTICES relation. The OBJECT A/V relation contains the bounding box that contains the object. The FACES relation is an unary relation listing the faces of the object and giving for each one a name or identifier, a type indicating the kind of face, a pointer to a spatial data structure describing the face at the next level of the hierarchy, and a pointer to a matrix representing a transformation by which the points in the face can be transformed into the object coordinates system. The EDGES relation lists all the object edges. Each tuple in the relation consists of an edge identifier, a pointer to the spatial data structure describing the edge, and a transformation matrix. The VERTICES relation lists the vertices of the object, giving for each one a name or identifier, a pointer to its corresponding SDS, and a matrix representing the transformation by which the vertex can be transformed into the object coordinate system.

At the next level, there are three different types of entities: object faces, object

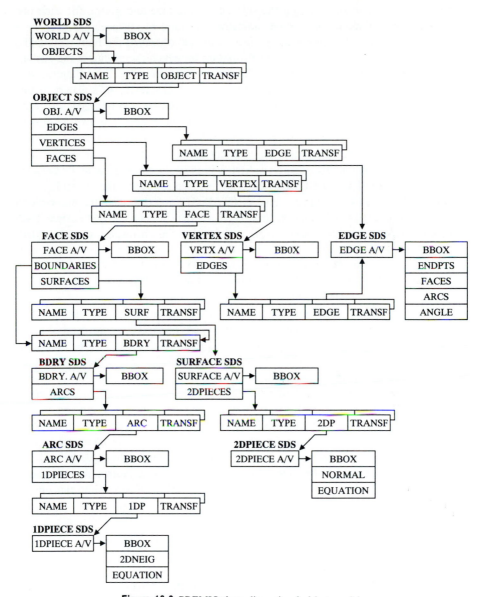

Figure 18.9 PREMIO three-dimensional object model.

edges, and object vertices. The FACE SDS consists of three relations: a FACE A/V relation, a BOUNDARIES relation, and a SURFACES relation. The BOUNDARIES relation is an unary relation listing the boundaries in the face and giving for each one a boundary name or identifier, a type indicating the kind of boundary, a pointer to a spatial data structure describing the boundary at the next level of the hierarchy, and a pointer to a matrix representing a transformation by which the points in the boundary can be transformed into the face coordinates system. The SURFACES relation is a unary relation listing the surfaces in the face and giving for each one a surface name or identifier, a type indicating the kind of surface, a pointer to a spatial data structure describing the surface at the next level of the hierarchy, and a pointer to a matrix representing a transformation by which the points in the surface can be transformed into the face coordinates system. An edge is the intersection of two object faces. Although it is a single entity at this level, it is a part of two separate boundaries, one for each adjacent face. The entities that represent this edge in the two separate boundaries are called arcs. Thus each edge has two associated arcs. The EDGE SDS consists of one relation: the EDGE A/V relation. It contains the edge bounding box, pointers to its endpoints SDS, pointers to its adjacent faces SDS, pointers to the corresponding arc SDS's in these faces, and the angle between the adjacent faces. The VERTEX SDS consist of two relations: the VERTEX A/V relation and the EDGES relation. The VERTEX A/V relation contains the location of the vertex. The EDGES relation lists the edges that intersect at that vertex. Each tuple contains an edge identifier, a type, a pointer to the corresponding edge SDS, and a transformation matrix.

At the fourth level are the BOUNDARY SDS and the SURFACE SDS. A BOUNDARY SDS consists of two relations: a BOUNDARY A/V relation and an ARCS relation. The BOUNDARY A/V relation contains the bounding box of the boundary. The ARCS relation is an unary relation listing the arcs in the boundary and giving for each one an arc name or identifier, a type indicating the kind of arc, a pointer to a spatial data structure describing the arc at the next level of the hierarchy, and a pointer to a matrix representing a transformation by which the points in the arc can be transformed into the boundary coordinates system.

A SURFACE SDS consists of two relations: a SURFACE A/V relation and a 2D-PIECES relation. The SURFACE A/V relation contains the bounding box of the surface. The 2D-PIECES relation is a unary relation listing the two-dimensional pieces in the surface and giving for each one a piece name or identifier, a type indicating the kind of piece, a pointer to a spatial data structure describing the piece at the next level of the hierarchy, and a pointer to a matrix representing a transformation by which the points in the piece can be transformed into the surface coordinates system.

At the next level are the 2D-PIECE SDS and the ARC SDS. A 2D-PIECE SDS consists of only one relation: the 2D-PIECES A/V relation. It contains the bounding box of the piece, a normal code indicating whether the surface is concave or convex, and its parametric equation.

An ARC SDS consists of two relations: an ARC A/V relation and a 1D-PIECES relation. The 1D-PIECES relation is an unary relation listing the one-dimensional pieces in the arc and giving for each one a piece name or identifier, a type indicating the kind of piece, a pointer to a spatial data structure describing the piece at the next level of the hierarchy, and a pointer to a matrix representing a transformation by which the points in the piece can be transformed into the arc coordinates system. At the last level is the 1D-PIECE SDS. A 1D-PIECE SDS consists of one relation: the 1D-PIECES A/V relation. It contains the bounding box of the piece, a two-dimensional neighborhood indicating the position of the face surface with respect to the piece, and its parametric equation.

The surface-edge-vertex representation is designed to be general purpose and to allow easy access to a variety of information about an object. Thus it can be used in many different matching procedures. Because it is a relational model, it is especially well suited for the relational-distance approach to matching covered later in this chapter.

18.3.4 Sticks, Plates, and Blobs

The surface-edge-vertex model is a very precise model of an object. Once the correspondences have been found between a surface-edge-vertex model and the image(s) of an object, the pose of the object can be determined from the point correspondences. This assumes that the images to be analyzed are images of the exact parts that were modeled. Sometimes a recognition task requires only a rough estimate of the identity or structure of an object. This rough matching may be all that is required for the task at hand or it may be the first phase of a more complex sequence in which a more detailed model to try is chosen on the basis of the results of the rough match. The models described in this section are meant to be rough models of three-dimensional objects that can be used in the rough-matching phase. They are called *sticks, plates, and blobs* models and were originally described in Shapiro et al. (1984).

Complex manmade objects, such as office furniture and industrial tools, are made from parts. The parts can have flat or curved surfaces, and they exist in a large variety. Instead of trying to describe each part precisely, as in the surface-edge-vertex models, each part can be classified as a *stick*, a *plate*, or a *blob*. Sticks are long, thin parts that have only one significant dimension. Plates are flatish, wide parts with two nearly flat surfaces connected by a thin edge between them. Plates have two significant dimensions. Blobs are parts that have three significant dimensions. All three kinds of parts are near-convex, so a stick cannot bend very much, the surfaces of a plate cannot fold very much, and a blob can be bumpy but cannot have large concavities. Figure 18.10 shows several examples of sticks, plates, and blobs.

To use sticks, plates, and blobs in a computer model of an object, we must define them more precisely. Formally, a stick is a 4-tuple $ST = (En, I, Cm, L)$, where En is the set of two endpoints of the stick, I is the set of interior points, Cm is its center of mass, and L is its length. Since straight-line segments have each of the components of a stick, we can represent all sticks by straight-line segments in the models.

A plate is a 4-tuple $PL = (Eg, S, Cm, A)$, where Eg is the set of edge points, $S = \{S_1, S_2\}$ is the set of surface points of the plate partitioned into the two surfaces, Cm is the center of mass, and A is the area. We can represent all plates by circles in the models.

A blob is a triple $BL = (S, Cm, V)$, where S is the set of surface points, CM is the center of mass, and V is the volume of the blob. Blobs can be represented as spheres in the model.

Sticks

Plates

Blobs

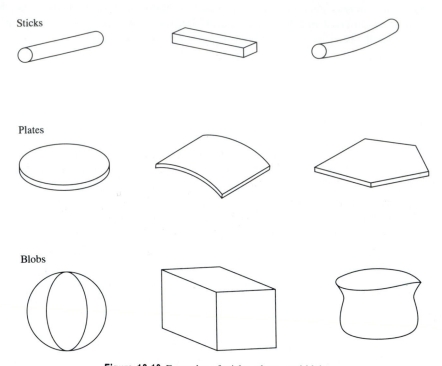

Figure 18.10 Examples of sticks, plates, and blobs.

The models describe how the sticks, plates, and blobs are put together. These descriptions are also rough; they cannot specify the physical points where two parts join. The stick has two logical endpoints, a logical set of interior points, and a logical center of mass that can be specified as connection points. The plate has a set of edge points, a set of surface points, and a center of mass. The blob has a set of surface points and a center of mass. Only this minimal information can be used in the object models.

The models described in Shapiro et al. (1984) were relational models that employed the same spatial data structure (SDS) described above. The SDS for a sticks-plates-and-blobs model consists of an attribute-value table and five other relations. The unary SIMPLE PARTS relation is a list of the parts of the object. Each part is represented by its own SDS that consists only of an attribute/value table with attributes TYPE (stick, plate, or blob), RELATIVE LENGTH, RELATIVE AREA, and RELATIVE VOLUME. The length, area, and volume values may be real numbers or may be marked "don't care" when they are unimportant or inappropriate.

The CONNECTS/SUPPORTS relation contains some of the most important information on the structure of the object. It consists of 10-tuples of the form $(s_1, s_2, SUPPORTS, HOW, vl_1, vh_1, vl_2, vh_2, vl_3, vh_3)$. The components s_1 and s_2 are simple parts; SUPPORTS is true if s_1 supports s_2 and false otherwise; and HOW describes the connection type of s_1 and s_2. The values in the HOW field are elements of the set { end-end, end-interior, end-center, end-edge, interior-center, center-center }. The pairs (vl_1, vh_1), (vl_2, vh_2), and (vl_3, vh_3) hold the low and high values for the allowed angle ranges for the (at most) three angles that can be specified for a binary connection.

The other four relations express constraints. The TRIPLE CONSTRAINT relation has 6-tuples of the form $(s_1, s_2, s_3, SAME, vl, vh)$, where simple part s_2 touches both s_1 and s_3; SAME is true if s_1 and s_3 touch s_2 on the same end (or surface) of s_2 and false otherwise; and vl and vh specify the permissible low and high values for the constrained angle. This angle is the one subtended by the centers of mass of s_1 and s_3 at the center of mass of s_2. The PARALLEL relation and the PERPENDICULAR relation have pairs of the form (s_1, s_2), where simple parts s_1 and s_2 are parallel (or perpendicular) in the model. Figure 18.11 illustrates the sticks-plates-and-blobs model of a prototype chair object. All chairs with similar relations should match this model, regardless of the exact shapes of the parts.

Since the sticks-plates-and-blobs model is relational like the surface-edge-vertex model, some form of relational matching, such as the relational-distance matching described later, is appropriate. In general, sticks will project to long, thin regions of the image; plates will project to compact regions; and blobs will project to one or more connected regions. Thus matching of the three-dimensional models to two-dimensional images under perspective projection is possible (Mulgaonkar, Shapiro, and Haralick, 1984).

18.3.5 Generalized Cylinder Representation

A *generalized cylinder* is a volumetric primitive defined by a space curve axis and a cross-section function at each point of the axis. The cross section is swept along the axis, creating a solid. For example, an actual cylinder is a generalized cylinder whose axis is a straight-line segment and whose cross section is a circle of constant radius. A cone is a generalized cylinder whose axis is a straight-line segment and whose cross section is a circle whose radius starts out zero at one endpoint of the axis and grows to its maximum at the other endpoint. A rectangular solid is a generalized cylinder whose axis is a straight line segment and whose cross section is a constant rectangle. A torus is a generalized cylinder whose axis is a circle and whose cross section is a constant circle. In many applications of generalized cylinders, the cross section and axis are polygonal lines, polynomials, or conics (Nevatia and Binford, 1977; Soroka, Anderson, and Bajcesy, 1981; Brooks, Grenier, and Binford, 1979; Brooks, 1981). Shani and Ballard (1984) discuss the use of cubic uniform B splines. A generalized cylinder representation is a representation that uses generalized cylinders (Binford, 1971) as its primitives.

Whereas the surface-edge-vertex model is very precise and the sticks-plates-and-blobs model is very rough, the generalized cylinder model is somewhere in between. Many objects can be exactly defined by a generalized cylinder model, but for others only an approximation is possible.

A generalized cylinder model of an object must include descriptions of the generalized cylinders and the spatial relationships among them. Nevatia and Binford

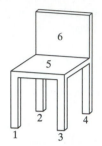

ATTRIBUTE-VALUE TABLE

BASE SUPPORTS	4
TOP TYPE	2
NO. STICKS	3
NO. PLATES	2
NO. BLOBS	0
NO. UPRIGHTS	5
HORIZONTALS	1
SLANTEDS	0
NO. LEVELS	3
TOP PCS. POS.	2

SIMPLE-PARTS Relation

SIMPT	TYPE	LENGTH	AREA	VOLUME
1	1	1.0	0.0	0.0
2	1	1.0	0.0	0.0
3	1	1.0	0.0	0.0
4	1	1.0	0.0	0.0
5	2	1.0	1.0	0.0
6	2	1.0	1.0	0.0

CONNECTS-SUPPORTS Relation

SP1	SP2	SUPPORTS	HOW
1	5	True	12
2	5	True	12
3	5	True	12
4	5	True	12
5	6	True	23

TRIPLES Relation

SP1	SP2	SP3	SAME
1	5	2	True
1	5	3	True
1	5	4	True
1	5	6	False
2	5	3	True
2	5	4	True
2	5	6	False
3	5	4	True
3	5	6	False
4	5	6	False

PARALLEL Relation

SP1	SP2
1	2
1	3
1	4
2	3
2	4
3	4

PERPENDICULAR Relation

SP1	SP2
1	5
2	5
3	5
4	5
5	6

Figure 18.11 Full relational structure of the sticks-plates-and-blobs model of a chair object. The attribute-value table contains global attributes; the simple-parts relation lists the parts and their attributes; the connects-supports relation gives connections between pairs of parts; the triples relation specifies connections between three parts at a time; and the parallel and perpendicular relations list pairs of parts that are parallel or perpendicular, respectively.

(1977) worked with three-dimensional models consisting of generalized cylinders with normal cross sections for primitives, plus connectivity relations and global properties. Cylinders were described by length of axis, average cross-section width, ratio of the two, and cone angle. Global properties of an object include number of pieces (cylinders), number of elongated pieces, and symmetry of the connections. Marr and Nishihara (1977) envisioned hierarchical generalized cylinder models. Each level in the hierarchy represents a closer approximation to the actual object. For example, a person might be modeled very roughly as a stick figure (as shown in Fig. 18.12) consisting of cylinders for the head, torso, arms, and legs. At the next level of the hierarchy, the torso might be divided into a neck and lower torso; the arms into three cylinders for upper arm, lower arm, and hand; and the legs similarly. At the next level, the hands might be broken into a main piece and five fingers, and one level deeper, the fingers might be broken into three pieces and the thumb into

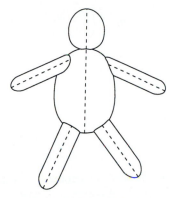

Figure 18.12 Rough generalized cylinder model of a person. The dotted lines are the axes of the cylinders.

two. Marr and Nishihara also used the connectivity relationship, specifying for two touching cylinders the position at which they touch, the inclination angle, and the girdle angle describing the rotation of one about the other.

A three-dimensional generalized cylinder can project to two different kinds of two-dimensional regions on an image: ribbons and ellipses. A *ribbon* is the projection of the long portion of the cylinder, and an *ellipse* is the projection of the cross section. Of course, the cross section is not always circular, so its projection is not always elliptical, and some generalized cylinders are completely symmetric, so they have no longer or shorter parts. For those that do, algorithms have been developed to find the ribbons in images of the modeled objects. These algorithms generally look for long regions that can support the notion of an axis. Mohan and Nevatia (1989) have proposed a new method for locating and recognizing ribbons in complex images by using an approach based on perceptual organization and symmetry. Generalized cylinders have been used in the ACRONYM system (Brooks, 1981) described in the next chapter; in the SUCCESSOR system (Binford, 1989), which is the successor to ACRONYM; and in various other works (Soroka, 1979; Agin and Binford, 1976; Rao and Nevatia, 1989).

18.3.6 Superquadric Representation

Superquadrics are models originally developed for computer graphics (Barr, 1981; Miller, 1988) and proposed for use in computer vision by Pentland (1986). Superquadrics can intuitively be thought of as lumps of clay that can be deformed and glued together into object models. Mathematically superquadrics form a parameterized family of shapes. A *superquadric* is a closed surface spanned by a vector whose x-, y-, and z-components are specified as functions of the angles η and ω via the spherical product of two two-dimensional parametrized curves

$$h = \begin{pmatrix} h_1(\eta) \\ h_2(\eta) \end{pmatrix}$$

and

$$m = \begin{pmatrix} m_1(\omega) \\ m_2(\omega) \end{pmatrix}$$

that come from one of three basic forms. The spherical product of h with m (Barr, 1981) scaled by the vector $\begin{pmatrix} a_1 \\ a_2 \\ a_3 \end{pmatrix}$ is defined by

$$h \otimes m = \begin{bmatrix} a_1 h_1(\eta) m_1(\omega) \\ a_2 h_1(\eta) m_2(\omega) \\ a_3 h_2(\omega) \end{bmatrix}$$

Superquadrics in canonical position are generated by taking the scaled spherical product of curves coming from the three basic forms

$$\begin{pmatrix} \cos^\epsilon \xi \\ \sin^\epsilon \xi \end{pmatrix}, \quad \begin{pmatrix} \sec^\epsilon \xi \\ \tan^\epsilon \xi \end{pmatrix}, \quad \begin{pmatrix} a + \cos^\epsilon \xi \\ \sin^\epsilon \xi \end{pmatrix}$$

For example, with scaling vector $\begin{pmatrix} a_1 \\ a_2 \\ a_3 \end{pmatrix}$ the spherical product

$$\begin{pmatrix} \cos^{\epsilon_1} \eta \\ \sin^{\epsilon_1} \eta \end{pmatrix} \otimes \begin{pmatrix} \cos^{\epsilon_2} \omega \\ \sin^{\epsilon_2} \omega \end{pmatrix} = \begin{bmatrix} a_1 \cos^{\epsilon_1}(\eta) \cos^{\epsilon_2}(\omega) \\ a_2 \cos^{\epsilon_1}(\eta) \sin^{\epsilon_2}(\omega) \\ a_3 \sin^{\epsilon_1}(\eta) \end{bmatrix}$$

for $-\frac{\pi}{2} \le \eta \le \frac{\pi}{2}$ and $-\pi \le \omega < \pi$ generates a surface that is a superellipsoid having the implicit form

$$\left[\left(\frac{x}{a_1} \right)^{\frac{2}{\epsilon_2}} + \left(\frac{y}{a_2} \right)^{\frac{2}{\epsilon_2}} \right]^{\frac{\epsilon_2}{\epsilon_1}} + \left(\frac{z}{a_3} \right)^{\frac{2}{\epsilon_1}} = 1$$

The parameters a_1, a_2, and a_3 relate to the length of the axes of the superellipse in the x-, y-, and z-directions, respectively. The parameters ϵ_1 and ϵ_2 represent the squareness in the latitude and longitude planes. Superellipsoids can be used to model cylinders ($\epsilon_1 < 1$ and $\epsilon_2 = 1$), ellipsoids ($\epsilon_1 = 1, \epsilon_2 = 1$), and parallelopipeds ($\epsilon_1 < 1$ and $\epsilon_2 < 1$).

The spherical product

$$\begin{pmatrix} \sec^{\epsilon_1} \eta \\ \tan^{\epsilon_1} \eta \end{pmatrix} \otimes \begin{pmatrix} \cos^{\epsilon_2} \omega \\ \sin^{\epsilon_2} \omega \end{pmatrix} = \begin{pmatrix} a_1 \sec^{\epsilon_1} \eta \cos^{\epsilon_2} \omega \\ a_2 \sec^{\epsilon_1} \eta \sin^{\epsilon_2} \omega \\ a_3 \tan^{\epsilon_1} \eta \end{pmatrix}, \quad \begin{array}{l} \frac{-\pi}{2} \le \eta \le \frac{\pi}{2} \\ -\pi \le \omega < \pi \end{array}$$

generates a surface that is a superhyperboloid of one piece whose implicit equation is

$$\left[\left(\frac{x}{a_1} \right)^{\frac{2}{\epsilon_2}} + \left(\frac{y}{a_2} \right)^{\frac{2}{\epsilon_2}} \right]^{\frac{\epsilon_2}{\epsilon_1}} - \left(\frac{z}{a_3} \right)^{\frac{2}{\epsilon_1}} = 1$$

The spherical product of

$$
\begin{pmatrix} \sec^{\epsilon_1} \eta \\ \tan^{\epsilon_1} \eta \end{pmatrix} \otimes \begin{pmatrix} \sec^{\epsilon_2} \omega \\ \tan^{\epsilon_2} \omega \end{pmatrix} = \begin{pmatrix} a_1 \sec^{\epsilon_1} \eta \sec^{\epsilon_2} \omega \\ a_2 \sec^{\epsilon_1} \eta \tan^{\epsilon_2} \omega \\ a_3 \tan^{\epsilon_1} \eta \end{pmatrix}, \quad \begin{array}{l} -\frac{\pi}{2} \le \eta \le \frac{\pi}{2} \\ -\frac{\pi}{2} \le \omega \le \frac{\pi}{2} \text{(piece 1)} \\ \frac{\pi}{2} \le \omega \le \frac{3\pi}{2} \text{(piece 2)} \end{array}
$$

generates a superhyperboloid of two sheets having implicit equation

$$
\left[\left(\frac{x}{a_1}\right)^{\frac{2}{\epsilon_2}} - \left(\frac{y}{a_2}\right)^{\frac{2}{\epsilon_2}} \right]^{\frac{\epsilon_2}{\epsilon_1}} - \left(\frac{z}{a_3}\right)^{\frac{2}{\epsilon_1}} = 1
$$

The spherical product

$$
\begin{pmatrix} a + \cos^{\epsilon_1} \eta \\ \sin^{\epsilon_1} \eta \end{pmatrix} \otimes \begin{pmatrix} \cos^{\epsilon_2} \omega \\ \sin^{\epsilon_2} \omega \end{pmatrix} = \begin{pmatrix} a_1(a + \cos^{\epsilon_1} \eta) \cos^{\epsilon_2} \omega \\ a_2(a + \cos^{\epsilon_1} \eta) \sin^{\epsilon_2} \omega \\ a_3 \sin^{\epsilon_1} \end{pmatrix}, \quad \begin{array}{l} -\pi \le \eta \le \pi \\ -\pi \le \omega \le \pi \end{array}
$$

generates a supertoroid having implicit equation

$$
\left(\left[\left(\frac{x}{a_1}\right)^{\frac{2}{\epsilon_2}} + \left(\frac{y}{a_2}\right)^{\frac{2}{\epsilon_2}} \right]^{\frac{\epsilon_2}{2}} - a \right)^{\frac{2}{\epsilon_1}} + \left(\frac{z}{a_3}\right)^{\frac{2}{\epsilon_1}} = 1
$$

The power of the superquadric representation lies not in its ability to model perfect geometric shapes but in its ability to model deformed geometric shapes. Gupta, Bogoni, and Bajcsy (1989) defined two deformations: *tapering* and *bending*. Linear tapering along the z-axis is given by the transformation

$$
x' = \left(\frac{k_x}{a_3} z + 1\right) x
$$

$$
y' = \left(\frac{k_y}{a_3} z + 1\right) y
$$

$$
z' = z
$$

where k_x and k_y ($-1 \le k_x, k_y \le 1$) are the tapering parameters with respect to the x- and y-planes, respectively, relative to the z-direction. The bending deformation is defined by the transformation

$$
x' = x + \cos(\alpha)(R - r)
$$

$$
y' = y + \sin(\alpha)(R - r)
$$

$$
z' = \sin(\gamma)(\frac{1}{k} - r)
$$

where k is the curvature, r is the projection of the x- and y-components onto the bending plane $z - r$ given by

$$r = \cos\left[\alpha - \tan^{-1}\left(\frac{y}{x}\right)\right]\sqrt{x^2 + y^2}$$

R is the transformation of r given by

$$R = k^{-1} - \cos(\gamma)(k^{-1} - r)$$

and γ is the bending angle

$$\gamma = zk^{-1}$$

Superquadric models are mainly for use with range data, since it is unclear how to match them to gray tone images, and several procedures for recovering the parameters of a superquadric fit to a surface have been proposed (Pentland, 1986; Bajcsy and Solina, 1987). Figure 18.13 illustrates the fitting procedures described by Ferrie, Lagarde, and Whaite (1989).

18.3.7   Octree Representation

Octree encoding is a geometric modeling technique used in computer vision, robotics, and computer graphics to represent arbitrary three-dimensional objects. An *octree* is a hierarchical 8-ary tree structure. Each node in the tree corresponds to a cubic region of the universe. The *label* of a node is either *full*, if the cube is completely enclosed by the three-dimensional object; *empty*, if the cube contains

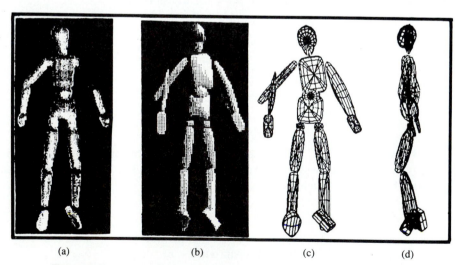

(a)   (b)   (c)   (d)

Figure 18.13 Range data image of (a) a doll, (b) its superquadric fit, and (c) and (d) two wire frame views of the superquadric fit (c, d). (From Ferrie, Lagarde, and Whaite, 1989.)

no part of the object; or *partial*, if the cube partly intersects the object. These labels correspond to the labels "black," "white," and "gray" that are often used in quadtrees, the two-dimensional version of octrees. A node with label *full* or *empty* has no children. A node with label *partial* has eight children, representing the partition of the cube into octants.

A three-dimensional object can be represented by a $2^n \times 2^n \times 2^n$ three-dimensional array for some integer n. The elements of the array are called *voxels* and have a value of 1 (full) or 0 (empty), indicating the presence or absence of the object. The octree encoding of the object is equivalent to the three-dimensional array representation, but will generally require much less space. Figure 18.14 gives a simple example of an object and its octree encoding, using the octant numbering scheme of Jackins and Tanimoto (1980). Octrees were originally proposed by Hunter (1978). Jackins and Tanimoto gave algorithms for 90° rotation and translation of objects represented by octrees, and described a Pascal program for octree manipulation. Meagher (1982) gave linear time algorithms for Boolean operations (union, intersection, and difference), geometric operations (translation, scaling, and rotation), N-dimensional interference detection, and display with hidden surface removal. Octrees are discussed in detail in Samet (1990).

18.3.8 The Extended Gaussian Image

Another way of thinking about a three-dimensional object is as a collection of surface normals, one at each point of the surface of the object. If a surface is planar, all the points on that surface map to the same surface normal. If a convex surface has

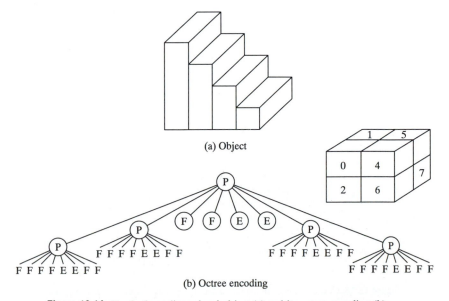

(a) Object

(b) Octree encoding

Figure 18.14 Simple three-dimensional object (a) and its octree encoding (b).

positive curvature everywhere, every point has a distinct surface normal. The set of surface normals of an object can be mapped to a unit sphere (called the Gaussian sphere) by placing the tail of the normal vector at the center of the sphere and allowing the head to pierce the surface of the sphere. The resultant set of points on the Gaussian sphere is called the *Gaussian image* of the object. Figure 18.15 illustrates these concepts.

For planar objects the Gaussian image is not invertible and does not represent the object precisely enough to use in most vision tasks. The Gaussian image of a planar object can be made more useful by mapping each planar surface to a pair consisting of the point on the Gaussian sphere pierced by the normal to the plane and a weight reflecting the area of the planar surface on the object. This idea can be extended to curved objects as follows:

Let δO be a small surface patch of the object and δS be the corresponding surface patch on the Gaussian sphere. The *Gaussian curvature K* is defined by

$$K = \lim_{\delta O \to 0} \frac{\delta S}{\delta O} = \frac{dS}{dO}$$

The *extended Gaussian image* is then given by

$$G(\xi, \eta) = \frac{1}{K(u, v)}$$

where (ξ, η) is a point on the Gaussian sphere that corresponds to the point (u, v) on the object surface. A planar region, whose Gaussian curvature is zero, corresponds to a point mass of the extended Gaussian image proportional to the area of the planar region.

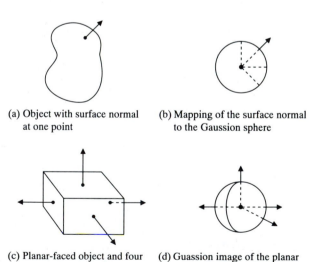

(a) Object with surface normal
at one point

(b) Mapping of the surface normal
to the Gaussion sphere

(c) Planar-faced object and four
of its surface normals

(d) Guassion image of the planar
object (including only the
four surface normals shown)

Figure 18.15 Concepts associated with the Gaussian image of a three-dimensional object.

The extended Gaussian image was proposed by Horn (1979, 1984) and has been used in recognition algorithms by Dane and Bajcsy (1981), Ikeuchi (1981), and Horn and Ikeuchi (1984). It is covered extensively in Horn's (1986) book.

18.3.9 View-Class Representation

All the object representations we have just discussed emphasize the three-dimensional nature of the objects, but ignore the problem of recognizing an object from a two-dimensional image taken from an arbitrary viewpoint. Most objects look different when seen from different viewpoints. A cylinder that projects to a ribbon (see above) in one set of viewpoints also projects to an ellipse in another set of viewpoints. In general, we can partition the space of viewpoints into a finite set of *view classes*, each representing a set of viewpoints that share some property (Korn and Dyer, 1987). The property may be that the same surfaces of the object are visible in an image taken from that set of viewpoints, the same line segments are visible, or the relational distances between relational structures extracted from line drawings at each of the viewpoints are similar enough. (See the following section on relational distance.) Figure 18.16 shows the view classes of a cube defined by grouping together those viewpoints that produce line drawings that are topologically isomorphic. The main point is that once the correct view class has been determined for an object, the matching to determine the correspondences necessary for pose determination is a highly constrained two-dimensional kind of matching.

The introduction of the view-class concept is generally credited to Koenderink and Van Doorn (1979). They proposed the use of a graph structure that has become known as the *aspect graph*. The aspect graph of an object is a graph structure in which (1) each node of the graph represents a topologically distinct view of the object as seen from some maximal connected cell of viewpoint space, (2) there is a node for each such view of the object, (3) each arc represents a *visual event* that occurs at the transition from one cell of viewpoint space to a neighboring cell, and (4) there is an arc for each such transition.

Algorithms to compute the view classes (or aspect graph) of an object can be formulated in a variety of ways. Currently the most general and widely used approach is to tessellate the Gaussian sphere, compute the view of the object for

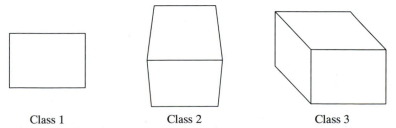

Class 1   Class 2   Class 3

Figure 18.16 The three view classes of a cube defined by grouping together viewpoints that produce topologically isomorphic line drawings.

each facet of the tessellation, and group together viewpoints that see topologically equivalent views. The resulting partition of viewpoint space is approximate, but no particular assumptions about the object geometry are required to create it. An example of this approach is given by Ikeuchi (1987), who uses the definition that the same faces of an object are visible in each member of a view class. A more common definition is that used by Chakravarty and Freeman (1982), who defined *characteristic views* as sets of viewpoints producing topologically isomorphic line drawings and represented each characteristic view by a vector containing the number of junctions of each type.

Several algorithms have been developed to compute the aspect graph from an exact partition of viewpoint space, but they are currently available only for restricted classes of object geometry. Solutions for polyhedra under the Gaussian sphere model of viewpoint space have been described by Plantinga and Dyer (1986), Gigus and Malik (1988), and Gigus, Canny, and Seidel (1988). Solutions for polyhedra under

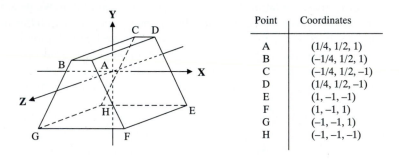

Point	Coordinates
A	(1/4, 1/2, 1)
B	(−1/4, 1/2, 1)
C	(−1/4, 1/2, −1)
D	(1/4, 1/2, −1)
E	(1, −1, −1)
F	(1, −1, 1)
G	(−1, −1, 1)
H	(−1, −1, −1)

(a) Truncated wedge and coordinates of its vertices

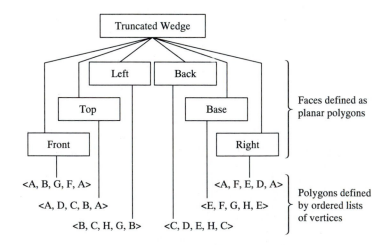

(b) Faces and vertices used in object definition

Figure 18.17 Surface-edge-vertex representation of a wedge.

the model of viewpoint space as three-dimensional space have been described by Plantinga and Dyer (1986) and Stewman and Bowyer (1987, 1988). A solution for solids of revolution under the Gaussian sphere model of viewpoint space has been described by Eggert and Bowyer (1989).

Figure 18.17 depicts the surface-edge-vertex representation of a simple polyhedral object, and Fig. 18.18 shows the aspect graph of the object. This aspect graph was constructed under the model of viewpoint space as all of three-dimensional space. The node of the aspect graph indicated by the arrow represents a view that

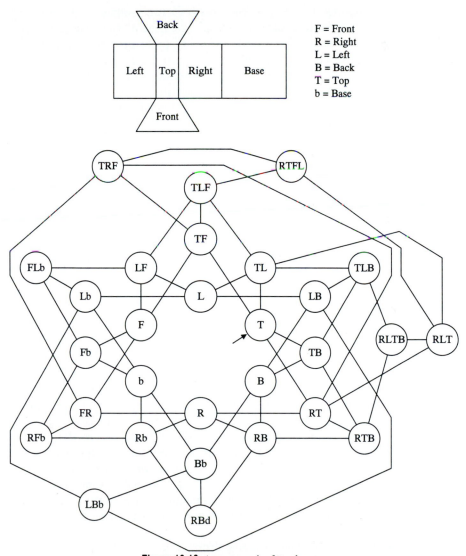

Figure 18.18 Aspect graph of a cube.

exists only for a finite-extent cell of space located just above the object. Aspect graphs created by using the Gaussian sphere model of viewpoint space do not include such views.

Note that different nodes of an aspect graph may represent topologically equivalent views of an object. Grouping together the different nodes that see topologically equivalent views essentially results in the *view classes* of the object.

18.4   General Frameworks for Matching

Matching means finding a correspondence between two entities. The consistent-labeling procedures given in Chapter 17 are all examples of matching algorithms. These algorithms are designed for a wide class of problems that fit a particular model. Thus the consistent-labeling problem and corresponding procedures constitute a general framework for matching. The statistical pattern recognition model of Chapter 3 is another general framework. Although everybody's model matching procedure is somewhat unique, there are several other approaches that we can classify as general. In this section we describe some frameworks for matching.

18.4.1   Relational-Distance Approach to Matching

In the previous chapter we defined an inexact version of the consistent-labeling problem that, unlike the exact problem, allowed for the accumulation of some relational error. This idea leads to the concept of a relational distance that can compare two structures and determine their relational similarity. Here we discuss the relational distance as a framework for matching.

Relational-Distance Definition

A *relational description* D_X is a sequence of relations $D_X = \{R_1, \ldots, R_I\}$, where for each $i = 1, \ldots, I$, there exists a positive integer n_i with $R_i \subseteq X^{n_i}$ for some set X. Intuitively X is a set of the parts of the entity being described, and the relations R_i indicate various relationships among the parts. A relational description is a data structure that may be used to describe two-dimensional shape models, three-dimensional object models, regions on an image, and so on. In the next chapter we discuss the use of relational descriptions in knowledge-based vision. Here we are concerned with defining a distance measure for pairs of relational descriptions.

Let $D_A = \{R_1, \ldots, R_I\}$ be a relational description with part set A and $D_B = \{S_1, \ldots, S_I\}$ a relational description with part set B. We will assume that $\mid A \mid = \mid B \mid$; if this is not the case, we will add enough dummy parts to the smaller set to make it so. The assumption is made in order to guarantee that the relational distance is a metric.

Let f be any one-one, onto mapping from A to B. For any $R \subseteq A^N$, N a positive integer, the *composition* $R \circ f$ of relation R with function f is given by

$$R \circ f = \{(b_1, \ldots, b_N) \in B^N \mid there\ exists\ (a_1, \ldots, a_N) \in R$$

$$with\ f(a_n) = b_n,\ n = 1, \ldots, N\}$$

This composition operator, which is the same one used in the relational homomorphisms of Chapter 17, takes N-tuples of R and maps them, component by component, into N-tuples of B^N.

The function f maps parts from set A to parts from set B. The *structural error* of f for the ith pair of corresponding relations (R_i and S_i) in D_A and D_B is given by

$$E_S^i(f) = \mid R_i \circ f - S_i \mid + \mid S_i \circ f^{-1} - R_i \mid$$

The structural error indicates how many tuples in R_i are not mapped by f to tuples in S_i and how many tuples in S_i are not mapped by f^{-1} to tuples in R_i. The structural error is expressed with respect to only one pair of corresponding relations.

The *total error* of f with respect to D_A and D_B is the sum of the structural errors for each pair of corresponding relations. That is,

$$E(f) = \sum_{i=1}^{I} E_S^i(f)$$

The total error gives a quantitative idea of the difference between the two relational descriptions D_A and D_B with respect to the mapping f.

The *relational distance* $GD(D_A, D_B)$ between D_A and D_B is then given by

$$GD(D_A, D_B) = \min_{\substack{1-1 \\ f: A \to B \\ onto}} E(f)$$

That is, the relational distance is the minimal total error obtained for any one-one, onto mapping f from A to B. We call a mapping f that minimizes total error a *best mapping* from D_A to D_B. If there is more than one best mapping, one can be arbitrarily selected as the designated best mapping. More than one best mapping will occur when the relational descriptions involve certain kinds of symmetries.

It is also possible to think of an N-ary relation $R \subseteq A^N$ as a bit vector. The bit vector has a position for each possible N-tuple $(a_1, \ldots, a_N) \in A^N$. Those positions that represent N-tuples of R have value 1, and the rest have value 0. The composition operation $R \circ f$ is achieved by a permutation of the bit vector of R, resulting in the new bit vector $R \circ f$. The structural error of a permutation $f : A \to B$ with respect to relations $R \subseteq A^N$ and $S \subseteq B^N$ is merely the number of 1-bits in the bit vector $(R \circ f) \oplus S$, where $\oplus$ stands for the exclusive-or operation. The total error and relational-distance definitions remain the same.

Relational-Distance Examples

Figure 18.19 shows two digraphs, each having four nodes. A best mapping from $A = \{1,2,3,4\}$ to $B = \{a,b,c,d\}$ is $\{f(1) = a, f(2) = b, f(3) = c, f(4) = d\}$. For this mapping we have

$$
\begin{aligned}
|R \circ f - S| &= |\{(1,2)(2,3)(3,4)(4,2)\} \circ f - \{(a,b)(c,b)(d,b)\}| \\
&= |\{(a,b)(b,c)(c,d)(d,b)\} - \{(a,b)(c,b)(d,b)\}| \\
&= |\{(b,c)(c,d)\}| \\
&= 2 \\
|S \circ f^{-1} - R| &= |\{(a,b)(c,b)(d,b)\} \circ f^{-1} - \{(1,2)(2,3)(3,4)(4,2)\}| \\
&= |\{(1,2)(3,2)(4,2)\} - \{(1,2)(2,3)(3,4)(4,2)\}| \\
&= |\{(3,2)\}| \\
&= 1 \\
E(f) &= |R \circ f - S| + |S \circ f^{-1} - R| \\
&= \quad\quad 2 \quad\quad + \quad\quad 1 \\
&= 3
\end{aligned}
$$

Since f is a best mapping, the relational distance is also 3.

Figure 18.20 gives a set of object models M_1, M_2, M_3, and M_4. The primitives are sticks, plates, and blobs, as described above. Two relations are shown in the figure: the connection relation (R_1, S_1) and the parallel relation (R_2, S_2). Both are binary relations over the set of primitives. Consider the first two models, M_1 and M_2. The best mapping f maps primitive 1 to $1'$, 2 to $2'$, and 3 to $3'$. Under this mapping the connection relations are isomorphic. The parallel relationship $(2,3)$ in model M_1 does not hold between $2'$ and $3'$ in model M_2. Thus the relational distance between M_1 and M_2 is exactly 1. Now consider models M_1 and M_3. The best mapping maps 1 to $1''$, 2 to $2''$, 3 to $3''$, and a dummy primitive to $4''$. Under this mapping, the parallel relations are now isomorphic, but there is one more connection in M_3 than in M_2. Again the relational distance is exactly 1.

Finally, consider models M_3 and M_4. The best mapping maps $1''$ to 1^*, $2''$ to 2^*, $3''$ to 3^*, $4''$ to 4^*, 5_d to 5^*, and 6_d to 6^*. (5_d and 6_d are dummy primitives.)

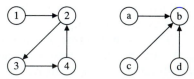

Figure 18.19 Two digraphs whose relational distance is 3.

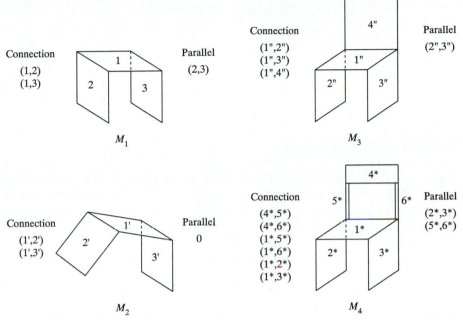

Figure 18.20 Four object models. The relational distance of model M_1 to M_2 and M_1 to M_3 is 1. The relational distance of model M_3 to M_4 is 6.

For this mapping we have

$$\begin{aligned}
|R_1 \circ f - S_1| &= |\{(1'',2'')(1'',3'')(1'',4'')\} \circ f - \\
&\quad \{(4^*,5^*)(4^*,6^*)(1^*,5^*)(1^*,6^*)(1^*,2^*)(1^*,3^*)\}| \\
&= |\{(1^*,2^*)(1^*,3^*)(1^*,4^*)\} - \\
&\quad \{(4^*,5^*)(4^*,6^*)(1^*,5^*)(1^*,6^*)(1^*,2^*)(1^*,3^*)\}| \\
&= |\{(1^*,4^*)\}| \\
&= 1 \\
|S_1 \circ f^{-1} R_1| &= |\{(4^*,5^*)(4^*,6^*)(1^*,5^*)(1^*,6^*)(1^*,2^*)(1^*,3^*)\} \circ f^{-1} - \\
&\quad \{(1'',2'')(1'',3'')(1'',4'')\}| \\
&= |\{(4'',5_d)(4'',6_d)(1'',5_d)(1'',6_d)(1'',2'')(1'',3'')\} - \\
&\quad \{1'',2'')(1'',3'')(1'',4'')\}| \\
&= |\{(4'',5_d)(4'',6_d)(1'',5_d)(1'',6_d)\}| \\
&= 4 \\
|R_2 \circ f - S_2| &= |\{(2'',3'')\} \circ f - \{(2^*,3^*)(5^*,6^*)\}| \\
&= |\{(2^*,3^*)\} - \{(2^*,3^*)(5^*,6^*)\}| \\
&= |\phi| \\
&= 0
\end{aligned}$$

$$|S_2 \circ f^{-1} - R_2| = |\{(2^*,3^*)(5^*,6^*)\} \circ f^{-1} - \{(2'',3'')\}|$$
$$= |\{(2'',3'')(5_d,6_d)\} - \{(2'',3'')\}|$$
$$= |\{(5_d,6_d)\}|$$
$$= 1$$
$$E_s^1(f) = 1 + 4 = 5$$
$$E_s^2(f) = 0 + 1 = 1$$
$$E(f) = 6$$

Relational Distance as a Metric

The relational distance can be used to determine the similarity of an unknown object to an object model. It can also be used to compare object models for the purpose of grouping models in a large database. Model grouping will be discussed in Section 18.4. Here we show that GD is a metric, which makes the grouping possible. The metric proof was originally given in Shapiro and Haralick (1985).

Let f be a one-one, onto function from A to B. We say that f is a *relational isomorphism* if $E(f) = 0$. In this case D_A and D_B are said to be isomorphic. Intuitively this means that the parts of D_A perfectly match the parts of D_B with respect to all required attributes and relationships. The following theorem establishes the metric property of GD.

Theorem: Let GD be the relational-distance measure, and let D_A, D_B, and D_C be arbitrary relational descriptions. Then

1. $GD(D_A,D_B) = 0$ if and only if D_A and D_B are isomorphic;
2. $GD(D_A,D_B) = GD(D_B,D_A)$;
3. $GD(D_A,D_B) \leq GD(D_A,D_C) + GD(D_C,D_B)$.

Proof:
1. If f is an isomorphism between D_A and D_B, then $E(f) = 0$. If $GD(D_A,D_B) = 0$, then there exists one-one, onto f with $E(f) = 0$. Thus f is an isomorphism between D_A and D_B.
2. $GD(D_A,D_B)$

$$= \min_f \sum_{i=1}^{I} |R_i \circ f - S_i| + |S_i \circ f^{-1} - R_i|$$

$$= \min_{f^{-1}} \sum_{i=1}^{I} |R_i \circ f^{-1} - S_i| + |S_i \circ (f^{-1})^{-1} - R_i|$$

$$= \min_{f^{-1}} \sum_{i=1}^{I} |S_i \circ f - R_i| + |R_i \circ f^{-1} - S_i|$$

$$= \min_f \sum_{i=1}^{I} |S_i \circ f - R_i| + |R_i \circ f^{-1} - S_i|$$

$$= GD(D_B,D_A)$$

3. Let $D_A = \{R_1, \ldots, R_I\}$, $D_B = \{S_1, \ldots, S_I\}$, and $D_C = \{T_1, \ldots, T_I\}$, where for each $i = 1, \ldots, I$, $R_i \subseteq A^{n_i}$, $S_i \subseteq B^{n_i}$, and $T_i \subseteq C^{n_i}$. Let $f_1 \subseteq A \times C$ be one-one, onto and the f_1 that minimizes $GD(D_A, D_C)$. Let $f_2 \subseteq C \times B$ be one-one, single valued and the f_2 that minimizes $GD(D_C, D_B)$.

Let $f : A \to B = f_1 \circ f_2$. Then f is one-one and onto and produces some error $E(f)$ with respect to D_A and D_B.

Let $x \in R_i \circ f_1 \circ f_2 - S_i$. Then $x \in R_i \circ f_1 \circ f_2$ and $x \notin S_i$. Since $x \in R_i \circ f_1 \circ f_2$ and f_2 is one-one and onto, there exists a unique $y \in R_i \circ f_1$ such that $\{y\} = \{x\} \circ f_2^{-1}$ and $\{x\} = \{y\} \circ f_2$. If $y \notin T_i$, then $y \in R_i \circ f_1 - T_i$. If $y \in T_i$, then $\{x\} = \{y\} \circ f_2$ is an element of $T_i \circ f_2$. Hence $x \in T_i \circ f_2 - S_i$.

Since for each $x \in R_i \circ f_1 \circ f_2 - S_i$ either $x \in T_i \circ f_2 - S_i$ or $y = x \circ f_2^{-1} \in R_i \circ f_1 - T_i$, we have

$$| R_i \circ f_1 \circ f_2 - S_i | \leq | R_i \circ f_1 - T_i | + | T_i \circ f_2 - S_i |$$

Thus

$$\sum_{i=1}^{I} | R_i \circ f_1 \circ f_2 - S_i | \leq \sum_{i=1}^{I} | R_i \circ f_1 - T_i | + \sum_{i=1}^{I} | T_i \circ f_2 - S_i |$$

Similarly we can show that

$$\sum_{i=1}^{I} | S_i \circ f_2^{-1} \circ f_1^{-1} - S_i | \leq \sum_{i=1}^{I} | T_i \circ f_1^{-1} - R_i | + \sum_{i=1}^{I} | S_i \circ f_2^{-1} - T_i |$$

Adding, we get

$$\sum_{i=1}^{I} | R_i \circ f_1 \circ f_2 - S_i | + | S_i \circ f_2^{-1} \circ f_1^{-1} - S_i |$$

$$\leq \sum_{i=1}^{I} | R_i \circ f_1 - T_i | + | T_i \circ f_1^{-1} - R_i | + \sum_{i=1}^{I} | T_i \circ f_2 - S_i | + | S_i \circ f_2^{-1} - T_i |$$

which says

$$E(f) \text{wrt } D_A \text{ and } D_B \leq GD(D_A, D_c) + GD(D_C, D_B)$$

But

$$GD(D_A, D_B) \leq E(f)$$

So

$$GD(D_A, D_B) \leq GD(D_A, D_C) + GD(D_C, D_B)$$

Thus the relational distance of two relational descriptions is a metric up to isomorphism.

Attributed Relational Descriptions and Relational Distance

The relational descriptions defined above describe relationships among parts, but not properties of parts, properties of the whole, or properties of these relationships. However, it is easy to extend both the concept of relational description and the definition of relational distance to include them. Intuitively an m-tuple of attributes added to an n-tuple of parts produces an $n + m$-tuple that specifies a relationship plus the properties of that relationship. If $n = 1$ and $m > 0$, each tuple lists a part and its properties. If $n = 0$, $m > 0$, and the relation has only one tuple, this is a property vector describing the global properties of the object. Formally the definitions change to the following:

Let X be a set of parts of object O_X and P be a set of property values. Generally we can assume P is the set of real numbers. An *attributed relation* over part set X with property value set P is a subset of $X^n \times P^m$ for some nonnegative integers n and m. An *attributed relational description* D_X is a sequence of attributed relations $D_X = \{R_1, \ldots, R_I\}$, where for each $i = 1, \ldots, I$, there exists a nonnegative integer n_i, a nonnegative integer m_i (where $n_i + m_i > 0$), and a property value set P_i with $R_i \subseteq X^{n_i} \times P_i^{m_i}$. For example, a binary parts connection relation $R \subseteq X^2$ can be extended to an attributed relation $R' \subseteq X^2 \times \Re$, $\Re$ the set of real numbers, where an attributed pair (x_1, x_2, a) specifies that part x_1 connects to part x_2 at angle a.

Consider an attributed relation $R \subseteq A^n \times P^m$ over some part set A and property value set P. Let $r \in R$ be an $n + m$-tuple having n parts followed by m property values. Let $S \subseteq B^n \times P^m$ be a second attributed relation over part set B and property value set P. Let $f : A \to B$ be a one-one, onto mapping from A to B. We define the composition $r \circ f$ of attributed tuple r with f by

$$r \circ f = \{(b_1, \ldots, b_n, p_1, \ldots, p_m) \in B^n \times P^m$$
$$| \text{ there exists } (a_1, \ldots, a_n, p_1, \ldots, p_m)$$
$$\in R \text{ with } f(a_i) = b_i, i = 1, \ldots, n\}$$

Assume that if $(b_1, \ldots, b_n, p_1 \ldots, p_m) \in S$ and $(b_1, \ldots, b_n, q_1, \ldots, q_m) \in S$, then $p_1 = q_1, \ldots, p_m = q_m$. That is, each n-tuple of parts has only one m-tuple of properties. The error of a tuple $t = (b_1, \ldots, b_n, p_1, \ldots, p_m)$ with respect to a relation $S \subseteq B^n \times P^m$ is given by

$$e(t,S) = \begin{cases} \text{norm\_dis}[(p_1, \ldots, p_m), (q_1, \ldots, q_m)] & \text{if } (b_1, \ldots, b_n, q_1, \ldots, q_m) \in S \\ 1 & \text{otherwise} \end{cases}$$

where norm_dis returns the Euclidean distance (or any other desired distance) between two vectors, normalized by dividing by some maximum possible distance. Thus $e(t,S)$ is a quantity between 0 and 1. Now we can extend the definition of the structural error of f for the ith pair of corresponding relations (R_i and S_i) to

$$E_s^i(f) = \sum_{r \in R_i} e(r \circ f, S_i) + \sum_{s \in S_i} e(s \circ f^{-1}, R_i)$$

Total error and relational distance are defined as given above.

18.4.2 Ordered Structural Matching

The relational distance can be determined by using a tree-search procedure with some kind of discrete or continuous relaxation, as discussed in Chapter 17. Because the tree search may require exponential computation time, it is not suitable for vision tasks that require a real-time response. In many two-dimensional computer vision problems, the spatial arrangement of the primitives allows the definition of an *ordering* on the primitives that greatly reduces the complexity of the search.

Suppose that we wish to compare an object represented by description D_A to another object represented by description D_B. Suppose that the primitive set A has the ordering $< a_1, a_2, \ldots, a_s >$ and that the primitive set B has the ordering $< b_1, b_2, \ldots, b_t >$. Suppose that during the matching process it is hypothesized that primitive a_i maps to primitive b_j. The ordering tells us that one of the following conditions holds:

1. a_{i+1} maps to b_{j+1}.
2. a_{i+1} maps to b_{j+k}, $k > 1$, and no primitive of D_A maps to any of b_{j+1}, b_{j+2}, $\ldots$, b_{j+k-1}.
3. a_{i+1} maps to no primitive of D_B,

where a_{i+1} is the next primitive (circularly) after a_i in the ordering. Thus, once a_i is mapped to b_j, the ordering can be used to find the correspondences between all the other primitives in polynomial time. Ordered structural matching was used in Shapiro, MacDonald, and Sternberg (1987) for two-dimensional shape matching using shape primitives extracted by operations of mathematical morphology. Syntactic pattern recognition algorithms that represent object models by grammars, that convert an object to a string of symbols, and that parse the string according to the grammar are also examples of ordered structural matching. See Bunke and Sanfeliu (1990).

18.4.3 Hypothesizing and Testing with Viewpoint Consistency Constraint

Another way to make relational matching more efficient is to find a limited number of correspondences between model and image; use those correspondences to *hypothesize* a transformation matrix that describes the position and orientation of the object with respect to the camera; *test* the hypothesis by projecting the model to the image plane via the transformation; and evaluate the goodness of the fit of the transformed model to the image. This hypothesize-and-test paradigm is made possible by the *viewpoint consistency constraint* (Lowe, 1987) that states:

> The locations of all projected model features in an image must be consistent with projection from a single viewpoint.

Thus, neglecting any effects of noise or mismatch, if the first few features matched dictate a particular viewpoint, then the remainder of the image must also match to the model under the same viewpoint, that is, the same pose parameters. This constraint has been used in a number of vision systems, including Lowe's own SCERPO system (Chapter 19), the range data work of Faugeras (1984) and of Grimson and Lozano-Perez (1984), the optimal-matching search work of Ben-Arie and Meiri (1987), and the ACRONYM system (Brooks, 1981).

As Lowe points out, though the viewpoint consistency constraint sounds very simple, it is not so simple in practice. The following problems must be solved in order for this technique to be used successfully.

1. How should the initial features to be matched be selected from the set of features that are or can be extracted from the image?
2. How many points of correspondence should be established before a transformation is hypothesized?
3. How can hidden lines be removed in the testing step when the object is three-dimensional?
4. How should the match of the projection of the model to the image be evaluated?

There are no generally agreed-upon answers to these questions, only some suggestions. With respect to problem 1, Lowe suggested the use of perceptual organization in feature selection. He advocated a grouping process based on proximity, collinearity, and parallelness of line segments. Bolles and Cain (1982) analyzed the set of objects to be matched in a preprocessing step (see above) to determine focus features and nearby related features that could be used to determine the correct object and viewpoint.

Problem 2 requires an error analysis and is related to the discussion in Chapter 14. Solving problem 3 requires a very simple hidden-line-removal algorithm or a parallel architecture. Problem 4 has been ignored by most vision researchers. Lowe (1987) suggested a probabilistic model for solving this problem. In comparing projected model edges to image edges, he considers two different kinds of matching error. The first type of error, which comes from inaccuracies in prediction, causes an unrelated image edge to appear close to a predicted model edge. Lowe modeled this type of error by assuming a random distribution of possibly matching image edges and calculating the probability that a random edge could match the predicted edge. The second type of error comes about when there are two closely competing matches in the image. Lowe suggested examining all the potential matches for each particular prediction in order to determine the ambiguity from which the probablility of selecting the wrong match can be calculated. Another verification analysis model, based on robust statistics, was suggested in Chapter 14.

18.4.4 View-Class Matching

When a three-dimensional object is represented by a view-class model, the matching can be divided into two stages: (1) determining the view class of the object and (2) determining the precise viewpoint within that view class.

Determining View Class

Depending on the complexity of the object and on the definition of view class used, there may be only a few view classes or hundreds. When the number of view classes is large, it is especially necessary to rule out most of the view classes rapidly before attempting any of the more detailed matching needed to determine the pose of the object. What is needed is a robust classification procedure. Several possible techniques have been explored in CAD-based vision systems. Ikeuchi (1987), working with range data, precomputes features of the visible surfaces for each view class and constructs a decision tree whose traversal can indicate the proper view class. The trees constructed in this way are not guaranteed to work if the data are not perfect. However, with proper training data all the techniques of statistical pattern recognition are available for robust decision tree construction (see Chapter 4).

Lu and Shapiro (1989) used an accumulator-based method. In the Lu and Shapiro system, a view class is represented by a hierarchical relational structure called the *relational pyramid,* in which primitives appear on the lowest level and each of the other levels represents relationships among entities from lower levels. In this particular implementation, the level-1 primitives were straight- and curved-line segments, the level-2 relations were junctions and loops, and the level-3 relations were adjacency, collinearity, and parallelness of junctions and loop-inside-loop relationships.

The full relational pyramid structure is for use in detailed matching for determining the exact pose of the object after the view class is identified. For rapid view-class identification, relational summaries were derived from the relational pyramids. If the relational pyramid has a relation R with c tuples $\{[(N_1, t_{1,j}), \ldots, (N_n, t_{n,j})] \mid j = 1, \ldots, c\}$ (where each $t_{i,j}$ is a tuple from a lower level of the pyramid and N_i is the name of the relation that tuple $t_{i,j}$ comes from), the summary has a corresponding relation R with a single tuple $[(N_1, \ldots, N_n), c]$ representing those c tuples. For example, if the collinear relation has four tuples of the form $[(FORK, f), (ARROW, a)]$, then the collinear summary relation has one tuple $[(FORK, ARROW), 4]$, indicating that there are four collinearity relationships between a fork junction and an arrow junction in line drawings of this view class.

An index structure allows direct access, given a summary tuple of the form $[(N_1, \ldots, N_n), c]$, to a list of all view classes that have this tuple in their summary structures. The on-line system keeps an evidence accumulator for each view class, initialized to zero. For exact matching, the system traverses the summary structure derived from the unknown view, and for each tuple in the summary structure, it adds one to the accumulators of all the view classes on the list attached to that tuple in the index. The view class or classes with maximal evidence are selected. For inexact matching, when considering summary tuple $[(N_1, \ldots, N_n), c]$, the system adds $e^{-k^2/2}$ to the accumulators of those view classes on lists attached to $[(N_1, \ldots, N_n), c + k]$ and $[(N_1, \ldots, N_n), c - k]$ for $k = 0, \ldots, K$, where K is the maximum amount of deviation allowed.

In this system, features and relationships detected in the unknown view of the object are used to accumulate evidence for each possible view class. The relationships used in this study were topological relationships among straight- and curved-line segments, but any kind of relationship could theoretically be used. Again the important point is to make the algorithm robust by using some kind of training data or theoretical analysis so that the reliability of the features to be used in evidence accumulation is known and the evidence accumulation procedure can make use of this knowledge.

Pose Determination within View Class

Once the view class has been determined, feature correspondences must be found to determine the pose. Because the view class has been identified, the exact set of line segments (or surfaces in three-dimensions) that should be visible in that view class and their spatial arrangement in that view class are known. Thus the matching algorithm in this phase can be very tightly constrained. However, it still must take into account that missing or extra features may occur in the image as the result of illumination or other environmental effects that were not taken into account in the models. There are two possible approaches. One is to use a general-purpose matching procedure for each view class, with the particular relationships and constraints of that view class used to prune the search for a solution. The relational pyramid structure discussed earlier is a hierarchical, relational structure that can constrain the feature matching at each possible level of the pyramid. A strategy that tries to use higher-level features first and propagate the results to lower levels of the pyramid can result in a fast match. The second possibility is to develop a customized procedure for each view class. A preprocessing program analyzes each view class, possibly using training data or theoretical analysis to determine reliable features, and then selects a sequence of features to look for in the matching. This approach has been implemented by Ikeuchi and Hong (1989).

18.4.5   Affine-Invariant Matching

Affine-invariant matching, originally proposed by Hummel and Wolfson (1988), makes use of local features and their relationships in an unusual way. We will describe the method using point features, but line and curve segments are also feasible (Lamdan, Schwartz, and Wolfson, 1988). A set of interest points of the object, determined on the basis of sharp convexities and deep concavities along the boundary of the objects, is first computed. The exact technique for computing the interest points depends on the class of objects and does not affect the method.

The interest points of flat objects can be modeled as a set $\mathcal{M} = \{(x_m, y_m, z_0)\}_{m=1}^M$ of points all lying in the $z = z_0$ plane. The perspective projection produces the observed image data points $O = \{(u_n, v_n)\}_{n=1}^N$, where

$$u_n = f\frac{r_{11}x_m + r_{12}y_m + r_{13}z_0 + t_1}{r_{31}x_m + r_{32}y_m + r_{33}z_0 + t_3}$$

$$v_n = f\frac{r_{21}x_m + r_{22}y_m + r_{23}z_0 + t_2}{r_{31}x_m + r_{32}y_m + r_{33}z_0 + t_3}$$

$\begin{pmatrix} r_{11} & r_{12} & r_{13} \\ r_{21} & r_{22} & r_{23} \\ r_{31} & r_{32} & r_{33} \end{pmatrix}$ is a rotation matrix relating the model reference frame to the

camera reference frame, $\begin{pmatrix} t_1 \\ t_2 \\ t_3 \end{pmatrix}$ is the translation of the object reference frame to

the camera reference frame, and f is the distance the image plane is in front of the center of perspectivity.

When the translation t_3 in the z-direction is large compared with $r_{31}x_m + r_{32}y_m$ for all m, then (u_m, v_m) can be written as

$$\begin{pmatrix} u_m \\ v_m \end{pmatrix} = \begin{pmatrix} r_{11} & r_{12} \\ r_{21} & r_{22} \end{pmatrix} \begin{pmatrix} x_m \\ y_m \end{pmatrix} + \begin{pmatrix} b_1 \\ b_2 \end{pmatrix}$$

In this case there exists an affine two-dimensional correspondence between two different images of the same planar object: Each point w in the model corresponds to a transformed point $Aw + b$ in the image, where A is a 2×2 (scaling, rotation, and skewing) matrix, and b is a two-dimensional (translation) vector. We assume that most of the model points are recognized as distinctive points in the image and that the image may contain other distinctive points that are unrelated to any model point. The problem is to recognize the object in the image and to find the affine transformation between the set of image points and the set of model points, so that the position and orientation of the object can be determined.

Affine Transformation of Points in a Plane

It is well known that a necessary and sufficient condition to define a plane uniquely is a set of three noncollinear points in space. Consequently the affine transformation of the plane is also uniquely defined by the transformation of three noncollinear points. Moreover, there is a unique map of any noncollinear triplet (here called a basis) in the plane to another noncollinear triplet; this mapping is defined by the affine transformation of the plane that contains the original triplet.

The most important observation is that for each noncollinear basis triplet, the coordinates of all other points in the plane, given in the coordinate system of the basis triplet, are affine invariant. If a, b, and c are three noncollinear points in a plane, each represented as a 2×1 vector, then any other point v, also represented by a 2×1 vector, with affine coordinates (ξ, η) with respect to basis $< a,b,c >$, will still have coordinates (ξ, η) if the entire plane undergoes the affine transformation T, assuming that the same triplet of transformed points $< Ta, Tb, Tc >$ is chosen as the basis.

The mathematical representation of a generic point v, in terms of its affine coordinates (ξ, η) and the basis triplet $< a,b,c >$ that defines the plane, is given by the following equation:

$$\begin{pmatrix} \xi \\ \eta \end{pmatrix} = (a - c \quad b - c)^{-1} (v - c)$$

Hence, given the affine-invariant coordinates and the basis, one may compute the given point by

$$v = \xi(a - c) + \eta(b - c) + c \tag{18.1}$$

Notice that if point v and its basis are transformed by T, its new affine coordinates are

$$
\begin{aligned}
\begin{pmatrix} \xi \\ \eta \end{pmatrix} &= [T(a - c) \quad T(b - c)]^{-1} \quad T(v - c) \\
&= [T(a - c \quad b - c)]^{-1} \quad T(v - c) \\
&= (a - c \quad b - c)^{-1} \quad T^{-1} \quad T(v - c) \\
&= (a - c \quad b - c)^{-1} \quad (v - c)
\end{aligned}
\tag{18.2}
$$

which clearly shows that its transformed coordinates are affine invariant.

The Hummel-Wolfson-Lamdan Matching Algorithm

Given that the affine transformation is uniquely defined as the transformation of three noncollinear points in the plane, one can try to match noncollinear triplets in the set of model interest points against noncollinear triplets in the set of scene interest points. The original algorithm (Hummel and Wolfson, 1988) consists of two major steps: a preprocessing step and a recognition step. The first step converts the model interest points into an affine-invariant model representation. The second step, the matching proper, performs the same basic task, but now for the observed image points, and tries to match model against image by using the affine representation. The following discussion describes these two steps in detail.

Preprocessing: In this off-line step, all possible combinations of noncollinear triplets from the model are used as possible bases for the planar transformation. That is, for each ordered set of three noncollinear points selected from the set of model interest points, the affine coordinates of each of the remaining points with respect to these three points as the basis are computed by using Eq. (18.1). Each time a pair of affine-transformed coordinates is computed, these values are quantized and used as an entry to a hash table, where the basis triplet and the model from which these coordinates came are recorded. This is done for as many models as needed. If new models have to be added to the database, they can be processed independently, so there is no need to recompute the hash table.

Recognition: In the on-line recognition step, we are given a set of interest points that represent the projection of the object (or objects) in the image. Starting with any ordered noncollinear basis triplet of image points, the transformed coordinates of each of the remaining image points are computed, just as in the preprocessing stage. Then each affine-invariant coordinate votes for the closest *(model, basis triplet)* tuples to it. For simplicity, in the following description of this voting, we assume that there is only one model. Let $[\xi(m;i,j,k),\eta(m;i,j,k)]$ be the affine-invariant coordinates of the mth model point with respect to a model basis $<i,j,k>$; let $[\alpha(n;a,b,c),\beta(n;a,b,c)]$ be the affine-invariant coordinates of the nth image point with respect to an image basis $<a,b,c>$; and let q be the quantization function. Then for each image basis $<a,b,c>$, the Hummel-Wolfson voting produces the count $V(i,j,k;a,b,c)$ for basis model $<i,j,k>$ defined by

$$V(i,j,k;a,b,c) = \#\{ \; n \mid \text{for some } m, \; q[\alpha(n;a,b,c),\beta(n;a,b,c)] =$$

$$q[\xi(m;i,j,k),\eta(m;i,j,k)]\}$$

The hash table mechanism is used to establish closeness by identity of quantized value. Affine-invariant coordinates are taken to be close if their quantization is identical. If they are identical, they will hash to the same bucket. In operation the quantized affine-invariant coordinates of each image point with respect to image basis $<a,b,c>$ accesses the hash table, where it may pick up a pointer to a model basis having some model point whose affine coordinates hash to the same bucket, if there is such a one. The votes are then added into an accumulator associated with the model basis. When all the votes have been cast, the algorithm checks to see if any model basis triplet scored high enough. If no model basis triplet achieves a high enough score, it means that the image basis triplet selected in the set of image interest points does not correspond to any triplet in the set (or sets, in the case of more than one model) of model interest points. Another ordered basis triplet in the image is then used, and the procedure is repeated until a certain basis triplet scores high enough (a match is declared to be found) or until all the possible combinations of image interest points have been tried as bases and no pair has a sufficient score (no match is found).

Shortcomings of the Affine-Invariant Matching Technique

The affine-invariant matching technique is mathematically sound in the noiseless case. However, it has a number of shortcomings in practice:

1. If the three noncollinear points selected as a basis are not numerically stable with respect to the other points, the coordinates of the transformed points are not reliable.

2. On real images the coordinates of the detected interest points are noisy. This causes the wrong bin of the hash table to be accessed and produces unreliable results.

3. Partial object symmetries may cause bins representing incorrect transformations to have counts as high or nearly as high as the bin representing the correct transformation.

The first problem can be solved by only choosing basis triplets whose defining triangle has large enough area (Costa et al., 1989). The solution to problems 2 and

3 is a variant of the procedure that uses an explicit noise model and replaces the quantized bins used in hashing with a distance calculation. The improved method follows:

An Explicit Noise Model and Optimal Voting

Let $< i, j, k >$ represent a model basis and let $(\hat{\alpha}_n, \hat{\beta}_n), n = 1, \dots, N - 3$ represent the affine-invariant coordinates of the observed image points with respect to a given image basis, which in this discussion we hold fixed. The question we wish to answer is, what is the best way for the observations $(\hat{\alpha}_n, \hat{\beta}_n), n = 1, \dots, N - 3$ to vote for a model basis $< i, j, k >$? For the votes to be meaningful, they should be related to

$$P\{< i, j, k > \mid (\hat{\alpha}_n, \hat{\beta}_n), n = 1, \dots, N - 3\}$$

Now by the definition of conditional probability,

$$P\{< i, j, k > \mid (\hat{\alpha}_n, \hat{\beta}_n), n = 1, \dots, N - 3\} =$$

$$\frac{P\{(\hat{\alpha}_n, \hat{\beta}_n), n = 1, \dots, N - 3 \mid < i, j, k >\} P\{< i, j, k >\}}{\sum_{<i', j', k'>} P\{(\hat{\alpha}_n, \hat{\beta}_n), n = 1, \dots, N - 3 \mid < i', j', k' >\} P\{< i', j', k' >\}}$$

If the prior probabilities for a model basis are equal, we have

$$P\{< i, j, k > \mid (\hat{\alpha}_n, \hat{\beta}_n), n = 1, \dots, N - 3\} =$$

$$\frac{P\{(\hat{\alpha}_n, \hat{\beta}_n), n = 1, \dots, N - 3 \mid < i, j, k >\}}{\sum_{<i', j', k'>} P\{(\hat{\alpha}_n, \hat{\beta}_n), n = 1, \dots, N - 3 \mid < i', j', k' >\}}$$

Concentrating on the key term $P\{(\hat{\alpha}_n, \hat{\beta}_n), n = 1, \dots, N - 3 \mid < i, j, k >\}$, we assume that the observed affine-invariant coordinates for those image points that arise from model points or that are unrelated to model points are conditionally independent of the given model basis $< i, j, k >$. Hence

$$P\{(\hat{\alpha}_n, \hat{\beta}_n), n = 1, \dots, N - 3 \mid < i, j, k >\} = \prod_{n=1}^{N-3} P\{(\hat{\alpha}_n, \hat{\beta}_n) \mid < i, j, k >\}$$

Since we desire to relate the observed affine-invariant coordinates to a model point, if they can be so related, we let $B_{<i,j,k>}$ be the set of model points excluding the basis points i, j, and k, and m be a point in that set, that is, $m \in B_{<i,j,k>}$. We then define the terms of the product by

$$P\{(\hat{\alpha}_n, \hat{\beta}_n) \mid < i, j, k >\}$$

$$= \sum_{m \in B_{<i,j,k>}} P\{(\hat{\alpha}_n, \hat{\beta}_n), m \mid < i, j, k >\}$$

$$= \sum_{m \in B_{<i,j,k>}} P\{(\hat{\alpha}_n, \hat{\beta}_n) \mid m, < i, j, k >\} P\{m \mid < i, j, k >\}$$

We assume that the conditional probability of a model point m occurring in the image is independent of basis $<i,j,k>$. Thus

$$P\{(\hat{\alpha}_n,\hat{\beta}_n) \mid <i,j,k>\} = \sum_{m \in B_{<i,j,k>}} P\{(\hat{\alpha}_n,\hat{\beta}_n) \mid m, <i,j,k>\}P\{m\}$$

so that

$$P\{(\hat{\alpha}_n,\hat{\beta}_n), n = 1,\dots,N-3 \mid <i,j,k>\}$$
$$= \prod_{n=1}^{N-3} \sum_{m \in B_{<i,j,k>}} P\{(\hat{\alpha}_n,\hat{\beta}_n) \mid m, <i,j,k>\}P\{m\}$$

Hence

$$P\{<i,j,k> \mid (\hat{\alpha}_n,\hat{\beta}_n), n = 1,\dots,N-3\}$$
$$= \prod_{n=1}^{N-3} \sum_{m \in B_{<i,j,k>}} \frac{P\{(\hat{\alpha}_n,\hat{\beta}_n) \mid m, <i,j,k>\}P\{m\}}{\Delta} \quad (18.3)$$

where Δ is a normalizing constant to make the probability sum to 1 when summed over all bases.

In order to model the probability density $P\{(\hat{\alpha}_n,\hat{\beta}_n) \mid m, <i,j,k>\}$, two distinct cases must be taken into account: first, the case when the observed coordinates $(\hat{\alpha}_n,\hat{\beta}_n)$ arise from a point that appears in the model; and second, the case when the observed coordinates $(\hat{\alpha}_n,\hat{\beta}_n)$ arise from a point that does not appear in the model. In the second case the image point giving rise to affine coordinates $(\hat{\alpha}_n,\hat{\beta}_n)$ is an extraneous point. If we define a Bernoulli random variable, y, given by

$$y = \begin{cases} 1; & \text{if } (\hat{\alpha}_n,\hat{\beta}_n) \text{ comes from a point appearing in the model} \\ 0; & \text{otherwise} \end{cases}$$

the probability density given above can be written as

$$P\{(\hat{\alpha}_n,\hat{\beta}_n) \mid m, <i,j,k>\} = P\{(\hat{\alpha}_n,\hat{\beta}_n), y = 1 \mid m, <i,j,k>\}+$$
$$P\{(\hat{\alpha}_n,\hat{\beta}_n), y = 0 \mid m, <i,j,k>\} \quad (18.4)$$

Taking into account that neither y nor the probability of observing affine coordinates $(\hat{\alpha}_n,\hat{\beta}_n)$ that do not arise from the model depends on the model basis $<i,j,k>$ or the model point m, we can rewrite the two terms on the right-hand side of Eq. (18.4) as

$$P\{(\hat{\alpha}_n,\hat{\beta}_n), y = 1 \mid m, <i,j,k>\}$$
$$= P\{(\hat{\alpha}_n,\hat{\beta}_n) \mid y = 1, m, <i,j,k>\}P\{y = 1 \mid m, <i,j,k>\}$$
$$= P\{(\hat{\alpha}_n,\hat{\beta}_n) \mid y = 1, m, <i,j,k>\}P\{y = 1\}$$
$$(18.5)$$

and

$$P\{(\hat{\alpha}_n, \hat{\beta}_n), y = 0 \mid m, <i,j,k> \}$$
$$= P\{(\hat{\alpha}_n, \hat{\beta}_n) \mid y = 0, m, <i,j,k> \} P\{y = 0 \mid m, <i,j,k> \}$$
$$= P\{(\hat{\alpha}_n, \hat{\beta}_n) \mid y = 0\} P\{y = 0\}$$

(18.6)

Now let $P\{y = 1\} = q$. Consequently $P\{y = 0\} = 1 - q$. Also let

$$P\{(\hat{\alpha}_n, \hat{\beta}_n) \mid y = 1, m, <i,j,k> \} = D_1$$

and

$$P\{(\hat{\alpha}_n, \hat{\beta}_n) \mid y = 0\} = D_2$$

Equation (18.4) can then be written as

$$P\{(\hat{\alpha}_n, \hat{\beta}_n) \mid m, <i,j,k> \} = D_1 \, q + D_2 \, (1 - q) \qquad (18.7)$$

The quantity q is the probability that the observed affine coordinates $(\hat{\alpha}_n, \hat{\beta}_n)$ come from a point that appears in the model. Letting z be a variable denoting the number of model points that might appear in the image, we can write q as

$$q = \sum_z P\{y = 1, z\}$$

By definition of conditional probability, the following immediately results:

$$q = \sum_z P\{y = 1 \mid z\} P\{z\} \qquad (18.8)$$

Let r be the probability that a model point appears in the image, M be the number of model points, and N be the number of image points. Then in Eq. (18.8), $P\{y = 1|z\}$ is given by the following binomial distribution:

$$P\{y = 1 \mid z\} = \binom{M}{z} r^z (1 - r)^{M-z}$$

and $P\{z \mid <i,j,k> \} = z/N$. Rewriting Eq. (18.8), we obtain

$$q = \sum_{z=0}^{M} \frac{z}{N} \binom{M}{z} r^z (1 - r)^{M-z}$$

But $\sum_{z=0}^{M} z \binom{M}{z} r^z (1 - r)^{M-z}$ is the expected value of a binomial distribution having parameters M and r, and it is equal to Mr. Therefore q is finally given by

$$q = \frac{M \, r}{N} \qquad (18.9)$$

To determine r, we estimate the number L of model points that are not likely to appear in the image. Then r can be estimated by $r = M - L/M$.

The two densities D_1 and D_2 in Eq. (18.7) can be modeled as follows: D_2 denotes the probability of observing the affine coordinates $(\hat{\alpha}_n, \hat{\beta}_n)$ that do not arise from the model. Therefore we take it to be given by $D_2 = C$, where C is a constant related to the area of the affine plane in which we expect values.

Let (α_m, β_m) be the affine-invariant coordinates of the mth model point with respect to basis $<i, j, k>$. In this case we assume normal errors:

$$D_1 = \frac{1}{2\pi |\Sigma_m|^{\frac{1}{2}}} \, exp\left\{ -\frac{1}{2} \begin{pmatrix} \hat{\alpha}_n - \alpha_m \\ \hat{\beta}_n - \beta_m \end{pmatrix}' \Sigma_m^{-1} \begin{pmatrix} \hat{\alpha}_n - \alpha_m \\ \hat{\beta}_n - \beta_m \end{pmatrix} \right\} \tag{18.10}$$

To determine the covariance Σ_m, we note that the affine-invariant coordinates $(\hat{\alpha}, \hat{\beta})$ of a point (v_x, v_y) with respect to basis $<a, b, c>$ is given by

$$\hat{\alpha} = \frac{(v_x - c_x)(b_y - c_y) - (b_x - c_x)(v_y - c_y)}{(a_x - c_x)(b_y - c_y) - (b_x - c_x)(a_y - c_y)} \tag{18.11}$$

and

$$\hat{\beta} = \frac{(a_x - c_x)(v_y - c_y) - (v_x - c_x)(a_y - c_y)}{(a_x - c_x)(b_y - c_y) - (b_x - c_x)(a_y - c_y)} \tag{18.12}$$

indicating that each of them is a function of eight random variables $a_x, b_x, c_x, v_x,$ $a_y, b_y, c_y,$ and v_y. Thus we can write

$$\hat{\alpha} = f(x_1, x_2, x_3, x_4, x_5, x_6, x_7, x_8)$$

and

$$\hat{\beta} = g(x_1, x_2, x_3, x_4, x_5, x_6, x_7, x_8) \tag{18.13}$$

where each one of the eight random variables x_i can be expressed as

$$x_i = \bar{x}_i + \xi_i$$

where $\bar{x}_i$ is the true unknown value of x_i and ξ_i is a random perturbation (noise) added to $\bar{x}_i$. Assuming that the ξ_i are independent and identically distributed with zero mean and standard deviation σ_ξ, we write

$$E[\xi_i] = 0 \quad \text{and} \quad E[\xi_i^2] = \sigma_\xi^2 \tag{18.14}$$

Therefore Eq. (18.13) can be rewritten as

$$\hat{\alpha} = f(\bar{x}_i + \xi_i : i = 1, \dots, 8) \quad \text{and} \quad \hat{\beta} = g(\bar{x}_i + \xi_i : i = 1, \dots, 8)$$

If we linearize these functions by expanding them in a Taylor series and neglecting second- and higher-order terms, we obtain

$$\hat{\alpha} = f(\bar{x}_1, \dots, \bar{x}_8) + \sum_{i=1}^{8} \xi_i \frac{\partial}{\partial \bar{x}_i} f(\bar{x}_1, \dots, \bar{x}_8) \tag{18.15}$$

and

$$\hat{\beta} = g(\bar{x}_1, \ldots, \bar{x}_8) + \sum_{i=1}^{8} \xi_i \frac{\partial}{\partial \bar{x}_i} g(\bar{x}_1, \ldots, \bar{x}_8) \tag{18.16}$$

In these equations note that

$$f(\bar{x}_1, \ldots, \bar{x}_8) = \alpha \qquad \text{and} \qquad g(\bar{x}_1, \ldots, \bar{x}_8) = \beta$$

since the values $\bar{x}_i$ are the coordinates of the points without noise (model).

In order to find the covariance matrix associated with $\hat{\alpha}$ and $\hat{\beta}$, we need to find their first and second moments. From Eq. (18.15) we can calculate the expected value of $\hat{\alpha}$ as follows:

$$
\begin{aligned}
E[\hat{\alpha}] &= E[f(\bar{x}_1, \ldots, \bar{x}_8)] + E[\sum_{i=1}^{8} \xi_i \frac{\partial}{\partial \bar{x}_i} f(\bar{x}_1, \ldots, \bar{x}_8)] \\
&= E[f(\bar{x}_1, \ldots, \bar{x}_8)] + \sum_{i=1}^{8} E[\xi_i] \frac{\partial}{\partial \bar{x}_i} f(\bar{x}_1, \ldots, \bar{x}_8) \\
&= E[f(\bar{x}_1, \ldots, \bar{x}_8)] \\
&= E[\alpha] \\
&= \alpha
\end{aligned}
\tag{18.17}
$$

The variance of $\hat{\alpha}$ is given by

$$\sigma_{\hat{\alpha}}^2 = E[(\hat{\alpha} - E[\hat{\alpha}])^2] \tag{18.18}$$

From this point on, for simplicity, the functions $f(\bar{x}_1, \ldots, \bar{x}_8)$ and $g(\bar{x}_1, \ldots, \bar{x}_8)$ will be indicated in a vector notation by $f(\bar{X})$ and $g(\bar{X})$, respectively. Substituting Eqs. (18.15) and (18.17) into Eq. (18.18) and using Eq. (18.14), we can calculate the expression for $\sigma_{\hat{\alpha}}^2$ as follows:

$$
\begin{aligned}
\sigma_{\hat{\alpha}}^2 &= E\{[\sum_{i=1}^{8} \xi_i \frac{\partial}{\partial \bar{x}_i} f(\bar{X})]^2\} \\
&= E[\sum_{i=1}^{8} \xi_i \frac{\partial}{\partial \bar{x}_i} f(\bar{X}) \sum_{j=1}^{8} \xi_j \frac{\partial}{\partial \bar{x}_j} f(\bar{X})] \\
&= \sum_{i=1}^{8} \sum_{j=1}^{8} E[\xi_i \frac{\partial}{\partial \bar{x}_i} f(\bar{X}) \xi_j \frac{\partial}{\partial \bar{x}_j}] f(\bar{X}) \\
&= \sum_{i=1}^{8} \sum_{j=1}^{8} \frac{\partial}{\partial \bar{x}_i} f(\bar{X}) \frac{\partial}{\partial \bar{x}_j} f(\bar{X}) E[\xi_i \xi_j]
\end{aligned}
\tag{18.19}
$$

Assuming that the noise distribution for all point coordinates is the same, that is, they have the same mean (zero) and same standard deviation σ_ξ, and knowing that

$$E[\xi_i \xi_j] = \begin{cases} \sigma_\xi^2 & ; \quad i = j \\ 0 & ; \quad i \neq j \end{cases} \tag{18.20}$$

since ξ_i are independent random variables, we can rewrite Eq. (18.19) as

$$\sigma_{\hat{\alpha}}^2 = \sum_{i=1}^{8} \left(\frac{\partial}{\partial \bar{x}_i} f(\bar{X}) \right)^2 E[\xi_i^2]$$

Hence

$$\sigma_{\hat{\alpha}}^2 = \sigma_\xi^2 \sum_{i=1}^{8} \left(\frac{\partial}{\partial \bar{x}_i} f(\bar{X}) \right)^2$$

Following the same procedure for $\hat{\beta}$, we can show that the expected value and variance of $\hat{\beta}$ are given by

$$E[\hat{\beta}] = \beta \tag{18.21}$$

and

$$\sigma_{\hat{\beta}}^2 = \sigma_\xi^2 \sum_{i=1}^{8} \left(\frac{\partial}{\partial \bar{x}_i} g(\bar{X}) \right)^2 \tag{18.22}$$

The computation of the covariance of the two random variables $\hat{\alpha}$ and $\hat{\beta}$ is also similar and is given below. The covariance of $\hat{\alpha}$ and $\hat{\beta}$ is given by

$$\sigma_{\hat{\alpha},\hat{\beta}}^2 = E\{ [\hat{\alpha} - E(\hat{\alpha})] [\hat{\beta} - E(\hat{\beta})] \} \tag{18.23}$$

Substituting Eqs. (18.13), (18.14), (18.15), and (18.19) into Eq. (18.23) yields

$$\sigma_{\hat{\alpha},\hat{\beta}}^2 = E\{ [\sum_{i=1}^{8} \xi_i \frac{\partial}{\partial \bar{x}_i} f(\bar{X})][\sum_{j=1}^{8} \xi_j \frac{\partial}{\partial \bar{x}_j} g(\bar{X})] \}$$

$$= E[\sum_{i=1}^{8} \sum_{j=1}^{8} \xi_i \xi_j \frac{\partial}{\partial \bar{x}_i} f(\bar{X}) \frac{\partial}{\partial \bar{x}_j} g(\bar{X})]$$

$$= \sum_{i=1}^{8} \sum_{j=1}^{8} \frac{\partial}{\partial \bar{x}_i} f(\bar{X}) \frac{\partial}{\partial \bar{x}_j} g(\bar{X}) E[\xi_i \xi_j]$$

$$= \sum_{i=1}^{8} \frac{\partial}{\partial \bar{x}_i} f(\bar{X}) \frac{\partial}{\partial \bar{x}_i} g(\bar{X}) E[\xi_i^2]$$

and hence

$$\sigma_{\hat{\alpha},\hat{\beta}}^2 = \sigma_\xi^2 \sum_{i=1}^{8} \frac{\partial}{\partial \bar{x}_i} f(\bar{X}) \frac{\partial}{\partial \bar{x}_i} g(\bar{X}) \qquad (18.24)$$

The covariance matrix Σ associated with $\hat{\alpha}$ and $\hat{\beta}$ is given by

$$\Sigma = \begin{pmatrix} \sigma_{\hat{\alpha}}^2 & \sigma_{\hat{\alpha},\hat{\beta}}^2 \\ \sigma_{\hat{\alpha},\hat{\beta}}^2 & \sigma_{\hat{\beta}}^2 \end{pmatrix} = \sigma_\xi^2 \begin{pmatrix} A & C \\ C & B \end{pmatrix} \qquad (18.25)$$

The values A, B, C, and D can be calculated by computing the partial derivatives of the functions defined in Eqs. (18.11) and (18.12), and they are given by

$$A = \frac{1}{D^2}[(b_y - c_y)^2 + (v_y - b_y)^2 + (v_x - c_x)^2 + (b_x - v_x)^2 +$$
$$(c_y - v_y)^2 + (c_x - b_x)^2] +$$
$$\frac{N_1^2}{D^4}[(b_y - a_y)^2 + (c_x - a_x)^2 + (a_x - b_x)^2 + (a_y - c_y)^2 +$$
$$(c_y - b_y)^2 + (b_x - c_x)^2] +$$
$$\frac{2N_1}{D^3}[(v_y - b_y)(b_y - a_y) + (v_x - c_x)(c_x - a_x) +$$
$$(a_x - b_x)(b_x - v_x) + (a_y - c_y)(c_y - v_y)]$$

$$B = \frac{1}{D^2}[(c_y - a_y)^2 + (a_y - v_y)^2 + (v_x - a_x)^2 + (a_x - c_x)^2 +$$
$$(v_y - c_y)^2 + (c_x - v_x)^2] +$$
$$\frac{N_2^2}{D^4}[(b_y - a_y)^2 + (c_x - a_x)^2 + (a_x - b_x)^2 + (a_y - c_y)^2 +$$
$$(c_y - b_y)^2 + (b_x - c_x)^2] +$$
$$\frac{2N_2}{D^3}[(b_y - a_y)(a_y - v_y) + (v_x - a_x)(a_x - b_x) +$$
$$(v_y - c_y)(c_y - b_y) + (b_x - c_x)(c_x - v_x)]$$

$$C = \frac{1}{D^2}[(b_y - c_y)(c_y - a_y) + (a_y - v_y)(v_y - b_y) +$$
$$(b_x - v_x)(v_x - a_x) + (a_x - c_x)(c_x - b_x)] +$$
$$\frac{N_1}{D^3}[(b_y - a_y)(a_y - v_y) + (v_x - a_x)(a_x - b_x) +$$
$$(v_y - c_y)(c_y - b_y) + (b_x - c_x)(c_x - v_x)] +$$
$$\frac{N_2}{D^3}[(v_y - b_y)(b_y - a_y) + (v_x - c_x)(c_x - a_x) +$$
$$(a_x - b_x)(b_x - v_x) + (a_y - c_y)(c_y - v_y)] +$$
$$\frac{N_1 N_2}{D^4}[(b_y - a_y)^2 + (c_x - a_x)^2 + (a_x - b_x)^2 + (a_y - c_y)^2 +$$
$$(c_y - b_y)^2 + (b_x - c_x)^2]$$

where

$$D = c_x a_y - a_y b_x + b_x c_y - c_y a_x + a_x b_y - b_y c_x$$

$$N_1 = v_x b_y - b_y c_x + c_x v_y - v_y b_x + b_x c_y - c_y v_x$$

$$N_2 = v_x c_y - c_y a_x + a_x v_y - v_y c_x + c_x a_y - a_y v_x$$

To use this optimal voting procedure, the voting accumulators must keep track of model point m, image point n, and basis triplet $< i, j, k >$. The actual vote V is a computed probability density:

$$V(< i, j, k >, m, n) = P\{(\hat{\alpha}_n, \hat{\beta}_n) | m, < i, j, k >\} P\{m\}$$

where $P\{m\} = 1/(M - 3)$. Notice that in this formulation there is no required quantization of the affine coordinates nor any explicit statement requiring that closeness of image affine coordinates with model affine coordinates be determined through a hashing function. However, to save computation, a hashing function can be used to retrieve an explicit set of linked lists of model bases and affine-invariant coordinates with respect to a model basis of all model points lying close enough to the affine-invariant coordinates of an image point. In this case the computation for $P\{(\hat{\alpha}_n, \hat{\beta}_n) | m, < i, j, k >\} = P\{(\hat{\alpha}_n, \hat{\beta}_n) | (\alpha_m, \beta_m), \Sigma_m, < i, j, k >\}$, as given by Eq. (18.3), is done only for the bases and model points on these linked lists. The criteria used for selecting the linked lists are as follows: Lists that are considered to contain model bases for which $P\{(\hat{\alpha}_n, \hat{\beta}_n) | m, < i, j, k >\}$ is not significantly different from zero must be associated with buckets in the hash table that overlap with a given elliptical cross section of the Gaussian density of Eq. (18.3). This cross section is chosen to be the one for which the value of the density is equal to 90% of its peak value. To speed the voting process, all the lists associated with the square neighborhood of buckets that encloses the selected ellipse are used for voting.

18.5 Model Database Organization

In many real application areas, such as navigation or industrial vision, a robot vision system must be able to recognize a large set of objects from their stored models. While many experimental systems perform recognition by sequentially trying each model in the database, this approach is not efficient for a system that (a) has a large database of models and (b) uses a fairly complex recognition algorithm. Thus an important problem for robot vision systems is to organize the database of models in a way that allows rapid access to the most likely candidate models. Several approaches to this problem have already been given here in other contexts. For example, the original affine-invariant matching technique just discussed stores (model, basis) pairs for all possible models in its hash table structure. If there are m models, each with an average of n interest points, the total number of such pairs stored in the database is

$$(n) \frac{m!}{(m - 4)!}$$

which can grow quite large.

If the object models can be characterized by their global attributes, then, like the view-class determination discussed earlier, the object classes can be selected by some type of decision tree or other statistical or syntactic classifier. Such classification may be very simple, but very helpful. For example, Nevatia and Binford (1977) used an indexing scheme to access generalized cylinder models likely to match an unknown object. Each model had a three-bit code describing its distinguished (usually biggest) piece. The encoded attributes were (1) connectivity (one end or both),

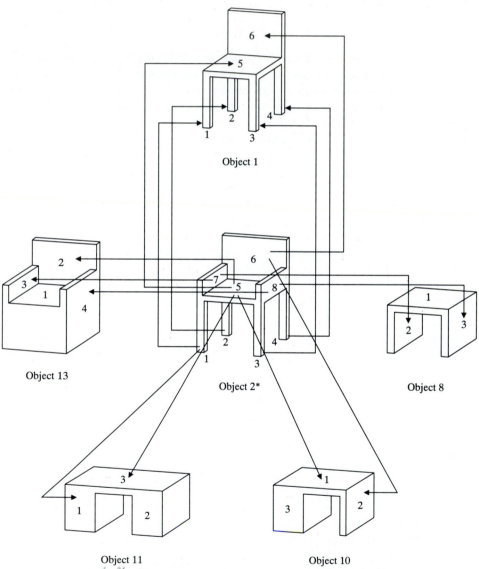

Figure 18.21 Cluster of object models whose representative is object 2. The arrows indicate the mappings from parts of object 2 to parts of the other objects obtained in the calculation of relational distance.

(2) type (long or wide), and (3) conical (true or false). Objects having the same code were grouped together, and the correct group was found before full relational matching was attempted. Schneier (1979) employed a structure called the graph of models, in which common primitives and relations were shared across and within models. In his matching process, primitives and relations index all models in which they occur, and models index all primitives and relations within them.

The representations and techniques described in this chapter were designed to be used in vision systems, computer programs that can recognize objects, determine their positions and orientations, and aid in inspection, manipulation, or assembly tasks. Design of vision systems involves the determination of the knowledge that the system will require for its particular task. In the next chapter we discuss the topic of knowledge-based vision.

Shapiro and Haralick (1982) developed a scheme for organizing relational models based on the relational-distance metric described in this chapter. The idea is to group similar relational models into clusters and to choose a representative for each cluster. An unknown object would be compared with the cluster representatives and then with the models in those clusters deemed similar enough. Clustering can be done by any clustering algorithm that can work with distances between objects (as opposed to points in n-dimensional space). The representative should be a relational description within the cluster that somehow best represents that cluster. Suppose that a cluster C has been constructed. For any relational description D_i in C, define the total distance of D_i with respect to C by

$$T(D_i, C) = \sum_{D \in C} GD(D_i, D)$$

The relational description D_b that satisfies

$$T(D_b, C) = \min_{D \in C} T(D, C)$$

is used as the representative of the cluster. If the clusters are large, this entire process can be repeated to create a hierarchical structure (or tree) of clusters and representatives. For a one-level clustering of furniture object models, Fig. 18.21 illustrates a cluster of similar objects and the mappings from the parts of the representative object to the parts of the other objects in the cluster that were obtained during the calculation of relational distance.

■ Exercises

18.1.   Let $S = \{(x, y) | f(x, y) = 1\}$ be a set of points in a binary image representing a two-dimensional shape. Let $S^* = \{x^*, y^*) | x^* = ax + b, y^* = cy + d, (x, y) \in S\}$ be a set of points obtained by "stretching" or "squeezing" set S. (a) Prove that $m_{00}(S^*) = m_{00}(S)$, that is, the $(0,0)$th normalized moment of S^* is equal to the $(0,0)$th moment of S. (b) Prove that $\bar{x}(S^*) = a\bar{x}(S) + b$. (c) Prove that

$$\sigma_x(S^*) = a\sigma_x(S)$$
$$\sigma_y(S^*) = c\sigma_y(S)$$

18.2.   Show that Hu's first three functions $\phi(1), \phi(2)$, and $\phi(3)$ are rotation invariant.

18.3.   Apply the Phillips and Rosenfeld method to find interest points in the sequence of
points   <   (141, 365), (142, 366), (140, 367), (141, 368), (142, 368), (143, 369),
(141, 370), (142, 371), (143, 372), (144, 373), (142, 374), (143, 375), (144, 375),
(142, 376), (141, 376), (143, 377), (143, 378), (143, 379), (145, 379), (144, 380),
(143, 381), (144, 382), (145, 382), (146, 382), (147, 381), (148, 382), (149, 383),
(149, 381), (150, 380), (151, 381), (152, 382), (153, 381), (154, 383), (155, 381),
(156, 382), (157, 383), (158, 382), (159, 381), (160, 380), (161, 381), (162, 382),
(163, 381) > for $k = 5$ and $t(k) = 1.5$.

18.4.   Show that an affine transformation takes lines into lines.

18.5.   Show that the ratio of the length of two lines is invariant under affine transformation.
How can this be used to help the matching process?

18.6.   Figure 18.22 shows three primary views of a three-dimensional object and one oblique
view. What three-dimensional features might be used to recognize the object and
determine its pose?

18.7.   Draw the complete three-dimensional wire frame representation, labeling each edge
for the object of the previous exercise.

18.8.   Extend your answer from the previous exercise to a full surface-edge-vertex representation.

18.9.   What does the stick-plates-and-blobs model of the object of Exercise 18.6 consist of?

18.10. Represent the object of Exercise 18.6 as a union of generalized cylinders. Is this
representation unique?

18.11. Assuming the origin and lengths of an object are as shown in Fig. 18.23, give an
octree encoding of the object.

18.12. Show the Gaussian image of the object of Exercise 18.11.

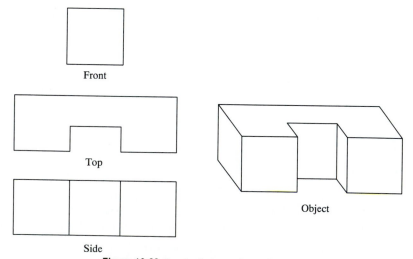

Front

Top

Side

Object

Figure 18.22 Standard views of a single object.

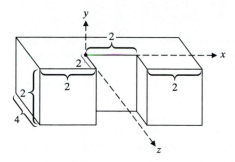

Figure 18.23 Dimensions for the object in Fig. 18.22.

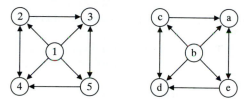

Figure 18.24 Two graphs.

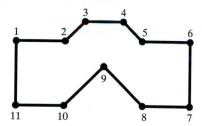

Figure 18.25 Two-dimensional polygonal shape. The coordinates of its vertices are:
1. (1,4) 2. (3,4) 3. (4,3) 4. (5,5) 5. (6,4) 6. (8,4) 7. (8,1) 8. (6,1)
9. (4.5,3) 10. (3,1) 11. (1,1)

18.13. Draw six different view classes of the object of Exercise 18.11.

18.14. What are the best mapping and relational distance between the two graphs of Fig. 18.24?

18.15. Construct a relational pyramid representation of the view class shown in Exercise 18.12 using the following relations:

Level 1	straight-line segments
Level 2	fork junctions, arrow junctions, L junctions, T junctions
Level 3	adjacency of two junctions (two junctions are adjacent if connected by a line segment)

18.16. For the two-dimensional shape of Fig. 18.25, using vertices 3, 7, and 11 as a basis, construct the affine representation of all the other points.

■ Bibliography

Abu-Mostafa, Y. S., and D. Psaltis, "Recognitive Aspects of Moment Invariants," *IEEE Transactions on Pattern Analysis and Machine Intelligence,* Vol. PAMI-6, 1984, pp. 698–706.

Aggarwal, J. K., et al., "Survey: Representation Methods for Three-Dimensional Objects," *Progress in Pattern Recognition,* 1981, pp. 377–391.

Agin, G. J., and T. O. Binford, "Computer Description of Curved Objects," *IEEE Transactions on Computers*, Vol. 25, 1976, pp. 439–440.

Alt, F. L., "Digital Pattern Recognition by Moments," *Journal of the ACM*, Vol. 11, 1962, pp. 240–258.

Anderson, R. L., "Real-Time Gray-Scale Video Processing Using a Moment-Generating Chip," *IEEE Journal of Robotics and Automation,* Vol. RA-1, 1985, pp. 70–85.

Baer, A., C. Eastman, and M. Henrion, "Geometric Modeling: A Survey," *Computer Aided Design,* 1979, pp. 1–20.

Bajcsy, R., and F. Solina, "Three-Dimensional Representations Revisited," *Proceedings of the First International Conference on Computer Vision,* London, 1987, pp. 231–240.

Bamieh, B., and R. J. P. De Figueiredo, "A General Moment–Invariants/Attributed–Graph Method for Three–Dimensional Object Recognition from a Single Image," *IEEE Journal of Robotics and Automation,* Vol. RA-2, 1986, pp. 31–41.

Barnhill, R. E., "A Survey of the Representation and Design of Surfaces," *IEEE Computer Graphics and Applications,* Vol. 3, 1983, pp. 9–16.

Barr, A.H., "Superquadrics and Angle Preserving Transformations," *IEEE Computer Graphics and Applications,* Vol. 1, 1981, pp. 11–23.

——, "Global and Local Deformations of Solid Primitives," *Computer Graphics,* Vol. 18, 1984, pp. 21–30.

Ben-Arie, J., and A. Z. Meiri, "3D Objects Recognition by Optimal Matching Search of Multinary Relation Graphs," *Computer Vision, Graphics, and Image Processing,* Vol. 37, 1987, pp. 345–361.

Besl, P. J., and R. J. Jain, "Three-Dimensional Object Recognition," *ACM Computing Survey,* Vol. 17, 1985, pp. 75–145.

Bhanu, B., "Representation and Shape Matching of 3-D Objects," *IEEE Transactions on Pattern Analysis and Machine Intelligence,* Vol. PAMI-6, 1984, pp. 340–351.

Binford, T. O., "Visual Perception by a Computer," *Proceedings of the IEEE Conference on Systems and Controls*, Miami, 1971.

——, "Spatial Understanding: The Successor System," *Proceedings of the DARPA Image Understanding Workshop*, Palo Alto, 1989, pp. 12–20.

Blum, H., "Biological Shape and Visual Science (Part I)," *Journal of Theoretical Biology*, Vol. 38, 1973, pp. 205–287.

Blum, H., and R. Nagel, "Shape Description Using Weighted Symmetric Axis Features," *Pattern Recognition*, Vol. 10, 1978, pp. 167–180.

Bolles, R. C., and R. A. Cain, "Recognizing and Locating Partially Visible Objects: The Local-Feature Focus Method," *International Journal of Robotics Research*, Vol. 1, 1982, pp. 57–82.

Brady, M., and H. Asada, "Smoothed Local Symmetries and Their Implementation," *International Journal of Robotics Research*, Vol. 3, 1984, pp. 36–61.

Brooks, R. A., "Symbolic Reasoning among 3D Models and 2D Images," *Artificial Intelligence*, Vol. 17, 1981, pp. 285–348.

Brooks, R. A., R. Grenier, and T. O. Binford, "The ACRONYM Model-Based Vision System," *Proceedings of the International Japanese Conference on Artificial Intelligence*, Tokyo, 1979, pp. 105–113.

Bunke, H., and A. Sanfeliu (eds.), *Syntactic and Structural Pattern Recognition: Theory and Applications*, World Scientific, Singapore, 1990.

Burns, J. B., A. R. Hanson, and E. M. Riseman, "Extracting Straight Lines," *IEEE Transactions on Pattern Analysis and Machine Intelligence*, Vol. PAMI-8, 1986, pp. 425–455.

Camps, O. I., L. G. Shapiro, and R. M. Haralick, "PREMIO: The Use of Prediction in a CAD-Model Based Vision System," *Technical Report* EE–ISL–89–01, Department of Electrical Engineering, University of Washington, Seattle, 1989.

Cederburg, R. L. T., "Chain–Link Coding and Segmentation for Raster Scan Devices," *Computer Graphics and Image Processing,* Vol. 10, 1979, pp. 224–234.

Chakravarty, I., and H. Freeman, "Characteristic Views as a Basis for Three-Dimensional Object Recognition," *Proceedings of SPIE 336 (Robot Vision)*, Arlington, VA, 1982, pp. 37–45.

Chellappa, R., and R. Bagdazian, "Fourier Coding of Image Boundaries," *IEEE Transactions on Pattern Analysis and Machine Intelligence,* Vol. PAMI-6, 1984, pp. 102–105.

Chin, R. T., and C. R. Dyer, "Model-Based Recognition in Robot Vision," *Computing Surveys,* Vol. 18, 1986, pp. 67–108.

Cohen, E., "Some Mathematical Tools for a Modeler's Workbench," *IEEE Computer Graphics and Applications,* Vol. 3, 1983, pp. 663–66.

Costa, M., et al., "Optimal Affine Invariant Point Matching," *Proceedings of the SPIE Conference on Applications of Artificial Intelligence*, Orlando, FL, 1989, pp. 515–530.

Crimmins, T. R., "A Complete Set of Fourier Descriptors for Two-Dimensional Shapes," *IEEE Transactions on Systems, Man, and Cybernetics,* Vol. SMC-12, 1982, pp. 848–855.

Cyganski, D., and J. A. Orr, "Object Recognition and Orientation Determination by Tensor Methods," in *Advances in Computer Vision and Image Processing,* T. Huang (ed.), JAI Press, Greenwich, CT, 1988, Vol. 3, pp. 101–144.

Dane, C., and R. Bajcsy, "Three-Dimensional Segmentation Using the Gaussian Image and Spatial Information," *Proceedings of the Pattern Recognition–Image Processing Conference,* Dallas, 1981, pp. 54–56.

Davis, L. S., "Shape Matching Using Relaxation Techniques," *IEEE Transactions on Pattern Analysis and Machine Intelligence*, Vol. PAMI-1, 1979, pp. 60–72.

Dekking, F. M., and P. J. Van Otterloo, "Fourier Coding and Reconstruction of Complicated Contours," *IEEE Transactions on Systems, Man, and Cybernetics,* Vol. SMC-16, 1986, pp. 395–404.

Dorst, L., "Length Estimators Compared," *Proceedings of the Fourth Scandinavian Conference on Image Analysis,* Trondheim, Norway, 1985, pp. 743–751.

Dorst, L., and A. W. M. Smeulders, "Best Linear Unbiased Estimators for Properties of Digitized Straight Lines," *IEEE Transactions on Pattern Analysis and Machine Intelligence,* Vol. PAMI-8, 1986, pp. 276–282, 676.

——, "Length Estimators for Digitized Contours," *Computer Vision, Graphics, and Image Processing,* Vol. 40, 1987, pp. 311–333.

Dudani, S. A., K. J. Breeding, and R. B. McGhee, "Aircraft Identification by Moment Invariants," *IEEE Transactions on Computers,* Vol. C-26, 1977, pp. 39–45.

Eggert, D., and K. W. Bowyer, "Computing the Orthographic Projection Aspect Graph of Solids of Revolution," *Proceedings of the IEEE Workshop on Interpretation of 3D Scenes,* Austin, TX, 1989, pp. 102–108.

Faber, T. L., and E. M. Stokely, "Orientation of 3D Structures in Medical Images," *IEEE Transactions on Pattern Analysis and Machine Intelligence,* Vol. 10, 1988, pp. 626–633.

Faugeras, O. D., "New Steps Toward a Flexible 3-D Vision System for Robotics," *Proceedings of the Seventh International Conference on Pattern Recognition,* Montreal, 1984, pp. 796–805.

Feng, H. F., and T. Pavlidis, "Decomposition of Polygons into Simpler Components: Feature Generation for Syntactic Pattern Recognition," *IEEE Transactions on Computers,* Vol. C-24, 1975, pp. 636–650.

Ferrie, F. P., J. Lagarde, and P. Whaite, "Darboux Frames, Snakes, and Super-Quadrics: Geometry from the Bottom Up," *Proceedings of the IEEE Workshop on Interpretation of 3D Scenes,* Austin, TX, 1989, pp. 170–176.

Freeman, H., "On the Encoding of Arbitrary Geometric Configurations," *IRE Transactions on Electronic Computers,* Vol. EC-10, 1961, pp. 260–268.

——, "Boundary Encoding and Processing," in *Picture Processing and Psychopictorics,* B.S. Lipkin and A. Rosenfeld (eds.), Academic Press, New York, 1970, pp. 241–266.

——, "Computer Processing of Line-Drawing Images," *Computing Surveys,* Vol. 6, 1974, pp. 57–97.

——, "Application of the Generalized Coding Scheme to Map Data Processing," *Proceedings of the IEEE Pattern Recognition and Image Processing Conference,* Chicago, 1978a, pp. 220–226.

——, "Shape Description via the Use of Critical Points," *Pattern Recognition,* Vol. 10, 1978b, pp. 159–166.

Gardner, M., "The Superellipse: A Curve That Lies Between the Ellipse and the Rectangle," *Scientific American,* September 1965, pp. 222–236.

Gigus, Z., J. Canny, and R. Seidel, "Efficiently Computing and Representing the Aspect Graphs of Polyhedral Objects," *Proceedings of the Second International Conference on Computer Vision,* Tampa, FL, 1988, pp. 30–39.

Gigus, Z., and J. Malik, "Computing the Aspect Graph for Line Drawings of Polyhedral Objects," *Proceedings of the IEEE Conference on Computer Vision and Pattern Recognition,* Ann Arbor, MI, 1988, pp. 654–661.

Gordon, W. J., "An Operator Calculus for Surface and Volume Modeling," *IEEE Computer Graphics and Applications,* Vol. 3, 1983, pp. 18–22.

Granlund, G. H., "Fourier Preprocessing for Hand Print Character Recognition," *IEEE Transactions on Computers,* Vol. C-21, 1972, pp. 195–201.

Grimson, W. E., and T. Lozano-Perez, "Model-Based Recognition and Localization from Sparse Range or Tactile Data," *International Journal of Robotics Research,* Vol. 3, 1984, pp. 3–35.

Groen, F. C. A., and P. W. Verbeek, "Freeman Code Probabilities of Object Boundary Quantized Contours," *Computer Graphics and Image Processing,* Vol. 7, 1978, pp. 391–402.

Gross, A. D., and T. E. Boult, "Error of Fit Measures for Recovering Parametric Solids," *Proceedings of the Second International Conference of IEEE,* Tampa, FL, 1988, pp. 690–694.

Gupta, A., L. Bogoni, and R. Bajcsy, "Quantitative and Qualitative Measures for the Evaluation of the Superquadric Model," *Proceedings of the IEEE Workshop on Interpretation of 3D Scenes,* Austin, TX, 1989, pp. 162–169.

Gupta, L., and M. D. Srinath, "Contour Sequence Moments for the Classification of Closed Planar Shapes," *Pattern Recognition,* Vol. 20, 1987, pp. 267–272.

——, "Invariant Planar Shape Recognition Using Dynamic Alignment," *Pattern Recognition,* Vol. 21, 1988, pp. 235–239.

Hanna, S. L., J. F. Abel, and D. P. Greenberg, "Intersection of Parametric Surfaces by Mean Look-Up Tables," *IEEE Computer Graphics and Applications,* Vol. 3, 1983, pp. 39–48.

Hanson, A. R., and E. M. Riseman, "VISIONS: A Computer System for Interpreting Scenes," in *Computer Vision Systems*, A. Hanson and E. Riseman (eds.), Academic Press, New York, 1978, pp. 303–333.

Horaud, P., and R. C. Bolles, "3DPO: A System for Matching 3-D Objects in Range Data," in *From Pixels to Predicates*, A. P. Pentland (ed.), Ablex, Norwood, NJ, 1986, pp. 359–370.

Horn, B. K. P., "Sequins and Quills—Representation for Surface Topography," MIT AI Lab Memo 536, May 1979.

——, "Extended Gaussian Images," *Proceedings of the IEEE*, Vol. 72, 1984, pp. 1671–686.

——, *Robot Vision*, MIT Press, Cambridge, MA, 1986.

Horn, B. K. P. and K. Ikeuchi, "The Mechanical Manipulation of Randomly Oriented Parts," *Scientific American*, August 1984, pp. 100–111.

Hsia, T. C., "A Note on Invariant Moments in Image Processing," *IEEE Transactions on Systems, Man, and Cybernetics,* Vol. SMC-11, 1981, pp. 831–834.

Hsu, Y. N., and H. H. Arsenault, "Optical Pattern Recognition Using Circular Harmonic Expansion," *Applied Optics,* Vol. 21, 1982, pp. 4016–019.

——, "Pattern Discrimination by Multiple Circular Harmonic Components," *Applied Optics,* Vol. 23, 1984, pp. 841–844.

Hsu, Y. N., H. H. Arsenault, and G. April, "Rotation–Invariant Digital Pattern Recognition Using Circular Harmonic Expansion," *Applied Optics,* Vol. 21, 1982, pp. 4012–019.

Hu, M. K., "Visual Pattern Recognition by Moment Invariants," *IRE Transactions on Information Theory*, Vol. IT-8, 1962, pp. 179–187.

Hummel, R., and H. Wolfson, "Affine Invariant Matching," *Proceedings of the DARPA Image Understanding Workshop,* Cambridge, MA, 1988, pp. 351–364.

Hunter, G. M., *Efficient Computation and Data Structures for Graphics*, Ph.D. diss., Princeton University, Princeton, NJ, 1978.

Ikeuchi, K., "Recognition of 3-D Objects Using the Extended Gaussian Image," *Proceedings of the International Joint Conference on Artificial Intelligence*, Vancouver, 1981, pp. 595–600.

——, "Generating an Interpretation Tree from a CAD Model for 3D-Object Recognition in Bin-Picking Tasks," *International Journal of Computer Vision*, Vol. 1, 1987, pp. 145–165.

Ikeuchi, K., and K. S. Hong, "Determining Linear Shape Change: Toward Automatic Generation of Object Recognition Programs," *Proceedings of the IEEE Conference on Computer Vision and Pattern Recognition,* San Diego, CA, 1989, pp. 450–457.

Jackins, C. L., and S. L. Tanimoto, "Oct-trees and Their Use in Representing Three-Dimensional Objects," *Computer Graphics and Image Processing*, Vol. 14, 1980, pp. 249–270.

Khotanzad, A., and Y. H. Hong, "Rotation Invariant Pattern Recognition Using Zernike Moments," *Proceedings of the Ninth International Conference on Pattern Recognition,* Rome, 1988, pp. 326–328.

Koenderink, J. J., and A. J. Van Doorn, "The Internal Representation of Solid Shape with Respect to Vision," *Biological Cybernetics*, Vol. 32, 1979, pp. 211–216.

Koparkar, P. A., and S. P. Mudur, "Computational Techniques for Processing Parametric Surfaces," *Computer Vision, Graphics, and Image Processing,* Vol. 28, 1984, pp. 303–322.

Korn, M. R., and C. R. Dyer, "3-D Multiview Object Representations for Model-Based Object Recognition," *Pattern Recognition*, Vol. 20, 1987, pp. 91–103.

Kuhl, F. P., and C. R. Giardina, "Elliptic Fourier Features of a Closed Contour," *Computer Graphics and Image Processing,* Vol. 18, 1982, pp. 236–258.

Lamdan, Y., J. Schwartz, and H. Wolfson, "Object Recognition by Affine Invariant Matching," *Proceedings of the IEEE Conference on Computer Vision and Pattern Recognition,* Ann Arbor, MI, 1988, pp. 335–344.

Lamdan, Y., and H. Wolfson, "Geometric Hashing: A General and Efficient Model-Based Recognition Scheme," *Proceedings of the Second International Conference on Computer Vision*, Tarpon Springs, MD, 1988, pp. 238–249.

Langridge, D. J., "Curve Encoding and the Detection of Discontinuities," *Computer Graphics and Image Processing,* Vol. 20, 1982, pp. 58–71.

Lee, D. T., "Visibility of a Simple Polygon," *Computer Vision, Graphics, and Image Processing,* Vol. 22, 1983, pp. 207–221.

Lee, Y. C., and K. F. J. Jea, "PAR: A CSG-Based Unique Representation Scheme for Rotational Parts," *IEEE Transactions on Systems, Man, and Cybernetics,* Vol. SMC-17, 1987, pp. 1039–49.

Lin, C. C., and R. Chellappa, "Classification of Partial 2D Shapes Using Fourier Descriptors," *IEEE Transactions on Pattern Analysis and Machine Intelligence,* Vol. PAMI-9, 1987, pp. 686–690.

Lowe, D. G., "The Viewpoint Consistency Constraint," *International Journal of Computer Vision*, Vol. 1, 1987, pp. 57–72.

Lu, H., and L. G. Shapiro, "Model-Based Vision Using Relational Summaries," *Proceedings of the SPIE Conference on Applications of Artificial Intelligence VII*, Orlando, IL, 1989, pp. 662–675.

Maitra, S., "Moment Invariants," *Proceedings of the IEEE,* Vol. 67, 1979, pp. 697–699.

Marr, D., and K. Nishihara, "Representation and Recognition of the Spatial Organization of Three-Dimensional Shapes," *Proceedings of the Royal Society of London*, B200, 1977, pp. 269–294.

Meagher, D., "Geometric Modeling Using Octree Encoding," *Computer Graphics and Image Processing*, Vol. 19, 1982, pp. 129–147.

Miller, J. R., "Analysis of Quadric-Surface-Based Solid Models," *IEEE Computer Graphics and Applications,* Vol. 8, 1988, pp. 28–42.

Minami, T., and K. Shinohara, "Encoding of Line Drawings with a Multiple Grid Chain Code," *IEEE Transactions on Pattern Analysis and Machine Intelligence,* Vol. PAMI-8, 1986, pp. 269–276.

Mohan, R., and R. Nevatia, "Perceptual Organization for Segmentation and Description," *Proceedings of the DARPA Image Understanding Workshop*, Palo Alto, 1989, pp. 415–424.

Mori, S., and M. Doh, "A Sequential Tracking Extraction of Shape Features and Its Constructive Description," *Computer Graphics and Image Processing,* Vol. 19, 1982, pp. 349–366.

Mulgaonkar, P. G., L. G. Shapiro, and R. M. Haralick, "Matching 'Sticks, Plates, and Blobs' Using Geometric and Relational Constraints," *Image and Vision Computing*, Vol. 2, 1984, pp. 85–98.

Nevatia, R., and T. O. Binford, "Description and Recognition of Curved Objects," *Artificial Intelligence*, Vol. 8, 1977, pp. 77–98.

Pavlidis, T., "Analysis of Set Patterns," *Pattern Recognition*, Vol. 1, 1968, pp. 165–178.

——, "Representations of Figures by Labelled Graphs," *Pattern Recognition,* Vol. 4, 1972, pp. 5–17.

——, "Algorithm for Shape Analysis of Contours and Waveforms," *Proceedings of the Fourth International Joint Conference on Pattern Recognition,* Kyoto, Japan, 1978a, pp. 70–85.

——, "A Review of Algorithms for Shape Analysis," *Computer Graphics and Image Processing,* Vol. 7, 1978b, pp. 243–258.

Pavlidis, T., and S. L. Horowitz, "Segmentation of Plane Curves," *IEEE Transactions on Computers*, Vol. C-23, 1974, pp. 860–870.

Pentland, A. P., "Perceptual Organization and the Representation of Natural Form," *Artificial Intelligence,* Vol. 28, 1986, pp. 29–73.

——, "Recognition by Parts," *Proceedings of the IEEE Computer Society International Conference on Computer Vision,* London, 1987, pp. 612–620.

Persoon, E., and K. S. Fu, "Shape Discrimination Using Fourier Descriptors," *IEEE Transactions on Systems, Man, and Cybernetics*, Vol. SMC-7, 1977, pp. 170–179.

Phillips, T., "A Shrinking Technique for Complex Object Decomposition," *Pattern Recognition Letters,* Vol. 3, 1985, pp. 271–277.

Phillips, T., and A. Rosenfeld, "A Method of Curve Partitioning Using Arc–Chord Distance," *Pattern Recognition Letters,* Vol. 5, 1987, pp. 285–288.

Plantinga, W. H., and C. R. Dyer, "An Algorithm for Construction of the Aspect Graph," *Proceedings of the IEEE Symposium on Foundations of Computer Science*, Toronto, 1986, pp. 123–131.

Proffitt, D., "Normalization of Discrete Planar Objects," *Pattern Recognition,* Vol. 15, 1982, pp. 137–143.

Proffitt, D., and D. Rosen, "Metrication Errors and Coding Efficiency of Chain-Encoding Schemes for the Representation of Lines and Edges," *Computer Graphics and Image Processing,* Vol. 10, 1979, pp. 318–332.

Rao, K., and R. Nevatia, "Descriptions of Complex Objects from Incomplete and Imperfect Data," *Proceeding of the DARPA Image Understanding Workshop*, Palo Alto, 1989, pp. 399–414.

Reddi, S. S., "Radial and Angular Moment Invariants for Image Identification," *IEEE Transactions on Pattern Analysis and Machine Intelligence,* Vol. PAMI-3, 1981, pp. 240–242.

Reeves, A. P., and R. W. Taylor, "Identification of Three–Dimensional Objects Using Range Information," *IEEE Transactions on Pattern Analysis and Machine Intelligence,* Vol. 11, 1989, pp. 403–410.

Reeves, A. P., et al., "Three-Dimensional Shape Analysis Using Moments and Fourier Descriptors," *IEEE Transactions on Pattern Analysis and Machine Intelligence*, Vol. 10, 1988, pp. 937–943.

Requicha, A. A. G., "Representations for Rigid Solids: Theory, Methods, and Systems," *Computing Surveys,* Vol. 12, 1980, pp. 437–464.

——, "Toward a Theory of Geometric Tolerancing," *The International Journal of Robotics Research,* Vol. 2, 1983, pp. 45–60.

Requicha, A. A. G., and H. B. Voelcker, "Solid Modeling: Current Status and Research Directions," *IEEE Computer Graphics and Applications,* Vol. 3, 1983, pp. 25–37.

Richard, C. W., and H. Hemami, "Identification of Three-Dimensional Objects Using Fourier Descriptors of the Boundary Curve,"*IEEE Transactions on Systems, Man, and Cybernetics*, Vol. SMC-4, 1974, pp. 371–377.

Roberts, L. G., "Machine Perception of Three-Dimensional Solids," in *Optical and Electro-Optical Information Processing*, Tippet et al. (eds.), MIT Press, Cambridge, MA, 1965, pp. 159–197.

Sadjadi, F., "Recognition of Complex Three-Dimensional Objects Using Three-Dimensional Moment Invariants," *SPIE Intelligent Robots and Computer Vision,* Vol. 521, 1984, pp. 87–90.

Samet, H., *Design and Analysis of Spatial Data Structures*, Addison-Wesley, Reading, MA, 1990.

Schmidt, R. A., "Fast Generation of Chain-Code Image Descriptors," *Proceedings of the IEEE International Conference on Robotics and Automation,* St. Louis, MO, 1985, pp. 868–872.

Schneier, M., "A Compact Relational Representation," Workshop on the Representation of Three-Dimensional Objects, University of Pennsylvania, Philadelphia, 1979.

Scholten, D. K., and S. G. Wilson, "Chain Coding with a Hexagonal Lattice," *IEEE Transactions on Pattern Analysis and Machine Intelligence,* Vol. PAMI-5, 1983, pp. 526–533.

Shafer, S. A., *Shadows and Silhouettes in Computer Vision,* Kluwer, Boston, MA, 1985.

Shani, U., and D. H. Ballard, "Splines as Embeddings for Generalized Cylinders," *Computer Vison, Graphics, and Image Processing,* Vol. 27, 1984, pp. 129–156.

Shapiro, L. G., "A Structural Model of Shape," *IEEE Transactions on Pattern Analysis and Machine Intelligence*, Vol. PAMI-2, 1980, pp. 111–126.

Shapiro, L. G., and R. M. Haralick, "Decomposition of Two-Dimensional Shapes by Graph-Theoretic Clustering," *IEEE Transactions on Pattern Analysis and Machine Intelligence,* Vol. PAMI-1, 1979, pp. 10–20.

——, "A Spatial Data Structure," *Geo-Processing*, Vol. 1, 1980, pp. 313–337.

——, "Organization of Relational Models for Scene Analysis," *IEEE Transactions on Pattern Analysis and Machine Intelligence*, Vol. PAMI-4, 1982, pp. 595–602.

——, "A Metric for Comparing Relational Descriptions," *IEEE Transactions on Pattern Analysis and Machine Intelligence,* Vol. PAMI-7, 1985, pp. 90–94.

Shapiro, L. G., R. MacDonald, and S. R. Sternberg, "Ordered Structural Shape Matching with Primitive Extraction by Mathematical Morphology," *Pattern Recognition*, Vol. 20, 1987, pp. 75–90.

Shapiro, L. G., et al., "Matching Three-Dimensional Objects Using a Relational Paradigm," *Pattern Recognition*, Vol. 17, 1984, pp. 385–405.

Soroka, B. I., "Generalized Cylinders from Parallel Slices," *Proceedings of the IEEE Conference on Pattern Recognition and Image Processing,* Chicago, 1979, pp. 421–426.

Soroka, B. I., R. L. Anderson, and R. K. Bajcsy, "Generalized Cylinders from Local Aggregation of Sections," *Pattern Recognition,* Vol. 13, 1981, pp. 353–363.

Sriraman, R., J. Kpolowitz, and S. Mohan, "Tree Searched Chain Coding for Subpixel Reconstruction of Planar Curves," *IEEE Transactions on Pattern Analysis and Machine Intelligence,* Vol. 11, 1989, pp. 95–104.

Stewman, J. H., and K. W. Bowyer, "Aspect Graphs for Planar–Face Convex Objects," *Proceedings of the IEEE Workshop on Computer Vision*, Miami Beach, FL, 1987, pp. 123–130.

——, "Creating the Perspective Projection Aspect Graph of Polyhedral Objects," *Proceedings of the Second International Conference on Computer Vision*, Tarpon Springs, MD, 1988, pp. 494–500.

Stracklee, J., and N. J. D. Nagelkerke, "On Closing the Fourier Descriptor Presentation," *IEEE Transactions on Pattern Analysis and Machine Intelligence,* Vol. PAMI-5, 1983, pp. 660–661.

Suk, M., and H. Kang, "New Measures of Similarity between Two Contours Based on Optimal Bivariate Transforms," *Computer Vision, Graphics, and Image Processing,* Vol. 26, 1984, pp. 168–182.

Teague, M. R., "Image Analysis via the General Theory of Moments," *Journal of the Optical Society of America,* Vol. 70, 1980, pp. 920–930.

Teh, C. H., and R. T. Chin, "On Digital Approximation of Moment Invariants," *Computer Vision, Graphics, and Image Processing,* Vol. 33, 1986, pp. 318–326.

——, "On Image Analysis by the Methods of Moments," *IEEE Transactions on Pattern Analysis and Machine Intelligence,* Vol. 10, 1988, pp. 496–513.

Voelcker, H. B., and A. A. G. Requicha, "Geometric Modeling of Mechanical Parts and Processes," *Computer,* December 1977, pp. 48–57.

Vossepoel, A. M., and A. W. M. Smeulders, "Vector Code Probability and Metrication Error in the Representation of Straight Lines of Finite Length," *Computer Graphics and Image Processing,* Vol. 20, 1982, pp. 347–364.

Wallace, T. P., and P. A. Wintz, "An Efficient Three–Dimensional Aircraft Recognition Algorithm Using Normalized Fourier Descriptors," *Computer Graphics and Image Processing,* Vol. 13, 1980, pp. 99–126.

Wang, Y. F., M. J. Magee, and J. K. Aggarwal, "Matching Three–Dimensional Objects Using Silhouettes," *IEEE Transactions on Pattern Analysis and Machine Intelligence,* Vol. PAMI-6, 1984, pp. 513–518.

Xie, S-E., and T. W. Calvert, "CSG-EESI: A New Solid Representation Scheme and a Conversion Expert System," *IEEE Transactions on Pattern Analysis and Machine Intelligence,* Vol. 10, 1988, pp. 221–234.

Yuille, A. L., "The Creation of Structure in Dynamic Shape," *Proceedings of the Second International Conference of IEEE,* Tampa, FL, 1988.

Zahn, C. T., and R. Z. Roskies, "Fourier Descriptors for Plane Closed Curves," *IEEE Transactions on Computers*, Vol. C-21, 1972, pp. 269–281.

19 KNOWLEDGE-BASED VISION

19.1 Introduction

A *knowledge-based vision* system uses domain knowledge to analyze images from that domain. The knowledge might be very general knowledge about three-dimensional objects or it might be extremely specific. A system that works with urban scenes might know about the appearance of houses, roads, and office buildings. A system that works with scenes of airports would know about runways, terminals, and hangars. A system that works with machined three-dimensional parts would know about the possible structures of those parts.

This chapter discusses knowledge-based vision from two points of view: concepts and systems. The main concepts covered are knowledge representations, control strategies, and information integration. The systems discussed illustrate various techniques and applications.

19.2 Knowledge Representations

Computer vision makes use of a large variety of knowledge representations, from simple vectors to complex structures. In this section we categorize these structures into five main types: feature vectors, relational structures, hierarchical structures, rules, and frames.

19.2.1 Feature Vectors

The simplest form of knowledge representation used in computer vision is the *feature vector*, commonly employed in statistical pattern recognition and useful in many kinds of vision systems. A feature vector is simply a tuple of measurements of features that have numeric (or sometimes symbolic) values and can be used to aid in

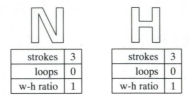

strokes	3
loops	0
w-h ratio	1

strokes	3
loops	0
w-h ratio	1

Figure 19.1 Two hand-printed characters and their feature vectors or attribute-value tables. The global features employed here are number of straight strokes, number of loops, and width-to-height ratio.

recognizing a pattern or an object. For example, in character recognition some useful numeric-valued features might be number of straight strokes, number of loops, and width-to-height ratio of the character. In general two-dimensional shape recognition, such features as moments (Alt, 1962) and Fourier descriptors (Zahn and Roskies, 1972) have been used to classify the shapes. In view-class-based recognition of three-dimensional objects from extracted line drawings, number of FORK, ARROW, L, and T junctions and number of loops may be among the features used to help rapidly determine the appropriate view class (Chakravarty and Freeman, 1982; Shapiro and Lu, 1988). When the names of the features are explicitly present, feature vectors are often referred to as *attribute-value tables* or as *property lists*.

The features used in a feature-vector representation of an object are the *global* features of the object. They characterize the entire object rather than any of its parts. Figure 19.1 gives the features and their values for two different handprinted characters, N and H. Notice that these global features do not aid in distinguishing between the two characters. More powerful global features could help, but global features alone are insufficient for recognition of complex objects and scenes. For this reason hierarchical and relational structures are widely used in computer vision.

19.2.2 Relational Structures

Complex objects and scenes are often composed of recognizable parts. A full description of a complex entity consists of (1) its global features, (2) the global features of each of its parts, and (3) the relationships among the parts. In terms similar to the relational formalisms of the previous two chapters, a *relational description* D_O of entity O is a pair (P, R), where P is the set of parts or *primitives* of O, and R is a set of named (or numbered) *relations* among the primitives. For each relation R_i in R, there is a nonnegative integer n_i indicating the number of primitives involved in the tuples of relation R_i and a second nonnegative integer m_i indicating the number of features or attributes involved. If A is the set of all possible attribute values, then $R_i \subseteq P^{n_i} \times A^{m_i}$. That is, each tuple of R_i represents an attributed relationship among n_i primitives and characterized by m_i attributes. When $n_i = 0$, $m_i > 0$, and $| R_i | = 1$, relation R_i is just a feature vector representing global attributes of O. When $n_i = 1$ and $m_i \geq 0$, relation R_i is a possibly attributed list of the primitives of O. When $n_i > 1$ and $m_i \geq 0$, relation R_i describes possibly attributed relationships among the primitives.

As a concrete example, suppose we wish to represent by a relational description a line drawing that has been extracted from a gray tone image. Suppose that the global attributes we wish to represent are the total number of line segments, density of line segments (average number of segments per unit area), and size (number of rows and number of columns) of the image from which the line drawing was extracted. These global attributes can be represented by a single-tuple relation Image_Properties with four components: Total_Lines, Density, Rows, and Columns. Now suppose we wish to keep track of the start and endpoint coordinates, the length, and the angle of each line segment. This information can be encoded as an attributed relation Line_Segments, whose tuples contain one primitive (Line_Segment) and the values of six attributes: Row_Start, Col_Start, Row_End, Col_End, Length, and Angle. Thus far we have represented only attributes and no relationships.

Many different relationships among straight line segments can be useful in image analysis. Lowe (1987) formed "perceptual groupings" of line segments to use in matching an image to a model. He chose three relationships to define the perceptual groupings: proximity, parallelism, and collinearity. One way to represent these relationships is as attributed binary relations. The tuples of the proximate relation would contain a pair of line segments and a real number indicating how close they come to touching. The parallel and collinear relations would contain tuples with a pair of line segments and a measure of their parallelness or collinearity. To illustrate these concepts, Fig. 19.2 shows a line drawing and a relational description employing the Image_Properties, Line_Segments, and Parallel relations. The real-valued quantities of density and parallelness were estimated for this example. Length was defined as (rounded) Euclidean distance between start and endpoint, and angle was defined as (rounded) $\tan^{-1}(\text{Row\_End} - \text{Row\_Start} / \text{Col\_End} - \text{Col\_Start})$.

Although binary relations are more common, higher-level relations are often more useful in matching, since they can constrain the match more effectively. The ternary relations used in the shape-matching example of Chapter 17 caused the search tree to be severely pruned. Junctions where several line segments meet are commonly used to organize the segments into small groups of two or more. In the pure relational model described here, there might be a separate relation for each junction type (FORK, ARROW, L, T, PSI, MULTI, etc.). If we relax the formalism slightly and define a *list* of primitives as a valid type, then a junctions relation can be constructed whose tuples each contain two components: a list of segments that meet or nearly meet at that junction and a measure of "junctionness" that expresses the probability that this configuration of lines is the result of a real three-dimensional junction.

19.2.3 Hierarchical Structures

Although it is frequently sufficient to describe an object by its parts and their relationships, more complex objects and scenes may have so many parts that the description becomes unwieldy. If an entity can be naturally broken into its major parts, the major parts into subparts, and so on, then a hierarchical structure may

Image_Properties

Total_Lines	Density	Rows	Columns
9	0.03	512	480

Line_Segments

Line_Segments	Row_Start	Col_Start	Row_End	Col_End	Length	Angle
1	124	120	124	400	280	0
2	124	120	250	190	144	290
3	124	120	249	122	125	270
4	125	400	250	471	14	300
5	250	190	250	469	279	0
6	249	120	378	191	147	290
7	250	190	378	191	128	270
8	250	470	375	469	125	270
9	378	191	375	471	280	1

Parallel

Segment1	Segment2	Parallelness
1	5	1.
1	9	.99
5	9	.99
3	7	1
3	8	1
7	8	1
2	4	.99
2	6	1

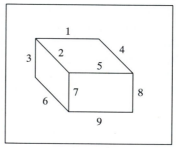

Figure 19.2 Relational description of a simple line drawing. The Image_Properties relation gives the global attributes of the image. The Line_Segments relation lists each line segment along with its start and endpoint, length, and angle. The Parallel relation is a binary relation giving pairs of near-parallel segments and a measure (here just estimated) of their parallelness.

be a better representation. The hierarchical structures we use have both a hierarchical and a relational component. Formally a *hierarchical, relational description* D_O of entity O is a pair (P, R), where P is a set of parts and R is a set of relations among the parts. The relations work the same as in the single-level structure, but the parts, instead of primitives, either can be *atomic parts* that can be broken down no further or can themselves be hierarchical, relational descriptions. Figure 19.3 gives an example of a hierarchical, relational structure representing an outdoor

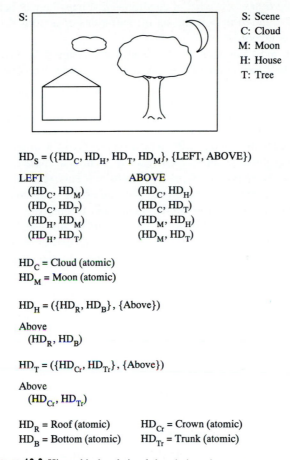

S:

S: Scene
C: Cloud
M: Moon
H: House
T: Tree

$HD_S = (\{HD_C, HD_H, HD_T, HD_M\}, \{LEFT, ABOVE\})$

LEFT
  (HD_C, HD_M)
  (HD_C, HD_T)
  (HD_H, HD_M)
  (HD_H, HD_T)

ABOVE
  (HD_C, HD_H)
  (HD_C, HD_T)
  (HD_M, HD_H)
  (HD_M, HD_T)

HD_C = Cloud (atomic)
HD_M = Moon (atomic)

$HD_H = (\{HD_R, HD_B\}, \{Above\})$

Above
  (HD_R, HD_B)

$HD_T = (\{HD_{Cr}, HD_{Tr}\}, \{Above\})$

Above
  (HD_{Cr}, HD_{Tr})

HD_R = Roof (atomic) HD_{Cr} = Crown (atomic)
HD_B = Bottom (atomic) HD_{Tr} = Trunk (atomic)

Figure 19.3 Hierarchical, relational description of an outdoor scene.

scene. The parts (objects) in the scene are a cloud, a house, a tree, and the moon. The relations LEFT and ABOVE define the spatial relationships among the objects. The cloud and the moon are atoms, but the house decomposes further into a roof and bottom, and the tree decomposes further into a crown and a trunk. These last entities are atoms. Although atoms do not have parts, they may have attributes. A feature vector is a good representation for an atom. Thus a hierarchical, relational structure can be thought of as a hierarchy of relational structures whose terminal entities are feature vectors.

19.2.4 Rules

Not all complex knowledge is hierarchical or relational. Rule-based systems encode knowledge in the form of *rules*. A rule has the form

$$< antecedents > \rightarrow < actions >$$

where $<antecedents>$ is a list of conditions that must be met before the rule can be applied and $<actions>$ is the set of operations or actions that are performed when the rule is invoked.

One of the most common kinds of rules encountered by vision students and researchers is the *production rule* associated with context-free grammars. The production rule

$$<expression> \rightarrow <expression> + <term>$$

used in a grammar describing arithmetic expressions in a programming language says that, during parsing, an $<expression>$ in a derivation may be replaced by the string $<expression> + <term>$. This rule is used to parse expressions involving the + operation. In the context of computer vision, syntactic pattern recognition systems use similar grammars to describe classes of patterns. The terminal symbols used by such *picture grammars* are usually two-dimensional entities such as line or curve segments or primitive two-dimensional shapes. For example, a grammar for representing chromosomes (Ledley et al., 1965) contains such productions as

$$<submedian\ chromosome> \rightarrow <arm\ pair><arm\ pair>$$
$$<arm\ pair> \rightarrow <side><arm\ pair>$$
$$<arm\ pair> \rightarrow <arm\ pair><side>$$
$$<arm\ pair> \rightarrow <arm><right\ part>$$
$$<arm\ pair> \rightarrow <left\ part><arm>$$
$$<left\ part> \rightarrow <arm>\ c$$
$$<right\ part> \rightarrow c\ <arm>$$
$$<arm> \rightarrow b\ <arm>$$
$$<arm> \rightarrow <arm>\ b$$
$$<arm> \rightarrow a$$
$$<side> \rightarrow b\ <side>$$
$$<side> \rightarrow <side>\ b$$
$$<side> \rightarrow b$$
$$<side> \rightarrow d$$

where b is a two-dimensional segment. The first rule says that a $<submedian\ chromosome>$ can be made up of an $<arm\ pair>$ followed by a second $<arm\ pair>$. The second rule says that an $<arm\ pair>$ consists of a $<side>$ followed by another $<arm\ pair>$. The third rule says that a $<side>$ can consist of a segment (b) followed by another $<side>$. The fourth rule says that a $<side>$ can also consist of just a segment (b). These rules are a subset of the entire grammar for chromosomes. The rules can be used to generate or recognize the shapes of the outer boundaries of the chromosomes they describe. Figure 19.4 shows the shape of a chromosome that can be generated from these rules.

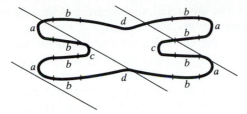

Figure 19.4 The shape of a chromosome that can be generated from the given rules. The derivation of the chromosome is:

> *<submedian chromosome>*
> $\overset{*}{\Rightarrow}$ *<arm pair><arm pair>*
> $\overset{*}{\Rightarrow}$ *<side><arm pair><side><arm pair>*
> $\overset{*}{\Rightarrow}$ *<side><arm><right><side><arm><right>*
> $\overset{*}{\Rightarrow}$ *<side><arm>c<arm><side><arm>c<arm>*
> $\overset{*}{\Rightarrow}$ *<side>b<arm>b c b<arm>b<side>b<arm>b c b<arm>b*
> $\overset{*}{\Rightarrow}$ *d b a b c b a b d b a b c b a b*

Rules can also be used in expert systems for vision and nonvision applications. In a vision expert system for analyzing aerial images of urban scenes, a rule such as

< if there are two adjacent corners > → < hypothesize a building >

may be used to tell the system that two corners connected by a straight line are a good cue for a rectangular-shaped building. At a lower level, rules may guide a knowledge-based segmentation of an image (Levine and Nazif, 1985; McKeown, Wilson, and McDermott, 1985). This will be discussed in more detail in the section on example systems. Finally, rules can be used as a control mechanism. This usage will be discussed in the section on control, where the rules are procedures called knowledge sources.

19.2.5 Frames and Schemas

Minsky (1975) introduced the *frame* as "a data-structure for representing a stereotyped situation." He envisioned a relational structure whose terminal nodes consisted of *slots*, which are attributes, and *fillers*, which are the values for those attributes. A filler can be atomic or can reference another frame. A filler can also be empty and waiting for a value or can contain a *default* value that is used until the slot is filled. A filler may be a variable or a symbolic expression that must be evaluated to obtain the value. In fact, a filler may be the name of a procedure that must be executed to produce the value. Thus frames are very flexible structures that could be used to implement any of the knowledge representations we have discussed so far, and more.

A *schema* in database terminology is a model or prototype. In a relational database system, the data are stored in relations, and the schema associated with a relation gives the names, types, and meanings of the components of the relation. In artificial intelligence the term *schema* either is a synonym for frame or means a general frame used as the prototype or model for an entity. Instances of the schema are more-specific frames that describe individual instances of the entity.

Frames and schemas have been used in vision systems to describe models of objects and of scene classes. The VISIONS system (Hanson and Riseman, 1978) uses schemas to define the expected objects and spatial relationships in road and house scenes. Their suburban-house scene schema includes slots for a house, a garage, a driveway, and a car. The ACRONYM system (Brooks, 1981) uses frames as nodes in an object graph. The frames describe classes and instances of objects. For example, the frame that describes a generalized cylinder model of an electric motor is as follows:

```
NODE: ELECTRIC_MOTOR_CONE
     CLASS:              SIMPLE_CONE
     SPINE:              Z0014
     SWEEPING_RULE:      CONSTANT_SWEEPING_RULE
     CROSS_SECTION:      Z0013
```

This node describes the top level of a hierarchical model of an electric motor. The node is implemented as a frame with four slots: CLASS, SPINE, SWEEPING_RULE, and CROSS_SECTION. The fillers for CLASS and SWEEP-ING_RULE are constants that have some meaning to the system. The fillers for SPINE and CROSS_SECTION are references to other nodes (frames) in the object graph. Brooks also allowed his fillers to be variables. The same variable (such as MOTOR_LENGTH) could be used as a filler in many different slots. When the variable was assigned a value, all the slots would have that same value. ACRONYM is described in more detail in the section on example systems.

Another good use of frames is in image-sequence matching. Since each image of a time sequence differs only a little from the preceding one, a frame that represents the features of one image may be used to predict the features that are expected in the next. The predictions may be used in a top-down matching strategy.

19.3   Control Strategies

The control strategy of a system dictates how the knowledge in that system will be used. We discuss the two major forms of control, hierarchical and heterarchical, and variants of each. Figure 19.5 illustrates their general structure.

19.3.1 Hierarchical Control

Hierarchical control is the most common control structure used in computer pro-gramming. It is defined by a predefined hierarchy or ordering of the procedures that perform the required analysis task. In the most common scenario, a main pro-

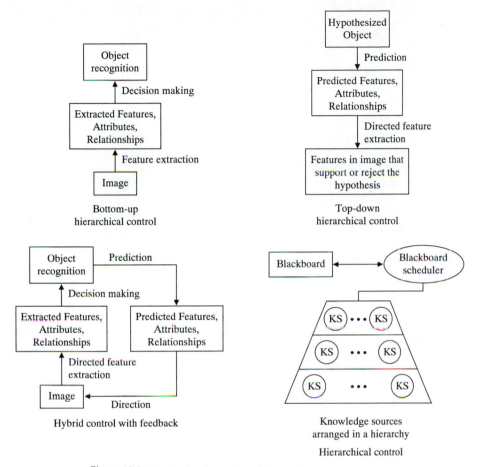

Figure 19.5 Four kinds of control used in machine vision systems.

cedure calls on a set of preprocessing routines that convert the original image to a form most suitable for extracting the primitive features. Next, a feature extraction routine locates the features of interest and constructs a symbolic description of these features and their attributes and relationships. Finally, a decision-making procedure performs some type of recognition task using the symbolic description. Each of these procedures has its own task and its own place in the hierarchy. They communicate with each other only through intermediate results.

There are two main kinds of hierarchical control: *bottom-up* and *top-down*. Bottom-up control is similar to the scenario described above. Features are extracted from the image and grouped in some way, with no knowledge of the structure of the object or scene. Only after a complete symbolic description has been constructed are object or scene models used in a matching procedure. Top-down control starts with a hypothesis that the image contains a particular object or can be categorized as a particular type of scene. This leads to further hypotheses about the parts of the object or scene that may be present. These hypotheses lead to expectations of certain kinds of primitives in certain relationships. Eventually the feature extraction routines are called on either to verify the existence of a primitive and report its location or to report failure.

Neither strict bottom-up control nor strict top-down control is flexible enough for analysis of complex images. A powerful variant is *hybrid control with feedback*. This strategy starts out with an initial segmentation of the image and extraction of a preliminary set of features and relationships. On the basis of this preliminary description, the identity of one or more objects is hypothesized. Now a top-down strategy can be used to verify or disprove the existence of these objects. Knowledge of these verified objects and of the expected class of scene allows new hypotheses to be generated. Furthermore, as objects or parts of objects are recognized, more information can be deduced that will help in the overall recognition process. For example, determining the size of a particular piece of an object can constrain the sizes of other pieces of the same object (Brooks, 1981). Matching several features of an object allows the estimation of pose parameters. Once the pose parameters have been determined, the corresponding transformation can be applied to the features of the model to predict the features that should appear in the image (Lowe, 1987). Now a verification step can be used to decide whether the match is correct. There are many different variants of the hybrid control with feedback idea. The SPAM system (McKeown, Wilson, and McDermott, 1985) described later is an excellent example.

19.3.2 Heterarchical Control

Whereas hierarchical control dictates the order in which different processes operate, heterarchical control lets the data themselves dictate the order in which things are done. In the heterarchical control model, knowledge is embodied in a set of procedures called *knowledge sources*. Each knowledge source can communicate with some or all of the other knowledge sources. The knowledge sources are supposed to work cooperatively to analyze the image. For example, finding two corners in the data could activate a knowledge source that would call on procedures to search for evidence of two adjoining corners to form a building. Hypothesizing the building would lead to the activation of another knowledge source to look for a street. If the street hypothesis were verified, the next action might be to look for more buildings along the same street. If not, an alternative approach might be taken. Although this idea could work in principle, in practice it leads to a system that is difficult to keep track of.

One result of trying to add some order to a heterarchy is the blackboard approach. A *blackboard* is a global database shared by a set of independent knowledge sources. Each knowledge source checks the blackboard to determine whether its preconditions have been met. If so, it executes and then posts its results on the blackboard, where they can be used by other knowledge sources. The blackboard has its own organization. In Erman et al. (1980) it was organized into levels corresponding to the different levels of the task (speech analysis in this case) being performed. In computer vision the levels might include the scene as a whole, the objects that form the scene, the surfaces and edges that form the objects, the regions and lines of the image that are projections of these surfaces and edges, and the pixels of the image that form the regions and lines. A *blackboard scheduler* controls the access to the blackboard by the knowledge sources. Like a process scheduler in an operating system, it determines the order of execution of competing knowledge sources.

19.4   Information Integration

A *hypothesis* is a proposition, a statement that can be either *true* or *false*. In image analysis the *hypothesize-and-test* paradigm is commonly used within any of the control structures. A hypothesis is generated on the basis of some initial evidence. It then must be verified either by additional evidence or by being judged consistent with other established hypotheses. For example, a shape recognition knowledge source may hypothesize that region 56 is a telephone on the basis of the shape of the region's boundary. A texture knowledge source may be asked to verify that the texture of region 56 is acceptable for telephones (telephones rarely have dots or stripes). A matching expert may be called on to look for the parts, such as the receiver, dial, or pushbuttons. A global constraint analysis procedure may be asked to check the validity of the label "telephone" in the context of the adjacent objects' labels.

The phrase "hypothesize-and-test" oversimplifies the problem. If testing can be accomplished by a hypothesis verification procedure that returns either *true* or *false*, then the testing results either in the establishment of the hypothesis or in its removal from further consideration. But image analysis is usually not a yes-or-no task. Instead, each knowledge source that tries to verify a hypothesis will produce a number indicating its *certainty* that the hypothesis is true. So the shape knowledge source that originally proposed the label "telephone" for region 56 may say that the certainty of this being a telephone shape is .86. The texture knowledge source may say that this is telephone texture with certainty .95. The matching expert may find the receiver, fail to find any evidence of dial or pushbuttons, and return a certainty of .5. Meanwhile the global analyzer may determine that the region surrounding the telephone is already labeled "desk," which is 100% compatible with label "telephone," so it returns certainty 1.0. Now how can all these different judgments be combined to decide whether the hypothesis that region 56 is a telephone is true, or better yet, to decide on the combined certainty that region 56 is a telephone? The area of research that deals with this question is *information integration*. Two main approaches to the information integration problem have been used in computer vision: the Bayesian belief network and the Dempster-Shafer theory of evidence. Use of context in information interpretation is sometimes possible with relaxation approaches (Rosenfeld, Hummel, and Zucker, 1976). We discuss both the Bayesian-belief-network and the theory-of-evidence approaches as well as the probabilistic approach to belief.

19.4.1 Bayesian Approach

The Bayesian approach to information integration is well illustrated by Pearl's (1987) model. Pearl defines a *Bayesian belief network* as a directed acyclic graph whose nodes represent propositional variables and whose arcs represent causal relationships. Suppose that the nodes of the graph are variables $X_1, X_2, \ldots, X_N$. Each propositional variable X_i has a finite set of possible values $\{x_i\}$. An arc (X_j, X_i) from node X_j to node X_i indicates that the value of X_j is a direct cause of the value of X_i. Suppose X_i has I possible values, and X_j has J possible values. For each pair of values (x_j, x_i), the strength of the causality is the conditional probability $P(x_i \mid x_j)$. These probabilities can be thought of as forming a $J \times I$ matrix associated with the arc.

The support set of a node is the set of the node's predecessors. Suppose that node X_i has support set S_i. Then $P(x_i \mid s_i)$ denotes the joint conditional probability of $X_i = x_i$, given the set of values $\{s_i\}$ for the support variables. The distribution corresponding to the entire network is then

$$P(x_1, x_2, \ldots, x_N) = \prod_{i=1}^{N} P(x_i \mid s_i)$$

The matrices of strengths on each of the causal links are static input to the network. As information propagates through the network, the *belief value* BEL(x) of each proposition $X = x$ changes. The dynamically changing belief value BEL(x) denotes the probability $P(x \mid e)$ of $X = x$, given all the evidence e received so far. Figure 19.6 illustrates the idea for a portion of a belief network to be used in image analysis. Nodes S, K, and T represent objects to be identified. Each object has two possible labels: Object S can be a desk or a workstation, object K can be a set of pushbuttons or a keyboard, and object T can be a telephone or a computer terminal. Labels of "desk" for S and "pushbuttons" for K tend to support the label of "telephone" for T, and labels of workstation for S and "keyboard" for K tend to support the label of "terminal" for T.

The model of belief propagation defines how beliefs in the network are updated as new evidence enters the system. Suppose that nodes U and V are predecessors of node X, and nodes Y and Z are successors of node X, as shown in Fig. 19.7. Let $P(x \mid u, v)$ be the fixed conditional-probability matrix that relates the variable x to its immediate parents u and v. Let $\pi_X(u)$ be the current strength of the *causal support* contributed by an incoming link to X. Let $\lambda_Y(x)$ be the current strength of the *diagnostic support* contributed by an outgoing link from X. Causal support represents evidence propagating forward from parents to their children, whereas diagnostic support represents feedback from children to their parents. Updating a node X thus involves updating not only its belief function but also its λ and π functions. Belief updating is given by the formula

$$\text{BEL}(x) = \alpha \lambda_Y(x) \lambda_Z(x) \sum_{u,v} P(x \mid u, v) \pi_X(u) \pi_X(v)$$

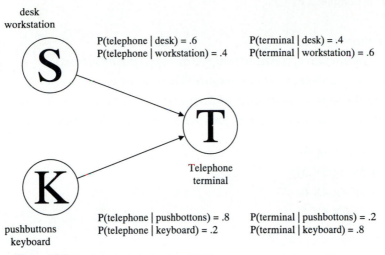

desk
workstation

P(telephone | desk) = .6   P(terminal | desk) = .4
P(telephone | workstation) = .4   P(terminal | workstation) = .6

Telephone
terminal

P(telephone | pushbottons) = .8   P(terminal | pushbottons) = .2
pushbuttons   P(telephone | keyboard) = .2   P(terminal | keyboard) = .8
keyboard

Figure 19.6 Portion of a belief network to be used in determining the identities of objects in an image.

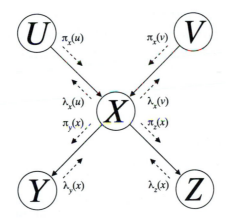

Figure 19.7 Propagation of beliefs through a piece of a belief network.

where α is a normalizing constant that makes $\sum_x \text{BEL}(x) = 1$. For example, in Fig. 19.6, suppose at some point in the analysis the causal support functions have the values

$$\pi_T(desk) = .9$$
$$\pi_T(workstation) = .4$$
$$\pi_T(pushbuttons) = .6$$
$$\pi_T(keyboard) = .5$$

and the values of the diagnostic functions λ_Y *(telephone)* and λ_Y *(terminal)* have been set to 1, since node T has no successors. Then the unnormalized belief function for node T is computed by

BEL(*telephone*) =

$P(telephone|desk,pushbuttons)\pi_T(desk)\pi_T(pushbuttons)+$
$P(telephone|desk,keyboard)\pi_T(desk)\pi_T(keyboard)+$
$P(telephone|workstation,pushbuttons)\pi_T(workstation)\pi_T(pushbuttons)+$
$P(telephone|workstation,keyboard)\pi_T(workstation)\pi_T(keyboard)$

$=$ $(.6)(.8)(.9)(.6)$
    $+ (.6)(.2)(.9)(.5)$
    $+ (.4)(.8)(.4)(.6)$
    $+ (.4)(.2)(.4)(.5)$

$=$ $.41$

BEL(*terminal*) =

$P(terminal|desk,pushbuttons)\pi_T(desk)\pi_T(pushbuttons)+$
$P(terminal|desk,keyboard)\pi_T(desk)\pi_T(keyboard)+$
$P(terminal|workstation,pushbuttons)\pi_T(workstation)\pi_T(pushbutton)+$
$P(terminal|workstation,keyboard)\pi_T(workstation)\pi_T(keyboard)$

$=$ $(.4)(.2)(.9)(.6)$
    $+ (.4)(.8)(.9)(.5)$
    $+ (.6)(.2)(.4)(.6)$
    $+ (.6)(.8)(.4)(.5)$

$=$ $.31$

Since BEL(*telephone*)+BEL(*terminal*) = .41+.31 = .72, and they are supposed
to sum to 1, we can normalize by multiplying by 1/.72=1.39. The normalized values
would be BEL *(telephone)* = *.57* and BEL *(terminal)* = *.43*. λ updating is given
by

$$\lambda_X(u) = \alpha \sum_v \left[\pi_X(v) \sum_x \lambda_Y(x)\lambda_Z(x)P(x \mid u,v) \right]$$

Finally, π updating is given by

$$\pi_Y(x) = \alpha\lambda_Z(x) \left[\sum_{u,v} P(x \mid u,v)\pi_X(u)\pi_X(v) \right]$$

In the belief network there will be some nodes (leaves) that have no successors
and some nodes (roots) that have no predecessors. The leaf nodes can receive no
diagnostic support. To accommodate for this, the λ values for such a node are all

set to 1. A root node receives no causal support. To handle this, Pearl introduces a dummy parent node U for each root node X. U is instantiated to 1, and the conditional probabilities $P(x \mid u)$ on the link $U \to X$ are set to the prior probabilities $P(x)$ of $X = x$.

In the course of evidence accumulation, a variable X may be instantiated to one of its possible values x'. In Pearl's model such a node is called an *evidence node*. The diagnostic support for an evidence node is computed by introducing a dummy child node Z with

$$\lambda_Z(x) = \begin{cases} 1 & \text{if } x = x' \\ 0 & \text{otherwise} \end{cases}$$

If X has children $Y_1, \ldots, Y_m$, each child receives the same message $\pi_{Y_i}(x) = \lambda_Z(x)$ from X.

The updating process described here can proceed asynchronously and eventually halts at a fixed point. At the fixed point, BEL(x) is the final measure of belief that $X = x$ and is equivalent to $P(x|e)$, where e is the total evidence that was given to the system.

19.4.2 Dempster-Shafer Theory

The Bayesian model allows for the expression of positive belief in a proposition, but it has no representation for disbelief. Dempster-Shafer theory (Garvey, Lowrance, and Fischler, 1981) is a formalism for information integration that allows both belief and disbelief.

Representation of Belief

In Shafer's belief model (Shafer, 1981), belief in a proposition is represented by the *belief interval*. This is the unit interval [0,1] further demarcated by two points j and k, $k \geq j$. Suppose that a belief interval describes proposition A. Then the subinterval [0,j) is called Belief(A); the subinterval (k,1] is called Disbelief(A); and the remainder, the subinterval [j,k], is called Uncertainty(A). Belief(A) is the degree to which current evidence supports the proposition A, Disbelief(A) is the degree to which current evidence supports the negation of A, and Uncertainty(A) is the degree to which we believe nothing one way or the other about proposition A. As new evidence is collected, the remaining uncertainty will decrease, and each piece of length that it loses will be given to Belief(A) or Disbelief(A). With that in mind, we can define two other measures: the Plausibility of A is the sum of Belief(A) and Uncertainty(A). This would be the maximum possible extent of Belief(A) if all of the remaining uncertainty were to turn eventually into belief: the extent to which the evidence does not actually refute the proposition. The opposite measure is Doubt(A), which is Uncertainty(A) plus Disbelief(A). Figure 19.8 illustrates these concepts.

A *frame of discernment* Θ is the set of mutually exclusive atomic propositions concerning a given subject area, exactly one of which corresponds to the truth. The

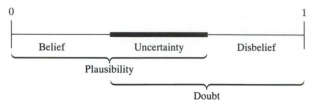

Figure 19.8 Belief interval.

subsets of Θ represent compound propositions. A *basic probability assignment* is a function m (for mass) defined on 2^Θ that satisfies the following conditions:

1. $0 \leq m(A) \leq 1$, for every $A \subseteq \Theta$
2. $m(\phi) = 0$
3. $\sum_{A \subseteq \Theta} m(A) = 1$

where ϕ represents the empty set. Intuitively $m(a)$ is the degree of belief committed exactly to subset A by function m. Each distinct piece of evidence e has an associated basic probability function m_e. So knowledge contributed by different knowledge sources is represented through the use of different functions.

A belief function Bel over a frame of discernment Θ is a function that assigns to each subset of Θ a belief. Bel must satisfy the following conditions:

1. $\mathrm{Bel}(\phi) = 0$
2. $\mathrm{Bel}(\Theta) = 1$
3. $A \subseteq B \subset \Theta$ implies $\mathrm{Bel}(A) \leq \mathrm{Bel}(B)$

Belief functions are easily constructed from basic probability assignment functions. If m is a basic probability assignment function on 2^Θ, then the belief function constructed on the basis of m is defined by

$$\mathrm{Bel}(A) = \sum_{B \subseteq A} m(B)$$

It is the probability that the true state of affairs implies A.

The entities that produce evidence for or against the propositions are knowledge sources (KS's), such as sensors. For each KS there is a set of propositions to which it can directly contribute belief, which is that knowledge source's frame of discernment. Each KS has a unit of probability that it can distribute across the propositions and their disjunction in its frame of discernment. One simple way of distributing the probability is for a knowledge source to assign some probability to each proposition in its frame of discernment and to assign some probability to the disjunction of all the propositions. As more evidence becomes available, this probability of the disjunction of all the propositions will move toward the proposition in the frame of discernment.

As an example, consider a knowledge source KS_1, which analyzes an image containing a single object and produces a message related to what it found on the image. Let Θ be the set of all possible objects that it may find on an image. For the particular image examined, suppose that KS_1 computes the following probabilities:

1. $m_1(\{CHAIR\}) = .3$
2. $m_1(\{TABLE\}) = .1$
3. $m_1(\{DESK\}) = .1$
4. $m_1(\{WINDOW\}) = .15$
5. $m_1(\{PERSON\}) = .05$
6. $m_1(\Theta) = .3$

The knowledge source has decided that the degrees of belief committed to the object in the image being a CHAIR, TABLE, DESK, WINDOW, or PERSON are .3, .1, .1, .15, and .05, respectively. The assignment of mass .3 to Θ means that with probability .3 the knowledge source knows that something in Θ has occurred, but doesn't know what. Of course {CHAIR, TABLE, DESK, WINDOW, PERSON } $\subseteq \Theta$, but Θ may contain other possibilities as well. The belief function Bel_{KS_1}, which can be constructed from the model just given, is defined by

$$
Bel_{KS_1}(A) = \begin{cases} \sum_{l \in A} m_1(\{l\}) & \text{if } A \neq \phi \\ 0 & \text{if } \{CHAIR, TABLE, DESK, \\ & \quad WINDOW, PERSON\} \cap A = \phi \\ 1 & \text{if } A = \Theta \end{cases}
$$

Belief Combination

Dempster's rule of combination is used to combine the judgments of two independent knowledge sources. Suppose m_1 is the function that assigns probabilities to subsets of Θ by the first knowledge source. Hence the belief function Bel_1 for the first knowledge source is given by

$$
Bel_1(A) = \sum_{B \subseteq A} m_1(B)
$$

Likewise, suppose m_2 is the function that assigns probabilities to subsets of Θ by the second knowledge source. The belief function Bel_2 for the second knowledge source is given by

$$
Bel_2(A) = \sum_{B \subseteq A} m_2(B)
$$

Dempster's rule of combination defines the function m_{12} by

$$
m_{12}(C) = \frac{\displaystyle\sum_{A \subseteq \Theta,\ B \subseteq \Theta,\ A \cap B = C} m_1(A)m_2(B)}{\displaystyle\sum_{A \subseteq \Theta,\ B \subseteq \Theta,\ A \cap B \neq \phi} m_1(A)m_2(B)}
$$

when $C \neq \phi$ and $m_{12}(\phi) = 0$. The belief function Bel_{12} for the combined evidence is then defined by

$$\text{Bel}_{12}(A) = \sum_{B \subseteq A} m_{12}(B)$$

The underlying rationale of Dempster's rule of combination is as follows: The chance that the meaning of the message is A from the first knowledge source is $m_1(A)$, and the chance that the meaning of the message is B from the second knowledge source is $m_2(B)$. Because the knowledge sources are independent, the chance that the message means $A \cap B$ is $m_1(A) m_2(B)$. But if $A \cap B = \phi$, the messages are contradictory. Contradictions do not describe any situation of reality. So the chance distribution of the combined message is conditioned on the meanings being noncontradictory.

Returning to our example of knowledge source KS_1, suppose that there is a second knowledge source KS_2 that shares the same frame of discernment as KS_1, but the evidence that forms the basis for its belief in the various propositions is acoustic rather than visual. Suppose the m-values corresponding to its belief function are

$$m_2(\{CHAIR\}) = .2$$
$$m_2(\{TABLE\}) = .05$$
$$m_2(\{DESK\}) = .25$$

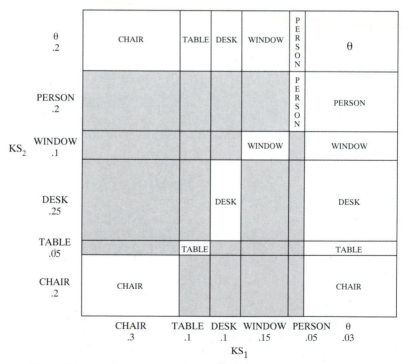

Figure 19.9 How beliefs from KS_1 and KS_2 are combined by Dempster's rule of combination. The white rectangles indicate the area representing the denominator of the combining equation.

$m_2(\{WINDOW\}) = .1$
$m_2(\{PERSON\}) = .2$
$m_2(\Theta) = .2$

To determine the combined m-values from these two sources, the two subdivided intervals from m_1 and m_2 are placed so as to define two contiguous sides of a square, which is then cut up into rectangles by lines drawn perpendicular to the two sides through the points that mark their subdivisions, as shown in Fig. 19.9. The areas of the rectangles thus formed are then normalized over the total area not committed to impossible intersections by either knowledge source. In Fig. 19.9 this would be equivalent to ignoring all the shaded area and renormalizing the remainder to be unit area. The resulting rectangles are the chance probabilities or, as Shafer likes to call them, "the m-values" committed to the intersections of propositions in the set KS_1 propositions $\times$ KS_2 propositions. The combined belief function is then constructed by using the equation $\text{Bel}(A) = \sum_{B \subseteq A} m(B)$.

In our example with knowledge sources KS_1 and KS_2, $m_{12}(\{CHAIR\})$ would represent the combined probability assigned to the label CHAIR. This would be computed as follows:

$$m_{12}(\{CHAIR\}) = \frac{\begin{matrix} m_1(\{CHAIR\})m_2(\{CHAIR\})+ \\ m_1(\{CHAIR\})m_2(\Theta)+ \\ m_1(\Theta)m_2(\{CHAIR\}) \end{matrix}}{\sum_{A \subseteq \Theta,\ B \subseteq \Theta,\ A \cap B \neq \phi} m_1(A)m_2(B)}$$

$$= \frac{(.2)(.3) + (.2)(.3) + (.2)(.3)}{\begin{matrix}(.3)(.2) + (.3)(.2) + (.1)(.05) + (.1)(.2) + (.1)(.25)+ \\ (.1)(.2) + (.15)(.1) + (.15)(.2) + (.05)(.2) + (.05)(.2)+ \\ (.3)(.2 + .05 + .25 + .1 + .2 + .2)\end{matrix}}$$

$$= \frac{.18}{.555}$$

$$= .32$$

One other belief-combining effect is mentioned by Garvey (Garvey, Lowrance, and Fischler, 1981). He points out that KS's contribute belief to the most precise proposition possible, even if it is a disjunction rather than a simple proposition. In this way the KS's can at least focus belief on subsets of possible propositions. Suppose, for example, that the simple propositions to which both KS's can contribute belief are $A, B,$ and C. Then the set of all disjunctions is $A \vee B$, $A \vee C$, $B \vee C$, and $A \vee B \vee C$. These disjunctions can be represented by the set unions $A \cup B$, $A \cup C$, $B \cup C$, and $A \cup B \cup C$, which is Θ. Since the conjunction of A and a disjunction that contains A is equal to A, the intersection of the belief contributed by KS_1 to Θ and the belief contributed by KS_2 to A becomes added belief in A.

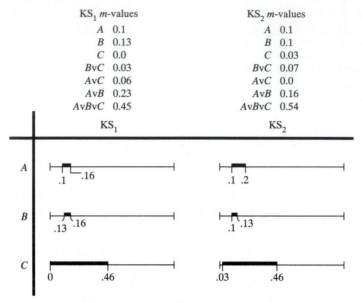

Figure 19.10 Belief intervals from both KS's.

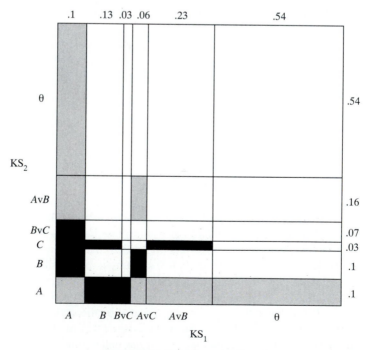

Figure 19.11 Belief in proposition *A*.

$$KS_1 \oplus KS_2$$

A	.1726
B	.217
C	.0555
BvC	.0561
AvC	.0324
AvB	.2474
	.781

Figure 19.12 Result of combined m-values.

$$KS_1 \oplus KS_2$$

A	.221
B	.278
C	.071
BvC	.072
AvC	.041
AvB	.316
	.999

Figure 19.13 Total areas renormalized to ignore useless areas.

The process in the more general case, when probability is distributed over all subsets of Θ, is illustrated in Figs. 19.10–19.13. Suppose that there are three propositions, $A, B,$ and C, that are exhaustive and mutually exclusive, as well as two knowledge sources, KS_1 and KS_2, capable of contributing belief to any one or combinations of the three. Figure 19.10 shows possible belief intervals for each proposition from both KS's at some point in the hypothetical investigation. (Judging by the amount of the uncertainty remaining, one may assume that the investigation is still in an early stage or the knowledge sources are having trouble collecting evidence.) Figure 19.11 showing the belief in proposition A illustrates that a proposition gains belief from the intersection of itself with itself, as well as from the intersection of itself with a disjunction that contains it and from the intersection of a pair of disjunctions that contain it. Figure 19.12 shows the results of the orthogonal sum $KS_1 \oplus KS_2$ before normalization. The total areas are normalized in Fig. 19.13 to ignore intersections of mutually exclusive propositions, such as $A \wedge B$, giving the final result.

19.5   A Probabilistic Basis for Evidential Reasoning

One difficulty with the Dempster-Shafer theory of evidence is the lack of an operational basis for the belief values it generates. Here we describe an operational probabilistic basis for belief: What we do not believe, we make no use of. Hence the degree of belief associated with a proposition in a body of evidence is related to the chance probability that an inference mechanism has access to the proposition for use in making an inference. In this model propositions in a body of evidence are treated as random propositions. A random proposition has two states: assertable and not assertable. A random proposition whose chance state is assertable is available for making inferences. A random proposition whose chance state is not assertable is not available for making inferences. An assertable proposition is considered to be true. A proposition whose chance state is not assertable is simply considered not to exist. The degree of belief in a proposition in a body of evidence is then the chance probability that it is assertable. The degree of belief in a proposition inferrable from a body of evidence is the conditional probability that a logical inference mechanism can semantically infer the proposition from the conjunction of chance suppositions in the body of evidence, given that the conjunction of assertions is not self-contradictory.

The operational probabilistic model for belief described here is simultaneously a generalization of Shafer's theory of evidence and a Bayesian approach to belief. To understand this, note that Shafer's frame of discernment contains a set of possibilities, exactly one of which corresponds to the truth. Our body of evidence contains a set of atomic propositions as well as possibly other well-formed formulas of propositions. There is no requirement that the propositions correspond to a set of possibilities, exactly one of which is true. Shafer uses Dempster's rule of combination, which is a generalization of Bernoulli's rule of combination as the mechanism to pool and combine evidence. We will demonstrate how Dempster's rule of combination is a specialization of the natural combination rule when belief in a proposition is considered as the chance probability that the proposition is semantically implied from a set of noncontradictory premises.

We begin our development by considering a model of the legal system, one of whose essential functions is to facilitate the establishment of evidence and the combining of evidence to help a jury decide what to believe. We conclude from this analysis that conditional probability of hypothesis, given evidence, cannot be used as the basis of a belief system calculus.

19.5.1 Legal Court Paradigm

There is an argument or question about h. To make a judgment about whether h is true or false, the court has two attorneys A and B, a jury to make the judgment, and a judge to administer and make sure that the jury is provided fair and relevant information. Attorney A is paid to uncover, organize, and present all evidence and inferences that are supportive of h. Attorney B is paid to uncover, organize, and present all evidence and inferences that are supportive of $\overline{h}$, the negation of h.

Evidence consists of physically observed facts and statements made by witnesses in response to the questions by the attorneys. The jury listens to all the evidence and all inferences made by the attorneys based on the evidence. In addition, the jury may make inferences from the evidence not stated by the attorneys. Then the jury collectively determines the weight of evidence in favor of h, and the weight of evidence in favor of $\bar{h}$ and makes a judgment of h or $\bar{h}$ accordingly.

What is happening here? The cases made by the two attorneys correspond to the establishment of a set of elementary propositions that constitute the evidence. With each elementary proposition in the evidence, the jury collectively associates an elementary or initial belief. In our model this initial degree of belief is the probability with which the elementary proposition is asserted and therefore available for use in an inference process.

Attorney A organizes an argument from a selected consistent subset of the body of evidence to show that h can be semantically implied from the body of evidence. Attorney B organizes an argument from a different selected subset of the body of evidence to show that $\bar{h}$ can be semantically implied from the body of evidence.

The attorneys' job is not just to present and select evidence but to present and select jury-believable evidence. Evidence, for example, that is secondhand is considered hearsay. Juries should not find it believable. Therefore it is legally inadmissible. Statements that cannot be inferred from the evidence is jury-unbelievable. Therefore the attorneys must present all the evidence relative to the inferences they would like to make.

Since h and $\bar{h}$ are contradictory, there must obviously be inconsistencies between the selected subsets of propositions in support of h and $\bar{h}$. Both subsets cannot be simultaneously believed. Indeed, the dependencies of each subset on the other influence the degree of belief in each subset and in the inferences that might be drawn from each subset.

A theory for evidential reasoning, if it follows the pattern of the legal paradigm, must then have the following five features:

1. Each piece of evidence is a proposition.
2. The question to be decided is a proposition.
3. Each proposition in the body of evidence has associated with it a measure of its assertability.
4. There must be a logic calculus for inferring propositions from conjunctions of other propositions.
5. There must be a belief calculus for computing a degree of belief for each proposition the inference calculus is able to infer. The degree of belief in an inferred proposition will depend on the measure of assertability for the proposition in the body of evidence.

It is natural that the logic calculus be based on symbolic logic. A proposition r can be inferred from proposition $q_1, \ldots, q_N$ if and only if r is semantically implied by the proposition $q_1, \ldots, q_N$. It is also natural that the degree of belief in a proposition q should be equal to the degree of belief in a proposition r if q and r are logically equivalent. The commitment to these two constraints has an important consequence: The mechanism that computes a measure of belief for any inferred proposition cannot use conditional probability as its degree of belief in the proposition $q \rightarrow r$. Belief in $q \rightarrow r$ must be equal to belief in $\bar{r} \rightarrow \bar{q}$, since the two are logically equivalent. Yet $P(r|q)$ is independent of $P(\bar{q}|\bar{r})$.

To ensure consistency in belief values, belief in $q \rightarrow r$ must be taken to be belief in $\overline{q} \vee r$, which is the logical definition of $q \rightarrow r$. How to compute belief in $\overline{q} \vee r$ is discussed in the next sections on the belief calculus.

19.5.2 Degree of Belief as Chance Probability of Being Inferred

In our development we use probability as the measure of belief of a proposition. The derived probability associated with belief has a frequentist interpretation, although the assertion probabilities on which the derived belief probabilities depend may have an objectivist frequentist interpretation or a subjectivist personal probability interpretation.

The use of probability as a means of belief is not new. Indeed Good (1952) defined the theory of probability as the logic of degrees of belief. Shafer (1981) disagrees with this view. Shafer argues that using probability commits one to have the sum of the degree of belief in a proposition h and the degree of belief in a proposition $\overline{h}$ total to 1, the total belief. This argument would be correct if there were only one way to use probability as a measure of degree of belief, and that way associated the probability of the truth or validity of a proposition h with the belief of h. But degree of belief in a proposition does not have to do with the validity of the proposition, with its truth or falseness. We maintain that degree of belief has to do with the chance probability with which the proposition can be inferred from its body of evidence. And in conformance with the law of probability, the chance probability that a proposition can be inferred plus the chance probability that it cannot be inferred does total to 1. But the chance probability that a proposition can be inferred from a body of evidence has little to do with the chance probability that the negation of a proposition can be inferred.

In order to understand the operational consequences of this point, we must make a distinction between a proposition, say q, and the assertion of the proposition q. In the logic of Boole, this is the distinction between the statement x and the statement $x = 1$, meaning x is true. In order to eliminate the logical difficulty of having to deal with statements of the form $(x = 1) = 1$, whose pattern of $(\ldots) = 1$ can continue on ad infinitum, Whitehead and Russel developed a symbolic logic that could have no such statements, since they considered the statement q itself as the assertion of q. But Whitehead and Russel were concerned with validity and not with belief. When we state a proposition, our statement serves only to call the proposition into existence. When we assert a proposition, then we can use it to make inferences of other propositions. Any proposition that has been inferred is considered asserted. Degree of belief makes the connection between the existence of a proposition and the degree to which it can be asserted.

19.5.3 Belief Calculus

The belief calculus is a set of rules from which the degree of belief in any inferred proposition can be computed from the measure of assertability of propositions in body of evidence. By our definition, the degree of belief in our inferred proposition is the conditional probability of the chance inference of the proposition, given a chance state of noncontradictory propositional assertions. The key to understanding the belief calculus then revolves around the meaning of chance state of propositional assertions and the way to compute its probability.

We found the belief calculus on the following stochastic model. Each proposition in the body of evidence can be in either the state of asserted or the state of nonasserted. The probability with which a proposition in the body of evidence can be asserted is what we have earlier called its measure of assertability. The body of evidence itself consists of the sets of its propositions, each associated with its probability of being asserted. A state of assertion for E is given by a subset S of E. The propositions in the subset S are considered to be asserted, and the propositions in the subset $E - S$ are considered to be nonasserted. We consider that the propositions in E are mutually independent insofar as their assertability is concerned. The probability for the chance state of assertion S can then be computed by

$$\prod_{q \in S} m(q) \prod_{r \in E-S} [1 - m(r)]$$

As there are $2^{\#E}$ subsets of E, there are $2^{\#E}$ chance states of assertion. Each chance state of assertion S contains propositions that are considered to be asserted. Let C be the collection of those states of assertion from which the contradiction cannot be inferred. Let H be the collection of those states of assertion from which the proposition h can be inferred. Then the degree of belief in h is defined by

$$Bel\ (h) = \frac{m(H \cap C)}{m(C)}$$

$$= \frac{\displaystyle\sum_{S \in H \cap C} \prod_{q \in S} m(q) \prod_{r \in E-S} [1 - m(r)]}{\displaystyle\sum_{S \in C} \prod_{s \in S} m(s) \prod_{r \in E-S} [1 - m(r)]}$$

We now present a variety of examples to illustrate this calculation concretely so that we can fully appreciate its meaning and consequences.

19.5.4 Examples

First we consider the case in which the body of evidence consists only of the propositions q and $q \to r$. There are four possible states of assertion:

q	$q \to r$	
A	A	$\Rightarrow r$
A	NA	
NA	A	
NA	NA	

where we use A to designate the state "asserted" and NA to designate the state "not asserted." Note that a proposition whose state is "not asserted" is not a proposition that is denied. A proposition that is denied is one whose negation is asserted. To be not asserted simply means that it is unavailable to the inference mechanism.

The only subset of $E = \{q, q \rightarrow r\}$ from which the inference r can be made is $S = \{q, q \rightarrow r\}$. Since no subset of E is contradictory, we can compute

$$Bel\ (r) = m(q)\ m(q \rightarrow r)$$

Next suppose that the question is the proposition q and the entire body of evidence consists of the conjunction qr; $E = \{qr\}$.

$$
\begin{array}{l}
qr \\
A \quad \Rightarrow q \\
NA
\end{array}
$$

Again, the only subset of $E = \{qr\}$ from which the inference r can be made is $S = \{qr\}$, and since S is noncontradictory, we have

$$Bel\ (r) = Bel\ (qr)$$

Now suppose that the question is the proposition $q \vee r$ and the entire body of evidence consists of the proposition q,r; $E = \{q,r\}$.

$$
\begin{array}{lll}
q & r & \\
A & A & \Rightarrow q \vee r \\
A & NA & \Rightarrow q \vee r \\
NA & A & \Rightarrow q \vee r \\
NA & NA &
\end{array}
$$

The subsets from which the inference $q \vee r$ can be made are $S_1 = \{q\}$, $S_2 = \{r\}$, and $S_3 = \{q,r\}$. Since each of these is noncontradictory, the chance probability that $q \vee r$ can be inferred is

$$Bel\ (q \vee r) = m(q) + m(r) - m(q)\ m(r)$$

This is not unlike Bernoulli's rule of combination.

Some evidence can be supportive of other evidence. Consider the case in which the evidence consists of $q,r,q \vee r$. The question is the proposition $q \vee r$. Again, there are no states of assertion that are contradictory, and the only state of assertion from which the inference $q \vee r$ cannot be drawn is $\emptyset$. Hence

$$
\begin{aligned}
Bel\ (q \vee r) &= 1 - [1 - m(q)][1 - m(r)][1 - m(q \vee r)] \\
&= m(q) + m(r) - m(q)\ m(r) + \\
&\quad \{1 - [m(q) + m(r) - m(q)\ m(r)]\}m(q \vee r)
\end{aligned}
$$

This is clearly greater than what the degree of belief in $q \vee r$ would be if the supporting evidence $q \vee r$ were not present.

The simplest case in which the body of evidence contains two conflicting propositions is when $E = \{q,\overline{q}\}$. Consider the case when the question is q. The subsets from which the inference q can be made is only the subset $\{q\}$. The noncontradictory subsets are $\{q\}$, $\{\overline{q}\}$, and $\emptyset$. Here we obtain

$$Bel\ (q) = \frac{m(q)[1 - m(\overline{q})]}{m(q)[1 - m(\overline{q})] + m(\overline{q})[1 - m(q)] + [1 - m(q)][r - m(\overline{q})]}$$
$$= \frac{m(q)[1 - m(\overline{q})]}{1 - m(q)m(\overline{q})}$$

This form is like Dempster's rule of combination. The numerator is the sum of products of the chance probability of noncontradictory assertions that semantically imply the proposition q, and the denominator is 1 minus the product of the chance probability of assertions that are contradictory. By rearranging, there results

$$Bel\ (q) = m(q) - \frac{m(q)m(\overline{q})[1 - m(q)]}{1 - m(q)\ m(\overline{q})}$$

Thus conflicting evidence can make a belief of an inferred proposition have a lower value than the probability of its assertability.

Next we consider a case in which there is a more complex conflict dependency among the propositions of the evidence. Suppose the body of evidence consists of q, r, and $q \rightarrow \overline{r}$. The question is the proposition $q \vee r$.

q	r	$q \rightarrow \overline{r}$	
A	A	A	$\Rightarrow f$
A	A	NA	$\Rightarrow q \vee r$
A	NA	A	$\Rightarrow q \vee r$
A	NA	NA	$\Rightarrow q \vee r$
NA	A	A	$\Rightarrow q \vee r$
NA	A	NA	$\Rightarrow q \vee r$
NA	NA	A	
NA	NA	NA	

There are five noncontradictory states of assertion from which $p \vee r$ can be inferred. They are $\{q,r\}$, $\{q, q \rightarrow \overline{r}\}$, $\{q\}$, $\{r, q \rightarrow \overline{r}\}$, $\{r\}$. The contradictory state is $\{q, r, q \rightarrow \overline{r}\}$. Hence we have

$$Bel\ (q \vee r) = \{m(q)m(r)[1 - m(q \rightarrow \overline{r})] + m(q)[1 - m(r)]m(q \rightarrow \overline{r})$$
$$+ m(q)[1 - m(r)][1 - m(q \rightarrow \overline{r})] + [1 - m(q)]m(r)m(q \rightarrow \overline{r})$$
$$+ [1 - m(q)]m(r)[1 - m(q \rightarrow \overline{r})]\} \Big/ [1 - m(q)m(r)m(q \rightarrow \overline{r})]$$

$$= \{m(q) + m(r) - m(q)m(r)[1 + m(q \rightarrow \overline{r})]\} \Big/ [1 - m(q)m(r)m(q \rightarrow \overline{r})]$$

$$= m(q) + m(r) - m(q)m(r) -$$
$$\frac{[1 - [m(q) + m(r) - m(q)m(r)]]m(q)m(r)m(q \rightarrow \overline{r})}{1 - m(q)m(r)m(q \rightarrow \overline{r})}$$

This shows again that when conflicting relevant evidence exists, the degree of belief in a proposition will be less than what it would be if the conflicting evidence did not exist.

For our next case of conflict, suppose the evidence $E = \{q, q \rightarrow r, q \rightarrow \bar{r}\}$ and the question is r.

q	$q \rightarrow r$	$q \rightarrow \bar{r}$	
A	A	A	$\Rightarrow f$
A	A	NA	$\Rightarrow r$
A	NA	A	
A	NA	NA	
NA	A	A	
NA	A	NA	
NA	NA	A	
NA	NA	NA	

Here we have

$$Bel\ (r) = \frac{m(q)\ m(q \rightarrow r)[1 - m(q \rightarrow \bar{r})]}{1 - m(q)\ m(q \rightarrow r)\ m(q \rightarrow \bar{r})}$$

From the form of this relation, we can see that if the degree of assertability for $q \rightarrow \bar{r}$ is 1, then the belief in r must be zero. And this happens even if the degree of belief in $q \rightarrow r$ is also 1. The largest that $Bel\ (r)$ can be is $Bel\ (q)Bel\ (q \rightarrow r)$.

For our final case of conflict, suppose $E = \{q, q \rightarrow r, s, s \rightarrow \bar{r}\}$ and the question is $q \rightarrow \bar{s}$.

q	$q \rightarrow r$	s	$s \rightarrow \bar{r}$	
A	A	A	A	$\Rightarrow f$
A	A	A	NA	
A	A	NA	A	$\Rightarrow g \rightarrow \bar{s}$
A	A	NA	NA	
A	NA	A	A	
A	NA	A	NA	
A	NA	NA	A	
A	NA	NA	NA	
NA	A	A	A	$\Rightarrow q \rightarrow \bar{s}$
NA	A	A	NA	
NA	A	NA	A	$\Rightarrow q \rightarrow \bar{s}$
NA	A	NA	NA	
NA	NA	A	A	
NA	NA	A	NA	
NA	NA	NA	A	
NA	NA	NA	NA	
A	NA	NA	A	

Here we have

$$Bel\ (q \rightarrow \bar{s}) = \frac{m(q \rightarrow r)m(s \rightarrow \bar{r})[1 - m(q)m(s)]}{1 - m(q \rightarrow r)m(s)m(q)m(s \rightarrow \bar{r})}$$

So if, for example, $m(s) = .9$, $m(q) = .2$, $m(q \rightarrow r) = .8$, and $m(s \rightarrow \bar{r}) = .9$, then

$$Bel \ (q \rightarrow \bar{s}) = \frac{.8(.9)[1 - .2(.9)]}{1 - (.8)(.9)(.2)(.9)}$$

$$= \frac{.5904}{.8704} = .6783$$

while if the probability of the assertability of q is raised to .7 so that $m(2) = .9$, $m(q) = .7$, $m(q \rightarrow r) = .8$, and $m(s \rightarrow r) = .9$, then

$$Bel \ (q \rightarrow \bar{s}) = \frac{.8(.9)[1 - .7(.9)]}{1 - .8(.9)(.7)(.9)}$$

$$= \frac{.2664}{.5464} = .4876$$

These conflict examples illustrate that the degree of belief in an inferred proposition is not just a function of the degree of the assertability of the proposition from which the inference stems. It is also a function of the context of those propositions in the body of evidence. The issue of contextual dependence immediately raises the question of contextual irrelevancy. If the evidence E can be divided into two mutually exclusive sets E_1 and E_2, and if

$$Bel(h|E_1, E_2) = Bel(h|E_1)$$

then with respect to h, E_2 is contextually irrelevant.

19.6   Example Systems

To explore knowledge-based vision further, we have selected a set of knowledge-based vision systems to look at in some detail. Each system attempts to use some form of knowledge to analyze images. The knowledge varies from highly scene-specific to general knowledge of perspective projections. The applications of these systems include such diverse ones as aerial image analysis, bin picking, and automatic mail sorting. Binford (1982) provides a survey of some knowledge-based systems.

19.6.1   VISIONS

Hanson and Riseman (1978) designed one of the early knowledge-based systems, called VISIONS (Visual Integration by Semantic Interpretation of Natural Scenes). The system has multiple levels of representation both for the information on a particular image and for the information contained in the long-term knowledge base. It has modular knowledge sources that transform information from one level of representation to another. It has a hierarchical strategy to control the employment of the knowledge sources, and it has a search tree that stores the history of the search process. The VISIONS system uses both declarative and procedural forms of knowledge and provides bottom-up and top-down paths for hypothesis development. Declarative knowledge is represented in the semantic network style. Three-dimensional shapes are represented in terms of B splines and Coon's surface patches.

VISIONS operates in the domain of natural outdoor scenes and therefore has to employ partial models. It does this through schema mechanisms that permit a concept to be described at three levels:

1. As an entity in and of itself having properties that do not relate to the context in which the entity may exist;

2. As an entity that is part of a schema and therefore carries information relative to the role it plays within the context of the schema;

3. As an entity that itself is a schema and thereby contains information about all its parts and their relationships.

VISIONS has an uncertainty calculus that has evolved from generalized confidence to Bayesian belief probabilities to Dempster-Shafer belief functions. Draper et al. (1989) discuss its schema system in detail. Beveridge et al. (1989) discuss its segmentation procedure.

19.6.2 ACRONYM

The ACRONYM system (Brooks, 1981, 1983) uses symbolic reasoning to aid in analyzing scenes. It performs iterations of prediction, description, and interpretation in a coordinated effort to describe the objects present in an image.

ACRONYM uses stored models whose primitives are generalized cones. A set of primitives, along with constraints on size, structure, and spatial relationships, defines a class of models. Additional constraints define subclasses and eventually specific instances of object models. The basic unit for object description is the frame. Each frame has a unique name and a set of slots. Slot fillers can be constants, variables, expressions, or references to other nodes. Frames are the nodes of a structure called the *object graph*. All models are stored in the object graph. The object graph is a hierarchical structure, representing objects from coarse detail to fine detail.

When the user specifies models, she can also specify algebraic constraint expressions. As the system receives feedback from the low-level processes working with the image, it builds up more constraints. Constraints are stored in the *restriction graph*, also a hierarchical structure. The base-restriction node of the restriction graph has an empty constraint set. Arcs are directed from less restrictive nodes to more restrictive nodes. Thus a hierarchy of stricter and stricter constraints is built up during the analysis of an image. Symbolic constraint manipulation is performed by a reasoning system based on the work of Bledsoe (1975).

Using knowledge from the object models and feedback from the low-level processes, ACRONYM constructs a *prediction graph*. The nodes of this structure represent either predictions of specific image features (such as shapes) or complete prediction graphs of higher-level features. Arcs indicate relationships between nodes. The restriction graph is attached to the prediction graph as follows: A least-restrictive restriction node is associated with the entire prediction graph. More-restrictive restriction nodes are attached to each node and arc in the prediction graph.

The entities that ACRONYM can predict are called *observables* and are defined recursively as image features or directly computable relations between observables. The most useful observables are invariant ones—those that remain constant and observable over the entire range of variation in model size, structure, and position. At the lowest level ACRONYM deals with two kinds of observable features: ribbons (the two-dimensional projections of the generalized cones) and ellipses (which describe the shapes generated by the ends of generalized cones).

The low-level descriptive processes of ACRONYM begin with the image, produce edge elements, perform edge linking, and finally construct a set of ribbons and ellipses under the direction of the predictor module. These observables and their relationships are stored in the *observation graph*. The interpreter module matches these descriptions to models, constructing an *interpretation graph*, and then uses measurements on the image features to constrain parameters of the models to be matched. The contraints are added to the restriction graph and are used by the predictor module to further direct the descriptive processes. Restriction nodes that are inconsistent cause elimination of the interpretation nodes that caused them. Additional iterations of prediction, description, and interpretation occur as finer and finer details are identified. The final version of the interpretation graph represents a match of a maximal subgraph of the prediction graph and constitutes an interpretation of the image.

The purpose of the constraint manipulation system is to determine whether a set of constraints is inconsistent, thus making certain hypothesized interpretations impossible. The ACRONYM system was the first computer vision system to employ a general symbolic constraint manipulation system for this purpose. Other systems have employed constraint consistency checks, but in a more restrictive setting.

ACRONYM has been tested on aerial images of scenes containing wide-bodied jets. It was given a generic model of wide-bodied passenger jets plus class specializations to L-1011s and Boeing-747s. The Boeing-747 class was further specialized to Boeing-747B and Boeing-747SP. Using extremely noisy data, ACRONYM was able to classify accurately those substructures it recognized as airplanes. However, in many cases the extracted airplane features were not close enough to what ACRONYM was expecting, and it failed to recognize them at all. Although many of the problems were due to low-to-mid-level processing, the remainder were caused by the rigidity of the models and matching process. ACRONYM is important for its intelligent use of models, constraints, and feedback.

19.6.3 SPAM

SPAM (McKeown, Wilson, and McDermott, 1985) is a knowledge-based system that uses map and domain-specific knowledge to interpret airport scenes. The system has three main components: an image/map database, a set of image-processing tools, and a rule-based system. The image/map database contains facts about the airport, represented in map space, not image space, and related to the image through a camera model. Facts include locations (latitude, longitude) and elevations of features and facilities to compute properties and relationships. The image-processing tools include a region-growing segmentation procedure, a road follower, procedures for linear feature extraction and three-dimensional junction analysis, and an interactive human segmentation system.

The idea behind the system is to perform an initial segmentation, hypothesize interpretations for regions, and verify or weaken hypotheses based on interpretations of nearby regions. The regions that form consistent cliques that model most of the airport end up with the highest confidences. Then such a clique can be used as a starting point from which the remainder of the regions can be classified.

The rules manipulate three kinds of primitives: region, fragments, and functional areas. Regions are from the image and are classified as either linear, compact, small blob, or large blob. Fragments are interpretations for regions. Linear regions can be interpreted as subclass runway, taxiway, or access road; compact regions can be interpreted as terminal buildings or hangars; small-blob regions can be interpreted as parking lots or parking aprons; and large-blob regions can be interpreted as tarmac or grassy areas. Functional areas represent distinct spatial subdivisions of the scene. A functional area contains features normally found in close proximity.

The initial segmentation gives lots of errors. There are missing and extra regions. Furthermore, objects and shadows cause regions inside what would normally be large, uniform areas. Thus much of the interpretation problem turns out to be a resegmentation problem.

There are five sequential phases to the procedure. The *build phase* selects regions and creates fragment interpretations. It calculates the initial confidences for the classes and subclasses into which a region may fall. The rules used by the build phase are basically classification rules. The *local evaluation phase* processes fragment interpretations. Its rules invoke image-processing tools for region enlargement, extension, joining, and merging. For example, two approximately collinear roads separated by a gap might be joined to form one roadway.

The *consistency phase* checks the consistency of the interpretations of neighboring regions. For example, one rule specifies that runways are oriented parallel to terminal buildings. Another says that taxiways intersect runways. A third says that access roads generally have curved segments. The *functional area phase* tries to find functional areas starting from the consistent cliques around such focal features as terminals, access roads, runways, and hangars. The functional area rules dictate when fragments can be grouped as a functional area. The last phase is a *global evaluation phase* to find the complete airport interpretation. It uses goal generation rules that recognize unlikely or impossible situations and prune the weakly consistent fragments that caused them. For example, a rule might dictate the removal of weak runway fragments that have much stronger support as taxiways or roads.

The system has successfully interpreted difficult airport scenes where the initial segmentations were poor. The rules for improving the segmentation make it much more powerful than ACRONYM, which could only interpret those parts of the image that had segmented well. However, the amount of scene-specific knowledge required could be a burden in constructing systems for other domains and makes a general-purpose system infeasible. Papers related to SPAM include McKeown and Denlinger (1984, 1988), McKeown and Pane (1985), and McKeown (1988).

19.6.4 MOSAIC

The goal of the MOSAIC image-understanding system (Herman and Kanade, 1986) was to build and incrementally improve a three-dimensional model of a complex urban scene, using a sequence of both monocular views and stereo pairs. The system did not have any specific models of the buildings it was looking at, but had built-in knowledge of buildings in general, particularly of how they would appear in aerial images. Three main tasks are executed by the system: stereo analysis, monocular analysis, and the construction and modification of the scene model.

The stereo-matching subsystem performs junction-based stereo matching by using a constraint satisfaction approach. It uses the (built-in) knowledge that the images come from vertical aerial photography, that L junctions appear on building roofs, and that all junctions on a single roof should have the same height. A cost function is developed on the basis of both the gray-level matches between potentially corresponding junctions and the three-dimensional consistency of matches between pairs of junctions. A network is constructed whose nodes are pairs of possibly-matching junctions. A beam search that extends only the best n paths at each stage tries to find a path through the network that minimizes a sum of costs. Then a triangulation procedure is used to construct a three-dimensional wire-frame representation of the scene as determined from the two views.

The monocular analysis is unusual in that, owing to the highly restrictive nature of the domain, it can produce a reasonable hypothesis of the three-dimensional structure of the scene from just one view. It first uses its knowledge of buildings in the feature-extraction phase. After initial lines and junctions are extracted, a graph is constructed whose nodes are junctions and whose links represent possible connections between pairs of junctions that are hypothesized to belong to the same building, on the basis of proximity and spatial relationships. The graph is then pruned by a verification technique that takes into acount the percentage of the hypothesized line connecting two junctions that can actually be accounted for by line segments in the image. Finally, original junction legs that may have been deleted are added back in and extraneous segments deleted. The result is a much improved line–junction structure. A three-dimensional wire frame is constructed from this structure by using a combination of ground–plane knowledge, camera geometry, and heuristics. Given a junction point from the image structure, the ray on which its corresponding three-dimensional point lies is known. In the absence of other data, MOSAIC selects an arbitrary depth for the point, thus fixing its three-dimensional coordinates. It then uses building heuristics to propagate the depth consistently within a single structure and even between structures if two horizontal lines are known to be at the same height.

The model-construction-and-modification module uses a relational structure called the *structure graph* to represent the current three-dimensional scene model. Nodes of the graph represent topological primitives (faces, edges, vertices, objects, and edge groups) and geometric primitives (planes, lines, and points). Primitives are *confirmed* if they come from one of the sequence of input images and *unconfirmed* if they are only hypothesized. Edges represent either part-of links or geometric constraint links that attach topological primitives to corresponding geometric primitives. The wire-frame description obtained from the first image or stereo pair is used to construct the first scene-model structure graph. The wire-frame description obtained from the most recent stereo or monocular analysis is represented as a separate image structure graph.

Updating the scene model is a process of matching the image structure graph to the model structure graph, merging matched vertices and edges from the image structure with those in the model structure, deleting inconsistent hypothesized elements from the model structure, and adding unmatched vertices and edges from the image structure to the model structure as new confirmed elements. Finally, a hypothesizing step is performed that uses knowledge of planar-faced objects and of urban scenes to add missing features and produce a complete surface model of the scene that can be displayed graphically.

MOSAIC was tested on a number of aerial images of urban scenes. The successful results demonstrated that a knowledge-based system in a very limited domain can be made quite powerful. Since many of the techniques were strongly dependent on the objects in the scene being mostly rectangular buildings with flat roofs, it would be difficult to apply MOSAIC to other domains.

19.6.5 Mulgaonkar's Hypothesis-Based Reasoning

Mulgaonkar (Mulgaonkar, Shapiro, and Haralick, 1986) developed a hypothesis-based reasoning system and used it for interpreting line drawings of perspective views of scenes. His reasoning system works with hypotheses that are relational tuples containing a relation name, a set of related entities, and the attributes of the relationship. For example, if SURFACE1 and SURFACE2 are planar surfaces in three-space, then (CONNECT, SURFACE1, SURFACE2, 45) represents the hypothesis that SURFACE1 and SURFACE2 meet at a 45° angle. The state of the world (scene) is the set of the "correct" hypotheses selected from the large space of possible hypotheses. It is the job of the interpretation system to input the entities in the observed instance of the world (the image) and the measurements made, and to determine the unknown state of the world.

Mulgaonkar uses rules called inference engines that operate on the relational tuples and measured attributes of the image and compute from them values for some of the unknown attributes. A hypothesis is considered inconsistent if two separate inference engines compute a value for the same attributes of the hypothesis and the two values are incompatible. For example, if one inference engine infers (CONNECT, SURFACE1, SURFACE2, 10°) and a second infers (CONNECT, SURFACE1, SURFACE2, 90°), then something is wrong.

The system reasons about the three-dimensional entities LINE3D, POINT3D, and PLANE3D. The attributes of LINE3D are its vanishing point on the image and the direction cosines of the line in three-space. The attributes of POINT3D are its coordinates in three-space. The attributes of PLANE3D are the normal to the plane, the equation of the vanishing trace of the plane in the image (once located), and the distance of the plane from the origin. The other three-dimensional quantity that the system reasons about is the focal length of the camera. On the image the system works with lines, points, arcs, and planes (lists of lines and areas whose three-dimensional counterparts occupy the same plane).

At the beginning of the reasoning process, none of the numerical three-dimensional attributes are known, unless available from other analyses such as a shape-from-shading or stereo process. As the reasoning proceeds, the inference engines compute values for three-dimensional attributes, and the appropriate slots are filled with numerical values. Inference engines can compute such quantities as vanishing points of two parallel lines, vanishing trace of a plane, focal length, and so on. They contain knowledge about the relationship of the perspective projections of lines, rectangles, and conics to the three-dimensional entities. For example, given the hypothesis that two lines are parallel and that their projections are not parallel, one inference engine determines the vanishing point on the image as the intersection of the projections. Given the hypothesis that two lines are coplanar and do not have a common vanishing point, a second inference engine computes the vanishing trace for the plane.

Most of the work in Mulgaonkar's research lay in developing the hypothesis-based reasoning system and the mathematics behind the inference engines. The actual applications to vision were on fairly simple scenes. The utility of such a system in a real application has yet to be proved.

19.6.6 SCERPO

The SCERPO (Spatial Correspondence, Evidential Reasoning, and Perceptual Organization) system (Lowe, 1987) performs three-dimensional object recognition from single two-dimensional images. The overall control structure of the system is to find a few initial matches between model and image features, determine a hypothesized transformation from model to image, apply the transformation to the model to obtain a two-dimensional line drawing (with hidden lines removed), and verify that the lines in the original image match well to the line drawing. This is a hierarchical control structure that works bottom up until a model and transformation are selected and then works top-down to verify.

SCERPO's models are three-dimensional wire-frame objects, and line segments are the only primitives. The features to be used in matching were chosen on the basis of being detectable in the absence of domain knowledge by a bottom-up analysis and yet of sufficient specificity that they could serve as indexing terms into a database of objects. Lowe chose three relationships among line segments as features to be used in matching: proximity, parallelism, and collinearity. Line segments in these relationships form perceptual groupings that humans can easily find amid noise segments, and the relationships are invariant or approximately invariant under perspective projection if the camera is far enough away from the object.

One novel aspect of Lowe's work was the idea of ranking the perceptual groupings found in the image before trying to use any of them in matching. Groupings that have a low probability of occurring accidentally, and therefore a high probability of actually matching a corresponding grouping in the object model, get high rankings, whereas groupings that have a high probability of occurring by accident are assigned low rankings. To this end Lowe has developed formulas that express (1) the expected number of endpoints within a radius r of an endpoint of a line segment of length l; (2) the expected number of line segments within a given separation s of a given line segment, and within a given angle θ of the angle of that line segment; and (3) the expected number of line segments within a given gap length g and separation s of a given line segment, and within a given angle θ of the angle of that line segment. The first formula allows SCERPO to calculate the significance of proximate relationships among line segments, the second is for parallel relationships, and the third is for collinear relationships.

Structures in the object models that give rise to perceptual groupings are calculated off line and stored in a database. In the on-line processing, SCERPO extracts line segments from the image, forms perceptual groupings, and ranks them as described above. It then iteratively performs matching of image grouping to model grouping, verification based on the computation of the transformation from matched image segments to model segments, and extension of the match to include more line segments.

SCERPO was tested on a number of common objects, such as a stapler and a disposable razor. The idea of ranking perceptual groupings to be used in matching is an important one that needs further exploration.

19.6.7 Ikeuchi's Model-Based Bin-Picking System

Ikeuchi (1987) designed and implemented a model-based vision system for bin-picking tasks, using range data obtained from dual photometric stereo. The system has two modes: compile and run. In compile mode a geometric modeler is used to generate views of an object under various view directions. For an object with n faces, each of the generated views is represented by a vector of the form $(X_1, X_2, \ldots, X_n)$, where X_i is equal to 1 if face i is visible in this view, and to zero otherwise. The views that have identical vectors become an *attitude group*, and each attitude group is represented by a representative attitude. A work model consisting of a set of attributes, such as inertial moments of faces, relative positions of faces, face shapes, edge information, and extended Gaussian image, is constructed for each of the representative attitudes. A procedure that was executed by humans in Ikeuchi (1987), but is now automatic, constructs an *interpretation tree* from the set of representative work models. The interpretation tree is a decision-tree classifier. Each node represents a decision to be made on the basis of the value of one of the model's attributes.

In run mode the system generates three edge maps, two needle maps, and a depth map of the scene being analyzed, using dual photometric stereo. The highest region in the depth map is passed to the interpretation tree to make its first decision. The interpretation tree dictates which features are to be extracted from the maps and how to compare them with the features in the work models. The unknown object at the top of the bin is first classified into one of the attitude groups, and then its exact pose is found. The object can be removed from the bin by a robot and the process repeated for the next object.

The interpretation tree used in this system is a simple method for precompiling knowledge that can be used to recognize an object rapidly (there is no backtracking in this system). However, if the interpretation tree is generated from CAD models or other perfect data, it is not guaranteed to recognize objects from noisy data. Ikeuchi attempted to solve this problem in later work (Ikeuchi and Kanade, 1987) by modeling sensors and predicting variations in the features that can be detected. The next system we will look at uses training data to solve this problem.

19.6.8 Jain's Evidence-Based Recognition System

Jain and Hoffman (1988) have constructed a three-dimensional object recognition system that uses range data for input and combines statistical and structural techniques to identify objects from a database of models. Instead of matching image structures to object models, they constructed a rulebase of evidence rules from training data. They then used the rules to identify an unknown view of an unknown object.

Since this system uses range data, it applies three-dimensional segmentation techniques to partition the data into surface patches. The patches are classified as planar, convex, or concave, and the edges that separate them are classified as *jump edges*, which have discontinuous depth, or *crease edges*, which have continuous depth and discontinuous normals. From this set of patches, an initial representation structure is constructed. The initial representation contains three different kinds of information. *Morphological* information is global information concerning the entire three-dimensional shape. The morphological information computed includes the perimeter of the set of nonbackground pixels, the number of connected background components, and the number of connected background holes. Since surface patches are the object primitives, patch information is local information about each primitive. *Patch* information includes the sense of the patch (planar, convex, or concave), its size, its span, and a piecewise linear fit of the boundary. Relational information describes the relationships between pairs of patches. The relational information computed for a pair of surface patches includes the type of their boundary edge, the normal angle between them, the minimum and maximum distances between them, the boundary angle, a linear fit of their boundary, and the jump gap between them. But segmentation of range data is not perfect, and noise in the data can cause extra patches. Jain and Hoffman solved this problem by constructing from the initial representation n modified representations, one for each object model in the database. The modified representation corresponding to object i uses knowledge of the boundary angles between faces of object i as constraints to force merging of patches with impossible angles between them.

The evidence-based recognition procedure uses morphological, patch, and relational evidence conditions. Each condition consists of a combination of features of the appropriate type with associated bounds on each feature. The conditions for recognition are stored in an evidence feature rulebase. Each rule in the rulebase has four parts: a list of features, a list of corresponding bounds for those features, a list of bounds on the number of occurrences of those features, and a vector of weights between -1.0 and 1.0. If there are n objects in the database, then each rule has a vector of n weights. The ith weight gives the support for the hypothesis that the object being identified is object i.

Suppose that there are r rules in the rulebase. An *evidence r-vector* for model i is a vector of r weights, one for each rule in the database. The jth weight is the weight that is found in rule j supporting the hypothesis that the unknown object being analyzed is object i. An *instance r-vector* is a vector of r entries, each entry having value 0 or value 1. The jth entry has value 1 if the modified representation of the unknown object corresponding to the ith object satisfies rule j, and 0 otherwise. Since there are n different modified representations, there will actually be n different instance r-vectors to be matched against the n different models.

In the process of matching the ith instance r-vector to the ith evidence r-vector, the components that represent positive evidence for object i are distinguished from those that represent negative evidence for object i. The similarity measure that is computed between the instance and the model is the difference between a measure of the support for the model and a measure of the refutation. The support measure is the product of the proportion of observed features that are positive evidence for object i and the proportion of object i's positive evidence that is observed. The refutation measure is the square of the proportion of observed features that are negative evidence for object i.

The recognition procedure calculates the similarity between the unknown object and each of the models. It classifies the unknown object as object i if the latter has a unique maximum similarity and if there is at least one instance of major evidence (weight=1) for object i. The rulebase used in the recognition procedure is generated automatically by an inductive algorithm. The technique has been tested for 5 to 10 object databases with several hundred experiments and achieved a high success rate for single-object images. Only a few trials have been run with multi-object images, and it is not clear if the method is really robust enough to work well in the multi-object case.

19.6.9   ABLS

A different approach to evidence-based recognition is espoused by Wang and Srihari (1988) in the ABLS system for locating address blocks on mail pieces. They have developed a general framework for object recognition in a complex environment, using diverse knowledge sources. As in the consistent-labeling problem, this system deals with a set of objects from the image and attempts to assign a label to each one. Instead of relational constraints, the system is provided with a set of knowledge sources, each of which is a rule with preconditions and actions. The knowledge sources are organized in a *dependency graph*, in which an arc from KS_1 to KS_2 indicates that KS_2 cannot execute until KS_1 has finished.

Each knowledge source in the system is assigned a *utility value*. The utility of a knowledge source is based on its efficiency, its effectiveness, its processing time, the proportion of the population of objects that it can deal with, and a special situation-adjustment factor. The control system is blackboard based with the following scheduling strategy:

- If only one object satisfies the termination criteria, that object is the address block. Succeed and quit.

- If no object satisfies the termination criteria, try another bottom-up segmentation tool.

- Or else apply an unused knowledge source to the candidate objects to generate more evidence.

The dependency graph dictates which knowledge sources are eligible to execute. The knowledge source that is selected is the one with the highest utility value. It calls on procedures from an image-processing/feature-extraction toolkit to compute a piece of evidence about the labels of the objects in the image. Evidence combination is performed via the Dempster-Shafer theory described earlier.

ABLS operates in a difficult domain of packages and letters with hand-printed, handwritten, and typewritten labels. In addition to the sought-for address block, there are the return address, the stamp, and various extraneous printing and graphics. Although a general model indicates where the address block should be in relation to the other objects, the rigid techniques of model-based vision do not apply. ABLS achieves a success rate of 81% overall and 98% if the segmentation is correct. Like ACRONYM and perhaps all high-level systems, the quality of the segmentation strongly affects the results.

19.7   Summary

In discussing the use of both simple and complex knowledge representations in knowledge-based vision systems, we have described feature vectors, relational and hierarchical structures, rules, frames, and schema. We have reviewed a variety of control strategies, including hierarchical and heterarchical ones, and have discussed mechanisms for information integration, including a Bayesian-belief-network approach, a theory-of-evidence approach, and a probabilistic basis for evidential reasoning. Finally, we have reviewed several knowledge-based vision systems. Interestingly, although such systems often employ a test-and-verify paradigm, relatively little work has been reported on the verification of their performance. In some sense the systems are toy systems. Also, although they work for mid- and high-level vision, they tend not to be applied at the low level. Developing a knowledge methodology by which all levels of vision can be integrated is an important research direction.

◼ Bibliography

Alt, F. L., "Digital Pattern Recognition by Moments," *Journal of the ACM,* Vol. 11, 1962, pp. 240–258.

Beveridge, J. R., et al., "Segmenting Images Using Localized Histograms and Region Merging," *International Journal of Computer Vision,* Vol. 2, 1989, pp. 311–347.

Binford, T. C., "Survey of Model-Based Image Analysis Systems," *International Journal of Robotics Research,* Vol. 1, 1982, pp. 18–64.

Bledsoe, W., "A New Method for Proving Certain Presburger Formulas," *Proceedings of the International Joint Conference on Artificial Intelligence,* Tbilisi, Georgia, USSR, 1975, pp. 15–21.

Brooks, R. A., "Symbolic Reasoning among 3-D Models and 2-D Images," *Artificial Intelligence,* Vol. 17, 1981, pp. 285–348.

———, "Model-Based Three-Dimensional Interpretations of Two-Dimensional Images," *IEEE Transactions on Pattern Analysis and Machine Intelligence,* Vol. PAMI-5, 1983, pp. 140–150.

Chakravarty, I., and H. Freeman, "Characteristic Views as a Basis for Three-Dimensional Object Recognition," *Proceedings of SPIE-336 (Robot Vision),* Arlington, VA, 1982, pp. 37–45.

Douglass, R. J., "Interpreting Three-Dimensional Scenes: A Model Building Approach," *Computer Graphics and Image Processing,* Vol. 17, 1981, pp. 91–113.

Draper, B. A., et al., "The Schema System," *International Journal of Computer Vision,* Vol. 2, 1989, pp. 209–250.

Erman, L. D., et al., "The Hearsay-II Speech-Understanding System: Integrating Knowledge to Resolve Uncertainty," *Computing Surveys,* Vol. 12, 1980, pp. 213–253.

Garvey, T. D., J. D. Lowrance, and M. A. Fischler, "An Intelligence Technique for Integrating Knowledge from Disparate Sources," *Proceedings of the International Joint Conference on Artificial Intelligence,* Vancouver, BC, 1981, pp. 319–325.

Good, I. J., "Rational Decisions," Journal Royal Statistical Society, Series B, Vol. 14, 1952, pp. 107–114.

Hanson, A. R., and E. M. Riseman, "VISIONS: A Computer System for Interpreting Scenes," in *Computer Vision Systems,* A.R. Hanson and E.M. Riseman (eds.), Academic Press, New York, 1978, pp. 303–334.

Herman, M., and T. Kanade, "Incremental Reconstruction of 3D Scenes from Multiple, Complex Images," *Artificial Intelligence,* Vol. 30, 1986, pp. 289–341.

Ikeuchi, K., "Generating an Interpretation Tree from a CAD Model for 3D-Object Recognition in Bin-Picking Tasks," *International Journal of Computer Vision,* Vol. 1, 1987, pp. 145–165.

Ikeuchi, K., and T. Kanade, "Modeling Sensor Detectability and Reliability in the Configuration Space for Model-Based Vision," *TR CMU-CS-87-144,* Carnegie Mellon University, Pittsburgh, 1987.

Jain, A. K., and R. Hoffman, "Evidence-Based Recognition of 3-D Objects," *IEEE Transactions on Pattern Analysis and Machine Intelligence,* Vol. 10, 1988, pp. 793–802.

Ledley, R. S., et al., "FIDAC: Film Input to Automatic Computer and Associated Sytax-Directed Pattern Recognition Programming System," in *Optical and Electro-Optical Information Processing,* J.T. Tippet (ed.), MIT Press, Cambridge, MA, 1965, p. 591.

Levine, M. D., and A. M. Nazif, "Rule-Based Image Segmentation: A Dynamic Control Strategy Approach," *Computer Vision, Graphics, and Image Processing,* Vol. 32, 1985, pp. 104–126.

Levine, M. D., and S. I. Shaheen, "A Modular Computer Vision System for Picture Segmentation and Interpretation," *IEEE Transactions on Pattern Analysis and Machine Intelligence,* Vol. PAMI-3, 1981, pp. 540–556.

Lowe, D. G., "Three-Dimensional Object Recognition from Single Two-Dimensional Images," *Artificial Intelligence,* Vol. 31, 1987, pp. 355–395.

McKeown, D. M., "Building Knowledge-Based Systems for Detecting Man-Made Structures from Remotely Sensed Imagery," *Philosophical Transactions of the Royal Society of London*, Vol. 324, 1988, pp. 423–435.

McKeown, D. M., and J. L. Denlinger, "Map-Guided Feature Extraction from Aerial Imagery," *Workshop on Computer Vision, Representation, and Control,* Annapolis, MD, 1984, pp. 205–213.

——, "Cooperative Methods for Road Tracking in Aerial Imagery," *Proceedings of the Computer Society Conference on Computer Vision and Pattern Recognition,* Ann Arbor, MI, 1988, pp. 662–672.

McKeown, D. M., and J. F. Pane, "Alignment and Connection of Fragmented Linear Features in Aerial Imagery," *Proceedings of the Conference on Computer Vision and Pattern Recognition,* San Francisco, 1985, pp. 55–61.

McKeown, D. M., A. H. Wilson, Jr., and J. McDermott, "Rule-Based Interpretation of Aerial Imagery," *IEEE Transactions on Pattern Analysis and Machine Intelligence,* Vol. PAMI-7, 1985, pp. 570–585.

Minsky, M., "A Framework for Representing Knowledge," in *The Psychology of Computer Vision,* P.H. Winston (ed.), McGraw-Hill, New York, 1975, pp. 211–277.

Mulgaonkar, P. G., L. G. Shapiro, and R. M. Haralick, "Shape from Perspective: A Rule-Based Approach," *Computer Vision, Graphics, and Image Processing,* Vol. 36, 1986, pp. 298–320.

Pearl, J., "Distributed Revision of Composite Beliefs," *Artificial Intelligence,* Vol. 33, 1987, pp. 173–215.

Rosenfeld, A., R. A. Hummel and S. W. Zucker, "Scene Labeling by Relaxation Operations," *IEEE Transactions on Systems, Man, and Cybernetics,* Vol. SMC-6, 1976, pp. 420–433.

Shafer, G., "Constructive Probability," *Synthese,* Vol. 48, 1981, pp. 1–60.

Shapiro, L. G., and H. Lu, "The Use of a Relational Pyramid Representation for View Classes in a CAD-to-Vision System," *Proceedings of the Ninth International Conference on Pattern Recognition,* 1988, pp. 379–381.

Wang, C. H., and S. N. Srihari, "A Framework for Object Recognition in a Visually Complex Environment and Its Application to Locating Address Blocks on Mail Pieces," *International Journal of Computer Vision,* Vol. 2, 1988, pp. 125–151.

Zahn, C. T., and R. Z. Roskies, "Fourier Descriptors for Plane Closed Curves," *IEEE Transactions on Computers,* Vol. C-21, 1972, pp. 269–281.

20 ACCURACY

20.1 Introduction

Accurately characterizing the performance of a vision algorithm or system is an important aspect of the successful employment of vision systems. In this chapter we briefly review what we can expect of positional inaccuracy due to quantizing error, how to estimate false alarm and misdetection rates for automated position inspection, experimental protocols for describing the experiments and analysis used to evaluate the performance of a given vision system, how to determine the repeatability and positional accuracy of a vision system, and how to assess the performance of a near-perfect vision system.

20.2 Mensuration Quantizing Error

Position on a digital grid has an inherent quantizing error due to the discrete spacing between pixel centers. In this section we discuss a quantizing model that will permit us to determine the position variance due to quantizing for a line endpoint, for the length of a line, for the centroid of a line, and for the centroid of a shape. Questions of this sort were answered by Ho (1982), who used a slightly different model from the one we assume and whose variance results differ somewhat. Following Ho, we consider a one-dimensional digital grid on which a line segment is laid.

Let B be the coordinate of the line's right endpoint. Let Δc be the spacing between pixel centers. Then B can be represented as $B = \Delta c(B^* - 1/2 + q)$, where $B^* = Ceiling(B/\Delta c - 1/2)$ and q is a uniform random variable, $0 \le q \le 1$. The relationship between all these parameters is illustrated in Figure 20.1. Let β^* represent the digital coordinate of the line's rightmost pixel. The quantizing model specifies the relationship between β^* and B. For any natural quantizing model, we would expect that the more B exceeds $\Delta c(B^* - 1/2)$, the greater the probability that the digital coordinate β^* will be B^*, and the more B is less than $\Delta c(B^* + 1/2)$,

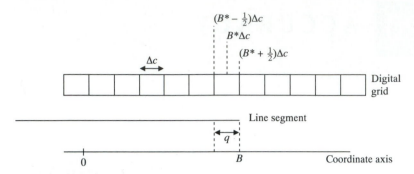

Figure 20.1 Relationship between the line segment end and the digital grid. The line segment whose right end has coordinate B has probability q of being quantized to B^* on the digital grid ($\Delta c B^*$ relative to the coordinate axis) and has probability $1 - q$ of being quantized to $B^* - 1$ on the digital grid [$\Delta c (B^* - 1)$ relative to the coordinate axis].

the greater the probability that the digital coordinate β^* will be $B^* - 1$. The model we use states

$$\beta^* = \begin{cases} B^* & \text{with probability } q \\ B^* - 1 & \text{with probability } 1 - q \end{cases}$$

Letting x be a random variable where

$$x = \begin{cases} 1 & \text{with probability } q \\ 0 & \text{with probability } 1 - q \end{cases}$$

we can restate the quantizing model:

$$\beta^* = B^* - 1 + x \tag{20.1}$$

Note that $E(x) = q$ and $E(x^2) = q$.

The first question to ask is, what is the mean and variance of the digital coordinate β^*?

$$E(\beta^*) = E(B^* - 1 + X)$$

$$= E\left(\frac{B}{\Delta c} + \frac{1}{2} - q - 1 + x\right)$$

$$= E_q\left[E_x\left(\frac{B}{\Delta c} - \frac{1}{2} + x - q \middle| q\right)\right]$$

$$= E_q\left(\frac{B}{\Delta c} - \frac{1}{2} - q + q\right)$$

$$= \frac{B}{\Delta c} - \frac{1}{2} \tag{20.2}$$

Hence an unbiased estimator $\hat{B}$ for the position B of the line's right endpoint is $\hat{B} = \Delta c(\beta^* + \frac{1}{2})$.

$$V(\beta^*) = E\left\{\left[\beta^* - \left(\frac{B}{\Delta c} - \frac{1}{2}\right)\right]^2\right\}$$

$$= E\left[\left(B^* - 1 + x - \frac{B}{\Delta c} + \frac{1}{2}\right)^2\right]$$

$$= E\left[\left(\frac{B}{\Delta c} + \frac{1}{2} - q - 1 + x - \frac{B}{\Delta c} + \frac{1}{2}\right)^2\right]$$

$$= E\left[(x - q)^2\right]$$

$$= E_q\left\{E_x\left[(x - q)^2 | q\right]\right\}$$

$$= E_q\left[E_x\left(x^2 - 2qx + q^2 | q\right)\right]$$

$$= E_q\left(q - 2q^2 + q^2\right) = E\left(q - q^2\right) = \frac{1}{2} - \frac{1}{3} = \frac{1}{6} \qquad (20.3)$$

This means that the variance for the estimator $\hat{B} = \Delta c(\beta^* + \frac{1}{2})$ is

$$V(\hat{B}) = V\left[\Delta c\left(\beta^* + \frac{1}{2}\right)\right]$$

$$= (\Delta c)^2 \, V\left(\beta^* + \frac{1}{2}\right)$$

$$= (\Delta c)^2 \, V(\beta^*)$$

$$= (\Delta c)^2 \, \frac{1}{6}$$

$$= \frac{(\Delta c)^2}{6} \qquad (20.4)$$

A line's left endpoint A can be handled in a similar way. We represent A by $A = \Delta c(A^* + 1/2 - s)$, where $A^* = Floor\,(A/\Delta c + 1/2)$ and s is a uniform random variable, $0 \le s \le 1$. Let α^* represent the digital coordinate of the line's leftmost pixel. The quantizing model states

$$\alpha^* = \begin{cases} A^* & \text{with probability } s \\ A^* + 1 & \text{with probability } 1 - s \end{cases}$$

Letting y be a random variable where

$$y = \begin{cases} 1 & \text{with probability } s \\ 0 & \text{with probability } 1 - s \end{cases}$$

we can restate the quantizing model:

$$\alpha^* = A^* + 1 - y \qquad (20.5)$$

We can now determine the mean and variance of α^*:

$$
\begin{aligned}
E\left(\alpha^{*}\right) &= E\left(A^{*}+1-y\right) \\
&= E\left(\frac{A}{\Delta c}-\frac{1}{2}+s+1-y\right) \\
&= E_{s}\left[E_{y}\left(\frac{A}{\Delta c}-\frac{1}{2}+s+1-y \mid s\right)\right] \\
&= E_{s}\left(\frac{A}{\Delta c}-\frac{1}{2}+1+s-s\right) \\
&= \frac{A}{\Delta c}+\frac{1}{2}
\end{aligned}
\tag{20.6}
$$

Hence an unbiased estimator $\hat{A}$ for the position A of the line's left endpoint is $\hat{A}=\Delta c(\alpha^{*}-1/2)$:

$$
\begin{aligned}
V(\alpha^{*}) &= E\left\{\left[\alpha^{*}-\left(\frac{A}{\Delta c}+\frac{1}{2}\right)\right]^{2}\right\} \\
&= E\left[\left(A^{*}+1-y-\frac{A}{\Delta c}-\frac{1}{2}\right)^{2}\right] \\
&= E\left[\left(\frac{A}{\Delta c}-\frac{1}{2}+s+1-y-\frac{A}{\Delta c}-\frac{1}{2}\right)^{2}\right] \\
&= E[(s-y)^{2}] \\
&= E_{s}[E_{y}(s^{2}-2sy+y^{2} \mid s)] \\
&= E_{s}(s^{2}-2s^{2}+s)=\frac{1}{6}
\end{aligned}
\tag{20.7}
$$

Now if $\hat{A}$ is an unbiased estimator of A, and $\hat{B}$ is an unbiased estimator of B, then $\hat{B}-\hat{A}$ will be an unbiased estimator for the length of the line segment:

$$
E(\hat{B}-\hat{A})=E(\hat{B})-E(\hat{A})=B-A
\tag{20.8}
$$

The variance of this estimator is also easily computed:

$$
V(\hat{B}-\hat{A})=V(\hat{B})+V(\hat{A})=\frac{(\Delta c)^{2}}{3}
\tag{20.9}
$$

An estimator for the centroid of the line is given naturally by $(\hat{A}+\hat{B})/2$, since

$$
E\left(\frac{\hat{A}+\hat{B}}{2}\right)=\frac{1}{2}[E(\hat{A})+E(\hat{B})]=\frac{A+B}{2}
$$

The variance of the centroid is

$$V\left(\frac{\hat{A} + \hat{B}}{2}\right) = \frac{1}{4}[V(\hat{A}) + V(\hat{B})]$$

$$= \frac{(\Delta c)^2}{12} \tag{20.10}$$

The information about the centroid of a line can be extended to an arbitrary shape by considering the shape to be composed of a set of N lines, line n beginning at A_n and ending at B_n. However, here we must assume that there is no systematic alignment of the shape with the row and column axes of the quantifying grid. The column centroid of such shape is given by

$$\bar{c} = \frac{\sum_{n=1}^{N}(B_n - A_n)\left(\frac{B_n + A_n}{2}\right)}{\sum_{n=1}^{N}(B_n - A_n)} \tag{20.11}$$

This suggests the estimator $\hat{c}$ defined by

$$\hat{c} = \frac{\sum_{n=1}^{N}(\hat{B}_n - \hat{A}_n)\left(\frac{\hat{A}_n + \hat{B}_n}{2}\right)}{\sum_{n=1}^{n}(\hat{B}_n - \hat{A}_n)} \tag{20.12}$$

Substituting the definitions for $\hat{A}_n$ and $\hat{B}_n$ in terms of the random variables q_n, s_n, x_n, and y_n results in

$$\hat{c} = \left(\frac{\sum_{n=1}^{N}[(B_n - A_n) + \Delta c(-q_n - s_n + x_n + y_n)]\left[\frac{A_n + B_n + \Delta c(-q_n + s_n + x_n - y_n)}{2}\right]}{\sum_{n=1}^{N}[(B_n - A_n) + \Delta c(-q_n - s_n + x_n + y_n)]}\right) \tag{20.13}$$

an expression bearing a close resemblance to the centroid $\bar{c}$.

We want to obtain the mean and variance of $\hat{c}$. But because the denominator is not statistically independent from the numerator, we cannot blindly proceed by taking the expectation of the numerator independent of the denominator. Instead we proceed by using a first-order approximation.

Let $M_n = B_n - A_n$, $P_n = (B_n + A_n)/2$, $W_n = \Delta c(-q_n - s_n + x_n + y_n)$, and $V_n = \Delta c/2(-q_n + s_n + x_n - y_n)$. Then we can rewrite $\hat{c}$ as

$$\hat{c}(M_1 + W_1, ..., M_N + W_N, P_1 + V_1, ..., P_N + V_N) = \frac{\sum_{n=1}^{N}(M_n + W_n)(P_n + V_n)}{\sum_{n=1}^{N}(M_n + W_n)} \tag{20.14}$$

Note that $\hat{c}(M_1, ..., M_N, P_1, ..., P_N) = \bar{c}$.

We approximate $\hat{c}(M_1 + W_1, ..., M_N + W_N, P_1 + V_1, ..., P_N + V_N)$ by expanding $\hat{c}$ in a Taylor series around the point $(M_1, ..., M_N, P_1, ..., P_N)$, and we consider only the first-order terms. We obtain

$$\hat{c}(M_1 + W_1, ..., M_N + W_N, P_1 + V_1, ..., P_N + V_N) \cong \bar{c} + \frac{\displaystyle\sum_{n=1}^{N}\left[(P_n - \bar{c})W_n + M_n V_n\right]}{\displaystyle\sum_{n=1}^{N} M_n}$$

$$(20.15)$$

Recognizing that $E(W_n) = E(V_n) = 0$, we easily obtain

$$E(\hat{c}) \cong \bar{c} \qquad (20.16)$$

Also,

$$V(\hat{c}) = E[(\hat{c} - \bar{c})^2]$$

$$= E\left[\left(\frac{\displaystyle\sum_{n=1}^{N}[(P_n - \bar{c})W_n + M_n V_n]}{\displaystyle\sum_{n=1}^{N} M_n}\right)^2\right]$$

$$= E\frac{\left[\displaystyle\sum_{n=1}^{N}[(P_n - \bar{c})W_n + M_n V_n]\sum_{m=1}^{N}[(P_m - \bar{c})W_m + M_m V_m]\right]}{\left(\displaystyle\sum_{n=1}^{N} M_n\right)^2}$$

$$V(\hat{c}) = \frac{\displaystyle\sum_{n=1}^{N}\sum_{m=1}^{N} E[(P_n - \bar{c})(P_m - \bar{c})W_n W_m + (P_n - \bar{c})M_m W_n V_m}{\left(\displaystyle\sum_{n=1}^{N} M_n\right)^2} \begin{array}{l} \\ + M_n(P_m - \bar{c})V_n W_m + M_n M_m V_n V_m] \end{array} \qquad (20.17)$$

Now

$$E(W_n W_m) = \begin{cases} 0 & n \neq m \\ \frac{(\Delta c)^2}{3} & n = m \end{cases}$$

$$E(W_n V_m) = 0 \qquad \text{for all } n, \ m$$

$$E(V_n V_m) = \begin{cases} 0 & n \neq m \\ \frac{(\Delta c)^2}{12} & n = m \end{cases}$$

After making the appropriate substitutions into Eq. (20.17), we obtain

$$V(\hat{c}) = \frac{\frac{(\Delta c)^2}{3} \sum_{n=1}^{N} \left[(P_n - \bar{c})^2 + \frac{M_n^2}{4} \right]}{\left(\sum_{n=1}^{N} M_n \right)^2} \tag{20.18}$$

For any shape that is symmetric about a vertical axis running through its column centroid, $P_n = \bar{c}$, $n = 1, ..., N$. In this case the variance of Eq. (20.18) reduces and becomes

$$V(\hat{c}) = \frac{\frac{(\Delta c)^2}{12} \sum_{n=1}^{N} M_n^2}{\left(\sum_{n=1}^{N} M_n \right)^2} \tag{20.19}$$

A disk is a shape that, regardless of its orientation, is symmetric about a vertical axis running through its column centroid. Let Δr be the spacing between rows. For the case of the disk, $\Delta r N$ represents its diameter and N represents the number of rows required to cover the disk. For a row that is n above the middle row of the disk, the chord length will be $2\Delta r \sqrt{(N/2)^2 - n^2}$ and the chord length squared will be $4\Delta r[(N/2)^2 - n^2]$. Likewise for a row that is n below the middle row. Hence when N is odd,

$$\sum_{n=1}^{N} M_n^2 = (\Delta r)^2 \left\{ N^2 + 8 \sum_{n=1}^{\frac{N-1}{2}} \left[\left(\frac{N}{2} \right)^2 - n^2 \right] \right\}$$

$$= (\Delta r)^2 \frac{2N^3 + N}{3} \tag{20.20}$$

And when N is even,

$$\sum_{n=1}^{N} M_n^2 = (\Delta r)^2 \left\{ 8 \sum_{n=1}^{\frac{N}{2}} \left[\left(\frac{N}{2} \right)^2 - n^2 \right] \right\}$$

$$= \frac{(\Delta r)^2 \{2N^3 - 3N^2 - 2N\}}{3} \tag{20.21}$$

Also, the area of the disk is easily given by

$$\Delta r \sum_{n=1}^{N} M_n$$

Since the disk has radius $\frac{N\Delta r}{2}$,

$$\Delta r \sum_{n=1}^{N} M_n = \pi \left(\frac{N\Delta r}{2} \right)^2 \tag{20.22}$$

from which we immediately obtain

$$\sum_{n=1}^{N} M_n = \frac{(\Delta r)^2 \pi N^2}{4\Delta r}$$

$$= \Delta r \frac{\pi N^2}{4} \tag{20.23}$$

Substituting Eqs. (20.20), (20.21), and (20.23) into Eq. (20.19) for the variance of $\hat{c}$, we obtain that the variance of the column centroid for a disk is

$$V(\hat{c}) = \begin{cases} (\frac{4\Delta c}{6\pi})^2 (\frac{2}{N} + \frac{1}{N^3}) & N \text{ odd} \\ (\frac{4\Delta c}{6\pi})^2 (\frac{2}{N} - \frac{3}{N^2} - \frac{2}{N^3}) & N \text{ even} \end{cases} \tag{20.24}$$

Thus to a first-order approximation we have

$$V(\hat{c}) = \left(\frac{4\Delta c}{6\pi} \right)^2 \frac{2}{N}$$

$$= \frac{.09(\Delta c)^2}{N} \tag{20.25}$$

And the standard deviation is then

$$\sigma_{\hat{c}} = \frac{.3\Delta c}{\sqrt{N}} \tag{20.26}$$

Similarly the standard deviation of the estimated row centroid $\hat{r}$ is given by

$$\sigma_{\hat{r}} = \frac{.3\Delta r}{\sqrt{Q}}$$

where Q is the number of columns it takes to cover the disks.

For Eq. (20.25) Ho gives a normalized form (his equation 20):

$$\frac{V(\hat{c})}{(\Delta c)^2} = \frac{\pi}{9}(4p^{-1} + p^{-3})$$

where p is the digital perimeter. Since the diameter of the disk is N, $p = \pi N$, so that to a first-order approximation Ho has

$$\frac{V(\hat{c})}{(\Delta c)^2} = \frac{4\pi}{9\pi N} = \frac{4}{9N}$$

$$= \frac{.4444}{N}$$

which is considerably higher than the results of Eq. (20.25).

A rectangle oriented parallel to the rows of an image is also symmetric about a vertical axis running through its column centroid. Suppose the rectangle occupies N rows and Q columns. Then

$$\sum_{n=1}^{N} M_n^2 = (\Delta c)^2 \sum_{n=1}^{N} Q^2$$

$$= (\Delta c)^2 Q^2 N \tag{20.27}$$

and

$$\sum_{n=1}^{N} M_n = \Delta c \sum_{n=1}^{N} Q$$

$$= \Delta c Q N \tag{20.28}$$

Hence, substituting Eqs. (20.27) and (20.28) into Eq. (20.19), we obtain

$$V(\hat{c}) = \frac{\frac{(\Delta c)^2}{3}(\Delta c)^2 Q^2 N}{(\Delta c Q N)^2}$$

$$= \frac{(\Delta c)^2}{12N} \tag{20.29}$$

And the standard deviation of the column centroid of a rectangle is

$$\sigma_{\hat{c}} = \frac{\Delta c}{2\sqrt{3N}} = \frac{.2886\Delta c}{\sqrt{N}} \tag{20.30}$$

Similarly,

$$\sigma_{\hat{r}} = \frac{.2886\Delta r}{\sqrt{Q}} \tag{20.31}$$

From the corresponding formulas for the disk and the rectangle, it appears that the rectangle centroid has the smaller standard deviation. However, their respective areas are not the same. Recall that the disk has area

$$\pi\left(\frac{N\Delta r}{2}\right)\left(\frac{Q\Delta c}{2}\right) = NQ\frac{\pi}{4}\,\Delta r\Delta c$$

where $N\Delta r = Q\Delta c$ is the diameter of the disk. If $N\Delta r = N\Delta c$ for the rectangle, the rectangle is a square whose area is

$$N\Delta r\,Q\Delta c = NQ\,\Delta r\Delta c$$

If we consider a square with the same area as the disk—that is, a square having $N\sqrt{\pi/4} = .8862\,N$ rows and $Q\sqrt{\pi/4} = .8862\,Q$ columns—then the standard deviation of the row and column centroid for the square would be

$$\sigma_{\hat{r}} = \frac{.2886\Delta r}{\sqrt{.8862 \ Q}} = \frac{.3066\Delta r}{\sqrt{Q}}$$

and

$$\sigma_{\hat{c}} = \frac{.2886\Delta r}{\sqrt{.8862 \ N}} = \frac{.3066\Delta c}{\sqrt{N}}$$

Both standard deviations are larger than the corresponding ones for the disk.

There is another reason why the positional accuracy for the centroid of a disk, which is its center, has a smaller standard deviation than the positional accuracy for the centroid of a square. Equation (20.19) for the variance of a shape holds only for shapes that are symmetric about a vertical axis through the shape's centroid. If the square were rotated, it would not necessarily be symmetric about a vertical axis through its centroid. Hence Eq. (20.18) must be used for the variance of the centroid. Its value is guaranteed to be greater than the one that assumes the symmetry. This means that the variance of the centroid of a square depends on its orientation, the variance being, in general, greater for orientations not aligned with the row and column axes. A rotated disk, on the other hand, retains its symmetry under rotation, so that the variance of its centroid is the same regardless of orientation.

Papers examining other aspects of how quantization affects accuracy include Klaasman (1975), Dorst and Smeulders (1986), and Dorst and Smeulders (1987), who discuss estimating properties of a digital line. Groen and Verbeek (1978), Ellis et al. (1979), Proffitt and Rosen (1979), and Koplowitz and Bruckstein (1989) discuss properties of digital planar shape, perimeter, and curves.

20.3   Automated Position Inspection: False-Alarm and Misdetection Rates

In the industrial position inspection task, an automated mechanism machines a part to given specifications or assembles an object from parts. Assembly involves placing one part on or into another. Inspection ensures that the machining or part placement is correct by measuring the relative position of critical points of the part or of one part with respect to another. Inspection, like machining or assembly, can also be automated.

The automated inspector consists of a machine that can identify critical points of an object and measure their position. The observed position is then compared with the specified position. If the difference between them is too large, the observed position does not pass inspection. In this section we describe and analyze a model that relates the placement characteristics of the machining or assembly mechanism to the accuracy of the inspection mechanism in order to derive the operating characteristics of the combined system in terms of false-alarm and misdetection rates.

We assume that the automated mechanism tries to machine a critical point to a position t or tries to place one part on another with relative position t, where

t is a known number. Unfortunately the actual position x is not t, because the automated machining or assembling makes some error. We assume that x has a Gaussian distribution with mean t and standard deviation σ_x.

It is not a requirement for a well-made part or assembly to be perfect in order for it to perform its function. An actual position x need only be close enough to the desired position t. How close is close enough is determined by how the part must function with other parts. Hence there is a tolerance interval, which we will assume is centered around position t. If the actual position x lies within the mechanical tolerance interval, the position in question is defined as good. If the actual position x lies outside the mechanical tolerance interval, the position in question is defined as bad. In our model the tolerance interval is $t \pm \alpha$. So if $|x - t| < \alpha$, the position is good, otherwise it is bad.

Unfortunately the actual position x is not known. The inspection system observes the actual position and measures it, but the measurement y that it produces is noisy. The measurement y is not equal to x. We assume that the conditional distribution of y given x is a Gaussian distribution with mean x and standard deviation σ_y.

The inspection system must use the observed measurement y to make a decision about whether or not the actual position x is in tolerance. It is natural for the inspection system to use an acceptance interval centered around the desired position t for this purpose. In our model the acceptance interval is $t \pm \beta$. So if $|y - t| < \beta$, the inspection system decides the position is good, otherwise it decides the position is bad.

Just as the machine that places the parts sometimes makes an error, and just as the measuring system that the automated inspector employs sometimes makes an error, so does the automated inspector itself also sometimes make an error. There are two kinds of error the automated inspector can make. A position could be within tolerance and yet not be accepted because the observed position y is not within the acceptance interval. In this case a good position is falsely called bad. Such an error is called a false alarm because the inspector thinks it has found a defect when in fact it has not. The second kind of error occurs when a position is in fact not within tolerance but is accepted because the observed position y is within the acceptance interval. In this case a bad position is missed and is incorrectly called good. Such an error is called a misdetection because the inspector that is looking for defects has missed one.

In terms of our model, the false-alarm rate is the conditional probability

$$P_F = P(|y - t| > \beta \mid |x - t| < \alpha)$$

and the misdetection rate is the conditional probability

$$P_M = P(|y - t| < \beta \mid |x - t| > \alpha)$$

The entire probability model is characterized by the five parameters t, σ_x, σ_y, α, and β. The problem we solve in this section is how to compute the false-alarm and misdetection probabilities in terms of these parameters.

20.3.1 Analysis

Let $P(x)$ denote the probability density function for the actual position x. Let $P(y|x)$ denote the conditional probability density function for the measurement y obtained when observing the position x. Then with the Gaussian distribution assumption,

$$P(x) = \frac{1}{\sqrt{2\pi}\sigma_x} e^{-\frac{1}{2}\left(\frac{x-t}{\sigma_x}\right)^2}$$

$$P(y|x) = \frac{1}{\sqrt{2\pi}\sigma_y} e^{-\frac{1}{2}\left(\frac{y-x}{\sigma_y}\right)^2}$$

Consider the conditional probability $P(|y - t| < \beta \mid |x - t| < \alpha)$, which is closely related to the false-alarm probability:

$$P(|y - t| < \beta \mid |x - t| < \alpha) = 1 - P(|y - t| > \beta \mid |x - t| < \alpha)$$

$$= 1 - P_F$$

Now

$$P(|y - t| < \beta \mid |x - t| < \alpha) = P(-\beta \le y - t \le \beta \mid -\alpha \le x - t \le \alpha)$$

$$= \frac{P(-\alpha \le x - t \le \alpha, \ -\beta \le y - t \le \beta)}{P(-\alpha \le x - t \le \alpha)}$$

$$= \frac{\displaystyle\int_{x=-\alpha+t}^{\alpha+t} \int_{y=-\beta+t}^{\beta+t} P(y|x)P(x)\, dy\, dx}{\displaystyle\int_{x=-\alpha+t}^{\alpha+t} P(x)\, dx}$$

$$= \frac{\displaystyle\int_{x=-\alpha+t}^{\alpha+t} \int_{y=-\beta+t}^{\beta+t} \frac{1}{\sqrt{2\pi}\sigma_y} e^{-\frac{1}{2}\left(\frac{y-x}{\sigma_y}\right)^2} \frac{1}{\sqrt{2\pi}\sigma_x} e^{-\frac{1}{2}\left(\frac{x-t}{\sigma_x}\right)^2} dy\, dx}{\displaystyle\int_{x=-\alpha+t}^{\alpha+t} \frac{1}{\sqrt{2\pi}\sigma_x} e^{-\frac{1}{2}\left(\frac{x-t}{\sigma_x}\right)^2} dx}$$

For the integral of the numerator, make a change of variable:

$$\begin{pmatrix} u \\ v \end{pmatrix} = \begin{pmatrix} \frac{1}{\sigma_x} & 0 \\ \frac{-1}{\sigma_y} & \frac{1}{\sigma_y} \end{pmatrix} \begin{pmatrix} x \\ y \end{pmatrix} - \begin{pmatrix} \frac{t}{\sigma_x} \\ 0 \end{pmatrix}$$

For the integral of the denominator, make a change of variable:

$$u = \frac{x - t}{\sigma_x}$$

Then we obtain

$$P(|y - t| < \beta \mid |x - t| < \alpha) = \frac{\displaystyle\int_{u=-\frac{\alpha}{\sigma_x}}^{\frac{\alpha}{\sigma_x}} \int_{v=\frac{-\beta-\sigma_x u}{\sigma_y}}^{\frac{\beta-\sigma_x u}{\sigma_y}} \frac{1}{2\pi} e^{-\frac{1}{2}(u^2+v^2)} dv \, du}{\displaystyle\int_{u=\frac{-\alpha}{\sigma_x}}^{\frac{\alpha}{\sigma_x}} \frac{1}{\sqrt{2\pi}} e^{-\frac{1}{2}u^2} du}$$

To simplify our notation, we let the integral

$$\int_{-\infty}^{z} \frac{1}{\sqrt{2\pi}} e^{-\frac{1}{2}u^2} du = \phi(z)$$

Hence

$$P(|y - t| < \beta \mid |x - t| < \alpha) = \frac{\displaystyle\int_{u=\frac{-\alpha}{\sigma_x}}^{\frac{\alpha}{\sigma_x}} \frac{1}{\sqrt{2\pi}} e^{-\frac{1}{2}u^2} \left[\phi\left(\frac{\beta-\sigma_x u}{\sigma_y}\right) - \phi\left(\frac{-\beta-\sigma_x u}{\sigma_y}\right) \right] du}{\phi\left(\frac{\alpha}{\sigma_x}\right) - \phi\left(\frac{-\alpha}{\sigma_x}\right)}$$

Noticing that $\phi(z) + \phi(-z) = 1$ so that

$$\phi\left(\frac{\beta - \sigma_x u}{\sigma_y}\right) - \phi\left(\frac{-\beta - \sigma_x u}{\sigma_y}\right) = \phi\left(\frac{\beta + \sigma_x u}{\sigma_y}\right) - \phi\left(\frac{-\beta + \sigma_x u}{\sigma_y}\right)$$

we can split the integral in the numerator into its negative half $\frac{-\alpha}{\sigma_x} \leq u \leq 0$ and its positive half $0 \leq u \leq \frac{\alpha}{\sigma_x}$ and then recombine it into twice the integral on the positive half. Hence

$$P(|y - t| < \beta \mid |x - t| < \alpha) = \frac{2\displaystyle\int_{0}^{\frac{\alpha}{\sigma_x}} \frac{1}{\sqrt{2\pi}} e^{\frac{-u^2}{2}} \left[\phi\left(\frac{\beta+\sigma_x u}{\sigma_y}\right) - \phi\left(\frac{-\beta+\sigma_x u}{\sigma_y}\right) \right] du}{\phi\left(\frac{\alpha}{\sigma_x}\right) - \phi\left(\frac{-\alpha}{\sigma_x}\right)}$$

and the false-alarm probability can be given by

$$
P_F = 1 - \frac{2 \int_0^{\frac{\alpha}{\sigma_x}} \frac{1}{\sqrt{2\pi}} e^{\frac{-u^2}{2}} \left[\phi \left(\frac{\beta + \sigma_x u}{\sigma_y} \right) - \phi \left(\frac{-\beta + \sigma_x u}{\sigma_y} \right) \right] du}{\phi \left(\frac{\alpha}{\sigma_x} \right) - \phi \left(\frac{-\alpha}{\sigma_x} \right)}
$$

Next we determine the misdetection probability $P_M = P(|t - y| < \beta \mid |t - x| > \alpha)$:

$$
P_M = \frac{P(|x - t| > \alpha, \ |y - t| < \beta)}{P(|x - t| > \alpha)}
$$

$$
= \frac{P(|y - t| < \beta) - P(|x - t| < \alpha, \ |y - t| < \beta)}{1 - P(|x - t| < \alpha)}
$$

But

$$
P(|y - t| < \beta) = P(-\beta + t \le y \le \beta + t)
$$

$$
= \int_{x=-\infty}^{\infty} \int_{y=-\beta+t}^{\beta+t} P(x, y) \, dy \, dx
$$

$$
= \int_{x=-\infty}^{\infty} \int_{y=-\beta+t}^{\beta+t} \frac{1}{\sqrt{2\pi}\sigma_x} e^{-\frac{1}{2}\left(\frac{x-t}{\sigma_x}\right)^2} \frac{1}{\sqrt{2\pi}\sigma_y} e^{-\frac{1}{2}\left(\frac{y-x}{\sigma_x}\right)^2} dy \, dx
$$

As before, we make a change of variable:

$$
\begin{pmatrix} u \\ v \end{pmatrix} = \begin{pmatrix} \frac{1}{\sigma_x} & 0 \\ \frac{-1}{\sigma_y} & \frac{1}{\sigma_y} \end{pmatrix} \begin{pmatrix} x \\ y \end{pmatrix} - \begin{pmatrix} \frac{t}{\sigma_x} \\ 0 \end{pmatrix}
$$

This yields

$$
P(|y - t| < \beta) = \int_{u=-\infty}^{\infty} \int_{v=\frac{-\beta-\sigma_x u}{\sigma_y}}^{\frac{\beta-\sigma_x u}{\sigma_y}} \frac{1}{2\pi} e^{-\frac{1}{2}(u^2 + v^2)} dv \, du
$$

Now we make another change of variable to rotate the axes in a way that separates the double integral. Let

$$
\begin{pmatrix} s \\ w \end{pmatrix} = \frac{1}{\sqrt{\sigma_x^2 + \sigma_y^2}} \begin{pmatrix} \sigma_y & -\sigma_x \\ \sigma_x & \sigma_y \end{pmatrix} \begin{pmatrix} u \\ v \end{pmatrix}
$$

This yields

$$P(|y - t| < \beta) = \int\limits_{s=-\infty}^{\infty} \int\limits_{w=\frac{-\beta}{\sqrt{\sigma_x^2+\sigma_y^2}}}^{\frac{\beta}{\sqrt{\sigma_x^2+\sigma_y^2}}} \frac{1}{2\pi} e^{-\frac{1}{2}(s^2+w^2)} ds\, dw$$

$$= \int\limits_{s=-\infty}^{\infty} \frac{1}{\sqrt{2\pi}} e^{-\frac{1}{2}s^2} ds \int\limits_{w=\frac{-\beta}{\sqrt{\sigma_x^2+\sigma_y^2}}}^{\frac{\beta}{\sqrt{\sigma_x^2+\sigma_y^2}}} \frac{1}{\sqrt{2\pi}} e^{-\frac{1}{2}w^2} dw$$

$$= \phi\left(\frac{\beta}{\sqrt{\sigma_x^2 + \sigma_y^2}}\right) - \phi\left(\frac{-\beta}{\sqrt{\sigma_x^2 + \sigma_y^2}}\right)$$

Finally, after making all the required substitutions, we obtain the equation for the misdetection probability:

$$P_M = \frac{\left\{ \begin{array}{c} \phi\left(\frac{\beta}{\sqrt{\sigma_x^2+\sigma_y^2}}\right) - \phi\left(\frac{-\beta}{\sqrt{\sigma_x^2+\sigma_y^2}}\right) \\[2mm] -2\int\limits_0^{\frac{\alpha}{\sigma_x}} \frac{1}{\sqrt{2\pi}} e^{-\frac{1}{2}u^2} \left[\phi\left(\frac{\beta+\sigma_x u}{\sigma_y}\right) - \phi\left(\frac{-\beta+\sigma_x u}{\sigma_y}\right) \right] du \end{array} \right\}}{1 - \left[\phi\left(\frac{\alpha}{\sigma_x}\right) - \phi\left(\frac{-\alpha}{\sigma_x}\right) \right]}$$

The quantities in the equations are all ratios: α/σ_x, β/σ_y, σ_x/σ_y, and $\beta/\sqrt{\sigma_x^2 + \sigma_y^2}$. This suggests that there is an inherent invariance of the false-alarm and misdetection probabilities to the scale of any one of the parameters σ_x, σ_y, α, and β. To simplify our results, we scale everything in terms of σ_y. We let parameters k_x, k_α, and k_β satisfy

$$\sigma_x = k_x \sigma_y$$

$$\alpha = k_\alpha \sigma_y$$

$$\beta = k_\beta \sigma_y$$

Then

$$\frac{\alpha}{\sigma_x} = \frac{k_\alpha}{k_x}$$

$$\frac{\beta}{\sigma_y} = k_\beta$$

$$\frac{\sigma_x}{\sigma_y} = k_x$$

$$\frac{\beta}{\sqrt{\sigma_x^2 + \sigma_y^2}} = \frac{k_\beta}{\sqrt{1 + k_x^2}}$$

To make the parameters more meaningful, we relate them back to the inspection situation. Given σ_x and α, the probability of the automated mechanism producing a critical point that is not within tolerance is $1 - [\phi(\frac{\alpha}{\sigma_x}) - \phi(\frac{-\alpha}{\sigma_x})]$. Call this the failure probability f. Then

$$f = 1 - \left[\phi\left(\frac{\alpha}{\sigma_x}\right) - \phi\left(\frac{-\alpha}{\sigma_x}\right)\right]$$

$$= 1 - \left[\phi\left(\frac{k_\alpha}{k_x}\right) - \phi\left(\frac{-k_\alpha}{k_x}\right)\right]$$

$$= 1 - \left\{\phi\left(\frac{k_\alpha}{k_x}\right) - \left[1 - \phi\left(\frac{k_\alpha}{k_x}\right)\right]\right\}$$

$$= 2 - 2\phi\left(\frac{k_\alpha}{k_x}\right)$$

Hence the failure probability determines the relationship between k_α and k_x:

$$k_\alpha = k_x \phi^{-1}\left(1 - \frac{f}{2}\right)$$

The inspector's measurement has a standard deviation of σ_y, and the tolerance interval for the machine has length 2α. We define the relative precision r of the measurement to be the ratio $\sigma_y / 2\alpha$:

$$r = \frac{\sigma_y}{2\alpha}$$

$$= \frac{\sigma_y}{2k_\alpha \sigma_y}$$

$$= \frac{1}{2k_\alpha}$$

Hence, given the relative precision r of the measuring instrument, the parameter k_α is determined by

$$k_\alpha = \frac{1}{2r}$$

Thus when the failure fraction and relative precision are given, the parameters k_α and k_x are determined. Once k_α and k_x are determined, the false-alarm probability and the misdetection probability can be computed as a function of the third parameter k_β. In effect this generates the operating curve of false-alarm rate as a function of misdetection rate.

20.3.2 Discussion

When k_β is large, the acceptance interval is relatively large, and we should expect that all good positions will be accepted. Hence the false-alarm rate will be small. On the other hand, bad positions will also be accepted, so there will be a high rate of misdetection. When k_β is small, the acceptance interval is relatively small, and we should expect that all bad positions will not be accepted. Hence the misdetection rate will be small. On the other hand, good positions will also not be accepted, so that there will be a high rate of false alarm. Figures 20.2 and 20.3 show this relationship between the misdetection and false-alarm rates for fixed values of failure rate and different values of relative precision.

Now suppose we fix both the failure rate and the misdetection rate. As the relative precision r gets better (r gets smaller, meaning the fraction of the tolerance interval represented by the standard deviation of the measurements gets smaller), we should also expect that the corresponding false-alarm rate will become smaller. Figures 20.2 and 20.3 show this relationship. Operating curves for smaller values of relative precision are uniformly below operating curves for larger values of relative precision.

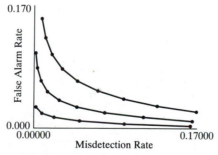

Figure 20.2 Three operating curves for a fixed failure rate of .05. The top operating curve is for a relative precision of .1; the middle, for a relative precision of .065; the bottom, for a relative precision of .05.

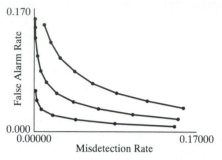

Figure 20.3 Three operating curves for a fixed failure rate of .01. The top operating curve is for a relative precision of .1; the middle, for a relative precision of .075; the bottom, for a relative precision of .05.

Finally, if we fix the relative precision and the misidentification rate and observe the false-alarm rate as the failure rate increases, we will discover that the false-alarm rate also increases (Fig. 20.4). Operating curves for larger failure rates are uniformly above operating curves for smaller failure rates.

To understand this, recall that for the failure rate to increase when the relative precision remains fixed, the tolerance interval must remain the same while the standard deviation of the actual position must get larger. If no change were made in the acceptance interval, this misidentification rate would decrease, since more of the defects have values far from the acceptance interval. To maintain a constant misidentification rate, the acceptance interval must increase some. If all other things were equal, an increase in the size of the acceptance interval would decrease the false-alarm rate. But here not everything else is equal. An increase in the standard deviation of the actual position causes a greater fraction of good positions to be near the boundary of the tolerance interval defining a good position. A noisy measurement of position near the boundary of the tolerance interval is more likely to fall outside the acceptance interval. Thus more good positions are not accepted, and the false-alarm rate increases.

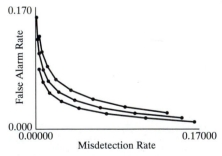

Figure 20.4 Three operating curves for a fixed relative precision of .075. The top operating curve is for a failure rate of .02; the middle, for a failure rate of .01; the bottom, for a failure rate of .005.

20.4   Experimental Protocol

Controlled experiments are an important component of computer vision, for the controlled experiment demonstrates that the algorithm designed by the computer vision researcher recognizes, locates, and measures what it is designed to do from image data. A properly designed scientific experiment provides evidence to accept or reject the hypothesis that the algorithm performs to a specified accuracy level. To set up such an experiment properly, so that it can be repeated and the evidence verified by another researcher, one must pay considerable attention to experimental protocol.

The experimental protocol states the quantity (or quantities) to be measured, the accuracy to which it is to be measured, and the population of scenes/images or artificially generated data on which the vision algorithm is to be applied. Then the protocol must give the experimental design and the data analysis plan.

The experimental design describes how a suitably random, independent, and representative set of images from the specified population is to be sampled, generated, or acquired. If the population includes, for example, a range of sizes of the object of interest, or if the object of interest can appear in a variety of situations, then the sampling mechanism must assume that a reasonable number of images are sampled, with the object appearing in each of the different sizes or appearing with sizes and orientations throughout its permissible range.

Similarly, if the object to be recognized or measured can appear in a variety of different lighting conditions that create a similar variety in shadowing, then the sampling must ensure that images are acquired with the lighting and shadowing varying throughout its permissible range. In general, the essential variation of the images in the population can be described by some number, say N variables. If the N variables $X_1, \ldots, X_N$ having to do with kind of lighting, light position, object position, object orientation, undesired object occlusion, environmental clutter, distortion, noise, and so on, have respective range sets $R_1, \ldots, R_N$, then the sampling design must ensure that the images sample the domain $R_1 \times R_2 \times \ldots \times R_N$ in a representative way. Since the number of images we are likely to sample is an infinitesimal fraction of the number of possibilities in $R_1 \times R_2 \times \ldots \times R_N$, this suggests that the experimental design may have to make judicious use of a Latin-square layout. If free parameters must be set in the vision algorithm, then they can also be part of the N variables.

The experiments must then be carried out for each image or data set in the sample. Suppose there are M different measurements $\hat{y}_1, \ldots, \hat{y}_M$ to be made on each image. These variables would be a subset of the N variables that might describe the images in the population of interest. There will be a difference between the true values $y_1, \ldots, y_M$ of the measured quantities and the measured values themselves. The accuracy criterion must state how the comparison between the true and measured values will be evaluated. Finally, the experimental data analysis plan must state how the hypothesis that the algorithm meets the specified requirement will be tested. It must indicate how the observed data (the true values and the corresponding measured values) will be analyzed. The plan must be detailed enough so that another researcher could carry out the analysis just as the designer of the plan would.

The plan must be supported by theoretically developed statistical analysis showing that an experiment carried out according to the experimental design and analyzed according to the data analysis plan will produce a statistical test itself having a given accuracy. That is, since the entire population was only sampled, the sampling fluctuation will introduce random fluctuation in the test results. For some fraction of experiments carried out according to the protocol, the hypothesis to be tested will be accepted, but the algorithm, in fact, if it were tried on the complete population of image variations, would not meet the specified requirements; and for some fraction of experiments carried out according to the protocol, the hypothesis to be tested will be rejected, but the algorithm, if in fact it were tried on the complete population of image variation, would meet the specified requirements. The specified size of these false- and missed-acceptance errors will dictate the number of images to include in the sample. This relation between sample size and false- and missed-acceptance rates of the test for the hypothesis must be determined on the basis of statistical theory. One would certainly expect the sample size to be large enough to keep the error rate below 20%.

For example, if the error rate of a vision algorithm is to be less than $1/1,000$, then in order to be about 85% sure that the performance meets specification, 10,000 tests will have to be run. If the vision algorithm performs incorrectly nine or fewer times, then we can assert that with 85% probability the vision algorithm meets specification (Haralick, 1989).

20.5 Determining the Repeatability of Vision Sensor Measuring Positions

Vision sensors may measure a position or location in one, two, or three dimensions. To determine the repeatability of the vision sensor (after it has been suitably warmed up and calibrated), an experiment must be performed in which some number of points are exposed to the sensor, each some number of times, and the repeatability is computed in terms of the degree to which the measured position for each point agrees with the corresponding mean measured position for each point. In what follows, we give a derivation that establishes an unbiased statistic of measurement repeatability and a relationship involving the number of samples that bound the relative uncertainty of this statistic.

20.5.1 The Model

Suppose that there are N points to be measured and each point is measured M times. Each point is K-dimensional. Let the mth measurement of the nth point be denoted by Y_{nm}. We assume that the measurements are independent and that the difference between the actual and measured positions are normally distributed with mean 0 and covariance $r^2 I$.

In this model the repeatability of the vision sensor is described by the standard deviation r, and the purpose of the experiment is to estimate r by using a large enough number of samples so that the estimate $\hat{r}$ of r is guaranteed to be sufficiently close to r.

20.5.2 Derivation

The natural statistic to compute from the experiment is the sum of the norms squared of the differences between the observed positions and the corresponding mean observed positions, summed over all points. Thus let

$$S^2 = \sum_{n=1}^{N} \sum_{m=1}^{M} (Y_{nm} - \hat{\mu}_n)' (Y_{nm} - \hat{\mu}_n)$$

where

$$\hat{\mu}_n = \frac{1}{M} \sum_{m=1}^{M} Y_{nm}$$

We need to determine the relationship between S^2 and r^2. Now

$$\sum_{n=1}^{N} \sum_{m=1}^{M} = (Y_{nm} - \hat{\mu}_n)'(Y_{nm} - \hat{\mu}_n)$$

$$= \sum_{n=1}^{N} \sum_{m=1}^{M} (Y_{nm} - \hat{\mu}_n)'(Y_{nm} - \hat{\mu}_n + \hat{\mu}_n - -\mu_n)'(Y_{nm} - \hat{\mu}_n + \hat{\mu}_n - \mu_n)$$

$$= \sum_{n=1}^{N} \sum_{m=1}^{M} (Y_{nm} - \hat{\mu}_n)'(Y_{nm} - \hat{\mu}_n)'(Y_{nm} - \hat{\mu}_n)$$

$$+ 2 \sum_{n=1}^{N} \sum_{m=1}^{M} (Y_{nm} - \hat{\mu}_n)'(Y \hat{\mu}_n - \mu_n) \sum_{n=1}^{N} \sum_{m=1}^{M} (\hat{\mu}_n - \mu_n)'(\hat{\mu}_n - \mu_n)$$

$$= \sum_{n=1}^{N} \sum_{m=1}^{M} (Y_{nm} - \hat{\mu}_n)'(Y_{nm} - \hat{\mu}_n) + M \sum_{n=1}^{N} (\hat{\mu}_n - \mu_n)'(\hat{\mu}_n - \mu_n)$$

Since $Y_{nm} - \mu_n$ are independent and normally distributed with mean 0 and covariance $\sigma_r^2 I$,

$$\sum_{n=1}^{N} \sum_{M=1}^{M} \frac{(Y_{nm} - \mu_n)'(Y_{nm} - \mu_n)}{r^2}$$

has a chi-square distribution with MNK degrees of freedom. Since $\hat{\mu}_n = \frac{1}{M} \sum_{m=1}^{M} Y_{nm}$, $\hat{\mu}_n - \mu_n$ are independent and normally distributed with mean 0 and covariance $\frac{r^2 I}{M}$.

Hence $\sum\limits_{n=1}^{N} \frac{(\hat{\mu}_n - \mu_n)'(\hat{\mu}_n - \mu_n)}{r^2/M}$ has a chi-square distribution with NK degrees of freedom. Because

$$\sum_{n=1}^{N}\sum_{m=1}^{M}(Y_{nm} - \mu_n)'(Y_{nm} - \mu_n) = \sum_{n=1}^{N}\sum_{m=1}^{M}(Y_{nm} - \hat{\mu}_n)'(Y_{nm} - \hat{\mu}_n)$$

$$+ M\sum_{n=1}^{N}(\hat{\mu}_n - \mu_n)'(\hat{\mu}_n - \mu_n)$$

the statistic $\sum\limits_{n=1}^{N}\sum\limits_{m=1}^{M}\frac{(Y_{nm} - \hat{\mu}_n)'(Y_{nm} - \hat{\mu}_n)}{r^2}$ must have a chi-square distribution with $MNK - NK = NK(M-1)$ degrees of freedom. Therefore

$$E[S^2] = r^2 NK(M-1)$$

which implies that the statistic

$$\hat{r}^2 = \frac{1}{NK(M-1)}\sum_{n=1}^{N}\sum_{m=1}^{M}(Y_{nm} - \hat{\mu}_n)'(Y_{nm} - \hat{\mu}_n)$$

has expected value r^2.

The next problem to be solved is, how many samples must be taken to ensure that $\hat{r}^2$ is sufficiently close to r^2? Noticing that

$$V\left[\sum_{n=1}^{N}\sum_{m=1}^{M}\frac{(Y_{nm} - \hat{\mu}_n)'(Y_{nm} - \hat{\mu}_n)}{r^2}\right] = 2NK(M-1)$$

we obtain that

$$V\left[\frac{\hat{r}^2 - r^2}{r^2}\right] = V\left[\frac{\hat{r}^2}{r^2}\right] = \frac{1}{[NK(M-1)]^2}2NK(M-1)$$

$$= \frac{2}{NK(M-1)}$$

Hence with probability close to 1,

$$\left|\frac{\hat{r}^2 - r^2}{r^2}\right|^2 \leq 9 \cdot \frac{2}{NK(M-1)}$$

To guarantee that for a specified relative error f,

$$\left|\frac{\hat{r}^2 - r^2}{r^2}\right| \leq f$$

we require only

$$\sqrt{\frac{18}{NK(M-1)}} \leq f$$

Hence

$$N(m-1) \geq \frac{18}{Kf^2}$$

For example, if $K = 2$ and $f = .2$, then $N(M-1) \geq 225$.

20.6   Determining the Positional Accuracy of Vision Sensors

Vision sensors may measure a position in one, two, or three dimensions. To determine the accuracy of the vision sensor (after it has been suitably calibrated), an experiment must be performed in which some number of points in known positions are exposed to the sensor, the measured positions are compared with the known positions, and the accuracy is computed in terms of the degree to which the actual and measured positions agree. The accuracy has two components: One is due to the bias of the measuring instrument about each point (a bias that may depend on position), and the other is due to random variation within the measuring instrument, which creates a randomness for the repeated measurement of the same point.

In performing the experiment, it is important for the position of the points to be independent. This suggests that positions of points must be chosen at random and must not follow a regular pattern. If there is any question about whether the positional accuracy changes as a function of position—for example, if there is greater accuracy in the center of the field of view than in its periphery—then the experiment must be independently repeated for each section of the field of view. If there is any question about whether the vision sensor's performance might change from day to day, then the experiment must be performed over a series of days.

There are two basic difficulties in determining how to do the required computation. The first is that point positions are never known perfectly, because the instrument that places each point into position is not perfect. Thus the nominal or mean positions are certainly known, but the actual positions themselves have to be considered as random variables with known variances. The second is that the statistic computed on the basis of the experiment is itself a random variable having a mean value and variance. For the positional accuracy determination experiment to have meaningful results, the instrument positioning the points must have a higher accuracy than the vision sensor; that is, the variances of the positions of the points being measured about their nominal positions must be much smaller than the variances of the measurements. Also, the number of points measured must be large enough so that the variance of the computed statistic of positional accuracy is sufficiently small. In what follows, we give the derivations that establish an unbiased statistic of measurement accuracy and a relationship involving the number of samples that bound the relative error of this statistic.

20.6.1   The Model

Suppose that there are N points to be measured and each point is measured μ times. Each point is K-dimensional. Denote the actual but unknown positions of these points by $\mu_1...\mu_N$, and their unknown expected positions (mean positions) by the given $\mu_1^*, ..., \mu_N^*$. We assume that the N points are independent and that the vector of the N deviations between their actual and nominal positions is normally distributed with mean 0 and covariance $\sigma_1^2 I$.

Let the mth measurement of the nth point be denoted by Y_{nm}. We assume that the measurements are independent and that the difference between the actual position μ_n and the measured position $Y_n m$ is normally distributed with mean $\mu_n + b_n$ and covariance $r^2 I$. The bias vector b_n itself is normally distributed with mean 0 and covariance $\sigma_b^2 I$.

In this model the positional accuracy of the vision sensor is described by $r^2 + \sigma_b^2$, and the purpose of the experiment is to estimate $r^2 + \sigma_b^2$ by using a large enough number of samples so that the unbiased estimate $\hat{\sigma}_2$ of $\hat{r}^2 + \hat{\sigma}_b^2$ is guaranteed to be sufficiently close to $r^2 + \sigma_b^2$.

20.6.2   Derivation

The natural statistic to compute from the experiment is the sum of the norms squared of the differences between the observed positions and the known expected positions. Thus let

$$S^2 = \sum_{n=1}^{N} (Y_n - \mu_n)' (Y_n - \mu_n)$$

We need to determine the relationship between S^2 and σ_1^2 and σ_2^2. Now

$$\sum_{n=1}^{N} (Y_n - \mu_n)' (Y_n - \mu_n) = \sum_{n=1}^{N} (Y_n - X_n + X_n - \mu_n)' (Y_n - X_n + X_n - \mu_n)$$

$$= \sum_{n=1}^{N} (Y_n - X_n)' (Y_n - X_n) + 2 \sum_{n=1}^{N} (Y_n - X_n)' (X_n - \mu_n)$$

$$+ \sum_{n=1}^{N} (X_n - \mu_n)' (X_n - \mu_n)$$

$$= \sum_{n=1}^{N} (Y_n - X_n)' (Y_n - X_n) + \sum_{n=1}^{N} (X_n - \mu_n)' (X_n - \mu_n)$$

As a consequence of the independence of points and the form of the assumed co-variances,

$$\sum_{n=1}^{N} \frac{(Y_n - X_n)'(Y_n - X_n)}{\sigma_2^2}$$

has a chi-square distribution with NK degrees of freedom (which we denote by χ_{NK}^2) and

$$\sum_{n=1}^{N} \frac{(X_n - \mu_n)'(X_n - \mu_n)}{\sigma_1^2}$$

has a χ_{NK}^2 distribution. Hence $\sum_{n=1}^{N}(Y_n - \mu_n)'(Y_n - \mu_n)$ has a $\sigma_2^2 \chi_{NK}^2 + \sigma_1^2 \chi_{NK}^2$ distribution. Since $E[\chi_{NK}^2] = NK$,

$$E\left[\sum_{n=1}^{N}(Y_n - \mu_n)'(Y_n - \mu_n)\right] = (\sigma_2^2 + \sigma_1^2)NK$$

This suggests that an appropriate statistic to estimate σ_2 is $\hat{\sigma}_2$, where

$$\hat{\sigma}_2^2 = \frac{1}{NK}\sum_{n=1}^{N}(Y_n - \mu_n)'(Y_n - \mu_n) - \sigma_1^2$$

since $E[\hat{\sigma}_2^2] = \sigma_2^2$.

Next we must determine how close we can expect $\hat{\sigma}_2^2$ to be to σ_2^2. The variance of $\hat{\sigma}_2^2$ is one appropriate measure:

$$V[\hat{\sigma}_2^2] = V\left[\frac{1}{NK}\sum_{n=1}^{N}(Y_n - \mu_n)'(Y_n - \mu_n) - \sigma_1^2\right]$$

$$= \frac{1}{(NK)^2}V\left[\sum_{n=1}^{N}(Y_n - \mu_n)'(Y_n - \mu_n)\right]$$

$$= \frac{1}{(NK)^2}V\left[\sum_{n=1}^{N}(Y_n - X_n)'(Y_n - X_n) + \sum_{n=1}^{N}(X_n - \mu_n)'(X_n - \mu_n)\right]$$

$$= \frac{1}{(NK)^2}\left\{\sigma_2^4 V\left[\sum_{n=1}^{N}\frac{(Y_n - X_n)'(Y_n - X_n)}{\sigma_2^2}\right]\right.$$

$$\left. + \sigma_1^4 V\left[\sum_{n=1}^{N}\frac{(X_n - \mu_n)'(X_n - \mu_n)}{\sigma_1^2}\right]\right\}$$

Since $V[\chi^2_{NK}] = 2NK$,

$$V[\hat{\sigma}^2_2] = \frac{1}{(NK)^2} \{\sigma^4_2 \, 2NK + \sigma^4_1 \, 2NK\}$$

$$= \frac{2(\sigma^4_1 + \sigma^4_2)}{NK}$$

Since $\sigma_1 < \sigma_2$, we certainly have

$$V[\hat{\sigma}^2_2] \leq \frac{4\sigma^4_2}{NK}$$

Because the variance of $\hat{\sigma}^2_2$ depends on σ^2_2, it is only natural to measure the closeness between $\hat{\sigma}^2_2$ and σ^2_2 in a relative way. From the expression for the variance of $\hat{\sigma}^2_2$, we immediately have

$$V\left[\frac{\hat{\sigma}^2_2}{\sigma^2_2}\right] \leq \frac{4}{NK}$$

From the expression for the expected value of $\hat{\sigma}^2_2$, we have

$$E\left[\frac{\hat{\sigma}^2_2}{\sigma^2_2}\right] = 1$$

Hence, by the Chebychev inequality, with probability close to 1,

$$\left|\frac{\hat{\sigma}^2_2}{\sigma^2_2} - 1\right|^2 \leq 9 \cdot \frac{4}{NK}$$

Ensuring that for a given relative error f,

$$\left|\frac{\hat{\sigma}^2_2}{\sigma^2_2} - 1\right| \leq f$$

then requires that

$$\sqrt{\frac{36}{NK}} \leq f$$

or
$$NK \geq \frac{36}{f^2}$$

For example, if the relative error f were .2, then $NK \geq 900$. If $K = 2$, then the number N of samples would be required to satisfy $N > 450$.

20.7   Performance Assessment of Near-Perfect Machines

Machines that are employed in recognition or defect inspection tasks are required to perform nearly flawlessly. Being sure that a machine meets performance specifications can require assessing its performance on a much larger sample than intuition might lead one to believe is necessary. And depending on what is meant by "meeting specifications," being sure that a machine meets specifications can require that its performance, which is sampled in a limited assessment test, be better than intuition might lead one to believe is necessary.

To set the stage for the explanation, consider the action of a computer vision recognition or inspection machine. It observes a part or document that it can either accept or reject. Rejection means that the machine is unable to carry out the recognition or inspection. Judgment is reserved. Acceptance means that the machine will make a judgment. This judgment can be either about the correct placement of an object or about the extent to which the observed object is defect free. The judgment can be either correct or incorrect. The fraction of time that the machine's judgment is incorrect is called the *error rate*.

The error rate is composed of two kinds of errors: false detection and misdetection. If the machine is an inspection machine, the observed part may be flawed or unflawed, and the machine's judgment may pronounce the part to be flawed or unflawed. The fraction of time that a part which is actually unflawed is judged to be flawed is called the *false-detection* or *false-alarm* rate. The fraction of time that a part which is actually flawed is judged to be unflawed is called the *misdetection rate*.

For a computer vision machine to be successful, it must be nearly perfect. The rejection, error, misdetection, and false-alarm rates must all be very small and close to zero. To determine whether a nearly perfect machine meets specifications, an acceptance test must be performed. In the acceptance test the machine is given a sample of N parts or documents to judge, and the resulting number of rejects, of machine judgment errors, of misdetections, or of false alarms is observed. The assessment of whether the machine meets specifications then requires an appropriate comparison of the observed number from the assessment test with the performance requirement. Because the assessment test observes only a finite sample, there is of necessity a difference between the observed performance on the test sample and the long-term performance on the total population. The issue of performance assessment then amounts to comparing the specifications and the observed performance so that the uncertainty due to sampling is precisely understood. The next section gives a derivation of the problem, which then leads to a description of how this comparison can be made. Full details can be found in Haralick (1989).

20.7.1 Derivation

Consider the case for false-alarm errors. The other cases are obviously similar. Let N, the sampling size, be the total number of parts observed, and let K be the number of false-alarm judgments observed to occur in the acceptance test of the N parts.

Suppose the machine performance specification indicates that its fraction of false alarms is to be less than f_o. The simplest intuitive way of making the comparison between f_o and K is to use K in the natural manner to estimate the true false-alarm rate f. The maximum likelihood estimate $\hat{f}$ of f based on K is $\hat{f} = K/N$. If the estimate $\hat{f}$ of f is less than f_o, we judge that the machine passes the acceptance test. If the estimate $\hat{f}$ is greater than f_o, we judge that the machine fails the test. The issue is, how sure are we, if we apply such a procedure, that the judgment we make about the machine's performance is correct? To resolve this issue, we must estimate the performance of our judgment. Let us start from the beginning.

To carry out the estimation, we suppose that, conditioned on the true error rate f, the machine's judgments are independent and identically distributed. Let X_n be a random variable, taking the value 1 for a false alarm and the value 0 otherwise, when the machine is judging the nth part. In the maximum-likelihood technique of estimation, we compute the estimate $\hat{f}$ as the value of f that maximizes

$$Prob \left(\sum_{n=1}^{N} X_n = K \mid f \right) = \binom{N}{K} f^K (1 - f)^{N-K}$$

Taking the partial derivative with respect to f and setting the derivative to zero results in $\hat{f} = K/N$, the natural estimate of f.

Suppose that we adopt the policy of accepting the machine if $\hat{f} \leq f_o$. To understand the consequences of this policy, consider the probability that the policy results in a correct acceptance decision. The probability that $f \leq f_o$, given that $\hat{f} \leq f_o$, needs to be computed:

$$Prob \, (f \leq f_o \mid \hat{f} \leq f_o) = \int_{f=0}^{f_o} Prob \, (f \mid \hat{f} \leq f_o) \, df$$

$$= \frac{\int_{f=0}^{f_o} Prob \, (\hat{f} \leq f_o \mid f) \, Prob \, (f) \, df}{Prob \, (\hat{f} \leq f_o)}$$

To make the mathematics simple, let f_o be constrained so that there is some integer K_o such that $f_o = K_o/N$. Then

$$Prob \, (f \leq f_o \mid \hat{f} \leq f_o) = \frac{\int_{f=0}^{f_o} Prob \left(\sum_{n=1}^{N} X_n \leq K_o \mid f \right) Prob \, (f) \, df}{\int_{f=0}^{1} Prob \left(\sum_{n=1}^{N} X_n \leq K_o \mid f \right) Prob(f) \, df}$$

The probability that the true value of f is less than or equal to f_o, given that the observed value $\hat{f}$ is less than f_o, will depend, in general, on the tester's prior probability function $Prob \, (f)$. So, depending on the acceptance tester's prior probability function $Prob \, (f)$, there will be some smallest number F, $0 \leq F \leq 1$, such that $Prob \, (f) = 0$ for all $f > F$. Here the support for the prior probability function is the interval $[0, F]$.

For example, an acceptance tester who has had successful experience with previous machines from the same manufacturer might have a prior probability function whose support is the interval $[0, 2f_o]$. An acceptance tester who has had no previous experience with the manufacturer might have a prior distribution whose support is the interval $[0, 10f_o]$. An acceptance tester who has had an unsuccessful experience with a previous machine from the same manufacturer might have a prior distribution for f whose support is the interval $[0, .5]$.

In each of these cases, we assume that neither we nor the tester knows anything more about the prior probability function than the interval of support $[0, F]$, where we assume that $F \geq f_o$, since if not, there would be no point in performing an acceptance test to establish something we already know. In this case we take *Prob* (f) to be that probability function defined on the interval $[0, F]$ having highest entropy. Such a *Prob* (f) is the uniform density on the interval $[0, F]$. Hence we take *Prob* $(f) = \frac{1}{F}$, $0 \leq f \leq F$. Therefore

$$
\begin{aligned}
\textit{Prob } (f \leq f_o \mid \hat{f} \leq f_o) &= \frac{\int\limits_{f=0}^{f_o} \sum\limits_{k=0}^{K_o} \binom{N}{k} f^k (1-f)^{N-k} df / F}{\int\limits_{f=0}^{F} \sum\limits_{k=0}^{K_o} \binom{N}{k} f^k (1-f)^{N-k} df / F} \\[2ex]
&= \frac{\sum\limits_{k=0}^{K_o} \binom{N}{k} B(k+1, N+1-k) I_{f_o}(k+1, N+1-k)}{\sum\limits_{k=0}^{K_o} \binom{N}{k} B(k+1, N+1-k) I_F(k+1, N+1-k)} \\[2ex]
&= \frac{\sum\limits_{k=0}^{K_o} I_{f_o}(k+1, N+1-k)/(N+1)}{\sum\limits_{k=0}^{K_o} I_F(k+1, N+1-k)/(N+1)} \\[2ex]
&= \frac{\sum\limits_{k=0}^{K_o} I_{f_o}(k+1, N+1-k)}{\sum\limits_{k=0}^{K_o} I_F(k+1, N+1-k)}
\end{aligned}
$$

where $I_{f_o}(k+1, N+1-k)$ is the incomplete beta ratio function.

Under the particular conditions we are interested in, $N \gg 100$, $f \ll 0.1$, and $k \ll N$. Hence $I_f(k+1, N+1-k) \approx I_f(k+1, N)$. This can be observed from the recurrence relation

$$
I_x(a, b) = x I_x(a-1, b) + (1-x) I_x(a, b-1)
$$

Now when $a + b > 6$ and $x \ll 1$, $I_x(a, b) \approx \phi(y)$, where

$$
y = \frac{3\left\{ (bx)^{\frac{1}{3}}(1 - 1/9b) - [a(1-x)]^{\frac{1}{3}}(1 - 1/9a) \right\}}{\left[\frac{(bx)^{\frac{2}{3}}}{b} + \frac{[a(1-x)]^{\frac{2}{3}}}{a} \right]^{\frac{1}{2}}}
$$

and ϕ is the cumulative normal $(0, 1)$ distribution function (Abramowitz and Stegun, 1972). From this approximation it follows that, when $f_o N > 1$, $x \ll 1$, $f_o mk \ll$

$1, \frac{N}{m} >> 1$, and $m > 1$; then

$$I_{f_o}(k+1\ N+1-k) \approx I_{f_o}(k+1,\ N)$$

$$\approx I_{mf_o}(k+1,\ \frac{N}{m})$$

$$\approx I_{mf_o}(k+1,\ \frac{N}{m}+1-k)$$

This means that instead of having to parametrize by f_o and N independently, we can create tables parametrized by the product $f_o N$.

For example, if $f_o = 0.0001, F \geq 10f_o$, and $N = 10^4$, then $K_o = 1$ and **Prob** $(f \leq f_o\ |\ \hat{f} \leq f_o) = \frac{1}{2}[0.6321 + 0.2642] = 0.4481$. If $f_o = .0001, F \geq 10f_o$, and $N = 2 \times 10^4$, then $K_o = 2$ and **Prob** $(f \leq f_o\ |\ \hat{f} \leq f_o) = \frac{1}{3}[0.8647 + 0.5940 + 0.3233] = 0.5940$. This means that with 2 or fewer observed false alarms out of 20,000 observations, the probability is only 0.5940 that the true false-alarm rate is less than 0.0001. It seems that such a policy does not provide very certain answers. Perhaps more observations would be helpful. If $N = 10^5$, then $K_o = 10$. In this case

$$\textbf{Prob } (f \leq f_o\ |\ \hat{f} \leq f_o) = \frac{1}{11}[10.0000 + 0.9995 + 0.9972 + 0.9897 + 0.9707$$

$$+ 0.9329 + 0.8699 + 0.7798 + 0.6672 + 0.5461 + 0.4170]$$

$$= 0.8332$$

Thus with 10 or fewer observed false alarms out of 100,000 observations, the probability is 0.8336 that the true false-alarm rate is less than 0.0001. This is certainly better, but depending on our own requirement for certainty, it may not be sure enough.

If we adopt a different policy, we can be more sure about our judgment of the true false-alarm rate. Suppose we desire to perform an acceptance test that guarantees the probability is α that the machine meets specification. In this case we adopt the policy that we accept the machine if $\hat{f} \leq f^*$ where f^* is chosen so that for the fixed probability $\alpha(f^*)$, **Prob** $(f \leq f_o\ |\ \hat{f} \leq f^*) = \alpha(f^*)$. This means we accept if $K \leq K^*$, where $K^* = Nf^*$. Proceeding as before to find K^*, we have

$$\alpha(K^*) = \textbf{Prob } (f \leq f_o\ |\ \hat{f} \leq f^*) = \frac{\int_{f=0}^{f_o} \sum_{k=0}^{K^*} \binom{N}{k} f^k (1-f)^{N-k} df}{\int_{f=0}^{F} \sum_{k=0}^{K^*} \binom{N}{k} f^k (1-f)^{N-k} df}$$

$$= \sum_{k=0}^{K^*} I_{f_o}(k+1, N+1-k)$$

$$\times \sum_{k=0}^{K^*} I_F(k+1, N+1-k)$$

$$= \sum_{k=0}^{K^*} \sum_{i=k+1}^{\infty} e^{-K_o} K_o^i / i! / \sum_{k=0}^{K^*} \sum_{i=k+1}^{\infty} e^{-K_o} K_o^i / i!$$

Then if $f_o = .0001, F \geq 10f_o, N = 10^5$, and $K^* = 8$, there results

$$\alpha (K^*) = \frac{1}{9}[1.000 + .9995 + .9972 + .9897 + .9707 + .9329 + .8699 + .7798 + .6672]$$

$$= .9119$$

So if $\hat{f} \leq 8/10^5$, the probability will be .9119 that the true false alarm rate is less than .0001.

In summary, we have obtained that $Prob$ $(f \leq f_o$ and $K \leq K^*) = \frac{1}{N+1} \sum_{k=0}^{K^*} I_{f_o}(k+1, N+1-k)$ and $Prob$ $(K \leq K^*) = \frac{K^*+1}{N+1}$. Since $Prob$ (f) is uniform, $Prob$ $(f \leq f_o) = f_o$. These three probabilities determine the missed-acceptance rate $Prob$ $(f \leq f_o \mid K > K^*)$; the false-acceptance rate $Prob$ $(K \leq K^* \mid f > f_o)$; the error rate $Prob$ $(f \leq f_o$ and $K > K^*) + Prob$ $(f > f_o$ and $K \leq K^*)$; the identification accuracy rate $Prob$ $(f \leq f_o$ and $K \leq K^*) + Prob$ $(f > f_o$ and $K > K^*)$; the acceptance capture rate $Prob$ $(K \leq K^* \mid f \leq f_o)$; and the capture certainty rate $P(f \leq f_o \mid K \leq K^*)$. Thus the complete operating characteristics of the acceptance policy we have discussed can be determined. Tables 20.1 and 20.2 show these relationships for $f_o N = 5$ and $f_o N = 10$. The calculations use the tables of the cumulative Poisson distribution as an approximation to the incomplete beta ratio function and assume that $F \geq 10f_o$.

Table 20.1   Missed-acceptance, false-acceptance, and acceptance capture rates for varying values of K^* when $f_o N = 10$.

Missed-acceptance rate $Prob\ K^*\ (K > K^* \mid f \leq f_o)$	False-acceptance rate $Prob\ (f > f_o \mid K \leq K^*)$	Acceptance capture rate $Prob\ (K \leq K^* \mid f \leq f_o)$
00 .8013	.0067	.1987
01 .6094	.0236	.3906
02 .4344	.0573	.5656
03 .2874	.1092	.7126
04 .1755	.1755	.8245
05 .0987	.2489	.8012
06 .0511	.3222	.9489
07 .0244	.3903	.9756
08 .0108	.4504	.9892
09 .0044	.5022	.9956
10 .0017	.5462	.9983
11 .0006	.5836	.9994
12 .0082	.6155	.9998

Table 20.2   Missed-acceptance, false-acceptance, and acceptance capture rates for varying values of K^* when $f_o N = 10$.

Missed-acceptance rate $Prob\ K^*\ (K > K^* \mid f \leq f_o)$	False-acceptance rate $Prob\ (f > f_o \mid K \leq K^*)$	Acceptance capture rate $Prob\ (K \leq K^* \mid f \leq f_o)$
00 .9000	.0000	.1000
01 .8000	.0003	.2000
02 .7003	.0011	.2997
03 .6014	.0034	.3986
04 .5043	.0086	.4957
05 .4111	.0185	.5889
06 .3241	.0344	.6759
07 .2461	.0576	.7539
08 .1794	.0882	.8206
09 .1252	.1252	.8748
10 .0835	.1668	.9165
11 .0532	.2110	.9467
12 .0323	.2556	.9677

20.7.2   Balancing the Acceptance Test

From the point of view of the buyer of a recognition or inspection machine, one wants to be very sure that the machine meets the specifications. This corresponds to an acceptance test using a small value of K^* and a consequent small false-acceptance rate. But as can be seen from Tables 20.1 and 20.2, a small false-acceptance rate means that the acceptance capture rate (the fraction of machines that are accepted, given that they do meet specification) is small. Obviously the seller of a recognition or inspection machine wants to be sure that the machines that do meet specification are in fact accepted by the buyer. To have a high acceptance capture rate, the seller wants K^* to be large.

If the buyer and seller balance their own self-interests exactly in a middle compromise, the operating point chosen for the acceptance test will be the one for which the false-acceptance rate (which the buyer wants to be small) equals the missed-acceptance rate (which the seller wants to be small).

In this case

$$Prob\ (K > K^* \mid f \leq f_o) = Prob\ (f > f_o \mid K \leq K^*).$$

Since

$$Prob\ (K > K^* \mid f \leq f_o) = 1 - Prob\ (K \leq K^* \mid f \leq f_o)$$

and

$$Prob\ (f > f_o \mid K \leq K^*) = 1 - Prob\ (f \leq f_o \mid K \leq K^*)$$

the equality of false-acceptance and missed-acceptance rates implies

$$Prob \ (K \leq K^* \mid f \leq f_o) = Prob \ (f \leq f_o \mid K \leq K^*)$$

But

$$Prob \ (K \leq K^* \mid f \leq f_o) = \frac{Prob \ (f \leq f_o \mid K \leq K^*) \ Prob \ (K \leq K^*)}{Prob \ (f \leq f_o)}$$

Hence the equality of the false-acceptance and missed-acceptance rates implies $Prob \ (f \leq f_o) = Prob \ (K \leq K^*)$. Using the uniform distribution assumption on f, we have $f_o = (K^* + 1)/(N + 1)$. From this we obtain $K^* = f_o(N + 1) - 1$. For large N this can be simplified to $K^* = f_o N - 1$. Choosing an operating point with a value of K^* less than $f_o N - 1$ favors the buyer, and choosing an operating point with a value of K^* greater than $f_o N - 1$ favors the seller.

The last issue we address is the relationship between the number of trials for the acceptance test and the false acceptance rate at the operating point in which the false-acceptance rate is equal to the missed-acceptance rate. Now the false-acceptance rate is given by

$$Prob \ (f > f_o \mid K \leq K^*) = 1 - \frac{1}{K^* + 1} \sum_{k=0}^{K^*} \sum_{i=k+1}^{\infty} e^{-K_o} K_o^i / i!$$

At the desired operating point, $K^* = f_o N - 1 = K_o - 1$. To illustrate the relationship, we simply compute $\frac{1}{Nf_o} \sum_{k=0}^{Nf_o - 1} \sum_{i=k+1}^{\infty} e^{-Nf_o}(Nf_o)^i / i!$ as a function of N. Since N occurs everywhere with f_o, we can compute the false-acceptance rate as a function of $K_o = Nf_o$ and obtain a little more generality. Table 20.3 shows that with $Nf_o = 5$, the false-acceptance rate is 0.1755. With $Nf_o = 10$, the false-acceptance rate is 0.1251. And with $Nf_o = 20$, the false-acceptance rate is 0.0888. This suggests that for the acceptance test to have certainty greater than 90%, $Nf_o \geq 15$. Haralick (1989) gives a set of tables for $Nf_0 = 1, \ldots, 15$ by 1 and $20, \ldots, 45$ by 5.

20.7.3 Lot Assessment

A related performance assessment problem is that of the automatic inspection of products by lots. In the usual lot inspection approach, a quality control inspector makes a complete inspection on a randomly chosen small sample from each lot. The reason for not inspecting all of the lot is, of course, cost. If more than a specified number of defective products are found in the sample, the entire lot is rejected.

Quality control by lots has a long-standing methodology (Juran, 1963; Pau, 1976). However, the established methodology assumes that the inspector is error free. Not only is it the case that quality control inspectors are not error free, but any automatic classification or mensuration system also is not error free. Kittler and Pau (1978, 1980) discuss a methodology for lot assessment appropriate for the case of the imperfect inspector.

Table 20.3 False-acceptance rate as a function of $K_\circ = Nf_\circ$, where $K^* = Nf_o - 1$ is chosen to equalize the false-acceptance and missed-acceptance rates.

$$K_\circ = Nf_\circ \quad 1 - \frac{1}{K_\circ} \sum_{k=0}^{K_\circ - 1} \sum_{i=k+1}^{\infty} e^{-K_\circ} K_\circ^i / i!$$

01	.3679
02	.2706
03	.2240
04	.1954
05	.1755
06	.1607
07	.1490
08	.1395
09	.1318
10	.1251
15	.1025
20	.0888
30	.0729
40	.0629

20.8 Summary

In this chapter we developed a mensuration quantizing error model that permits the computation of the variance of centroids due to the random error of quantization. We discussed the derivation of false-alarm and misdetection-rate relationships for automated position measuring and mechanisms, and the need for experimental protocol in vision experiments. Our perspective is similar to that of Petkovic et al. (1988). We also explained how to measure the repeatability and accuracy of positional measurements made by a vision system, and how to derive the test criterion on which a machine vision system can be tested to determine whether it meets the specifications the manufacturer has advertised.

Bibliography

Abromowitz, M., and I. A. Stegun, *Handbook of Mathematical Functions,* National Bureau of Standards, U.S. Government Printing Office, Washington, DC, 1972.

Berenstein, C. A., and D. Lavine, "On the Number of Digital Straight Line Segments," *IEEE Transactions on Pattern Analysis and Machine Intelligence,* Vol. PAMI10, 1988, pp. 880–887.

Chang, C. A., et al., "A Quality Control Issue in Using Computer Vision Inspection Systems," *Proceedings of Vision '86,* Detroit, MI, 1986, pp. 4.57–4.73.

Deng, T-C, and T.L. Bergen, "Accuracy of Position Estimation by Centroid," *SPIE Conference on Intelligent Robots and Computer Visions*, SPIE Vol. 848 1987, Cambridge, MA, pp. 141–150.

Dorst, L., and A. W. M. Smeulders, "Discrete Representation of Straight Lines," *IEEE Transactions on Pattern Analysis and Machine Intelligence,* Vol. PAMI-6, 1984, pp. 450–462.

——, "Length Estimators Compared," *Pattern Recognition in Practice II,* North-Holland, Amsterdam, 1985.

——, "Best Linear Unbiased Estimators for Properties of Digitized Straight Lines," *IEEE Transactions on Pattern Analysis and Machine Intelligence,* Vol. PAMI-8, 1986, pp. 276–282, 676.

——, "Length Estimator for Digitized Contours," *Computer Vision, Graphics, and Image Processing,* Vol. 40, 1987, pp. 311–333.

Ellis, T. J., et al., "Measurement of the Lengths of Digitized Curved Lines," *Computer Graphics and Image Processing,* Vol. 10, 1979, pp. 333–347.

Groen, F. C. A., and P. W. Verbeek, "Freeman-Code Probabilities of Object Boundary Quantized Contours," *Computer Graphics and Image Processing,* Vol. 7, 1978, pp. 391–402.

Haralick, R. M., "Performance Assessment of Near Perfect Machines, *Machine Vision Applications,* Vol. 2, 1989, pp. 1–16.

Havelock, D. I., "Geometric Precision in Noise-Free Digital Images," *IEEE Transactions on Pattern Analysis and Machine Intelligence,* Vol. 11, 1989, pp. 1065–74.

Ho, C.S., "Precision of Digital Vision Systems," *IEEE Transactions on Pattern Analysis and Machine Intelligence*, Vol. PAMI-5, Nov. 1983, pp. 593–691.

Juran, J. M., *Quality Control Handbook,* McGraw-Hill, New York, 1963.

Kittler, J., and L. F. Pau, "Small Sample Properties of a Pattern Recognition System in Lot Acceptance Sampling," *Proceedings of the Fourth International Conference on Pattern Recognition,* Kyoto, Japan, 1978, pp. 249–257.

——, "Automatic Inspection by Lots in the Presence of Classification Errors," *Pattern Recognition,* Vol. 12, 1980, pp. 237–241.

Klaasman, H., "Some Aspects of the Accuracy of the Approximated Position of a Straight Line on a Square Grid," *Computer Graphics and Image Processing,* Vol. 4, 1975, pp. 225–235.

Koplowitz, J., and A. M. Bruckstein, "Design of Perimeter Estimators for Digitized Shapes," *IEEE Transactions on Pattern Analysis and Machine Intelligence,* Vol. PAMI-11, 1989, pp. 611–622.

Kulpa, Z., "Area and Perimeter Measurements of Blobs in Discrete Binary Pictures," *Computer Graphics and Image Processing,* Vol. 6, 1977, pp. 431–451.

McIllroy, M. D., "A Note on Discrete Representation of Lines," *AT&T Technical Journal,* Vol. 64, 1984, pp. 481–490.

McVey, E. S., and J. W. Lee, "Some Accuracy and Resolution Aspects of Computer Vision Distance Measurements," *IEEE Transactions on Pattern Analysis and Machine Intelligence,* Vol. PAMI-4, 1982, pp. 646–649.

Pau, L. F., *Contrôle de Qualité Statistique,* Convention 290939, Bureau National de Métrologie, Paris, 1976.

——, *Contrôle de Qualité pour l'Instrumentation,* Chiron, Paris, 1978.

Petkovic, D., et al., "Verifying the Accuracy of Machine Vision Algorithms and Systems," Research Report RJ 6523 (63324), IBM Research Division, 1988.

Proffitt, D., and D. Rosen, "Metrication Errors and Coding Efficiency of Chain Encoding Schemes for the Representation of Lines and Edges," *Computer Graphics and Image Processing,* Vol. 19, 1979, pp. 318–332.

Vossepoel, A. M., and A. W. M. Smeulders, "Vector Code Probability and Metrication Error in the Representation of Straight Lines of Finite Length," *Computer Graphics and Image Processing,* Vol. 20, 1982, pp. 347–364.

21 GLOSSARY OF COMPUTER VISION TERMS

21.1 Image

1.   An *image* is a spatial representation of an object, of a two-dimensional or three-dimensional scene, or of another image. It can be real or it can be virtual, as in optics. In computer vision, "image" usually means recorded image, such as a video image, a digital image, or a picture. It may be abstractly thought of as a continuous function I of two variables defined on some bounded and usually rectangular region of a plane. The value of the image located at spatial coordinates (r,c) is denoted by $I(r,c)$. For optic or photographic sensors, $I(r,c)$ is typically proportional to the radiant energy received in the electromagnetic band to which the sensor or detector is sensitive in a small area around (r,c). For range finder sensors, $I(r,c)$ is a function of the line-of-sight distance from (r,c) to an object in the three-dimensional world. For a tactile sensor, $I(r,c)$ is proportional to the amount that the surface at and around (r,c) deforms the sensor. When the image is a map, $I(r,c)$ is an index or symbol associated with some category, such as color, thematic land-use category, soil type, or rock type. A recorded image may be in photographic, video signal, or digital format.

2.   A *video image* is an image in electronic signal format capable of being displayed on a cathode ray tube screen or monitor. The video signal can be generated from devices like a CCD camera, a vidicon, a flying spot scanner, a tactile sensor, a range sensor, or a frame buffer driving a digital-to-analog converter. Video images have two common formats. In the frame format, the video signal itself is a sequence of signals, the ith signal representing the ith line of the image. The ith signal is separated from the $(i + 1)$st signal by a horizontal sync or pulse. Each video frame is separated from the next video frame by a vertical sync pulse. In the interlaced format, the video signal is divided into two fields. The first field contains all the odd-numbered lines, and the second field contains all the even-numbered lines. As in the frame format, the ith line of the field is its ith signal, and it is separated from

the next line of the field by a horizontal sync pulse. Successive fields are separated by vertical sync pulses.

3.   The *gray level, gray shade, gray tone, gray tone intensity, image intensity, image density, brightness,* or *image value* is a number or value assigned to a position on an image. For optic or photographic sensors, the image intensity at (r,c) is proportional to the integrated output, reflectance, or transmittance of a small area, usually called a resolution cell or pixel, centered on the position (r,c). Its value can be related to transmittance; reflectance; a coordinate of the tristimulus, ICI, YIQ, or RGB color coordinate system; brightness; radiance; luminance; density; voltage; or current.

4.   *Resolution* is a generic term that describes how well a system, process, component, material, or image can reproduce an isolated object consisting of separate closely spaced objects or lines. The *limiting resolution, resolution limit,* or *spatial resolution* is described in terms of the smallest dimension of the target or object that can just be discriminated or observed. Resolution may be a function of object contrast and spatial position as well as of element shape (single point, number of points in a cluster, continuum, line, etc.).

5.   A *resolution cell* is the smallest, most elementary areal constituent having an associated image intensity in a digital image. A resolution cell is referenced by its spatial coordinates, which are the center coordinates of its area. The resolution cell or spatial formations of resolution cells constitute the basic unit for low-level processing of digital image data. Resolution cells usually have areas that are square, rectangular, or hexagonal.

6.   *Acutance* is a measure of the sharpness of edges in a photograph or an image. It is defined for any edge by the average squared rate of change of the image intensity across the edge divided by the total image intensity difference from one side of the edge to the other side.

7.   The *contrast* of an object against its background can be measured by (1) its *contrast ratio,* which is the ratio of the higher of object transmittance or background transmittance to the lower of object transmittance or background transmittance; (2) its *contrast difference,* which is the difference between the higher density of object or background and the lower density of object or background; and (3) its *contrast modulation,* which is the difference between the darker of object or background image intensity and the lighter of the two divided by the sum of object image intensity and background image intensity.

8.   A *pixel, picture element,* or *pel* is a pair whose first member is a resolution cell or (row, column) spatial position and whose second member is the image intensity value or vector of image values associated with the spatial position.

9.   A *voxel,* short for volume element, is an ordered pair whose first component is a (row, column, slice) location of a rectangular parallelepiped volume and whose second component is the vector of properties in the rectangular parallelepiped volume.

10.   An *edgel,* short for edge element, is a triplet whose first component is the (row, column) location of a pixel, whose second component is the position and orientation of an edge running through the pixel, and whose third component is the strength of the edge.

11.   *Raster scan order* refers to the sequence of pixel locations obtained by scanning the spatial domain of an image in a left-to-right scan of each image row with the rows taken in a top-to-bottom ordering. Frame format video images are images scanned in raster scan order.

12.   A *range image* is an image in which each pixel value is a function of the distance between the pixel and the object surface patch imaged on the pixel. Depending on the sensor and preprocessing used to create the range image, the distance can be the distance between the image plane and the ranged surface patch, the line-of-sight distance between the pixel and its corresponding ranged surface patch, or some function of these distances and the pixel's position.

13.   A *digital image, digitized image,* or *digital picture function* is an image in digital format obtained by partitioning the area of the image into a finite two-dimensional array of small, uniformly shaped, mutually exclusive regions called resolution cells and assigning a representative image value to each such spatial region. A digital image may be abstractly thought of as a function whose domain is the finite two-dimensional set of resolution cells and whose range is the set of possible image intensities.

14.   *Rangel* is the range data element produced by a range sensor. It is a pair whose first member is a row-column spatial position and whose second member is the range value or a vector whose first component is the range value and whose second component is the image intensity value.

15.   A *depth map* or *range map* is a digital range image in which the range value in each pixel's position is the distance between the image plane and the ranged surface patch corresponding to the pixel.

16.   An *orientation map* is a digital image in which each pixel contains the three-dimensional orientation vector of the normal to the three-dimensional surface patch corresponding to the pixel position.

17.   A *multi-image set* or *multiband image* is a set of registered images, each related to the same subject but taken at different times, from different positions, with different lighting, with different sensors, at different electromagnetic frequencies, with different polarizations, or from different sections of the subject. Although there is a high degree of information redundancy between images in a multi-image set, each image usually has some information not available in any one or combination of the other images in the set. If the multi-image set has N images, then each resolution cell is associated with an N-tuple of image values.

18.   A *multispectral image* is a multiband image in which each band is an image taken at the same time but sensitive in a different part of the electromagnetic spectrum.

19.   A *time-varying image, multitemporal image, dynamic imagery,* or *image time sequence* is a multi-image set in which each successive image in the set is taken of the same scene at a successive time. Between successive snapshots, the objects in the scene may move or change and the sensor may move.

20.   A *binary image* is an image in which each pixel takes either the value 0 or the value 1.

21.   A *gray scale image* or a *gray level image* is an image in which each pixel has a value in a range larger than just 0 or 1. Gray scale images typically have

values in the range 0 to 63, 0 to 255, or 0 to 1,023, corresponding to 6-bit, 8-bit, or 10-bit digitizations.

22.   A *symbolic image* is an image in which the value of each pixel is an index or a symbol.

23.   A *histogram* or *image histogram* is a function h defined on the set of image intensity values to the nonnegative integers. The value $h(k)$ is given by the number of pixels in the image having image intensity k. For images having a large gray tone range, the image will often be quantized before being histogrammed or will be quantized on the fly during the histogramming process.

21.2   Photometry and Illumination

24.   *Luminous flux* is radiant power evaluated according to its capacity to produce visual sensation. *Luminous intensity* in a given direction is measured in terms of luminous flux per steradian. The unit of luminous intensity is the *candela*. The luminance of a blackbody radiator at the temperature of solidification of platinum is 60 candelas per square centimeter. The unit of luminous flux is the *lumen*. The luminous flux emitted by a uniform point light source of luminous intensity of one candela in one steradian solid angle is one *lumen*.

25.   The *illumination* at a point on a surface is the luminous flux incident on an infinitesimal element of the surface centered at the given point divided by area of the surface element. The unit of illumination is the *lux* or meter candle. The lux is equal to one lumen per square meter. Another unit of illumination is the *foot candle*, and it is equal to one lumen per square foot. The illumination at a point on a surface due to a point source of light is proportional to the luminous intensity of the source in the direction of the surface point and to the cosine of the angle between this direction and the surface normal direction. It is inversely proportional to the square of the distance between the surface point and the source.

26.   The *illuminance* in a given direction at a surface point is the luminous intensity in that direction of an infinitesimal surface element containing the given point divided by the area of the orthogonal projection of the element on a plane perpendicular to the given direction.

27.   The *radiance* of an object is a measure of the power per unit foreshortened surface area per unit solid angle radiated or reflected by the object about a specified direction. Radiance can be a function of the viewing angle and the spectral wavelength and bandwidth.

28.   The *radiant intensity* of a point object is a measure of the radiant power per steradian radiated or reflected by the object. Radiant intensity can be a function of the viewing angle and the spectral wavelength and bandwidth.

29.   *Irradiance* is the power per unit area of radiant energy incident on a surface.

30.   The *reflectance*, the *reflection coefficient*, or the *bidirectional reflectance distribution function* of a surface is the ratio of the radiant power per unit area per steradian reflected by the surface to the radiant power per unit area incident on

the surface. The reflectance can be a function of the incident angle of the radiance, the viewing angle of the sensor, and the spectral wavelength and bandwidth.

31. A *reflectance image* or *reflectance map* is a digital image in which the value in each pixel's position is proportional to the reflectance of the surface patch imaged at the pixel's position for a given illumination and viewing direction.

32. A *Lambertian surface* is a uniformly diffusing surface. It appears as a matte surface, and it has a reflectance function that is a constant. The reflectance function of a Lambertian surface does not depend on the viewing angle, and therefore a planar surface having a Lambertian reflectance appears equally bright from all viewing angles. For a Lambertian surface, the luminous intensity per unit area in a given direction of the reflected light varies as the cosine of the angle between the direction and the surface normal direction.

33. *Backlighting* refers to an illumination arrangement in which the light source is on the opposite side of the object from the camera. Backlighting tends to produce images that are black-and-white silhouettes.

34. *Frontlighting* refers to an illumination arrangement in which the light source is on the same side of the object as the camera.

35. *Ambient light* refers to the light that is present in the environment around a machine vision system and that is generated from sources outside the system. From the point of view of the machine vision system, ambient light is unplanned light that might adversely affect the image processing. Care is usually taken to minimize its effect.

36. The r, g, b *chromaticity* coordinates of a multispectral pixel, where red brightness is R, green brightness is G, and blue brightness is B, is given by

$$r = \frac{R}{R + G + B}$$
$$g = \frac{G}{R + G + B}$$
$$b = \frac{B}{R + G + B}$$

37. The H, S *hue saturation* coordinates of a multispectral pixel whose chromaticity coordinates are r, g, b is given by

$$H = \begin{cases} \theta \\ 360 - \theta \end{cases} \text{ if } \begin{array}{c} b < g \\ b > g \end{array}$$

where

$$\theta = \arccos \frac{2r - g - b}{\sqrt{6} \left[(r - \tfrac{1}{3})^2 + (g - \tfrac{1}{3})^2 + (b - \tfrac{1}{3})^2 \right]^{1/2}}$$

$$S = 1 - 3 \min \{r, g, b\}$$

38. The *YIQ coordinate* used in NTSC color TV transmissions is related to the RGB coordinates by the following linear transformation:

$$\begin{pmatrix} Y \\ I \\ Q \end{pmatrix} = \begin{pmatrix} .299 & .587 & .144 \\ .596 & -.274 & -.322 \\ .211 & -.523 & .312 \end{pmatrix} \begin{pmatrix} R \\ G \\ B \end{pmatrix}$$

21.3   Photogrammetry

39.   *Analytic photogrammetry* refers to the analytic mathematical techniques that permit the inference of geometric relations between points and lines in the two-dimensional perspective projection image space and the three-dimensional object space.

40.   *Digital photogrammetry* refers to the computer processing of perspective projection digital images with analytic photogrammetry techniques as well as other computer techniques for the automatic interpretation of scene or image content.

41.   *Relative orientation* in analytic photogrammetry is the relative position and orientation of one common reference frame with respect to another. When two common reference frames are in known relative orientation, the rays emanating from the same object point located on each camera's image will intersect exactly at a point in three-dimensional space.

42.   *Exterior orientation* or *outer orientation* refers to the position and orientation of a camera reference frame with respect to a world reference frame, computed on the basis of corresponding 3D and 2D perspective projection points.

43.   *Absolute orientation* in analytic photogrammetry is the rotation and translation transformation or transformations by which one or more camera reference frames can be made to correspond to a world reference frame, computed on the basis of corresponding 3D points.

44.   The *optic axis* or *principal axis* is the straight line that passes through the centers of curvature of lens surfaces.

45.   The *principal point* is the point on the image that is the intersection of the image plane with the optic axis.

46.   The *center of perspectivity* of a perspective projection is the common point where all rays meet.

47.   The *principal distance* is the distance between the center of perspectivity and the image projection plane. For a fixed image, it is sometimes called the *camera constant* or *Gaussian focal length*.

48.   The *inner orientation* or *internal orientation* is given by the triple (u_0, v_0, f) whose (u_0, v_0) is the position of the principal point in the measurement image plane coordinate system, and whose f is the principal distance. Inner orientation may also include the values of the free parameters that describe the lens distortion.

49.   A *perspective projection* is defined in terms of a projection plane and a center of perspectivity. Imaging sensors whose ideal model is the pinhole camera generate perspective projection images. There are three commonly used frameworks for defining the relationship between a three-dimensional point and its two-dimensional perspective projection. All take the projection plane to be perpendicular to the z-axis. In the first framework, the center of perspectivity is taken to be the origin and the projection plane is a distance f from the origin on the positive z-axis. In this case the projection (P_x, P_y) of the point (x, y, z) is given by

$$P_x = \frac{fx}{z}, \qquad P_y = \frac{fy}{z}$$

In the second framework, the center of perspectivity is taken to be $(0, 0, -f)$ and the projection plane passes through the origin. In this case the projection (P_x, P_y) of the point (x, y, z), $z > 0$, is given by

$$P_x = \frac{fx}{f + z}, \qquad P_y = \frac{fy}{f + z}$$

In the third framework, the center of perspectivity is taken to be $(0, 0, f)$ and the projection plane passes through the origin. In this case the projection (P_x, P_y) of the point (x, y, z), $z < 0$, is given by

$$P_x = \frac{fx}{f - z}, \qquad P_y = \frac{fy}{f - z}$$

50.   A *parallel* or *orthographic projection* onto a plane perpendicular to the z-axis of a point (x, y, z) produces the projected point (P_x, P_y) defined by

$$P_x = sx, \qquad P_y = sy$$

where s is the scale factor of the projection.

51.   The *parallax* is the observed positional difference of a projected three-dimensional point on a pair of two-dimensional perspective images. The difference in position is caused by a shift in the position of the perspective centers and optic axis orientation. That portion of the parallax in the direction of the x-axis is called the *x-parallax*. That portion of the parallax in the direction of the y-axis is called the *y-parallax*. For a pair of stereo images in which the line joining the centers of perspectivity is parallel to the x-axis, the parallel will be entirely in the direction of the x-axis when the two image planes are in the same orientation. Should one image plane be tilted with respect to the other, the y-parallax will not be zero.

52.   A *vanishing point* is the point in the two-dimensional perspective projection image plane where a system of three-dimensional parallel lines converge. The vanishing points of all systems of three-dimensional parallel lines parallel to a given plane will lie along a corresponding line in the two-dimensional perspective projection image plane called the vanishing line for the given plane.

21.4   Image Operators

53.   An *image operator, image transform,* or *image transform operator* is a function that takes an image for its input and produces an image for its output. The domain of a transform operator is often called the spatial or space domain. The range of the transform operator is often called the transform domain. Some image transform operators have spatial and transform domains of entirely different geometry or character; the image in the spatial domain may appear entirely different and have a different interpretation from the image in the transform domain. Specific examples of such image transforms include Fourier, sine, cosine, slant, Haar, Hadamard, Mellin, Karhunen-Loeve, and Hough transforms. Image operators that have spatial and transform domains of similar geometry or character include point operators, neighborhood operators, and spatial filters.

54.   The *discrete Fourier transform* $\hat{I}$ of a digital image I represents the image in terms of a linear combination of complex exponentials. The Fourier transform $\hat{I}$ is defined by

$$\hat{I}(w_r, w_c) = \frac{1}{RC} \sum_{r=0}^{R-1} \sum_{c=0}^{C-1} I(r, c) e^{-2j\pi\left(\frac{rw_r}{R} + \frac{cw_c}{C}\right)}$$

$\hat{I}(w_r, w_c)$ is the coefficient of the complex exponential $e^{2j\pi\left(\frac{rw_r}{R} + \frac{cw_c}{C}\right)}$ in the linear combination representing I, as can immediately be seen from the corresponding relation

$$I(r, c) = \sum_{w_r=0}^{R} \sum_{w_c=0}^{C} \hat{I}(w_r, w_c) e^{2j\pi\left(\frac{rw_r}{R} + \frac{cw_c}{C}\right)}$$

which is called the *inverse discrete Fourier transform.* The variables w_r and w_c have the interpretation of being row and column *spatial frequencies.*

21.5   Point Operators

55.   A *point operator* is an image operator in which the output image value at each pixel position depends only on the input image value at the corresponding pixel position.

56.   *Thresholding* is an image point operation that produces a binary image from a gray scale image. A binary-1 is produced on the output image whenever a pixel value on the input image is above a specified minimum threshold level. A binary-0 is produced otherwise. Alternatively, thresholding can produce a binary-1 on the output image whenever a pixel value on the input image is below a specified maximum threshold level. A binary-0 is produced otherwise.

57.   *Level slicing* or *density slicing* is a point operation that employs two thresholds and produces a binary image. A binary-1 is produced on the output image

wherever a pixel value on the input image lies between the specified minimum and maximum threshold levels. A binary-0 is produced otherwise.

58.   *Multilevel thresholding* is a point operator employing two or more thresholds. Pixel values that are in the interval between two successive threshold values are assigned an index associated with the interval.

59.   *Contrast stretching* refers to any monotonically increasing point operator whose effect is to increase or enhance the visibility of an image's detail.

60.   *Quantizing* is a monotonically increasing point operator by which each image intensity value in a digital image is assigned a new value from a given finite set of quantized values. The quantized image has fewer distinct gray levels but may make better use of the dynamic range. Thus quantizing often enhances the image's appearance. There are four often-used methods of quantizing: equal-interval, equal-probability, minimum-variance, and histogram hyperbolization quantizing. In each method the range of image values from the maximum to the minimum value is divided into contiguous intervals, and each image value is assigned either the mean value or the index of the quantized class to which it belongs.

61.   In *equal-interval* or *linear quantizing,* the range of image values from maximum to minimum is divided into contiguous intervals, each of equal length, and each image value is assigned to the quantized class that corresponds to the interval within which it lies.

62.   In *equal-probability quantizing,* the range of image values is divided into contiguous intervals such that after the image values are assigned to their quantized class, there is an equal frequency of occurrence for each quantized value in the quantized digital image; equal-probability quantizing is sometimes referred to as *histogram equalization.*

63.   In *minimum-variance quantizing,* the range of image values is divided into contiguous intervals such that the weighted sum of the variances of the quantized intervals is minimized. The weights are usually chosen to be the quantized class probabilities that are computed as the proportional areas on the image that have values in the quantizing intervals.

64.   In *histogram hyperbolization quantizing,* the range of image values is divided into contiguous intervals and each image value is assigned to the mean of its quantized class. The division is done in such a way that the quantized image has a uniform perceived brightness. Histogram hyperbolization takes into account the nonlinearity of the human eye–brain combination.

65.   *Masking* is a point operator applied to a two-band image. One image band is a binary image B and is called the mask image band. The second image band I is called the image to be masked. Masking produces a resulting image J whose pixels take the value 0 wherever the mask image has value 0 and whose pixels take the value of the image I wherever the mask image has value 1. That is,

$$J(r, c) = \begin{cases} 0 & \text{if } B(r,c) = 0 \\ I(r,c) & \text{if } B(r,c) = 1 \end{cases}$$

66.   *Change detection* is the process by which two registered images may be compared, pixel by pixel, and a binary-1 value given to the output pixel whenever

corresponding pixels on the input images have significantly different enough gray levels. Corresponding pixels on the input images that do not have significantly different enough gray levels generate a binary-0 value on the output image. A change detection operator is a point operator.

21.6   Spatial Operators

67.   *Two-dimensional signal processing* refers to that area of image processing in which the one-dimensional signal-processing techniques of noise filtering, restoration, data compression, and detection have been generalized to two dimensions and thereby made applicable to image data.

68.   A *neighborhood operator* is an image operator in which the output image value at each pixel position depends only on the input image values in a neighborhood containing or surrounding the corresponding input pixel position.

69.   A *spatial filter* is an image operator in which the spatial and transform domains have similar geometries and in which the image output value at each pixel depends on more than one pixel value in the input image. Usually, but not always, the image output value has its highest dependence on the image input values in some neighborhood centered in the corresponding pixel in the input image.

70.   A *linear spatial filter* is a spatial filter for which the image intensity at coordinates (r, c) in the output image is some weighted average or linear combination of the image intensities located in a particular spatial pattern around coordinates (r, c) of the input image. A linear spatial filter is often used to change the spatial frequency characteristics of the image. For example, a linear spatial filter that emphasizes high spatial frequencies will tend to sharpen the edges in an image. A linear spatial filter that emphasizes low spatial frequencies will tend to blur the image and reduce salt and pepper noise. When the purpose of the filter is to enhance neighborhoods having certain shapes, the operation is sometimes called mask matching.

71.   The *kernel* of a linear spatial filter is a function defined on the domain of the spatial pattern of the filter and whose value at each pixel of the domain is the weight or coefficient of the linear combination that defines the spatial linear filter.

72.   A *box filter* is a linear spatial smoothing filter in which each pixel in the filtered image is the equally weighted average of the pixels in a rectangular window centered at its spatial position in the input image.

73.   A *Gaussian filter* is a linear spatial smoothing filter whose kernel is given by the two-dimensional Gaussian

$$k(r, c) = \frac{1}{2\pi} e^{-\frac{1}{2}\left(\frac{r^2}{\sigma_r^2} + \frac{c^2}{\sigma_c^2}\right)}$$

Filtering an image with a Gaussian filter will smooth the image.

74.   *Convolving* an image I with a kernel k having support or domain K produces a convolved image, denoted by $I * k$, that is defined by

$$(I * k)(r, c) = \sum_{(i,j) \in K} I(r - i, c - j)k(i, j)$$

Convolution is a linear operator.

75.   A two-dimensional filter is called a *separable filter* if the convolution can be decomposed into two successive one-dimensional convolutions, one convolution operating on the image row by row, and the other operating on the image column by column.

76.   *Correlating* an image I with a kernel k having support or domain K produces a correlated image J defined by

$$J(r, c) = \sum_{(i,j) \in K} I(r + i, c + j)k(i, j)$$

Correlation is a linear operator.

77.   A *high-pass filter* is a linear spatial filter that attenuates the low spatial frequencies of an image and accentuates the high spatial frequencies. It is typically used to enhance small details, edges, and lines.

78.   A *low-pass filter* is a linear spatial filter that attenuates the high spatial frequencies of an image and accentuates the low spatial frequencies. It is typically used to suppress small undesired details, eliminate noise, enhance coarse image features, or smooth the image.

79.   A *bandpass filter* is a linear spatial filter that attenuates spatial frequencies outside the band and accentuates those within the band. It is typically used to enhance details of the image whose spatial size characteristics are related to the spatial frequencies within the band.

80.   A *pyramid* or *image pyramid* is a sequence of copies of an image in which both sample density and resolution are decreased in regular steps. The bottom level of the pyramid is the original image. Each successive level is obtained from the previous level by a filtering operator followed by a sampling operator. There are three kinds of pyramids: Gaussian, morphological, and Laplacian.

81.   In the *Gaussian image pyramid*, the resolution is decreased by successive convolutions of the image at the previous level of the pyramid with a Gaussianlike kernel. After the low-pass Gaussian convolutions, the sample density is typically decreased by sampling every other pixel.

82.   In the *morphological image pyramid*, each successive level is obtained by an opening and closing operation on the previous level, followed by sampling.

83.   In the *Laplacian image pyramid*, each layer is obtained by taking the Laplacian of the corresponding level on the Gaussian pyramid. The Laplacian convolution kernel here is typically defined as the kernel obtained by taking the Laplacian of a Gaussian having an appropriately chosen value for its standard deviation. It can be rapidly implemented by taking the difference between two successive layers in the Gaussian pyramid.

84.   An *edge operator* or *step edge operator* is a neighborbood operation that determines the extent to which each pixel's neighborhood can be partitioned by a

simple arc passing through the pixel where pixels in the neighborhood on one side of the arc have one predominant value, and those on the other side have a different predominant value. Some edge operators can also produce a direction that is the predominant tangent direction of the arc as it passes through the pixel.

There are four classes of edge operators: *gradient, Laplacian, zero crossing, and morphological edge*. The gradient operators compute some quantity related to the magnitude of the slope of the underlying image gray tone intensity surface of which the observed image pixel values are a noisy discretized sample. The Laplacian operators compute some quantity related to the Laplacian of the underlying image gray tone intensity surface. The zero-crossing operators determine whether or not the digital Laplacian or the estimated second directional derivative has a zero crossing within the pixel. The morphological edge operators compute a quantity related to the residues of an erosion and/or dilation operation.

85.   An *edge image* is an image in which each pixel is labeled as "edge" or "nonedge." In addition to this basic labeling, pixels in an edge image may carry additional information, such as edge direction, contrast, or strength.

86.   *Edge linking* refers to the process by which neighboring edge-labeled pixels can be aggregated to constitute a chain or sequence of edge pixels.

87.   *Boundary detection* or *boundary delineation* refers to any process that determines a chain of pixels separating one image region from a neighboring image region.

88.   An *occluding edge* is an image edge that arises from a range or depth discontinuity. This typically happens where one object surface projects to a pixel on one side of the edge and another object surface that is some distance behind the first one projects to a pixel on the other side of the edge. Step edges in depth maps are always occluding edges.

89.   A pair of straight edges are said to be *antiparallel* if there are no edges between them and the edges have opposite contrast.

90.   *Homomorphic filtering* is a filtering process in which the filter is applied to the logarithm of the image and the output image is obtained by exponentiating the filtered logarithm image.

91.   A *median filter* is a nonlinear neighborhood image-smoothing spatial filter in which the value of an output pixel is the median value of all the input pixels in the supporting neighborhood of the filter about the given pixel's position. Median filters are used to smooth and remove noise from images.

92.   *Image smoothing* refers to any spatial filtering producing an output image that spatially simplifies and approximates the input image. Image smoothing suppresses small image details and enhances large or coarse image structures.

93.   A *scale space image* is an image in which each pixel's value is a function indicating for each standard deviation σ the value at the pixel's position of the convolution of the image with a Gaussian kernel having standard deviation σ.

94.   *Scale space structure* refers to that analysis of a scale space image in which each pixel's value is a function specifying for each possible standard deviation σ whether the pixel contains a zero crossing of some combination of a fixed order of spatial partial derivatives evaluated at the pixel's position.

21.7   Morphological Operators

95.   *Mathematical morphology* refers to an area of image processing concerned with the analysis of shape. The basic morphological operations consist of dilating, eroding, opening, and closing an image with a structuring element.

96.   The *structuring element* of a morphological operator is a function defined on the domain of the spatial pattern of the operator and whose value at each pixel of the domain is the weight or coefficient employed by the operator at that pixel position. The structuring element of a morphological operator has a role in morphology exactly analogous to the role of the kernel in a convolution operation.

97.   *Dilating* an image I by a structuring element s having support or domain S produces a dilated image denoted by $I \oplus s$ that is defined by

$$(I \oplus s)(r, \, c) = \max_{(i,j) \in S} \{I(r - i, \, c - j) + s(i, \, j)\}$$

Dilating is a commutative, associative, translation-invariant, and increasing operation. Dilating is the dual operation to eroding.

98.   *Eroding* an image I by a structuring element s having support or domain S produces an eroded image denoted by $I \ominus s$ that is defined by

$$(I \ominus s)(r, \, c) = \min_{(i,j) \in S} \{I(r + i, \, c + j) - s(i, \, j)\}$$

Eroding is a translation-invariant and increasing operation. It is the dual operation to dilating.

99.   *Opening* an image I with a structuring element s produces an opened image denoted by $I \circ s$ that is defined by

$$I \circ s = (I \ominus s) \oplus s$$

Opening is an increasing, antiextensive, and idempotent operation. It is the dual operation to closing. Opening an image with a disk-shaped structuring element smooths the contour, breaks narrow isthmuses, and eliminates islands and capes smaller in size or width than the disk structuring element.

100.   *Closing* an image I with a structuring element s produces a closed image denoted by $I \bullet s$ that is defined by

$$I \bullet s = (I \oplus s) \ominus s$$

Closing is an increasing, extensive, and idempotent operation. It is the dual operation to opening. Closing an image with a disk-shaped structuring element smooths the contours; fuses narrow breaks and long, thin gulfs; eliminates holes smaller in size than the disk structuring element; and fills gaps on the contour.

101.   A *thinning operator* is a symbolic image neighborhood operator that deletes, in some symmetric way, all the interior border pixels of a region that do not disconnect the region. Successive applications of a thinning operator reduce a region to a set of arcs that constitutes a *skeleton* of the region.

102.   A *thickening operator* is a symbolic image neighborhood operator that in some symmetric way aggregates all background pixels near enough to a region into the region.

21.8   Hough Transform

103.   The *discrete Radon transform* $R : Q \rightarrow [0, \infty)$ of a function $I : X \rightarrow [0, \infty)$ relative to a functional form $F : X \times Q \rightarrow [0, \infty)$ is defined by

$$R(q) = \sum_{\{x \in X | F(x,q)=0\}} I(x)$$

104.   The *Hough transform* $H : \mathcal{B} \rightarrow [0, \infty)$ of a function $I : X \rightarrow [0, \infty)$ relative to a functional form $F : X \times Q \rightarrow [0, \infty)$, where $\mathcal{B}$ is a partition of the parameter space Q, is defined by

$$H(B) = \sum_{\{x \in X | \text{ for some } q \in B, F(x, q)=0\}} I(x), \qquad B \in \mathcal{B}$$

When I has the form $I(x) = 1$, for every $x \in X_0$, where $X_0 \subseteq X$ and $I(x) = 0$ elsewhere, the Hough transform of X_0, relative to a functional form $F : X \times Q \rightarrow [0, \infty)$, has the simple form defined by

$$H(B) = \#\{x \in X_0 | \text{ for some } q \in B, F(x, q) = 0\}$$

The Hough transform can aid in the detection of image arcs of a given shape or form or in the detection of three-dimensional object shapes. Each shape or form has some free parameters that, when specified precisely, define the arc, shape, or form. The shape having free parameter q corresponds to the set $\{x \in X | F(x, q) = 0\}$. The free parameters consitute the transform domain or the parameter space of the Hough transform. Depending on the information available to the Hough transform, each neighborhood of the image or object surface being transformed will map to a point or a set of points in the Hough parameter space. The Hough transform discretizes the Hough parameter space into bins, and counts for each bin how many neighborhoods on the image or object surface have a transformed point lying in the volume assigned to the bin.

105.   The *Gaussian sphere* refers to a unit sphere and its associated spherical coordinate system. The quantities usually represented on the Gaussian sphere are orientation vectors.

106.   *Gradient space* is a two-dimensional space whose axes represent the first-order partial derivatives of a surface of the form $z = f(x, y)$. Each point in gradient space corresponds to the orientation of a possible surface normal.

107.   The *extended Gaussian image* or *orientation histogram* of a three-dimensional object is a two-dimensional histogram or Hough transform of the surface normal orientations of the object. It is computed by tessellating the surface of a sphere into cells and assigning to each cell a value that estimates the total area of the object's surface having a surface normal orientation that falls within the cell.

21.9   Digital Geometry

108.   The *Euclidean distance* between two points $p = (p_1, ..., p_N)$ and $q = (q_1, ..., q_N)$ is defined by

$$d(p, \, q) = \sqrt{\sum_{n=1}^{N} (p_n - q_n)^2}$$

109.   The *block or city block distance* between two points $p = (p_1, ..., p_N)$ and $q = (q_1, ..., q_N)$ is defined by

$$d(p, \, q) = \sum_{n=1}^{N} |p_n - q_n|$$

110.   The *square or max distance* between two points $p = (p_1, ..., p_N)$ and $q = (q_1, ..., q_N)$ is defined by

$$d(p, \, q) = \max_{n=1,...,N} |p_n - q_n|$$

111.   The *distance transform* of a binary image is an image having in each pixel's position its distance from the nearest binary-0 pixel of the input image. Distance can be city block distance, Euclidean distance, or square distance.

112.   A *figure F* or a *subimage F* in a continuous or digital image I is any function F whose domain is some subset A of the set of spatial coordinates or resolution cells, whose range is the set G of image intensities, and that is defined by $F(r, \, c) = I(r, \, c)$ for any $(r, \, c)$ belonging to A.

113.   A *region R* of an image is any subset of resolution cells in the spatial domain of the image.

114.   A neighboring pair of pixels are said to be *4-connected* if they share a common side. A neighboring pair of pixels are said to be *8-connected* if they share a common side or a common corner.

115.   A region R is *connected* if there is a path between any two resolution cells contained in R. More precisely, R is 4-connected (8-connected) if for each pair of resolution cells $(r, \, c)$ and $(u, \, v)$ belonging to R, there exists some sequence $< (a_1, \, b_1), \, (a_2, \, b_3), ..., (a_m, \, b_m) >$ of resolution cells belonging to R such that $(r, \, c) = (a_1, \, b_1)$, $(u, \, v) = (a_m, \, b_m)$; and $(a_i, \, b_i)$ is 4-connected (8-connected) to $(a_{i+1}, \, b_{i+1})$, $i = 1, 2, ..., m - 1$.

116.   A *blob* or *connected component* is a maximal-sized connected region.

117.   A *digital straight-line segment* between resolution cells $(r_1, \, c_1)$ and $(r_2, \, c_2)$ is that subset of all pixels such that some part of the line segment joining $(r_1, \, c_1)$ and $(r_2, \, c_2)$ has a nonempty intersection with the pixel's area.

118.   A region R is *convex* if for every pair of resolution cells in R, R contains the digital straight-line segment that joins the pair of resolution cells.

119.   A pixel is an *interior border pixel* of a region R if the pixel belongs to R and neighbors a pixel outside of R.

120.   A pixel is an *exterior border pixel* of a region R if the pixel does not belong to R and neighbors a pixel belonging to R.

121.   A pixel is an *interior pixel* of a region R if every pixel it neighbors belongs to R.

122.   A *simple boundary* is an oriented closed curve that does not touch or cross itself. Pixels that are on the inside of a simple boundary constitute a connected region having no holes.

123.   The *bounding contour* of a region R consists of the simple boundary that surrounds the pixels of R.

124.   A set of pixels H constitutes a *hole* of a region R if H is a maximal connected set of pixels that do not belong to R but are surrounded by R.

125.   The *border* or *boundary* of a connected region R consists of its bounding contour and the (possibly empty) set of simple boundaries, each of which surrounds the pixels belonging to some hole of R.

126.   *Boundary following* refers to the sequential procedure by which the chain of the boundary pixels of a region can be determined.

127.   A *concurve* is a continuous curve, usually representing a blob boundary, consisting of a connected chain of simply described arcs.

128.   The *minimum-perimeter olygon* of a digital curve C is the polygon of shortest length whose digitization is C. The shape of the minimum-perimeter polygon is often similar to the general perceived shape of the digital curve.

129.   *Contour tracing* is a searching or traversing process by which the bounding contour of a blob can be identified.

21.10   Two-Dimensional Shape Description

130.   The *perimeter* of a connected region R is the length of the bounding contour of R.

131.   The *area* A of a region R is defined by

$$A = [\#R] \cdot s$$

where s is the scale factor that specifies the area of a pixel.

132.   The *centroid* $(\bar{r}, \bar{c})$ of a region R is the center of mass of the region. It is the mean (row, column) position for all pixels in the region and is given by

$$\bar{r} = \frac{1}{\#R} \sum_{(r,c) \in R} r, \qquad \bar{c} = \frac{1}{\#R} \sum_{(r,c) \in R} c$$

133.   The (j,k)th *moment* M_{jk} of a digital shape S is given by

$$M_{jk} = \sum_{(r,c) \in S} r^j c^k$$

The center of gravity $(\bar{r}, \bar{c})$ of S can be expressed in terms of the moments of S:

$$\bar{r} = \frac{M_{10}}{M_{00}}, \qquad \bar{c} = \frac{M_{01}}{M_{00}}$$

134.   The (j,k)th *central moment* μ_{jk} of a digital shape S is given by

$$\mu_{jk} = \sum_{(r,c) \in S} (r - \bar{r})^j (c - \bar{c})^k$$

135. (j,k)th *normalized central moment* of S is given by

$$\eta_{jk} = \frac{\mu_{jk}}{\mu_{00}^\gamma}, \quad \text{where} \quad \gamma = \frac{j+k}{2} + 1$$

136.   *Rotation invariant moments* of S are given by

$$\phi\,(1) = \eta_{20} + \eta_{02}$$
$$\phi\,(2) = (\eta_{20} - \eta_{02})^2 + 4\eta_{11}^2$$
$$\phi\,(3) = (\eta_{30} - 3\eta_{12})^2 + (3\eta_{21} - \eta_{03})^2$$
$$\phi\,(4) = (\eta_{30} + \eta_{12})^2 + (\eta_{21} + \eta_{03})^2$$
$$\phi\,(5) = (\eta_{30} - 3\eta_{12})(\eta_{30} + \eta_{12})[(\eta_{30} + \eta_{12})^2 - 3(\eta_{21} + \eta_{03})^2]$$
$$\qquad + (3\eta_{21} - \eta_{03})(\eta_{21} + \eta_{03})[3(\eta_{30} + \eta_{12})^2 - (\eta_{21} + \eta_{03})^2]$$
$$\phi\,(6) = (\eta_{20} - \eta_{02})[(\eta_{30} + \eta_{12})^2 - (\eta_{21} + \eta_{03})^2]$$
$$\qquad + 4\eta_{11}(\eta_{30} + \eta_{12})(\eta_{21} + \eta_{03})$$
$$\phi\,(7) = (3\eta_{21} - \eta_{03})(\eta_{30} + \eta_{12})[(\eta_{30} + \eta_{12})^2 - 3(\eta_{21} + \eta_{03})^2]$$
$$\qquad - (\eta_{30} - 3\eta_{12})(\eta_{21} + \eta_{03})[3(\eta_{30} + \eta_{12})^2 - (\eta_{21} + \eta_{03})^2]$$

137.   The *Euler number* of a region is the number of its connected components minus the number of its holes.

138.   The *compactness* of a blob can be measured by the length of its perimeter squared divided by its area or, alternatively, by the standard deviation of the radii from the centroid to the boundary divided by the mean radius. The classical measure of perimeter squared divided by area has the disadvantage that in the digital domain it takes its smallest value not for a digital circle but for a digital octagon or diamond, depending on whether 8-connectivity or 4-connectivity is used in calculating the perimeter.

139.   The *bounding rectangle* of a region R is a rectangle that circumscribes R. Its sides are aligned with the row and column directions, its leftmost side aligning with the lowest-numbered column of R, its rightmost side aligning with the highest-numbered column of R, its topmost side aligning with the lowest-numbered row of R, and its bottommost side aligning with the highest-numbered row of R.

140.   An *extremal pixel* of R is a pixel having from among all pixels in R one of the following properties:

   a. An extremal row coordinate value r and an extremal column coordinate value taken from among all the column positions c such that $(r, c) \in R$;
   b. An extremal column coordinate value c and an extremal row coordinate value taken from among all the row positions r such that $(r, c) \in R$.

A region may have as many as eight distinct extremal points, each of which must be lying on the bounding rectangle of the region. Extremal pixels can be used to represent the areal extent of a region and to infer the dominant axis length and orientation of the region.

141.   The *second-moment matrix*

$$\Sigma = \begin{pmatrix} \mu_{rr} & \mu_{rc} \\ \mu_{rc} & \mu_{cc} \end{pmatrix}$$

of a region R is defined by

$$\mu_{rc} = \frac{s_r^2}{A} \sum_{(r,c)\in R} (r - \bar{r})^2$$

$$\mu_{rc} = \frac{s_r s_c}{A} \sum_{(r,c)\in R} (r - \bar{r})(c - \bar{c})$$

$$\mu_{cc} = \frac{s_c^2}{A} \sum_{(r,c)\in R} (c - \bar{c})^2$$

where A is the area of the region, s_r is the row scale factor, s_c is the column scale factor, $\bar{r}$ is the row centroid, and $\bar{c}$ is the column centroid.

142.   The *elongation* or *elongatedness* of a blob or connected region can be measured in a variety of ways. One technique is to use the ratio of the length of the maximum-length chord in the blob to the length of its maximum-length perpendicular chord. A second technique is to use the square root of the ratio of the largest to smallest eigenvalue of the second-central-moment matrix of the blob. A third technique is to use the ratio of the largest distance between an opposing pair of extremal points to the next-to-largest distance between an opposing pair of extremal points.

143.   The *symmetric axis* or *medial axis* of a blob is a subset of blob pixels that are the centers of maximal lines, squares, or disks contained in the blob. Associated with each pixel that is part of a symmetric axis may also be additional information, such as the size of the maximal line or square or the radius of the maximal disk of which it is the center.

144.   The *connected component operator* has as input a binary image and produces as output an image in which each binary-1 pixel is given a unique label of the maximally connected component of pixels having a binary-1 value to which it belongs.

145.   In *connected component analysis* or *blob analysis,* the position and shape properties of each connected component are measured. Typical shape properties include area, perimeter, number of holes, bounding rectangle, extremal points, centroid, second moments, and orientation derived from second moments or extremal points. The connected components are then identified or classified by a decision rule on the basis of their measured properties.

146.   *Signature analysis* of a binary image analyzes the image in terms of its projections. Projections can be vertical, horizontal, diagonal, circular, radial,

spiral, or general. The analysis consists of computing the projections, segmenting each projection, and taking property measurements of each projection segment. Signature analysis may also use the projection segmentation to induce a segmentation of the image.

147. A *binary image projection* is the histogram of the gray scale image produced by masking a projection index image with the given binary image. Each pixel of the *projection index image* contains a number that is the index of the projection bin to which the pixel belongs. A histogram of the masked projection index image then contains in projection bin i the number of pixels that on the binary image have binary value 1 and on the projection index image have index value i.

21.11   Curve and Image Data Structures

148. The *Fourier descriptors* of a closed planar curve are the coefficients of the Fourier series of the spatial positions of the curve as a function of arc length. Typically the low-frequency coefficients are the ones of greatest interest.

149. *Iterative endpoint curve fitting* refers to an iterative process of segmenting a curve into a set of piecewise linear segments that approximate the curve. The process begins by constructing a straight line between the endpoints of the curve. If the farthest distance between the curve and the straight line is less than a specified tolerance, then the approximation is considered to be suitable and the curve segment is divided no further. If the farthest distance between the curve and the straight line is greater than a specified tolerance, then the approximation is considered not to be suitable. The curve is then divided into two segments at this farthest distance point, and the straight-line fitting process independently continues on each segment.

150. The *chain code* representation of a digital arc or blob boundary is a sequence in which each element is a symbol representing the vector joining two neighboring pixels of the arc or boundary. The most common chain code uses the symbols 0 to 7 to represent the vectors $(0,\ 1)$, $(-1,\ 1)$, $(-1,\ 0)$, $(-1,\ -1)$, $(0,\ -1)$, $(1,\ -1)$, $(1,\ 0)$, and $(1,\ 1)$ of row and column coordinates, which can join two neighboring pixels. More-complex chain codes have more symbols and can represent the vector joining two more-distant pixels that define the beginning and ending of a digital straight-line segment that is part of a digital arc or blob boundary.

151. A *quadtree* is a tree data structure that represents an image. Each node of the quadtree represents a square subset of the image's spatial domain. The root node of the quadtree represents the spatial domain of the entire image. If all the pixels of the spatial domain subset represented by a node have the same value, then the value of the node is the value of the pixels in the subset. Such a node is called a pure node. If the node is a mixed or impure node, then the square represented by the node is partitioned into four quadrants and the node has four children nodes, one child node for each quadrant. If the image being represented is a binary image, then the corresponding quadtree is called a *binary quadtree*. If the image being

represented is a gray scale image, then the corresponding quadtree is called a *gray scale quadtree*.

152.   A *curve pyramid* consists of a sequence of symbolic images representing curves at multiple resolutions. The main operation in a bottom-up construction consists of locally connecting the short segments of the curve into longer ones. If the segments are described by binary "curve relations" of the labeled sides of a square cell, the concatenation of the short segments can be achieved formally by taking the transitive closure of the curve relations. Overlapping pyramids are necessary if the resulting pyramid is to have the "length reduction property": Long curves with many segments survive to high levels, whereas short curves disappear after a few reduction steps.

153.   An *octree* is a datum in a tree data structure that represents a function defined as a three-dimensional space volume. Each node of the octree represents a cube subset of the volume. The root node of the quadtree represents the entire volume. If all the voxels represented by a node have the same function values, then the value of the node is their function value. Such a node is called a pure node. If the node is a mixed or impure node, then the cube represented by the node is partitioned into eight volume octants and the node has eight children, one child for each octant. If the function is binary, then the corresponding octree is called a *binary octree*. The binary octree is useful for representing three-dimensional volumes. If the function being represented is nonbinary, such as a real or integer-valued function, then the corresponding octree is called a *gray scale octree*.

154.   *Run-length encoding* is a way to represent binary images compactly. There are a variety of run-length encoding formats. Each has a way of representing the starting column position of a maximally long horizontal string of binary-1 valued pixels as well as the number of pixels in a run. Many vision systems that recognize objects from their binary images use run-length encoding to reduce the volume of data to be processed.

155.   A *generalized cone* or *generalized cylinder* is a data structure for volumetric representation of a three-dimensional object. The volume is generated by sweeping an arbitrarily shaped cross section along a three-dimensional curve called the *generalized cone axis* or *generalized cylinder spine*. The axis passes through the centroids of the cross sections and is at a fixed angle (usually orthogonal) to them. The cross-sectional shape is permitted to have some free parameters, such as size or elongation. These values are specified for each axis point by the *cross-section function* or the *sweeping rule*.

156.   *Constructive solid geometry* is a mechanism for representing three-dimensional volumes by a constructive process that begins with simple-shaped volumes, which are then combined and subtracted from each other by the set of operations consisting of union, intersection, and set difference.

157.   A *boundary surface description* or *boundary representation* of a three-dimensional object or volume is a representation that contains each of the surface boundaries of the volume. Each surface boundary is represented in terms of simply described pieces, each of which has its own arc boundary. Each arc it-

self is represented in terms of simply described pieces that begin and terminate at endpoints or vertices.

158.   A *superquadric* is a closed surface spanned by a vector whose x-, y-, and z-components are specified as functions of the angles η and ω via the spherical product of two two-dimensional parametrized curves

$$h = \begin{pmatrix} h_1(\eta) \\ h_2(\eta) \end{pmatrix}$$

and

$$m = \begin{pmatrix} m_1(\omega) \\ m_2(\omega) \end{pmatrix}$$

that come from one of three basic trigonometric forms. The *spherical product* of h with m scaled by the vector $\begin{pmatrix} a_1 \\ a_2 \\ a_3 \end{pmatrix}$ is defined by

$$h \otimes m = \begin{pmatrix} a_1 h_1(\eta) m_1(\omega) \\ a_2 h_1(\eta) m_2(\omega) \\ a_3 h_2(\omega) \end{pmatrix}$$

159.   A *strip tree* is a binary tree data structure for hierarchically representing a planar arc segment. The root node of the tree represents the minimal-sized rectangle that bounds the arc. Each nonterminal node of the tree splits the arc, which its bounding rectangle approximates into two continuous pieces. Each of its two children nodes then contains the minimal-sized rectangle that bounds the arc segment piece belonging to it. Terminal nodes of the tree have bounding rectangles that are sufficiently close to the arc segment they contain.

160.   The *primal sketch* is a data structure for representing gray level intensity changes, their geometric distribution, and their organization in each image neighborhood. Primitives of the primal sketch include zero crossings, blobs, terminations and discontinuities, edge segments, virtual lines, groups, curvilinear organizations, and boundaries.

161.   In *facet model* image processing, the digital image's pixel values are regarded as noisy discretized sampled observations of its underlying and unknown gray tone intensity surface. Any operation to be performed on the image is defined in terms of this surface. Thus in order to do any processing, the underlying intensity surface must be estimated. This requires a model that describes what the general form of the surface would be in any image neighborhood if there were no noise. To estimate the surface from the neighborhood around a pixel then amounts to estimating the free parameters of the general form. Useful image-processing operations that can then be performed by using the facet-model-processing approach include gradient edge detection, zero-crossing edge detection, image segmentation, line detection, corner detection, three-dimensional shape estimation from shading, and determination of optic flow.

21.12   Texture

162.   A *discrete tonal feature* on a digital image is a connected set of resolution cells, all of which have the same or almost the same image intensity.

163.   *Texture* is concerned with the spatial distribution of the image intensities and discrete tonal features. When a small area of the image has little variation of discrete tonal features, the dominant property of that area is gray tone. When a small area has wide variation of discrete tonal features, the dominant property of that area is texture. Three things are crucial in this distinction: (1) the size of the small areas, (2) the relative sizes of the discrete tonal features, and (3) the number of distinguishable discrete tonal features. Texture can be described in terms of dimensions of uniformity, density, coarseness, roughness, regularity, intensity, and directionality.

164.   A *texel,* short for texture element, is a triplet whose first component is the (row, column) location of a small neighborhood, whose second component is the size of the neighborhood, and whose third component is the vector of texture properties.

165.   The *gray level dependence matrix* or *gray level co-occurrence matrix* characterizes the microtexture of an image region by measuring the dependence between pairs of gray levels arising from pixels in a specified spatial relation. For gray level pair (i, j), the gray level dependence matrix P for region R of image I has value $P(i, j)$, where $P(i, j)$ is the number of pairs of pixels in the region having the desired spatial relation, where the first pixel has gray level i, and the second pixel has gray level j. If S designates the set of all pairs of pixels in the desired spatial relation, then $S \subseteq R \times R$ and

$$P(i, j) = \#\{[(r_1, c_1), (r_2, c_2)] \in S \mid I(r_1, c_1) = i \text{ and } I(r_2, c_2) = j\}$$

The gray level dependence matrix P can be normalized. One normalized form that produces a joint probability is given by

$$P_1(i, j) = \frac{P(i, j)}{\#S}$$

A second normalized form that produces a conditional probability is given by

$$P_2(i, j) = \frac{P(i, j)}{\sum\limits_{j} P(i, j)}$$

166.   *Statistical texture measures* include the moments of the gray levels of the given region, typically the variance, the slewness, and the kurtosis.

167.   The *gray level difference histogram* at a distance d of an image I is the histogram of values $\langle |I(r,c) - I(r',c')| \quad : \quad (r - r')^2 + (c - c')^2 = d^2 \rangle$.

168.   A *structural texture description* is given by a set of primitives and by placement rules that govern the stochastic spatial relation between them.

169.   *Fourier-related texture descriptions* include the power spectrum and autocorrelation function.

21.13 Segmentation

170. *Image segmentation* is a process that typically partitions the spatial domain of an image into mutually exclusive subsets called regions. Each region is uniform and homogeneous with respect to some property, such as tone or texture, and its property value differs in some significant way from that of each neighboring region. An image segmentation process that uses image intensity as a property value produces regions that are called discrete tonal features.

171. *Region growing* refers to a sequential image segmentation process in which pixels are successively added to incomplete regions or initiate new regions when it is not appropriate to make them part of any of the existing incomplete regions. There are three basic kinds of region growing: tracking, aggregation, and merging.

172. In *region tracking*, the image is scanned in raster scan order. The similarity of each pixel is compared with the regions to which the already processed 4- or 8-connected neighboring pixels belong. If one of these already-processed neighboring pixels belongs to a region of sufficient similarity to the current pixel, then the pixel is added to the region. If the regions to which all the already-processed pixels belong are dissimilar from the current pixel, then the current pixel initiates a new region.

173. In *region aggregation,* seed pixels are first found that serve as prototype pixels for the regions in the desired segmentation. Then in a sequential fashion pixels having neighbors in any incomplete region join themselves into the region if they are similar enough. The aggregation process continues until all pixels are part of some region.

174. In *region merging,* the neighboring regions of an initial segmentation are successively merged if they have similar enough properties. After each merging iteration, properties of the new regions are recomputed. The merging iterations continue until the properties of each pair of neighboring regions are sufficiently different from each other.

175. *Contour filling* is the process by which all pixels inside a blob defined by its bounding contour or contours are marked with the same unique label.

21.14 Matching

176. *Template matching* is an operation that can be used to find out how well a template subimage matches a window of a given image. The degree of matching is often determined by translating the template subimage all over the given image and for each position evaluating the cross-correlation or the sum of the squared or absolute image intensity differences of corresponding pixels. Template matching can also be used to best match an observed measurement pattern with a prototype pattern.

177.   *Matched filtering* is a template-matching operation done by using the magnitude of the cross-correlation function to measure the degree of matching.

178.   *Image matching* refers to the process of determining the pixel-by-pixel, arc-by-arc, or region-by-region correspondence between two images taken of the same scene but with different sensors, different lighting, or a different viewing angle. Image matching can be used in the spectral/temporal pattern classification of remote sensing or in determining corresponding points for stereo, tracking, change analysis, and motion analysis. In one group of approaches, subimages of one image are translated over a second image. For each translation the differences between appropriately transformed gray tone intensities and/or edges are measured. In this signal level approach, the unit being operated with is the pixel, since the measured difference is between values of pixels in two images.

179.   In *symbolic registration* or *symbolic matching,* higher-level units are worked with. For example, the scene can be segmented and a region matching then performed by using segment features, such as area, position, perimeter^2/area, orientation, length-to-width ratio, area/area of minimum bounding rectangle, area/area of bounding ellipse, gray tone intensity or color of segment, and number of corresponding neighbors. Because this matching uses a higher-level unit than the pixel, it is called symbolic matching or symbolic registration.

180.   In *feature-point matching,* selected points of each image are first determined on the basis of the distinctive image values in a neighborhood or on the basis of an intersection between two feature lines. The location of each point can be determined to subpixel precision. After the locations of distinctive points are determined, a correspondence process associates as many selected points of one image as possible with selected points on the second image. The correspondence is based on similarity of the feature characteristics of the points.

181.   A *structural description* is a relational representation of a two-dimensional or three-dimensional entity. It consists of a set of primitives, each having its own attribute description, and a set of named relations composed of tuples whose components are primitives that stand in the relation specified by the relation name.

A function $h : A \rightarrow B$ is a *relational homomorphism* from N-ary relation $R \subseteq A^N$ to N-ary relation $S \subseteq B^N$ if

$$R \circ h \subseteq S$$

where $R \circ h = \{(b_1, ..., b_N) \in B^N |$ for some $(a_1, ..., a_N) \in R, \quad b_n = h(a_n), \quad n = 1, ..., N\}.$

182.   The relation R is said to *match* relation S if there exists a relational homomorphism satisfying

$$R \circ h = S \quad \text{and} \quad S \circ h^{-1} = R$$

183.   *Relational matching* refers to the process by which it is determined whether two relations match or do not match. *Structural matching* is a matching that establishes a correspondence or homomorphism from the primitives of one

structural description to the primitives of a second structural description. In the ideal match a tuple of primitives that stand in a given relation in the first description will have its corresponding tuple of primitives stand in the same given relation in the second structural description.

184.   The *local feature focus method* is a model-based object recognition and location technique in which one feature, referred to as the focus feature, on the image is found and is used along with the object model to predict what other nearby features might be. After locating a set of features, a relational matching is performed to infer a consistent correspondence between all the located image features and the object features. Once a consistent correspondence has been found, the object position and orientation can be hypothesized. The hypothesized object position and orientation are then verified by template matching.

185.   *Relaxation* refers to any computational mechanism that employs a set of locally interacting parallel processes, one associated with each image unit, that in an iterative fashion update each unit's current labeling in order to achieve a globally consistent interpretation of the image data. In *discrete relaxation,* the assessment of each unit's current state consists of that subset of labels not yet ruled out. In *probabilistic relaxation,* the assessment of each unit's current state is a probability function associating with each possible label a probability of its being the correct state.

21.15   Localization

186.   An *interest operator* is a neighborhood operator that is designed to locate, with high spatial accuracy, pixel or subpixel positions whose central neighborhoods have distinctive gray tone patterns. Such neighborhoods are typically those whose autocorrelation functions fall off rapidly. Interest operators are usually used to mark pixels on a pair of images taken of the same scene but with some shift of either camera position or object from one image to another. The marked pixels are then the candidate pixels that are input to an algorithm that establishes correspondences between the selected pixels.

187.   An *area-of-interest operator* is an operator that delineates regions of an image that have potentially interesting patterns. These delineated *regions of interest* are the ones that must be further processed.

188.   *Screening* is an operation of selecting photographs or images containing areas of potential interest from those in a set of photographs, some or most of which contain no interesting areas.

189.   In *area analysis* or *region analysis* the area of the image containing the objects or entities to be processed is located by some simple algorithm and a more complex processing algorithm is applied only in the located area. This strategy of processing can often increase execution speed. The algorithm locating the area to be processed is called the *focus-of-attention mechanism*.

21.16   General Image Processing

190.   *Preprocessing* is an operation applied before pattern identification is performed. Preprocessing produces, for the categories of interest, pattern features that tend to be invariant under changes such as translation, rotation, scale, illumination level, and noise. In essence, preprocessing converts the measurement patterns to a form that allows a simplification in the decision rule. Preprocessing can bring into registration; bring into congruence; remove noise; enhance images; segment target patterns; and detect, center, and normalize objects of interest.

191.   *Registering* or *registration* is the translation or translation-and-rotation alignment process by which two images of like geometries and of the same set of objects are positioned coincident with each other so that corresponding points of the imaged scene appear in the same position on the registered images. In this manner, corresponding image values can be made to represent the sensor output for the same object point over the full image frame.

192.   *Congruencing* is the geometric warping process by which two images of different geometries but of the same set of objects are spatially transformed so that the size, shape, position, and orientation of any object on one image are made to be the same as the size, shape, position, and orientation of that object on the other image.

193.   *Image compression* is an operation that preserves all or most of the information in the image and that reduces the amount of memory needed to store the image or the time needed to transmit it.

194.   *Image restoration* is a process by which a degraded image is restored, as clearly or as best as possible, to its ideal condition. Perfect image restoration is possible only to the extent that the degradation transform is mathematically invertible. Common forms of restoration include *inverse filtering, Wiener filtering,* and *constrained least-squares filtering.*

195.   *Image reconstruction* refers to the process of reconstructing an image from a set of its projections. The projections may be taken along a set of parallel rays, in which case they are called *parallel projections,* or they may be taken along a set of rays emanating from a point, in which case they are called *fan beam projections.* The most commonly employed reconstruction techniques are the *filtered back projection* and the *algebraic reconstruction* techniques. Image reconstruction techniques are important in computerized tomography, nuclear medicine, and ultrasonic imaging.

196.   *Image enhancement* is any one of a group of operations that improve the detectability of objects. These operations include, but are not limited to, contrast stretching, edge enhancement, spatial filtering, noise suppression, image smoothing, and image sharpening.

197.   *Image processing* or *picture processing* encompasses all the various operations that can be applied to image data. These include, but are not limited to, image compression, image restoration, image enhancement, preprocessing, quantization, spatial filtering, matching, and recognition techniques.

198.   *Interactive image processing* is carried out by an operator or an analyst at a console with a means of accessing, preprocessing, feature extracting, classifying, identifying, and displaying the original or processed imagery for subjective evaluation and further interactions.

21.17   Vision

199.   *Structured light* refers to a technique of projecting a carefully designed light pattern on a scene and viewing the scene from a different direction. Usually the pattern consists of successive planes of light at different positions and orientations. Those pixels that image a surface patch lit by a known light pattern have sufficient information to determine the three-dimensional coordinates of the surface patch, since the light pattern is designed so that the line of sight passing through the pixel and the lens will intersect the known light pattern in a unique point. For stereo-matching purposes, the structured light pattern may be "unstructured" in the sense of being a texture pattern or consisting of random stipples.

200.   *Light striping* refers to a simple form of structured lighting in which the light pattern consists of successive planes of light that are all parallel.

201.   *Stereopsis* refers to the capability of determining the depth of a three-dimensional point by observing the point on two perspective projection images taken from different positions.

202.   A *stereo image pair* refers to two perspective projection images taken of the same scene from slightly different positions. The common area appearing in both images of the stereo pair is usually 40% to 80% of the total image area.

203.   A point p on one image and a point q on a second image are said to form a *corresponding point pair* (p, q) if p and q are each a different sensor projection of the same three-dimensional point. The *visual correspondence* problem consists of matching all pairs of corresponding points from two images of the same scene.

204.   *Disparity* or *stereo disparity* refers to the difference in positions of the images of the same three-dimensional point in two perspective projection images taken from different positions.

205.   *Stereo matching* refers to the matching process by which corresponding points on a stereo image pair are identified.

206.   *Triangulation* refers to the process of determining the (x, y, z) coordinates of a three-dimensional point from the observed position of two perspective projections of the point. The centers of perspectivity and the perspective projection planes are assumed known.

207.   The *epipolar axis* of a stereo image pair is the line passing through the centers of perspectivity of the image.

208.   The two *epipoles* of a stereo image pair consist of one point on each of the perspective projection planes determined as the intersection of the image plane with the epipolar axis. For stereo image pairs having parallel perspective projection image planes, the epipoles are infinitely far to the left and right.

209.   An *epipolar ray* is the line segment between an epipole and a point on the perspective projection image plane.

210.   An *epipolar line* on one stereo image corresponding to a given point in another stereo image is the perspective projection on the first stereo image of the three-dimensional ray that is the inverse perspective projection of the given point from the other stereo image.

211.   An *epipolar plane* relative to a pair of stereo images is any plane determined by an observed three-dimensional point and the positions in three-dimensional space of its perspective projections on the left and right stereo images. Every epipolar plane contains the epipolar axis, and every plane that contains the epipolar axis is an epipolar plane.

212.   An *occluding boundary* is a boundary appearing on an image owing to a discontinuity in the range or depth of an object in the observed scene.

213.   For each fixed viewing position and point source of illumination, the *reflectance map* is a function defined on gradient space that specifies a surface's reflectivity. Thus at each possible orientation of the surface normal (as encoded by the surface's first-order partial derivatives), the surface's reflectance map specifies the reflectivity of the surface.

214.   *Shape from shading* refers to the capability of determining the three-dimensional shape characteristics of an object from the gray tone shading manifested by the object's surface on a perspective or orthographic projection image, that is, from its reflectance map.

215.   *Local shading analysis* refers to the capability of inferring the shape and predominant tilt of a section of an object's surface by the image intensities in a local neighborhood of a perspective or parallel projection view.

216.   *Photometric stereo* refers to the capability of determining surface orientation by means of the shading variations present on two or more images taken of the same scene from the same position and orientation but with the light source in different positions.

217.   The *motion field, image flow,* or *image flow field* is an image in which the value of each pixel is the projected translational velocity arising from a 3D surface point of an object in motion relative to the camera. Each projected translational velocity vector is called an *optic flow vector*.

218.   An *optic flow* or *optical flow image* is an image in which the value of each pixel is the estimated projected translational velocity arising from a surface point of an object in motion relative to the camera. Because the projected velocity may not be estimable at each pixel, some pixels in an optic flow image may have no optic flow information.

219.   The *focus of expansion* of a motion field image arising from a moving camera and a stationary scene is that point on the image where the optic flow is zero and such that the optic flows of the neighboring points are directed away from it. In cases of relative motion toward the camera, there will be exactly one focus of expansion point in such a motion field image. In a motion field image arising from an object in relative motion to the camera, the focus of expansion is that point on the image having all the optic flow vectors arising from the moving object directed away from it.

220.   The *focus of contraction* of a motion field image arising from a moving camera and a stationary scene is that point on the image where the optic flow is zero and such that the optic flows of the neighboring points are directed toward it. In cases of relative motion away from the camera, there will be exactly one focus of contraction point in such a motion field image. In a motion field image arising from an object in motion relative to the camera, the focus of contraction is that point on the image having all the optic flow vectors arising from the moving object directed toward it.

221.   *Structure from motion* refers to the capability of determining a moving object's shape characteristics, and its position and velocity as well, from a sequence of two or more images taken of the moving object. Equivalently, if the camera is in motion, structure from motion refers to the capability of determining an object's shape characteristics, as well as its position and the camera's velocity, from a sequence of two or more images taken of the object by the moving camera. One fundamental kind of structure-from-motion problem is to determine the fixed position of M three-dimensional points from a time sequence of N views each containing the two-dimensional perspective projection of the M three-dimensional points.

222.   *Passive navigation* refers to the determination of the motion of a camera from a time-varying image sequence.

223.   *Dynamic scene analysis* refers to the analysis of time-varying imagery. The purpose of the analysis may be to track moving objects, determine the motions of the objects, recognize the objects, determine the spatial positions of the objects at the time each image was obtained, or determine a shape description or characterization of one or more objects.

224.   *Shape from texture* refers to the capability of determining the three-dimensional shape characteristics of a homogeneously textured surface from the texture density variations manifested by the surface on a perspective or orthographic projection image.

225.   *Surface reconstruction* refers to the process by which a three-dimensional surface is analytically described by its boundary representation on the basis of processing a stereo image pair, a range map, or a time-varying image sequence of the observed surface.

226.   *Shape from contour* or *shape from shape* refers to the capability of inferring the three-dimensional shape of an object from a two-dimensional perspective projection view of a set of regularly marked contours on the object's surface.

227.   The $2\frac{1}{2}D$ *sketch* is a multiband image, each pixel providing information about the depth and surface orientation of the surface projected on it as well as indicating the existence of nearby depth and surface discontinuities.

228.   The *intrinsic scene characteristics* for each object surface point include its depth from the image focal plane, its surface orientation, its reflectance, and its incident illumination.

229.   An *intrinsic image* is a multiband image in which each pixel contains the predominant intrinsic scene characteristics of the surface patch projecting to its position. Hence each pixel of the intrinsic image specifies depth, surface orientation, reflectance, and incident illumination for the surface patch projecting to its position.

230.   *Inverse optics* refers to the capability of inferring the three-dimensional position and/or the surface normal of each point on an object's surface and/or the surface shape from one or more perspective projection images of the object. Included in the techniques of inverse optics are stereo, photometric stereo, shape from shading, shape from texture, and structure from motion. Inverse optics techniques can be thought of as techniques that invert the perspective projection process and therefore belong to the reconstructionist school of applied physics computer vision.

231.   The *blocks world* refers to a world in which all objects have simple surfaces. The most common kind of objects in the blocks world are polyhedral objects.

232.   A *view aspect* is a maximal connected region of viewpoint space; having the property that when looking at a given object's center from any point of a view aspect, the resulting views are topologically identical.

233.   An *aspect graph* is a graph in which the nodes are the view aspects and the arcs connect adjacent view aspects. The views represented by the nodes in the graph are all the stable views, characteristic views, and principal views.

234.   *Three-dimensional vision* refers to the capability of a machine vision system to infer some three-dimensional characteristic of an object or object feature, such as its position, dimensions, orientation, or motion, or some three-dimensional characteristic of a point or an object surface, such as its three-dimensional position or surface normal orientation.

235.   *Automatic visual inspection* or *automatic vision inspection* refers to an inspection process that uses sensor-producing image data and techniques from image processing, pattern recognition, or computer vision to measure and/or interpret the imaged objects in order to determine whether they have been manufactured within permitted tolerances. Automatic visual inspection systems usually integrate the technologies of material handling, illumination, image acquisition, and special-purpose computer hardware, along with the appropriate image analysis algorithms, into a system intended to be of practical use in the factory. Benefits from using automated visual inspection can include more accurate, reliable, repeatable, and complete quality assurance at a lower price and a higher speed of inspection than possible by manual labor.

236.   A *machine vision system* is a system capable of acquiring one or more images of an object; of processing, analyzing, and measuring various characteristics of the acquired images; and of interpreting the results of the measurements in such a way that some useful decision can be made about the object. Functions of machine vision systems include locating, inspecting, gauging, identifying, recognizing, counting, and motion estimating.

237.   *Visual fixturing* or *visual pose determination* refers to the capability of inferring the position and orientation of a known object by using a suitable object model and one or more cameras, range sensors, or triangulation-based vision sensors.

238.   *Optical gauging* or *visual gauging* refers to the ability to measure specific positions or dimensions of a manufactured object by using noncontact light-sensitive sensors to compare these measurements with preselected tolerance limits

for quality inspection and sorting decisions. Gauging has wide application in manufacturing, because it can determine the diameters of holes, openings, or cutouts, the widths of shafts, components, gaps, wires, or rods; and the relative locations of holes, folds, features, components, openings, or breaks.

239.   A gauging system can be either a *fixed-inspection* or a *flexible-inspection* system. The fixed-inspection system holds the part in a precision test fixture and has one or more sensors to take the required measurements. Flexible inspection utilizes sensors that are moved about the part being inspected, the motion being done along a programmed path trajectory. Flexible-inspection gauging is sometimes called *robotic gauging*.

240.   A vision procedure is said to be *robust* if small changes in the assumed model on which the procedure or technique was developed produce only small changes in the result. Small fractions of the data that do not fit the assumed model, and in fact are very far from fitting it, constitute a small change in the assumed model. Data not fitting an assumed model may be due to rounding or quantizing errors, gross errors, or the fact that the model itself is only an idealized approximation of reality.

241.   *Computer vision, image understanding,* or *scene analysis* is the combination of image processing, pattern recognition, and artificial intelligence technologies that focuses on the computer analysis of one or more images, taken with a single/multiband sensor or in time sequence. The analysis recognizes, locates the position and orientation of, and provides a sufficiently detailed symbolic description or recognition of those imaged objects deemed to be of interest in the three-dimensional environment. The computer vision process often uses geometric modeling and complex-knowledge representations in an expectation- or model-based matching or searching methodology. The searching can include bottom-up, top-down, blackboard, hierarchical, and heterarchical control strategies.

242.   The *bottom-up control strategy* is an approach to problem solving that is *data driven*. It employs no object models in its early stages and uses only general knowledge about the world being sensed. In a computer vision system using a bottom-up control strategy, the observed image data are interpreted and aggregated. The interpretations and aggregations are then successively manipulated and aggregated until a sufficiently high-level description of the scene has been generated.

243.   The *top-down control strategy* is an approach to problem solving that is *goal directed* or *expectation directed*. A form of solution is generated or hypothesized. Assuming the hypothesis is true and using the information in the knowledge database, the inference mechanism then infers, if possible, some consistent set of values for the unknown variables or parameters. If a consistent set can be inferred, then the problem has been solved. If a consistent set cannot be inferred, then a new form of solution is generated or hypothesized. In a computer vision system using a top-down control strategy, the number or types of objects being sensed in the image are usually highly constrained, and information about the objects, relationships between objects, and object parts are all known. The system hypothesizes that the image shows a particular set of objects, infers values for parameters, and then tests to verify that the hypothesis is consistent with the observed data.

244.   A *hierarchical control strategy* is an approach to problem solving in which the given problem is solved by dividing it into a set of subproblems, each of which encapsulates an important or major aspect of the original problem. Then each subproblem is successively divided into more-detailed subproblems. The refinement continues until the most-refined subproblems can be solved directly.

245.   In the *heterarchical control model,* knowledge is embodied in a set of procedures called knowledge sources. Each knowledge source can communicate with some or all of the other knowledge sources. Whereas hierarchical control dictates the order in which different processes operate, heterarchical control lets the data themselves dictate the order in which things are done.

246.   A *blackboard control strategy* is an approach to problem solving in which the various components of the inference mechanism communicate with one another through a common working data storage area called the blackboard. When the blackboard has sufficient data to permit one component of the inference mechanism to make a deduction, the inference mechanism goes to work and writes its results on the blackboard, where it becomes available for the other components of the inference mechanism. In this manner the inferred constraints are successively propagated, and the required search is more limited.

247.   *Model-based computer vision* is a computer vision process that employs an explicit model of the object to be recognized. Recognition proceeds in a top-down manner by matching the object data structure inferred from the observed image to the model data structure.

248.   *Knowledge-based computer vision* refers to a computer vision process that has an image-processing component, a reasoning or inference component, and a knowledge database component. The knowledge database stores information about the environment being imaged. The image-processing component extracts primitive point, line, curve, and region information from an observed image. The reasoning or inference component is typically rule based. It integrates the information produced by the image-processing component with the information in the knowledge database and reasons about what hypothesis should be generated next, what hypothesis should be validated next, what new information can be inferred from what has already been established, and what new primitives the image-processing component should extract next.

21.18   Pattern Recognition

249.   *Pattern recognition* techniques can be used to construct decision rules that enable one to identify units on the basis of their measurement patterns. Pattern recognition techniques can also be employed to cluster units having similar enough measurement patterns. In *statistical pattern recognition,* the measurement patterns have the form of N-tuples or vectors. In *syntactic pattern recognition,* the measurement patterns have the form of sentences from the language of a phrase structure grammar. In *structural pattern recognition,* the measurements do not have the form

of an N-tuple or a vector. Rather, the unit being measured is encoded in terms of its parts and their relationships and properties.

250.   *Pictorial pattern recognition* refers to techniques that treat the image as a pattern and either categorize the image or produce a description of it.

251.   The *unit* is the entity that is observed and whose measured properties constitute the measurement pattern. The simplest and most practical unit to observe and measure in the pattern recognition of image data is often the pixel (the gray tone intensity or the gray tone intensity N-tuple in a particular resolution cell). This is what makes pictorial pattern recognition so difficult, because the objects requiring analysis or identification are not single pixels but are often complex spatial formations of pixels.

252.   A *measurement pattern* or *pattern* is the data structure of the measurements resulting from observing a unit.

253.   A *measurement N-tuple* or *measurement vector* is the ordered N-tuple of measurements obtained from a unit under observation. Each component of the N-tuple is a measurement of a particular quality, property, feature, or characteristic of the unit. In image pattern recognition, the units are usually picture elements or simple formations of picture elements, and the measurement N-tuples are the corresponding gray tone intensities, gray tone intensity N-tuples, or properties of formations of gray tone intensities.

254.   The *Cartesian product* of two sets A and B, denoted by $A \times B$, is the set of all ordered pairs in which the first component of the pair is some element from the first set and the second component is some element from the second set. The Cartesian product of N sets can be defined inductively.

255.   *Measurement space* is a set large enough to include the set of all possible measurement patterns that could be obtained by observing some set of units.

256.   The *range set R_i* for the ith sensor, which produces the ith image in the multi-image set, is the set of all measurements that can be produced by the ith sensor. Simply, it is the set of all gray tone intensities that could possibly exist in the ith image. When the units are the pixels, measurement space M is the Cartesian product of the range sets of the sensors: $M = R_1 \times R_2 \times \ldots \times R_n$.

257.   Each unit is assumed to be of one and only one given type. The set of types is called the set of *pattern classes* or *categories C,* each type being a particular category.

258.   A *feature, feature pattern, feature N-tuple, feature vector,* or *pattern feature* is an N-tuple or vector whose components are functions of the initial measurement pattern variables or some subset of them. Feature N-tuples or vectors are designed to contain a high amount of information relative to the discrimination between units of the types of categories in the given category set. Sometimes the features are predetermined; at other times they are determined when the pattern discrimination problem is being solved. In image pattern recognition, features often contain information relative to gray tone intensity, texture, or region shape.

259.   *Feature space* is the set of all possible feature N-tuples.

260.   *Feature selection* is the process by which the features to be used in the pattern recognition problem are determined. Sometimes feature selection is called *property selection*.

261.   *Feature extraction* is the process by which an initial measurement pattern or some subset of measurement patterns is transformed to a new pattern feature. Sometimes feature extraction is called *property extraction*.

The word *pattern* can be used in three distinct senses: as measurement pattern; as feature pattern; and as the dependency pattern or patterns of relationships among the components of any measurement or feature N-tuple derived from units of a particular category and that are unique to those N-tuples, that is, they are dependencies that do not occur in any other category.

262.   A *classifier* is a device or process that sorts patterns into categories or classes.

263.   The *compactness hypothesis* states that the pattern measurements of a given class are nearer to other pattern measurements in the class than to the pattern measurements of other classes.

264.   The region of space occupied by pattern measurements from the same class or category is called a *class region*.

265.   Two classes or categories are said to be *separable* if their class regions do not overlap. If for every class region there exists a hyperplane that separates it from all other class regions, the classes are said to be *linearly separable*.

266.   A *prototype pattern* or *reference pattern* is the observable or characteristic measurement or feature pattern derived from units of a particular category. A category is said to have a prototype pattern only if the characteristic pattern is highly representative of the N-tuples obtained from units of that category.

267.   A *data sequence* $S_d = < d_1, d_2, \ldots, d_j >$ is a sequence of patterns derived from the measurement patterns or features of some sequence of observed units: d_1 is the pattern associated with the first unit; d_2 is the pattern associated with the second unit; and d_j is the pattern associated with the jth unit.

268.   A *decision rule* f usually assigns one and only one category to each observed unit on the basis of the sequence of measurement patterns in the data sequence S_d or on the basis of the corresponding sequence of feature patterns.

269.   A *simple decision rule* is a decision rule that assigns a category to a unit solely on the basis of the measurements or features associated with the unit. Hence the units are treated independently, and the decision rule f may be thought of as a function that assigns one and only one category to each pattern in measurement space or to each feature in feature space.

270.   A *hierarchical decision rule* is a decision rule in tree form. In binary trees, each nonterminal node of the tree contains a simple decision rule that classifies patterns as belonging to its left child or to its right child. Each terminal node of the tree contains the assigned class or category of the observed unit.

271.   A *compound decision rule* is a decision rule that assigns a unit to a category on the basis of some ontrivial subsequence of measurement patterns in the data sequence or in the corresponding sequence of feature patterns.

272.   Provision can be made for a decision rule to *reserve judgment* or to *defer assignment* if the pattern is too close to the category boundary in measurement or feature space. With this provision a deferred assignment is an assignment to the category of "reserved judgment."

273.   A *category identification sequence* or *ground truth* $S_c = < c_1, c_2, \ldots, c_j >$ is a sequence of category identifications obtained from some sequence of observed units: c_1 is the category identification of the first unit: c_2 is the category identification of the second unit; and c_j is the category identification of the *j*th unit.

274.   A *training sequence* is a set of two sequences: (1) the data sequence and (2) a corresponding category identification sequence. A training datum is the pair consisting of a pattern in the data sequence and the corresponding category identification in the category identification sequence. The training sequence is used to estimate the category conditional probability distributions from which the decision rule is constructed, or it may be used to estimate the decision rule itself.

275.   A *training procedure* is a procedure that uses the training sequence to construct a decision rule. It may operate by passing through the entire training sequence only one time and constructing the decision rule in a manner independent of the order in which the training data occur in the training sequence. It may operate iteratively, in which case it passes through the training sequence many times, and after handling each training datum, it modifies or updates the decision rule. Such iterative training procedures may be affected by the order in which the training data occur in the training sequence.

276.   A *window-training procedure* is an iterative training procedure in which each adjustment of the decision rule is made only when the training datum falls within a specified window, a subset of the pattern measurement space. Usually this window contains the decision boundary.

277.   An *error-correcting training procedure* is an iterative sequential training procedure in which, at each iteration, the decision rule is adjusted in response to a misclassification of a training datum.

278.   A classifier is said to *learn* if its iterative training procedure increases the classification performance accuracy of the classifier after each few iterations.

279.   The *conditional probability* of a measurement or feature *N*-tuple *d* given category *c* is usually denoted by $P_c(d)$ or by $P(d \mid c)$ and is defined as the relative frequency or proportion of times the *N*-tuple *d* is derived from a unit whose true category identification is *c*.

280.   A *distribution-free* or *nonparametric decision rule* is one that makes no assumptions about the functional form of the conditional probability distribution of the patterns given the categories.

281.   A simple *maximum-likelihood decision rule* is one that treats the units independently and assigns a unit *u* having pattern measurements or features *d* to that category *c* whose units are the most probable ones to have given rise to pattern or feature vector *d*, that is, such that the conditional probability of *d*, given *c*, is highest.

282.   A simple *Bayes decision rule* is one that treats the units independently and assigns a unit u having pattern measurements or features d to the category c whose conditional probability, given d, is highest.

283.   Let $< u_1, u_2, \ldots, u_j >$ be a sequence of units with corresponding data sequence $< d_1, d_2, \ldots, d_j >$ and known category identification sequence $< c_1, c_2, \ldots, c_j >$. A simple *nearest neighbor decision rule* is one that treats the units independently and assigns a unit u of unknown identification and with pattern measurements or features d to category c_j, where d_j is that pattern closest to d by some given metric or distance function.

284.   A *discriminant function* $f_i(d)$ is a scalar function whose domain is usually measurement space and whose range is usually the real numbers. When $f_i(d) \geq f_k(d)$, for $k = 1, 2, \ldots, K$, then the decision rule assigns the ith category to the unit giving rise to pattern d.

285.   A *linear discriminant function* f is a discriminant function of the form $f(d) = \sum_{j=1}^{n} a_j \delta_j + a_o$, where $d = (\delta_1, \delta_2, \ldots, \delta_n)$ represents the measurement pattern.

286.   A *quadratic discriminant function* f is a discriminant function of the form $f(d) = \sum_{i=1}^{n} \sum_{j=i}^{n} a_{ij} \delta_i \delta_j + \sum_{j=1}^{n} a_j \delta_j + a_o$.

287.   A *decision boundary* between the ith and kth categories is a subset H of patterns in measurement space M defined by

$$H = \{d \in M \mid f_i(d) = f_k(d)\}$$

where f_i and f_k are the discriminant functions for the ith and kth categories.

288.   A *hyperplane decision boundary* is the special name given to decision boundaries arising from the use of linear discriminant functions.

289.   A *linear decision rule* is a simple statistical pattern recognition decision rule that usually treats the units independently and makes the category assignments using linear discriminant functions. The decision boundaries obtained from linear decision rules are hyperplanes.

290.   The *pattern discrimination* problem is concerned with how to construct the decision rule that assigns a unit to a particular category on the basis of the measurement or feature patterns in the data sequence.

291.   *Pattern identification* is the process in which a decision rule is applied. If $S_u = < u_1, u_2, \ldots, u_j >$ is the sequence of units to be observed and identified, and if $S_d = < d_1, d_2, \ldots, d_j >$ is the corresponding data sequence of patterns, then the pattern identification process produces a category identification sequence $S_c = < c_1, c_2, \ldots, c_j >$, where c_i is the category in C to which the decision rule assigns unit u_i on the basis of the j patterns in S_d. In general, each category in S_c can be assigned by the decision rule as a function of all the patterns in S_d. Sometimes pattern identification is called *pattern classification* or *classification*.

292.   A *perceptron* or *neural network* is an interconnected network of non-linear units or processing elements capable of learning and self-organizing. The response of a unit or a processing element is a nonlinear monotonic function of a weighted sum of the inputs to the processing elements. The weights, called *synaptic*

weights, are modified by a learning or reinforcement algorithm. Typical nonlinear processing functions are sgn(x), $1/(1 + e^{-x})$, and tanh(x). When each processing element contributes one component to the output response vector and its inputs are selected from the components of the input pattern vector, the perceptron is called a *simple perceptron.* Processing units whose output only indirectly influences the components of the output response vector are called *hidden units.*

293. An *error-corrective reinforcement* for a perceptron is a learning or training algorithm in which the change in the synaptic weights is a function of the degree to which the output of the processing unit is not what it is desired to be. Error-corrective reinforcement algorithms are also called *error back-propagation* algorithms.

294. A *forward-coupled perceptron* is a perceptron in which the processing units are layered. The inputs to the processing units in layer n come from the outputs of processing units in layers prior to layer n. Single-layered perceptrons can create linear decision surfaces. Two-layered perceptrons can create convex decision regions. Three-layered perceptrons can create almost arbitrarily shaped decision regions.

295. A *series-coupled perceptron* is a forward-coupled perceptron in which the inputs to the processing center in layer n come from the outputs of processing units in layer $n - 1$.

296. A *back-coupled perceptron* is a perceptron that is not forward coupled. That is, there is some processor in layer n in which output feeds back and is the input to a processor in some layer prior to layer n.

297. A *cluster* is a homogeneous group of units that are very "like" one another. "Likeness" between units is usually determined by the association, similarity, or distance between the measurement patterns associated with the units.

298. A *cluster assignment function* is a function that assigns each observed unit to a cluster on the basis of the measurement patterns in the data sequence or on the basis of their corresponding features. Sometimes the units are treated independently. In this case the cluster assignment function can be considered as a transformation from measurement space to the set of clusters.

299. The *pattern classification* problem is concerned with constructing the cluster assignment function that groups similar units. Pattern classification is synonymous with *numerical taxonomy* or *clustering.*

300. The *cluster identification* process is the process in which the cluster assignment function is applied to the sequence of observed units, thereby yielding a cluster identification sequence.

301. A *misidentification, misdetection,* or *type I error* occurs for category c_i if a unit whose true category identification is c_i is assigned by the decision rule to category c_k, $k \neq i$. A misidentification error is often called an *error of omission.*

302. A *false identification, false alarm,* or *type II error* occurs for category c_i if a unit whose true category identification is c_k, $k \neq i$, is assigned by the decision rule to category c_i. A false-identification error is often called an *error of commission.*

303.   A *prediction sequence, test sequence,* or *generalization sequence* is a set of two sequences: (1) a data sequence (whose corresponding true category identification sequence may be considered as unknown to the decision rule) and (2) a corresponding category identification sequence determined by the decision rule assignment. By comparing the category identification sequence determined by the decision rule assignment with that determined by the ground truth, the misidentification and false-identification rates for each category may be estimated.

304.   A *confusion matrix* or *contingency table* is an array of probabilities whose rows and columns are both similarly designated by category label and that indicates the probability of correct identification for each category, as well as the probability of type I and type II errors. The (ith, kth) element P_{ik} is the probability that a unit has true category identification c_i and is assigned by the decision rule to category c_k.

305.   A unit is said to be *detected* if the decision rule is able to assign it as belonging only to some given subset A of categories from the set C of categories. To detect a unit does not imply that the decision rule is able to identify the unit as specifically belonging to one particular category.

306.   A unit is said to be *recognized, identified, classified, categorized,* or *sorted* if the decision rule is able to assign it to some category from the set of given categories. In some applications there may be a definite distinction between recognize and identify. In these applications, for a unit to be recognized, the decision rule must be able to assign it to a type of category that includes many subcategories. For a unit to be identified, the decision rule must be able to assign it not only to a type of category but also to a subcategory of the category type. For example, a small area ground patch that may be recognized as containing trees may be specifically identified as containing apple trees.

307.   A unit is said to be *located* if specific coordinates can be given for the unit's physical location.

308.   *Accuracy* refers to the degree of closeness an estimate has to the true value of what it is estimating.

309.   *Precision* refers to the degree of closeness an estimate has to its expected value.

310.   The *receiver operating characteristic* or the *receiver operating curve* of a pattern classifier is a function of its misdetection rate against its false-alarm rate.

311.   The *leave-K-out* method of evaluating a pattern classifier divides the training set into L mutually exclusive subsets, each having K patterns. The classifier is successively trained by using $L-1$ of the subsets and tested on the Lth subset. The evaluation is then made on the accumulated performance tests of the experiments, where in each experiment K patterns were omitted from the training set and then used in the testing set. Performance estimates obtained by using the leave-K-out method are unbiased. However, for small K the estimates will have high variance.

312.   The *resubstitution method* of evaluating a pattern classifier uses the same set for training and testing. Performance estimates obtained by using the resubstitution method are always biased high.

21.19   Index of Terms

INDEX